Handbook of Experimental Pharmacology

Volume 163/II

Springer
Berlin
Heidelberg
New York
Hong Kong
London
Milan
Paris
Tokyo

Angiotensin

Vol. II

Contributors

W.M. Aartsen, G. Aguilera, M. Azizi, A. Blume, M. Burnier, A.M. Capponi, R. Corti, J. Culman, M.J.A.P. Daemen, D.A. Denton, J. Díez, H. Drexler, J. Fielitz, J.W. Fischer, M. de Gasparo, P. Gohlke, T.L. Goodfriend, O. Grisk, T.F. Lüscher, M. Maillard, M.L. Mathai, M.J. McKinley, J. Ménard, T. Münzel, B.J. Oldfield, F. Qadri, V. Regitz-Zagrosek, R. Rettig, E.M. Richards, M.F. Rossier, F. Ruschitzka, B. Schieffer, E.L. Schiffrin, B.A. Schölkens, W. Schulz, J.F.M. Smits, C. Sumners, N. Tsilimingas, A. Warnholtz, R.S. Weisinger, M. Wendt, G. Wiemer, P. Wohlfart

Editors

Thomas Unger and Bernward A. Schölkens

Springer

Professor
Thomas Unger
Center for Cardiovascular Research (CCR)
Institut für Pharmakologie und Toxikologie
Charité Campus Mitte
Charité – Universitätsmedizin Berlin
Hessische Strasse 3-4
10115 Berlin, Germany
e-mail: Thomas.Unger@charite.de

Professor
Bernward A. Schölkens
Aventis Pharma Deutschland GmbH
Drug Innovation & Approval
Bldg. H 831, Room C 320
65926 Frankfurt/Main, Germany
e-mail: Bernward.Schoelkens@aventis.com

With 48 Figures and 12 Tables

ISSN 0171-2004

ISBN 3-540-40641-7 Springer-Verlag Berlin Heidelberg New York

Library of Congress Cataloging-in-Publication Data
Angiotensin / contributors, E.M. Abdel-Rahman ... [et al.] ; editors, Thomas Unger and Bernward A. Schölkens. p. ; cm. – (Handbook of experimental pharmacology ; v. 163) Includes bibliographical references and indexes.
ISBN 3-540-40640-9 (hard : alk. paper) – ISBN 3-540-40641-7 (hard : v. 2 : alk. paper)
1. Angiotensins. I. Abdel-Rahman, E. M. II. Unger, Th. III. Schölkens, Bernward A., 1943- . IV. Series. [DNLM: 1. Angiotensins–physiology. 2. Receptors, Angiotensin–physiology. 3. Renin-Angiotensin System–genetics. 4. Renin-Angiotensin System–physiology. 5. Angiotensin-Converting Enzyme Inhibitors–pharmacokinetics. W1 HASIL v. 163 2004 / WG 106 A58778 2004]
OP905.H3 vol. 163 [QP572.A54] 615'.1s–dc22 [616.1'32] 2003060602

Springer-Verlag is a part of Springer Science+Business Media
springeronline.com

Printed in Germany

Cover design: design & production GmbH, Heidelberg
Typesetting: Stürtz AG, 97080 Würzburg

Printed on acid-free paper 27/3150 hs – 5 4 3 2 1 0

Preface

The 1974 volume on angiotensin edited by Irvine H. Page and F. Merlin Bumpus expanded the *Handbook of Experimental Pharmacology* series. In the preface, the editors of the first edition commented on their subject matters as follows:

"...Initially, it seemed that the action of angiotensin was relatively simple but this proved grossly misleading... Even after two decades [since angiotensin was identified as the major effector peptide of the renin–angiotensin system (editors' note)] the multiplicity of its actions appears not to have been fully discovered. To call attention to its many functions is one of the purposes of this book."

Thirty years later, this statement still holds true. Nevertheless, this new edition of the volume on angiotensin attempts to provide an updated account of the knowledge and findings accumulated since the complexity of angiotensin was so accurately recognized.

Certainly, the editors of the first volume on angiotensin would have been gratified by the wealth of new data on their subject flooding the literature since 1974, adding to the complexity of actions of angiotensin peptides they had predicted. This, of course, does not make our present-day task of understanding the multiple facets of the renin–angiotensin system any easier. However, it justifies our current endeavor to take a new look at this venerable system that still offers so many hidden miracles to be unveiled and described and, after all, has proved in the last 30 years to be of utmost clinical importance.

It is indeed the advent of the inhibitors of the renin–angiotensin system, notably the angiotensin-converting enzyme inhibitors and the angiotensin AT_1 receptor antagonists, which has helped enormously in gaining deeper insights into the system, notwithstanding the fact that these compounds have become some of the most successful drugs ever developed, not only to control hypertension but also to protect target organs like the kidney, heart, blood vessels, and brain, and, most importantly, to reduce cardiovascular mortality.

Naturally, the focus of any published work on medical science changes over the years and between editions. In 1974, much emphasis was laid on the biochemistry of the renin–angiotensin system, since the major discoveries had been made in this area. The interaction of renin and converting enzyme with their substrates, newly developed assays to analyze the various components of the system, and angiotensin analogs and their structure–function relationships were in the center of interest.

In the years that followed, attention gradually shifted to other areas, such as the genetics of the renin–angiotensin system, angiotensin receptors, their regulation, signaling pathways and various functions, and the inhibitors of the renin–angiotensin system with their mechanisms of action and their clinical use.

Moreover, whereas 30 years ago angiotensin II was still predominantly seen as a regulator of blood pressure and body volume, this peptide and its active fragments together with another major effector molecule of the system, the adrenal steroid aldosterone, are now considered to play an important role in a variety of (patho-) physiological functions that may be as diversified as vascular growth and atheroma formation, renal protein handling and glomerulosclerosis, cardiac left ventricular hypertrophy, fibrosis and postinfarction remodeling, or central osmoregulation and neuroregeneration. And along with the immense progress in biological sciences that we have witnessed during the last decades, the renin–angiotensin–aldosterone system has been connected to a number of biological phenomena such as cellular differentiation, neuroplasticity, and apoptosis.

This shift in scientific interest and research activities is reflected in the present volume, although we strongly felt that the fundamental knowledge on the system accumulated and substantiated over the last 100 years should never be omitted but be present as an undercurrent to help our understanding and, even more importantly, to put our temporary knowledge of today into a historical perspective.

The editors of the 1974 volume finished their preface stating, "...Books today are expensive and time-consuming to read..."

Again, their statement holds true for the two volumes of this new edition. Today, electronic media provide us with virtually any information including, of course, what has been written on angiotensin to date, and a host of review articles has been published on almost every aspect of the renin–angiotensin–aldosterone system. However, as with Page and Bumpus in their day, we are convinced that even in our time, there is still a place for books of this kind which invite the scholar to in-depth reading of what acknowledged experts have compiled as the essentials in their field.

When we asked the authors if they were willing to contribute to this edition, we did so with a certain degree of apprehension for the above-mentioned reasons but were overwhelmed by their unanimous positive response to our request. We would like to thank all authors for their efforts to make this volume a solid source of comprehensive information. We would also like to express our gratitude to our secretaries, Miranda Schröder and Undine Schelle, as well as to Sibylle Melzer and Ellen Scheibe and also to Susanne Dathe, our partner at Springer-Verlag, for their continuous, invaluable support, which enabled us to achieve our task.

Berlin and Frankfurt, January 2004

Thomas Unger
Bernward A. Schölkens

List of Contributors

(Addresses stated at the beginning of respective chapters.)

List of Contents

Part 4. Tissues

Vascular

Brain/Nervous System

Heart

Kidney and Adrenal Gland

Part 5. Inhibition of the Renin-Angiotensin System

ACE Inhibitors

AT_1 Antagonists

NEP/ACE Inhibitors

Part 6. Clinical Qutlook

Contents of Companion Volume 163/I

Angiotensin

Part 1. General Aspects

Part 2. Genetics

Part 3. ANG Receptors

AT_1

AT_2

Part 4
Tissues
Vascular

Angiotensin II and Oxidative Stress

N. Tsilimingas[1] · A. Warnholtz[2] · M. Wendt[2] · T. Münzel[2]

[1] Cardiovascular Surgery, Hamburg, Germany
[2] Medizinische Klinik III, Schwerpunkt Kardiologie/Angiologie, Universitätsklinikum Hamburg-Eppendorf, Martinistraße 52, 20246 Hamburg, Germany
e-mail: muenzel@ukc.uni-hamburg.de

Abstract There is an increasing body of evidence that oxidative stress markedly contributes to endothelial dysfunction and to poor prognosis in patients with coronary artery disease and hypertension. Among many stimuli for oxidative stress in vascular tissue, the role of angiotensin II has been extensively studied. When given acutely, angiotensin II stimulates the release of nitric oxide and activates a nonphagocytic NAD(P)H oxidase, leading to formation of superoxide and of the nitric oxide/superoxide reaction product peroxynitrite. Peroxynitrite in turn may cause tyrosine nitration of prostacyclin synthase and/or oxidation of zinc–thiolate complexes of the nitric oxide synthase associated with inhibition or uncoupling of these enzymes. Further, angiotensin II causes endothelial dysfunction, hypertension, increases the endothelin expression, and inhibition of the cGMP-cGK-I signaling cascade, all of which are secondary to angiotensin II-stimulated production of reactive oxygen species.

Keywords Angiotensin II · NADPH oxidase · Endothelium · Peroxynitrite · Superoxide

Abbreviations

ACE	Angiotensin converting enzyme
ADMA	Asymmetric dimethylarginine
AT-1	Angiotensin type 1 receptor
BH_4	Tetrahydrobiopterin
cGK	cGMP-dependent protein kinase
cGMP	Cyclic guanosine monophosphate
DDAH	Dimethylarginine dimethyl-aminohydrolase
H_2O_2	Hydrogen peroxide
LDL	Low-density lipoprotein
L-NAME	N^G-nitro-L-arginine methyl ester
L-NNA	N^G-nitro-L-arginine
NADH	Reduced nicotinamide adenine dinucleotide
NADPH	Reduced nicotinamide adenine dinucleotide phosphate
NOS III	Nitric oxide synthase
nox	Nonphagocytic NAD(P)H oxidase
$O_2^{\cdot -}$	Superoxide
$ONOO^-$	Peroxynitrite
OH	Hydroxyl radical
PDE 1A1	Phosphodiesterase type 1A1
PKC	Protein kinase C
PRMT	Protein arginine methyltransferase
P-VASP	Phosphorylated vasodilatory stimulated phosphoprotein
RAAS	Renin-angiotensin-aldosterone system
ROS	Reactive oxygen species
sGC	Soluble guanylyl cyclase
SOD	Superoxide dismutase

1 Introduction

In the last decade, a great deal has been learned about the critical role of reactive oxygen species (ROS) in the pathophysiology of vascular disease. The effects of angiotensin II on ROS production and its functional consequences for endothelial function, hypertension, endothelin expression and activation of superoxide ($O_2^{\cdot -}$) producing enzymes such as the NAD(P)H oxidase, have been studied extensively. In this review, mechanisms by which angiotensin II-mediated ROS production affects vascular function will be discussed.

2 Endothelial Dysfunction, Oxidative Stress and Prognosis

Traditionally, the role of the endothelium was thought to be primarily that of a selective barrier to the diffusion of macromolecules from the blood lumen into interstitial space. During the past 20 years, numerous additional roles for the endothelium have been defined such as regulation of vascular tone, modulation of inflammation, promotion as well as inhibition of vascular growth, and modulation of platelet aggregation and coagulation. Endothelial dysfunction is characteristic of patients with apparent coronary atherosclerosis or patients with cardiovascular risk factors such as hypercholesterolemia, hypertension, diabetes and chronic smoking. Recent studies indicate that assessment of endothelial function in both the coronary and peripheral circulation provide important prognostic information concerning future cardiovascular events (Al Suwaidi et al. 2001; Heitzer et al. 2001). Examination of patients with mild coronary artery disease and endothelial dysfunction, at the level of coronary conductance and vessel resistance, demonstrated a greater incidence of cardiovascular events compared to patients with better endothelial function (Schachinger et al. 2000). More recent studies indicate that endothelial dysfunction in patients with angiographically normal coronary arteries was predictive of subsequent cardiovascular events (Halcox et al. 2002). Although the mechanisms underlying endothelial dysfunction may be multifactorial, there is a growing body of evidence that increased production of free radicals may considerably contribute to this phenomenon. In patients with coronary artery disease, the extent of vitamin C-induced improvement of endothelial dysfunction was a strong and independent predictor of subsequent cardiovascular events (Heitzer et al. 2001). This indicates that high oxidative stress in vascular tissue not only contributes to endothelial dysfunction, but also decisively determines the prognosis in patients with cardiovascular risk factors.

3 Oxidative Stress

3.1 Reactive Oxygen Species

ROS are molecules that are initially derived from oxygen but have undergone univalent reduction, so that they readily react with other biological products. ROSs include $O_2^{\cdot -}$, hydrogen peroxide (H_2O_2), hydroxyl radical (OH), nitric oxide (NO) and the NO/$O_2^{\cdot -}$ reaction product peroxynitrite. Of these, $O_2^{\cdot -}$, H_2O_2, NO and $ONOO^-$ are formed in response to angiotensin II treatment and modify the activity of signaling proteins and enzymes. In contrast to $O_2^{\cdot -}$, NO, OH,

H_2O_2, $ONOO^-$ and hypochlorous acid are not free radicals per se, but have oxidizing effects that markedly contribute to oxidative stress. Superoxide is dismuted by superoxide dismutase (SOD) to the more stable radical H_2O_2. Catalase and glutathionperoxidase are important scavengers of H_2O_2, leading to the formation of water. The rapid bimolecular reaction between NO and $O_2^{\cdot -}$ to yield $ONOO^-$ is more than three times faster than the enzymatic dismutation of $O_2^{\cdot -}$ catalyzed by SOD. Thus, $ONOO^-$ formation represents a major potential pathway for NO reactivity, which depends on rates of tissue $O_2^{\cdot -}$ production. $ONOO^-$ has a cytotoxic potential about 1,000 times higher than that of H_2O_2, and has a low stimulatory capacity of the target enzyme guanylyl cyclase. $ONOO^-$ will also protonate to peroxynitrous acid (ONOOH), to yield an oxidant with the reactivity of $^{\cdot}OH$, via metal independent mechanisms. Peroxynitrite in pure form will cause oxidative damage to protein, lipid, carbohydrate, DNA, subcellular organelles, and cell systems. Peroxynitrite may also cause tyrosine nitration of prostacyclin synthase (Zou et al. 1999) and oxidation of the zinc–thiolate complex of NOS (Zou et al. 2002) leading to uncoupling or inhibition of targeted enzymes.

3.2 Angiotensin and ROS Production

3.2.1 Endothelium

Incubation of cultured rat aortic endothelial cells with angiotensin II stimulates release of NO (Pueyo et al. 1998). These actions are mediated by the AT_1 receptor as confirmed by their inhibition with the AT_1 receptor antagonist losartan. For example, angiotensin II-stimulated cGMP production by reporter cells was prevented by a specific calmodulin antagonist, suggesting that angiotensin II stimulated endothelial calmodulin-dependent NOS (Pueyo et al. 1998). In porcine pulmonary endothelial cells, there is evidence that angiotensin IV rather than angiotensin II contributes to NO release, since an angiotensin IV antagonist blocked both angiotensin II- and angiotensin IV-induced NO production (Hill-Kapturczak et al. 1999). Angiotensin II also stimulates the production of $O_2^{\cdot -}$, which is mediated by the AT_1 receptor (Sohn et al. 2000). There is also evidence that stimulation of the AT_2 subtype functionally antagonizes AT_1 receptor-induced $O_2^{\cdot -}$production, involving a tyrosine phosphatase pathway (Sohn et al. 2000). The simultaneous release of NO and $O_2^{\cdot -}$, in response to angiotensin II stimulation, has been recently demonstrated in vitro (Pueyo et al. 1998) as well as in humans (Dijkhorst-Oei et al. 1999). In cultured endothelial cells, angiotensin II stimulated the formation of $ONOO^-$, as indicated by the increase in luminol-dependent chemiluminescence and by the inhibitory effects of SOD and NOS III inhibitors on chemiluminescence signals (Pueyo et al. 1998). By using venous occlusion plethysmography, Tonk Rabelinks' group showed that in a NO-free system (achieved by N^G-monomethyl-L-arginine treatment), angiotensin II-induced vasoconstriction was significantly enhanced in the forearm circulation of healthy volunteers. Likewise, angiotensin II-induced vasoconstriction was greatly diminished in response to concomitant treatment with vitamin C, suggesting that angiotensin II simultaneously stimulates NO and $O_2^{\cdot-}$ production in humans (Dijkhorst-Oei et al. 1999).

3.2.2 Vascular Smooth Muscle

Incubation of cultured smooth muscle cells with angiotensin II leads to a marked increase in the production of $O_2^{\cdot-}$. The likely $O_2^{\cdot-}$ source was identified as a NADH/NADPH driven oxidase (Griendling et al. 1994). This assumption was based on inhibitor experiments where diphenylene iodonium (inhibitor of flavin-containing oxidases, DPI) and quinacrine inhibited NADH and NADPH-stimulated $O_2^{\cdot-}$ production (Griendling et al. 1994). Further studies demonstrated that $O_2^{\cdot-}$ production induced by angiotensin II was inhibited by *N*-acetylcysteine and by the free radical scavenger tiron (Laursen et al. 1997). Angiotensin II-induced formation of the $O_2^{\cdot-}$dismutation product H_2O_2 can be detected within minutes (Ushio-Fukai et al. 1996). Both antisense against the $p22^{phox}$ NADPH oxidase subunit and overexpression of catalase prevent angiotensin II-stimulated formation of H_2O_2 (Ushio-Fukai et al. 1996). Intracellular signaling mechanisms by which angiotensin II stimulates ROS formation is still under debate, and may involve arachidonic acid made by phospholipase A_2, or indirectly via phospholipase D-mediated degradation of phosphatidylcholine to phosphatidic acid (Griendling et al. 1994).

Most of the experiments to detect $O_2^{\cdot-}$ production in homogenates use the chemiluminescence substance lucigenin. Lucigenin has been demonstrated to undergo redox cycling when used in high concentrations (250 μM), and in particular when NADH (Janiszewski et al. 2002) is used as a substrate. These limitations do not occur when low concentrations of lucigenin (5 μM) (Li et al. 1998) and NADPH as substrate are used (Griendling et al. 2000). Recent studies using electron paramagnetic resonance to detect $O_2^{\cdot-}$ produced by cultured smooth muscle cells, in response to angiotensin II stimulation, revealed that NADPH is the major substrate for oxidase (Sorescu et al. 2001). In cultured endothelial cells, both NADPH and NADH were equally effective in stimulating superoxide production (Somers et al. 2000). Thus, the question whether NADH and/or NADPH are the preferred substrate for this enzyme is still under debate.

4
Effects of Angiotensin II

4.1
Activation of Oxidases

4.1.1
Effects of Angiotensin II on the Activity and Expression of the Nonphagocytic NAD(P)H Oxidase

Although vascular NAD(P)H oxidase is quite similar to the phagocytic multicomponent NAD(P)H oxidase, there are several distinct features that allow differentiation between both oxidases (Griendling et al. 2000). Activation occurs within seconds with the phagocytic oxidase, and within minutes or hours with the nonphagocytic oxidase (see also Table 1). The $O_2^{\cdot-}$-producing capacity of the phagocytic oxidase is in the micromolar range, and $O_2^{\cdot-}$ is being released in a burst-like fashion. In contrast, a steady release of low amounts of $O_2^{\cdot-}$, in the nanomolar range, is produced by vascular NADPH oxidase. Endothelial and adventitial cells, as well as inflammatory cells such as neutrophils and macrophages, express NAD(P)H oxidase consisting of the flavocytochrome b_{558} subunits gp91^{phox} and p22^{phox}, the cytosolic factors p47^{phox} and p67^{phox}, and the small GTPase rac1 (Griendling et al. 2000). In contrast to endothelial, adventitial and inflammatory cells, smooth muscle cells lack gp91^{phox}; however, recent studies identified the existence of two gp91^{phox} homologs, namely nox1 and nox4 (Lambeth et al. 2000; Suh et al. 1999). Because gp91^{phox} and nox harbor electron transfer moieties of the enzyme and thus serve as the catalytic component, regulation of these subunits is of utmost importance to the ultimate functioning of the enzyme.

In vitro, incubation of endothelial and smooth muscle cells with angiotensin II resulted in increases in activity and expression of NADPH oxidase subunits

Table 1 Characteristics of phagocytic and vascular NAD(P)H oxidases

	Phagocytic oxidase	Vascular oxidase
Substrate K_m	NADPH (30 μM)	NADPH (50 μM) NADH (10 μM)
Kinetics of activation	Seconds	Minutes to hours
Capacity	130 nmol $O_2^{\cdot-}$/min*mg cell protein	40 nmol $O_2^{\cdot-}$/min*mg cell protein
Fate of released $O_2^{\cdot-}$	Extracellular	Intracellular and/or extracellular
Cellular location	Phagosomes	Plasma membrane/microsomes
Subunit structure	gp91^{phox} p22^{phox} p47^{phox} p67^{phox} rac-1/rac-2	Endothelial cells and fibroblasts: gp91^{phox}, p22^{phox}, p47^{phox} p67^{phox}, rac-1/ rac-2 Vascular smooth muscle cells: nox1, nox4, p22^{phox}, p47^{phox}, p67^{phox}, rac-1/ rac-2

such as gp91^{phox}, p22^{phox}, p47^{phox}, p67^{phox}, nox1 and nox4 (Touyz et al. 2002; Wingler et al. 2001). Antisense directed against p22^{phox} and nox-1 resulted in an inhibition of angiotensin II-induced ROS formation (Lassegue et al. 2001), while antisense against nox4 resulted in a decrease in basal $O_2^{\cdot-}$ production (Szocs et al. 2002). In vivo treatment with angiotensin II increased expressions of p22^{phox} (three- to fourfold), gp91^{phox} (threefold), p47 and p67^{phox} in endothelial and smooth muscle cells, and the adventitia, respectively (Cifuentes et al. 2000; Di Wang et al. 1999; Mollnau et al. 2002; Pagano et al. 1998; Pagano et al. 1997). Likewise, in experimental hypercholesterolemia, increased activity of the NAD(P)H oxidase was linked with increased AT_1 receptor expression, and the activity of the enzyme was reduced by in vivo AT_1 receptor blockade (Warnholtz et al. 1999). Evidence for a role of the RAAS in the regulation of gp91^{phox} isoform expression in smooth muscle cells was provided by Mollnau et al. (2002) and Wingler et al. (2001), who demonstrated increased expression of nox1 and nox4 in aortas from hypertensive, angiotensin II-infused animals and in transgenic rats, overexpressing the Ren2 gene.

Further insight into the role of NAD(P)H oxidase in vascular disease was provided by knockout experiments. Infusion of angiotensin II into wild-type mice increased $O_2^{\cdot-}$ production, blood pressure, media viscosity, gp91^{phox} expression and nitrotyrosine content (footprint for $ONOO^-$-induced vascular damage) in vascular tissue, all of which were significantly reduced in gp91^{phox} knockout animals (Wang et al. 2001). Infusion of angiotensin II caused significantly greater increases in vascular $O_2^{\cdot-}$ production in aortas from wild-type mice as compared to aortas from p47^{phox} knockout animals (Brandes et al. 2002). Likewise, infusion of angiotensin II at subpressor doses caused myocardial hypertrophy in wild-type, but not in gp91^{phox} knockout mice, suggesting an involvement of myocardial NAD(P)H oxidase in mediating myocardial hypertrophy independent of changes in blood pressure (Bendall et al. 2002). These data clearly indicate a crucial role for the NAD(P)H oxidase in mediating angiotensin II-induced increases in oxidative stress, in vascular and myocardial tissue.

4.1.2 Angiotensin II and the Nitric Oxide Synthase

As mentioned above, angiotensin II stimulates the simultaneous release of NO and $O_2^{\cdot-}$leading to $ONOO^-$ formation (Pueyo et al. 1998). Long-term incubation of pulmonary endothelial cells with angiotensin II markedly increased NOS III expression at the mRNA and protein levels, with a maximum after 8 h (+300%) (Olson et al. 1997). These effects were mediated by the AT-1 subtype, since pretreatment with losartan abolished the effects on NOS III expression. Despite the marked increases in NOS III expression, angiotensin II treatment usually led to significant endothelial dysfunction (Mollnau et al. 2002). The discrepancy between the increases in NOS III expression and vascular $O_2^{\cdot-}$ production may indicate that NOS III is itself a significant $O_2^{\cdot-}$ source, under these conditions. Recent results from in vivo studies in angiotensin II-treated, hypertensive animals confirm this concept. At both the RNA and protein levels, NOS III enzyme was up-regulated more than twofold in the myocardium (Tambascia et al. 2001) and vessels (Mollnau et al. 2002; Sullivan et al. 2002) of angiotensin II-treated animals as compared to controls. Nevertheless, there was considerable endothelial dysfunction as well as a marked decrease in vascular NO bioavailability in angiotensin II-treated animals, indicating that vascular $O_2^{\cdot -}$from hypertensive animals may either overwhelm nitric oxide production of up-regulated NOS III, or that the up-regulated NOS III itself is uncoupled, thereby contributing to $O_2^{\cdot -}$ production.

Conditions leading to NOS III uncoupling, such as BH_4 deficiency (NOS III cofactor) (Xia et al. 1998) and intracellular L-arginine (NOS III substrate) (Gorren et al. 1998) depletion, have been characterized. In BH_4 deficiency, electrons flowing from the NOS III reductase domain to the oxygenase domain are diverted to molecular oxygen, rather than to L-arginine, resulting in production of $O_2^{\cdot -}$ rather than NO. What conditions are responsible for NOS III uncoupling in vessels from angiotensin II-infused animals? In vitro angiotensin II treatment increased the vascular formation of the $NO/O_2^{\cdot -}$ reaction product, $ONOO^-$ (Pueyo et al. 1998). Peroxynitrite in turn has been shown to rapidly oxidize the active NOS III cofactor BH_4 to inactive molecules, e.g., BH_2 switching NOS III from a NO to an $O_2^{\cdot -}$-producing enzyme (Laursen et al. 2001). An alternative explanation whereby $ONOO^-$ may cause NOS III uncoupling was recently reported by Zou et al (2002). Using cultured endothelial cells, $ONOO^-$ was shown to directly cause oxidation of the zinc–thiolate complex within the enzyme, all of which may favor NOS III monomer over dimer formation, leading to NOS III uncoupling. Likewise, Jennifer Pollock's group demonstrated increased NOS III expression in vessels from angiotensin II-infused animals (Sullivan et al. 2002). Interestingly, a considerable portion of the increased NOS III protein was located in the cytosolic fraction. Since NOS III in its activated state is a membrane-associated enzyme (targeted to caveolae and golgi membranes via *N*-myristoylation and palmitoylation processes), redistribution toward the cytosol and therefore the altered subcellular location of NOS III may also contribute to endothelial dysfunction and NOS III uncoupling in this particular animal model.

As pointed out above, another mechanism leading to NOS III uncoupling is depletion of intracellular L-arginine. Recent in vitro and in vivo studies demonstrate that in almost all situations where oxidative stress is encountered in vascular tissue, intracellular and plasma concentrations of asymmetric dimethylarginines (ADMA) are increased (Boger et al. 2000a; Boger et al. 1998). This phenomenon may be explained either by an increase in the activity of methylating enzymes such as S-adenosylmethionine-dependent methyltransferases (PRMT Type I) (Boger et al. 2000b), or a decrease in the activity of ADMA demethylating enzymes such as dimethylarginine dimethylaminohydrolase (DDAH); all of which result in increased intracellular production of ADMA. In concentrations reached by stimulation of methyltransferases or inhibition of DDAH, ADMA has been shown to significantly inhibit NOS III activity. Interestingly, the activity of both enzymes regulating intracellular ADMA concentrations has been reported to be redox-sensitive. Oxidative stress increases the activity of methylating enzyme such as PRMT I (Boger et al. 2000b), while decreasing the activity of demethylating DDAH (Ito et al. 1999). These observations may explain why in angiotensin II hypertension, L-arginine is able to improve endothelial dysfunction in experimental animals (Pucci et al. 1995), even when intracellular L-arginine levels are not decreased (the so-called L-arginine paradox). Recent studies with patients diagnosed with essential hypertension revealed increased plasma levels of ADMA (Surdacki et al. 1999). Interestingly, treatment with ACE-inhibitors, AT_1 receptor blockers, but not beta-blockers, reduced plasma ADMA levels. However, all drugs were able to normalize blood pressure, pointing to a role for the RAAS in mediating increased plasma ADMA levels in hypertensive patients (Ito et al. 2001).

How can we assess NOS III uncoupling in vascular tissue? Pritchard et al. previously demonstrated that NOS III inhibition in endothelial cells cultured with L-NNA increased steady state $O_2^{\cdot-}$ levels (Vasquez-Vivar et al. 1998). These findings indicate that a large portion of baseline $O_2^{\cdot-}$ production is scavenged due to its interaction with NO. After incubation of endothelial cells with native LDL, $O_2^{\cdot-}$ levels increased markedly, a phenomenon which largely was blocked by L-NAME(Pritchard et al. 1995). Reduction of steady state $O_2^{\cdot-}$ levels by means of NOS III inhibition identified NOS III as an important $O_2^{\cdot-}$ source. The assumption that NOS III is uncoupled in angiotensin II hypertension was strengthened recently by experiments with the NOS inhibitor L-NNA. Vascular $O_2^{\cdot-}$ production was assessed using lucigenin- and coelenterazine-derived chemiluminescence. In control vessels with an intact endothelium, NOS-inhibition with L-NNA increased vascular $O_2^{\cdot-}$, indicating that basal production of endothelium-derived nitric oxide quenches the baseline chemiluminescence signal. In contrast, incubation of aortas, from angiotensin II-treated animals, with L-NNA markedly reduced the chemiluminescence signal, thus indicating that NOS III is an important $O_2^{\cdot-}$ source (Mollnau et al. 2002). It seems that in most situations where high oxidative stress is encountered, NOS III is in an uncoupled state as observed in the experimental animal (Hink et al. 2001; Laursen et al. 2001; Oelze et al. 2000) and in patients with hypercholesterolemia (Stroes et al. 1997) and diabetes (Heitzer et al. 2000b) and in chronic smokers (Heitzer et al. 2000a).

4.1.3
Angiotensin II and the Xanthine Oxidase

Xanthine oxidase is an oxidoreductase that catalyzes the oxidation of hypoxanthine and xanthine in purine metabolism. This enzyme exists in two interconvertible forms, either as xanthine dehydrogenase or xanthine oxidase. Recent studies with double-transgenic rats (dTGR) harboring human renin and angiotensinogen genes provide evidence for angiotensin II involvement in the regulation of xanthine oxidase-mediated $O_2^{\cdot-}$ production (Mervaala et al. 2001). Incubation of vascular tissue with SOD and the xanthine oxidase inhibitor oxypurinol, improved endothelial dysfunction. Increased production of 8-iso-prostaglandin F2α, decreased NO production, increased xanthine oxidase activity in the kidney, and endothelial dysfunction were corrected by in vivo treatment with the angiotensin II (type 1) receptor blocker valsartan, pointing to a role for angiotensin II xanthine oxidase activation in this particular animal model.

4.2 Effects of Angiotensin II-Induced Hypertension on cGMP, sGC and cGK-I Signaling

Vessels from animals with angiotensin II-evoked hypertension show not only reduced vasodilation to endothelium-dependent vasodilators, but also to endothelium-independent vasodilators such as SNP and NTG (Rajagopalan et al. 1996). Reduced nitrovasodilator responses, in vessels from angiotensin-infused hypertensive animals, may be explained by the reduced expression of both sGC subunits (Mollnau et al. 2002). This observation complements reports from other animal models of hypertension, where the expression of one or both sGC subunits (α_1 and β_1) and/or NO-dependent sGC activity were significantly decreased (Jacke et al. 2000; Kloss et al. 2000; Lopez-Farre et al. 2002; Ruetten et al. 1999). In all these animal models, endothelial dysfunction was associated with enhanced vascular ROS formation. Normalization of blood pressure by hydralazine-treatment (Bauersachs et al. 1998), vitamin C treatment (Marques et al. 2001), in vivo PKC-inhibition, Ca^{2+}-antagonist or chronic ACE-inhibitor treatment (Jacke et al. 2000), normalized or even enhanced sGC expression. The question remains whether the high blood pressure or the observed increases in ROS production within smooth muscle cells represents the crucial stimulus for a decrease in sGC expression. Georg Kojda's group recently showed that incubation of vascular tissue with the sGC inhibitor LY83583, a redox-cycler that generates intracellular $O_2^{\cdot-}$, mimicked the changes observed in vessels from cholesterol-fed animals: that is, sGCβ_1 subunit expression was up-regulated. However, in contrast to hypercholesterolemic animals, NO-dependent activity of sGC, in aortic cytosol, was abolished (Laber et al. 2002). Thus, it appears that the observed decrease in sGC expression, in vessels from angiotensin II-treated animals, was mainly elicited by high blood pressure and not linked to increased $O_2^{\cdot-}$ formation. Infusion of angiotensin II did not modify cGK I expression. However, cGK I activity, as assessed by phosphorylation of the cGK-I substrate, the vasodilatory stimulated phosphoprotein (P-VASP), was markedly reduced, suggesting a signaling defect upstream of cGK I (Mollnau et al. 2002). This was supported by the fact that in vivo treatment with chelerythrine reduced oxidative stress and improved both endothelial dysfunction and P-VASP (M. Oelze, unpublished observation). Angiotensin II increased the activity and expression of the PDE1A1 isoform in cultured smooth muscle cells, providing evidence that increased cGMP metabolism may partly contribute to endothelial dysfunction and decreased cGK I activity, as observed in angiotensin II-induced hypertension (Kim et al. 2001).

4.3 Key Role for Protein Kinase C in Angiotensin II Induced $O_2^{\cdot-}$ Production

Recent in vitro studies revealed that angiotensin II down-regulates nox4, while markedly up-regulating the nox1 isoform (Lassegue et al. 2001). Up-regulation of nox1 was also observed in response to the phorbol ester, phorbol myristate acetate (a direct activator of PKC). Angiotensin II-induced up-regulation of nox1 was inhibited by specific inhibitors of PKC, suggesting a crucial role for PKC in the up-regulation of the nox1 message (Gorlach et al. 2000). Antisense nox1 mRNA completely inhibited angiotensin II-induced $O_2^{\cdot-}$ production, supporting a role for nox1 in redox signaling in vascular smooth muscle cells (Lassegue et al. 2001). The results obtained from our recent in vivo experiments support this concept. Using a nonradioactive PKC assay, we found a twofold increase in whole vessel homogenates from angiotensin II-treated animals, as compared with controls. While nox4 expression was only slightly modified by angiotensin II treatment, dramatic increases in the expressions of nox1 (~sevenfold), $gp91^{phox}$ (threefold) and $p22^{phox}$ (threefold) were observed. In vitro and in vivo treatment with the PKC inhibitor chelerythrine markedly reduced angiotensin II-stimulated $O_2^{\cdot-}$ production and prevented up-regulation of nox1 (Mollnau et al. 2002). An interesting aspect of this particular study was that the in vivo PKC inhibition also prevented the up-regulation of $p22^{phox}$, suggesting that PKC may function as a general regulator of oxidase function by regulating both activity and expression. Interestingly, recent in vitro studies indicate that $O_2^{\cdot-}$ is a strong stimulus for PKC activation (Nishikawa et al. 2000). Thus, PKC increases oxidative stress in angiotensin II-treated vascular tissue by stimulating NADPH-oxidase mediated $O_2^{\cdot-}$ production, and by its activation in response to oxidative stress in a positive feedback fashion.

4.4 Angiotensin II, Oxidative Stress, Endothelin Expression and Enhanced Sensitivity to Vasoconstrictors

Vasoconstriction and hypertension result from angiotensin II action on the AT_1 receptor and its subsequent activation of second messenger pathways, including PKC. Incubation of endothelial (Chua et al. 1993) and smooth muscle cells (Sung et al. 1994) with angiotensin II stimulates preproendothelin expression via activation of the AT_1 receptor and then PKC. Infusion of angiotensin II for 5 days leads to a marked increase in sensitivity to vasoconstrictors such as phenylephrine, KCl and serotonin, while the sensitivity to endothelin-1 is significantly reduced (Rajagopalan et al. 1997). In vivo treatment with an ET_a receptor antagonist normalized hypercontractile responses and prevented desensitization of the vasculature to endothelin-1. Endothelin expression, as assessed by immunostaining, was increased throughout the vascular wall (Rajagopalan et al. 1997). Superoxide is a strong stimulus for preproendothelin expression in endothelial (Kahler et al. 2000) but also smooth muscle cells (Kahler et al. 2001). Endothelin itself also stimulates vascular $O_2^{\cdot-}$ production (Wedgwood et al. 2001). Increased production of endothelin within smooth muscle and/or endothelial cells may prime PKC, which in turn may mediate hypersensitivity to a variety of vasoconstrictors and activate the NADPH oxidase. Recent studies also indicate that vasoconstriction induced by angiotensin II, but not by endothelin-1, phenylephrine or potassium chloride, is mediated via stimulation of vascular $O_2^{\cdot-}$ production. TEMPOL, a potent SOD mimetic, markedly attenuated the maximal constriction in response to angiotensin II, in spontaneously hypertensive rats (Shastri et al. 2002).

4.5
Role of $O_2^{\cdot-}$ in Angiotensin II-Induced Hypertension

Angiotensin II stimulates vasoconstriction via AT_1 receptor and ROS production, which shorten the half-life of NO. To assess the contribution of $O_2^{\cdot-}$ in angiotensin II-induced hypertension, animals were treated with angiotensin II and the blood pressure lowering effects of liposome encapsulated SOD was tested (Laursen et al. 1997). As a reference constrictor, norepinephrine was used. Despite similar degrees of hypertension, angiotensin II, but not norepinephrine, caused a marked increase in vascular $O_2^{\cdot-}$ production. SOD treatment reduced blood pressure in angiotensin II- but not in norepinephrine-treated rats (Laursen et al. 1997). Likewise, SOD enhanced in vivo hypotensive responses to acetylcholine and in vitro responses to endothelium-dependent vasodilators, indicating that angiotensin II-induced hypertension is likely the result of increased NO degradation, secondary to the stimulatory effects of vascular $O_2^{\cdot-}$ production. To analyze the contribution of vascular NAD(P)H oxidase in angiotensin II-induced hypertension, Fukui et al. studied the time course of angiotensin II-induced hypertension and the expression of the NADPH oxidase subunit $p22^{phox}$. Blood pressure began to rise within 3 days of angiotensin II treatment, and remained elevated for up to 14 days (Fukui et al. 1997). Expression of $p22^{phox}$ was significantly increased on day 3 and peaked on day 5, after pump implantation. SOD treatment lowered blood pressure and inhibited expression of $p22^{phox}$, indicating that oxidative stress may be crucial for enzyme expression. Further evidence for the role of $O_2^{\cdot-}$ in the pressor response of angiotensin II was provided by Wang et al. (2002). The authors demonstrate that angiotensin II-induced simultaneously increased blood pressure and blunted vascular $O_2^{\cdot-}$ production in vessels from mice overexpressing human SOD. (Wang et al. 2002).

Similar findings were obtained from animals with renovascular hypertension. In vessels from hypertensive animals, marked endothelial dysfunction was associated with increased NAD(P)H oxidase-mediated superoxide production. In vitro treatment with SOD and the PKC-inhibitor calphostin C, reduced superoxide production and improved endothelial function, suggesting an involvement of PKC in activating the oxidase in this particular renin-dependent hypertension (Heitzer et al. 1999).

5 Conclusions

Based on the above findings, we propose that ROS play a pivotal role in mediating endothelial dysfunction under angiotensin II treatment: Angiotensin II activates NAD(P)H oxidases in endothelial and smooth muscle cells and within the adventitia, partly via PKC activation. NAD(P)H oxidase-derived $O_2^{\cdot-}$ may com-

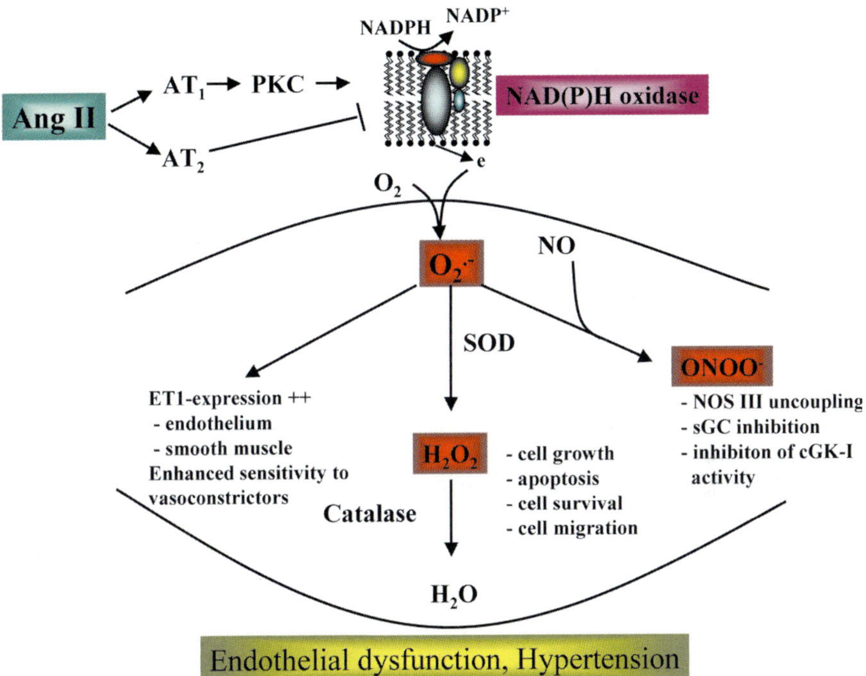

Fig. 1 Scheme depicting mechanisms underlying angiotensin II (*Ang II*)-induced endothelial dysfunction and hypertension. Stimulation of the AT_1 receptor subtype activates NAD(P)H oxidase in a protein kinase C (*PKC*)-dependent fashion. NAD(P)H oxidase activation increases superoxide ($O_2^{\cdot-}$) production within endothelial and smooth muscle cells and within the adventitia, while stimulation of the AT_2 subtype leads to an inhibition of $O_2^{\cdot-}$ formation. Superoxide may react with nitric oxide (*NO*) to produce the reactive intermediate peroxynitrite (*ONOO–*). ONOO– may cause NOS III uncoupling, inhibition of the activity of the soluble guanylyl cyclase (*sGC*) and of the cGMP-dependent protein kinase-I. Superoxide dismutase (*SOD*) dismutes $O_2^{\cdot-}$ to hydrogen peroxide (H_2O_2), which has been shown to be a potent stimulus for cell proliferation, apoptosis, cell migration and cell survival. Superoxide increases the expression of preproendothelin within endothelial cells and smooth muscle cells, leading to enhanced endothelin-mediated vasoconstriction, but also leading to hypersensitivity of the vasculature to vasoconstricting agonists such as norepinephrine and serotonin. All these events may contribute at least in part to endothelial dysfunction and hypertension in response to angiotensin II treatment

bine with NO to form the highly reactive intermediate $ONOO^-$. Peroxynitrite oxidizes the NOS III cofactor BH_4 to BH_2, or may cause oxidation of the zinc-thiolate complex of NOS III. Superoxide may also stimulate and/or inhibit L-arginine methylating or ADMA demethylating enzymes, leading to increased intracellular concentrations of ADMA. Intracellular BH_4 depletion, oxidation of the zinc-thiolate complex of NOS III, and/or increased ADMA concentrations will lead to NOS III uncoupling, which may further increase oxidative stress within vascular tissue. Superoxide also stimulates preproendothelin expression in endothelial and smooth muscle cells, leading to enhanced endothelin-mediated constriction and to a hypersensitivity of the vasculature to vasoconstricting agonists such as norepinephrine and serotonin. It is conceivable that all these mechanisms contribute to endothelial dysfunction and hypertension, in response to angiotensin II treatment (see Fig. 1).

Acknowledgements. This study was supported by the Deutsche Forschungsgemeinschaft (Mu 1079/4-1).

References

Al Suwaidi J, Higano ST, Holmes DR Jr, Lerman A (2001) Pathophysiology, diagnosis, and current management strategies for chest pain in patients with normal findings on angiography. Mayo Clin Proc 76:813–822

Bauersachs J, Bouloumie A, Fraccarollo D, Hu K, Busse R, Ertl G (1998) Hydralazine prevents endothelial dysfunction, but not the increase in superoxide production in nitric oxide-deficient hypertension. Eur J Pharmacol 362:77–81

Bendall JK, Cave AC, Heymes C, Gall N, Shah AM (2002) Pivotal role of a gp91(phox)-containing NADPH oxidase in angiotensin II-induced cardiac hypertrophy in mice. Circulation 105:293–296

Boger RH, Bode-Boger SM, Szuba A, Tsao PS, Chan JR, Tangphao O, Blaschke TF, Cooke JP (1998) Asymmetric dimethylarginine (ADMA): a novel risk factor for endothelial dysfunction: its role in hypercholesterolemia. Circulation 98:1842–1847

Boger RH, Bode-Boger SM, Sydow K, Heistad DD, Lentz SR (2000a) Plasma concentration of asymmetric dimethylarginine, an endogenous inhibitor of nitric oxide synthase, is elevated in monkeys with hyperhomocyst(e)inemia or hypercholesterolemia. Arterioscler Thromb Vasc Biol 20:1557–1564

Boger RH, Sydow K, Borlak J, Thum T, Lenzen H, Schubert B, Tsikas D, Bode-Boger SM (2000b) LDL cholesterol upregulates synthesis of asymmetrical dimethylarginine in human endothelial cells: involvement of S-adenosylmethionine-dependent methyltransferases. Circ Res 87:99–105

Brandes RP, Miller FJ, Beer S, Haendeler J, Hoffmann J, Ha T, Holland SM, Gorlach A, Busse R (2002) The vascular NADPH oxidase subunit p47phox is involved in redox-mediated gene expression. Free Radic Biol Med 32:1116–1122

Chua BH, Chua CC, Diglio CA, Siu BB (1993) Regulation of endothelin-1 mRNA by angiotensin II in rat heart endothelial cells. Biochim Biophys Acta 1178:201–206

Cifuentes ME, Rey FE, Carretero OA, Pagano PJ (2000) Upregulation of p67(phox) and gp91(phox) in aortas from angiotensin II-infused mice. Am J Physiol Heart Circ Physiol 279:H2234–H2240

Di Wang H, Hope S, Du Y, Quinn MT, Cayatte A, Pagano PJ, Cohen RA (1999) Paracrine role of adventitial superoxide anion in mediating spontaneous tone of the isolated rat aorta in angiotensin II-induced hypertension. Hypertension 33:1225–1232

Dijkhorst-Oei LT, Stroes ES, Koomans HA, Rabelink TJ (1999) Acute simultaneous stimulation of nitric oxide and oxygen radicals by angiotensin II in humans in vivo. J Cardiovasc Pharmacol 33:420–424

Fukui T, Ishizaka N, Rajagopalan S, Laursen JB, Capers QT, Taylor WR, Harrison DG, de Leon H, Wilcox JN, Griendling KK (1997) p22phox mRNA expression and NADPH oxidase activity are increased in aortas from hypertensive rats. Circ Res 80:45–51

Gorlach A, Brandes RP, Nguyen K, Amidi M, Dehghani F, Busse R (2000) A gp91phox containing NADPH oxidase selectively expressed in endothelial cells is a major source of oxygen radical generation in the arterial wall. Circ Res 87:26–32

Gorren AC, Mayer B (1998) The versatile and complex enzymology of nitric oxide synthase. Biochemistry (Mosc) 63:734–743

Griendling KK, Minieri CA, Ollerenshaw JD, Alexander RW (1994) Angiotensin II stimulates NADH and NADPH oxidase activity in cultured vascular smooth muscle cells. Circ Res 74:1141–1148

Griendling KK, Sorescu D, Ushio-Fukai M (2000) NAD(P)H oxidase: role in cardiovascular biology and disease. Circ Res 86:494–501

Halcox JP, Schenke WH, Zalos G, Mincemoyer R, Prasad A, Waclawiw MA, Nour KR, Quyyumi AA (2002) Prognostic value of coronary vascular endothelial dysfunction. Circulation 106:653–658

Heitzer T, Wenzel U, Hink U, Krollner D, Skatchkov M, Stahl RA, MacHarzina R, Brasen JH, Meinertz T, Munzel T (1999) Increased NAD(P)H oxidase-mediated superoxide production in renovascular hypertension: evidence for an involvement of protein kinase C. Kidney Int 55:252–260

Heitzer T, Brockhoff C, Mayer B, Warnholtz A, Mollnau H, Henne S, Meinertz T, Munzel T (2000a) Tetrahydrobiopterin improves endothelium-dependent vasodilation in chronic smokers: evidence for a dysfunctional nitric oxide synthase. Circ Res 86:E36–E41

Heitzer T, Krohn K, Albers S, Meinertz T (2000b) Tetrahydrobiopterin improves endothelium-dependent vasodilation by increasing nitric oxide activity in patients with type II diabetes mellitus. Diabetologia 43:1435–1438

Heitzer T, Schlinzig T, Krohn K, Meinertz T, Munzel T (2001) Endothelial dysfunction, oxidative stress, and risk of cardiovascular events in patients with coronary artery disease. Circulation 104:2673–2678

Hill-Kapturczak N, Kapturczak MH, Block ER, Patel JM, Malinski T, Madsen KM, Tisher CC (1999) Angiotensin II-stimulated nitric oxide release from porcine pulmonary endothelium is mediated by angiotensin IV. J Am Soc Nephrol 10:481–491

Hink U, Li H, Mollnau H, Oelze M, Matheis E, Hartmann M, Skatchkov M, Thaiss F, Stahl RA, Warnholtz A, Meinertz T, Griendling K, Harrison DG, Forstermann U, Munzel T (2001) Mechanisms underlying endothelial dysfunction in diabetes mellitus. Circ Res 88:E14–E22

Ito A, Tsao PS, Adimoolam S, Kimoto M, Ogawa T, Cooke JP (1999) Novel mechanism for endothelial dysfunction: dysregulation of dimethylarginine dimethylaminohydrolase. Circulation 99:3092–3095

Ito A, Egashira K, Narishige T, Muramatsu K, Takeshita A (2001) Renin-angiotensin system is involved in the mechanism of increased serum asymmetric dimethylarginine in essential hypertension. Jpn Circ J 65:775–778

Jacke K, Witte K, Huser L, Behrends S, Lemmer B (2000) Contribution of the renin-angiotensin system to subsensitivity of soluble guanylyl cyclase in TGR(mREN2)27 rats. Eur J Pharmacol 403:27–35

Janiszewski M, Souza HP, Liu X, Pedro MA, Zweier JL, Laurindo FR (2002) Overestimation of NADH-driven vascular oxidase activity due to lucigenin artifacts. Free Radic Biol Med 32:446–453

Kahler J, Mendel S, Weckmuller J, Orzechowski HD, Mittmann C, Koster R, Paul M, Meinertz T, Munzel T (2000) Oxidative stress increases synthesis of big endothelin-1 by activation of the endothelin-1 promoter. J Mol Cell Cardiol 32:1429–1437

Kahler J, Ewert A, Weckmuller J, Stobbe S, Mittmann C, Koster R, Paul M, Meinertz T, Munzel T (2001) Oxidative stress increases endothelin-1 synthesis in human coronary artery smooth muscle cells. J Cardiovasc Pharmacol 38:49–57

Kim D, Rybalkin SD, Pi X, Wang Y, Zhang C, Munzel T, Beavo JA, Berk BC, Yan C (2001) Upregulation of phosphodiesterase 1A1 expression is associated with the development of nitrate tolerance. Circulation 104:2338–2343

Kloss S, Bouloumie A, Mulsch A (2000) Aging and chronic hypertension decrease expression of rat aortic soluble guanylyl cyclase. Hypertension 35:43–47

Laber U, Kober T, Schmitz V, Schrammel A, Meyer W, Mayer B, Weber M, Kojda G (2002) Effect of hypercholesterolemia on expression and function of vascular soluble guanylyl cyclase. Circulation 105:855–860

Lambeth JD, Cheng G, Arnold RS, Edens WA (2000) Novel homologs of gp91phox. Trends Biochem Sci 25:459–461

Lassegue B, Sorescu D, Szocs K, Yin Q, Akers M, Zhang Y, Grant SL, Lambeth JD, Griendling KK (2001) Novel gp91(phox) homologues in vascular smooth muscle cells: nox1 mediates angiotensin II-induced superoxide formation and redox-sensitive signaling pathways. Circ Res 88:888–894

Laursen JB, Rajagopalan S, Galis Z, Tarpey M, Freeman BA, Harrison DG (1997) Role of superoxide in angiotensin II-induced but not catecholamine-induced hypertension. Circulation 95:588–593

Laursen JB, Somers M, Kurz S, McCann L, Warnholtz A, Freeman BA, Tarpey M, Fukai T, Harrison DG (2001) Endothelial regulation of vasomotion in apoE-deficient mice: implications for interactions between peroxynitrite and tetrahydrobiopterin. Circulation 103:1282–1288

Li Y, Zhu H, Kuppusamy P, Roubaud V, Zweier JL, Trush MA (1998) Validation of lucigenin (bis-N-methylacridinium) as a chemilumigenic probe for detecting superoxide anion radical production by enzymatic and cellular systems. J Biol Chem 273:2015–2023

Lopez-Farre A, Rodriguez-Feo JA, Garcia-Colis E, Gomez J, Lopez-Blaya A, Fortes J, de Andres R, Rico L, Casado S (2002) Reduction of the soluble cyclic GMP vasorelaxing system in the vascular wall of stroke-prone spontaneously hypertensive rats: effect of the alpha1 receptor blocker doxazosin. J Hypertens 20:463–470

Marques M, Millas I, Jimenez A, Garcia-Colis E, Rodriguez-Feo JA, Velasco S, Barrientos A, Casado S, Lopez-Farre A (2001) Alteration of the soluble guanylate cyclase system in the vascular wall of lead-induced hypertension in rats. J Am Soc Nephrol 12:2594–2600

Mervaala EM, Cheng ZJ, Tikkanen I, Lapatto R, Nurminen K, Vapaatalo H, Muller DN, Fiebeler A, Ganten U, Ganten D, Luft FC (2001) Endothelial dysfunction and xanthine oxidoreductase activity in rats with human renin and angiotensinogen genes. Hypertension 37:414–418

Mollnau H, Wendt M, Szocs K, Lassegue B, Schulz E, Oelze M, Li H, Bodenschatz M, August M, Kleschyov AL, Tsilimingas N, Walter U, Forstermann U, Meinertz T, Griendling K, Munzel T (2002) Effects of angiotensin II infusion on the expression and function of NAD(P)H oxidase and components of nitric oxide/cGMP signaling. Circ Res 90:E58–E65

Nishikawa T, Edelstein D, Du XL, Yamagishi S, Matsumura T, Kaneda Y, Yorek MA, Beebe D, Oates PJ, Hammes HP, Giardino I, Brownlee M (2000) Normalizing mitochondrial superoxide production blocks three pathways of hyperglycaemic damage. Nature 404:787–790

Oelze M, Mollnau H, Hoffmann N, Warnholtz A, Bodenschatz M, Smolenski A, Walter U, Skatchkov M, Meinertz T, Munzel T (2000) Vasodilator-stimulated phosphoprotein serine 239 phosphorylation as a sensitive monitor of defective nitric oxide/cGMP signaling and endothelial dysfunction. Circ Res 87:999–1005

Olson SC, Dowds TA, Pino PA, Barry MT, Burke-Wolin T (1997) ANG II stimulates endothelial nitric oxide synthase expression in bovine pulmonary artery endothelium. Am J Physiol 273:L315–L321

Pagano PJ, Clark JK, Cifuentes-Pagano ME, Clark SM, Callis GM, Quinn MT (1997) Localization of a constitutively active, phagocyte-like NADPH oxidase in rabbit aortic adventitia: enhancement by angiotensin II. Proc Natl Acad Sci U S A 94:14483–14488

Pagano PJ, Chanock SJ, Siwik DA, Colucci WS, Clark JK (1998) Angiotensin II induces p67phox mRNA expression and NADPH oxidase superoxide generation in rabbit aortic adventitial fibroblasts. Hypertension 32:331–337

Pritchard KA Jr, Groszek L, Smalley DM, Sessa WC, Wu M, Villalon P, Wolin MS, Stemerman MB (1995) Native low-density lipoprotein increases endothelial cell nitric oxide synthase generation of superoxide anion. Circ Res 77:510–518

Pucci ML, Dick LB, Miller KB, Smith CJ, Nasjletti A (1995) Enhanced responses to L-arginine in aortic rings from rats with angiotensin-dependent hypertension. J Pharmacol Exp Ther 274:1–7

Pueyo ME, Arnal JF, Rami J, Michel JB (1998) Angiotensin II stimulates the production of NO and peroxynitrite in endothelial cells. Am J Physiol 274:C214–C220

Rajagopalan S, Kurz S, Munzel T, Tarpey M, Freeman BA, Griendling KK, Harrison DG (1996) Angiotensin II-mediated hypertension in the rat increases vascular superoxide production via membrane NADH/NADPH oxidase activation. Contribution to alterations of vasomotor tone. J Clin Invest 97:1916–1923

Rajagopalan S, Laursen JB, Borthayre A, Kurz S, Keiser J, Haleen S, Giaid A, Harrison DG (1997) Role for endothelin-1 in angiotensin II-mediated hypertension. Hypertension 30:29–34

Ruetten H, Zabel U, Linz W, Schmidt HH (1999) Downregulation of soluble guanylyl cyclase in young and aging spontaneously hypertensive rats. Circ Res 85:534–541

Schachinger V, Britten MB, Zeiher AM (2000) Prognostic impact of coronary vasodilator dysfunction on adverse long-term outcome of coronary heart disease. Circulation 101:1899–1906

Shastri S, Gopalakrishnan V, Poduri R, DiWang H (2002) Tempol selectively attenuates angiotensin II evoked vasoconstrictor responses in spontaneously hypertensive rats. J Hypertens 20:1381–1391

Sohn HY, Raff U, Hoffmann A, Gloe T, Heermeier K, Galle J, Pohl U (2000) Differential role of angiotensin II receptor subtypes on endothelial superoxide formation. Br J Pharmacol 131:667–672

Somers MJ, Burchfield JS, Harrison DG (2000) Evidence for a NADH/NADPH oxidase in human umbilical vein endothelial cells using electron spin resonance. Antioxid Redox Signal 2:779–787

Sorescu D, Somers MJ, Lassegue B, Grant S, Harrison DG, Griendling KK (2001) Electron spin resonance characterization of the NAD(P)H oxidase in vascular smooth muscle cells. Free Radic Biol Med 30:603–612

Stroes E, Kastelein J, Cosentino F, Erkelens W, Wever R, Koomans H, Luscher T, Rabelink T (1997) Tetrahydrobiopterin restores endothelial function in hypercholesterolemia. J Clin Invest 99:41–46

Suh YA, Arnold RS, Lassegue B, Shi J, Xu X, Sorescu D, Chung AB, Griendling KK, Lambeth JD (1999) Cell transformation by the superoxide-generating oxidase Mox1. Nature 401:79–82

Sullivan JC, Pollock DM, Pollock JS (2002) Altered nitric oxide synthase 3 distribution in mesenteric arteries of hypertensive rats. Hypertension 39:597–602

Sung CP, Arleth AJ, Storer BL, Ohlstein EH (1994) Angiotensin type 1 receptors mediate smooth muscle proliferation and endothelin biosynthesis in rat vascular smooth muscle. J Pharmacol Exp Ther 271:429–437

Surdacki A, Nowicki M, Sandmann J, Tsikas D, Boeger RH, Bode-Boeger SM, Kruszelnicka-Kwiatkowska O, Kokot F, Dubiel JS, Froelich JC (1999) Reduced urinary excretion of nitric oxide metabolites and increased plasma levels of asymmetric dimethylarginine in men with essential hypertension. J Cardiovasc Pharmacol 33:652–658

Szocs K, Lassegue B, Sorescu D, Hilenski LL, Valppu L, Couse TL, Wilcox JN, Quinn MT, Lambeth JD, Griendling KK (2002) Upregulation of Nox-based NAD(P)H oxidases in restenosis after carotid injury. Arterioscler Thromb Vasc Biol 22:21–27

Tambascia RC, Fonseca PM, Corat PD, Moreno H Jr, Saad MJ, Franchini KG (2001) Expression and distribution of NOS1 and NOS3 in the myocardium of angiotensin II-infused rats. Hypertension 37:1423–1428

Touyz RM, Chen X, Tabet F, Yao G, He G, Quinn MT, Pagano PJ, Schiffrin EL (2002) Expression of a functionally active gp91phox-containing neutrophil-type NAD(P)H oxidase in smooth muscle cells from human resistance arteries: regulation by angiotensin II. Circ Res 90:1205–1213

Ushio-Fukai M, Zafari AM, Fukui T, Ishizaka N, Griendling KK (1996) p22phox is a critical component of the superoxide-generating NADH/NADPH oxidase system and regulates angiotensin II-induced hypertrophy in vascular smooth muscle cells. J Biol Chem 271:23317–23321

Vasquez-Vivar J, Kalyanaraman B, Martasek P, Hogg N, Masters BS, Karoui H, Tordo P, Pritchard KA Jr (1998) Superoxide generation by endothelial nitric oxide synthase: the influence of cofactors. Proc Natl Acad Sci U S A 95:9220–9225

Wang HD, Xu S, Johns DG, Du Y, Quinn MT, Cayatte AJ, Cohen RA (2001) Role of NADPH oxidase in the vascular hypertrophic and oxidative stress response to angiotensin II in mice. Circ Res 88:947–953

Wang HD, Johns DG, Xu S, Cohen RA (2002) Role of superoxide anion in regulating pressor and vascular hypertrophic response to angiotensin II. Am J Physiol Heart Circ Physiol 282:H1697–H1702

Warnholtz A, Nickenig G, Schulz E, Macharzina R, Brasen JH, Skatchkov M, Heitzer T, Stasch JP, Griendling KK, Harrison DG, Bohm M, Meinertz T, Munzel T (1999) Increased NADH-oxidase-mediated superoxide production in the early stages of atherosclerosis: evidence for involvement of the renin-angiotensin system. Circulation 99:2027–2033

Wedgwood S, Dettman RW, Black SM (2001) ET-1 stimulates pulmonary arterial smooth muscle cell proliferation via induction of reactive oxygen species. Am J Physiol Lung Cell Mol Physiol 281:L1058–L1067

Wingler K, Wunsch S, Kreutz R, Rothermund L, Paul M, Schmidt HH (2001) Upregulation of the vascular NAD(P)H-oxidase isoforms Nox1 and Nox4 by the renin-angiotensin system in vitro and in vivo. Free Radic Biol Med 31:1456–1464

Xia Y, Tsai AL, Berka V, Zweier JL (1998) Superoxide generation from endothelial nitric-oxide synthase. A Ca2+/calmodulin-dependent and tetrahydrobiopterin regulatory process. J Biol Chem 273:25804–25808

Zou M, Yesilkaya A, Ullrich V (1999) Peroxynitrite inactivates prostacyclin synthase by heme-thiolate-catalyzed tyrosine nitration. Drug Metab Rev 31:343–349

Zou MH, Shi C, Cohen RA (2002) Oxidation of the zinc-thiolate complex and uncoupling of endothelial nitric oxide synthase by peroxynitrite. J Clin Invest 109:817–826

Angiotensin II and Atherosclerosis

H. Drexler · B. Schieffer

Abteilung Kardiologie und Angiologie, Medizinische Hochschule Hannover,
Carl Neuberg Strasse 1, 30625 Hannover, Germany
e-mail: Schieffer.Bernhard@MH-Hannover.de

Abstract Atherosclerosis is a chronic inflammatory disease perpetuated by a variety of pro-inflammatory mediators. Fatal endpoints of this inflammatory disease are myocardial infarction, stroke or sudden death, which together remain the major cause of morbidity and mortality in industrialized countries. Accumulated evidence obtained from pathological observations, state-of-the-art imaging and large-scale clinical trials showed that these fatal events are mainly endpoints of pathological vascular remodeling processes, suggesting a close interaction between inflammation and vascular remodeling. In this regard, the renin-angiotensin system (RAS) was—at least partially—accused of contributing to the development and/or progression of atherosclerosis. The Heart Outcome and Prevention Evaluation (HOPE) study convincingly demonstrated that clinical blockade of the RAS may reduce fatal endpoints of severe atherosclerosis, which is characterized by the pathological triad of inflammation, RAS and remodeling: this suggests a role for the RAS in plaque development and stability. Since experimental and clinical evidence point to inflammatory mediators such as interferon-γ, interleukin-1β, interleukin-18 and -10, and interleukin-6 (1) as promotors/predictors of cardiovascular events, we will discuss the clinical basis and the potential molecular and cellular interactions between the RAS and cytokines, and their potential impact on remodeling processes at the atherosclerotic plaques. In particular, we will focus on the interaction of angiotensin II (ANG II) and cytokines such as interleukin 6 (IL-6), and their potential impact on plaque development and stability.

Keywords Angiotensin II · IL-6 · Matrixmetalloproteinases · Plasminogen activator inhibitor 1 · C reactive protein · Atherosclerosis · Acute coronary syndrome · Myocardial infarction · Renin-angiotensin system · Angiotensin converting enzyme

Abbreviations

ANG II	Angiotensin II
MMP	Matrixmetalloproteinases
PAI-1	Plasminogen activator inhibitor 1
C-RP	C reactive protein
RAS	Renin-angiotensin system
ACE	Angiotensin-converting enzyme

1 Introduction: Role of Inflammation in Atherosclerosis

Inflammation is a hallmark of atherosclerosis (Ross 1993, 1999), and atherosclerotic plaques are chronic inflammatory lesions composed of dysfunctional endothelium, smooth muscle cells, lipid-laden macrophages, and T lymphocytes (van der Wall et al. 1994). These lipid-laden activated macrophages and T-lymphocytes stimulate adjacent cells to erode the collagen and elastin framework that forms the plaque's cap (Libby 1995; Davies and Thomas 1985). Mediators of inflammation like interferons, interleukin 1β, interleukin 6 (IL-6) and tumor necrosis factor, are thought to be involved in the development of unstable pla-

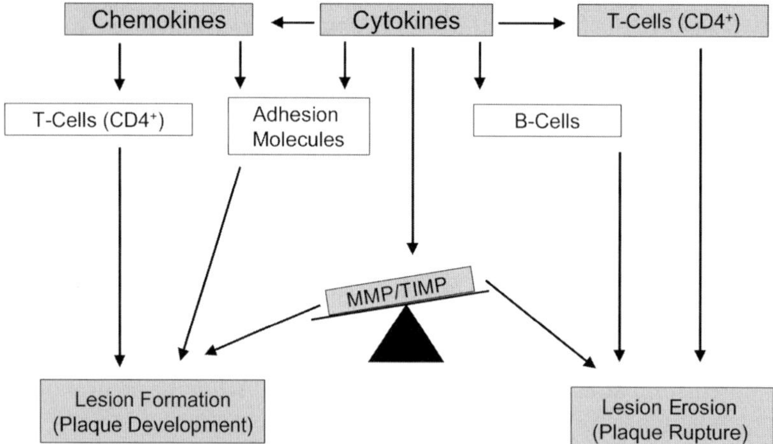

Fig. 1 Schematic showing how cytokines such as IL-6 may contribute to plaque destabilization by enhancing metalloprotease (*MMP*) activity and suppressing their *TIMP* inhibitors (tissue inhibitors of metalloprotease). This latter is mediated by maturation of T and B cells and also involves the expression and presentation of chemokines such as MCP-1 and its CCR2 receptor

ques (Fig. 1). Myocardial infarction is one fatal endpoint of progressive atherosclerosis, and is thought to be a result of pathological remodeling processes (Ross 1999).

2 Renin-Angiotensin-System and Plaque Stability

Blockade of the RAS has been extensively investigated in the treatment of hypertension and heart failure. Recently, it was suggested that blockade of the RAS may also be beneficial in preventing or reducing atherosclerosis and its sequelae, and that drugs that interfere with this system may also be useful in patients without hypertension or heart failure.

The impact of an activated renin-angiotensin system (RAS) on the development of acute myocardial infarction, as a clinical endpoint of plaque instability, was first demonstrated in epidemiological studies in the early 1990s. Alderman and colleagues demonstrated that an elevated renin-sodium profile (a surrogate marker of RAS activation) is associated with a fivefold higher risk of myocardial infarctions as compared to those with lower profiles (Aldermann et al. 1991). In addition, genetic data indicated that the deletion (D/D) polymorphism in the angiotensin-converting enzyme (*ACE*) gene is associated with higher tissue and plasma levels of ACE, whereas the insertion polymorphism *ACE-I/I* genotype is associated with lower ACE levels, and the ACE-I/D level with intermediate levels (Cambien et al. 1992). Cambien and colleagues reported an increased prevalence of ACE-DD genotype in a population of subjects with prior myocardial infarctions, as compared to those without previous infarctions; neither group had other significant CV risk factors (Cambien et al. 1994).

Although the results of these studies remain controversial, some observations indicate that genetically defined polymorphisms of RAS components may be associated with increased risk of myocardial infarction (Danser et al. 1995).

The results of these epidemiological observations stimulated a variety of in vitro studies and clinical studies investigating the potential mechanisms by which ANG II may contribute to and how chronic RAS inhibition may prevent the development of acute coronary syndrome. Earlier studies focused primarily on the role of the RAS in hypertension, but it is now well established that the RAS can affect cardiac and vascular structure and function via numerous complex pathways (Rolland et al. 1993). RAS has traditionally been described as a classic endocrine system, in which renin of renal origin acts on angiotensinogen of hepatic origin to produce angiotensin I in the plasma. Angiotensin I in turn is converted by pulmonary endothelial angiotensin-converting enzyme (ACE) to angiotensin II, considered the primary mediator of physiological actions of the RAS and its detrimental effects in various disease states (Jackson 2001). Recently, the importance of local tissue RASs, of other angiotensin peptides in addition to angiotensin II such as angiotensin III, IV and angiotensin (1–7) (Dzau 2001; Tallant and Ferrario 1999), and of the interaction with other systems such as the endogenous kallikrein-kininogen-kinin system have been established. Since most actions of angiotensin II are exerted through angiotensin type 1 (AT1) receptors, other specific cell surface receptors, including AT2, AT4 and angiotensin (1–7), are involved in the actions of angiotensin peptides and are discussed elsewhere in this book.

The major effects of angiotensin II (Jackson 2001) include vasoconstriction; enhancement of peripheral noradrenergic transmission; increased central nervous sympathetic discharge; release of aldosterone; decreased urinary Na+; increased urinary K+ excretion; migration, proliferation, and hypertrophy of vascular smooth muscle cells; hypertrophy of cardiac myocytes; proliferation and increased synthetic capacity of fibroblasts leading to increased cardiac and vascular extracellular matrix formation; release of thromboxane A2; and increased production of matrix metalloproteinases. Other effects are stimulation of plasminogen activator inhibitor-1 (PAI-1) synthesis, activation of endothelial NADP/NADPH oxidase leading to increased generation of superoxide anion, and activation and release of mediators of inflammation. Important cardiac and vascular consequences of these effects include an increase in blood pressure, changes in preload and afterload, cardiac hypertrophy and remodeling; hypertrophy and remodeling of blood vessels, decreased nitric oxide (NO) activity, increased oxidative stress, endothelial dysfunction, increased oxidation of low-density lipoprotein (LDL) cholesterol, thrombosis, and plaque growth and rupture (Dzau 2001; Jackson 2001).

Some recent experimental studies pointed to the potential interaction of ANG II with inflammatory cells (Ross 1999; Alexander 1994). ANG II was shown to enhance the migration and differentiation of blood-derived monocytes to macrophages, then into atherosclerotic plaques (Hernandez-Presa et al. 1997), a mechanism thought to be essential for both the progression of atherosclerosis and the development of acute coronary syndrome (Ross 1999).

In various animal models, it was conclusively shown that the RAS contributes to plaque development. In fact, infusion of angiotensin II in apoE-deficient mice exacerbated experimental atherosclerosis and blockade of AT1 receptors by losartan blunted atherosclerotic plaque formation in monkeys (Strawn et al. 2000; Weiss et al. 2001). Molecular and cellular mechanisms, however, remained to be determined.

3 Angiotensin II, Superoxide Anions and Inflammation

Reactive oxygen species—generated by the membrane-bound NADPH oxidase system and stimulated by ANG II via the AT1 receptor—seem to be a pivotal step in the cross-link of inflammatory processes that contribute to atherosclerotic plaque formation. Griendling and colleagues first demonstrated that ANG II stimulates the generation of reactive oxygen species in vascular cells and macrophages (Greindling and Alexander 1997; Greindling et al. 1994), which are known activators for cytoplasmic signaling cascades such as the NF-kB, MAP-kinases or JAK/STAT cascade (Chakraborti and Chakraborti 1998) (Fig. 2).

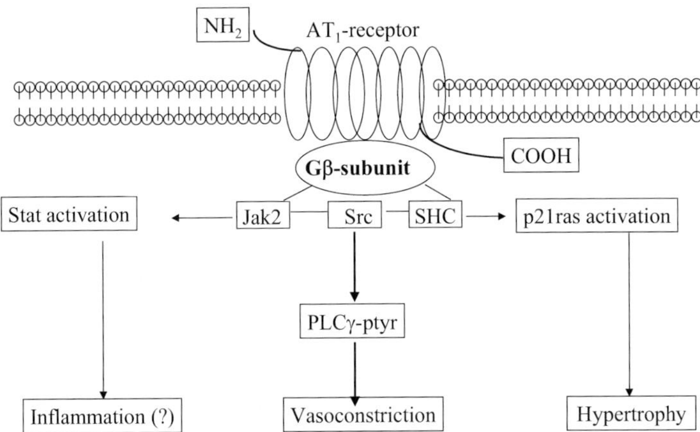

Fig. 2 Model cascade of angiotensin II (ANG II) signal transduction leading to vascular smooth muscle cell hypertrophy, vasoconstriction and inflammation. Upon binding to its G protein-coupled AT1 receptor, ANG II activates via tyrosine phosphorylation second messenger system such as the JAK/STAT cascade, connector proteins such as SHC, and subsequently p21ras

Together, these mechanisms may enhance oxidative stress within the vascular wall and lead to activation of redox-sensitive genes, such as those for pro-inflammatory cytokines (Northermann et al. 1989; Kishimoto et al. 1995). IL-6 transcription was shown to be regulated by a redox-sensitive mechanism (Fuhrman et al. 1994; Han et al. 1999; Schieffer et al. 1998). A pivotal cellular signaling intermediate, involved in ANG II-dependent pro-atherogenic potency, may be oxygen free radicals (Fig. 3).

ANG II is capable of stimulating superoxide anions via the NADH/NADPH oxidase system, which involves a free-radical-forming enzyme formerly exclusively known for leukocytes (Greindling and Alexander 1997; Greindling et al. 1994). Superoxide anions are known activators of signaling systems such as the JAK/STAT or NF-kB system (Schieffer et al. 1998; Pagano et al. 1997). These observations suggest that ANG II may—via redox-sensitive mechanisms—activate IL-6 synthesis and release. Recent observations indicated that pro-inflammatory eicosanoids like leukotriene B4 and thromboxanes are involved in AT1 receptor-dependent NADPH oxidase activation. The latter not only links inflammation with the RAS, but also pro-thrombotic mechanisms. Pro-inflammatory eicosanoids are to elevated in patients with unstable angina and do play a critical role with their vasoconstrictive and mitogenic properties. A major urinary metabolite of thromboxane A_2 synthesized from extrarenal sources is 11-dehydro thromboxane B_2 (TXB2), which is elevated in patients with unstable angina, and a major portion of this metabolite is believed to come from activated platelets. Eikelboom and colleagues (Eikelboom et al. 2002) reported that patients enrolled in the HOPE trial, whose urinary 11-dehydro TXB2 levels were in the highest quartile, had an odds ratio of 2 for developing myocardial infarction, and 3.5 for cardiovascular-related death compared with those subjects in the

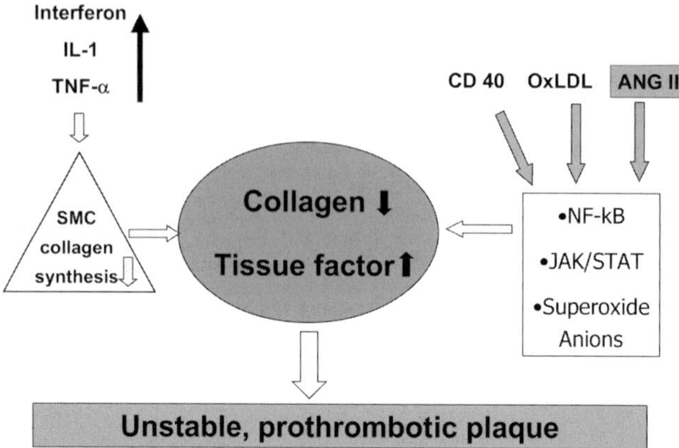

Fig. 3 Interactions and contributions of inflammatory mediators oxidized lipoprotein and ANG II to plaque destabilization. This involves signaling mechanisms such as *NF-kB*, the *JAK/STAT* cascade and superoxide anions. Together, these mediators reduce collagen concentration and smooth muscle cell content in the atherosclerotic lesion and simultaneously enhance the concentration of tissue factor. This molecular and cellular mimicry leads to destabilization of the fibrous cap and promotes atherosclerotic plaque instability

lowest quartile. The deaths occurred in spite of the prophylactic measure of 100 mg of acetylsalicylic acid; the underlying mechanism remained to be determined.

When ANG II-induced effects are blocked by chronic ACE inhibition, macrophage recruitment into the vessel wall was abolished, in both a rabbit model of atherosclerosis and apoE-deficient mice (Hernandez-Presa et al. 1997; Fuhmann et al 1994). Blockade of the AT1 receptor by losartan prevented the accumulation of oxidative reactants in atherosclerotic vessel walls and reduced the extent of atherosclerotic lesion in the apo E-deficient animal model and monkeys (Keidar et al. 1997; Kalra et al. 2002). Thus, the interaction between reactive oxygen species, inflammatory cells and the RAS seem to be important not only for development of acute coronary syndrome, but also for the progression of atherosclerosis (Fig. 4). With regard to development of atherosclerotic lesions, evidence from other animal models, including rodents and primates, showed that ACE inhibition may reduce the extent of vascular lesions (Chobanian et al. 1990; Hayek et al. 1995; Aberg and Ferrer 1990). Additional mechanisms by which the RAS, via ANG II, may encourage the development of atherosclerosis and involve the activation of thrombosis pathways via PAI-1 (Ridker et al. 1993) and the stimulation of pro-inflammatory cytokines. PAI-1 serum levels in formerly healthy volunteers of the Physician Health Study were shown to be elevated and associated with a higher risk of myocardial infarction (Ridker and Vaughan 1995).

Recent evidence suggests that tumor necrosis factor (TNFα) and interleukin-1β are expressed in the shoulder region of atherosclerotic plaques (Mach et al. 1997; Fiotti et al. 1999). Both cytokines are known stimulators of extracellular matrix degrading metalloproteinases, e.g., MMP-1, MMP-2,MMP-3 and MMP-9

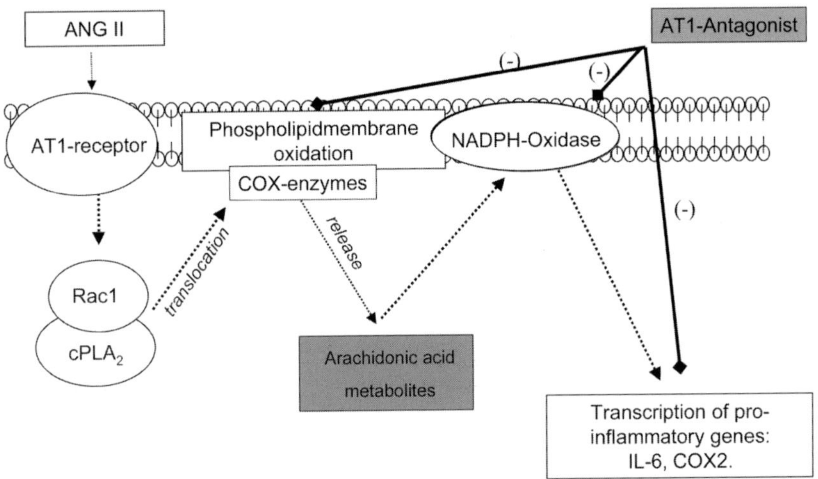

Fig. 4 Scheme of angiotensin II-induced cytokine synthesis and release. Via AT1 receptor activation, ANG II induces the release of arachidonic acid from the plasma membrane, which in turn activates NADPH oxidase via arachidonic acid metabolites. Superoxide anions generated by NADPH oxidase will subsequently stimulate synthesis and release of pro-inflammatory cytokines such as IL-6 and COX2. This action can be blunted by AT1 receptor antagonists

(for summary see Rabbani and Topol 1999). However, their potential interaction with neurohumoral systems such as the RAS and their overall impact on the progression of atherosclerosis or development of acute coronary syndrome remains to be determined. Unpublished experiments from our laboratory demonstrated that ANG II does not stimulate TNFα or Interleukin 1β expression in human coronary vascular smooth muscle cells or macrophages. However, ANG II is capable of inducing TNFα in myocytes (Kalra et al. 2002).

Serum levels of interleukin 6 are elevated in patients with unstable angina (Biassuci et al. 1996; Marx et al. 1997) and have been implicated in the onset of acute coronary syndrome (Marx et al. 1997). Similar relationships have not been documented for other cytokines such as TNFα or IL1-β. Thus, from a clinical point of view, the pathophysiological role of IL-6 is of particular interest. The question arises of whether IL-6 is only a marker or in fact a mediator.

IL-6 is involved in a variety of physiological functions, including the stimulation of acute phase protein synthesis (α2-macroglobulin and C-RP leading to the activation of complement cascades), the induction of pro-thrombotic factors (e.g., PAI-1) and the stimulation of matrix degrading enzymes, such as MMP-1 and MMP-9 (Kishimoto et al. 1995; Solis-Herruzo et al. 1996).

A recent report by Rekhter and co-workers demonstrated that the delicate balance of plaque stability is controlled, on one hand, by intrinsic properties of the tissue and, on the other hand, by external forces to which the plaque is subjected (Rekhter et al. 2000). Based on this evidence, increased lipid content and decreased collagen content have been long suspected to decrease the mechanical

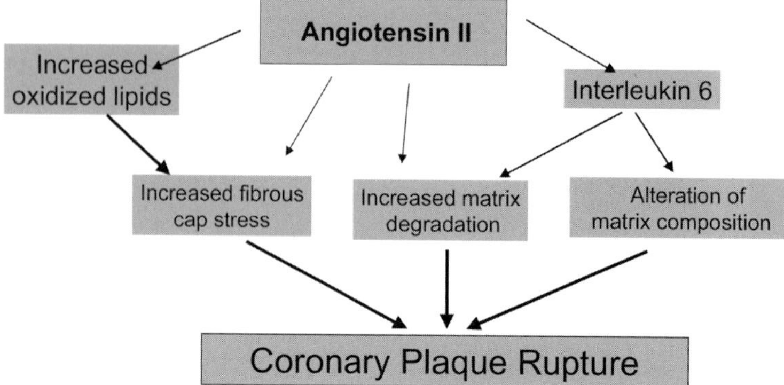

Fig. 5 Model cascade leading to coronary plaque rupture—role of angiotensin II. Summary of potential intermediates activated by angiotensin II via interleukin 6. ANG II induces IL-6 independently of lipid peroxidation and vasoconstriction, most probably via superoxide anions. In contrast, MMP expression and activation is induced in an IL-6-dependent manner in vitro. Together, these factors initiate the process of plaque destabilization and may contribute to the development of an acute coronary syndrome

strength of plaques (Fig. 5) (Richardson et al. 1989; Davies et al. 1994). Thus, potential plaque-rupturing factors may include a powerful vasoconstrictor, e.g., angiotensin II, responsible for an increase in local circumferential stress, and pro-inflammatory cytokines such as IL-6, stimulating extracellular matrix decomposition. Rekhter et al. demonstrated that collagen loss was focal a feature that may have important consequences for formation of weak spots in plaques, (Rekhter et al. 2000) which are prone to rupture: vulnerable plaques [36]. Collagen loss was also found to occur predominantly in areas rich in macrophage-derived foam cells (CD68 positive cells) (Galis et al. 1994); this is in agreement with previous observations that human atheroma tends to rupture in areas with inflammatory infiltrates (Gamos et am/ 1994; van der Wal et al. 1994; Shah et al. 1995).

In this regard, Diet and co-workers first demonstrated that the ANG II-forming protease ACE is expressed in human atherosclerotic plaques (Diet et al. 1996). The authors demonstrated that in early- and intermediate-stage atherosclerotic lesions, ACE was predominantly expressed in lipid-laden macrophages (similarly to pro-inflammatory cytokines), whereas in advanced lesions, ACE was predominantly localized throughout the microvasculature of plaques (Diet et al. 1996). Potter and co-workers further demonstrated that lipid-laden macrophages contain ANG II in a primate model of atherosclerosis (Potter et al. 1998). In humans, at least two major enzymes—ACE and chymase—are involved in the conversion of ANG-I to ANG-II, and may contribute to ANG II formation in coronary arteries. Further investigations, in normal and atheromatous coronary arterial tissue from patients dying of noncardiovascular diseases, demonstrated that only ACE but not chymase was colocalized with ANG-II in the intima of stable arthrosclerotic plaques (Ohishi et al. 1999). These findings suggest that ACE is likely the primary source of ANG II in atherosclerotic human coronary arteries (Shieffer et al. 2000). Interestingly, Hoshida and colleagues demonstrated that tissue ACE activity is selectively up-regulated in patients with acute coronary syndromes, but serum ACE activity is not Hoshida et al. 2001). These observations suggest that tissue ACE activity may be an important regulator of ANG II formation at atherosclerotic lesions (Dzau 2001).

Recent evidence suggests that IL-6 is increased in patients with acute coronary syndrome (Biassuci et al. 1996). In this regard, our laboratory demonstrated that ANG II and ACE are expressed predominantly in areas of clustered macrophages, in atherosclerotic coronary segments (Diet et al. 1996; Potter et al. 1998; Ohishi et al. 1999). Since both ANG II and IL-6 may interact and also may be involved in the development of acute coronary syndrome, we investigated the in vitro and in vivo interactions between ANG II and IL-6. Our results demonstrated that ANG II induces the synthesis and release of IL-6 in smooth muscle cells and human macrophages (Schieffer et al. 2000), and immunohistological analysis further showed that ANG II, the AT1 receptor, and ACE are expressed at strategically relevant sites of human coronary atherosclerotic plaques, that is, at the shoulder of macrophage-rich atherosclerotic plaques. Furthermore, ANG II was detected in close proximity to potential plaque rupture sites in coronary artery sections from patients who died from acute myocardial infarctions. Colocalization of components of the RAS with IL-6 was observed in stable coronary plaques and atherectomy tissues (Schieffer et al. 2000). Together, these findings suggest that the RAS may contribute to inflammatory processes within the arthrosclerotic vascular wall, and thereby may contribute to the development of acute coronary syndrome (Schieffer et al. 2000).

However, is the in vitro interaction and the in vivo colocalization of ANG II and IL-6 of clinical importance for the stability of atherosclerotic lesion? As a mediator of inflammation, IL-6 stimulates a variety of intracellular signaling mechanisms, including the traditional cytokine signaling cascade of JAK-kinases and STAT transcription factors (signal transducers and activators of transcription) (Kishimoto et al. 1995) (Fig. 1). Via this signaling cascade, IL-6 mediates its physiological functions, including macrophage differentiation, B-cell maturation, acute phase protein and smooth muscle cell proliferation. As indicated above, cytokine-stimulated smooth muscle cells synthesize and release enzymes responsible for extracellular matrix degradation, which may potentially destabilize the fibrous caps of plaques (Rabbani and Topol 1999) and may play an important role for pro-inflammatory cytokines (e.g., IL-6) in the development of acute coronary syndrome. Finally, IL-6 regulates expressions of adhesion molecules and other cytokines, (e.g., IL-1β and TNFα), which together may enhance an inflammatory reaction at atherosclerotic plaques.

Since both ANG II and IL-6 activate identical cellular signaling events, we investigated whether ANG II induces the release of PAI-1 and C-RP via activation of the JAK/STAT cascade. Preliminary results demonstrated that PAI-1 and C-RP were released when smooth muscle cells were stimulated with ANG II (Han et al. 1999). Preliminary results from ANG II-stimulated, IL-6-deficient murine smooth muscle cells revealed that depletion of IL-6 blunted ANG II-induced metalloprotease expression and activity. Based on these findings, we suggest a model in which ANG II, via IL-6, may promote the development of acute coronary syndrome via induction of pro-thrombotic, complement activation and actions of matrix degrading factors (e.g., PAI-1, CRP, metalloproteases). Accumulation of these factors contribute to the subsequent pro-atherogenic potency of ANG II (see Fig. 4).

Thus it is suggested that IL-6 induction by ANG II is redox-sensitive and involves the aforementioned signaling cascades. In-fact, recent observations reported that blockade of superoxide anion generation by ANG II abolished ANG II-induced IL-6 release in vitro [47]. In summary, generation of superoxide anions via the AT1 receptor seems to be pivotal for the pro-inflammatory and pro-atherogenic properties of ANG II.

4
Impact of Angiotensin II Inhibition in Patients with Coronary Artery Disease

The most conclusive evidence for the beneficial effects of angiotensin II (ANG II) inhibition in patients with atherosclerotic vascular disease, is demonstrated by large morbidity and mortality trials. Initial trials consisted of patients with reduced left ventricular ejection fraction (LVEF), with or without heart failure, and demonstrated that long-term therapy with ACE inhibitors reduced the risk for MI and cardiovascular death (Pfeffer et al. 1992; Yusuf et al. 1992; Rutherford et al. 1994; Kober et al. 1992; AIRE Study Investigators 1993; Flather et al. 2000). More recently, the HOPE trial (Yusuf et al. 2000) demonstrated a similar effect in patients with cardiovascular disease, but with preserved LVEF. This randomized trial involved over 9,000 patients with a mean age of 55 years, presenting with CAD, prior stroke, peripheral arterial disease, or diabetes. These patients were then randomly assigned to either 10 mg/d of ramipril or placebo. After 4.5 years, the clinical study reported a statistically significant reduction in the composite primary endpoint of MI, stroke, or death from cardiovascular causes. In addition, the risk of stroke, MI, revascularization procedures, heart failure, and new diabetes was reduced. These beneficial effects of treatment were attained with only a modest reduction in blood pressure (a 3-mmHg reduction in systolic and a 1.8-mmHg reduction in diastolic blood pressure).

The only ACE inhibitor trial that failed to demonstrate a treatment benefit is the QUIET trial (Pitt et al. 2001)), which randomized 1,750 patients with CAD to treatment with 20 mg/d of quinapril or placebo. After a mean of 27 months, there were no significant differences in major ischemic events between the treatment and control groups. However, this study had several important limitations such as the lack of statistical reliability due to small sample size and the enrollment of low-risk study subjects (i.e., with a low incidence of major ischemic events); the quinapril dosage; and the choice of a composite primary outcome, which included so-called soft events.

The long-term effects of ACE inhibitor therapy on MI risk, in large randomized trials, support a beneficial action of these agents in atherosclerosis. The Perindopril Protection Against Recurrent Stroke Study (PROGRESS) (PROGRESS Collaborative Group 2001), conducted in 6,105 patients with prior stroke or transient ischemic attack, reported a 28% reduction in recurrent stroke and a 26% reduction in major ischemic events after 5 years of treatment with an ACE inhibitor-based regimen (4 mg/d of perindopril with or without the addition of the diuretic indapamide). This beneficial effect of treatment is at least in part due to blood pressure lowering, but also to the antiatherosclerotic actions of ACE inhibitors.

These large morbidity and mortality trials clearly support an important role for ACE inhibitors in the treatment of atherosclerotic vascular diseases, although several questions do remain. Thus, it is uncertain whether the benefits seen with ramipril in the HOPE trial extend to other ACE inhibitors and to low-risk patients with CAD and with preserved LV function (e.g., younger individuals with less extensive disease than the HOPE study participants and those treated aggressively with lipid-lowering agents). Also whether ACE inhibitors are superior to other drugs in preventing atherosclerotic complications in hypertension remains in question.

Some of these questions will be answered by ongoing clinical trails. The Prevention of Events with Angiotensin-Converting Enzyme (PEACE) (Pfeffer et al. 2001) trial with over 8,000 participants and the European Trial of Reduction of Cardiac Events with Perindopril in Stable Coronary Artery Disease (EUROPA) (Davies et al. 1994) with over 10,000 patients, are evaluating trandolapril and perindopril in CAD with preserved LV function. The Ischemic Management with Accupril Post-Bypass Graft via Inhibition of Converting Enzyme (IMAGINE) trial, which enrolled over 2,000 patients, is testing the effect of early quinapril treatment, post-bypass graft surgery. The Antihypertensive and Lipid-Lowering Treatment to Prevent Heart Attack Trial (ALLHAT), which has completed the randomization of over 40,000 study participants, is comparing different blood pressure-lowering drugs, including lisinopril (Pepine 1996).

5 AT1 Blocker or ACE Inhibitors in Patients with Coronary Artery Disease?

The angiotensin II receptor blockers (AT1 antagonists) available for clinical use all bind to the AT1 receptor with high affinity. AT1 antagonists reduce the activity of AT1 receptor-mediated actions of angiotensin II more effectively than do ACE inhibitors. ACE inhibitors reduce biosynthesis of angiotensin II produced by ACE action on angiotensin I, but do not inhibit alternate non-ACE angiotensin II-generating pathways such as those of chymase, cathepsin G, or tonin. In contrast to ACE inhibitors, AT1-antagonists indirectly activate AT2 receptors, although the importance of AT2-mediated effects is not clearly defined. Recent evidence suggests that AT2 receptors may exert antiproliferative, pro-apoptotic, and vasodilatory actions and may also have a modest effect on promoting bradykinin release. ACE inhibitors more effectively increase angiotensin (1–7) levels more than AT1-antagonists, and this may result in additional beneficial cardiac and vascular effects. Lastly, ACE inhibitors increase the levels of various ACE substrates that are not angiotensin peptides, including bradykinin. Increased bradykinin levels may also contribute to the beneficial cardiovascular effects of ACE inhibitors. Whether or not these distinct pharmacological differences between AT1 antagonists and ACE inhibitors result in significant differences in therapeutic outcomes is unknown at present.

In vivo and in vitro studies demonstrate that AT1 antagonists inhibit most of the biological actions of angiotensin II, leading to decreased vascular smooth muscle cell contraction, decreased pressor responses, improved renal blood flow, and decreased cellular hypertrophy and hyperplasia (Sun et al. 2001; Ferrario 2002). Animal models treated with AT1 antagonists demonstrated decreased cardiac and arterial medial hypertrophy. Reduced extent of atherosclerotic lesions, in response to AT1 receptor blockade treatment, was also observed in several animal models of atherosclerosis (Hayek et al. 1995; Aberg and Ferrer 1990; Strawn et al. 2000).

The clinical effects of AT1 antagonists on vascular functions and structures in normotensive individuals has been evaluated in several trials. The Brachial Artery Normalization of Forearm Function (BANFF) study compared the effects of daily administration of 20 mg of quinapril, 10 mg of enalapril, 50 mg of losartan, or 5 mg of amlodipine on endothelial function, as assessed by brachial artery flow-mediated vasodilation in 80 patients with CAD. After 2 weeks of treatment, losartan had no significant effect, whereas quinapril improved endothelial function, and this effect was related to the presence of an insertion allele of the ACE genotype (Chakraborti and Chakraborti 1998). In addition, our group (Hornig et al. 2001) studied 35 patients with CAD and reported that 4 weeks of treatment with 100 mg/d of losartan had a similar beneficial effect on endothelial function as 10 mg/d of ramipril. The different findings from these two studies may be related to the higher losartan dose, the length of therapy, and the assessment of endothelium-mediated vasodilation in the radial artery (as opposed to the more sensitive method of measuring brachial artery dilation). The Losartan Intervention for Endpoint reduction in hypertension study (LIFE) (Hornig et al. 2001; Dahlöf et al. 2002) randomized over 9,000 patients (55–80 years of age) with moderate to severe hypertension and left ventricular hypertrophy to either losartan or atenolol treatment. Mean follow-up time was 4.8 years. Even though the blood pressure lowering was similar in both groups (a 30-mmHg decrease in systolic and a 17-mmHg decrease in diastolic blood pressure), the composite endpoint of death, MI or stroke was significantly lower in the losartan group. The greatest benefit was the 25% reduction in the incidence of stroke. Trends toward fewer all-cause deaths, cardiovascular deaths, and infarctions were observed in the losartan group; however, these observed differences were statistically insignificant. No significant differences were noted in the incidences of angina, heart failure, revascularization, and resuscitated cardiac arrest.

Whether AT1 receptor blockade is as effective as ACE blockade in patients with MI or chronic atherosclerotic vascular disease was studied in the OPTIMAAL trial (Dickstein and Kjekshus 2002). With over 5,000 patients enrolled in the study, the results showed that 50 mg of losartan had no beneficial effect with respect to mortality and clinical events, as compared to 150 mg of captopril. The investigators therefore suggest that at the moment, first-line treatment of patients with documented atherosclerosis should be ACE inhibitors. However, ongoing large-scale clinical trials will compare AT1 antagonists against ACE inhibitors, and evaluate AT1 antagonists versus placebo in ACE-intolerant patients. The combination therapy of AT1 receptor antagonist and ACE inhibitor will be investigated. Combined therapy is theoretically appealing and was shown, in the Randomized Evaluation Of strategies for Left Ventricular Dysfunction (RESOLVD) study (McKelvie et al. 1999) to be more beneficial than either single treatment in improving the neurohormonal balance and left ventricular remodeling. The Valsartan in Acute MI (VALIANT) trial (Cohn and Tognoni 2001) was designed to randomize over 14,000 patients with heart failure and/or LVEF with less than 40% within 10 days after MI to treatment with valsartan alone or in combination with captopril (Cohn and Tognoni 2001). The Ongoing Telmisartan Alone and the Ramipril Global Endpoint Trial (ONTARGET) (Yusuf 2002) will enroll approximately 23,000 patients 55 years or older with a history of CAD, stroke, or peripheral arterial disease: however, subjects will be without heart failure or known low LVEF. The Telmisartan Randomized Assessment Study in ACE Inhibitor-Intolerant Patients with Cardiovascular Disease (TRANSCEND) trial (Yusuf 2002) will compare telmisartan to placebo in 5,000 patients who cannot tolerate ACE inhibitors. The ONTARGET and TRANSCEND trials are designed to test the effectiveness of AT1-antagonists and/or combined AT1-/ACE-inhibitor treatment in reducing atherosclerotic events. Based on these studies, the current recommended action for first-line treatment of patients with documented atherosclerosis remains ACE inhibitors.

6 Conclusion

Components of the RAS are expressed in the shoulder region of coronary atherosclerotic plaques, areas prone to increased risk of plaque rupture. In vitro observations suggest a close interaction between ANG II and IL-6, which involves the generation of pro-inflammatory eicosanoids and superoxide anions by the NADPH oxidase system. This interaction of ANG II and IL-6 may play a pivotal role in vessel wall inflammation, thrombosis and remodeling, by stimulating acute phase proteins, pro-thrombotic factors such as PAI-1 and metalloproteases. Thus, the induction of IL-6 by ANG II may be *one* important mechanism by which ANG II could be involved in the development of atherosclerosis and acute coronary syndrome.

Acknowledgements. Studies summarized in this manuscript were supported by DFG-Sonderforschungsbereich 566/B9 as well as the Leduque Foundation.

References

Aberg G, Ferrer P (1990) Effects of Captopril on atherosclerosis in cynomolgus monkeys. J Cardiovasc Pharmacol 15 [Suppl I]:S65–S72

Acute Infarction Ramipril Efficacy (AIRE) Study Investigators (1993) Effect of ramipril on mortality and morbidity of survivors of acute myocardial infarction with clinical evidence of heart failure. Lancet 342:821–828

Aldermann MH, Madhavan S, Ooi WL, Cohen H, Sealy JE, Laragh JH (1991) Association of the renin-sodium profile with the risk of myocardial infarction in patients with hypertension. N Engl J Med 324:1098–1104

Alexander RW (1994) Inflammation and coronary artery disease. N Engl J Med. 331: 468–469

Biassuci L, Vitelli A, Liuzzo G, Altamura S, Caliguri G, Monaco C, Rebuzzi A, Ciliberto G, Maseri A (1996) Elevated levels of interleukin-6 in unstable angina. Circulation 94:874–877

Cambien F, Poirier O, Lecerf L, Evans A, Cambou JP, Arvelier D, Luc G, Bard JM, Bara R, Richard S, Tiret L, Amouyel P, Alhenc-Gelas F, Sourbrier F (1992) Deletion polymorphism in the gene for angiotensin-converting enzyme is a potent risk factor for myocardial infarction. Nature 359:641–644

Cambien F, Costerousse O, Tirret L, Poirier O, Lecerf L, Gonzales MF, Evans A, Arveilier D, Cambou JP, Luc G, Rakotovao R, Ducimetiere P, Sourbrier F, Alhenc-Gelas F (1994) Plasma levels and gene polymorphism of angiotensin converting enzyme in relation to myocardial infarction. Circulation 90:669–676

Cashin-Hemphill L, Holmvang G, Chan RC et al (1999) Angiotensin-converting enzyme inhibition as antiatherosclerotic therapy: no answer yet. Quinapril Ischemic Event Trial. Am J Cardiol 83:43–47

Chakraborti S, Chakraborti T (1998) Oxidant-mediated activation of mitogen-activated protein kinases and nuclear transcription factors in the cardiovascular system: a brief overview. Cell Signal 10:675–683

Chobanian AV, Haudenschild CC, Nickerson C, Drago R (1990) Antiatherogenic effect of captopril in Watanabe heritable hyperlipidemic rabbits. Hypertension 15:327–331

Cohn JN, Tognoni G, for the Valsartan Heart Failure Trial Investigators (2001) A randomised trial of the angiotensin-receptor blocker valsartan in chronic heart failure. N Engl J Med 345:1667–1675

Dahlöf B, Devereux RB, Kjeldsen SE et al (2002) Cardiovascular morbidity and in the Losartan Intervention For Endpoint reduction in hypertension study (LIFE): a randomised trial against atenolol. Lancet 359:995–1003

Danser AH, Schalekamp MA, Bax WA, van den Brink AM, Saxena PR, Riegger GA, Schunkert H (1995) Angiotensin-converting enzyme in the human heart. Effect of the deletion/insertion polymorphism. Circulation 92:1387–1388

Davies MJ, Thomas AC (1985) Plaque fissuring: the cause of acute myocardial infarction, sudden ischemic death, and crescendo angina. Br Heart J 53:363–373

Davies MJ, Woolf N, Rowles P, Richardson PD (1994) Lipid and cellular constituents of unstable human aortic plaques. Basic Res Cardiol 89:33–39

Dickstein K, Kjekshus K for the OPTIMAAL Study Group (2002) Effects of losartan and captopril on mortality and morbidity in high-risk patients after acute myocardial infarction: the OPTIMAAL randomised trial. Lancet 360:752–760

Diet F, Pratt R, Berry GJ, Momose N, Gibbons G, Dzau VJ (1996) Increased accumulation of tissue ACE in human atherosclerotic coronary artery disease. Circulation 94:2576–2767

Dzau VJ (2001) Tissue angiotensin and pathobiology of vascular disease—a unifying hypothesis. Hypertension 37:1047–1052

Dzau VJ, Bernstein K, Celermeier D et al (2001) The relevance of tissue angiotensin-converting enzyme: manifestations in mechanistic and endpoint data. Am J Cardiol 88 [Suppl]:1L–20L

Eikelboom JW, Hirsh J, Weitz JI et al (2002) Aspirin-resistant thromboxane biosynthesis and the risk of myocardial infarction, stroke, or cardiovascular death in patients at high risk for cardiovascular events. Circulation 105:1650–1655

Ferrario CM (2002) Use of angiotensin II receptor blockers in animal models of atherosclerosis. Am J Hypertens 15:9S–13S

Fiotti N, Giansante C, Ponte E, Delbello C, Calabrese S, Zacchi T, Dobrina A, Guarnieri G (1999) Atherosclerosis and inflammation. Patterns of cytokine regulation in patients with peripheral arterial disease. Atherosclerosis 145:51–60

Flather MC, Yusuf S, Køber L et al (2000) Long-term ACE-inhibitor therapy in patients with heart failure or left-ventricular dysfunction: a systematic overview of data from individual patients. Lancet 355:1575–1581

Fuhrman B, Oiknine J, Aviram M (1994) Iron induces lipid peroxidation in cultured macrophages, increases their ability to oxidatively modify LDL and affect their secretory properties. Atherosclerosis 111:65–78

Galis ZS (1999) Metalloproteases in remodeling of vascular extracellular matrix. Fibrin Proteol 13:54–63

Galis ZS, Sukhova GK, Lark MW, Libby P (1994) Increased expression of matrix metalloproteinases and matrix degrading activity in vulnerable regions of human atherosclerotic plaques. J Clin Invest 94:2493–2503

Griendling KK, Alexander RW (1997) Oxidative stress and cardiovascular disease. Circulation 96:3264–3265

Griendling KK, Minieri CA, Ollerenshaw JD, Alexander RW (1994) Angiotensin II stimulates NADH and NADPH oxidase activity in cultured vascular smooth muscle cells. Circ Res 74:1141–1148

Han Y, Runge M, Brasier A (1999) Angiotensin II induces interleukin-6 transcription in vascular smooth muscle cells through pleiotropic activation of nuclear factor-kB transcription factor. Circ Res 84:695–703

Hayek T, Keidar S, Mei-Yi, Oikine J, Breslow J (1995) Effect of angiotensin converting enzyme inhibitors on LDL lipid peroxidation and atherosclerosis progression in apoE deficient mice. Circulation 92 [Suppl I]:I–625

Hernandez-Presa M, Bustos C, Ortega M, Tunon J, Renedo G, Ruiz-Ortega M, Egidio J (1997) Angiotensin-converting enzyme Inhibition prevents arterial nuclear factor-kB activation, monocyte chemoattactant proetein-1 expression and macrophage infiltration in a rabbit model of early accelerated atherosclerosis. Circulation 95:1532–1541

Hornig B, Landmesser U, Kohler C et al (2001) Comparative effect of ACE inhibition and angiotensin II type 1 receptor antagonism on bioavailability of nitric oxide in patients with coronary artery disease: role of superoxide dismutase. Circulation 103:799–805

Hoshida S, Nishida M, Yamashita N et al (1997) Vascular angiotensin-converting enzyme activity in cholesterol-fed rabbits: effects of enalapril. Atherosclerosis 130:53–59

Hoshida S, Kato J, Nishino M, Egami Y, Takeda T, Kawabata M, Tanouchi J, Yamada Y, Kamada T (2001) Increased angiotensin-converting enzyme activity in coronary artery specimens from patients with acute coronary syndrome. Circulation103:630–633

Jackson EK (2001) Renin and angiotensin. In: Hardman JG, Limbird LE, Gilman AG (eds) Goodman & Gilman's the pharmacological basis of therapeutics. McGraw-Hill, New York

Kalra D, Sivusabvasubramanian N, Mann DL (2002) Angiotensin II induces tumor necrosis factor biosynthesis in the adult mammalian heart through a protein kinase C-dependent pathway. Circulation 105:2198–2205

Keidar S, Attias J, Smith J, Breslow JL, Hayek T (1997) The angiotensin II receptor antagonist, losartan, inhibits LDL lipid peroxidation and atherosclerosis in apolipoprotein E deficient mice. Biochem Biophys Res Commun 236:622–625

Kishimoto T, Akira S, Narazaki M, Taga T (1995) IL-6 family of cytokines and gp130. Blood 86:1243–1254

Kober L, Torp-Pedersen C, Carlsen JE for the Trandolapril Cardiac Evaluation (TRACE) Study Group (1992) A clinical trial of the angiotensin-converting-enzyme inhibitor trandolapril in patients with left ventricular dysfunction after myocardial infarction. N Engl J Med 333:1670–1676

Libby P (1995) Molecular bases of the acute coronary syndrome. Circulation 21:2844–2850

Lindholm LH, Ibsen H, Dahlöf B et al (2002) Cardiovascular morbidity and mortality in patients with diabetes in the Losartan Intervention For Endpoint reduction in hypertension study (LIFE): a randomised trial against atenolol. Lancet 359:1004–1010

Lonn E, Yusuf S, Dzavik V, for the SECURE Investigators (2001) Effects of ramipril and of vitamin E on atherosclerosis: results of the prospective randomized Study to Evaluate Carotid Ultrasound changes in patients treated with Ramipril and vitamin E (SECURE). Circulation 103:919–925

Luchtefeld M., Drexler H, Schieffer B (2003) 5-Lipoxygenase is involved in the angiotensin II-induced NAD(P)H-oxidase activation. Biochem Biophys Res Commun 308:668–672

Mach F, Schonbeck U, Sukhova GK, Bourcier T, Bonnefoy JY, Pober JS, Libby P (1997) Functional CD40 ligand is expressed on human vascular endothelial cells, smooth muscle cells, and macrophages: implications for CD40-CD40 ligand signaling in atherosclerosis. Proc Natl Acad Sci 94:1931–1936

Marx N, Neumann FJ, Ott I, Gawaz M, Koch W, Pikau T, Schömig A (1997) Induction of cytokine expression in leucocytes in acute myocardial infarction. J Am Coll Cardiol 30:165–170

MacMahon S, Sharpe N, Gamble G et al.(2000) Randomized, placebo-controlled trial of the angiotensin converting enzyme inhibitor, ramipril, in patients with coronary or occlusive arterial disease. J Am Coll Cardiol 36:438–443

McKelvie RS, Yusuf S, Pericak D et al (1999) Comparison of candesartan, enalapril, and their combination in congestive heart failure: randomized evaluation of strategies for left ventricular dysfunction (RESOLVD) pilot study. Circulation 100:1056–1064

Northermann W, Braciak TA, Hattori M, Lee F, Fey GH (1989) Structure of the rat IL-6 gene and its expression in macrophage-derived cells. J Biol Chem 264:16072–16082

Ohishi M, Ueda M, Rakugi H, Naruko T, Kolima A, Okamura A, Higaki J, Ogihara T (1999) Relative localization of angiotensin-converting enzyme, chymase and angiotensin II in human coronary atherosclerotic lesion. J Hypertens 17:547–553

Pagano PJ, Clark JK, Cifuentes-Pagano ME, Clark SM, Callis GM, Quinn MT (1997) Localization of a constitutively active, phagocyte-like NADPH oxidase in rabbit aortic adventitia: enhancement by angiotensin II. Proc Natl Acad Sci U S A 94:14483–14488

Pepine CJ (1996) Ongoing clinical trials of angiotensin-converting enzyme inhibitors for treatment of coronary artery disease in patients with preserved left ventricular function. J Am Coll Cardiol 27:1048–1052

Pfeffer M, Braunwald E, Moye L, Basta L, Brown EJ, Cuddy TE, Davis BR, Geltmann EM, Goldman S, Flaker GC, Klein M, Lamas G, Packer M, Rouleau J, Rutherford J, Wertheimer JH, Hawkins CM (1992) Effect of captopril on mortality and morbidity in patients with left ventricular dysfunction after myocardial infarction: results of the survival and ventricular enlargement trial. N Engl J Med 327:669–677

Pfeffer MA, Domanski M, Verter J et al (2001) The continuation of the Prevention of Events With Angiotensin Converting Enzyme Inhibition (PEACE) Trial. Am Heart J 142:375–377

Pitt B, O'Neill B, Feldman R et al. (2001) The Quinapril Ischemic Event Trial (QUIET): evaluation of chronic ACE inhibitor therapy in patients with ischemic heart disease and preserved left ventricular function. Am J Cardiol 87:1058–1063

Potter DD, Sobey CG, Tompkins PK, Rossen JD and Heistad DD (1998) Evidence that macrophages in atherosclerotic lesions contain angiotensin II. Circulation 98:800–807

Powell JS, Clozel JP, Muller RK et al (1989) Inhibitors of angiotensin-converting enzyme prevent myointimal proliferation after vascular injury. Science 245:186–188

PROGRESS Collaborative Group (2001) Randomised trial of a perindopril-based blood-pressure-lowering regimen among 6105 individuals with previous stroke or transient ischaemic attack. Lancet 358:1033–1041

Rabbani R, Topol EJ (1999) Strategies to achieve coronary artery plaque stabilization. Cardiovasc Res 41:401–417

Rekhter MD, Hicks GW, Brammer DW, Hallak H, Kindt E, Chen J, Rosebury WS, Anderson MK, Kuipers PJ, Ryan MJ (2000) Hypercholesterolemia causes mechanical weakening of rabbit atheroma: local collagen loss as a prerequisite of plaque rupture. Circ Res 86:101–108

Richardson PD, Davies MJ, Born GV (1989) Influence of plaque configuration and stress distribution on fissuring of coronary atherosclerotic plaques. Lancet 2:941–944

Ridker PM, Vaughan DE (1995) Hemostatic factors and the risk of myocardial infarction. N Engl J Med 10;333:389–390

Ridker PM, Gaboury CJL, Conlin PR, Seely EW, Williams GH, Vaughan DE (1993) Stimulation of plasminogen activator inhibitor in vivo by infusion of angiotensin II: evidence of a potential interaction between the renin-angiotensin system and fibrinolytic function. Circulation 87:1969–1973

Rolland PH, Carpiot P, Friggi A et al (1993) Effects of angiotensin-converting enzyme inhibition with perindopril on hemodynamics, arterial structure and wall rheology in the hindquarters of atherosclerotic mini-pigs. Am J Cardiol 71:22E–27E

Ross R (1993) Pathogenesis of atherosclerosis: a perspective for the 1990s. Nature 362:801–809

Ross R (1999) Atherosclerosis–an inflammatory disease. N Engl J Med 340:115–126

Rutherford JD, Pfeffer MA, Moye LA et al (1994) Effects of captopril on ischemic events after myocardial infarction. Results of the Survival and Ventricular Enlargement trial. SAVE Investigators. Circulation 90:1731–1738

Schieffer B, Kowert A, Brauer N, Wengler A, Schieffer E, Gutzke S, Harringer W, Haverich A, Drexler H (1998) Irbesartan inhibits angiotensin II-induced pro-inflammatory and prothrombotic factors in human coronary arteries: role of the Jak/STAT cascade. Eur Heart J [Suppl II]:322

Schieffer B, Luchtefeld M, Hilfiker A, Drexler H (2000) Activation of the JAK/STAT cascade by angiotensin II depends on the NAD(P)H oxidase. Circ Res 87:1195–1201

Schieffer B, Schieffer E, Hilfiker-Kleiner D, Hilfiker A, Kovanen PT, Kaartinen M, Nussberger J, Harringer W, Drexler H (2000) Expression of ANGIOTENSIN II and Interleukin 6 in human coronary atherosclerotic plaques: Potential implications for inflammation and plaque instability. Circulation 101:1372–1378

Schjiffrin EL, Bae Park J, Intengan HD et al (2000) Correction of arterial structure and endothelial dysfunction in human essential hypertension by the angiotensin receptor antagonist losartan. Circulation 101:1653–1659

Shah PK, Falk E, Badimon JJ, Fernandez-Ortiz A, Mailhac A, Villareal-Levy G, Fallon JT, Regnstrom J, Fuster V (1995) Human monocyte-derived macrophages induce collagen breakdown in fibrous caps of atherosclerotic plaques. Potential role of matrix-degrading metalloproteinases and implications for plaque rupture. Circulation 92:1565–1569

Solis-Herruzo JA, Rippe A, Schrum LW, de la Torre P, Immaculada G, Jeffrey J, Munoz-Yague T, Brenner D (1996) Interleukin 6 increases ra metalloproteinases-13 gene expression through stimulation of activator protein 1 transcription factor in cultured fibroblasts. J Biol Chem 274:30919–30926

The SOLVD Investigators (1992) Effect of enalapril on mortality and the development of heart failure in asymptomatic patients with reduced left ventricular ejection fraction. N Engl J Med 327:568–574

Strawn WB, Chappell MC, Dean RH, Kivlighn S, Ferrario CM (2000) Inhibition of early atherogenesis by losartan in monkeys with diet-induced hypercholesterolemia. Circulation 101:1586–1589

Sun YP, Zhu BQ, Browne AE et al (2001) Comparative effects of ACE inhibitors and an angiotensin receptor blocker on atherosclerosis and vascular function. J Cardiovasc Pharmacol Ther 6:175–181

Tallant EA, Ferrario CM (1999) Antiproliferative actions of angiotensin-(1–7) in vascular smooth muscle. Hypertension 34:950–957

Van der Wal AC, Becker AE, van der Loos CM, Das PK (1994) Site of intimal rupture or erosion of thrombosed coronary atherosclerotic plaques is characterized by an inflammatory process irrespective of the dominant plaque morphology. Circulation 89:36–44

Weiss D, Kools JJ, Taylor WR (2001) Angiotensin II-induced hypertension accelerates the development of atherosclerosis in apoE-deficient mice. Circulation 103:448–454

Yusuf S (2002) From the HOPE to the ONTARGET and the TRANSCEND studies: challenges in improving prognosis. Am J Cardiol 89 [Suppl]:18A–26A

Yusuf S, Pepine CJ, Garces C et al (1992) Effect of enalapril on myocardial infarction and unstable angina in patients with low ejection fractions. Lancet 340:1173–1178

Yusuf S, Sleight P, Pogue J, Bosch J, Davies R, Dagenais G (2000) Effects of an angiotensin-converting-enzyme inhibitor, ramipril, on cardiovascular events in high-risk patients. The Heart Outcomes Prevention Evaluation Study Investigators. N Engl J Med 342:145–153

Angiotensin II and Vascular Extracellular Matrix

J. W. Fischer

Molecular Pharmacology, Institute of Pharmacology and Clinical Pharmacology,
Moorenstraße 5, 40225 Düsseldorf, Germany
e-mail: jens.fischer@uni-duesseldorf.de

Abstract Extracellular matrix (ECM) is an important structural and functional constituent of both healthy and diseased blood vessels. Vascular cells are embedded in the multicomponent network of the ECM that contains collagens, other glycoproteins, proteoglycans and carbohydrates. The interaction between embedded cells and ECM is mediated through specific transmembrane receptors such as integrins or syndecans. Upon activation integrins initiate intracellular signaling cascades and establish a connection to the cytoskeleton. Production of ECM by vascular smooth muscle cells, which are the main source of ECM in blood vessels, is regulated by a variety of growth factors, cytokines and peptide hormones, among them angiotensin II (ANG II). ANG II regulates the composition of the vascular ECM by a variety of mechanisms, including increased expression of particular ECM genes, which is mediated either directly or via the stimulation of autocrine growth factor release. In addition, ANG II induces plasminogen activator inhibitor I (PAI1), which is involved in the fine tuning of proteolytic ECM turnover. This review focuses on the regulation of fibronectin, collagen and proteoglycan expression by ANG II in VSMC and blood vessels and on the modulation of ECM degradation by matrix metalloproteinases. Furthermore, the modulation of focal adhesion signaling through the convergence with ANG II-induced tyrosine phosphorylation is discussed.

Keywords Angiotensin II · AT_1 receptor · Smooth muscle cells · Collagen · Proteoglycans · Fibronectin · Matrix metalloproteinase · Focal adhesions

Abbreviations

ANG II	Angiotensin II
ECM	Extracellular matrix
EGF	Epidermal growth factor
FAK	Focal adhesion kinase
FN	Fibronectin
GAG	Glycosaminoglycan
GPCR	G protein-coupled receptor
HA	Hyaluronic acid
MAPK	Mitogen-activated protein kinase
MMP	Matrix metalloproteinase
PAI-1	Plasmin activator inhibitor 1
PDGF	Platelet derived growth factor
PI3K	Phosphatidylinositol 3-kinase
PKC	Protein kinase C
PLC	Phospholipase C
Pyk-2	Proline-rich tyrosine kinase 2
RGD	Arginine, glycine, aspartate motif
TGFβ1	Transforming growth factor beta 1
TIMP	Tissue inhibitor of MMP
TN-C	Tenascin
uPA	Urokinase plasmin activator

1 Introduction

Increased activity of the renin angiotensin aldosterone system (RAAS) is associated with higher risks of cardiovascular morbidity and mortality, caused by elevation of arterial blood pressure and by blood pressure-independent mechanisms (Alderman et al. 1991). Trophic effects of angiotensin II (ANG II) on the vasculature include cellular hypertrophy, increased migration and mitosis of vascular cells. Increased accumulation of vascular extracellular matrix (ECM) is involved in these adverse pressure-independent effects of ANG II. The peri- and extracellular portions of the arterial wall are composed of ECM, which has a strong influence on the phenotype of vascular cells. The ECM of blood vessels is elaborated by vascular cells, and forms a complex network comprising collagens, glycoproteins, proteoglycans and hyaluronic acid (HA). Many ECM molecules bind to each other, creating a multicomponent scaffold that interacts with embedded cells. Vascular cells express integrins, CD44 and other ECM receptors that enable cells to form contacts with the ECM, and transduce signals from the ECM to the inside of the cells (for review see Giancotti and Ruoslahti 1999). In situ, vascular smooth muscle cells (VSMC) are surrounded by basement membrane comprising mainly collagen I, -III, -IV, -VI, laminin and perlecan. The basement membrane is believed to augment the differentiated phenotype of VSMC, which is characterized by contractility, low synthetic and proliferative activity (Thyberg and Hultgardh-Nilsson 1994). The ECM surrounding VSMC can undergo rapid and extensive remodeling during development of vascular disease. ECM remodeling entails degradation of old components, synthesis of new components and expression of new cell surface integrin receptors. The ECM is capable of determining the responsiveness of vascular cells to growth factors, peptide hormones and cytokines (Raines et al. 2000). In addition, ECM (a) modulates mechanical properties of the vessel wall, i.e., vascular stiffness and atherosclerotic plaque stability; (b) sequesters growth factors, i.e., transforming growth factor beta-1 (TGFβ1) and fibroblast growth factor-2 (FGF-2); and (c) participates in lipid accumulation and modification.

This chapter focuses on VSMC—the main source of vascular ECM—and the effect of ANG II on ECM composition, ECM-mediated signals and ECM-mediated changes in the phenotype of VSMC. ANG II mediates the majority of its actions via two G-protein-coupled receptors (GPCRs), angiotensin II type 1 receptor (AT_1 receptor) and angiotensin II type 2 receptor (AT_2 receptor). As most known effects of ANG II on ECM expression, function and signaling are mediated by AT_1 receptors, this review discusses mainly AT_1 receptor-mediated effects in vasculature and VSMC. AT_2 receptor-mediated effects on ECM in VSMC will be only briefly summarized (see Sect. 3.1, "AT_2 Receptor and Vascular EMC"). ANG II has a profound impact on the expression and degradation of vascular ECM and ECM-mediated signaling, which are thought to contribute significantly to the long-term effect of ANG II on vascular remodeling.

2
Focal Adhesion Complex and Cellular Phenotype

It has long been known that ECM contact is essential for adhesion, growth and survival of adhesive cells in connective tissues. Cells interact with the ECM via specific receptors. The best-characterized, and likely the most important group, are integrin heterodimers (Aplin et al. 1998); however, other ECM receptors are also potent modulators of cellular function and phenotype, such as the HA-receptors, CD-44 and the receptor for HA-mediated motility (RHAMM) (Turley et al. 2002). Transformed cells lose the integrin dependency for cell cycle progression and survival, which is fundamental to their propensity for metastasis. As with most cells in adult tissues, VSMC are permanently engaged with the surrounding ECM and form stable contacts with the ECM (focal adhesions). However, in order to proliferate and migrate during tissue remodeling, cells need to break existing focal adhesions and establish new contacts. This leads to local changes in the activity of downstream signaling events, such as mitogen-activated protein kinase (MAPK) activity, which converge with the signaling of receptor tyrosine kinases or GPCR. The phenotype of a cell at a given time reflects the integration of a variety of signals initiated by extracellular ligands of integrins and other adhesion receptors, tyrosine kinase receptors, GPCRs, and additional mediators such as nitric oxide, reactive oxygen species or mechanical factors. Structural components of the focal adhesion complex are briefly introduced. Integrins are heterodimeric transmembrane receptors comprising alpha and beta subunits that mediate cell adhesion to the ECM at focal adhesions. Eight beta subunits and 16 alpha subunits have been cloned in mammals, forming at least 23 defined $\alpha\beta$-heterodimers (for review see Aplin et al. 1998; Giancotti and Ruoslahti 1999). The integrin heterodimers have relatively broad substrate specificities, often allowing the ligation of a particular heterodimer by several ECM molecules. Often, it is the arginine-glycine-aspartate (RGD) sequence within the protein core of ECM molecules that is recognized by integrins (Fig. 3). Upon interactions with their respective ligands, integrins cluster and enable cells to form focal contacts with the surrounding ECM. Integrins are devoid of enzymatic activity and associate with adapter proteins such as vinculin, talin, paxillin, and $p130^{cas}$, to form a multimolecular platform, the focal adhesion complex, which allows initiation of signaling cascades that transduce information from the ECM into the cell (Fig. 3). Adaptor molecules of the focal adhesion complex connect to actin-containing filaments of the cytoskeleton. Furthermore, integrins recruit various tyrosine kinases to the focal adhesion complex, including focal adhesion kinase (FAK), integrin linked kinase (ILK), Src-family kinases and proline-rich tyrosine kinase 2 (Pyk-2). Through these kinases, integrins initiate signaling cascades that support mitogenic growth factor signaling, control cell cycle progression and inhibit apoptosis. Another important function of integrins is the control of cell spreading and migration, which is in part mediated through the Rho-family of small guanine nucleotide-binding proteins, Rho, Rac and Cdc42 (Keely et al. 1997). From the above-mentioned interactions between cells and ECM, it can be concluded that any factor that alters the composition of ECM or interferes with the signals generated by the ECM receptors will thereby have a profound impact on cell behavior, phenotype and gene expression. These mechanisms of interference will be further discussed, with respect to ANG II.

3
Modulation of ECM-Mediated Signals by ANG II

There are at least three avenues via which ECM-mediated signals can be modulated by ANG II. Firstly, ANG II, via AT_1 or AT_2 receptors, can induce or repress ECM gene expression; this can be either a direct effect or one mediated indirectly through the autocrine release of growth factors. Secondly, ANG II regulates degradation of ECM components, which terminates existing ECM signals, and generates new signals through either the formation of bioactive ECM fragments or the exposure of cryptic sites within ECM molecules. Thirdly, ANG II receptor-mediated signals interfere with focal adhesion signaling that is initiated by ECM–integrin interaction. See Fig. 1.

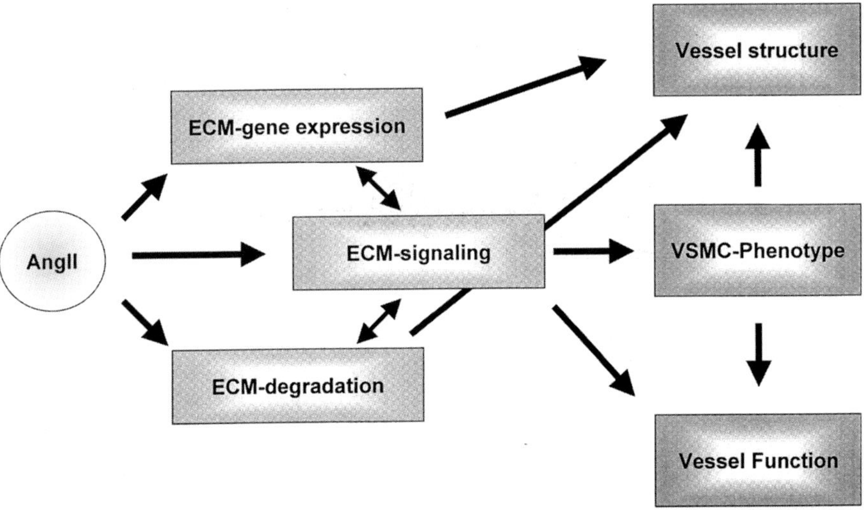

Fig. 1 Interrelationship between ANG II, ECM and vascular remodeling. ANG II induces the expression of a variety of ECM genes, inhibits ECM degradation and directly interferes with the signaling of focal adhesion complexes. Consequently, the composition of the ECM changes, which has direct effects on vessel structure, mechanical properties and ultimately vessel function. ANG II modulates the phenotype of vascular cells directly by interference with focal adhesion signaling and indirectly via changes in the composition of ECM. Alterations of the vascular phenotype, including the propensity of cells to proliferate and migrate, contribute to changes in vessel structure, function and progression of vascular disease

3.1 AT_2 Receptor and Vascular ECM

The AT_2 receptor is discussed, in detail, in the chapters by Balt and Pfaffendorf and by Wruck et al., of this volume. AT_2 receptors are abundantly expressed during embryonic development, but are less present after birth. It is, however, present in adult human vasculature, e.g., in renal and coronary arteries. The AT_2 receptor often counteracts with the AT_1 receptor and is involved in regeneration and repair programs (for review see de Gasparo et al. 2000; Stoll and Unger 2001). Knowledge about AT_2 receptor effects on ECM remodeling in the vessel wall and VSMC is limited, and thus will only be discussed briefly. It is possible that the functional significance of AT_2 receptors during ECM remodeling is underestimated, since many studies investigating ANG II and ECM were performed in cultured adult VSMC, which do not express AT_2 receptors. During vascular disease, expression of AT_2 receptors is induced and thus could potentially play a role in the control of ECM remodeling or ECM signaling (Akishita et al. 2000a; Nakajima et al. 1995). In support of this hypothesis are studies showing that experimental overexpression of AT_2 receptors in cultured VSMC induce collagen, fibronectin (FN) and proteoglycan expression (Chassagne et al. 2002; Mifune et al. 2000; Shimizu-Hirota et al. 2001). Moreover, in microvascular endothelial cells endogenously expressed AT_2 receptors mediate growth inhibition and induction of thrombospondin 1 (TSP-1) (Fischer et al. 2001a). Experiments using gene transfer of the AT_2 receptor into the rat balloon injury model demonstrated inhibition of neointimal hyperplasia by the AT_2 receptor (Nakajima et al. 1995). This is in agreement with known antiproliferative effects of the AT_2 receptor in cultured endothelial cells (Stoll et al. 1995), which has been confirmed in many other cell types, including VSMC. Most studies using AT_2 receptor-deficient mice suggest that AT_2 receptors inhibit hypertrophy and vascular fibrosis (Akishita et al. 2000b; Brede et al. 2001). In contrast, the one study in rats and another in AT_2 receptor-deficient mice revealed the opposite results, suggesting that AT_2 receptors promote hypertrophy and fibrosis (Levy et al. 1996; Senbonmatsu et al. 2000). These contrasting results might be related to differences in genetic background, since different AT_2 receptor-deficient mouse strains (FVB/N vs C57B6) were used. The FVB/N mice, in which the AT_2 receptor was shown to inhibit vascular fibrosis (Akishita et al. 2000b; Brede et al. 2001), are known to form an extensive neointima in response to vascular injury. In contrast C57B6 mice, in which the AT_2 receptor was reported to support vascular fibrosis (Senbonmatsu et al. 2000), are less responsive to arterial injury (Harmon et al. 2000). An alternative explanation for these contradictory results might be the differences in the function of AT_2 receptors in vessels of different origin. Furthermore, many glycoproteins and proteoglycans known to be involved in fibrotic remodeling and neointimal hyperplasia have yet to be characterized with respect to regulation by the AT_2 receptor. Thus, although experimental evidence suggests that the AT_2 receptor is involved in regulation of vascular ECM composition in vivo, a definitive conclusion is not yet possible as to whether it is inhibitory or stimulatory and which ECM components are regulated by the AT_2 receptor.

3.2
Modulation of Expression

ECM remodeling during vascular disease (i.e., atherosclerotic lesion development) involves the formation of significant amounts of new ECM components such as proteoglycans, FN, osteopontin, tenascin C (TN-C), vitronectin and TSP-1. These molecules are all important modulators of VSMC phenotype, and most of them are up-regulated in response to ANG II in VSMC (see Table 1). With respect to modulation of ECM expression, this review focuses on collagen, FN and proteoglycans, since there exists convincing evidence of significant roles for these ECM components in various vascular diseases and for interrelationships with the RAAS.

3.2.1
Fibronectin

Fibronectin (FN) is a glycoprotein of the vascular ECM (Ruoslahti 1988), which can be derived from the circulation or synthesized by cells of the vascular wall.

Table 1 The table displays a selection of representative studies on the regulation of individual ECM glycoproteins and proteoglycans by Ang II. The involved receptors are indicated as well as the involvement of autocrine factors

ECM component	Receptor subtype		Paracrine factors involved	Reference(s)
Elastin	AT_1	↓	ND	Tokimitsu et al. 1994
Tenascin	AT_1	↑	ND	Sharifi et al. 1992; Mackie et al. 1992
Fibronectin	AT_1	↑	TGF-β1	Sharifi et al. 1992; Matsubara et al. 2000
	AT_2[a]	↑	ND	Chassagne et al. 2002
Collagen types I, III	AT_1	↑	TGF-β1	Ford et al. 1999
	AT_2[a]	↑	ND	Mifune et al. 2000
Thrombospondin-1	AT_1	↑	ND (TGF-β1, PDGF-A)	Scott-Burden et al. 1990
Versican, biglycan, perlecan	AT_1	↑	TGF-β1	Shimizu-Hirota et al. 2001
	AT_2[a]	↑		
Syndecan-1	ND	↑	ND	Cizmeci-Smith et al. 1993
Osteopontin	ND	↑	ND	Giachelli et al. 1993

↓↑, Down- or up-regulation.
[a] Retroviral overexpression of AT-2 receptors in VSMC.
ND, not determined.

Cellular FN generated by vascular cells is distinct from plasma FN that is secreted by the liver, by virtue of the FN type III modules (ED-A, ED-B) generated by alternative splicing (for review see Kosmehl et al. 1996). FN is expressed at low levels in normal healthy arteries. However, it is markedly induced and accumulated during atherosclerosis and neointimal hyperplasia (Labat-Robert et al. 1985). Soluble dimeric FN molecules polymerize into a FN matrix in an integrin ($\alpha5\beta1$) dependent manner (Pickering et al. 2000). FN is a ligand to several integrin heterodimers such as $\alpha v\beta3$, $\alpha v\beta5$ and $\alpha5\beta1$. Among these integrins, only $\alpha5\beta1$ is specific for FN. The FN–integrin interactions induce signaling cascades that merge with those of growth factors and ANG II (see also Sect. 3.4, "Modulation of Focal Adhesion Signaling") and determine the response of VSMC to these stimuli.

Animal studies in rats along with in vitro studies of human and rat VSMC have shown that FN is induced by ANG II via the AT_1 receptor. After balloon injury Sprague-Dawley rats exhibited a marked up-regulation in FN, which was blocked by a specific AT_1 receptor antagonist (Kim et al. 1995). Furthermore, the AT_1 receptor induces FN also ANG II in uninjured vessels (Himeno et al. 1994). The effect of ANG II on FN was at least partially independent of blood pressure elevation. Recently, it was demonstrated that AT_1 receptors mediate FN expression, especially FN-EIIIA, via signaling involving phosphatidylinositol-specific phospholipase C (PI-PLC), protein kinase C (PKC), protein tyrosin phosphorylation and phosphatidylinositol 3 kinase (PI3K), eventually leading to activation of the AP1-like binding site of the FN promotor (Tamura et al. 1998, 2000). AT_1 receptors are also involved in stretch-induced FN expression

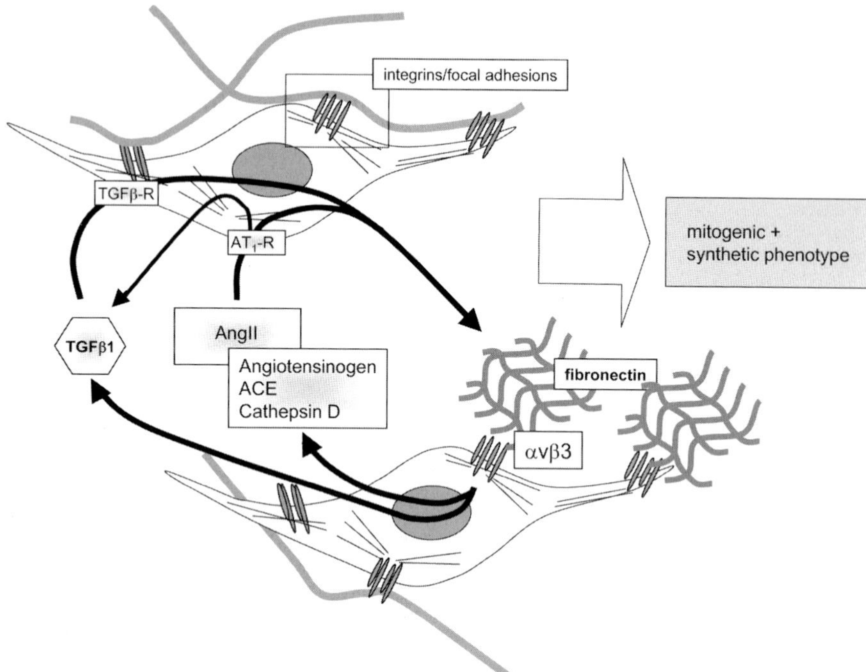

Fig. 2 Interdependency of tissue RAS and fibronectin expression. In VSMCs, FN induces, via $\alpha v\beta 3$-integrins, the expression of angiotensinogen, cathepsin D and ACE, leading to autocrine generation of ANG II and up-regulation of TGFβ1. In turn, ANG II and TGFβ1 induce additional FN expression. Furthermore, ANG II itself induces release of active TGFβ1 and increases the expression of TGFβ1-type 1 receptors (not shown). Through this mechanism, VSMCs shift toward the synthetic, proliferative and migratory phenotype

(Tamura et al. 2000). In addition, mechanical stretch induces angiotensin-converting enzyme (ACE) and AT_1 receptor expression, and ANG II generation in VSMC (Matsubara et al. 2000). Furthermore, cultured VSMC express angiotensinogen, cathepsin D, ACE and TGFβ1 in response to contact with FN, which results in increased ANG II production (Hu et al. 2000). The described interdependencies between mechanical stretch, FN expression, ANG II generation and autocrine TGFβ1 may create a positive feedback loop leading to sustained increase in FN expression and activity of the tissue renin-angiotensin system. See Fig. 2.

The functional significance of ANG II induced FN production in VSMC is confirmed by the phenotypic modulation of VSMC by FN: VSMC, in contact with FN, are more responsive to mitogenic and migratory stimuli (Hedin and Thyberg 1987). Furthermore, VSMC proliferation depends on FN matrix assembly, as inhibition of FN assembly was shown to almost completely inhibit SMC proliferation (Mercurius and Morla 1998). After balloon injury of the rat carotid artery, the luminal dedifferentiated VSMC up-regulate the FN-specific integrin, $\alpha 5\beta 1$, and rapidly assemble a polymeric FN matrix (Pickering et al. 2000). In vivo, in neointimal cells, FN is thought to provide a substrate for adhesion, to promote cell migration and proliferation and to prevent apoptosis. ANG II-induced FN is therefore likely to support the proliferative response of neointimal cells and to facilitate the migratory response of VSMC during vascular remodeling.

3.2.2 Collagen

Type I collagen is abundant in both normal adult arteries and remodeled arteries in vascular disease. The polymerization process of collagen, in contrast to FN fibril formation, is independent of interactions with cells and integrins. The size of collagen type I fibrils, the degree of lateral fusion and the three-dimensional orientation within the ECM are, however, subject to modulation by proteoglycans (i.e., decorin), type III collagen and other ECM components. VSMC adhere to collagen mainly via $\alpha2\beta1$, but also $\alpha1\beta1$ and $\alpha3\beta1$ integrins (Yamamoto and Yamamoto 1994; Gardner 1996). Functional consequences of collagen accumulation in the vessel wall are complex and can vary among vascular diseases such as hypertension, restenosis and atherosclerosis. Depending on the fibrillar organization of collagen, different consequences have been described, ranging from promotion to inhibition of proliferation and migration. Functional aspects of collagen accumulation in the vessel wall and its interrelationships with the RAS are discussed in following paragraphs.

The extent of collagen type I and III deposition in the vascular wall and the three-dimensional organization of collagen determine mechanical properties of blood vessels. One clinically important aspect is arterial stiffness, which—if it is increased—is associated with higher cardiovascular morbidity and mortality in patients (O'Rourke and Frohlich 1999). Increased vascular stiffness contributes to the development of increased systolic blood pressure, pulse pressure and aortic pulse wave velocity, which then expands the vascular load on the heart followed by adverse left ventricle remodeling and impaired cardiac function (Lakatta 1993). In chronic hypertension, increased vascular collagen accumulation and vascular stiffness are observed (Ooshima et al. 1974). Many in vitro (Kato et al. 1991) and animal studies (Fakhouri et al. 2001) suggest that ANG II, via the AT_1 receptor, increases collagen expression by VSMC and collagen accumulation in vessel walls. In human VSMC, paracrine release of TGFβ1 mediates collagen synthesis and expression, in response to AT_1 receptor activation (Ford et al. 1999). The significance of AT_1 receptor-driven collagen expression is further underlined by the association between aortic stiffness and AT_1 receptor polymorphisms in hypertensive patients (Benetos et al. 1996).

In vascular lesions, collagen accumulation can be both detrimental and beneficial. Collagen is an important constituent of vascular lesions, serving as a scaffold for the binding of other ECM components such as FN and proteoglycans. Extensive collagen accumulation will increase neointimal thickness and the extent of stenosis. On the other hand, high collagen content of the fibrous cap, which is determined by the balance of collagen expression and degradation, will increase atherosclerotic plaque stability (Plutzky 1999). Recently, this view was experimentally confirmed by the inhibition of TGFβ1 resulting in decreased plaque stability in primary atherosclerosis in mice (Lutgens et al. 2002; Mallat et al. 2001). Therefore, with respect to the atherosclerotic plaque, increased expression of collagen in response to ANG II would not necessarily be detrimental to the individual. Another interesting consequence of ANG II-mediated stimulation of collagen deposition in the vascular wall is the phenotypic modulation of VSMC. It is well known that polymeric collagen inhibits VSMC proliferation induced by serum, platelet derived growth factor (PDGF) BB and other growth factors (Raines et al. 2000). It was shown that the growth-inhibiting effects of fibrillar collagen are mediated by cyclin-dependent kinase inhibitors p27 and p21 (Koyama et al. 1996). In contrast, monomeric collagen supports VSMC growth (Koyama et al. 1996). It seems to be a contradiction that ANG II should support via the AT_1 receptor proliferation (Su et al. 1998) and at the same time increase synthesis of collagen type I, which is antiproliferative in its polymerized form. However, ANG II often causes vascular and cellular hypertrophy rather than hyperplasia (Geisterfer et al. 1988). Therefore, it can be hypothesized that the induction of a fibrillar collagen matrix could balance mitogenic stimulation by ANG II and other factors in the vessel wall.

In summary, AT_1 receptor-induced collagen accumulation in adult arteries contributes to increased wall stiffness, which is associated with elevated cardiovascular morbidity and mortality. ANG II-driven collagen and ECM accumulation will adversely affect atherosclerotic plaques and restenotic lesions by increasing neointimal volume. On the other hand, collagen deposition in the fibrous cap can stabilize the plaque. How the phenotype of VSMC will be affected by ANG II-mediated stimulation of collagen expression depends on fibrillar organization, the extent of proteolysis and the presence of other ECM molecules. In conclusion, regulation of collagen expression and accumulation by ANG II is of functional and clinical significance since collagen regulates SMC phenotype, contributes significantly to the ECM of vascular lesions and determines mechanical properties of the vessel wall.

3.2.3 Proteoglycans

Proteoglycans consist of a protein core with one or more glycosaminoglycan (GAG) chains attached. Vascular proteoglycans are differentiated into extracellular and pericellular proteoglycans. The main extracellular proteoglycans produced by VSMC are chondroitin/dermatan sulfate proteoglycans (CS/DS-PGs) versican, biglycan, decorin and the heparan sulfate proteoglycan (HS-PG) perlecan (Evanko et al. 1998; Yao et al. 1994). Important pericellular proteoglycans in VSMC are HS-PGs syndecan 1–4 and glypicans. The former are transmembrane molecules that are capable of signaling and function as coreceptors for FGF-2 (Kiefer and Barr 1990). In the following paragraph, the functional significance of proteoglycans and associated molecules for VSMC phenotype is summarized. Proteoglycans do not directly interact with integrins; however, they may interrupt integrin–ECM interactions by displacing integrin ligands and modulate the polymerization of integrin ligands such as collagens and FN. Biglycan and versican are modulators of elastic fiber formation (Merrilees et al. 2002; Reinboth et al. 2002). Decorin is a regulator of collagen fibril formation (Danielson et al. 1997; Fischer et al. 2000; Scott and Haigh 1988) and an inhibitor of TGFβ1 activity in VSMC (Fischer et al. 2001b; Yamaguchi et al. 1990). Versican contains glomerular G1 and G3 domains, both of which are believed to promote proliferation (Yang et al. 1999; Zhang et al. 1998). Biglycan and versican bind low-density lipoproteins (LDL) and oxidized LDL (oxLDL), resulting in increased lipid accumulation in the vessel wall (O'Brien et al. 1998). Interestingly, versican and biglycan are also induced by oxLDL in VSMC, which is probably a positive feedback loop aggravating lipid accumulation (Chang et al. 2000). Taken together, proteoglycans are believed to play an important role during vascular remodeling and the progression of human restenotic and atherosclerotic lesions, by providing a pro-proliferative and pro-migratory environment and by sequestering LDL particles within the ECM. PG synthesis is increased in large arteries of hypertensive rats (Walker-Caprioglio et al. 1992). Furthermore, ANG II stimulates CS/DS-PG production in cultured rat VSMC (Bailey et al. 1994) and in the vasculature of Spraque-Dawley rats in vivo (Simon et al. 1994). It has also been suggested that ANG II regulates the extent of sulfation in GAG side chains in VSMC (Castro et al. 1999). With respect to specific gene regulation, ANG II induces the expression of syndecan-1 (Cizmeci-Smith et al. 1993), versican, biglycan and perlecan via the AT_1 receptor in rat VSMC (Shimizu-Hirota et al. 2001). This ANG II effect seems to involve transactivation of the EGF receptor and activation of MAPK/ERK kinase (MEK) (Shimizu-Hirota et al. 2001). Recently, it was demonstrated that ANG II induces proteoglycans with a higher affinity to LDL in VSMC, by modulating the composition and sulfation of the GAG chains (Figueroa and Vijayagopal 2002). This mechanism could be of relevance in vivo since it could support lipid accumulation and foam cell formation in the vessel wall. These in vitro results are supported by the observation that a AT_1 receptor antagonist reduces biglycan and decorin expression in the kidney of hypertensive rats (Sasamura et al. 2001). In summary, ANG II, via the AT_1 receptor, regulates PG expression in VSMC in vitro and in vivo. This regulation is likely biologically relevant since even a subtle shift in PG expression by ANG II can significantly alter the properties of vascular ECM, as detailed above.

3.2.4 Induction of Autocrine Factors

In VSMC, ANG II induces autocrine release of TGFβ1 (Gibbons et al. 1992; Itoh et al. 1993) through activation of the AP-1 complex (Morishita et al. 1998). Additionally, PDGF A-chain and FGF-2 are released in response to ANG II (Itoh et al. 1993; Naftilan et al. 1989). Recent experiments in mice have implicated the

Krüppel-like zinc-finger transcription factor 5 (KLF-5) in the expression of PDGF-A and probably also TGFβ1 in response to ANG II (Shindo et al. 2002). Furthermore, the study conducted by Shindo et al. (2002) demonstrated that the induction of PDGF-A chain and TGFβ1 is essential for hyperplastic and fibrotic vascular remodeling in response to arterial injury in heterozygote KLF-5-deficient mice. Autocrine release of these growth factors contributes to the migratory, proliferative or hypertrophic responses of VSMC to ANG II, as shown in numerous in vivo and in vitro studies (for review see Berk 2001). With respect to ECM modulation, TGFβ1 induction is particularly important since it inhibits ECM degradation by matrix metalloproteinases (MMPs), via up-regulation of tissue inhibitors of MMPs (TIMPs) and plasminogen activator inhibitor-1 (PAI-1, see Sect. 3.3, "Inhibition of EMC Turnover"), and down-regulation of MMPs. TGFβ1 is also a potent inducer of ECM-expression (Roberts et al. 1992). Vascular proteoglycans such as biglycan, versican and perlecan are induced by TGFβ1, as well as collagen, FN and many other ECM molecules (Iozzo 1997; Schönherr et al. 1991). The importance of autocrine release of growth factors in response to ANG II for ECM remodeling ANG II was recently demonstrated: in vascular VSMC, ANG II induces, via the AT_1 receptor and autocrine release of TGFβ1, collagen type-1, FN and proteoglycans (Ford et al. 1999; Matsubara et al. 2000; Shimizu-Hirota et al. 2001). Autocrine release of the PDGF-A chain might be of importance for ECM expression as well, since PDGF-A elevates expression of vascular proteoglycans versican and biglycan in VSMC (Schönherr et al. 1991; Schönherr and Wight 1993). In contrast, autocrine release of FGF-2 is thought to be less important with respect to ECM.

Additionally, cyclooxygenase (COX) 2 is induced by ANG II in VSMC, resulting in increased synthesis of prostacyclin and prostaglandin E2 and E1 (Ohnaka et al. 2000). These eicosanoids are important modulators of VSMC proliferation and migration (for review see Schrör and Weber 1997; Weber and Schrör 1999). Knowledge of the role of prostaglandin derivatives on ECM synthesis and degradation in SMC is still sparse, but a few studies reported modulation of ECM expression and degradation in SMC by eicosanoids. For example, prostaglandin E2 induces the expression of the HA-binding proteins and tumor necrosis factor stimulated protein-6, which is significant for the formation of pericellular ECM coats and migration (Fujimoto et al. 2002). Prostacyclin suppresses proteoglycan synthesis in VSMC (Koh et al. 1993), while prostaglandins E1 and E2 suppress procollagen synthesis in VSMC (Fitzsimmons et al. 1999). In ciliary SMC, prostaglandins stimulated the release of MMP-1, -2, -3, and -9 (Weinreb et al. 1997). Taken together, induction of growth factors and other mediators (i.e., cyclooxygenase products) provides ANG II with a powerful mechanism to enhance or extend its actions on ECM components, which would otherwise not be directly affected by AT-1- or AT-2-mediated signaling.

3.3 Inhibition of ECM Turnover

In whole tissues and organs, ECM degradation balances new ECM synthesis thereby controlling ECM accumulation and composition. A shift in this balance will determine the extent of vascular fibrosis, neointimal expansion and plaque stability. On the cellular and subcellular levels, spatial and local coordination of ECM degradation is required for cell migration and proliferation (reviewed in Stetler-Stevenson and Yu 2001). Proteolytic cleavage of ECM has been implicated in angiogenesis, wound healing and metastasis, and in promoting VSMC migration and proliferation (for review see Stetler-Stevenson et al. 1993). Therefore, it is evident that modulation of ECM degradation is an important pathway through which ANG II can influence cellular phenotype and ECM accumulation during vascular remodeling. Clinical evidence has demonstrated that serum levels of MMP-1 (collagenase 1) and MMP-9 are decreased in hypertensive patients (Laviades et al. 1998; Li-Saw-Hee et al. 2000). Antihypertensive treatment with ACE inhibitors enhanced collagen degradation, increased MMP-1 levels and decreased TIMP-1 levels (Laviades et al. 1998). ACE inhibitors were effective in normalizing increased collagen synthesis in hypertensive patients (Diez et al. 1995). A similar increase in collagen degradation, due to pharmacological ACE inhibition, has been demonstrated in SHR rats (Brilla et al. 1996). These data suggest that, during hypertension, the balance between collagen synthesis and degradation is shifted toward a fibrogenic state that may result in tissue fibrosis, and that RAAS is an important mediator of this shift.

Plasmin activates many of the MMPs, like MMP-1, -8, -3, -10, -7 and -9, in the vascular wall (for review see Murphy and Knauper 1997). Therefore, serum-derived plasmin is a key activator of MMPs and ECM turnover in vasculature. In addition, plasmin itself participates in the degradation of fibronectin, laminin and proteoglycans (Kenagy et al. 2002; Mignatti and Rifkin 1993). The activators of plasmin, urokinase plasminogen activator (uPA) and tissue plasminogen activator (tPA), are serine proteases capable of ECM degradation. Urokinase PA interacts with its receptor on the cell membrane of vascular cells, and catalyzes the activation of plasminogen at the cell membrane. In addition, uPA–uPA receptor complexes induce intracellular signaling and modulate cell proliferation and migration (Dumler et al. 1999). PAI-1 is the main inhibitor of uPA-mediated plasmin activation, and thus inhibits fibrinolysis and MMP activation and interfers with the uPA–uPA receptor-mediated signaling. PAI-1 resides within the pericellular ECM, bound to vitronectin (Preissner et al. 1990). In various vascular diseases, elevated PAI-1 levels are associated with decreased fibrinolysis and increased ECM accumulation: PAI-1 is elevated in atherosclerotic plaques (Schneiderman et al. 1992), increased in the serum of patients developing restenosis (Huber et al. 1992), and is positively correlated to the risk of myocardial infarction (Eriksson et al. 1995; Hamsten et al. 1985).

Since VSMC are capable of synthesizing both the activators and inhibitors of ECM degradation, the coordination of ECM turnover will depend on differential regulation of the various components. ANG II acts by strongly inducing PAI-1 expression, which in turn leads to decreased activation of MMPs by plasmin. ANG II infusion induces rapid elevations in plasma PAI-1 concentration (Ridker et al. 1993). In the rat balloon injury model, PAI-1 expression is rapidly induced within 3 h after injury, returns to baseline at 2 days and strongly increases thereafter. The second induction between 2 and 7 days was abolished by ACE-inhibition (Hamdan et al. 1996). TGFβ1 and thrombin are also strong inducers of PAI-1 expression. In human vascular SMC, the AT_1 receptor mediates PAI-1 induction and suppression of tPA and MMP-2 secretion (Papakonstantinou et al. 2001; Takeda et al. 2001; van Leeuwen et al. 1994). PAI-1 induction by the AT_1 receptor is mediated via activation of the PAI-1 promotor and prolongation of mRNA half-life, and is dependent on EGF receptor transactivation and MEK/ERK and RhoA activation (Takeda et al. 2001). Another study suggested that PKC and AP-1 activation are also involved (Ahn et al. 2001). In line with these in vitro studies, pharmacological blockade of AT-1 receptors was shown to prevent PAI-1 induction in the rat aorta, independent of lowering blood pressure (Nakamura et al. 2000). Thus, ANG II promotes fibrosis of the arterial wall through PAI-1 induction, which inhibits MMP activation, resulting in decreased ECM turnover. With respect to the RAAS, recent evidence suggests that aldosterone promotes fibrosis as well. Aldosterone and ANG II were shown to synergistically up-regulate PAI-1 in VSMC and EC (Brown et al. 2000a). In a rat model of glomerulosclerosis, aldosterone antagonism decreased PAI-1 levels (Brown et al. 2000b). In humans, aldosterone levels correlate with PAI-1 antigen concentrations (Brown et al. 1998; Sawathiparnich et al. 2002). In summary, experimental and clinical evidence suggests an inhibitory role of the RAAS in ECM turnover. In addition to up-regulation of ECM expression (see also Sect. 3.2, “Modulation of Expression”), ANG II inhibits MMP activity via PAI-1 and TGFβ1, thereby disturbing the balance between ECM synthesis and degradation, eventually leading to ECM accumulation and tissue fibrosis.

3.4
Modulation of Focal Adhesion Signaling

Integrin ligation by ECM molecules induces clustering of integrins and formation of focal adhesion complexes. Subsequently, FAK and ILK become activated and initiate intracellular signaling cascades. Downstream of the activated focal adhesion complex, activation of Src, Ras, MAPKs, PI3K and small GTPases occurs (for review see Giancotti and Ruoslahti 1999). Most of these mediators of integrin signaling are also involved in growth factor and ANG II signaling. Recently, a number of studies showed that growth factor mediated signaling is strongly influenced by ECM composition and integrin signaling (Schwartz and Baron 1999). It can be assumed that integrins modulate, in a similar way, the response of VSMC to ANG II. In support of this hypothesis is the observation that stimulation of DNA synthesis by ANG II was more pronounced in cells plated on FN or collagen (Sudhir et al. 1993). Furthermore, phosphorylation levels of

MAPK/ERK kinase (MEK) and extracellular signal-regulated kinase (ERK), in response to ANG II, are strongly elevated if VSMC were plated on FN (Tamura et al. 2001), suggesting that integrin ligation and ANG II stimulation act synergistically on MEK and ERK activation. In human VSMC, collagen and FN support, likely via the transcription factor egr-1 (early growth response gene 1), AT_1 receptor-mediated expression of PDGF-A , which then mediates DNA synthesis and mitosis (Ling et al. 1999). This effect was specific for FN and collagen, since on plastic and laminin, PDGF-A was not induced by ANG II, suggesting that the autocrine production of growth factors in response to ANG II is dependent on ECM as well.

ANG II-mediated signaling is partially dependent on Ca^{++}-sensitive transactivation of epidermal growth factor receptors (EGF-R) (Eguchi et al. 1998; Murasawa et al. 1998; Touyz et al. 2002). Proteolytic activity of MMPs and elastases can expose new RGD-binding sites for $\beta3$ integrins in VSMC (Jones et al. 1997). Subsequently, $\beta3$ integrin ligation induces TN-C expression, which promotes via $\alpha v\beta3$ integrins EGF-R signaling, thereby strongly enhancing EGF-induced proliferation (Jones et al. 1997). Interestingly, TN-C is induced by ANG II in VSMC (Mackie et al. 1992). Thus, a positive feedback loop might exist, since ANG II signaling is partially dependent on EGF receptor transactivation, EGF-R signaling is enhanced by TN-C-induced integrin signaling and TN-C is induced by ANG II.

ANG II also directly interferes with focal adhesion signaling. FAK phosphorylation and activation, as well as paxillin phosphorylation are initiated in response to AT_1 receptor activation (Leduc and Meloche 1995; Turner et al. 1995). These are rapid responses to ANG II and are followed by translocation of FAK to focal adhesions (Okuda et al. 1995; Polte et al. 1994). In addition, $p130^{cas}$ is phosphorylated in response to ANG II (Sayeski et al. 1998) and serves as an adapter molecule providing SH3 and SH2 domains. The SH3 domain of $p130^{cas}$ mediates c-src binding , whereas the SH2 domain mediates binding of FAK. $p130^{cas}$ also recruits paxillin and tensin, thereby playing a crucial role in integrin-mediated signaling, cell adhesion and cytoskeletal organization.

Since the AT_1 receptor is a GPCR, activation of intermediate kinases must occur to mediate phosphorylation of FAK and other constituents of the focal adhesion complex, in response to ANG II. A tyrosine kinase that is likely to play an important role is Pyk-2, which is closely related to FAK. Upon AT_1 receptor stimulation, Pyk-2 is translocated to focal adhesions and activated. The translocation is dependent on the cytoskeleton. In response to ANG II, Pyk-2 forms a complex with the upstream regulators of the ERK pathway, src, Shc and Grb2 (Rocic et al. 2001b; Sabri et al. 1998), which leads subsequently to ERK activation (Eguchi et al. 1999). Pyk-2 activates src, which is of importance since c-src stimulates bundling of actin filaments and assembly of focal adhesion in VSMC (Ishida et al. 1999). Effects of c-src on actin filaments involve phosphorylation of $p130^{cas}$, paxillin and tensin, which mediates cross linking of actin filaments. Pyk-2 also mediates FAK-phosphorylation in response to AT_1-receptor activation, as shown by experimental down-regulation of Pyk-2 expression in VSMC

(Rocic et al. 2001b; Rocic and Lucchesi 2001a). Furthermore, the PI3-K/Akt–p70S6 pathway is activated downstream of Pyk-2 in response to ANG II (Rocic and Lucchesi 2001a).

Integrin signaling also utilizes Pyk-2-mediated tyrosine phosphorylation, leading to an association between Pyk-2 and p130 ^{Cas}, PI3K and paxillin, and activation of c-src (Kaplan et al. 1995; Lakkakorpi et al. 1999). Litvak et al. (2000) demonstrated that Pyk-2 translocates to the focal adhesion sites after stimulation of both GPCR (histamine type 1 receptor) and adhesion to FN (Litvak et al. 2000). The G protein induced activation of Pyk-2 was PKC dependent and seems to occur in a similar fashion upon AT_1 receptor stimulation in VSMC (Frank et al. 2002). In addition, clustering of integrins and integrin-dependent Pyk-2 translocation to focal adhesions are PKC dependent. The observation that Pyk-2 is activated by both AT_1 receptors and integrins suggests that Pyk2 might be an important proximal point of convergence between AT_1 receptor- and integrin-induced signals. Convergence of AT_1 receptor and integrin signaling at the level

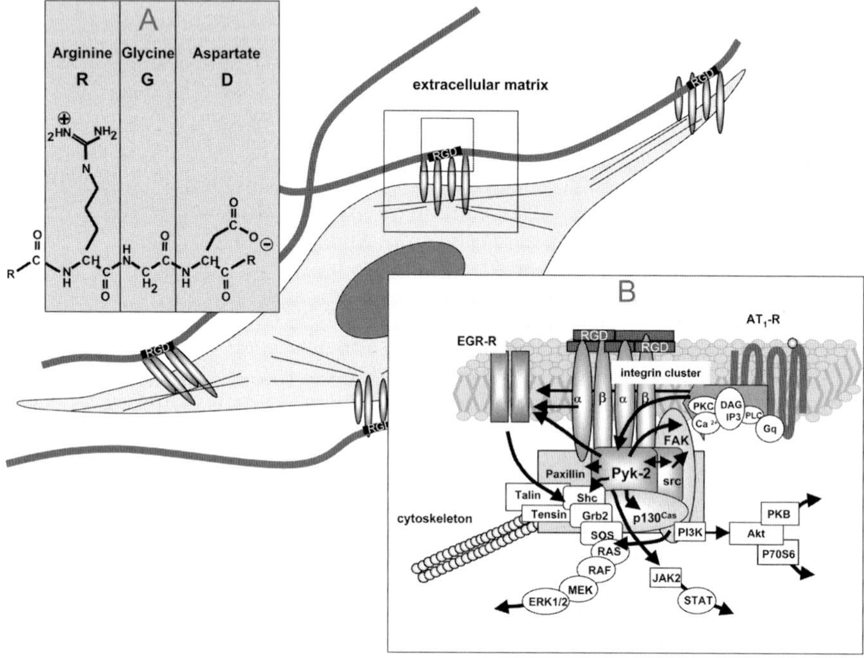

Fig. 3A, B Convergence of integrin and AT_1 receptor signaling at focal adhesions. VSMCs form focal contacts with the surrounding ECM. The focal contacts are mediated by integrin $\alpha\beta$ heterodimers that recognize arginine-glycine-aspartate (*RGD*) motifs within the core proteins of ECM components (**A**). Cytoplasmic tails of integrins recruit a multimolecular complex (focal adhesion complex) of adapter molecules and kinases that establish firm contacts with the cytoskeleton and initiate signaling cascades that determine cellular phenotype. Possible interrelationships between the signaling of AT_1 receptors, EGF receptors and integrins are presented schematically (**B**). Emphasized is the role of Pyk-2 as a possible point of convergence, mediating FAK and Src activation, EGF receptor transactivation, and initiation of the Ras/Raf, Jak/Stat and PI3K/Akt pathways. (*FAK* focal adhesion kinase; *Pyk-2* proline-rich protein kinase 2; *PI3K* phosphatidylinositol 3-kinase)

of Pyk-2, and possible interrelationships between the signaling pathways of integrins and AT_1 receptors are depicted in Fig. 3.

Some recent studies started to demonstrate the functional and pathophysiological significance of modulating Pyk-2 activity by integrins and ANG II. Pyk-2 phosphorylation, in response to ANG II, is increased in VSMC derived from SHR compared to those from WKY (Rocic et al. 2002). Pyk-2 has been reported to modulate VSMC migration in response to ANG II (Blaschke et al. 2002). In addition, EGF-R might be a target for Pyk-2 in response to ANG II, resulting in transactivation of EGF-R and ERK1/2 activation (Ginnan and Singer 2002). Furthermore, it was shown that AT_1 receptor-mediated activation of JAK2 (janus kinase 2) is dependent on Ca^{++}, PKCδ and their target kinase Pyk-2 (Frank et al. 2002) in VSMC.

Taken together, ANG II interferes with FA signaling by initiating the recruitment, phosphorylation and activation of several key molecules of the focal adhesion complex, among them FAK, Paxillin, tensin, c-src, p130 cas and Pyk-2. All these molecules are critically involved in integrin signaling. Activation of Pyk-2 might be of particular importance as a point of convergence, as it is directly activated in response to AT_1 receptors and after integrin ligation. Downstream of Pyk-2 are the PI-3 kinase/Akt pathways, the ERK/MAPK pathways, and possibly the JAK/STAT pathway, which regulate VSMC proliferation, migration and protein synthesis.

4 Summary and Conclusion

Angiotensin II increases the expression of a variety of ECM molecules, either directly or via the autocrine release of growth factors, inhibits proteolytic removal of ECM, and directly interferes with the signaling of focal adhesions. These actions provide ANG II with a repertoire of pathways that will lead to changes in ECM composition and subsequently to long-term structural and mechanical alterations of the vessel wall. VSMC within the altered ECM will be subjected to changes in integrin signaling, leading to phenotypic modulation, and the responsiveness to other stimuli such as growth factors and ligands of GPCRs. The remarkable success of pharmacological modulation of the RAAS is without doubt not only due to the acute effects on hemodynamic parameters, but also to the inhibition of detrimental changes in ECM composition, cellular phenotype and vessel structure. The pharmacological interference with the RAAS is therefore also an example of successful therapeutic ECM modulation and should encourage further research on targeted modulation of vascular ECM.

References

Ahn JD, Morishita R, Kaneda Y, Lee KU, Park JY, Jeon YJ, Song HS, Lee IK (2001) Transcription factor decoy for activator protein-1 (AP-1) inhibits high glucose- and angiotensin II-induced type 1 plasminogen activator inhibitor (PAI-1) gene expression in cultured human vascular smooth muscle cells. Diabetologia 44:713–720

Akishita M, Horiuchi M, Yamada H, Zhang L, Shirakami G, Tamura K, Ouchi Y, Dzau VJ (2000a) Inflammation influences vascular remodeling through AT2 receptor expression and signaling. Physiol Genomics 2:13–20

Akishita M, Iwai M, Wu L, Zhang L, Ouchi Y, Dzau VJ, Horiuchi M (2000b) Inhibitory effect of angiotensin II type 2 receptor on coronary arterial remodeling after aortic banding in mice. Circulation 102:1684–1689

Alderman MH, Madhavan S, Ooi WL, Cohen H, Sealey JE, Laragh JH (1991) Association of the renin-sodium profile with the risk of myocardial infarction in patients with hypertension. N Engl J Med 324:1098–1104

Aplin AE, Howe A, Alahari SK, Juliano RL (1998) Signal transduction and signal modulation by cell adhesion receptors: the role of integrins, cadherins, immunoglobulin-cell adhesion molecules, and selectins. Pharmacol Rev 50:197–263

Bailey WL, LaFleur DW, Forrester JS, Fagin JA, Sharifi BG (1994) Stimulation of rat vascular smooth muscle cell glycosaminoglycan production by angiotensin II. Atherosclerosis 111:55–64

Benetos A, Gautier S, Ricard S, Topouchian J, Asmar R, Poirier O, Larosa E, Guize L, Safar M, Soubrier F, Cambien F (1996) Influence of angiotensin-converting enzyme and angiotensin II type 1 receptor gene polymorphisms on aortic stiffness in normotensive and hypertensive patients. Circulation 94:698–703

Berk BC (2001) Vascular smooth muscle growth: autocrine growth mechanisms. Physiol Rev 81:999–1030

Blaschke F, Stawowy P, Kappert K, Goetze S, Kintscher U, Wollert-Wulf B, Fleck E, Graf K (2002) Angiotensin II-augmented migration of VSMCs towards PDGF-BB involves Pyk2 and ERK 1/2 activation. Basic Res Cardiol 97:334–342

Brede M, Hadamek K, Meinel L, Wiesmann F, Peters J, Engelhardt S, Simm A, Haase A, Lohse MJ, Hein L (2001) Vascular hypertrophy and increased P70S6 kinase in mice lacking the angiotensin II AT(2) receptor. Circulation 104:2602–2607

Brilla CG, Matsubara L, Weber KT (1996) Advanced hypertensive heart disease in spontaneously hypertensive rats. Lisinopril-mediated regression of myocardial fibrosis. Hypertension 28:269–275

Brown NJ, Agirbasli MA, Williams GH, Litchfield WR, Vaughan DE (1998) Effect of activation and inhibition of the renin-angiotensin system on plasma PAI-1. Hypertension 32:965–971

Brown NJ, Kim KS, Chen YQ, Blevins LS, Nadeau JH, Meranze SG, Vaughan DE (2000a) Synergistic effect of adrenal steroids and angiotensin II on plasminogen activator inhibitor-1 production. J Clin Endocrinol Metab 85:336–344

Brown NJ, Nakamura S, Ma L, Nakamura I, Donnert E, Freeman M, Vaughan DE, Fogo AB (2000b) Aldosterone modulates plasminogen activator inhibitor-1 and glomerulosclerosis in vivo. Kidney Int 58:1219–1227

Castro CM, Cruzado MC, Miatello RM, Risler NR (1999) Proteoglycan production by vascular smooth muscle cells from resistance arteries of hypertensive rats. Hypertension 34:893–896

Chang MY, Potter-Perigo S, Tsoi C, Chait A, Wight TN (2000) Oxidized low-density lipoproteins regulate synthesis of monkey aortic smooth muscle cell proteoglycans that have enhanced native low-density lipoprotein binding properties. J Biol Chem 275:4766–4773

Chassagne C, Adamy C, Ratajczak P, Gingras B, Teiger E, Planus E, Oliviero P, Rappaport L, Samuel JL, Meloche S (2002) Angiotensin II AT(2) receptor inhibits smooth muscle cell migration via fibronectin cell production and binding. Am J Physiol Cell Physiol 282:C654–C664

Cizmeci-Smith G, Stahl RC, Showalter LJ, Carey DJ (1993) Differential expression of transmembrane proteoglycans in vascular smooth muscle cells. J Biol Chem 268:18740–18747

Danielson KG, Baribault H, Holmes DF, Graham H, Kadler KE, Iozzo RV (1997) Targeted disruption of decorin leads to abnormal collagen fibril morphology and skin fragility. J Cell Biol 136:729–743

De Gasparo M, Catt KJ, Inagami T, Wright JW, Unger T (2000) International union of pharmacology. XXIII. The angiotensin II receptors. Pharmacol Rev 52:415–472

Diez J, Laviades C, Mayor G, Gil MJ, Monreal I (1995) Increased serum concentrations of procollagen peptides in essential hypertension. Relation to cardiac alterations. Circulation 91:1450–1456

Dumler I, Kopmann A, Weis A, Mayboroda OA, Wagner K, Gulba DC, Haller H (1999) Urokinase activates the Jak/Stat signal transduction pathway in human vascular endothelial cells. Arterioscler Thromb Vasc Biol 19:290–297

Eguchi S, Iwasaki H, Inagami T, Numaguchi K, Yamakawa T, Motley ED, Owada KM, Marumo F, Hirata Y (1999) Involvement of PYK2 in angiotensin II signaling of vascular smooth muscle cells. Hypertension 33:201–206

Eguchi S, Numaguchi K, Iwasaki H, Matsumoto T, Yamakawa T, Utsunomiya H, Motley ED, Kawakatsu H, Owada KM, Hirata Y, Marumo F, Inagami T (1998) Calcium-dependent epidermal growth factor receptor transactivation mediates the angiotensin II-induced mitogen-activated protein kinase activation in vascular smooth muscle cells. J Biol Chem 273:8890–8896

Eriksson P, Kallin B, van't Hooft FM, Bavenholm P, Hamsten A (1995) Allele-specific increase in basal transcription of the plasminogen-activator inhibitor 1 gene is associated with myocardial infarction. Proc Natl Acad Sci U S A 92:1851–1855

Evanko S, Raines EW, Ross R, Gold LI, Wight TN (1998) Proteoglycan distribution in lesions of atherosclerosis depends on lesion severity, structural characteristics and the proximity of PDGF and TGF-β1. Am J Pathol 152:533–546

Fakhouri F, Placier S, Ardaillou R, Dussaule JC, Chatziantoniou C (2001) Angiotensin II activates collagen type I gene in the renal cortex and aorta of transgenic mice through interaction with endothelin and TGF-beta. J Am Soc Nephrol 12:2701–2710

Figueroa JE, Vijayagopal P (2002) Angiotensin II stimulates synthesis of vascular smooth muscle cell proteoglycans with enhanced low-density lipoprotein binding properties. Atherosclerosis 162:261–268

Fischer JW, Kinsella MG, Clowes MM, Lara S, Clowes AW, Wight TN (2000) Local expression of bovine decorin by cell-mediated gene transfer reduces neointimal formation after balloon injury in rats. Circ Res 86:676–683

Fischer JW, Kinsella MG, Levkau B, Clowes AW, Wight TN (2001b) Retroviral overexpression of decorin differentially affects the response of arterial smooth muscle cells to growth factors. Arterioscler Thromb Vasc Biol 21:777–784

Fischer JW, Stoll M, Hahn AW, Unger T (2001a) Differential regulation of thrombospondin-1 and fibronectin by angiotensin II receptor subtypes in cultured endothelial cells. Cardiovasc Res 51:784–791

Fitzsimmons C, Proudfoot D, Bowyer DE (1999) Monocyte prostaglandins inhibit procollagen secretion by human vascular smooth muscle cells: implications for plaque stability. Atherosclerosis 142:287–293

Ford CM, Li S, Pickering JG (1999) Angiotensin II stimulates collagen synthesis in human vascular smooth muscle cells. Involvement of the AT(1) receptor, transforming growth factor-beta, and tyrosine phosphorylation. Arterioscler Thromb Vasc Biol 19:1843–1851

Frank GD, Saito S, Motley ED, Sasaki T, Ohba M, Kuroki T, Inagami T, Eguchi S (2002) Requirement of Ca(2+) and PKCdelta for Janus kinase 2 activation by angiotensin II: involvement of PYK2. Mol Endocrinol 16:367–377

Fujimoto T, Savani RC, Watari M, Day AJ, Strauss JF 3rd (2002) Induction of the hyaluronic acid-binding protein, tumor necrosis factor-stimulated gene-6, in cervical smooth muscle cells by tumor necrosis factor-alpha and prostaglandin E(2). Am J Pathol 160:1495–1502

Geisterfer AA, Peach MJ, Owens GK (1988) Angiotensin II induces hypertrophy, not hyperplasia, of cultured rat aortic smooth muscle cells. Circ Res 62:749–756

Giachelli CM, Bae N, Almeida M, Denhardt DT, Alpers CE, Schwartz SM (1993) Osteopontin is elevated during neointima formation in rat arteries and is a novel component of human atherosclerotic plaque. J Clin Invest 92:1686–1696

Giancotti FG, Ruoslahti E (1999) Integrin signaling. Science 285:1028–1032

Gibbons GH, Pratt RE, Dzau VJ (1992) Vascular smooth muscle cell hypertrophy vs. hyperplasia. Autocrine transforming growth factor-beta 1 expression determines growth response to angiotensin II. J Clin Invest 90:456–461

Ginnan R, Singer HA (2002) CaM kinase II-dependent activation of tyrosine kinases and ERK1/2 in vascular smooth muscle. Am J Physiol Cell Physiol 282:C754–C761

Hamdan AD, Quist WC, Gagne JB, Feener EP (1996) Angiotensin-converting enzyme inhibition suppresses plasminogen activator inhibitor-1 expression in the neointima of balloon-injured rat aorta. Circulation 93:1073–1078

Hamsten A, Wiman B, de Faire U, Blomback M (1985) Increased plasma levels of a rapid inhibitor of tissue plasminogen activator in young survivors of myocardial infarction. N Engl J Med 313:1557–1563

Harmon KJ, Couper LL, Lindner V (2000) Strain-dependent vascular remodeling phenotypes in inbred mice. Am J Pathol 156:1741–1748

Hedin U, Thyberg J (1987) Plasma fibronectin promotes modulation of arterial smooth-muscle cells from contractile to synthetic phenotype. Differentiation 33:239–246

Himeno H, Crawford DC, Hosoi M, Chobanian AV, Brecher P (1994) Angiotensin II alters aortic fibronectin independently of hypertension. Hypertension 23:823–826

Hu WY, Fukuda N, Satoh C, Jian T, Kubo A, Nakayama M, Kishioka H, Kanmatsuse K (2000) Phenotypic modulation by fibronectin enhances the angiotensin II-generating system in cultured vascular smooth muscle cells. Arterioscler Thromb Vasc Biol 20:1500–1505

Huber K, Jorg M, Probst P, Schuster E, Lang I, Kaindl F, Binder BR (1992) A decrease in plasminogen activator inhibitor-1 activity after successful percutaneous transluminal coronary angioplasty is associated with a significantly reduced risk for coronary restenosis. Thromb Haemost 67:209–213

Iozzo RV, Pillarisetti J, Sharma B, Murdoch AD, Danielson KG, Uitto J, Mauviel A (1997) Structural and functional characterization of the human perlecan gene promoter. J Biol Chem 272:5219–5228

Ishida T, Ishida M, Suero J, Takahashi M, Berk BC (1999) Agonist-stimulated cytoskeletal reorganization and signal transduction at focal adhesions in vascular smooth muscle cells require c-Src. J Clin Invest 103:789–797

Itoh H, Mukoyama M, Pratt RE, Gibbons GH, Dzau VJ (1993) Multiple autocrine growth factors modulate vascular smooth muscle cell growth response to angiotensin II. J Clin Invest 91:2268–2274

Jones PL, Crack J, Rabinovitch M (1997) Regulation of tenascin-C, a vascular smooth muscle cell survival factor that interacts with the alpha v beta 3 integrin to promote epidermal growth factor receptor phosphorylation and growth. J Cell Biol 139:279–293

Kaplan KB, Swedlow JR, Morgan DO, Varmus HE (1995) c-Src enhances the spreading of src-/- fibroblasts on fibronectin by a kinase-independent mechanism. Genes Dev 9:1505–1517

Kato H, Suzuki H, Tajima S, Ogata Y, Tominaga T, Sato A, Saruta T (1991) Angiotensin II stimulates collagen synthesis in cultured vascular smooth muscle cells. J Hypertens 9:17–22

Keely PJ, Westwick JK, Whitehead IP, Der CJ, Parise LV (1997) Cdc42 and Rac1 induce integrin-mediated cell motility and invasiveness through PI(3)K. Nature 390:632–636

Kenagy RD, Fischer JW, Davies MG, Berceli SA, Hawkins SM, Wight TN, Clowes AW (2002) Increased plasmin and serine proteinase activity during flow-induced intimal atrophy in baboon PTFE grafts. Arterioscler Thromb Vasc Biol 22:400–404

Kiefer MC, Stephans JC, Crawford K, Okino K, Barr PJ (1990) Ligand-affinity cloning and structure of a cell surface heparan sulfate proteoglycan that binds basic fibroblast growth factor. Proc Natl Acad Sci U S A 87:6985–6989

Kim S, Kawamura M, Wanibuchi H, Ohta K, Hamaguchi A, Omura T, Yukimura T, Miura K, Iwao H (1995) Angiotensin II type 1 receptor blockade inhibits the expression of immediate-early genes and fibronectin in rat injured artery. Circulation 92:88–95

Koh E, Morimoto S, Jiang B, Inoue T, Nabata T, Kitano S, Yasuda O, Fukuo K, Ogihara T (1993) Effects of beraprost sodium, a stable analogue of prostacyclin, on hyperplasia, hypertrophy and glycosaminoglycan synthesis of rat aortic smooth muscle cells. Artery 20:242–252

Kosmehl H, Berndt A, Katenkamp D (1996) Molecular variants of fibronectin and laminin: structure, physiological occurrence and histopathological aspects. Virchows Arch 429:311–322

Koyama H, Raines EW, Bornfeldt KE, Roberts JM, Ross R (1996) Fibrillar collagen inhibits arterial smooth muscle proliferation through regulation of Cdk2 inhibitors. Cell 87:1069–1078

Labat-Robert J, Szendroi M, Godeau G, Robert L (1985) Comparative distribution patterns of type I and III collagens and fibronectin in human arteriosclerotic aorta. Pathol Biol (Paris) 33:261–265

Lakatta EG (1993) Cardiovascular regulatory mechanisms in advanced age. Physiol Rev 73:413–467

Lakkakorpi PT, Nakamura I, Nagy RM, Parsons JT, Rodan GA, Duong LT (1999) Stable association of PYK2 and p130(Cas) in osteoclasts and their co-localization in the sealing zone. J Biol Chem 274:4900–4907

Laviades C, Varo N, Fernandez J, Mayor G, Gil MJ, Monreal I, Diez J (1998) Abnormalities of the extracellular degradation of collagen type I in essential hypertension. Circulation 98:535–540

Leduc I, Meloche S (1995) Angiotensin II stimulates tyrosine phosphorylation of the focal adhesion-associated protein paxillin in aortic smooth muscle cells. J Biol Chem 270:4401–4404

Levy BI, Benessiano J, Henrion D, Caputo L, Heymes C, Duriez M, Poitevin P, Samuel JL (1996) Chronic blockade of AT2-subtype receptors prevents the effect of angiotensin II on the rat vascular structure. J Clin Invest 98:418–425

Ling S, Dai A, Ma YH, Wilson E, Chatterjee K, Ives HE, Sudhir K (1999) Matrix-dependent gene expression of egr-1 and PDGF A regulate angiotensin II-induced proliferation in human vascular smooth muscle cells. Hypertension 34:1141–1146

Li-Saw-Hee FL, Edmunds E, Blann AD, Beevers DG, Lip GY (2000) Matrix metalloproteinase-9 and tissue inhibitor metalloproteinase-1 levels in essential hypertension. Relationship to left ventricular mass and anti-hypertensive therapy. Int J Cardiol 75:43–47

Litvak V, Tian D, Shaul YD, Lev S (2000) Targeting of PYK2 to focal adhesions as a cellular mechanism for convergence between integrins and G protein-coupled receptor signaling cascades. J Biol Chem 275:32736–32746

Lutgens E, Gijbels M, Smook M, Heeringa P, Gotwals P, Koteliansky VE, Daemen MJ (2002) Transforming growth factor-beta mediates balance between inflammation and fibrosis during plaque progression. Arterioscler Thromb Vasc Biol 22:975–982

Mackie EJ, Scott-Burden T, Hahn AW, Kern F, Bernhardt J, Regenass S, Weller A, Buhler FR (1992) Expression of tenascin by vascular smooth muscle cells. Alterations in hypertensive rats and stimulation by angiotensin II. Am J Pathol 141:377–388

Mallat Z, Gojova A, Marchiol-Fournigault C, Esposito B, Kamate C, Merval R, Fradelizi D, Tedgui A (2001) Inhibition of transforming growth factor-beta signaling accelerates atherosclerosis and induces an unstable plaque phenotype in mice. Circ Res 89:930–934

Matsubara H, Moriguchi Y, Mori Y, Masaki H, Tsutsumi Y, Shibasaki Y, Uchiyama-Tanaka Y, Fujiyama S, Koyama Y, Nose-Fujiyama A, Iba S, Tateishi E, Iwasaka T (2000) Transactivation of EGF receptor induced by angiotensin II regulates fibronectin and TGF-beta gene expression via transcriptional and post-transcriptional mechanisms. Mol Cell Biochem 212:187–201

Mercurius KO, Morla AO (1998) Inhibition of vascular smooth muscle cell growth by inhibition of fibronectin matrix assembly. Circ Res 82:548–556

Merrilees MJ, Lemire JM, Fischer JW, Kinsella MG, Braun KR, Clowes AW, Wight TN (2002) Retrovirally mediated overexpression of versican v3 by arterial smooth muscle cells induces tropoelastin synthesis and elastic fiber formation in vitro and in neointima after vascular injury. Circ Res 90:481–487

Mifune M, Sasamura H, Shimizu-Hirota R, Miyazaki H, Saruta T (2000) Angiotensin II type 2 receptors stimulate collagen synthesis in cultured vascular smooth muscle cells. Hypertension 36:845–850

Mignatti P, Rifkin DB (1993) Biology and biochemistry of proteinases in tumor invasion. Physiol Rev 73:161–195

Morishita R, Gibbons GH, Horiuchi M, Kaneda Y, Ogihara T, Dzau VJ (1998) Role of AP-1 complex in angiotensin II-mediated transforming growth factor-beta expression and growth of smooth muscle cells: using decoy approach against AP-1 binding site. Biochem Biophys Res Commun 243:361–367

Murasawa S, Mori Y, Nozawa Y, Gotoh N, Shibuya M, Masaki H, Maruyama K, Tsutsumi Y, Moriguchi Y, Shibazaki Y, Tanaka Y, Iwasaka T, Inada M, Matsubara H (1998) Angiotensin II type 1 receptor-induced extracellular signal-regulated protein kinase activation is mediated by Ca2+/calmodulin-dependent transactivation of epidermal growth factor receptor. Circ Res 82:1338–1348

Murphy G, Knauper V (1997) Relating matrix metalloproteinase structure to function: why the "hemopexin" domain? Matrix Biol 15:511–518

Naftilan AJ, Pratt RE, Dzau VJ (1989) Induction of platelet-derived growth factor A-chain and c-myc gene expressions by angiotensin II in cultured rat vascular smooth muscle cells. J Clin Invest 83:1419–14124

Nakajima M, Hutchinson HG, Fujinaga M, Hayashida W, Morishita R, Zhang L, Horiuchi M, Pratt RE, Dzau VJ (1995) The angiotensin II type 2 (AT2) receptor antagonizes the growth effects of the AT1 receptor: gain-of-function study using gene transfer. Proc Natl Acad Sci U S A 92:10663–10667

Nakamura S, Nakamura I, Ma L, Vaughan DE, Fogo AB (2000) Plasminogen activator inhibitor-1 expression is regulated by the angiotensin type 1 receptor in vivo. Kidney Int 58:251–259

O'Brien KD, Olin KL, Alpers CE, Chiu W, Ferguson M, Hudkins K, Wight TN, Chait A (1998) A comparison of apolipoprotein and proteoglycan deposits in human coronary atherosclerotic plaques: co-localization of biglycan with apolipoproteins. Circulation 98:519–527

Ohnaka K, Numaguchi K, Yamakawa T, Inagami T (2000) Induction of cyclooxygenase-2 by angiotensin II in cultured rat vascular smooth muscle cells. Hypertension 35:68–75

Okuda M, Kawahara Y, Nakayama I, Hoshijima M, Yokoyama M (1995) Angiotensin II transduces its signal to focal adhesions via angiotensin II type 1 receptors in vascular smooth muscle cells. FEBS Lett 368:343–347

Ooshima A, Fuller GC, Cardinale GJ, Spector S, Udenfriend S (1974) Increased collagen synthesis in blood vessels of hypertensive rats and its reversal by antihypertensive agents. Proc Natl Acad Sci U S A 71:3019–3023

O'Rourke M, Frohlich ED (1999) Pulse pressure: Is this a clinically useful risk factor? Hypertension 34:372–374

Papakonstantinou E, Roth M, Kokkas B, Papadopoulos C, Karakiulakis G (2001) Losartan inhibits the angiotensin II-induced modifications on fibrinolysis and matrix deposition by primary human vascular smooth muscle cells. J Cardiovasc Pharmacol 38:715–728

Pickering JG, Chow LH, Li S, Rogers KA, Rocnik EF, Zhong R, Chan BM (2000) Alpha5beta1 integrin expression and luminal edge fibronectin matrix assembly by smooth muscle cells after arterial injury. Am J Pathol 156:453–465

Plutzky J (1999) Atherosclerotic plaque rupture: emerging insights and opportunities. Am J Cardiol 84:15J–20J

Polte TR, Naftilan AJ, Hanks SK (1994) Focal adhesion kinase is abundant in developing blood vessels and elevation of its phosphotyrosine content in vascular smooth muscle cells is a rapid response to angiotensin II. J Cell Biochem 55:106–119

Preissner KT, Grulich-Henn J, Ehrlich HJ, Declerck P, Justus C, Collen D, Pannekoek H, Muller-Berghaus G (1990) Structural requirements for the extracellular interaction of plasminogen activator inhibitor 1 with endothelial cell matrix-associated vitronectin. J Biol Chem 265:18490–18498

Raines EW, Koyama H, Carragher NO (2000) The extracellular matrix dynamically regulates smooth muscle cell responsiveness to PDGF. Ann N Y Acad Sci 902:39–51; discussion 51–52

Reinboth B, Hanssen E, Cleary EG, Gibson MA (2002) Molecular interactions of biglycan and decorin with elastic fiber components: biglycan forms a ternary complex with tropoelastin and microfibril-associated glycoprotein 1. J Biol Chem 277:3950–3957

Ridker PM, Gaboury CL, Conlin PR, Seely EW, Williams GH, Vaughan DE (1993) Stimulation of plasminogen activator inhibitor in vivo by infusion of angiotensin II. Evidence of a potential interaction between the renin-angiotensin system and fibrinolytic function. Circulation 87:1969–1973

Roberts AB, McCune BK, Sporn MB (1992) TGF-beta: regulation of extracellular matrix. Kidney Int 41:557–559

Rocic P, Lucchesi PA (2001a) Down-regulation by antisense oligonucleotides establishes a role for the proline-rich tyrosine kinase PYK2 in angiotensin II-induced signaling in vascular smooth muscle. J Biol Chem 276:21902–21906

Rocic P, Govindarajan G, Sabri A, Lucchesi PA (2001b) A role for PYK2 in regulation of ERK1/2 MAP kinases and PI 3-kinase by ANG II in vascular smooth muscle. Am J Physiol Cell Physiol 280:C90–C99

Rocic P, Griffin TM, McRae CN, Lucchesi PA (2002) Altered PYK2 phosphorylation by ANG II in hypertensive vascular smooth muscle. Am J Physiol Heart Circ Physiol 282:H457–H465

Ruoslahti E (1988) Fibronectin and its receptors. Annu Rev Biochem 57:375–413

Sabri A, Govindarajan G, Griffin TM, Byron KL, Samarel AM, Lucchesi PA (1998) Calcium- and protein kinase C-dependent activation of the tyrosine kinase PYK2 by angiotensin II in vascular smooth muscle. Circ Res 83:841–851

Sasamura H, Shimizu-Hirota R, Nakaya H, Saruta T (2001) Effects of AT1 receptor antagonist on proteoglycan gene expression in hypertensive rats. Hypertens Res 24:165–172

Sawathiparnich P, Kumar S, Vaughan DE, Brown NJ (2002) Spironolactone abolishes the relationship between aldosterone and plasminogen activator inhibitor-1 in humans. J Clin Endocrinol Metab 87:448–452

Sayeski PP, Ali MS, Harp JB, Marrero MB, Bernstein KE (1998) Phosphorylation of p130Cas by angiotensin II is dependent on c-Src, intracellular Ca2+, and protein kinase C. Circ Res 82:1279–1288

Schneiderman J, Sawdey MS, Keeton MR, Bordin GM, Bernstein EF, Dilley RB, Loskutoff DJ (1992) Increased type 1 plasminogen activator inhibitor gene expression in atherosclerotic human arteries. Proc Natl Acad Sci U S A 89:6998–7002

Schönherr E, Järveläinen HT, Sandell LJ, Wight TN (1991) Effects of platelet-derived growth factor and transforming growth factor-beta 1 on the synthesis of a large versican-like chondroitin sulfate proteoglycan by arterial smooth muscle cells. J Biol Chem 266:17640–17647

Schönherr E, Järveläinen HT, Kinsella MG, Sandell LJ, Wight TN (1993) Platelet-derived growth factor and transforming growth factor-b1 differentially affect the synthesis of biglycan and decorin by monkey arterial smooth muscle cells. Arterioscler Thromb 13:1026–1036

Schrör K, Weber AA (1997) Roles of vasodilatory prostaglandins in mitogenesis of vascular smooth muscle cells. Agents Actions Suppl 48:63–91

Schwartz MA, Baron V (1999) Interactions between mitogenic stimuli, or, a thousand and one connections. Curr Opin Cell Biol 11:197–202

Scott JE, Haigh M (1988) Identification of specific binding sites for keratan sulphate proteoglycans and chondroitin-dermatan sulphate proteoglycans on collagen fibrils in cornea by the use of cupromeronic blue in 'critical- electrolyte-concentration' techniques [published erratum appears in Biochem J 255:1061]. Biochem J 253:607–610

Scott-Burden T, Resink TJ, Hahn AW, Buhler FR (1990) Induction of thrombospondin expression in vascular smooth muscle cells by angiotensin II. J Cardiovasc Pharmacol 16:S17–S20

Senbonmatsu T, Ichihara S, Price E Jr, Gaffney FA, Inagami T (2000) Evidence for angiotensin II type 2 receptor-mediated cardiac myocyte enlargement during in vivo pressure overload. J Clin Invest 106:R25–R29

Sharifi BG, LaFleur DW, Pirola CJ, Forrester JS, Fagin JA (1992) Angiotensin II regulates tenascin gene expression in vascular smooth muscle cells. J Biol Chem 267:23910–23915

Shimizu-Hirota R, Sasamura H, Mifune M, Nakaya H, Kuroda M, Hayashi M, Saruta T (2001) Regulation of vascular proteoglycan synthesis by angiotensin II type 1 and type 2 receptors. J Am Soc Nephrol 12:2609–2615

Shindo T, Manabe I, Fukushima Y, Tobe K, Aizawa K, Miyamoto S, Kawai-Kowase K, Moriyama N, Imai Y, Kawakami H, Nishimatsu H, Ishikawa T, Suzuki T, Morita H, Maemura K, Sata M, Hirata Y, Komukai M, Kagechika H, Kadowaki T, Kurabayashi M, Nagai R (2002) Kruppel-like zinc-finger transcription factor KLF5/BTEB2 is a target for angiotensin II signaling and an essential regulator of cardiovascular remodeling. Nat Med 8:856–863

Simon G, Abraham G, Altman S (1994) Stimulation of vascular glycosaminoglycan synthesis by subpressor angiotensin II in rats. Hypertension 23:I148-I151

Stetler-Stevenson WG, Liotta LA, Kleiner DE Jr (1993) Extracellular matrix 6: role of matrix metalloproteinases in tumor invasion and metastasis. FASEB J 7:1434–1441

Stetler-Stevenson WG, Yu AE (2001) Proteases in invasion: matrix metalloproteinases. Semin Cancer Biol 11:143–152

Stoll M, Steckelings UM, Paul M, Bottari SP, Metzger R, Unger T (1995) The angiotensin AT2-receptor mediates inhibition of cell proliferation in coronary endothelial cells. J Clin Invest 95:651–657

Stoll M, Unger T (2001) Angiotensin and its AT2 receptor: new insights into an old system. Regul Pept 99:175–182

Su EJ, Lombardi DM, Siegal J, Schwartz SM (1998) Angiotensin II induces vascular smooth muscle cell replication independent of blood pressure. Hypertension 31:1331–1337

Sudhir K, Wilson E, Chatterjee K, Ives HE (1993) Mechanical strain and collagen potentiate mitogenic activity of angiotensin II in rat vascular smooth muscle cells. J Clin Invest 92:3003–3007

Takeda K, Ichiki T, Tokunou T, Iino N, Fujii S, Kitabatake A, Shimokawa H, Takeshita A (2001) Critical role of Rho-kinase and MEK/ERK pathways for angiotensin II-induced plasminogen activator inhibitor type-1 gene expression. Arterioscler Thromb Vasc Biol 21:868–873

Tamura K, Chen YE, Lopez-Ilasaca M, Daviet L, Tamura N, Ishigami T, Akishita M, Takasaki I, Tokita Y, Pratt RE, Horiuchi M, Dzau VJ, Umemura S (2000) Molecular mechanism of fibronectin gene activation by cyclic stretch in vascular smooth muscle cells. J Biol Chem 275:34619–34627

Tamura K, Nyui N, Tamura N, Fujita T, Kihara M, Toya Y, Takasaki I, Takagi N, Ishii M, Oda K, Horiuchi M, Umemura S (1998) Mechanism of angiotensin II-mediated regulation of fibronectin gene in rat vascular smooth muscle cells. J Biol Chem 273:26487–26496

Tamura K, Okazaki M, Tamura M, Kanegae K, Okuda H, Abe H, Nakashima Y (2001) Synergistic interaction of integrin and angiotensin II in activation of extracellular signal-regulated kinase pathways in vascular smooth muscle cells. J Cardiovasc Pharmacol 38 [Suppl 1]:S59–S62

Thyberg J, Hultgardh-Nilsson A (1994) Fibronectin and the basement membrane components laminin and collagen type IV influence the phenotypic properties of subcultured rat aortic smooth muscle cells differently. Cell Tissue Res 276:263–271

Tokimitsu I, Kato H, Wachi H, Tajima S (1994) Elastin synthesis is inhibited by angiotensin II but not by platelet-derived growth factor in arterial smooth muscle cells. Biochim Biophys Acta 1207:68–73

Touyz RM, Wu XH, He G, Salomon S, Schiffrin EL (2002) Increased angiotensin II-mediated Src signaling via epidermal growth factor receptor transactivation is associated with decreased C-terminal Src kinase activity in vascular smooth muscle cells from spontaneously hypertensive rats. Hypertension 39:479–485

Turley EA, Noble PW, Bourguignon LY (2002) Signaling properties of hyaluronan receptors. J Biol Chem 277:4589–4592

Turner CE, Pietras KM, Taylor DS, Molloy CJ (1995) Angiotensin II stimulation of rapid paxillin tyrosine phosphorylation correlates with the formation of focal adhesions in rat aortic smooth muscle cells. J Cell Sci 108:333–342

Van Leeuwen RT, Kol A, Andreotti F, Kluft C, Maseri A, Sperti G (1994) Angiotensin II increases plasminogen activator inhibitor type 1 and tissue-type plasminogen activator messenger RNA in cultured rat aortic smooth muscle cells. Circulation 90:362–368

Walker-Caprioglio HM, Koob TJ, McGuffee LJ (1992) Proteoglycan synthesis in normotensive and spontaneously hypertensive rat arteries in vitro. Matrix 12:308–320

Weber AA, Schrör K (1999) The significance of platelet-derived growth factors for proliferation of vascular smooth muscle cells. Platelets 10:77–96

Weinreb RN, Kashiwagi K, Kashiwagi F, Tsukahara S, Lindsey JD (1997) Prostaglandins increase matrix metalloproteinase release from human ciliary smooth muscle cells. Invest Ophthalmol Vis Sci 38:2772–2780

Yamaguchi Y, Mann DM, Ruoslahti E (1990) Negative regulation of transforming growth factor-beta by the proteoglycan decorin. Nature 346:281–284

Yang BL, Zhang Y, Cao L, Yang BB (1999) Cell adhesion and proliferation mediated through the G1 domain of versican. J Cell Biochem 72:210–220

Yao LY, Moody C, Schonherr E, Wight TN, Sandell LJ (1994) Identification of the proteoglycan versican in aorta and smooth muscle cells by DNA sequence analysis, in situ hybridization and immunohistochemistry. Matrix Biol 14:213–225

Zhang Y, Cao L, Yang BL, Yang BB (1998) The G3 domain of versican enhances cell proliferation via epidermal growth factor-like motifs. J Biol Chem 273:21342–21351

Renin-Angiotensin Inhibitors and Vascular Effects

E. L. Schiffrin

IRCM 110 Pine Avenue West, Montréal, Québec, H2W 1R7, Canada
e-mail: ernesto.schiffrin@ircm.qc.ca

Abstract Angiotensin II induces vasoconstriction, endothelial dysfunction, vascular inflammation and structural changes that are associated with cell proliferation and hypertrophy as well as fibrosis of the media of vessels. Similar changes occur in experimental and human hypertension. Large and small arteries are remodeled in hypertension, and their structure, function and mechanics are altered. These changes participate in the mechanisms of elevation of blood pressure and in the pathophysiology of its complications. For this reason it has been thought that renin-angiotensin blockade with either angiotensin-converting enzyme (ACE) inhibitors or angiotensin receptor blockers could prevent or regress toward normal the vascular changes present in experimental or human hypertension. Changes found in small resistance arteries may be the first manifestation of target organ damage in hypertensive patients. Endothelial dysfunction is often found in vessels from experimental animals and in many patients with elevated blood pressure. Interruption of the renin-angiotensin system may correct many of these abnormalities. Both ACE inhibitors and angiotensin AT_1 receptor blockers have been shown to improve vascular structure and endothelial dysfunction in experimental hypertension and in human essential hypertension. In patients, whereas renin-angiotensin blockade corrected vascular changes, the same level of blood pressure lowering with a beta-blocker failed to favorably impact the vasculature. Improved outcomes in clinical trials using ACE inhibitors and angiotensin AT_1 receptor antagonists may be a consequence in part of the vascular protective effects offered by these agents.

Keywords Converting enzyme inhibitors · AT_1 receptor antagonists · Endothelium · Resistance arteries · Remodeling

1 Introduction

Previous chapters have detailed the effects of angiotensin II on smooth muscle contraction, generation of oxidative stress, endothelial dysfunction, inflammation, growth and proliferation. We have recently reviewed the cellular mechanisms that at the level of smooth muscle cells lead to the pathophysiological consequences that angiotensin II has on blood vessels (Touyz and Schiffrin 2000). Angiotensin II will not only induce vasoconstriction, but also activation of growth pathways, stimulation of production of reactive oxygen species, of mediators of inflammation (NFkappaB, AP-1) that increase the expression of cytokines, chemokines and adhesion molecules and lead to vascular damage associated with angiotensin II infusion. Many of these changes in blood vessels found in response to angiotensin II are similar to the remodeling documented in experimental animals and in humans with high blood pressure.

Large conduit arteries in hypertensive experimental animals exhibit increased lumen diameter and a thickened wall, with fibrosis of the media and enhanced stiffness of the wall (Laurent 1995; Safar et al. 1996). In addition, endothelial function of large arteries has been shown to be impaired in different models of experimental hypertension (Lockette et al. 1986, Lüscher et al. 1986). The structure and function of small arteries of the mesenteric circulation, the heart and the kidney (Deng and Schiffrin 1992; Li and Schiffrin 1995; Li et al. 1996; Thybo et al. 1994), and in the brain of spontaneously hypertensive rats (SHRs) (Baumbach and Heistad 1989; Thybo et al. 1994) present structural abnormalities with some similarity to those induced by angiotensin II infusion in rats (Griffin et al. 1991) and also resemble those described in gluteal subcutaneous small arteries of hypertensive humans (Aalkjaer et al. 1987; Heagerty et al. 1993; Korsgaard et al. 1993; Park and Schiffrin 2001; Schiffrin 1992; Schiffrin and Deng 1999; Schiffrin et al. 1993). Indeed, the first manifestation of target organ damage present in mild hypertension in humans before development of cardiac hypertrophy, increased carotid intima-media thickness or microalbuminuria and nephroangiosclerosis is the remodeling of small arteries, demonstrated by biopsy studies of subcutaneous vessels (Park and Schiffrin 2001). Resistance arteries play a critical role in determining peripheral resistance (Christensen and Mulvany 1993; Folkow 1995; Mulvany and Aalkjaer 1990), but may also participate in triggering complications of high blood pressure such as stroke and nephroangiosclerosis, and, increasingly recognized, also in myocardial ischemia (Brush et al. 1988; Hasdai D et al. 1997; Kinlay et al. 1998). Whereas large arteries in hypertension present two types of changes [atherosclerosis, a disease of the intima triggered by inflammation and endothelial dysfunction and associated with dyslipidemia, and arteriosclerosis, that is, thickening of the vessel wall with fibrosis of the media, and increased lumen diameter (outward hypertrophic remodeling)], small arteries of hypertensive patients exhibit a reduction in the lumen and external diameter, normal or increased media thickness and increased media-to-lumen ratio but normal media cross-section (or volume of the media per unit length) (Heagerty et al. 1993; Korsgaard et al. 1993; Park and Schiffrin 2001; Schiffrin 1992; Schiffrin and Deng 1999; Schiffrin et al. 1993). The latter is what has recently been defined as "inward eutrophic remodeling" (Mulvany et al. 1996). These vessels have the same number of smooth muscle cells and little evidence of cell hypertrophy (Korsgaard et al. 1993), although some studies in experimental animals have reported either hyperplasia or cell hypertrophy or both. In humans, cells are restructured around a smaller vessel lumen. It is unknown how this rearrangement occurs. It is possible that apoptosis offsets growth to result in eutrophic remodeling. Indeed, angiotensin II-induced growth may be associated with secondarily enhanced apoptotic rates in the muscular media (Diep et al. 1999). Another mechanism may be chronic vasoconstriction, recently shown experimentally to induce inward eutrophic remodeling (Bakker et al. 2002). Angiotensin II and other vasoconstrictors act via a complex array of signaling mechanisms inside the cell, including MAP kinases and others, to elicit the constrictor, growth, motility and inflammatory changes that may contribute to remodeling (Touyz and Schiffrin 2000). Eutrophic remodeling found in resistance arteries may also result in part from changes in cell adhesion molecules (Intengan and Schiffrin 2000, 2001; Intengan et al. 1999a) or extracellular matrix deposition or the spatial arrangement of fibrillar material. We have recently shown in both spontaneously hypertensive rats (SHRs) (Intengan et al. 1999a; Sharifi et al. 1998) and in human resistance arteries from hypertensive patients (Intengan et al. 1999b) that collagen deposition is enhanced, and we believe that this is a fundamental aspect of vascular remodeling associated with hypertension, with reorganization of extracellular fibrillar elements, or changes in expression or topographic distribution of cell adhesion molecules, resulting in changes in cell-extracellular matrix attachment. Interestingly, collagen production by smooth muscle cells is stimulated by angiotensin II acting on AT_1 receptors, which generates the hope that it may be possible to prevent or regress vascular fibrosis with renin-angiotensin blockade, as has been recently demonstrated both in vivo in experimental animals and in vitro on smooth muscle cells in culture (Intengan et al. 1999a; Schiffrin et al. 1998; Sharifi et al. 1998; Touyz et al. 2001; Vacher et al. 1995). In secondary forms of hypertension (Rizzoni et al. 1996, 2000) and perhaps in essential hypertension as blood pressure elevation becomes more severe, eutrophic remodeling may be replaced by hypertrophic remodeling, with increased media cross-sectional area, although this transition has not been demonstrated. On the other hand, in angiotensin II-infused rats and in SHRs, some measure of hypertrophic remodeling as a consequence of activation of growth pathways may be found (Deng and Schiffrin 1992; Diep et al. 2000; Griffin et al. 1991).

Similarly to what occurs in response to angiotensin II infusion into experimental animals (Rajagopalan et al. 1996), the function of small blood vessels is also altered in hypertensive patients. The media stress developed in response to angiotensin II is increased or sometimes normal, whereas responses to most vasoconstrictors are reduced (Aalkjaer et al. 1987; Schiffrin et al. 1992, 1993). Interestingly, under some experimental conditions, if a state of mild hyperaldosteronism is induced by angiotensin II infusion, angiotensin II binding is enhanced rather than down-regulated (Schiffrin et al. 1984, 1985). This effect appears to be the result of mineralocorticoid action (Schiffrin et al. 1984, 1985; Ullian et al. 1992, 1996) and could be a mechanism for augmented vascular responsiveness to angiotensin II in hypertension. The structural abnormalities of resistance arteries, however, also contribute to enhance vasoconstrictor responses (Folkow 1982), participating in the mechanisms that lead to an elevated vascular tone in hypertension.

Endothelial function is impaired in hypertension: vasodilatory responses to acetylcholine in precontracted vessels are blunted both in hypertensive animals (Lockette et al. 1986; Lüscher et al. 1986) and humans (Deng et al. 1995; Panza et al. 1990), although normal responses have been reported in some studies (Cockroft et al. 1994). In our own studies, 40% of mild hypertensive patients may present impaired response to acetylcholine (Park and Schiffrin 2001). Endothelial dysfunction may be due to deficient generation of nitric oxide or increased degradation of nitric oxide, the latter perhaps a consequence of enhanced oxidative stress in the vascular wall (Tschudi et al. 1996), which can be induced by angiotensin II through stimulation of NADPH oxidase (Cai and Harrison 2000; Rajagopalan et al. 1996). Alterations in generation of endothelium-derived hyperpolarizing factors (EDHFs) has also been blamed for abnormal endothelial responses to acetylcholine in hypertension (Taddei et al. 2001). In some experimental animals, endothelium-dependent relaxation is blunted because acetylcholine induces contractions via generation of a cyclooxygenase-derived endothelium-dependent contracting factor or EDCF (Deng et al. 1995; Diedrich et al. 1990). The agent involved may be an endoperoxide or other cyclooxygenase-derived products. Another endothelium-derived vasoconstrictor that potentially plays an important role in abnormal endothelial function, and in elevated blood pressure and its consequences is endothelin-1 (Schiffrin 1999). Angiotensin II has been shown to induce the expression of endothelin in experimental animals (Rajagopalan et al. 1997) and could be involved in the enhanced production of endothelin by small arteries, which has been shown to occur in moderate to severe hypertension (Schiffrin et al. 1997).

2
Effect of Renin-Angiotensin Blockade on the Vasculature of Experimental Animals

2.1
ACE Inhibitors

Treatment of SHRs with various ACE inhibitors (Deng and Schiffrin 1993; Dohi et al. 1994; Harrap et al. 1990; Li and Schiffrin 1996; Rizzoni et al. 1995, 1998; Sharifi et al. 1998; Thybo et al. 1994) has been shown to improve impaired endothelial function and result in regression of vascular remodeling, including the deposition of excess collagen (Intengan et al. 1999a; Schiffrin et al. 1998; Sharifi et al. 1998). Effects of ACE inhibitors may be the result of inhibition of angiotensin II generation or of accumulation of bradykinin, both of which could also beneficially influence endothelial function (angiotensin II reduction by attenuation of oxidative stress and decreased scavenging of nitric oxide, bradykinin via stimulation of nitric oxide synthase).

2.2
Angiotensin Receptor Blockers

SHRs treated with angiotensin II receptor antagonists (Ledingham et al. 2000; Li et al. 1997; Rizzoni et al. 1998; Shaw et al. 1995) were demonstrated to exhibit correction of abnormal endothelial function and regression of vascular remodeling and the excess collagen deposition in the media of blood vessels (Schiffrin et al. 1998). Angiotensin receptor antagonists such as losartan, irbesartan, valsartan or others probably act mostly via blockade of angiotensin II-induced remodeling, which is mediated in large measure by increased production of reactive oxygen species (Touyz et al. 2002). As well, the absence of blockade of AT_2 receptors stimulated by the elevated angiotensin II concentrations found under treatment with AT_1 antagonists may stimulate nitric oxide production, which has anti-growth and pro-apoptotic effects, and also would correct the nitric oxide deficiency. The effect of angiotensin II on nitric oxide may involve stimulation of bradykinin via AT_2 receptors, which in turn would stimulate nitric oxide synthase activity (Gohlke et al. 1998).

3
Effects of Renin-Angiotensin Blockade on the Vasculature in Essential Human Hypertension

3.1
ACE Inhibitors

We have studied the effects of antihypertensive treatment on blood vessels of hypertensive patients by performing biopsies of gluteal subcutaneous tissue and

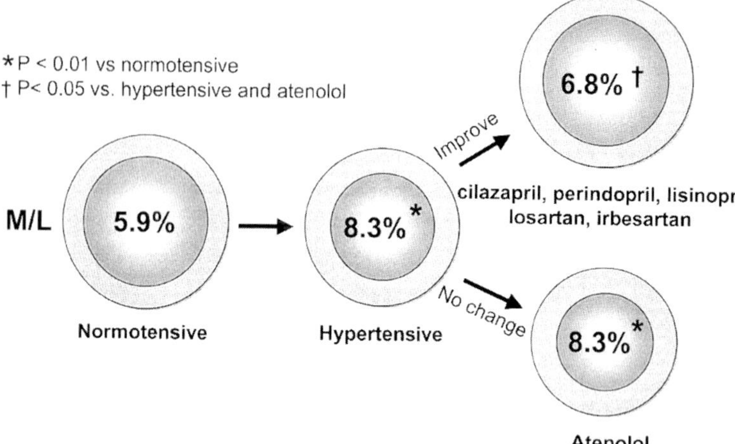

Fig. 1 Diagram demonstrating the structure of small (resistance) arteries of hypertensive patients as demonstrated by the media-to-lumen ratio and the eutrophic remodeling present of gluteal subcutaneous small arteries from hypertensive patients. Improvement of media-to-lumen ratio of small arteries was observed in patients treated with the ACE inhibitors cilazapril for 1 (Schiffrin et al. 1994) or 2 years (Schiffrin et al. 1995), perindopril for 1 year (Thybo et al. 1995), or lisinopril for 3 years (Rizzoni et al. 1997), or with the angiotensin receptor blocker losartan (Schiffrin et al. 2000) or irbesartan (Schiffrin et al. 2002). In contrast, patients treated with the beta-blocker atenolol for 1 (Schiffrin et al. 2000, 2002; Thybo et al. 1995) or 2 years (Schiffrin and Deng 1995) experienced no improvement. Similar improvement was noted on endothelial dysfunction with the ACE inhibitors (Rizzoni et al. 1997; Schiffrin and Deng 1995; Schiffrin et al. 1994) or with the angiotensin receptor blockers (Schiffrin et al. 2000, 2002; Ghiadoni et al. 2000) but not with the beta-blocker

dissecting out small arteries that were then mounted on a myograph and investigated (Deng and Schiffrin 1999; Deng et al. 1995; Mulvany et al. 1996; Park and Schiffrin 2001; Schiffrin and Deng 1999; Schiffrin et al. 1992, 1993). We first showed that the angiotensin-converting enzyme inhibitor cilazapril corrected the structure of these small arteries dissected from gluteal subcutaneous biopsies in hypertensive patients (Fig. 1), whereas the beta-blocker atenolol had no effect despite identical blood pressure lowering (Schiffrin et al. 1994, 1995). Associated with the regression of vascular remodeling, we were able to report as well improved endothelial function (Schiffrin and Deng 1995; Schiffrin et al. 1994). The angiotensin-converting enzyme inhibitor perindopril compared to atenolol in a 1-year trial later showed identical results on vascular structure of small arteries from hypertensive patients (Thybo et al. 1995). Another study examined the effect of the ACE inhibitor lisinopril on hypertensive patients with left ventricular hypertrophy (Rizzoni et al. 1997). Vessels from hypertensive patients were compared at the end of the 3 years of treatment to those of a different group of untreated hypertensive subjects. Improved structure and endothelial function of resistance arteries were demonstrated.

3.2 Angiotensin Receptor Blockers

More recently we evaluated the effects of the AT_1 receptor antagonist losartan (Fig. 1) compared to atenolol (Schiffrin et al. 2000). Treatment for 1 year with losartan resulted in correction of small artery structure and endothelial dysfunction in the losartan-treated patients, whereas there was no change in the atenolol-treated group, even though blood pressure control was identical. A decrease in the stiffness of wall components of small arteries could also be demonstrated in the losartan-treated group of patients, but was not found in the patients treated with atenolol (Park et al. 2000). In our and other studies (Schiffrin and Deng 1996; Schiffrin et al. 1994,1995,1996), patients on atenolol have never derived vascular benefits from treatment with the beta-blocker atenolol. Thus, since beta-blockade and renin-angiotensin blockade induced the same blood pressure lowering in hypertensive subjects, the effects of either angiotensin receptor blockers or ACE inhibitors appear quite selective, and beta-blockers a good negative control. Although patients in these studies were randomly assigned to treatment with the beta-blocker, we speculated that it could be possible that, since groups in each study were small, this could have resulted in some assignment bias. For this reason, we conducted a study in which we switched patients who had been in the course of other studies randomly assigned to be part of an atenolol group, to the angiotensin receptor blocker irbesartan (Schiffrin et al. 2002). Patients were treated for 1 year and blood pressure was reduced in this group to the same level as that achieved under atenolol. In this study, patients who exhibited altered structure of small arteries despite good blood pressure control for 1 year under atenolol had improved structure after 1 year of treatment with irbesartan. Maximal acetylcholine-induced endothelium-dependent relaxation was impaired when patients were on atenolol and was unchanged from before starting treatment, whereas it was normalized by irbesartan. Thus, while beta-blocker treatment did not correct the vascular structure or endothelial dysfunction in these hypertensive patients, switching the patients to treatment with irbesartan with identical control of blood pressure resulted in normalization of structure of blood vessels and endothelial function. Another study reported that hypertensive patients treated with irbesartan for 2 months exhibited improved distensibility of the carotid and reduced cross-sectional area of the radial artery (Benetos et al. 2000), which can be presumed to be favorable effects on these larger vessels, one a conduit artery, the other a large muscular artery. The angiotensin receptor blocker candesartan was shown to improve bradykinin-induced dilatation of the forearm circulation of hypertensive patients, which suggests that altered EDHF production was corrected by candesartan in this particular study (Ghiadoni et al. 2000).

4 Pathophysiological Relevance of Evidence Obtained with ACE Inhibitors and Angiotensin Receptor Blockers in Hypertensive Humans

It is important to determine whether the effects of ACE inhibitors and angiotensin receptor blockers reported above have a pathophysiological impact and alter the natural course of the disease. In favor of pathophysiological relevance of the results just reviewed are the following: (a) changes demonstrated on gluteal subcutaneous small arteries in hypertensive humans are similar to those in hypertensive rat vessels from more physiologically important vascular beds (coronary, renal and cerebral small arteries) (Intengan and Schiffrin 2000, 2001); (b) treatment with the angiotensin-converting enzyme inhibitors enalapril (Motz and Strauer 1996) and perindopril improved coronary reserve (Schwartzkopff et al. 2000), and perindopril also decreased pericoronary fibrosis in small vessels in the heart, as demonstrated by endomyocardial biopsy (Schwartzkopff et al. 2000); (c) there is a reasonably good correlation between endothelial function of small arteries obtained from gluteal subcutaneous biopsies and brachial artery flow-mediated dilatation, which is an endothelium-dependent response (Park et al. 2001). Brachial artery flow-mediated dilatation correlates with coronary vasomotion (Anderson et al. 1995), and endothelial dysfunction of the coronary vasculature has been shown to be predictive of cardiovascular events (Heitzer T et al. 2001; Schächinger et al. 1999), which underscores the relevance of findings using the gluteal subcutaneous small artery model in humans.

5 Conclusion

The data reviewed demonstrates that regression toward normal of the structural and functional changes present in small arteries of hypertensive humans can be achieved using agents that interrupt the renin-angiotensin system. For equal degree of blood pressure lowering, drugs that normalize the altered structure and endothelial dysfunction of small arteries, such as ACE inhibitors or angiotensin receptor blockers, may improve outcomes in hypertension to a greater degree than agents that do not, such as beta blockers. A recent systematic review of effects of beta-blockers in hypertensive patients aged more than 60 years has demonstrated that in ten trials (involving 16,164 patients), diuretic therapy was superior to β-blockade with regard to all endpoints (Messerli et al. 1998). Beta-blocker therapy was ineffective in preventing coronary heart disease, cardiovascular mortality, and all-cause mortality, although it did reduce the odds for cerebrovascular events. We speculate that the absence of benefit demonstrated may result from the fact that beta-blockers do not provide vascular protection, as our studies and those of other investigators have demonstrated (Schiffrin et al. 1994, 1995, 1996, 2000, 2002; Thybo et al. 1995). In the Captopril Prevention Project (Hansson et al. 1999a), cardiovascular mortality was lower with captopril than with conventional therapy, but the rate of fatal and nonfatal infarction was similar. However, fatal and nonfatal strokes were more common with captopril, which may have resulted from the fact that patients allocated to captopril had higher blood pressure before and during the study. STOP-2 (Hansson et al. 1999b; Lindholm et al. 2000) and HOPE (The Heart Outcomes Prevention Evaluation Investigators 2000) using ACE inhibitors, and the recent publication of the LIFE study (Dalhof et al. 2002; Lindholm et al. 2002) as well as several trials with diabetic renal patients (Brenner et al. 2001; Lewis et al. 2001; Viberti et al. 2002) using angiotensin receptor blockers, suggest that blockade of the renin-angiotensin system may offer cardiovascular and renal protection beyond blood pressure control. Whether benefits beyond blood pressure lowering found in some of these trials depend in part on the vascular protective actions of agents that interrupt the renin-angiotensin system, however, has yet to be conclusively demonstrated. Nevertheless, it does appear that blockade of the renin-angiotensin system may indeed have beneficial effects beyond blood pressure reduction that could depend to a large extent on the vascular protective action of these agents.

Acknowledgements. The work of the author reported here was supported by grants 13570 and 37917, and a Group grant to the Multidisciplinary Research Group on Hypertension, all from the Canadian Institutes of Health Research (or formerly the Medical Research Council of Canada), and by grants from Bristol-Myers Squibb and Sanofi-Synthélabo, Hoffmann-LaRoche and Merck-Frosst.

References

Aalkjaer C, Heagerty AM, Petersen KK, Swales JD, Mulvany MJ (1987) Evidence for increased media thickness, increased neuronal amine uptake, and depressed excitation-contraction coupling in isolated resistance vessels from essential hypertensives. Circ Res 61:181–186

Anderson TJ, Uehata A, Gerhard MD, Meredith IT, Knab S, Delagrange D, Lieberman EH, Ganz P, Creager MA, Yeung AC (1995) Close relation of endothelial function in the human coronary and peripheral circulations. J Am Coll Cardiol 26:1235–1241

Bakker ENTP, Van der Meulen ET, Van den Berg BM, Everts V, Spaan JAE, VanBavel E (2002) Inward remodeling follows chronic vasoconstriction in isolated resistance arteries. J Vasc Res 39:12–20

Baumbach GL, Heistad DD (1989) Remodeling of cerebral arterioles in chronic hypertension. Hypertension 13:968–972

Benetos A, Gautier S, Lafleche A, Topouchian J, Frangin, G, Girerd X, Sissmann J, Safar ME (2000) Blockade of angiotensin II type 1 receptors: effect on carotid and radial artery structure and function in hypertensive humans. J Vasc Res 37:8–15

Brenner BM, Cooper ME, de Zeeuw D, Keane WF, Mitch WE, Parving H-H, Remuzzi G, Snapinn SM, Zhang Z, Shahinfar S (2001) Effects of losartan on renal and cardiovascular outcomes in patients with type 2 diabetes and nephropathy. N Engl J Med 345:861–869

Brush JE, Cannon RO, Schenke WH, Bonow RO, Leon MB, Maron BJ, Epstein SE (1988) Angina due to coronary microvascular disease in hypertensive patients without left ventricular hypertrophy. N Engl J Med 319:1302–1307

Cai H, Harrison DG (2000) Endothelial dysfunction in cardiovascular diseases—the role of oxidant stress. Circ Res 87:840–844

Christensen KL, Mulvany MJ (1993) Mesenteric arcade arteries contribute substantially to vascular resistance in conscious rats. J Vasc Res 30:73–79

Cockroft JR, Chowienczyk PJ, Benjamin N, Ritter JM (1994) Preserved endothelium-dependent vasodilatation in patients with essential hypertension New Engl J Med 330:1036–1040

Dahlof B, Devereux RB, Kjeldsen SE, Julius S, Beevers G, Faire U, Fyhrquist F, Ibsen H, Kristiansson K, Lederballe-Pedersen O, Lindholm LH, Nieminen MS, Omvik P, Oparil S, Wedel H, the LIFE Study Group (2002) Cardiovascular morbidity and mortality in the Losartan Intervention For Endpoint reduction in hypertension study (LIFE): a randomised trial against atenolol. Lancet 359:995–1003

Deng LY, Schiffrin EL (1992) Effects of endothelin-1 and vasopressin on small arteries of spontaneously hypertensive rats. Am J Hypertens 5:817–822

Deng LY, Schiffrin EL (1993) Effect of antihypertensive treatment on response to endothelin of resistance arteries of hypertensive rats. J Cardiovasc Pharmacol 21:725–731

Deng LY, Li J-S, Schiffrin EL (1995) Endothelium-dependent relaxation of small arteries from essential hypertensive patients. Clin Sci 88:611–622

Diedrich DA, Yang Z, Bühler FR, Lüscher TF (1990) Impaired endothelium-dependent relaxations in hypertensive small arteries involve the cyclooxygenase pathway. Am J Physiol (Heart Circ Physiol) 258:H445–H451

Diep Q, Li JS, Schiffrin EL (1999) In vivo study of the role of AT_1 and AT_2 angiotensin receptors in apoptosis in rat blood vessels. Hypertension 34:617–624

Dohi Y, Criscione L, Pfeiffer K, Lüscher TF (1994) Angiotensin blockade or calcium antagonists improve endothelial dysfunction in hypertension: studies in perfused mesenteric resistance arteries. J Cardiovasc Pharmacol 24:372–379

Folkow B (1982) Physiological aspects of primary hypertension. Physiol Rev 62:347–504

Folkow B (1995) Hypertensive structural changes in systemic precapillary resistance vessels: how important are they for in vivo haemodynamics? J Hypertens 13:1546–1559

Ghiadoni L, Virdis A, Magagna A, Taddei S, Salvetti A (2000) Effect of the angiotensin II type 1 receptor blocker candesartan on endothelial function in patients with essential hypertension. Hypertension 35:501–506

Gohlke P, Pees C, Unger T (1998) AT2 receptor stimulation increases aortic cyclic GMP in SHRSP by a kinin-dependent mechanism. Hypertension 31:349–355

Griffin SA, Brown WC, MacPherson F, McGrath JC, Wilson, VG, Korsgaard N, Mulvany MJ, Lever AF (1991) Angiotensin II causes vascular hypertrophy in part by a non-pressor mechanism. Hypertension 17:626–635

Hansson L, Lindholm LH, Niskanen L, Lanke J, Hedner T, Niklason A, Luomanmaki K, Dahlof B, De Faire U, Morlin C, Karlberg BE, Wester PO, Bjorck JE (1999a) Effect of angiotensin converting enzyme inhibition compared with conventional therapy on cardiovascular morbidity and mortality in hypertension: the Captopril Prevention Project (CAPPP) randomised trial. Lancet 353:611–616

Hansson L, Lindholm LH, Ekbom T, Dahlöf B, Lanke J, Scherstén B, Wester PO, Hedner T, De Faire U (1999b) Randomised trial of old and new antihypertensive drugs in elderly patients: cardiovascular mortality and morbidity in the Swedish Trial in Old Patients with Hypertension-2 study. Lancet 354:1751–1756

Harrap SB, Van der Merwe WM, Griffin SA, McPherson F, Lever AF (1990) Brief angiotensin converting enzyme inhibitor treatment in young spontaneously hypertensive rats reduces blood pressure long-term. Hypertension 16:603–614

Hasdai D, Gibbons RJ, Holmes DRJ, Higano ST, Lerman A (1997) Coronary endothelial dysfunction is associated with myocardial perfusion defects. Circulation 96:3390–3395

Heagerty AM, Aalkjaer C, Bund SJ, Korsgaard N, Mulvany MJ (1993) Small artery structure in hypertension: Dual processes of remodelling and growth. Hypertension 21:391–397

Heitzer T, Schlinzig T, Krohn K, Meinertz T, Münzel T (2001) Endothelial dysfunction, oxidative stress, and risk of cardiovascular events in patients with coronary artery disease. Circulation 104:2673–2678

Intengan HD, Schiffrin EL (2000) Structure and mechanical properties of resistance arteries in hypertension. Hypertension 36:312–318

Intengan HD, Schiffrin EL (2001) Vascular remodeling in hypertension: roles of apoptosis, inflammation and fibrosis. Hypertension 38:581–587

Intengan HD, Thibault G, Li JS, Schiffrin EL (1999a) Resistance artery mechanics, structure, and extracellular components in spontaneously hypertensive rats effects of angiotensin receptor antagonism and converting enzyme inhibition. Circulation 100:2267–2275

Intengan HD, Deng LY, Li JS, Schiffrin EL (1999b) Mechanics and composition of human subcutaneous resistance arteries in essential hypertension. Hypertension 33 (part II):366–372

Kinlay S, Selwyn AP, Libby P, Ganz P (1998) Inflammation, the endothelium, and the acute coronary syndromes. J Cardiovasc Pharmacol 32:S62–S66

Korsgaard N, Aalkjaer C, Heagerty AM, Izzard AS, Mulvany MJ (1993) Histology of subcutaneous small arteries from patients with essential hypertension. Hypertension 22:523–526

Ledingham JM, Phelan EL, Cross MA, Laverty R (2000) Prevention of increases in blood pressure and left ventricular mass and remodeling of resistance arteries in young New Zealand genetically hypertensive rats: the effects of chronic treatment with valsartan, enalapril and felodipine. J Vasc Res 37:134–145

Lewis EJ, Hunsicker LF, Clarke WR, Berl T, Pohl MA, Lewis JB, Ritz E, Atkins RC, Rohde R, Raz I (2001) Renoprotective effect of the angiotensin receptor antagonist irbesartan in patients with nephropathy due to type 2 diabetes N Engl J Med 345:851–860

Li J-S, Schiffrin EL (1996) Effect of calcium channel blockade or angiotensin converting enzyme inhibition on structure of coronary, renal and other small arteries in SHR. J Cardiovasc Pharmacol 28:68–74

Li J-S, Sharifi MA, Schiffrin EL (1997) Effect of AT_1 angiotensin receptor blockade on structure and function of small arteries in SHR. J Cardiovasc Pharmacol 30:75–83

Lindholm LH, Hansson L, Ekbom T, Dahlof B, Lanke J, Linjer E, Schersten B, Wester PO, Hedner T, De Faire U (2000) Comparison of antihypertensive treatments in preventing cardiovascular events in elderly diabetic patients: results from the Swedish Trial in Old Patients with Hypertension-2. J Hypertens 18:1671–1675

Lindholm LH, Ibsen H, Dahlof B, Devereux RB, Beevers G, De Faire U, Fyhrquist F, Julius S, Kjeldsen SE, Kristiansson K, Lederballe-Pedersen O, Nieminen MS, Omvik P, Oparil S, Wedel H, Aurup P, Edelman J, Snapinn S, The LIFE Study Group (2002) Cardiovascular morbidity and mortality in patients with diabetes in the Losartan Intervention For Endpoint reduction in hypertension study (LIFE): a randomised trial against atenolol. Lancet 359:1004–1010

Laurent S (1995) Arterial wall hypertrophy and stiffness in essential hypertensive patients. Hypertension 26:355–362

Lockette W, Otsuka Y, Carretero O (1986) The loss of endothelium-dependent vascular relaxation in hypertension. Hypertension 8 [Suppl 2]:II61–II66

Lüscher TF, Vanhoutte PM (1986) Endothelium-dependent contractions to acetylcholine in the aorta of the spontaneously hypertensive rat. Hypertension 8:344–348

Messerli FH, Grossman E, Goldbourt U (1998) Are β-blockers efficacious as first-line therapy for hypertension in the elderly? JAMA 279:1903–1907

Motz W, Strauer BE (1996) Improvement of coronary flow reserve after long-term therapy with enalapril. Hypertension 27:1031–1038

Mulvany MJ, Aalkjaer C (1990) Structure and function of small arteries. Physiol Rev 70:921–971

Mulvany MJ, Baumbach GL, Aalkjaer C, Heagerty AM, Korsgaard N, Schiffrin EL, Heistad DD (1996) Vascular remodeling. Letter to the Editor. Hypertension 27:505–506

Panza JA, Quyyumi AA, Brush JE, Epstein SE (1990) Abnormal endothelium dependent vascular relaxation in patients with essential hypertension. N Engl J Med 323:22–27

Park JB, Schiffrin EL (2001) Small artery remodeling is the most prevalent (earliest?) form of target organ damage in mild essential hypertension. J Hypertens 19:921–930

Park JB, Intengan HD, Schiffrin EL (2000) Reduction of resistance artery stiffness by treatment with the AT_1 receptor antagonist losartan in essential hypertension. J Renin Angiotens System 1:40–45

Park JB, Charbonneau F, Schiffrin EL (2001) Correlation of endothelial function in large and small arteries in human essential hypertension. J Hypertens 19:415–420

Parving HH, Lehnert H, Bröchner-Mortensen J, Gomis R, Andersen S, Arner P (2001) The effect of irbesartan on the development of diabetic nephropathy in patients with type 2 diabetes. N Engl J Med 345:870–878

Rajagopalan S, Laursen JB, Borthayre A, Kurz S, Keiser J, Haleen S, Giaid A, Harrison DG (1997) Role for endothelin-1 in angiotensin II-mediated hypertension. Hypertension 30:29–34

Rizzoni D, Castellano M, Porteri E, Bettoni G, Muiesan ML, Cinelli A, Rosei EA (1995) Effects of low and high doses of fosinopril on the structure and function of resistance arteries. Hypertension 26:118–123

Rizzoni D, Porteri E, Castellano M, Bettoni G, Muiesan ML, Muiesan P, Giulini SM, Agabiti-Rosei E (1996) Vascular hypertrophy and remodeling in secondary hypertension. Hypertension 28:785–790

Rizzoni D, Muiesan ML, Porteri E, Castellano M, Zulli R, Bettoni G, Salvetti M, Monteduro C, Agabiti-Rosei E (1997) Effects of long-term antihypertensive treatment with lisinopril on resistance arteries in hypertensive patients with left ventricular hypertrophy. J Hypertens 15:197–204

Rizzoni D, Porteri E, Bettoni G, Piccoli A, Castellano M, Muiesan ML, Pasini G, Guelfi D, Agabiti-Rosei E (1998) Effects of candesartan cilexetil and enalapril on structural alterations and endothelial function in small resistance arteries of spontaneously hypertensive rats. J Cardiovasc Pharmacol 32:798–806

Rizzoni D, Porteri E, Guefi D, Piccoli A, Castellano M, Pasini G, Muiesan ML, Mulvany MJ, Agabiti-Rosei E (2000) Cellular hypertrophy in subcutaneous small arteries of patients with renovascular hypertension. Hypertension 35:931–935

Safar ME, Girerd X, Laurent S (1996) Structural changes of large conduit arteries in hypertension. J Hypertens 14:545–555

Schächinger V, Britten MB, Elsner M, Walter DH, Scharrer I, Zeiher AM (1999) A positive family history of premature coronary artery disease is associated with impaired endothelium-dependent coronary blood flow regulation. Circulation 100:1502–1508

Schiffrin EL (1992) Reactivity of small blood vessels in hypertension. Relationship with structural changes. Hypertension19 [Suppl II]:II1–II9

Schiffrin EL (1999) Role of endothelin-1 in hypertension. Hypertension 34:876–881

Schiffrin EL, Deng LY (1995) Comparison of effects of angiotensin converting enzyme inhibition and beta blockade on function of small arteries from hypertensive patients. Hypertension 25:699–703

Schiffrin EL, Deng LY (1996) Structure and function of resistance arteries of hypertensive patients treated with a β-blocker or a calcium channel antagonist. J Hypertens 14:1247–1255

Schiffrin EL, Deng LY (1999) Relationship of small artery structure with systolic, diastolic and pulse pressure in essential hypertension. J Hypertens 17:381–387

Schiffrin EL, Gutkowska J, Genest J (1984) Effect of angiotensin II and deoxycorticosterone infusion on vascular angiotensin II receptors in rats. Am J Physiol (Heart Circ Physiol15) 246:H608–H614

Schiffrin EL, Franks DJ, Gutkowska J (1985) Effect of aldosterone on vascular angiotensin II receptors in the rat. Can J Physiol Pharmacol 63:1522–1527

Schiffrin EL, Deng LY, Larochelle P (1992) Blunted effects of endothelin upon small subcutaneous resistance arteries of mild essential hypertensive patients. J Hypertens 10:437–444

Schiffrin EL, Deng LY, Larochelle P (1993) Morphology of resistance arteries and comparison of effects of vasoconstrictors in mild essential hypertensive patients. Clin Invest Med 16:177–186

Schiffrin EL, Deng LY, Larochelle P (1994) Effects of a beta blocker or a converting enzyme inhibitor on resistance arteries in essential hypertension. Hypertension 23:83–91

Schiffrin EL, Deng LY, Larochelle P (1995) Progressive improvement in the structure of resistance arteries of hypertensive patients after 2 years of treatment with an angiotensin converting enzyme inhibitor. Comparison with effects of a beta blocker. Am J Hypertens 8:229–236

Schiffrin EL, Deng LY, Sventek P, Day R (1997) Enhanced expression of endothelin-1 gene in resistance arteries in severe human essential hypertension. J Hypertens 15:57–63

Schiffrin EL, Li JS, Sharifi AM (1998) AT_1 angiotensin receptor blockade and angiotensin converting enzyme inhibition: effects on vascular remodeling and endothelial dysfunction in SHR. In: Dhalla NS, Zahradka P, Dixon IMC, Beamish RE (eds) Angiotensin II receptor blockade: physiological and clinical implications. Kluwer Academic Publishers, Norwell, MA, pp 33–40

Schiffrin EL, Park JB, Intengan HD, Touyz RM (2000) Correction of arterial structure and endothelial dysfunction in human essential hypertension by the angiotensin antagonist losartan. Circulation 101:1653–1659

Schiffrin EL, Park JB, Pu Q (2002) Effect of crossing over hypertensive patients from a beta-blocker to an angiotensin receptor antagonist on resistance artery structure and on endothelial function. J Hypertens 20:71–78

Schwartzkopff B, Brehm M, Mundhenke M, Strauer BE (2000) Repair of coronary arterioles after treatment with perindopril in hypertensive heart disease. Hypertension 36:220–225

Sharifi AM, Li JS, Endemann D, Schiffrin EL (1998) Comparison of effects of the angiotensin converting enzyme inhibitor enalapril and the calcium channel antagonist amlodipine on small artery structure and composition, and on endothelial dysfunction in SHR. J Hypertens 16:457–466

Shaw LM, George PR, Oldham AA, Heagerty AM (1995) A comparison of the effect of angiotensin converting enzyme inhibition and angiotensin II receptor antagonism on the structural changes associated with hypertension in rat small arteries. J Hypertens 13:1135–1143

Taddei S, Virdis A, Ghiadoni L, Sudano I, Salvetti A (2001) Endothelial dysfunction in hypertension. J Cardiovasc Pharmacol 38:S11–S14

The Heart Outcomes Prevention Evaluation Study Investigators (2000) Effects of an angiotensin-converting-enzyme inhibitor, ramipril, on cardiovascular events in high-risk patients. N Engl J Med 342:145–154

Thybo NK, Korsgaard N, Eriksen S, Christensen KL, Mulvany MJ (1994) Dose-dependent effects of perindopril on blood pressure and small-artery structure. Hypertension 23:659–666

Thybo NK, Stephens N, Cooper A, Aalkjaer C, Heagerty AM, Mulvany MJ (1995) Effect of antihypertensive treatment on small arteries of patients with previously untreated essential hypertension. Hypertension 25:474–481

Touyz RM, Schiffrin EL (2000) Signal transduction mechanisms mediating the physiological and pathophysiological actions of angiotensin II in vascular smooth muscle cells. Pharmacol Rev 52:639–672

Touyz RM, Tolloczko B, Schiffrin EL (1994) Mesenteric vascular smooth muscle cells from SHR display increased Ca2+ responses to angiotensin II but not to endothelin-1. J Hypertens12:663–673

Touyz RM, He G, El Mabrouk M, Schiffrin EL (2001) p38 MAP kinase regulates vascular smooth muscle cell collagen synthesis by angiotensin II in SHR but not in WKY. Hypertension 37:574–580

Touyz RM, Chen X, Tabet F, Yao G, He G, Quinn MT, Pagano PJ, Schiffrin EL (2002) Expression of a functionally active gp91phox-containing neutrophil-type NAD(P)H oxidase in smooth muscle cells from human resistance arteries: regulation by angiotensin II. Circ Res 90:1205–1213

Tschudi MR, Mesaros S, Lüscher TF, Malinski T (1996) Direct in situ measurement of nitric oxide in mesenteric resistance arteries. Hypertension 27:32–35

Ullian ME, Schelling JR, Linas SL (1992) Aldosterone enhances angiotensin II receptor binding and inositol phosphate responses. Hypertension 20:67–73

Ullian ME, Walsh LG, Morinelli TA (1996) Potentiation of angiotensin II action by corticosteroids in vascular tissue. Cardiovasc Res 32:266–273

Vacher E, Fornes P, Richer C, Bruneval P, Nisato D, Giudicelli J-F (1995) Early and late haemodynamic and morphological effects of angiotensin II subtype 1 receptor blockade during genetic hypertension development. J Hypertens 13:675–682

Viberti G, Wheeldon NM, for the MicroAlbuminuria Reduction With VALsartan (MARVAL) Study Investigators (2002) Microalbuminuria reduction with valsartan in patients with type 2 diabetes mellitus: a blood pressure-independent effect. Circulation 106:672–678

Part 4
Tissues

Brain/Nervous System

Angiotensin Pathways and Brain Function

F. Qadri

Institute of Experimental and Clinical Pharmacology and Toxicology,
University Clinic Schleswig Holstein Campus Lübeck, Lübeck, Germany
e-mail: qadri@medinf.mu-luebeck.de

Abstract The renin-angiotensin system (RAS) was initially identified as a circulating humoral system with the effector peptide, angiotensin II (ANG II). The existence of a brain RAS as one of the various tissue RASs is fully established. ANG-containing pathways within brain areas are involved in central blood pressure regulation, pituitary hormone release and in the maintenance of body fluid homeostasis. In addition, ANG II and its main biologically active degradative product ANG III play a pivotal role in drinking behavior, osmo- and thermoregulation and natriuresis. Systemic stimuli that may activate central angiotensinergic pathways include plasma hypertonicity and circulating hormones such as ANG II. In this chapter, the specific sites where ANG II/III is acting within polysynaptic neural pathways mediating homeostatic functions are reviewed.

Keywords Brain ANG pathways · Blood pressure · Body fluid homeostasis · Drinking behavior · AVP · Natriuresis · Osmoregulation

The renin-angiotensin system (RAS) was initially identified as a circulating humoral system with the effector peptide, angiotensin II (ANG II), generated by an enzymatic cascade. Angiotensinogen, which is synthesized in the liver, is cleaved by renin, a product of the juxtaglomerular cells of the kidney, to form ANG I, which in turn is cleaved by angiotensin-converting enzyme (ACE) to form ANG II. Apart from the production of ANG II in plasma and peripheral organs such as kidney, adrenal gland, vasculature, heart and ovaries, all components of RAS have also been described in the brain (Ganten et al. 1984; Saavedra 1992).

Central ANG II interacts with the autonomic control of the cardiovascular system to influence blood pressure, regulates body fluid homeostasis, modulates neuroendocrine systems and possibly interacts with cognitive functions (Phillips 1987; Unger et al. 1988; Wright and Harding 1992). The classic actions of ANG II are mediated through the AT_1 receptor, whereas the AT_2 receptor can offset or counteract some of the effects mediated by the AT_1 receptor, e.g., cell proliferation, water intake and blood pressure (Lucius et al. 1999; Unger 1999). Mostly, central ANG II-induced effects are similar to those of the systemic peptide on peripheral target organs. Systemic ANG II affects the brain functions through AT_1 receptors located in the circumventricular organs devoid of blood–brain barrier (BBB) and neuronally derived ANG II acts at many sites in the CNS located behind the BBB (Bunnemann et al. 1993; McKinley et al. 2001; Song et al. 1991).

1
Angiotensin Pathways and Central Cardiovascular Regulation

Blood pressure increases when ANG II is given i.v. or injected into the lateral brain ventricle. This response involves activation of the sympathetic nervous system (SNS), release of arginine vasopressin (AVP) and inhibition of the baroreceptor reflex (BRR) (Unger et al. 1983, 1985). The activation of SNS occurs via fibers from the hypothalamic paraventricular nucleus (PVN) to the ventrolateral medullary sympathetic nuclei in the brainstem. These nuclei project through the spinal cord to the intermediate lateral nucleus, and from there via the sympathetic chain to terminate on blood vessels. AVP is a strong vasoconstrictory neurohormone. It is synthesized in the hypothalamic PVN and supraoptic nucleus (SON) and stored in the neurohypophysis. Its release from the neurohypophysis is stimulated by the activation of AT_1 receptors. BRR inhibition is obtained when brain ANG II acts on the first synapse of the vagus in the nucleus tractus solitarii (NTS) in the brainstem (Casto and Phillips 1985).

1.1
Activation of Sympathetic Nervous System and Inhibition of Baroreceptor Reflex

The regulation of arterial blood pressure and heart rate is the result and outcome of feedback or compensatory control systems that operate over short and extended timeframes. The BRR is a relatively high-gain control system that provides for the regulation of arterial blood pressure over a timeframe of seconds to minutes (short-term mechanisms in arterial blood pressure regulation). The rapidity of this regulatory system is afforded by feedback loops operating through the autonomic nervous system. On the other hand, arterial blood pressure is also under the influence of neurohormones and sympathetic activity, for example, salt and water balance, which are regulated by the RAS from the kidney. The effectiveness of this regulatory system occurs over periods of hours and days (long-term mechanisms in arterial blood pressure regulation) (Averill and Diz 2000; Dampney et al. 2002).

The medulla oblongata contains neuronal circuits such as the BRR arc capable of generating vasomotor sympathetic and cardiovascular tone, which play an important role in the short-term regulation of arterial blood pressure. The arterial baroreceptors are located in the walls of the carotid sinus and aortic arch. Their adequate stimulus is stretch. Studies in conscious animals using the method of immediate early gene expression in combination with neurochemical tracing and immunohistochemistry, the central pathways and neurotransmitters that subserve the BRR arc, are investigated (Horiuchi et al. 1999; Li and Dampney 1994). The BRR arc in the medulla oblongata is composed of the following sequential arrangements of components. Peripheral afferent fibers arising from arterial and atrial/ventricular baroreceptors make their initial synapse upon the first-order neurons of the nucleus of the solitary tract (NTS). NTS neurons may have axonal projections to vagal preganglionic neurons in the nucleus ambiguus (Amb) and/or dorsal motor nucleus of the vagus. NTS also innervates GABAergic neurons in the caudal ventrolateral medulla (CVLM), also termed the depressor region of the ventrolateral medulla (VLM), since L-glutamate injected into this nucleus lowered blood pressure. CVLM, in turns, innervates neurons in the rostral ventrolateral medulla (RVLM), which sends axonal projections to the sympathetic thoracic segment of the spinal cord. RVLM is termed the rostral pressor region of the VLM, since L-glutamate given into this nucleus increases blood pressure and sympathetic nerve activity.

The presence of AT_1 receptors within the NTS and throughout the vagal sensory and motor systems has been shown by many authors (Diz et al. 1987; Lenkei et al. 1997, 1998; Lewis et al. 1986). In addition to the anatomical evidence, electrophysiological and pharmacological studies support the concept of functional ANG II receptors in the NTS (Andresen and Mendelowitz 1996; Averill et al. 1987; Barnes et al. 1993; Casto and Phillips 1986; Rettig et al. 1986). Endogenous (Campagnole-Santos et al. 1988; Matsumura et al. 1998) or exogenously (Casto and Phillips 1986; Michelini and Bonagamba 1998) administered ANG II at the level of the NTS inhibits the baroreceptor control of heart rate. A low dose of ANG II and ANG (1–7) microinjected into the NTS elicited depressor and bradycardic responses in rats (Campagnole-Santos et al. 1989; Diz et al. 1984; Rettig et al. 1986). The hypotensive and bradycardic responses were the combined effects of sympathetic inhibition and parasympathetic stimulation. On the contrary, high doses of ANG II evoked pressor and tachycardic responses (Casto and Phillips 1984; Mosqueda-Garcia et al. 1990; Rettig et al. 1986). These effects of ANG II were mediated via the AT_1 receptor subtype found in the NTS (Fow et al. 1994). Thus, it appears that ANG II and its metabolite ANG (1–7) mimic baroreceptor afferent stimulation of the NTS.

The neurons within the CVLM and RVLM are an important switchboard of the BRR arc. The presympathetic neurons in the RVLM play a pivotal role in the tonic and reflex control of sympathetic vasomotor activity and thus resting arterial pressure. Topical application of ANG II to the RVLM increased blood pressure and this effect was prevented by prior application of the ANG II antagonist, SarThran (Andreatta et al. 1988). Subsequent studies showed that microinjection of ANG II into the RVLM increased sympathetic nerve activity and arterial pressure and these effects were blocked by the broad-acting ANG II analog antagonists, SarThran and Sar^1,Ile^8-ANG II (Chan et al. 1991, 1994). Later studies have shown that pressor and sympathoexcitatory effects of exogenously applied ANG II were blocked by the AT_1 receptor selective antagonist losartan (Averill et al. 1994). Although experiments employing exogenous administration of ANG II or angiotensin peptides in the RVLM have been valuable in determining the possible effect of peptides and the receptors at which angiotensin peptides may act, the most interesting and valuable data come from studies where the involvement of endogenous angiotensin peptides within the RVLM have been studied. This has been achieved by microinjecting the broad-acting (SarThran) and later the specific angiotensin receptor antagonist (losartan). In the early studies, unilateral as well as bilateral microinjection of SarThran in the RVLM reduced baseline blood pressure and sympathetic nerve activity (Muratani et al. 1993; Sasaki and Dampney 1990). These initial data provide evidence for the concept that ANG II found endogenously in the RVLM may be important in determining the activity of vasomotor neurons, and hence the amount of vasomotor outflow responsible for maintaining a normal or basal blood pressure. Attempts to define the specific angiotensin receptor subtypes responsible for the tonic excitation of RVLM vasomotor neurons have led to some unexpected findings. It has been reported that blockade of either AT_1 or AT_2 receptors in the RVLM does not reduce baseline blood pressure or sympathetic nerve activity (Fontes et al. 1997; Hirooka et al. 1997; Lin et al. 1997). Thus, it seems that endogenous ANG II may also act at non-AT_1 and non-AT_2 receptors to stimulate vasomotor neurons. Several authors have suggested that the naturally occurring angiotensin peptide responsible for the tonic activation of sympathoexcitatory RVLM neurons may be other than ANG II, for example it may be ANG I (Lima et al. 1996), ANG (1–7) (Fontes et al. 1994; Lima et al. 1996) or ANG (3–8) (ANG IV) (Sved and Ito 1999). To date, it is uncertain which of the naturally occurring angiotensin peptides are involved in the sympathoexcitatory action of neurons in the RVLM.

Although the neurons in the CVLM play a key role in the maintenance of blood pressure, modest attention has been paid to it as an important site of an action for angiotensin peptides. Microinjection of ANG II or ANG (1–7) into the CVLM caused dose-dependent decreases in blood pressure and sympathetic nerve activity (Muratani et al. 1991; Sasaki and Dampney 1990; Silva et al. 1993), whereas SarThran increased blood pressure and renal sympathetic nerve activity (Muratani et al. 1993; Sesoko et al. 1995). These data suggest that either ANG II or ANG (1–7) found endogenously in the CVLM stimulates CVLM neurons, which inhibit vasomotor neurons of the RVLM, since inhibition of the RVLM abolished the pressor response evoked by the blockade of ANG II receptors in the CVLM (Muratani et al. 1993). Nevertheless, these findings provide strong evidence for the notion that tonic stimulation of angiotensin receptors on CVLM neurons contributes to the inhibitory influence of these neurons upon presynaptic neurons of the RVLM.

Another short-term feedback regulation of the cardiovascular system is the chemoreceptor reflex. The chemoreceptors are stimulated primarily by a decrease in the oxygen partial pressure of arterial blood. Their stimulation reflexly evokes both an increase in ventilation (increases oxygen uptake) and sympathetically mediated vasoconstriction (reduces oxygen consumption) in most vascular beds, excluding the heart and brain (Dampney et al. 2002). Like baroreceptor primary afferent fibers, chemoreceptor primary afferent fibers terminate in the NTS. However, in contrast to the baroreceptor reflex pathways, chemoreceptor signals are transmitted to the RVLM presympathetic neurons via a direct excitatory glutamatergic synapse (Guyenet and Koshiya 1995). Whether ANG II is involved in this pathway has to be clarified.

Acute emotional and threatening stimuli can also elicit a marked cardiovascular response. The defense or alerting response (acute stress) is characterized by an increase in arterial pressure, heart rate and skeletal muscle blood flow, accompanied by vasoconstriction in the splanchnic, renal and cutaneous vascular beds (Hilton 1975). This response is appropriate for a living being that may need to defend itself from a threatening situation. Such a response can be regarded as a part of "feed-forward" and not a part of "feed-back" regulatory mechanisms (Hilton 1975). After a long time spent in search of the so called defense area, recently the dorsomedial hypothalamic nucleus (DMH) has been characterized and identified as the brain region that plays a key role in integrating the cardiovascular response to acute stress. Activation of DMH neurons, by local injection of either an excitatory amino acid or GABA receptor antagonists, resulted in a cardiovascular response that is similar to the defense or alerting reaction. In addition, the neuroendocrine, gastrointestinal and behavioral changes were also similar to those evoked by an acute emotional stress (DiMicco et al. 1996). Recently it has been shown that the vasomotor and cardiac responses evoked from the DMH are mediated by descending pathways that are dependent and independent, respectively, of synaptic transmission within the RVLM (Fontes et al. 2001). There are major inputs to the NTS from many supramedullary nuclei, including those that play an important role in mediating cardiovascular responses to acute stress. It has been shown that neurons in the DMH project directly to the NTS and a high proportion of these have collateralized projections also to the RVLM (Fontes et al. 2001). In addition, electrical stimulation of the hypothalamic defense area modulates the BRR (Spyer 1994).

It seems that the NTS is the initial information-receiving center from the periphery and brain submedullary centers in the stress pathway. There are major inputs to the NTS from many supramedullary nuclei, including those that are believed to play important roles in mediating cardiovascular responses to acute stress (short-term feed-forward regulation). The cardiovascular changes that result from the stimulation of baroreceptors or chemoreceptors (short-term feedback regulation) and occur after acute or emotional stress such as body exercise, defense reactions such as fight or flee, or air jet stress (short-term feed-forward regulation) are regulated by the same BRR pathway and may be under the control of the same neuropeptide, i.e., angiotensin and other neurotransmitters (Dampney et al. 2002).

1.2 Arginine Vasopressin Release

Arginine vasopressin (AVP) is released from the posterior pituitary into the blood circulation. Circulating AVP is synthesized in the hypothalamic magnocellular neurons originating from the SON and PVN (Bargmann and Scharrer 1951). The release of AVP from the posterior pituitary constitutes one of the several mechanisms that operate to maintain cardiovascular and hydromineral homeostasis.

AVP release induced by ANG II and ANG III is one of the mechanisms by which these neuropeptides control volume homeostasis under conditions of hypovolemia, by reducing renal water loss and increasing blood pressure (Renaud and Bourque 1991). The administration of ANG II into the lateral or third cerebral ventricles dose-dependently stimulates the secretion of AVP (Andersson et al. 1972). Increases in circulating ANG II have been shown to result in large increases in the concentration of both oxytocin (OXY) and AVP from the posterior pituitary, effects that are abolished by the destruction of the SFO (Ferguson and Kasting 1986, 1988; Knepel et al. 1982). Both systemic and local application of ANG have also been shown to result in AT_1 receptor-mediated increases in neuronal activity of antidromically identified OXY and AVP neurons (Ferguson and Renaud 1986; Li and Ferguson 1993). In addition, the ablation of tissue in the median preoptic nucleus (MnPO) located at the anterior wall of the third cerebral ventricle also prevents an ANG-induced AVP release (Bealer et al. 1979). The region plays an important role in the ANG-induced cardiovascular changes and modulates the signals coming from the periphery. This region is rich in ANG receptors and there are direct neural inputs from this region to the vasopressin-secreting cells in the hypothalamic PVN and SON. Administration of ANG II blockers into the MnPO prevents the stimulation of magnocellular neurons in the PVN via stimulation of the SFO (Tanaka et al. 1987; Wilkin et al. 1989). Presumably, an angiotensinergic synapse in the MnPO has a key role in AVP secretion. In addition, the forebrain ANG pathways, the NTS and the A1 cells of the CVLM, regions rich in ANG receptors, also send direct neural inputs to the hypothalamic PVN and SON, and it is possible that they may also be the sites at which ANG acts as neurotransmitter to stimulate AVP secretion (Allen et al. 2000; Wilkin et al. 1989).

The release of AVP into the blood circulation is mediated via stimulation of periventricular and hypothalamic AT_1 receptors (Veltmar et al. 1992; Qadri et al. 1993). In addition, AT_1 receptor-mediated increase in blood pressure, AVP release and drinking response were shown to be blocked by ICV injection of antisense oligonucleotides to AT_1 receptor mRNA (Gyurko et al. 1993; Meng et al.

1994). The importance of the hypothalamic PVN in the integration and transmission of effects induced by peripheral or central ANG II is increasingly recognized (Stadler et al. 1992; Swanson and Sawchenko 1983; Wright and Harding 1992). Using an in vivo microdialysis technique, it has been shown that ANG II and ANG III are released within the PVN in response to local hyperosmotic stimulation and dehydration (Harding et al. 1992; Qadri et al. 1994). In addition, local microinjections of *C*-terminal ANG II and ANG III, but not shorter fragments, can induce AVP release and drinking response via AT_1 receptors in the PVN. On the other hand, the N-terminal ANG (1–7) was less potent in stimulating AVP release compared to the *C*-terminal ANG II and III and had no influence on drinking (Qadri et al. 1998a).

Blood-borne signals such as plasma hypernatremia act on osmo/sodium or ANG receptors found in the circumventricular organs such as the subfornical organ (SFO) and the organum vasculosum lamina terminalis (OVLT), regions known to be involved in osmoregulation. Stimulation of these receptors in CVO activates neural pathways that project to the hypothalamic PVN and SON, which results in increased release of AVP into the circulation and increased mean arterial blood pressure (Ferguson and Casting 1986; Gutman et al. 1988; Miselis 1982; Qadri et al. 1998a). Intracarotid injections of hyperosmolar saline (0.6 M NaCl solution) increased the firing rate of neurosecretory cells in the PVN. The firing rate of 60% of cells stimulated by hyperosmolar saline was blocked by local application of hexamethonium into the PVN, of 80% after atropine pretreatment and of 40% after the ANG antagonist saralasin (Akaishi and Negoro 1983). In an in vivo study, a non-pressure-associated and moderate increase in osmolality in the CSF (0.2 and 0.3 M saline, ICV) stimulates AVP release, which is mediated through the AT_1 and cholinergic receptors found in the periventricular and hypothalamic PVN, whereas pressure-associated and hyperosmolar saline (0.6 M, ICV)-induced AVP was facilitated via AT_2 receptors found in the periventricular and hypothalamic PVN (Hogarty et al. 1994; Höhle et al. 1996; Qadri et al. 1998b). Thus, all these data suggest that changes in the plasma ANG concentration and plasma osmolality (plasma Na+ concentrations) induced an increase in blood pressure and AVP release via angiotensinergic and cholinergic pathways in the brain.

2 Angiotensin Pathways and Regulation of Thirst

Thirst is a sensation aroused by a need for water, and relief from it sought by drinking water. Increased sodium appetite indicates a need for sodium, and relief will be sought by consuming salt or salty foods. The natural stimuli for thirst can be divided into two types: hypovolumic and osmotic (Fitzsimons 1972). The hypovolumic stimulus occurs when there is volume loss such as from slow dehydration over a period of time, massive vasodilation or rapid hemorrhage. The second natural stimulus for thirst is an increase in plasma osmotic pressure. Loss of cell water is detected by osmoreceptors (possibly sodium-sensitive receptors) located mainly in the hypothalamus, and their stimulation gives rise to thirst. Hypovolemia is detected by stretch receptors found in the walls of the heart and vasculature (Gauer and Henry 1963; Smith 1957). During hypovolemia ANG II levels in plasma increase due to low pressure in the renal artery that stimulates renin release. The brain is then informed about the changes in volume by inputs from the heart and baroreceptors located on the blood vessels relayed through the nucleus tractus solitarii (NTS) in the dorsal part of the medulla. The neural inputs stimulate the release of stored ANG II in the crucial areas of the brain that arouse thirst sensation. These brain areas are the circumventricular organs devoid of a BBB such as the SFO and the OVLT, including the median preoptic nucleus (MnPO), which lies within the BBB. The region most vulnerable and vital to induce thirst is the MnPO, since destruction of the SFO does not produce a lack of drinking behavior; however, destroying the MnPO produces chronic lack of thirst (adipsia) (Johnson and Buggy 1978).

Circulating ANG II derived from renal renin contributes to hypovolumic thirst and it also plays a role in increased sodium appetite along with the mineralocorticoids and other hormones. The idea that the kidney might contribute to thirst, although not through renin secretion, was first proposed by Linazasoro et al. (1954) and later Fitzsimons (1964) proposed and proved the involvement of the renal renin-angiotensin system in certain types of drinking behavior in rats that underwent caval obstruction. Obstruction of the abdominal vena cava just above the renal veins mimic the circulatory effects of severe hypovolemia. Within 30 min of caval obstruction, there was an increase in water intake followed much later by an increase in sodium intake, a marked fall in urine flow and electrolyte excretion (Fitzsimons 1969a). It has been found that increased drinking in response to caval obstruction is caused by ANG II, since ANG II caused increased drinking following intravenous infusion into water-replete rats (Fitzsimons and Simons 1969). Coincidentally during the investigation of norepinephrine-induced eating in rats, it was observed that an intrahypothalamic injection of ANG II caused drinking (Booth 1968) and later it was shown that injection of small amounts of ANG II in the anterior hypothalamus and preoptic region caused dose-dependent increases in drinking in water-replete rats (Epstein et al. 1970). In addition, ANG II injected into the brain areas has also been shown to induce salt intake (Buggy and Fisher 1974). These pioneering studies in search of mechanisms involved in the urge to drink water and eat salt pointed out that increased circulating ANG II caused increased drinking and salt eating by acting on accessible regions in the CNS. This indicates that ANG II triggers some neuronal circuits that induce the sensation of thirst and the compelling behavioral drive to find water and drink.

The systemic ANG II administration causes water-replete animals to drink water, although the response is less intensive compared to the vigorous, short-latency burst of drinking induced by ANG II injected into the brain (Abraham et al. 1975; Epstein et al. 1970). Pharmacological experiments have been performed using different techniques to study the effects of ANG II on drinking behavior and to map the central angiotensinergic pathways involved in drinking.

Ablation of the SFO blocks the water intake induced by systemic infusion of ANG II (Buggy and Fisher 1976; Simpson et al. 1978). Interestingly, drinking induced by ICV injection is not inhibited by ablation of the SFO. This led to the conclusion that ANG II acts at another region of the brain that lies behind the BBB to stimulate drinking. On the other hand, ablation of the anteroventral wall of the third ventricle (AV3V) of rat brain abolishes drinking in response to ANG II given ICV (Buggy and Fisher 1976), suggesting a region that lies within the AV3V region is crucial in ICV ANG II-induced drinking. Later it was demonstrated that the MnPO, which is a major component of the AV3V region, mediates drinking in response to ANG II given ICV. Furthermore, direct microinjection of ANG II into the MnPO induced drinking (ÓNeill and Broody 1987). Microinjections of ANG II into different ventral forebrain regions such as the anterior hypothalamus, hypothalamic PVN and lateral hypothalamic area (LHA) also induced an increase in water intake (do Vale et al. 1997; Qadri et al. 1998a; Tanaka et al. 2001). These findings on tracing the pathways involved in ANG II-induced drinking following hypovolemia or central ANG II injection were supported by additional studies using different techniques. Immunohistochemical staining revealed that these regions are rich in angiotensinergic nerve terminals (Lind et al. 1985) and AT_1 receptors (Lenkei et al. 1997). Immunohistochemical identification of the protein products of inducible transcription factors or immediate early genes such as c-fos or c-jun has been used to map the pattern of activation of neurons in the basal forebrain regions following different types of stimuli that induce drinking and sodium appetite, including ANG II (Hughes and Dragunow 1995; Lebrun et al. 1995; Blume et al. 1999). ICV injection of ANG II caused not only an intense c-fos expression but also other transcriptional factors such as FosB, c-Jun, Krox-20/24 and JunD in the CVO such as the SFO and OVLT, including the MnPO (Blume et al. 1998). There were also high levels of expression in the SON, PVN and lateral division of the central nucleus of amygdala and the bed nucleus of the stria terminalis (BNST) (Blume et al. 1997; Herbert et al. 1992). ANG II-induced Fos immunoreactivity was abolished by pretreatment with AT_1 receptor antagonist losartan (Lebrun et al. 1995; Rowland et al. 1994). ICV injections of antisense oligonucleotides complementary in nucleotide sequence to the mRNA encoding the first four amino acids of angiotensinogen to block central production of angiotensinogen in the rat brain greatly reduced the water intake in response to ICV injection of renin, suggesting a probable reduction in the levels of the renin substrate angiotensinogen in the brain regions critical in mediating the drinking response (Sinnayah et al. 1997).

ANG II given i.v. to conscious rats also resulted in an increased c-fos expression in forebrain, hypothalamic and medullary sites such as SFO and OVLT, BNST, SON, PVN, central nucleus of amygdala, area postrema (AP) and NTS (Badoer and McKinley 1997; McKinley et al. 1992; Oldfield et al. 1994;). Fos staining of neurons in SFO and OVLT following i.v. infusion of ANG II is consistent with the role of ANG II in hemorrhage, since these regions also show increased c-fos expression after hemorrhage. Hemorrhage also resulted in c-fos expression in the SON, PVN, AP and NTS (Badoer et al. 1993a, 1993b). From these data, an angiotensinergic thirst pathway could be proposed as follows: neurons in the SFO that are stimulated by circulating ANG II following hypovolemia have afferent axonal connections with neurons in the MnPO that subserve thirst. MnPO, the major switch gear in the central thirst pathway, is further connected with the hypothalamic AVP synthesizing and releasing regions SON and PVN. An increase in plasma osmotic pressure, which induces thirst and stimulates the release of AVP from the posterior pituitary into the blood circulation is detected by ANG II neurons in the OVLT and SFO. Increased plasma osmolality induces thirst and is mediated through AT_1 receptors (Hogarty et al. 1994). Losartan, an AT_1 receptor antagonist, abolished the drinking and AVP response of rats to a hyperosmotic stimulus (Hogarty et al. 1994; Qadri et al. 1998a). In addition, it has been shown that the levels of ANG II and ANG III are increased in the PVN of water-depleted rats and after an increase in plasma osmolality (Wright and Harding 1994; Qadri et al. 1994). It is hypothesized that an increase in plasma osmotic pressure is detected by ANG II neurons in the SFO and OVLT and is relayed via angiotensinergic fibers to the PVN. In the PVN, ANG II as a neurotransmitter or neuromodulator is released (Wright and Harding 1994; Qadri et al. 1994) and binds to its receptors to stimulate the thirst circuits and AVP release.

For a long time, it was thought that ANG II (1–8) is the only active peptide which influences arterial blood pressure, AVP release, stimulates drinking behavior and salt appetite in the animals. The identification of biological activity for the heptapeptide, ANG III (des-Asp^1-ANG II) has changed this view (Blair-West et al. 1971). ANG II and ANG III have been shown to stimulate equipotently water intake and AVP release when applied ICV or microinjected directly into the PVN (Wright et al. 1985; Qadri et al. 1998a). As demonstrated for the osmotically induced AVP release, the central pathway for drinking following water deprivation appears to involve endogenous ANG II, acting on AT_1 receptors, and cholinergic mechanisms, since pretreatment with losartan and the muscarinic receptor antagonist, atropine, each reduced water intake after 24 h water deprivation (Stauss and Unger 1990).

Thus, from the above data a neural pathway involved in drinking could be proposed as follows: neurons within the CVO, including MnPO, detect changes in the chemical composition of plasma (hypo/hypernatremia) and fluid volume. Their efferent neural pathways become important in driving subsequent physiological responses. The SFO and OVLT share a number of efferent brain target regions. These include the adjacent MnPO, with which they are both connected and which sends efferent fibers to the hypothalamic PVN. The SFO and OVLT send their efferent projections indirectly via the MnPO and directly to the hypothalamic PVN and SON. These projections mediate their influences on neurohypophysial hormone secretion in response to thirst stimuli (McKinley et al. 1999; Miselis 1981; Thrasher and Keil 1987).

3 Angiotensin Pathways and Sodium Excretion (Natriuresis)

Intracerebroventricular (ICV) administration of ANG II causes a large increase in the sodium excretion by the kidney (Andersson et al. 1972; Brooks and Malvin 1982; McKinley et al. 1994; Unger et al. 1989a, b). The natriuretic response following ICV injection of ANG II is most probably a synergistic effect of several effects that appear parallel or simultaneously to induce sodium excretion such as suppression of renal sympathetic nerve activity (May and McAllen 1997), reduction in plasma renin levels (Malayan et al. 1979; May and McAllen 1997), and an increase in arterial blood pressure and plasma AVP levels (Malayan et al. 1979). On the other hand, ICV administration of hypertonic saline stimulates many effects similar to ICV ANG II (via central AT_1 receptors) such as an increase in pressor response and AVP release (Qadri et al. 1998b), reduction in renin secretion (Eriksson and Fyhrquist 1976; McKinley and Mathai 1996) and renal nerve activity (May and McAllen 1997), and natriuresis (He et al. 1989; Rohmeiss et al. 1995a; Sjoquist et al. 1986). The responses to hypertonic saline can be blocked by AT_1 receptor blockers. All these data provide evidence for the central angiotensinergic influence on a wide range of homeostatic functions relating to osmoregulation (Hogarty et al. 1994; Mathai et al. 1998; Rohmeiss et al. 1995a; Qadri et al. 1998b).

The peripheral application of hyperosmotic saline stimulates AVP release, which is mediated through central AT_1 receptors (Hogarty et al. 1994). Centrally applied hyperosmotic saline engenders similar behavioral (drinking) and endocrine (AVP release) effects as observed with ANG II (see above) and could be blocked by an AT_1 receptor blocker. In addition, AVP release induced by hyperosmotic saline also involved central muscarinic cholinergic mechanisms. The hyperosmotic saline-induced non-pressure-associated AVP release is mediated through a periventricular and hypothalamic angiotensinergic and cholinergic mechanisms, whereas pressure-associated AVP release is mediated through the hypothalamic cholinergic mechanism (Qadri et al. 1998b). In case of hyperosmotic saline, the induced non-pressure-associated natriuresis, but not the pressure-associated natriuresis is mediated via the periventricular angiotensinergic system (Rohmeiss et al. 1995a). As shown for the osmotically induced AVP release and natriuresis, the central pathway for drinking following water deprivation appears to involve angiotensinergic and cholinergic mechanisms, since treatment with losartan and the muscarine receptor antagonist atropine each reduced water intake after 24 h water deprivation (Stauss and Unger 1990).

The signals and pathways to the kidney that mediate the central action of ANG II or hyperosmotic saline are emerging. Ablation of the lamina terminalis results in severe hypernatremia. This is probably due to a negative fluid balance as a result of disordered water intake and AVP release and also due to impaired natriuretic mechanisms (Park et al. 1985). The natriuresis induced by ICV hyperosmotic saline was attenuated by blockade of AT_1 receptors in the SFO (Rohmeiss et al. 1995b). One of the first data sets providing evidence that the

SFO is the relay station in mediating the non-pressure- and pressure-associated natriuresis following central osmoreceptor stimulation. Recently, it has been demonstrated that the MnPO is involved in body fluid regulation not only by controlling AVP release and water intake, but also by modulating central sympathetic outflow, which regulates body fluid balance through an effect on the kidney (Yasuda et al. 2000). In this study, the renal sympathetic nerve activity (RSNA) and mean arterial blood pressure were elevated by the injection of hyperosmotic saline into the third ventricle, and were attenuated by microinjection of lidocaine into the MnPO. In another recently published study, it was demonstrated that there is an interaction between AT_1 and AT_2 receptors of the PVN and the septal area in the control of ANG-induced physiological responses in terms of water and sodium homeostasis and mean arterial blood pressure modulation (Camargo et al. 2002). In addition, alpha-adrenergic pathways involving the PVN are important for the water and sodium excretion and pressor responses induced by angiotensinergic activation of the medial septal area (Camargo and Saad 2001). Together, these data suggest common central angiotensinergic, cholinergic and adrenergic pathways between osmotically induced AVP release, natriuresis and drinking in response to osmotic stimulation involving the lamina terminalis and hypothalamic MnPO and PVN.

Taken together, blood-borne signals such as plasma hypernatremia act on osmoreceptors (sodium and/or angiotensin receptors) found in the lamina terminalis, specifically in the SFO. Stimulation of osmoreceptors in the lamina terminalis (SFO) activates angiotensinergic pathway(s) through the MnPO that project to the hypothalamic PVN, which results in increased release of AVP into the circulation, an increase in the RSNA and an increase in mean arterial blood pressure; all these effects concomitantly participate in the natriuresis/osmoregulation.

It is of great interest to point out that there is an endogenous antagonisms of ANG II-induced central natriuretic effects (Unger et al. 1990). Atrial natriuretic peptide (ANP) was identified as a functionally antagonistic circulating hormone involved in ANG II-induced body fluid regulation. Both ANG II and ANP have been localized in close vicinity in forebrain areas involved in the central fluid and electrolyte regulation. In addition, pretreatment of rats with ANP given ICV dose-dependently antagonized the central ANG II-induced natriuretic effects (Rohmeiss et al. 1989, 1991). These are one of the first data sets supporting the notion of a functional antagonism between ANG II and ANP in the brain.

Is it Angiotensin II or III ? Des-Asp^1-ANG II, or ANG III, was once believed to be an inactive catabolic product of ANG II through N-terminal degradation. The identification of biological activity for this heptapeptide in 1971 (Blair-West et al. 1971) drastically changed this view, and it was the beginning of evaluations of its physiological functions. However, among the main bioactive peptides of the brain RAS, ANG II and ANG III exhibit the same affinity for AT_1 and AT_2 receptors. Both peptides, injected ICV, cause similar increases in AVP release and blood pressure. An increasingly large number of data provide evidence that ANG III and not ANG II plays an important role in central regulation of cardiovascular and body fluid volume homeostasis (Reaux et al. 2001; Yang et al. 1995).

4 Conclusion

All the data that has been reviewed here show that angiotensin is generated in the brain and the angiotensinergic neuronal pathways play an important role in the regulation and control of body fluid homeostasis. Homeostatic responses that may be influenced by angiotensinergic pathways include increased arterial blood pressure, AVP release into the circulation, thirst and renal sodium excretion. Systemic stimuli that activate central angiotensinergic pathways include plasma hypernatremia and circulating ANG II/ANG III. Although at present a great deal of knowledge has been accumulated regarding the involvement and importance of angiotensin and angiotensinergic pathways in brain functions, several gaps in our knowledge still exist and await further elucidation.

References

Abraham SF, Baker RM, Blaine EH, Denton DA, McKinley MJ (1975) Water drinking induced in sheep by angiotensin—a physiological or pharmacological effect? J Comp Physiol Psychol 88:503–518

Akaishi T, Negoro H (1983) Effects of microelectrophoretically applied acetylcholine- and angiotensin-antagonists on the paraventricular neurosecretory cells excited by osmotic stimuli. Neurosci Lett 36:157–161

Allen AM, Zhuo J, Mendelsohn FA (2000) Localization and function of angiotensin AT1 receptors. Am J Hypertens 13:31S–38S

Andersson B, Eriksson L, Fernandez O, Kolmodin CG, Oltner R (1972) Centrally mediated effects of sodium and angiotensin II on arterial blood pressure and fluid balance. Acta Physiol Scand 85:398–407

Andreatta SH, Averill DB, Santos RA, Ferrario CM (1988) The ventrolateral medulla. A new site of action of the renin-angiotensin system. Hypertension 11:1163–1166

Andresen MC, Mendelowitz D (1996) Sensory afferent neurotransmission in caudal nucleus tractus solitarius—common denominators. Chem Senses 21:387–395

Averill DB, Diz DI (2000) Angiotensin peptides and baroreflex control of sympathetic outflow: pathways and mechanisms of the medulla oblongata. Brain Res Bull 51:119–128

Averill DB, Diz DI, Barnes KL, Ferrario CM (1987) Pressor responses of angiotensin II microinjected into the dorsomedial medulla of the dog. Brain Res 414:294–300

Averill DB, Tsuchihashi T, Khosla MC, Ferrario CM (1994) Losartan, nonpeptide angiotensin II-type 1 receptor antagonist, attenuates pressor and sympathoexcitatory responses evoked by angiotensin II and L-glutamate in rostral ventrolateral medulla. Brain Res 665:245–252

Badoer E, McKinlay D (1997) Effect of intravenous angiotensin II on Fos distribution and drinking behavior in rabbits. Am J Physiol 272:R1515–R1524

Bargmann W, Scharrer E (1951) The site of origin of the hormones of the posterior pituitary. Am Sci 39:255–259

Badoer E, Oldfield BJ, McKinley MJ (1993a) Haemorrhage-induced production of Fos in neurons of the lamina terminalis: role of endogenous angiotensin II. Neurosci Lett 159:151–154

Badoer E, McKinley MJ, Oldfield BJ, McAllen RM (1993b) A comparison of hypotensive and non-hypotensive hemorrhage on Fos expression in spinally projecting neurons of the paraventricular nucleus and rostral ventrolateral medulla. Brain Res 610:216–223

Barnes KL, McQueeney AJ, Ferrario CM (1993) Receptor subtype that mediates the neuronal effects of angiotensin II in the rat dorsal medulla. Brain Res Bull 31:195–200

Bealer SL, Phillips MI, Johnson AK, Schmid PG (1979) Anteroventral third ventricle lesions reduce antidiuretic responses to angiotensin II. Am J Physiol 236:E610–E615

Blair-West JR, Coghlin JP, Denton DA, Funder JW, Scoggins BA, Wright RD (1971) Effect of the heptapeptide (2–8) and hexapeptide (3–8) fragments of angiotensin II on aldosterone secretion. J Clin Endocrinol Metab 32:575–578

Blair-West JR, Denton DA, McKinley MJ, Weisinger RS (1997) Central infusion of the AT1 receptor antagonist losartan inhibits thirst but not sodium appetite in cattle. Am J Physiol 272:R1940–R1945

Blume A, Lebrun CJ, Herdegen T, Braaavo R, Linz W, Mollenhoff E, Unger T (1997) Increased brain transcription factor expression by angiotensin in genetic hypertension. Hypertension 29:592–598

Blume A, Seifert K, Lebrun CJ, Mollenhoff E, Gass P, Unger T, Herdegen T (1998) Differential time course of angiotensin-induced AP-1 and Krox proteins in the rat lamina terminalis and hypothalamus. Neurosci Lett 241:87–90

Blume A, Herdegen T, Unger T (1999) Angiotensin peptides and inducible transcription factors. J Mol Med 77:339–357

Blume A, Neumann C, Dorenkamp M, Culman J, Unger T (2002) Involvement of adrenoceptors in the angiotensin II-induced expression of inducible transcription factors in the rat forebrain and hypothalamus. Neuropharmacology 42:281–288

Booth DA (1968) Mechanism of action of norepinephrine in eliciting an eating response on injection into the rat hypothalamus. J Pharmacol Exp Ther 160:336–348

Brooks VL, Malvin RL (1982) Intracerebroventricular infusions of angiotensin II increases sodium excretion. Proc Soc Exp Biol Med 169:532–537

Buggy J, Fisher AE (1974) Evidence for a dual central role for angiotensin in water and sodium intake. Nature 250:733–735

Buggy J, Fisher AE (1976) Anteroventral third ventricle site of action for angiotensin induced thirst. Pharmacol Biochem Behav 4:651–660

Bunnemann B, Metzger R, Fuxe K, Ganten D (1993) Regional expression of angiotensinogen mRNA in the brain of one-week-old, adult and old male rats. Brain Res Dev Brain Res 73:41–45

Camargo LA, Saad WA (2001) Role of the alpha(1)- and alpha(2)-adrenoceptors of the paraventricular nucleus on the water and salt intake, renal excretion, and arterial pressure induced by angiotensin II injection into the medial septal area. Brain Res Bull 4:595–602

Camargo LA, Saad WA, Simoes S, Santos TA, Saad WA (2002) Interaction between paraventricular nucleus and septal area in the control of physiological responses induced by angiotensin II. Braz J Med Biol Res 35:1017–1023

Campagnole-Santos MJ, Diz DI, Ferrario CM (1988) Baroreceptor reflex modulation by angiotensin II at the nucleus tractus solitarii. Hypertension 11:1167–1171

Campagnole-Santos MJ, Diz DI, Santos RA, Khosla MC, Brosnihan KB, Ferrario CM (1989) Cardiovascular effects of angiotensin-(1–7) injected into the dorsal medulla of rats. Am J Physiol 257:H324–H329

Casto R, Phillips MI (1984) A role for central angiotensin in regulation of blood pressure at the nucleus tractus solitarius. Clin Exp Hypertens 6:1933–1937

Casto R, Phillips MI (1985) Neuropeptide action in nucleus tractus solitarius: angiotensin specificity and hypertensive rats. Am J Physiol 249:R341–R347
Casto R, Phillips MI (1986) Angiotensin II attenuates baroreflexes at nucleus tractus solitarius of rats. Am J Physiol 250:R193–R198
Chan RK, Chan YS, Wong TM (1991) Responses of cardiovascular neurons in the rostral ventrolateral medulla of the normotensive Wistar Kyoto and spontaneously hypertensive rats to iontophoretic application of angiotensin II. Brain Res 556:145–150
Chan RK, Chan YS, Wong TM (1994) Effects of [Sar1, Ile8]-angiotensin II on rostral ventrolateral medulla neurons and blood pressure in spontaneously hypertensive rats. Neuroscience 63:267–277
Dampney RA, Coleman MJ, Fontes MA, Hirooka Y, Horiuchi J, Li YW, Polson JW, Potts PD, Tagawa T (2002) Central mechanisms underlying short- and long-term regulation of the cardiovascular system. Clin Exp Pharmacol Physiol 29:261–268
DiMicco JA, Stotz-Potter EH, Monroe AJ, Morin SM (1996) Role of the dorsomedial hypothalamus in the cardiovascular response to stress. Clin Exp Pharmacol Physiol 23:171–176
Diz DI, Barnes KL, Ferrario CM (1984) Hypotensive actions of microinjections of angiotensin II into the dorsal motor nucleus of the vagus. J Hypertens Suppl 2:S53–S56
Diz DI, Barnes KL, Ferrario CM (1987) Functional characteristics of neuropeptides in the dorsal medulla oblongata and vagus nerve. Fed Proc 46:30–35
Do Vale CF, Camargo GM, Saad WA, Menani JV, Renzi A, Luiz AC, Cerri PS, Camargo LA (1997) Effect of ibotenate lesions of the ventromedial hypothalamus on the water and salt intake induced by activation of the median preoptic nucleus in sodium-depleted rats. J Auton Nerv Syst 66:19–25
Epstein AN, Fitzsimons JT, Rolls BJ (1970) Drinking induced by injection of angiotensin into the rain of the rat. J Physiol 210:457–474
Eriksson L, Fyhrquist F (1976) Plasma renin activity following central infusion of angiotensin II and altered CSF sodium concentration in the conscious goat. Acta Physiol Scand 98:209–216
Ferguson AV, Kasting NW (1986) Electrical stimulation in subfornical organ increases plasma vasopressin concentrations in the conscious rat. Am J Physiol 251:R425–R428
Ferguson AV, Kasting NW (1988) Angiotensin acts at the subfornical organ to increase plasma oxytocin concentrations in the rat. Regul Pept 23:343–352
Fitzsimons JT (1964) Drinking caused by constriction of the inferior vena cava in the rat. Nature 204:479–480
Fitzsimons JT (1969) The role of a renal thirst factor in drinking induced by extracellular stimuli. J Physiol 201:349–368
Fitzsimons JT (1972) Thirst. Physiol Rev 52:468–561
Fitzsimons JT, Simons BJ (1969) The effect on drinking in the rat of intravenous infusion of angiotensin, given alone or in combination with other stimuli of thirst. J Physiol 203:45–57
Fontes MA, Silva LC, Campagnole-Santos MJ, Khosla MC, Guertzenstein PG, Santos RA (1994) Evidence that angiotensin-(1–7) plays a role in the central control of blood pressure at the ventro-lateral medulla acting through specific receptors. Brain Res 665:175–180
Fontes MA, Pinge MC, Naves V, Campagnole-Santos MJ, Lopes OU, Khosla MC, Santos RA (1997) Cardiovascular effects produced by microinjection of angiotensins and angiotensin antagonists into the ventrolateral medulla of freely moving rats. Brain Res 750:305–310
Fontes MA, Tagawa T, Polson JW, Cavanagh SJ, Dampney RA (2001) Descending pathways mediating cardiovascular response from dorsomedial hypothalamic nucleus. Am J Physiol Heart Circ Physiol 280:H2891–H2901
Fow JE, Averill DB, Barnes KL (1994) Mechanisms of angiotensin-induced hypotension and bradycardia in the medial solitary tract nucleus. Am J Physiol 267:H259–H266

Ganten D, Lang RE, Lehmann E, Unger T (1984) Brain angiotensin: on the way to becoming a well-studied neuropeptide system. Biochem Pharmacol 33:3523–3528

Gauer OH, Henry JP (1963) Circulatory basis of fluid volume control. Physiol Rev 43:423–481

Gutman MB, Ciriello J, Mogenson GJ (1988) Effects of plasma angiotensin II and hypernatremia on subfornical organ neurons. Am J Physiol 254:R746–R754

Guyenet PG, Koshiya N (1995) Working model of the sympathetic chemoreflex in rats. Clin Exp Hypertens 17:167–179

Gyurko R, Wielbo D, Phillips MI (1993) Antisense inhibition of AT1 receptor mRNA and angiotensinogen mRNA in the brain of spontaneously hypertensive rats reduces hypertension of neurogenic origin. Regul Pept 49:167–174

Harding JW, Jensen LL, Hanesworth JM, Roberts KA, Page TA, Wright JW (1992) Release of angiotensins in paraventricular nucleus of rat in response to physiological and chemical stimuli. Am J Physiol 262:F17–F23

He XR, Zhang JF, Yao T (1989) Intracerebroventricular administration of hypertonic saline inhibits the reabsorption of water and sodium in the proximal tubule. Sheng Li Xue Bao 41:421–427

Herbert J, Forsling ML, Howes SR, Stacey PM, Shiers HM (1992) Regional expression of c-fos antigen in the basal forebrain following intraventricular infusions of angiotensin and its modulation by drinking either water or saline. Neuroscience 51:867–882

Hilton SM (1975) Ways of viewing the central nervous control of the circulation: old and new. Brain Res 87:213–219

Hirooka Y, Potts PD, Dampney RA (1997) Role of angiotensin II receptor subtypes in mediating the sympathoexcitatory effects of exogenous and endogenous angiotensin peptides in the rostral ventrolateral medulla of the rabbit. Brain Res 772:107–114

Hogarty DC, Tran DN, Phillips MI (1994) Involvement of angiotensin receptor subtypes in osmotically induced release of vasopressin. Brain Res 637:126–132

Hohle S, Culman J, Boser M, Qadri F, Unger T (1996) Effect of angiotensin AT2 and muscarinic receptor blockade on osmotically induced vasopressin release. Eur J Pharmacol 300:119–123

Hughes P, Dragunow M (1995) Induction of immediate-early genes and the control of neurotransmitter-regulated gene expression within the nervous system. Pharmacol Rev 47:133–178

Johnson AK, Buggy J (1978) Periventricular preoptic-hypothalamus is vital for thirst and normal water economy. Am J Physiol 234:R122–R129

Knepel W, Nutto D, Meyer DK (1982) Effect of transection of subfornical organ efferent projections on vasopressin release induced by angiotensin or isoprenaline in the rat. Brain Res 248:180–184

Lebrun CJ, Blume A, Herdegen T, Seifert K, Bravo R, Unger T (1995) Angiotensin II induces a complex activation of transcription factors in the rat brain: expression of Fos, Jun and Krox proteins. Neuroscience 65:93–99

Lenkei Z, Palkovits M, Corvol P, Llorens-Cortes C (1997) Expression of angiotensin type-1 (AT1) and type-2 (AT2) receptor mRNAs in the adult rat brain: a functional neuroanatomical review. Front Neuroendocrinol 18:383–439

Lenkei Z, Palkovits M, Corvol P, Llorens-Cortes C (1998) Distribution of angiotensin type-1 receptor messenger RNA expression in the adult rat brain. Neuroscience 82:827–841

Lewis SJ, Allen AM, Verberne AJ, Figdor R, Jarrott B, Mendelsohn FA (1986) Angiotensin II receptor binding in the rat nucleus tractus solitarii is reduced after unilateral nodose ganglionectomy or vagotomy. Eur J Pharmacol 125:305–307

Li YW, Dampney RA (1994) Expression of Fos-like protein in brain following sustained hypertension and hypotension in conscious rabbits. Neuroscience 61:613–634

Li Z, Ferguson AV (1993) Subfornical organ efferents to paraventricular nucleus utilize angiotensin as a neurotransmitter. Am J Physiol 265:R302–R309

Lima DX, Fontes MA, Oliveira RC, Campagnole-Santos MJ, Khosla MC, Santos RA (1996) Pressor action of angiotensin I at the ventrolateral medulla: effect of selective angiotensin blockade. Immunopharmacology 33:305–307

Lin KS, Chan JY, Chan SH (1997) Involvement of AT2 receptors at NRVL in tonic baroreflex suppression by endogenous angiotensins. Am J Physiol 272:H2204–H2210

Linazasoro JM, Jimenez Diaz C, Castro Mendoza H (1954) The kidney and thirst regulation. Bull Inst Med Res Madrid 7:53–61

Lind RW, Swanson LW, Bruhn TO, Ganten D (1985) The distribution of angiotensin II-immunoreactive cells and fibers in the paraventriculo-hypophysial system of the rat. Brain Res 338:81–89

Lucius R, Gallinat S, Busche S, Rosenstiel P, Unger T (1999) Beyond blood pressure: new roles for angiotensin II. Cell Mol Life Sci 56:1008–1019

Malayan SA, Keil LC, Ramsay DJ, Reid IA (1979) Mechanism of suppression of plasma renin activity by centrally administered angiotensin II. Endocrinology 104:672–675

Matsumura K, Averill DB, Ferrario CM (1998) Angiotensin II acts at AT1 receptors in the nucleus of the solitary tract to attenuate the baroreceptor reflex. Am J Physiol 275:R1611–R1619

May CN, McAllen RM (1997) Baroreceptor-independent renal nerve inhibition by intracerebroventricular angiotensin II in conscious sheep. Am J Physiol 273:R560–R567

McKinley MJ, Badoer E, Oldfield BJ (1992) Intravenous angiotensin II induces Fos-immunoreactivity in circumventricular organs of the lamina terminalis. Brain Res 594:295–300

McKinley MJ, Evered M, Mathai M, Coghlan JP (1994) Effects of central losartan on plasma renin and centrally mediated natriuresis. Kidney Int 46:1479–1482

McKinley MJ, Mathai ML (1996) Centrally administered losartan inhibits the reduction in plasma renin concentration caused by intracerebroventricular hypertonic saline in Na-depleted sheep. Regul Pept 66:37–40

McKinley MJ, Gerstberger R, Mathai ML, Oldfield BJ, Schmid H (1999) The lamina terminalis and its role in fluid and electrolyte homeostasis. J Clin Neurosci 6:289–301

McKinley MJ, Allen AM, Mathai ML, May C, McAllen RM, Oldfield BJ, Weisinger RS (2001) Brain angiotensin and body fluid homeostasis. Jpn J Physiol 51:281–289

Meng H, Wielbo D, Gyurko R, Phillips MI (1994) Antisense oligonucleotide to AT1 receptor mRNA inhibits central angiotensin induced thirst and vasopressin. Regul Pept 54:543–551

Michelini LC, Bonagamba LG (1990) Angiotensin II as a modulator of baroreceptor reflexes in the brainstem of conscious rats. Hypertension 15 [Suppl 2]:I45–I50

Miselis RR (1981) The efferent projections of the subfornical organ of the rat: a circumventricular organ within a neural network subserving water balance. Brain Res 230:1–23

Miselis RR (1982) The subfornical organ's neural connections and their role in water balance. Peptides 3:501–502

Mosqueda-Garcia R, Tseng CJ, Appalsamy M, Robertson D (1990) Cardiovascular effects of microinjection of angiotensin II in the brainstem of renal hypertensive rats. J Pharmacol Exp Ther 255:374–381

Muratani H, Averill DB, Ferrario CM (1991) Effect of angiotensin II in ventrolateral medulla of spontaneously hypertensive rats. Am J Physiol 260:R977–R984

Muratani H, Ferrario CM, Averill DB (1993) Ventrolateral medulla in spontaneously hypertensive rats: role of angiotensin II. Am J Physiol 264:R388–R395

O'Neill TP, Broody MJ (1987) Role for the median preoptic nucleus in centrally evoked pressor responses. 252:R1165–R1172

Oldfield BJ, Badoer E, Hards DK, McKinley MJ (1994) Fos production in retrogradely labelled neurons of the lamina terminalis following intravenous infusion of either hypertonic saline or angiotensin II. Neuroscience 60:255–262

Park RG, Congiu M, Denton DA, McKinley MJ (1985) Natriuresis induced by arginine vasopressin infusion in sheep. Am J Physiol 249:F799–F805

Phillips MI (1987) Functions of angiotensin in the central nervous system. Annu Rev Physiol 49:413–435

Qadri F, Culman J, Veltmar A, Maas K, Rascher W, Unger T (1993) Angiotensin II-induced vasopressin release is mediated through alpha-1 adrenoceptors and angiotensin II AT1 receptors in the supraoptic nucleus. J Pharmacol Exp Ther 267:567–574

Qadri F, Edling O, Wolf A, Gohlke P, Culman J, Unger T (1994) Release of angiotensin in the paraventricular nucleus in response to hyperosmotic stimulation in conscious rats: a microdialysis study. Brain Res 637:45–49

Qadri F, Wolf A, Waldmann T, Rascher W, Unger T (1998a) Sensitivity of hypothalamic paraventricular nucleus to C- and N-terminal angiotensin fragments: vasopressin release and drinking. J Neuroendocrinol 10:275–281

Qadri F, Waldmann T, Wolf A, Hohle S, Rascher W, Unger T (1998b) Differential contribution of angiotensinergic and cholinergic receptors in the hypothalamic paraventricular nucleus to osmotically induced AVP release. J Pharmacol Exp Ther 285:1012–1018

Reaux A, Fournie-Zaluski MC, Llorens-Cortes C (2001) Angiotensin III: a central regulator of vasopressin release and blood pressure. Trends Endocrinol Metab 12:157–162

Renaud LP, Bourque CW (1991) Neurophysiology and neuropharmacology of hypothalamic magnocellular neurons secreting vasopressin and oxytocin. Prog Neurobiol 36:131–169

Rettig R, Healy DP, Printz MP (1986) Cardiovascular effects of microinjections of angiotensin II into the nucleus tractus solitarii. Brain Res 364:233–240

Rohmeiss P, Demmert G, Rettig R, Unger T (1989) Centrally administered arterial natriuretic factor inhibits central angiotensin-induced natriuresis. Brain Res 502:198–203

Rohmeiss P, Demmert G, Unger T (1991) Central effects of atrial natriuretic factor on renal excretion. Am J Physiol 261:F354–F359

Rohmeiss P, Beyer C, Nagy E, Tschope C, Hohle S, Strauch M, Unger T (1995a) NaCl injections in brain induce natriuresis and blood pressure responses sensitive to ANG II AT1 receptors. Am J Physiol 269:F282–F288

Rohmeiss P, Beyer C, Hocher B, Qadri F, Gretz N, Strauch M, Unger T (1995b) Osmotically induced natriuresis and blood pressure response involves angiotensin AT1 receptors in the subfornical organ. J Hypertens 13:1399–1404

Rowland NE, Li BH, Rozelle AK, Smith GC (1994) Comparison of fos-like immunoreactivity induced in rat brain by central injection of angiotensin II and carbachol. Am J Physiol 267:R792–R798

Saavedra JM (1992) Brain and pituitary angiotensin. Endocr Rev 13:329–380

Sasaki S, Dampney RA (1990) Tonic cardiovascular effects of angiotensin II in the ventrolateral medulla. Hypertension 15:274–283

Sesoko S, Muratani H, Takishita S, Teruya H, Kawazoe N, Fukiyama K (1995) Modulation of baroreflex function by angiotensin II endogenous to the caudal ventrolateral medulla. Brain Res 671:38–44

Silva LC, Fontes MA, Campagnole-Santos MJ, Khosla MC, Campos RR Jr, Guertzenstein PG, Santos RA (1993) Cardiovascular effects produced by micro-injection of angiotensin-(1–7) on vasopressor and vasodepressor sites of the ventrolateral medulla. Brain Res 613:321–325

Simpson JB, Mangiapane ML, Dellmann HD (1978) Central receptor sites for angiotensin-induced drinking: a critical review. Fed Proc 37:2676–2682

Sinnayah P, Kachab E, Haralambidis J, Coghlan JP, McKinley MJ (1997) Effects of angiotensinogen antisense oligonucleotides on fluid intake in response to different dipsogenic stimuli in the rat. Brain Res Mol Brain Res 50:43–50

Sjoquist M, Goransson A, Hansell P, Isaksson B, Ulfendahl HR (1986) Redistribution of glomerular filtration and renal plasma flow in CNS-induced natriuresis. Acta Physiol Scand 127:491–497

Smith HW (1957) Salt and water volume receptors: an exercise in physiological apologetics. Am J Med 23:623–652

Song K, Allen AM, Paxinos G, Mendelsohn FA (1991) Angiotensin II receptor subtypes in rat brain. Clin Exp Pharmacol Physiol 18:93–96

Spyer KM (1994) Annual review prize lecture. Central nervous mechanisms contributing to cardiovascular control. J Physiol 474:1–19

Stadler T, Veltmar A, Qadri F, Unger T (1992) Angiotensin II evokes noradrenaline release from the paraventricular nucleus in conscious rats. Brain Res 569:117–122

Stauss H, Unger T (1990) Role of angiotensin in the drinking response after water deprivation. Naunyn-Schmiedeberg's Arch Pharmacol 343 [Suppl]:R106

Swanson LW, Sawchenko PE (1983) Hypothalamic integration: organization of the paraventricular and supraoptic nuclei. Annu Rev Neurosci 6:269–324

Tanaka J, Saito H, Kaba H (1987) Subfornical organ and hypothalamic paraventricular nucleus connections with median preoptic nucleus neurons: an electrophysiological study in the rat. Exp Brain Res 68:579–585

Tanaka J, Hori K, Nomura M (2001) Dipsogenic response induced by angiotensinergic pathways from the lateral hypothalamic area to the subfornical organ in rats. Behav Brain Res 118:111–116

Thrasher TN, Keil LC (1987) Regulation of drinking and vasopressin secretion: role of organum vasculosum laminae terminalis. Am J Physiol 253:R108–R120

Unger T (1999) The angiotensin type 2 receptor: variations on an enigmatic theme. J Hypertens 17:1775–1786

Unger T, Rascher W, Schuster C, Pavlovitch R, Schomig A, Dietz R, Ganten D (1981) Central blood pressure effects of substance P and angiotensin II: role of the sympathetic nervous system and vasopressin. Eur J Pharmacol 71:33–42

Unger T, Becker H, Petty M, Demmert G, Schneider B, Ganten D, Lang RE (1985) Differential effects of central angiotensin II and substance P on sympathetic nerve activity in conscious rats. Implications for cardiovascular adaptation to behavioral responses. Cir Res 56:563–575

Unger T, Badoer E, Ganten D, Lang RE, Rettig R (1988) Brain angiotensin: pathways and pharmacology. Circulation 77:I40–154

Unger T, Horst PJ, Bauer M, Demmert G, Rettig R, Rohmeiss P (1989a) Natriuretic action of angiotensin II in conscious rats. Brain Res 486:33–38

Unger T, Gohlke P, Kotrba M, Rettig R, Rohmeiss P (1989b) Angiotensin II and atrial natriuretic peptide in the brain: effects on volume and Na+ balance. Resuscitation 18:309–319

Unger T, Badoer E, Gareis C, Girchev R, Kotrba M, Qadri F, Rettig R, Rohmeiss P (1990) Atrial natriuretic peptide (ANP) as a neuropeptide: interaction with angiotensin II on volume control and renal sodium handling. Br J Clin Pharmacol 30 [Suppl 1]:83S–88S

Veltmar A, Culman J, Qadri F, Rascher W, Unger T (1992) Involvement of adrenergic and angiotensinergic receptors in the paraventricular nucleus in the angiotensin II-induced vasopressin release. J Pharmacol Exp Ther 263:1253–1260

Wilkin LD, Mitchell LD, Ganten D, Johnson AK (1989) The supraoptic nucleus: afferents from areas involved in control of body fluid homeostasis. Neuroscience 28:573–584

Wright JW, Harding JW (1992) Regulatory role of brain angiotensins in the control of physiological and behavioral responses. Brain Res Brain Res Rev 17:227–262

Wright JW, Morseth SL, Abhold RH, Harding JW (1985) Pressor action and dipsogenicity induced by angiotensin II and III in rats. Am J Physiol 249:R514–R521

Wright JW, Harding JW (1994) Brain angiotensin receptor subtypes in the control of physiological and behavioral responses. Neurosci Biobehav Rev 18:21–53

Yang CCH, Kuo TBJ, Chan SHH, Chan JYH (1995) Neurobiology of central angiotensin III and dipsogenesis. Biol Signals 4:59–71

Yasuda Y, Honda K, Negoro H, Higuchi T, Goto Y, Fukuda S (2000) The contribution of the median preoptic nucleus to renal sympathetic nerve activity increased by intracerebroventricular injection of hypertonic saline in the rat. Brain Res 867:107–114

Involvement of the Renin Angiotensin System in the Regulation of the Hypothalamic Pituitary Adrenal Axis

G. Aguilera

Section on Endocrine Physiology, DEB, NICHDNIH Bldg 10 Rm 10N 262,
10 Center Drive, MSC 1862, Bethesda, MD 20892-61862, USA
e-mail: Greti_Aguilera@nih.gov

Abstract The presence of type 1 Ang II (AT_1) receptors at sites critical for activation of the HPA axis, such as parvocellular corticotropin releasing hormone (CRH) neurons in the hypothalamic paraventricular nucleus (PVN), brain nuclei with connections to the PVN, and pituitary corticotrophs provide evidence for direct regulatory effects of the peptide on the HPA axis activity. Central AT_1 receptors modulate CRH expression and possibly vasopressin in parvocellular neurons of the PVN. Intracerebroventricular injection of low doses of Ang II increase CRH mRNA and plasma ACTH. The latter effect is at least in part mediated by CRH release, since it is prevented by CRH antiserum. However, in a number of conditions plasma ACTH and glucocorticoid responses to stress are unaffected by central AT_1 receptor blockade, indicating that Ang II is not essential for activation of the HPA axis during stress. Ang II could stimulate CRH secretion directly through interaction with receptors in the PVN or median eminence, or indirectly by stimulating afferent neural pathways from the circumventricular organs. AT_1 receptors are present in the pituitary corticotroph and Ang II stimulates ACTH secretion in vitro. However, there is little evidence for a major role of Ang II directly regulating pituitary ACTH secretion and it is more likely that the pituitary effects of the peptide relate to corticotroph differentiation.

Keywords Hypothalamus · Hypothalamic paraventricular nucleus · Corticotrophin releasing hormone · Ang II receptors · Pituitary corticotroph · Adrenal · Stress

1 Introduction

Activation of the hypothalamic pituitary adrenal axis (HPA) is essential for survival of the organism when confronted with alterations of the external or internal environment (Munck et al. 1984). The end result of HPA axis activation is secretion of adrenal glucocorticoids, which influence metabolism, neurotransmitter synthesis, immune system and signaling of other hormones such as thyroid hormone, catecholamines and vasoactive peptides (Munck et al. 1984; Whithworth 1987; Wong et al. 1995). The production of adrenal glucocorticoids is controlled by the anterior pituitary peptide, ACTH. The regulation of ACTH synthesis and secretion is multifactorial, involving the stimulatory effect of the neuropeptides, corticotropin-releasing hormone (CRH) and vasopressin (VP), produced by parvocellular neurons of the hypothalamic paraventricular nucleus (PVN) and negative feedback by glucocorticoids (Aguilera 1994; Antoni 1986; Whitnall 1993). In addition, a number of neurotransmitters and peptides, including angiotensin II (Ang II) have been implicated in the regulation of ACTH secretion by acting directly on pituitary corticotrophs and/or indirectly by controlling the expression and secretion of CRH and VP. A normal component of the stress response in humans and experimental animals is activation of the renin-angiotensin system, as a result of sympathetic activation (Golin et al. 1988). This leads to increases in circulating Ang II, which contribute to cardiovascular adaptation during stress and potentially to modulation of HPA axis activity. Components of the renin angiotensin system, including AT_1 and AT_2 receptors, have been identified in a number of sites related to the stress response, such as the pituitary corticotroph, parvocellular neurons of the PVN, and brain nuclei with afferent connections to the PVN (Aguilera et al. 1995b; Saavedra 1992; Song et al. 1992). This chapter will discuss current knowledge on the involvement of Ang II on the regulation of HPA axis activity at central and peripheral levels.

2 Central Regulation of the HPA Axis by Ang II

A large body of evidence supports the involvement of Ang II in the central regulation of the HPA axis (Aguilera et al. 1995b; Ganong and Murakami 1987). As discussed below, morphological and functional studies have revealed the presence of Ang II receptors and other components of the renin-angiotensin system in areas of the brain controlling HPA axis activity, and provided evidence for modulatory effects of central Ang II on the HPA axis.

2.1 Central Ang II Receptors and HPA Axis Activity

A prominent site of expression of Ang II receptors in the brain is the hypothalamic PVN (Fig. 1H and M). This nucleus is the site of production of CRH and

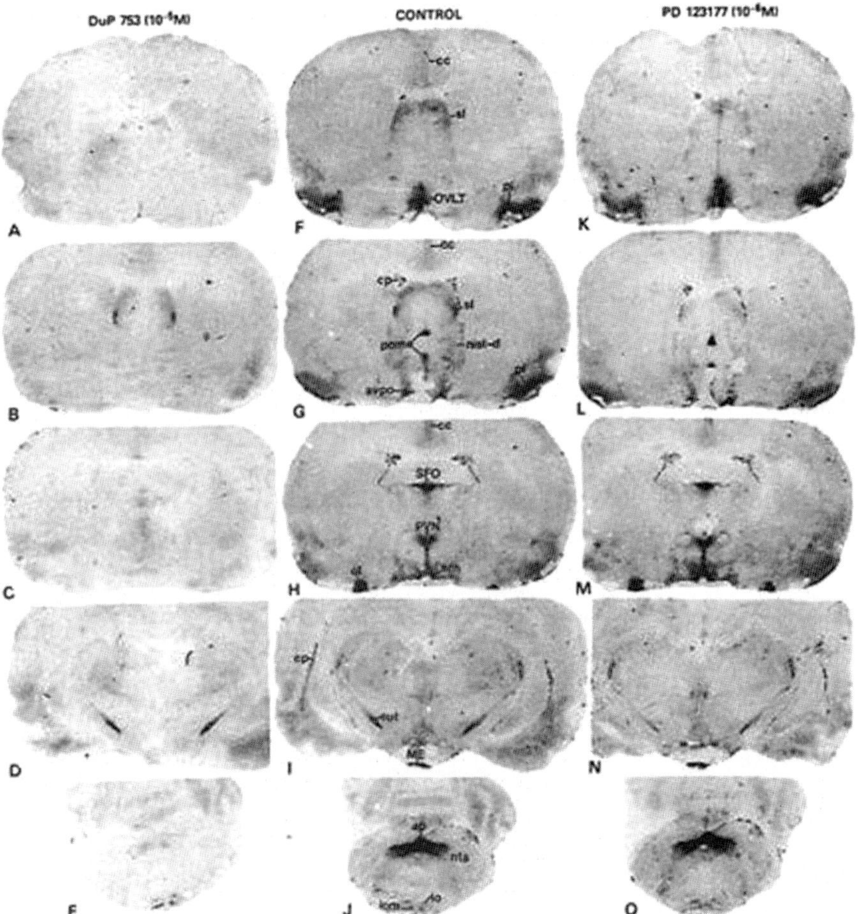

Fig. 1A–O Topographic distribution of Ang II receptor subtypes in the rat brain determined by binding autoradiography. Serial coronal sections of rat brain were incubated with ^{125}I[Sar^1, Ile^8]Ang II in the absence (**F–J**, total binding) or in the presence of saturating concentrations of AT_1-specific antagonist, Dup 753 (losartan) (**A–E**, AT_2 receptors), or the AT_2 antagonist, PD 123177 (**K–O**, AT_1 receptors). Abundant AT_1 receptors are present in the PVN, a key hypothalamic center regulating the HPA axis, and areas with connections to the PVN, including circumventricular organs (*OVLT, SFO, ME* and *AP*). AT_2 receptors are present in the lateral septum, subiculum and inferior olive but not in areas relevant to HPA axis activity (**A–E**). *cc*, cingular cortex; *sl*, lateral septum; *OVLT*, organum vasculoso of the lamina terminalis; *cp*, pirifom cortex; *SFO*, subfornical organ; *cp* choroid plexus; *POM* nucleus medianus; *nist* bed nucleus of the stria terminalis; *avpo* anteroventral preoptic area; *SFO*, subfornical organ; *PVN*, hypothalamic paraventricular nucleus; *of*, olfactory nucleus; *pf*, pirifom cortex; *sub*, subiculum; *ME*, median eminence; *ap*, area postrema; *nts*, nucleus of the solitary tract; *io*, inferior olive

VP, as well as a relay center for autonomic responses to stress (Cunningham and Sawchenko 1988; Swanson and Kuypers 1980). Ang II receptors are also present in a number of areas with connections to the PVN such as the circumventricular organs, the nucleus of the solitary tract in the brain stem and the median eminence (Millan et al. 1991; Saavedra 1992, Song et al. 1992) (Fig. 1H–O). Thus, Ang II can influence PVN function directly by interacting with receptors in PVN, or indirectly through release of other neurotransmitters by acting on presynaptic Ang II receptors. Ang II receptors in the PVN and other areas with afferent connections to the PVN are of type 1 (AT_1) (Bunneman et al. 1992; Millan et al. 1991; Saavedra 1999; Song et al. 1992). In the rat, AT_1 receptors at the later locations have been identified as belonging to the AT_{1A} subtype (Lenkei et al. 1997; Millan et al. 1991; Saavedra 1999) (Fig. 1). As shown in Fig. 1A–E, no AT_2 receptors are present in these areas. In the PVN, receptors are mainly located in parvocellular neurons expressing CRH or CRH and VP, cells that are critical for activation of the HPA axis during the stress response. Studies using double staining in situ hybridization techniques with a ^{35}S-labelled AT_1 receptor cRNA probe and digoxigenin-labeled cRNA probes for CRH, VP, oxytocin or TRH have shown that AT_1 receptor mRNA is located only in cells containing CRH mRNA (Aguilera et al. 1995c; Lenkei et al. 1995) (Fig. 2A). It is noteworthy that in both reports only small vasopressinergic neurons, with parvocellular characteristics, displayed significant AT_1 receptor staining (Fig. 2B). In contrast to the clear localization of AT_1 receptors in parvocellular neurons, the expression of these receptors in vasopressinergic magnocellular neurons is controversial. While in situ hybridization studies have failed to find AT_1 receptors in magnocellular neurons (Aguilera et al. 1995; Lenkei et al. 1995), some immunohisto-

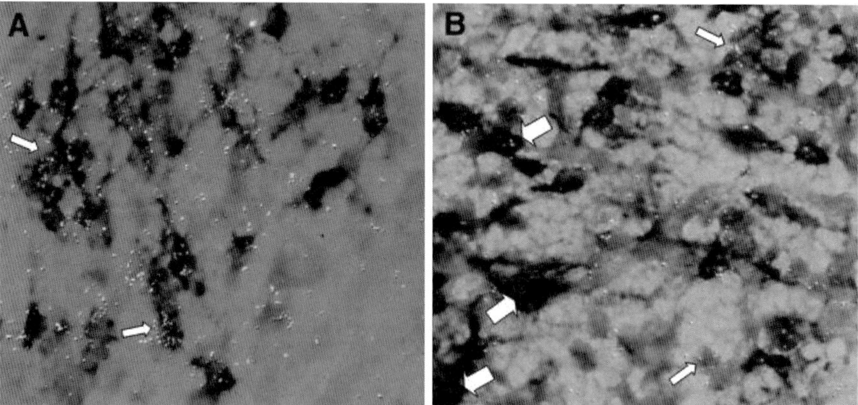

Fig. 2A, B Cellular localization of AT_1 receptor mRNA in the PVN determined by double labeling in situ hybridization using ^{35}S-labeled AT_1 receptor cRNA probes (*grains*), and digoxigenin-labeled CRH (**A**) and VP (**B**) cRNA probes. Thin *arrows* in **A** indicate representative CRH cells overlaid by AT_1 receptor transcripts (*grains*). Parvicellular vasopressinergic neurons containing AT_1 receptor transcripts are shown by the *thin arrows* in **B**. The *thick arrows* in **B** show magnocellular vasopressinergic neurons voided of AT_1 receptor mRNA

chemical studies have detected immunoreactivity in this region (Pfister et al. 1997). Ang II has well-recognized effects of Ang II in the central regulation of VP secretion (Phillips 1987; Unger et al. 1985). Although the presence of low levels of AT_1 receptor expression in magnocellular neurons cannot be ruled out, it is likely that at least part of the effects of Ang II in the PVN are indirect. This could be mediated through catecholamine release by activation of Ang II receptors located in afferent terminals innervating PVN neurons (Unger et al. 1985; Veltmar et al. 1992), or through glial cells, which contain abundant AT_1 receptors (Raizada et al. 1987).

The presence of high levels of AT_1 receptors as well as AT_1 receptor mRNA in the periventricular area indicates the presence of receptors in parvicellular perikarya other than CRH. This region of the PVN contains mainly TRH, somatostatin, and dopamine-producing cells (Swanson et al. 1987). The lack of significant colocalization of AT_1 receptor mRNA and TRH mRNA in the periventricular area renders it unlikely that AT_1 receptors are in TRH cells. On the other hand, it is likely that AT_1 receptors in the periventricular area are associated with dopaminergic neurons, as suggested by the ability of central administration of Ang II to increase dopaminergic turnover in the hypothalamus and to decrease prolactin secretion (Fuxe et al. 1980). Central administration of Ang II has been shown to reduce plasma levels of growth hormone (Steele et al. 1982), and it is possible that AT_1 receptors located in somatostatin cells mediate this effect.

2.2
Regulation of Ang II Receptors in the PVN

The presence of AT_1 receptor mRNA in cells containing CRH mRNA provides strong evidence for a role of Ang II in the regulation of CRH neurons, and suggests that these receptors mediate the increases in CRH mRNA observed following central administration of Ang II (Sumimoto et al. 1991). In situ hybridization and binding autoradiography studies have revealed that AT_1 receptor levels in the PVN are glucocorticoid dependent (Aguilera et al. 1995a). Withdrawal of circulating glucocorticoids by surgical adrenalectomy in the rat decreases AT_1 receptor expression in the PVN, while glucocorticoid administration has the opposite effect. Several stress paradigms have been shown to increase AT_1 receptor expression in the PVN in response to stress, with a pattern of distribution identical to that of CRH mRNA, suggesting that induction occurs in parvocellular neurons (Aguilera et al. 1995a; Armando et al. 2001; Leong et al. 2002) (Fig. 3A). Adrenalectomy with or without glucocorticoid replacement prevents stress-induced increases in AT_1 receptor expression, indicating that the increase in receptors is the result of the increases in circulating glucocorticoids following stress (Fig. 3B). These changes in Ang II receptor expression in the PVN during stress strongly suggest that Ang II has a modulatory role in PVN function during adaptation to stress (Aguilera et al. 1995a; Leong et al. 2002). However, the level of AT_1 receptors in the PVN have been shown to increase during stress, ir-

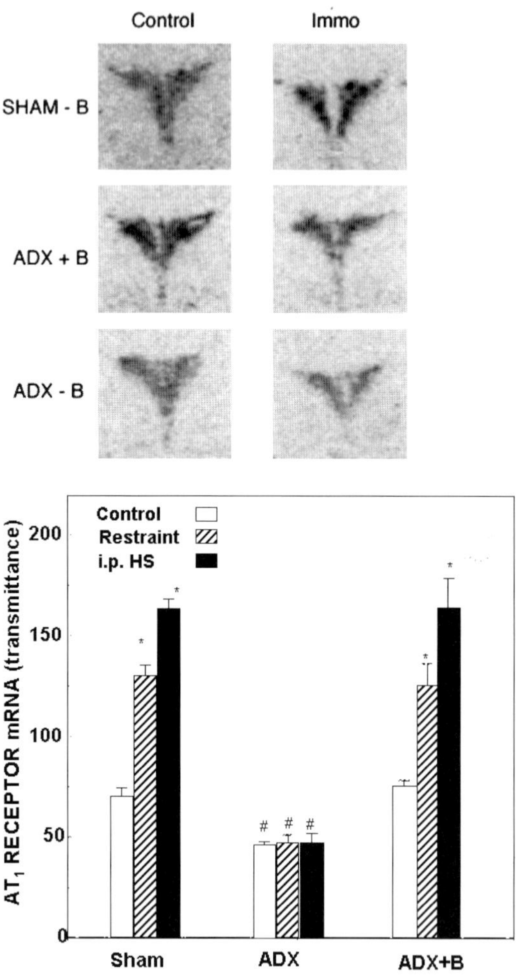

Fig. 3A, B Changes in AT_1 receptor mRNA expression in the PVN 4 h after 1 h restraint stress or a single i.p. hypertonic saline injection in 4-day sham operated rats (*SHAM-B*) or adrenalectomized rats, with (*ADX+B*) or without (*ADX–B*) glucocorticoid replacement. **A** Representative images of AT_1 receptor mRNA in situ hybridization in PVN sections from controls and rats subjected to restraint stress. *Bars* in **B** represent the mean and SE of the transmittance values obtained from the film images in five rats per experimental group. *, $p<0.001$ vs respective controls; #, $p<0.001$ vs sham operated or ADX+B rats

respective of CRH expression and ACTH responsiveness associated with the stressor, suggesting that the changes are a consequence of the glucocorticoid surge rather than a determinant of the HPA axis responsiveness. For example, AT_1 receptors in the PVN have been shown to increase equally in somatosensory stress paradigms associated with increased HPA responsiveness and during osmotic stimulation, a condition which is accompanied by decreases in CRH mRNA and ACTH responsiveness (Aguilera et al. 1995a). Although no AT_2 receptors have been described in areas of the brain involved in HPA axis regulation (Lenkei et al. 1997; Saavedra 1999), recent studies have shown increases in AT_1 receptors in the brain and activation of the HPA axis in knock-out mice for the AT_2 receptor (Armando et al. 2002). This indicates that AT_2 receptors can influence the expression of AT_1 receptors and suggests that cross-talk between the two receptor subtypes is critical for the angiotensinergic regulation of the HPA axis.

2.3 Central Ang II and HPA Axis Activity

Reports on the effect of central Ang II blockade on ACTH responses to stress are conflicting. While in sheep, ICV administration of a converting enzyme inhibitor attenuates ACTH responses to hemorrhage (Cameron et al. 1986), ACTH responses to ether (Buckner et al. 1986) or shaking stress (Hirasawa et al. 1990) in the rat are unchanged by central Ang II antagonists or converting enzyme inhibitors. Similarly, plasma ACTH and corticosterone responses to restraint stress were unaffected by central AT_1 receptor blockade with the non-peptide selective antagonist, losartan (Jezova et al. 1998). This inability to suppress central angiotensinergic activity to reduce the rise in plasma ACTH and corticosterone indicates that, at least in the rat, central Ang II is not required for the secretory response of the hypothalamus and pituitary to the latter acute stress paradigms. In contrast, peripheral administration of another non-peptide AT_1 receptor antagonist, candesartan, for 13 days was shown to abolish HPA responses to isolation stress, including the elevations in plasma ACTH and corticosterone as well as the increase in AT_1 receptor expression in the PVN (Armando et al. 2001). Since candesartan can cross the blood–brain barrier, it is not clear from the report whether the effects of long-term administration were due to blockade of peripheral or central AT_1 receptors.

Central AT_1 receptor blockade by icv administration of losartan attenuates CRH mRNA responses to acute immobilization, suggesting that endogenous Ang II in the brain is at least partially responsible for the increases in CRH mRNA in this stress paradigm (Jezova et al. 1998) (Fig. 4). A regulatory effect of Ang II on CRH mRNA in the PVN has also been suggested by studies showing increased CRH mRNA following ICV injection of Ang II (Sumimoto et al. 1991). Stress increases CRH receptor expression in the PVN but in contrast to CRH, this effect is not prevented by central AT_1 receptor blockade, suggesting that central Ang II is not involved in the regulation of hypothalamic CRH receptors (Jezova et al. 1998). The question of whether the stimulatory effect of ICV Ang II on CRH mRNA is directly mediated by AT_1 receptors in the PVN remains to be answered. The coexpression of AT_1 receptors in CRH cells of the PVN (Aguilera et al. 1995c; Lenkei et al. 1998), as well as the increase in Ang II binding and AT_1 receptor mRNA levels (Aguilera et al. 1995a; Leong et al. 2001) in the PVN observed following acute or chronic stress suggest a direct regulatory effect of Ang II on the CRH neuron. Overall, the evidence indicates that central Ang II is

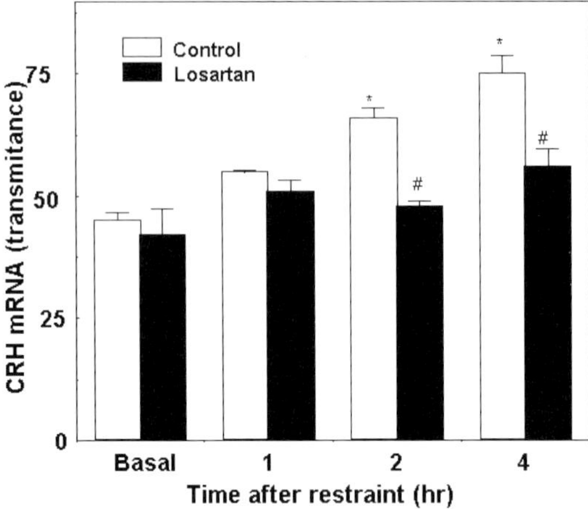

Fig. 4 Effect of acute immobilization on CRH mRNA in rats receiving an ICV injection of vehicle or 10 μg of the AT_1 receptor antagonist, losartan. Thirty minutes after ICV injection, rats were immobilized for 1 h and killed at the time points indicated. *Bars* represent the mean and SEM of the values measured by in situ hybridization in 6–16 rats per experimental group. *, $p<0.01$ vs basal; #, $p<0.05$ vs respective control; $p<0.01$ vs respective control

not involved in the acute activation of the HPA axis during stress. However, it does contribute to the regulation of CRH mRNA in the PVN and it may influence adaptive responses of the HPA axis to long-term stimulation.

3 Pituitary Effects of Angiotensin II

Several in vitro studies in rodents and primates have shown that Ang II has the potential to directly stimulate ACTH secretion in the pituitary corticotroph (Hauger et al. 1982; Aguilera et al. 1982, Spinedi and Negro-Villar 1983). Studies in isolated rat pituitary cells have shown that Ang II stimulates ACTH secretion in a dose-dependent manner and that the peptide potentiates the stimulatory effect of CRH (Aguilera et al. 1982; Spinedi and Negro-Vilar 1983). In pituitary cells from non-human primates, Ang II has no effect on its own, but it also enhances CRH-stimulated ACTH secretion (Millan et al. 1987). The pituitary effects of Ang II are mediated by AT_1 receptors present in the pituitary corticotroph of rodents and primates (Hauger et al. 1982; Zemel et al. 1990; Saavedra 1992). Studies using cell type enriched fractions of rat pituitary cells have shown that Ang receptors are located in corticotrophs and in lactotrophs but not in other cell types (Aguilera et al. 1983; Moreau et al. 1997). Pituitary Ang II receptors correspond to AT_1 receptors, with the majority belonging to the AT_{1B} subtype, as shown by in situ hybridization and RT-PCR in fractionated pituitary cells (Lenkey et al. 1999). In the rat, AT_1 receptors undergo marked down-regulation following estrogen administration (Chen and Printz 1982; Platia et al.

1986; Krishnamurthi et al. 1999). These changes probably reflect the effect of estrogen on AT_1 receptors in lactotrophs, which represent about 50% of the pituitary cell population. In primates, Ang II receptors are located only in corticotrophs and no information is available on their subtype (Millan et al. 1990). Although in AT_1 there is no evidence for an effect of glucocorticoids on anterior pituitary Ang II binding, it has been recently reported that restraint stress induces marked changes in pituitary AT_{1A} and AT_{1B} receptors (Leong et al. 2002). It is not possible to rule out that some of the changes in AT_1 receptor expression occurred in corticotrophs, but the magnitude of the effect suggests that they reflect mostly changes in lactotrophs, which constitute the larger proportion of AT_1-containing cells in the rat pituitary. This finding raises the possibility that AT_1 receptors in lactotrophs could mediate at least in part the increases in plasma prolactin, which are part of the stress response in the rat (Jurcovicova et al. 1990).

Since the affinity of pituitary Ang II receptors is in the nanomolar range (Chen and Printz 1983; Hauger et al. 1982), low levels of receptor occupancy would occur when exposed to circulating Ang II levels in the picomolar range. Experiments in vivo have shown than only high infusion levels of Ang II into the peripheral circulation increase plasma ACTH levels (Ganong and Murakami 1987) (Fig. 5). The stimulatory effect of Ang II on plasma ACTH can be prevented by administration of CRH antibody, suggesting that the effect of Ang II is indirect through stimulation of CRH release (Rivier et al. 1983). Supporting this possibility, peripheral injection of Ang II has been shown to induce rapid increases in CRH content in the ME followed by a decrease (Ganong and Murakami

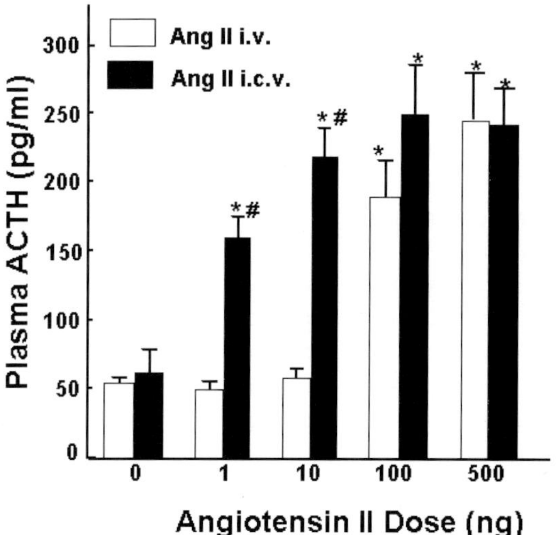

Fig. 5 Effect of central (*i.c.v.*) or peripheral (*i.v.*) administration of Ang II on plasma ACTH in conscious rats bearing jugular vein catheters. *Bars* are the mean and SE of values obtained in four to six rats per group. *, $p<0.001$ vs vehicle injected controls (0 dose); #, $p<0.01$ vs the respective IV dose

1987). In addition, in the rat, Ang II is locally synthesized in the pituitary by gonadotropes, from where it is stored and released together with LH. Thus, it is possible that locally released Ang II interact with receptors in corticotrophs and lactotrophs in a paracrine manner to modulate hormone secretion (Ganong and Murakami 1987).

Combination of in situ hybridization and immunohistochemistry studies have shown that only a subpopulation of corticotropes (about 25%) expresses AT_{1B} receptors (Moreau et al. 1997). Corticotropes are heterogeneous in size, shape, storage patterns, receptor expression and secretory responses (Childs 1992). Since in vitro Ang II can increase the percentage of cells that bind CRH and store ACTH (Childs 1992), it is possible that the primary effect of Ang II in the pituitary is corticotroph differentiation rather than stimulation of ACTH secretion. While there is evidence that Ang II stimulates ACTH secretion at the pituitary corticotroph level, further studies are required to determine the physiological importance of the direct effects of Ang II in the pituitary.

4 Adrenal Regulation by Ang II

In contrast to the key role of Ang II regulating mineralocorticoid secretion in the adrenal zona glomerulosa, the peptide does not appear to be a major regulator of glucocorticoid secretion. The functional involvement of Ang II in the zona fasciculata may be species dependent and it may not be directly related to acute regulation of steroidogenesis. For example, Ang II receptors show very low expression in the zona fasciculata of the rat, rendering unlikely that the peptide has any direct effect on glucocorticoid production in this species (Belloni et al. 1998). On the other hand, AT_1 receptors are present in the zona fasciculata of bovine adrenal gland and Ang II has been shown to increase cortisol production in cultured bovine adrenal fasciculata cells (Oualy et al. 1992). Low levels of Ang receptors have been also described in the adrenal fasciculata of human and non-human primates (Douglas et al. 1984). Concomitantly with binding affinities lower than those in the zona glomerulosa, in vitro cortisol responses to Ang II were less sensitive than aldosterone responses in adrenal glomerulosa cells (Douglas et al. 1984). Consistent with a minor effect of Ang II in the human adrenal fasciculata, i.v. infusion of Ang II has been reported to be ineffective in modifying plasma cortisol levels in humans (Calogero et al. 1991). In addition, no elevation in circulating glucocorticoids is associated with experimental or pathological conditions accompanied by activation of the renin-angiotensin system. While it is unlikely that Ang II plays a major role as a direct regulator of adrenal fasciculata steroidogenesis, it is plausible that the peptide influences adrenal fasciculate responsiveness indirectly, e.g., though modulation of adrenal medullary catecholamine production.

5 Summary and Conclusions

Evidence is accumulating to support the direct involvement of Ang II in the regulation of HPA axis activity both at the hypothalamic and pituitary levels. Type 1 Ang II (AT_1) receptors are present at sites critical for activation of the HPA axis, including parvocellular CRH neurons in the PVN, other brain nuclei with connections to the PVN, as well as pituitary corticotrophs. While AT_1 receptor expression in the PVN undergoes regulatory changes during alterations of HPA axis activity, these changes appear to result from altered circulating glucocorticoids rather than determine HPA axis responsiveness. Central AT_1 receptors modulate CRH expression and possibly vasopressin in parvocellular neurons. Intracerebroventricular injection of low doses of Ang II increase CRH mRNA and plasma ACTH, but in a number of conditions plasma ACTH and glucocorticoid responses to stress are unaffected by central AT_1 receptor blockade. The stimulatory effect of Ang II on ACTH secretion is mediated at least in part by CRH release, since CRH antiserum is able to block this action. Ang II could stimulate CRH secretion directly through interaction with receptors in the PVN or median eminence, or indirectly by stimulating afferent neural pathways from the circumventricular organs. AT_1 receptors are present in the pituitary corticotroph and Ang II stimulates ACTH secretion in vitro. However, there is little evidence for a major role of Ang II directly regulating pituitary ACTH secretion and it is more likely that the pituitary effects of the peptide relate to corticotroph differentiation.

References

Aguilera G (1994) Regulation of pituitary ACTH secretion during chronic stress. Frontier Neuroendocrinol 15:321–350

Aguilera G, Hyde CL, Catt KJ (1982) Angiotensin II receptors and prolactin release in pituitary lactotrophs. Endocrinology 111:1045–1050

Aguilera G, Harwood JP, Wilson JX, Morell J, Brown JH, Catt KJ (1983) Mechanisms of action of corticotropin-releasing factor and other regulators of corticotropin release in rat pituitary cells. J Biol Chem 258:8039–8045

Aguilera G, Kiss A, Luo X (1995a) Increased expression of type 1 angiotensin II receptors in the hypothalamic paraventricular nucleus following stress and glucocorticoid administration. J Neuroendocrinol 7:775–783

Aguilera G, Kiss A, Luo X, Sunar-Akbasak B (1995b) The renin-angiotensin system and the stress response. Ann N Y Acad Sci 771:173–186

Aguilera G, Young WS, Kiss A, Bathia A (1995c) Direct regulation of hypothalamic corticotropin-releasing-hormone neurons by angiotensin II. Neuroendocrinology 61:437–444

Antoni FA (1986) Hypothalamic control of adrenocorticotropin secretion: advances since the discovery of 41-residue corticotropin-releasing factor. Endocr Rev 7:351–378

Armando I, Carranza A, Nishimura Y, Hoe K-L, Barontini M, Terron, JA, Falcon-Neri A, Ito T, Juorio AV, Saavedra J (2001) Peripheral administration of an angiotensin II AT_1

receptor antagonist decreases the hypothalamic-pituitary adrenal response to isolation stress. Endocrinology 142:3880–3889

Armando I, Terron JA, Falcon-Neri A, Ito T, Hauser W, Inagami T, Saavedra JM (2002) Increased angiotensin II AT1 receptor expression in paraventricular nucleus and hypothalamic pituitary adrenal axis stimulation in AT2 receptor gene disrupted mice. Neuroendocrinology 76:137–147

Belloni AS, Andreis PG, Macchi V, Gottardo G, Malendowicz LK, Nussdorfer GG (1998) Distribution and functional significance of angiotensin-II AT1- and AT2-receptor subtypes in the rat adrenal gland. Endocr Res 24:1–15

Buckner FS, Chen FN, Wade CE, Ganong WF (1986) Centrally administered inhibitors of the generation and action of angiotensin II do not attenuate the increase in ACTH secretion produced by ether stress in rats. Neuroendocrinology 42:97-101

Bunnemann B, Iwai N, Metzger R, Fuxe K, Inagami T, Ganten D (1992) The distribution of angiotensin II receptor subtype mRNA in the rat brain. Neuroscience Letters 142:155–158

Calogero A, Fornito MC, Aliffi A, Vicari E, Mantero F, Polosa P, D'Agata R (1991) Role of peripherally infused angiotensin II on the human hypothalamic pituitary adrenal axis. Clin Endocrinol 34:183–186

Cameron VA, Espiner EA, Nicholls MG, MacFarlane MR, Sadler WA (1986) Intracerebroventricular captopril reduces plasma ACTH and vasopressin responses to hemorrhage stress. Life Sci 38:553-559

Chen FM, Printz MP (1983) Chronic estrogen treatment reduces angiotensin II receptors in the anterior pituitary. Endocrinology 113:1503–1510

Childs GV (1992) Structure-function correlates in the corticotropes of the anterior pituitary. Front Neuroendocrinol 13:271–317

Cunningham ET, Sawchenko PE (1988) Anatomical specificity of noradrenergic inputs to the paraventricular and supraoptic nuclei of the rat hypothalamus. J Comparative Neurol 274:60–76

Douglas JG, Brown GP, White C (1984) Angiotensin II receptors of human and primate adrenal fasciculata and glomerulosa: correlations of binding and steroidogenesis. Metabolism 33:685–688

Fuxe K, Anderson K, Ganten D, Hokfelt T, Encroth P (1980) Evidence for the existence of an angiotensin II-like immunoreactive central neuron system and its interactions with the central catecholaminergic pathways. In: Gross F, Vogel G (eds) Enzymatic release of vasoactive peptides. Raven press, New York, pp 161–170

Ganong WF, Murakami K (1987) The role of angiotensin II in the regulation of ACTH release. Ann N Y Acad Sci 512:176–186

Golin RMA, Gotoh E, Said SI, Ganong WF (1988) Pharmacological evidence that the sympathetic nervous system mediates the increase in renin secretion produced by immobilization and head up tilt in rats. Neuropharmacology 27:1209–1213

Hauger RL, Aguilera G, Baukal AJ, Catt KJ (1982) Characterization of angiotensin II receptors in the anterior pituitary gland. Mol Cell Endocrinol 25:203–212

Hirasawa R, Hashimoto K, Ota Z (1990) Role of central angiotensinergic mechanism in shaking stress-induced ACTH and catecholamine secretion. Brain Res 533:1–5

Jezova D, Ochedalski T, Kiss A, Aguilera G (1998) Brain angiotensin II modulates sympathoadrenal and hypothalamic pituitary adrenocortical activation during stress. J Neuroendocrinol 10:67–72

Jurcovicova J, Kvetnansky R, Dobrakovova M, Jezova D, Kiss A, Makara GB (1990) Prolactin response to immobilization stress and hemorrhage: the effect of hypothalamic deafferentations and posterior pituitary denervation. Endocrinology126:2527–2533

Krisnamurthi K, Verbalis JG, Zheng W, Wu Z, Clerch LB, Sandberg K (1999) Estrogen regulates angiotensin AT1 receptor expression via cytosolic proteins that bind to the 5′ leader sequence of the receptor mRNA. Endocrinology 140:5435–5438

Lenkei Z, Corvol P, Llorens-Cortes C (1995) Comparative expression of vasopressin and angiotensin type-1 receptor mRNA in rat hypothalamic nuclei: a double in situ hybridization study. Brain Res Mol Brain Res 34:135–142

Lenkei Z, Palkovits M, Corvol P, Llorens-Cortes C (1997) Expression of angiotensin type-1 (AT1) and type-2 (AT2) receptor mRNAs in the adult rat brain: a functional neuroanatomical review. Front Neuroendocrinol 8:383–439

Lenkei Z, Nuyt AM, Grouselle D, Corvol P, Llorens-Cortes C (1999) Identification of endocrine cell populations expressing the AT_{1B} subtype of angiotensin II receptors in the anterior pituitary. Endocrinology 140:472–477

Leong DS, Terron JA, Falcon Neri A, Armando I, Ito T, Johren O, Tonelli LH, Hoe K-L, Saavedra JM (2002) Restraint stress modulates brain, pituitary and adrenal expression of angiotensin II AT_{1A}, AT_{1B} and AT2 receptors. Neuroendocrinology 75:227–240

Millan MA, Abou-Samra A-B, Wynn PC, Catt KJ, Aguilera G (1987) Receptors and actions of corticotropin releasing factor in primate pituitary gland. J Clin Endocrinol Metab 64:1036–1041

Millan MA, Jacobowitz DM, Catt KJ, Aguilera G (1990) Distribution of angiotensin II receptors in the brain of non-human primates. Peptides 11:243–253

Millan MA, Jacobowitz DM, Aguilera G, Catt KJ (1991) Differential distribution of AT1 and AT2 angiotensin II receptor subtypes in the rat brain during development. Proc Natl Acad Sci U S A 88:11440–11444

Moreau C, Rasolojanahary R, Zamora AJ, Enjalbert A, Kordon C, Llorens-Cortes C (1997) Expression of angiotensin II receptor subtypes AT(1A) and AT(1B) in enriched fractions of dispersed rat pituitary cells. Neuroendocrinology 66:416–425

Munck A, Guyre PM, Holbrook NJ (1984) Physiological functions of glucocorticoids in stress and their relation to pharmacological actions. Endocrine Rev 5:25–44

Ouali R, Poulette S, Penhoat A, Saez JM (1992) Characterization and coupling of angiotensin-II receptor subtypes in cultured bovine adrenal fasciculata cells. J Steroid Biochem Mol Biol 43:271–280

Pfister J, Spengler C, Grouzmann E, Raizada MK, Felix D, Imboden H (1997) Intracellular staining of angiotensin receptors in the PVN and SON of the rat. Brain Res 754:307–310

Phillips MI (1987) Functions of angiotensin in the central nervous system. Annu Rev Physiol 49:413–435

Platia MP, Catt KJ, Aguilera G (1986) Effects of estradiol-17β on angiotensin II receptors and prolactin release in cultured pituitary cells. Endocrinology 119:2768–2772

Raizada MK, Phillips MI, Crews FT, Sumners C (1987) Distinct angiotensin II receptors in primary cultures of glial cells from rat brain. Proc Natl Acad Sci U S A 84:4655–4658

Rivier C, Vale W (1983) Effect of angiotensin II on ACTH release in vivo: role of corticotropin-releasing factor. Reg Peptides 7:253–258

Saavedra JM (1992) Brain and pituitary angiotensin. Endocr Rev 13:329–380

Saavedra JM (1999) Emerging features of brain angiotensin receptors. Regul Pept 85:31–45

Song K, Allen AM, Paxinos G, Mendelsohn FAO (1992) Mapping of angiotensin II receptor heterogeneity in rat brain. J Comp Neurol 316:467–484

Spinedi E, Negro-Vilar A (1983) Angiotensin II and ACTH release: site of action and potency relative to corticotropin releasing factor and vasopressin. Neuroendocrinology 17:446–453

Steele MK, McCann SD, Negro-Vilar A (1982) Modulation by dopamine and estradiol of the central effects of angiotensin II on pituitary hormone release. Endocrinology 111:722–729

Sumitomo T, Suda T, Nakano Y, Tozawa F, Yamada M, Demura H (1991) Angiotensin II increases the corticotropin-releasing factor messenger ribonucleic acid level in the rat hypothalamus. Endocrinology 128:2248–2252

Swanson LW, Kuypers HGLM (1980) The paraventricular nucleus of the hypothalamus: cytoarchitectonic subdivisions and organization of projections to the pituitary, dorsal vagal complex and spinal cord as demonstrated by retrograde fluorescence double labeling methods. J Comp Neurol 194:555–570

Swanson LW, Sawchenko PE, Lind RW, Rho JR (1987) The CRH motoneuron: differential peptide regulation in neurons with possible synaptic, paracrine, and endocrine outputs. Ann N Y Acad Sci 512:12–23

Unger T, Becker H, Petty M, Demmert G, Scheider B, Ganten D, Lang R (1985) Differential effects of central angiotensin II and substance P on sympathetic nerve activity in conscious rats: implications for cardiovascular adaptation to behavioral responses. Circ Res 56:563–575

Veltmar A, Culman J, Quadri F, Rascher W, Unger T (1992) Involvement of adrenergic and angiotensinergic receptors in the paraventricular nucleus in the angiotensin II-induced vasopressin release. J Pharm Exp Ther 263:1253–1260

Whitnall MH (1993) Regulation of the hypothalamic corticotropin-releasing hormone neurosecretory system. Prog Neurobiol 40:573–629

Whitworth JA (1987) Mechanisms of glucocorticoid induced hypertension. Kidney Int 31:1213–1224

Wong DL, Siddall B, Wang W (1995) Hormonal control of rat adrenal phenylethanolamine N-methyltransferase: enzyme activity, the final critical pathway. Neuropsycopharmacology 13:223–234

Zemel S, Millan MA, Feuillan P, Aguilera G (1990) Characterization and distribution of angiotensin II receptors in the primate fetus. J Clin Endocrinol Metab 71:1003–1007

Angiotensin Actions on the Brain Influencing Salt and Water Balance

M. J. McKinley · D. A. Denton · M. L. Mathai · B. J. Oldfield · R. S. Weisinger

Howard Florey Institute of Experimental Physiology and Medicine,
University of Melbourne, 3010 Melbourne, Australia
e-mail: mmck@hfi.unimelb.edu.au

Abstract Angiotensin (Ang) AT_1 receptors are located in many regions of the brain that influence fluid and electrolyte balance. Neurons that express AT_1 receptors in the subfornical organ and organum vasculosum of the lamina terminalis (OVLT), circumventricular organs that lack a blood-brain barrier, are stimulated by systemically administered Ang II to initiate water drinking, sodium appetite and vasopressin secretion. Intracerebroventricular (ICV) injection of Ang II has a potent dipsogenic effect and also stimulates vasopressin secretion, effects probably caused mainly by an action of centrally administered Ang II on the median preoptic nucleus. Central administration of drugs that block the AT_1 receptor, or prevent angiotensinogen production in the brain, inhibit drinking and vasopressin release in response to ICV injection of hypertonic saline, suggesting that angiotensin generated within the brain may have a role in body fluid homeostasis. Centrally administered Ang II also causes increased excretion of sodium by the kidney, and blockade of natriuretic responses by ICV injection of the AT_1 antagonist losartan suggests that a central angiotensinergic pathway may influence renal sodium excretion.

Keywords Subfornical organ · Organum vasculosum of the lamina terminalis (OVLT) · Circulating angiotensin · AT_1 receptor · Thirst · Water drinking · Vasopressin · Sodium appetite · Natriuresis · Brain angiotensin

Abbreviations

ACE	Angiotensin converting enzyme
Ang	Angiotensin
ANP	Atrial natriuretic peptide
AV3-V	Anteroventral third ventricle wall
ICV	Intracerebroventricular
MnPO	Median preoptic nucleus
OVLT	Organum vasculosum of the lamina terminalis

1 Introduction

Angiotensin (Ang) exerts multiple actions on the brain to influence fluid and electrolyte balance. These are in addition to its direct actions on the kidney and its stimulation of aldosterone secretion by the adrenal cortex to regulate sodium excretion. Initially, it was shown that circulating Ang II could stimulate the CNS to influence arterial pressure (Bickerton and Buckley 1961). Following this, it was shown that Ang II could stimulate water drinking in animals (Fitzsimons 1968), vasopressin secretion (Bonjour and Malvin 1970.) and sodium appetite (Buggy and Fisher 1974). There is also evidence that angiotensin may act on the brain to influence renal sodium excretion (Andersson et al. 1972).

When considering the actions of Ang on the brain, it is necessary to distinguish the central effects of circulating Ang II, from those that may result from the actions of Ang generated within the CNS. Circulating Ang II does not have ready passage across the blood-brain barrier (Schelling et al. 1976; Fei et al. 1982). Consequently, the brain sites at which blood-borne Ang can influence the central regulation of fluid and electrolyte homeostasis are limited to those sites that lack the blood-brain barrier, specifically the sensory circumventricular organs of the brain: the subfornical organ, organum vasculosum of the lamina terminalis (OVLT) and area postrema.

1.1 Angiotensin Receptors in Brain Regions That Regulate Body Fluid Homeostasis

Studies that initially used in vitro autoradiography, and more recently in situ hybridization and immunohistochemistry, have demonstrated that many regions of the brain that are known to have significant roles in the central regulation of fluid and electrolyte balance are sites rich in Ang receptor binding (Mendelsohn et al. 1984; Speth et al. 1986; McKinley et al. 1987; Mendelsohn et al. 1988). This binding represents AT_1 receptors, which are present in regions such as the lamina terminalis, the supraoptic and paraventricular nuclei of the hypothalamus, the bed nucleus of the stria terminalis, amygdala, the lateral parabrachial nucleus, the nucleus of the solitary tract and the caudal ventrolateral medulla (Song et al. 1991; Lenkei et al. 1997; Giles et al. 1998), which are all implicated in the regulation of body fluids.

While the AT_1 receptors in the subfornical organ and OVLT are influenced by circulating Ang II, those in the other aforementioned regions are not accessed by blood-borne Ang II. There is considerable evidence to show that angiotensinogen is synthesized within the brain (Stornetta et al. 1988) and that Ang peptides are generated within the brain (Ganten et al. 1983; Bunnemann et al. 1993). It is probable that Ang peptides are utilized as neurotransmitters or modulators within the CNS (Lind et al. 1985).

2 Angiotensin and Thirst

2.1 Peripheral Renin-Angiotensin System

It has been known for more than 30 years that Ang II is a potent dipsogenic agent. Pioneering studies by Fitzsimons, who studied water intake in bilaterally nephrectomized rats, established that a renal thirst factor, which he later showed to be renin, was necessary for water drinking in rats as a result of ligation of the inferior vena cava (Fitzsimons 1969). Subsequently, it was shown that systemic infusion of Ang II or renin was a stimulus to water drinking in many mammals, including rats, mice, dogs, sheep, goats, cows, monkeys and humans. In addition, peripherally administered Ang II is a dipsogenic agent in avian and reptilian species (Kobayashi et al. 1979). It has also been shown in an amphibian (the spadefoot toad) that Ang II has a central action to stimulate the toad to press its ventral skin to a moist surface and increase water uptake across the skin (Propper et al. 1995). A complete list of vertebrate species that have been shown to respond with a drinking response to peripheral administration of angiotensin has been provided by Fitzsimons (1998).

Administration of Ang II directly into the common carotid artery has been shown to be a more potent stimulus to water drinking in dogs and sheep than similar or greater amounts of intravenously infused Ang II. These results demonstrated that Ang II from the circulation probably acted directly on the brain to stimulate drinking behavior (Abraham et al. 1975; Reid et al. 1982). While numerous studies showed that systemically infused Ang II stimulates water drinking, the infusion rates needed to produce this effect often yielded blood concentrations of Ang II in excess of physiologically relevant concentrations. This appears to be the case in sheep and humans (Abraham et al. 1975; Phillips et al. 1985). However, in dogs and rats, blood Ang II levels associated with various physiological and pathophysiological conditions are sufficient to cause drinking (Hsiao et al. 1977; Ramsay et al. 1978; Fitzsimons and Kucharczyk 1978; Mann et al. 1980; Johnson et al. 1981).

An important consideration in regard to whether circulating concentrations of Ang II stimulate drinking is the concomitant inhibitory influence on drinking behavior of baroreceptor activation resulting from Ang II's pressor effect. If the pressor response is blocked or reduced by simultaneously administered captopril and vasodilator agents, the threshold dose of Ang II for water drinking could be reduced considerably (Robinson and Evered 1987; Evered et al. 1988). Thus, experiments that show lack of a dipsogenic response to systemically infused Ang II have to be regarded with caution because of the concomitant pressor response. In physiological conditions of hypovolemia or sodium depletion, in which Ang II levels rise, but no increase in arterial pressure occurs, the inhibitory influences from baroreceptors would not be a factor.

2.2 Effects of Peripherally Administered Angiotensin Antagonists on Drinking

In regard to the receptor subtype utilized by circulating Ang II to induce drinking, peripheral administration of the AT_1 antagonist losartan, but not the AT_2 antagonist PD 123177, prevented water drinking in response to subcutaneously administered Ang II (Fregly and Rowland 1991; Dourish et al. 1992). Dipsogenic responses to systemically infused Ang II can also be inhibited by centrally administered Ang antagonists. ICV administration of either the peptide analogue antagonist saralasin, or the specific AT_1 receptor antagonist losartan, inhibits water drinking induced by peripherally administered Ang II. Indeed, most studies of the effects of Ang receptor antagonists on water intake have employed the cerebroventricular route of delivery, and these will be discussed in a later section of this chapter. However, it should be pointed out that the Ang receptors that are directly stimulated by systemic Ang II are not necessarily the same Ang

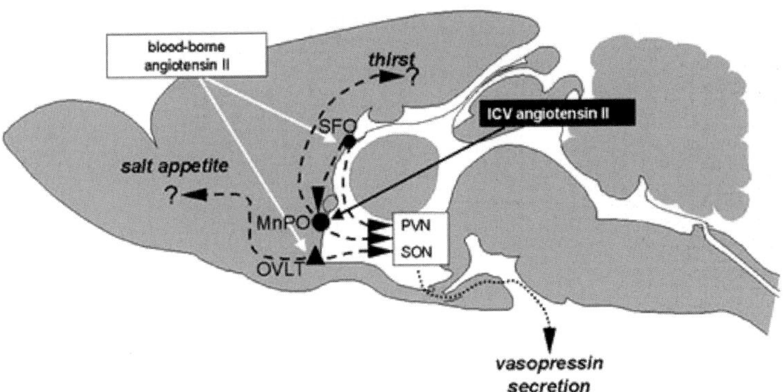

Fig. 1 Diagram of a mid-sagittal section of the rat brain showing some neural pathways (indicated by *dashed lines*) influencing salt and water balance that are stimulated by either circulating or intracerebroventricularly injected angiotensin II. The *question marks* indicate that the cortical sites subserving thirst and salt appetite are unknown. *ICV*, intracerebroventricular; *MnPO*, median preoptic nucleus; *OVLT*, organum vasculosum of the lamina terminalis; *PVN*, hypothalamic paraventricular nucleus; *SFO*, subfornical organ; *SON*, supraoptic nucleus

receptors that are being blocked by centrally administered Ang antagonists. This is because of the probable existence of a central neural pathway, driven by Ang II as a neurotransmitter in the median preoptic nucleus (MnPO), mediating drinking caused by circulating Ang II (Fig. 1) (Johnson et al. 1996). Nevertheless, the effectiveness of systemic AT_1 antagonists in blocking drinking responses clearly point to AT_1 receptors mediating water drinking caused by circulating Ang II. Peripherally administered AT_1 antagonists candesartan and telmisartan inhibit dipsogenic, vasopressin and pressor responses to centrally administered Ang II, showing that these AT_1 antagonists may gain access to Ang receptors for thirst inside the blood-brain barrier (Gohlke et al. 2001; Gohlke 2002).

2.3
Brain Site of the Dipsogenic Action of Circulating Angiotensin

The evidence that circulating Ang II is prevented from entering the interstitium of the brain by the blood-brain barrier (Schelling et al. 1976) leads to the question of how and where it can influence brain function to stimulate thirst. In the rat, it was initially shown that the subfornical organ is the central site for the dipsogenic actions of circulating Ang II (Fig. 1). Microinjection of very small quantities of Ang II into this circumventricular organ stimulates water drinking, and ablation of the subfornical organ prevents water drinking in response to systemically infused Ang II (Simpson and Routtenberg 1973; Simpson et al. 1978). Moreover, the subfornical organ contains some of the highest densities of AT_1 receptors in the brain (Lenkei et al. 1997; Allen et al. 2000) and along with other circumventricular organs is devoid of a blood-brain barrier because it has a fenestrated capillary endothelium (McKinley et al. 1991). Ablation of the subfornical organ in rats also prevents water drinking in response to stimuli that increase circulating levels of Ang II such as subcutaneous injection of the β-adrenergic agonist isoproterenol (Fitts 1994). The subfornical organ also mediates Ang II-induced drinking in the dog (Thrasher et al. 1982). In the sheep, however, ablation of the subfornical organ did not prevent the water drinking in response to systemically infused Ang II; it was suggested that the other circumventricular organ of the lamina terminalis, the organum vasculosum of the lamina terminalis (OVLT) may also play a role (McKinley et al. 1986). It should be pointed out that ablation of the subfornical organ has only a minor inhibitory effect, if any, on the drinking induced by intracerebroventricular (ICV) administration of Ang II (Lind and Johnson 1982), which must act mostly at another central site (discussed in Sect. 2.6, "Brain Sites for the Dipsogenic Action of the Centrally Administered Angiotensin" in this chapter). Atrial natriuretic peptide (ANP) injected into the subfornical organ inhibits Ang-induced drinking (Ehrlich and Fitts 1990), probably because it directly inhibits Ang-sensitive neurons in this site (Schmid and Simon 1992).

2.4 Effect of Angiotensin-Converting Enzyme Inhibitors on Drinking

The subfornical organ (and the OVLT) of the rat contains extremely high concentrations of angiotensin-converting enzyme (ACE), which enables Ang I derived from the circulation to be converted to Ang II locally within the subfornical organ and OVLT. This locally produced Ang II may be able to interact with the AT_1 receptors in the subfornical organ to mediate drinking behavior, and indeed there is evidence to support this view. Early studies of the effects of ACE inhibitors on water drinking in the rat reported that there was a paradoxical enhancement of drinking when the prototype ACE inhibitor SQ 20881 was administered peripherally to rats in conditions where the renin-angiotensin system is stimulated (Lehr et al. 1974). Subsequent studies using other ACE inhibitors such as captopril and enalapril also reported enhanced water drinking by rats when these agents were administered orally or subcutaneously (Schiffrin and Genest 1982; Elfont and Fitzsimons 1983; Weisinger et al. 1988). The probable explanation of the water drinking caused by systemic administration of ACE inhibitors is that the peripheral doses of ACE inhibitors are sufficient to block the conversion of circulating Ang I to Ang II, but not sufficient to block the very high concentrations of ACE within the subfornical organ and OVLT. Thus, circulating levels of renin and Ang I become greatly elevated due to lack of feedback inhibition of Ang II on the secretion of renin by the kidney. These high circulating levels of Ang I can reach the interstitium of the subfornical organ and are converted there to Ang II, with subsequent stimulation of AT_1 receptors in the subfornical organ and water drinking. Evidence to support this view comes from observations that much greater peripheral doses of ACE inhibitors or direct injection of ACE inhibitor into the subfornical organ that will block ACE and therefore local production of Ang II in the subfornical organ and OVLT prevent such water drinking (Thunhorst et al. 1989). Studies utilizing *c-fos* expression as an indicator of neuronal activation show that dipsogenic doses of captopril or enalapril cause many neurons in the subfornical organ and OVLT to be activated, while this effect, and the enhanced drinking, is blocked by peripherally administered AT_1 antagonist or by much higher peripheral doses of the ACE inhibitors (McKinley et al. 1997).

2.5 Dipsogenic Effects of Centrally Administered Angiotensin Peptides

Numerous studies have shown that microinjection of Ang II into the lateral or third cerebral ventricle is a potent stimulus to water drinking in mammals (Epstein et al. 1970; Severs et al. 1971; Andersson et al. 1972; Abraham et al. 1975; Fitzsimons and Kucharczyk 1978). Microinjection of Ang II into more specific sites such as the anteroventral third ventricle wall (AV3V) region, median preoptic nucleus (O'Neill and Brody 1987), lateral preoptic region (Fitzsimons and Kucharczyk 1978), lateral hypothalamus (Tanaka et al. 2001), subfornical organ (Simpson and Routtenberg 1971) or hypothalamic paraventricular nucleus (Jensen et al. 1992) has been reported to also stimulate drinking. Other components of the renin-angiotensin system - renin, the decapeptide Ang I, Ang-tetradecapeptide, and the heptapeptide Ang 2-8 (Ang III) also stimulate drinking behavior in the rat when injected centrally; however, the hexapeptide Ang 3-8 is a much less effective dipsogen (Fitzsimons 1971). Centrally administered Ang III is equally as dipsogenic as Ang II in the rat and baboon (Wright et al. 1985; Blair-West et al. 2001), but less so in the sheep (Weisinger et al. 1996). Bestatin, an inhibitor of aminopeptidase N, increases the half-life of Ang III in brain and has been shown to cause water drinking when administered intracerebroventricularly. It also potentiates the drinking response to centrally administered Ang II and Ang III. It was suggested that central conversion of Ang II to Ang III is necessary for Ang-induced drinking, and that in the brain, Ang III is a major endogenous dipsogenic peptide (Wright et al. 1988).

2.6 Brain Sites for the Dipsogenic Action of Centrally Administered Angiotensin

As mentioned in the previous paragraph, injections of Ang II into various brain regions cause water drinking in rats. Johnson and Epstein (1975) pointed out that for the sites in the brain where microinjections of Ang were dipsogenic, the injector needle usually had to traverse a ventricular space, and they suggested that much of the drinking stimulated by central injection of Ang was due to leakage of Ang into the ventricular space. It was also shown that when Ang was injected centrally, if it was prevented from reaching the (AV3V) region by a cold cream plug placed in the anterior part of the third ventricle, the injection of Ang was not dipsogenic. Because ablation of the AV3V region, but not the subfornical organ, abolished the dipsogenic effect of ICV injection of Ang II, they suggested that the AV3V rather than subfornical organ was the central site at which ICV Ang II induced drinking (Buggy and Fisher 1975; Buggy and Johnson 1977), a proposal also put forward by Andersson et al. (1975) as a result of studies of periventricular lesions in goats. The entire anterior wall of the third ventricle, which includes the median preoptic nucleus, is rich in AT_1 receptors (Song et al. 1992; Lenkei et al. 1997). Neurochemical ablation of the MnPO abolishes drinking in response to ICV Ang II (Cunningham et al. 1992) and direct microinjection of Ang II into the MnPO stimulates drinking (O'Neill and Brody 1987). Studies of immediate early gene expression also show that the MnPO is the main site in the AV3V that is stimulated by ICV injection of Ang II (Herbert et al. 1992; Lebrun et al. 1995; McKinley et al. 1995). Thus it likely that the MnPO is the major site in the AV3V region at which centrally administered Ang II acts to stimulate water drinking (Fig. 1).

2.7 Brain Angiotensin and Water Drinking

The effectiveness of centrally administered Ang peptides in eliciting drinking raises the question as to whether Ang endogenous to the brain is a dipsogenic agent. All components of a renin angiotensin system, renin, angiotensinogen, ACE, Ang I, II, and III, and Ang receptors are present in several parts of the brain (Ganten et al. 1983; Brownfield et al. 1983; Stornetta et al. 1988; Bunnemann et al. 1993; Dzau et al. 1996). An extensive angiotensinergic system of neurons and fibers has been mapped immunohistochemically and Ang immunoreactivity has been shown to be present in vesicles in synaptic profiles in rat brain (Lind et al. 1985; Oldfield et al. 1989). It has been proposed that some neural inputs to the MnPO utilize Ang as a neurotransmitter (Johnson et al. 1996). Evidence that Ang that is intrinsic to the brain participates in the generation of thirst comes from studies in which various pharmacological agents have been used to block receptors that may respond to brain Ang or by inhibiting the production of Ang within the brain.

2.8 Effect of Centrally Administered Angiotensin Inhibitors on Drinking

Several classes of pharmacological agents have been used centrally to block water drinking associated with Ang. These include both nonselective peptide antagonists, selective AT_1 and AT_2 receptor antagonists, ACE inhibitors and antisense oligonucleotides.

2.8.1 Peptide Antagonists

ICV administration of saralasin (Sar^1-Ala^8 Ang II) was initially used to block drinking in response to either peripherally or centrally administered Ang II (Abraham et al. 1976). While some investigators reported that ICV administration of saralasin could block drinking in response to water deprivation (Malvin et al. 1977), the majority of reports did not (Abraham et al. 1976; Lee et al.1981). Peptide Ang-analogue antagonists used in combination with muscarinic blockade suppressed water drinking in response to a dehydrative stimulus (Hoffmann et al. 1978). Because blood levels of Ang II increase in dehydrated animals, these experiments do not delineate whether it was centrally or peripherally derived Ang that was blocked.

2.8.2 Non-peptide Antagonists

ICV administration of AT_1 receptor antagonists (losartan, valsartan), at doses that do not escape into the periphery, have been consistently shown to exert a powerful block on water drinking in response to centrally administered Ang II (Fregly and Rowland 1991; Beresford and Fitzsimons 1992; McKinley et al. 1996), consistent with the presence of high densities of AT_1 receptors in the MnPO, a brain region crucial for Ang-stimulated drinking. Some centrally administered AT_2 antagonists have also been reported to block drinking in response to ICV Ang II (Rowland et al. 1992; Cooney and Fitzsimons 1993). However, the effective molar doses were higher than for AT_1 antagonists, and several investigators did not observe inhibition of the Ang-stimulated drinking response with PD123319, a widely used AT_2 antagonist (Beresford and Fitzsimons 1992; McKinley et al. 1996; Blair-West et al. 1997).

In addition to its blockade of drinking in response to centrally administered Ang, ICV losartan has been shown to block the water drinking that is stimulated by centrally administered hypertonic saline in several species (Blair-West et al. 1993; Mathai et al. 1996). It did not block drinking to intravenous infusion of hypertonic NaCl (Mathai et al. 1996). ICV losartan also inhibited the water intake stimulated by the administration of the hormone relaxin by either a peripheral or central route (Sinnayah et al. 1998; McKinley et al. 2001), as did ICV saralasin (Summerlee and Robertson 1995). It is unlikely that circulating Ang II levels were increased by the dipsogenic stimuli used in these experiments; therefore it seems likely that central stimulation with hypertonic saline or relaxin activates brain angiotensinergic pathways that participate in thirst mechanisms. This is in parallel with the role that such angiotensinergic pathways may play in the pressor response or vasopressin secretion stimulated by centrally administered hypertonic saline or relaxin (Rohmeiss et al. 1996; Geddes et al. 1994).

2.8.3 Antisense Oligonucleotides

Further support for the idea that brain-derived Ang II mediates dipsogenic responses comes from results using antisense oligonucleotides directed against angiotensinogen production in the brain. ICV administration of an 18mer antisense oligonucleotide complementary to the translation start codon region of angiotensinogen mRNA for 24 h caused a reduction in the drinking response to centrally administered hog renin in rats. Equivalent doses of mismatch or scrambled oligonucleotides did not, suggesting that less angiotensinogen was available to be cleaved to Ang peptides in the brain (Sinnayah et al. 1998). As well, such antisense treatment also caused a large reduction in the drinking response to subcutaneously administered isoproterenol, but not to several other dipsogenic stimuli that included water deprivation and ICV carbachol (Sinnayah et al. 1998). These results show that antisense treatment did not nonspecifically depress behavior, and also that circulating Ang II, which mediates isoproterenol-induced drinking, utilizes Ang that is synthesized in the brain in the neural pathways subserving its dipsogenic action. It has also been shown that ICV treatment with an antisense molecule directed against synthesis of the AT_1 receptor reduces the dipsogenic response to ICV Ang II (Sakai et al. 1994; Meng et al. 1994).

3 Sodium Appetite

3.1 Effect of Peripherally or Centrally Administered Angiotensin on Sodium Appetite

The administration of Ang II either directly into the brain of the rat, or peripherally, causes an increase in the appetite for NaCl (Buggy and Fisher 1974; Findlay and Epstein 1980; Bryant et al. 1980). Peripheral administration of renin or isoproterenol, which increase the blood levels of endogenous Ang II, do not initially stimulate a salt appetite; however, after repeated administration over days, a salt hunger is observed (Bryant et al. 1980). In other species that have been studied such as the sheep, mice and baboon, ICV administration of renin or Ang II increases the intake of hypertonic salt solution (Coghlan et al. 1981; Denton et al. 1990; Blair-West et al. 1998), although in the sheep there was evidence that a sodium deficit secondary to the natriuretic effect of ICV Ang II may have been the stimulus for the increased salt intake (Coghlan et al. 1981; Weisinger et al. 1986). The intake of NaCl in response to centrally administered Ang II, is significantly potentiated by simultaneously administered aldosterone (Fluharty and Epstein 1983; Sakai 1986), leading to the proposal that brain Ang acts in synergy with aldosterone to mediate the hunger for salt that may occur in sodium-depleted animals. Such a synergistic action between Ang and aldosterone has also been reported to occur in the pigeon (Massi and Epstein 1990) and baboon (Shade et al. 2002).

3.2
Effect of ACE Inhibitors and Angiotensin Antagonists on Salt Appetite

Similar to the enhancement of water drinking that occurs in the rat following peripheral treatment with captopril, the intake of NaCl in Na-replete (Fregly 1980; Evered and Robinson 1983; Elfont et al. 1984), Na-depleted (Moe et al. 1984; Weisinger et al. 1988), adrenalectomized (Elfont and Fitzsimons 1981) or hypovolemic (Stricker 1983) rats is increased by peripherally administered captopril or enalapril at 0.05-0.5 μg/kg. Treatment of rats with higher doses of captopril or enalapril (50-100 mg/kg) caused a reduction of NaCl intake in Na-depleted or adrenalectomized rats (Elfont and Fitzsimons 1981; Di Nicolantonio et al. 1982; Moe et al. 1984; Weisinger et al. 1988). These data have led to the suggestion that low doses of captopril, which may block peripheral but not central ACE, cause high levels of Ang I peripherally, which are then converted to Ang II centrally, because insufficient amounts of the ACE inhibitors reach the brain. On the other hand, the higher doses of captopril or enalapril block both peripheral and central ACE, and therefore the inhibition of Ang II formation in the brain inhibits salt appetite (Evered et al. 1980; Moe et al. 1984).

Further support for a physiological role of brain Ang II in the salt appetite of Na-depleted rats came from studies using the peptide antagonist, sarile, which blocks both AT_1 and AT_2 receptors. Sakai et al. (1990) observed that peripherally administered Ang II did not stimulate a sodium appetite, whether administered alone or in combination with mineralocorticoid treatment, whereas centrally administered Ang II was effective in this regard. They also showed that inhibition of Ang receptors by peripheral administration of the Ang antagonist sarile did not inhibit salt intake in Na-depleted rats, whereas centrally administered sarile was effective. This prompted these investigators to propose that brain Ang rather than circulating Ang mediated salt appetite (Sakai et al. 1990).

In double transgenic mice that expressed the human renin gene and also expressed the human angiotensinogen gene under the control of a neuron-specific promoter synapsin I, there was an increase in their preference for salt (Morimoto et al. 2002a), suggesting that neurally generated Ang in the brain plays a role in the generation of sodium appetite. When the human angiotensinogen gene was placed under the control of a glial-specific promoter (glial fibrillary acid protein) in transgenic mice expressing human renin, the salt preference of these animals was also elevated (Morimoto et al. 2002b). These data indicate that centrally generated Ang in both glia and neurons may have a role in sodium appetite in mice.

Studies with ACE inhibitors in ruminants, however, showed that circulating Ang II may be an important factor in the genesis of salt appetite in sodium-depleted animals. In both the sheep and cow, systemically administered captopril did cause a large inhibition of Na intake in Na-depleted animals, and it was found that intravenously infused Ang II at relatively low doses restored the Na appetite of these ACE-treated Na-depleted animals. Centrally infused Ang II did not restore the sodium appetite of such animals. It was suggested that blood-borne Ang II acting on Ang receptors outside the blood-brain barrier, such as those in circumventricular organs, stimulated a hunger for salt in these animals (Weisinger et al. 1987; Blair-West et al. 1988). Following these studies, systemically infused Ang II has been shown to be effective in restoring the sodium appetite of captopril-treated rats that had been sodium depleted by furosemide administration or adrenalectomy (Thunhorst et al. 1994; Weisinger et al. 1996; Schloorlemmer et al. 2001), suggesting that the rat, like the ruminant species, utilizes circulating Ang II as a signal to the brain to initiate a sodium appetite.

3.3
Brain Regions Mediating Influences of Angiotensin on Sodium Appetite

Following suggestions that circulating Ang II may act on circumventricular organs to stimulate sodium appetite, it was shown in the rat that ablation of the subfornical organ did cause a reduction in salt intake in salt-deficient rats (Weisinger et al. 1990; Thunhorst et al. 1990). However, in these rats, the Ang-dependent enhanced salt intake that occurs in response to the combination of sodium depletion and a low dose of captopril was not affected by ablation of the subfornical organ (Weisinger et al. 1990). As well, while microinjection of Ang II directly into the subfornical organ rapidly stimulates water drinking, it does not normally stimulate a salt appetite (Fitts et al. 2000). On the other hand, the other circumventricular organ of the lamina terminalis, the OVLT, may play an important role in Ang-mediated sodium appetite (Fig. 1). Injection of Ang II into the region of the OVLT stimulates an increase in salt intake in rats (Fitts et al. 2000), and ablation of the ventral lamina terminalis, which includes the OVLT, depresses the salt intake of sodium-depleted rats (Chiaraviglio 1984, Fitts et al. 1990) and also reduces the enhancement of salt appetite caused by treatment with oral doses of captopril (Fitts et al. 1990).

Ablation of the bed nucleus of the stria terminalis or the central nucleus of the amygdala also prevents increased sodium intake in sodium-depleted rats, suggestive that these regions have a physiological role in Ang-mediated sodium appetite (Galaverna et al. 1992; Johnson et al.1999). Several of the above-mentioned brain regions are also implicated by studies of immediate early gene expression in the rat brain in response to sodium depletion. Expression of *c-fos* in rats depleted of sodium chloride by either peritoneal dialysis or by treatment with the diuretic, furosemide, indicates that an increase in activity of neurons in the subfornical organ, OVLT, the central nucleus of the amygdala and bed nucleus of the stria terminalis, and the hypothalamic paraventricular and supraoptic nuclei occurs (Vivas et al. 1996; Thunhorst et al. 1998; Rowland et al. 1996). These regions are also activated by systemically infused Ang II (McKinley et al. 1992).

4 Vasopressin Secretion

4.1 Effect of Systemically Infused Angiotensin on Vasopressin Secretion

Included in the spectrum of responses that circulating Ang II is able to elicit via an action on the central nervous system, is secretion of the antidiuretic hormone vasopressin. Intravenously infused Ang II or renin was shown to increase the blood levels of vasopressin in dogs (Bonjour and Malvin 1970; Mouw et al. 1971; Reid et al. 1982; Ramsay et al. 1978; Thrasher 1985) by some but not all investigators who studied this question (Shade and Share 1975; Cadnapathornchai et al. 1975; Claybough et al. 1972).

Some of the investigators who did observe systemic infusions of Ang II at rates of 5-40 ng/kg/min to stimulate vasopressin secretion did not observe such an effect at higher infusion rates (Bonjour and Malvin 1970). As pointed out in several of the above publications, one of the factors that comes into play with systemic infusions of Ang II is the inhibitory influence that baroreceptor activation may have on vasopressin secretion that would be a consequence of the concomitant increase in arterial pressure. Other influences may also affect Ang II-induced vasopressin secretion. For instance, it has been observed that intravenous infusion of Ang II at 20 ng/kg/min progressively increases plasma vasopressin concentration over 60 min of infusion if dogs are not allowed to drink water. If dogs are allowed access to fluid and drink water in response to this rate of infusion, no increase in plasma vasopressin levels occurs (Ramsay et al. 1988). In some studies investigating the effects of systemically infused Ang II on vasopressin secretion, dogs were either water loaded or anesthetized (Cadnapathornchai et al. 1975; Shade and Share 1975), and these factors may have influenced the release of vasopressin.

Several studies have consistently shown that intravenous infusion of Ang II at 20 ng/kg/min in conscious dogs will stimulate increases in plasma vasopressin levels (Reid et al. 1982; Thrasher 1985; Ramsay et al. 1988). Reid et al. (1982) considered that the blood levels of Ang II required to stimulate vasopressin secretion were probably supraphysiological. Their observation that infusion of Ang II into carotid artery was a more effective stimulus to vasopressin secretion than the same dose administered into the femoral vein indicated that it was likely that Ang II acted directly on the brain to stimulate vasopressin secretion, a conclusion also reached by Mouw et al. (1971).

The initial studies in human subjects (uremic patients) were unable to show that intravenously infused Ang II stimulated the release of vasopressin (Hammer et al. 1980). However, more recently it has been shown that intravenously infused Ang II at 4-16 ng/kg/min increased plasma vasopressin concentrations in normal males (Chiodera et al. 1998), a response blocked by losartan, indicating that circulating Ang II acts on AT_1 receptors to stimulate vasopressin secretion (Chiodera et al. 1998).

4.2 Role of the Subfornical Organ and OVLT

For circulating Ang II to stimulate the secretion of vasopressin, it would be necessary for it to act on Ang II receptors that are on neurons accessible from the blood-stream, and connected (either directly or polysynaptically) to the vasopressin-secreting neurons of the hypothalamic supraoptic and paraventricular nuclei. Both the subfornical organ and OVLT have direct efferent connections to the supraoptic and paraventricular nuclei (Miselis et al. 1979). The two circumventricular organs may also connect to the neurosecretory cells of the supraoptic and paraventricular nuclei via a synaptic relay in the MnPO (Tanaka et al. 1997; Oldfield et al. 1990). As previously noted, neurons within the subfornical organ are rich in Ang II AT_1 receptors (Lenkei et al. 1997; Allen et al. 2000) and electrophysiological evidence in rats has shown that neurons within the subfornical organ that have efferent connections to the supraoptic and paraventricular nuclei are directly activated by blood-borne Ang II (Ferguson and Renaud 1986, Gutman et al. 1988; Tanaka et al. 1985). It is also likely that blood-borne Ang II may stimulate the indirect pathway from the subfornical organ to the paraventricular nucleus via a relay in the MnPO (Tanaka et al. 1987).

Further evidence that Ang receptors in the subfornical organ of the rat relay signals to the vasopressin-containing neurons of the hypothalamus comes from the study of lesions of the subfornical organ in this species. Ablation of the subfornical organ or transection of its efferent fibers in the rat caused an attenuation in the vasopressin released in response to systemic infusion of Ang II (Mangiapane et al. 1984; Knepel et al. 1982). Consistent with the concept of the subfornical organ being a site for circulating Ang II to influence vasopressin secretion is evidence from experiments investigating *c-fos* expression. Intravenously infused Ang II activates many neurons in the subfornical organ and the OVLT of conscious rats, as indicated by increased Fos immunoreactivity in these CVOs (McKinley et al. 1992). A significant proportion of the neurons in the subfornical organ that are activated by circulating Ang II have direct efferent projections to the supraoptic nucleus (Oldfield et al. 1994). These experiments also showed that neurons in the OVLT that project to the supraoptic nucleus are also stimulated by blood-borne Ang II. The fraction connecting to the supraoptic nucleus was less than that in the subfornical organ.

In regard to neurons in the OVLT mediating Ang II-stimulated vasopressin release, ablation of this CVO prevents vasopressin secretion in response to circulating Ang II in the dog (Thrasher 1985). It was proposed that the reason that OVLT lesions abolished Ang II-stimulated AVP secretion may have been because it removes a major facilitatory input to the median preoptic nucleus. This could lead to the threshold for excitatory input from the subfornical organ to the median preoptic nucleus being increased, with this latter site relaying signals on to vasopressin secreting cells (Thrasher 1985).

ANP has an inhibitory effect on Ang-stimulated vasopressin release (Antunes-Rodrigues et al. 1985). This may be the result of the action of ANP directly inhibiting Ang-sensitive neurons in the subfornical organ (Schmid and Simon 1992) and possibly other regions of the brain as well.

4.3
Effects of Centrally Administered Angiotensin on Vasopressin Secretion

While the effects of systemically infused Ang II on vasopressin secretion were somewhat equivocal, this is not the case with the effect of ICV administration of Ang II on vasopressin release and urine output. Many investigators in a variety of species have demonstrated that ICV infusion of Ang II or renin causes a marked reduction in urine output, renal free water clearance and increased urine osmolality in water-loaded animals (Severs et al. 1971; Hoffman et al. 1979; Andersson et al. 1972; Malayan and Reid 1976). This is almost certainly due to release of vasopressin because plasma AVP levels increase following an ICV infusion of Ang II or renin (Keil et al. 1975; Malayan et al. 1979; Yamaguchi et al. 1980). The effect of ICV infusion of Ang II to stimulate vasopressin release was inhibited by central administration of an aminopeptidase A inhibitor EC33, which prevents the conversion of Ang II to Ang III (Zini et al. 1996). These researchers also showed that ICV administration of an aminopeptidase N inhibitor, EC27, which prevented Ang III degradation, increased plasma vasopressin levels, this effect being inhibited by concomitant administration of saralasin. They proposed that Ang III is an important effector peptide for brain Ang mechanisms stimulating vasopressin secretion (Zini et al. 1996).

4.4
Site of Action of ICV Angiotensin to Stimulate Vasopressin Secretion

In regard to the brain sites at which ICV Ang II acts to stimulate vasopressin secretion, it seems likely that the MnPO is of particular importance. Ablation of the AV3V region blocks antidiuresis and vasopressin release in response to ICV infusion of Ang II in goats and rats, even though the vasopressin secreting neurons of the supraoptic and paraventricular nuclei are intact (Andersson et al. 1975; Bealer et al. 1979). This AV3V region encompasses a considerable part of the MnPO, which is a site of high-density AT_1 receptors, and ablation of the MnPO alone also prevents vasopressin secretion in response to ICV infusion of Ang II (Mangiapane et al. 1983). Neurons in the MnPO are directly sensitive to iontophoretically applied Ang II (Tanaka et al. 1987) and a tetrodotoxin-sensitive inward current in response to Ang II in was noted in voltage clamp recordings (Bai and Renaud 1998). ICV Ang II has been shown to strongly activate neurons in the MnPO, as shown by increased Fos immunoreactivity (McKinley et al. 1995) and neurons within this nucleus have efferent projections to vasopressin-secreting neurons of the supraoptic nucleus (Miselis 1981). It is unlikely that circulating Ang II has access to the MnPO because of the presence of the blood-brain barrier. Therefore Ang II that is generated within the brain is the likely endogenous ligand for AT_1 receptors in this nucleus that influences vasopressin release. Microinjection of losartan or cholinergic antagonists into the hypothalamic paraventricular nucleus also suppressed vasopressin release in response to non-pressor doses of ICV hypertonic saline, suggesting the paraventricular nucleus is also a site at which brain Ang mediates vasopressin release (Qadri et al. 1998).

4.5 Effect of Centrally Administered Inhibitors on Vasopressin Secretion

Initial studies of Ang antagonists used ICV administration of the nonselective Ang receptor antagonist saralasin. In rats, ICV administration of 10-100 ng saralasin blocked the increase in plasma vasopressin that occurs in response to ICV injection of Ang II (Yamaguchi et al. 1980; Keil et al. 1983). Yamaguchi and colleagues also showed that centrally administered saralasin (1 μg) reduced the plasma vasopressin levels of rats that had been water deprived, made hypovolemic by subcutaneous injection of polyethylene glycol, or injected with hypertonic saline (Yamaguchi et al. 1980, 1982). However, ICV injection of sarile (0.5-10 μg) did not reduce the vasopressin response to 24-72 h water deprivation or to hypertonic saline administration in rats, although when combined with ICV atropine, a 30% reduction in vasopressin levels in response to 48 h water deprivation was observed, leading to speculation that while central Ang is of significance for AVP release, redundancy exists in the central integrative mechanisms controlling vasopressin secretion (Keil et al. 1983). In vitro studies of hypothalamic explants showed that the osmotic stimulation of AVP release was inhibited by the addition of saralasin to the bathing medium, suggesting central Ang mediation of osmotically stimulated vasopressin release (Sladek and Joynt 1980).

In the last decade, specific inhibition of AT_1 or AT_2 receptors has been achieved by the central administration of either non-peptide receptor antagonists or antisense oligonucleotides. In rats and sheep, central administration of losartan blocked vasopressin secretion in response to centrally administered Ang II (Hogarty et al. 1992; Mathai et al. 1998); however, it was reported that ICV administration of an AT_2 antagonist also inhibited this response (Hogarty et al. 1992). Further evidence to support a role for AT_1 receptors in vasopressin secretion comes from studies showing that central administration of an antisense oligonucleotide to the AT_1 mRNA blocked vasopressin secretion to ICV Ang II (Meng et al. 1994). In addition, mice in which the gene for the AT_1 receptor had been deleted failed to increase vasopressin levels in response to water deprivation (Morris et al. 1999). This result would be consistent with studies showing that ICV administration of losartan can inhibit vasopressin secretion in response to centrally administered hypertonic saline (Hogarty et al. 1984; Rohmeiss et al. 1996; Mathai et al. 1998) and suggests that an angiotensinergic synaptic relay is involved in the osmotic stimulation of vasopressin secretion. Transgenic rats in which expression of an antisense nucleotide sequence reduced glial angiotensinogen production in the brain by more than 90% show impaired vasopressin secretion (Schinke et al. 1999). This results in a moderate form of diabetes insipidus, suggesting that brain Ang plays a role in the tonic release of vasopressin in rats.

5
Sodium Excretion

Unlike circulating Ang II, which promotes sodium retention by the kidney secondarily to aldosterone secretion, brain Ang II appears to promote the excretion of electrolytes. The main data supporting this contention comes from studies showing that ICV administration of Ang II causes a rapid and large increase in renal Na excretion in rats, goats, sheep, and dogs (Severs et al. 1970; Andersson et al. 1972; Coghlan et al. 1981). The natriuresis stimulated by ICV injection of Ang II is associated with a reduction in plasma renin concentration (Malayan and Reid 1979), an increase in arterial pressure (Andersson et al. 1972), increased plasma vasopressin levels (McKinley et al. 1998) and a reduction in renal sympathetic nerve activity (May et al. 2000), all factors that promote a natriuresis. The natriuretic effect of centrally administered Ang II is blocked by losartan pretreatment, indicating AT_1 receptors mediate this effect. Ablation of tissue in the lamina terminalis disrupts the natriuretic response to ICV Ang II (Andersson et al. 1975), indicating the likelihood that it is the AT_1 receptors in the lamina terminalis that are responsible for initiating the response. As well as inhibiting the natriuretic response to centrally administered Ang II, pretreatment with ICV administered losartan also prevents the natriuretic response to ICV infusion of hypertonic saline (McKinley et al. 1994; Rohmeiss et al. 1996), indicating that a central angiotensinergic neural pathway may subserve brain natriuretic mechanisms. The natriuretic effect of ICV Ang II is blocked by concomitant ICV injection of atrial natriuretic factor (Rohmeiss et al. 1989), supporting the view that central atrial natriuretic factor is inhibitory to many of the actions of Ang in the brain.

Acknowledgements. The work of the authors is supported by an Institute Block Grant from the National Health and Medical Research Foundation of Australia, the Robert J. Jr and Helen C. Kleberg Foundation, and the G. Harold and Leila Y. Mathers Charitable Foundation.

References

Abraham SF, Baker RM, Blaine EH, Denton DA, McKinley MJ (1975) Water drinking induced in sheep by angiotensin-a physiological or pharmacological effect? J Comp Physiol Psychol 88:503-518

Abraham SF, Denton DA, McKinley MJ, Weisinger RS (1976) Effect of an angiotensin antagonist, Sar1-Ala8-Angiotensin II on physiological thirst. Pharmacol Biochem Behav 4:243-247

Allen AM, Oldfield BJ, Giles ME, Paxinos G, McKinley MJ, Mendelsohn FAO (2000) Localization of angiotensin receptors in the nervous system. In: Quirion R, Bjorklund A,

Hokfelt T (eds) Handbook of chemical neuroanatomy, Vol. 16. Elsevier Science, Amsterdam, p 79

Andersson B, Eriksson L, Fernandez O, Kolmodin CG, Oltner R (1972) Centrally mediated effects of sodium and angiotensin II on arterial blood pressure and fluid balance. Acta Physiol Scand 85:398-407

Andersson B, Leksell LG, Lishajko F (1975) Perturbations in fluid balance induced by medially placed forebrain lesions. Brain Res 99:261-275

Bai D, Renaud LP (1998) Ang II AT_1 receptors induce depolarization and inward current in rat median preoptic neurons in vitro. Am J Physiol 275:R632-R639

Bealer SL, Phillips MI, Johnson AK, Schmid PG (1979) Anteroventral third ventricle lesions reduce antidiuretic responses to angiotensin II. Am J Physiol 236:E610-E615

Bickerton RK, Buckley JP (1961) Evidence for a central mechanism in angiotensin-induced hypertension. Proc Soc Exp Biol Med 106:834-836

Blair-West JR, Denton DA, McKinley MJ, Weisinger RS (1988) Angiotensin-related sodium appetite and thirst in cattle. Am J Physiol 255:R205-R211

Blair-West JR, Burns P, Denton DA, Ferraro T, McBurnie MI, Tarjan E, Weisinger RS (1994) Thirst induced by increasing brain sodium concentration is mediated by brain angiotensin. Brain Res 637:335-338

Blair-West JR, Denton DA, McKinley MJ, Weisinger RS (1997) Central infusion of the AT_1 receptor antagonist losartan inhibits thirst but not sodium appetite in cattle. Am J Physiol 272:R1940-R1945

Blair-West JR, Carey KD, Denton DA, Weisinger RS, Shade RE (1998) Evidence that brain angiotensin II is involved in both thirst and sodium appetite in baboons. Am J Physiol 275:R1639-R1646

Blair-West JR, Carey KD, Denton DA, Madden LJ, Weisinger RS, Shade RE (2001) Possible contribution of brain angiotensin III to ingestive behaviors in baboons. Am J Physiol Regul Integr Comp Physiol 281:R1633-R1636

Bonjour J-P, Malvin RL (1970) Stimulation of ADH release by the renin-angiotensin system. Am J Physiol 218:1555-1559

Brownfield MS, Reid IA, Ganten D, Ganong WF (1982) Differential distribution of immunoreactive angiotensin and angiotensin-converting enzyme in rat brain. Neuroscience 7:1759-1769

Bryant RW, Epstein AN, Fitzsimons JT, Fluharty SJ (1980) Arousal of a specific and persistent sodium appetite in the rat with continuous intracerebroventricular infusion of angiotensin II. J Physiol 301:365-382

Buggy J, Fisher AE (1974) Evidence for a dual central role for angiotensin in water and sodium intake. Nature 250:733-735

Buggy J, Fisher AE (1976) Anteroventral third ventricle site of action of angiotensin induced thirst. Pharmacol Biochem Behav 4:651-660

Buggy J, Johnson AK (1978) Angiotensin-induced thirst: effects of third ventricle obstruction and periventricular ablation. Brain Res 149:117-128

Bunnemann B, Iwai N, Metzger R, Fuxe K, Inagami T, Ganten D (1992) The distribution of angiotensin II AT_1 receptor subtype mRNA in the rat brain. Neurosci Lett 142:155-158

Bunnemann B, Fuxe K, Ganten D (1993) The renin-angiotensin system in the brain: an update 1993. Regul Pept 46:487-509

Cadnapaphornchai P, Boykin J, Harbottle JA, McDonald KM, Schrier RW (1975) Effect of angiotensin II on renal water excretion. Am J Physiol 228:155-159

Chai SY, Mendelsohn FA, Paxinos G (1987) Angiotensin converting enzyme in rat brain visualized by quantitative in vitro autoradiography. Neuroscience 20:615-627

Chiaraviglio E, Perez Guaita MF (1984) Anterior third ventricle (A3 V) lesions and homeostasis regulation. J Physiol 79:446-452

Chiodera P, Volpi R, Caiazza A, Giuliani N, Magotti MG, Coiro V (1998) Arginine vasopressin and oxytocin responses to angiotensin II are mediated by AT_1 receptor subtype in normal men. Metabolism 47:893-896

Coghlan JP, Considine PJ, Denton DA, Fei DTW, Leksell LG, McKinley MJ, Muller AF, Tarjan E, Weisinger RS, Bradshaw R (1981) Sodium appetite in sheep induced by cerebral ventricular infusion of angiotensin: comparison with sodium deficiency. Science 214:195-197

Cooney AS, Fitzsimons JT (1993) The effect of the putative AT_2 agonist, p-aminophenylalanine6 angiotensin II, on thirst and sodium appetite in rats. Exp Physiol 78:767-774

Cunningham JT, Beltz T, Johnson RF, Johnson AK (1992) The effects of ibotenate lesions of the median preoptic nucleus on experimentally-induced and circadian drinking behavior in rats. Brain Res 580:325-330

Denton DA, Blair-West JR, McBurnie M, Osborne PG, Tarjan E, Williams RM, Weisinger RS (1990) Angiotensin and salt appetite of BALB/c mice. Am J Physiol 259:R729-R735

DiNicolantonio R, Hutchinson JS, Mendelsohn FA (1982) Exaggerated salt appetite of spontaneously hypertensive rats is decreased by central angiotensin-converting enzyme blockade. Nature 298:846-848

Dourish CT, Duggan JA, Banks RJ (1992) Drinking induced by subcutaneous injection of angiotensin II in the rat is blocked by the selective AT_1 receptor antagonist DuP 753 but not by the selective AT_2 receptor antagonist WL 19. Eur J Pharmacol 211:113-116

Dzau VJ, Ingelfinger J, Pratt RE, Ellison KE (1986) Identification of renin and angiotensinogen messenger RNA sequences in mouse and rat brains. Hypertension 8:544-548

Ehrlich KJ, Fitts DA (1990) Atrial natriuretic peptide in the subfornical organ reduces drinking induced by angiotensin or in response to water deprivation. Behav Neurosci 104:365-372

Elfont RM, Fitzsimons JT (1983) Renin dependence of captopril-induced drinking after ureteric ligation in the rat. J Physiol 343:17-30

Elfont RM, Fitzsimons JT (1985) The effect of captopril on sodium appetite in adrenalectomized and deoxycorticosterone-treated rats. J Physiol 365:1-12

Elfont RM, Epstein AN, Fitzsimons JT (1984) Involvement of the renin-angiotensin system in captopril-induced sodium appetite in the rat. J Physiol 354:11-27

Epstein AN, Fitzsimons JT, Rolls BJ (1970) Drinking induced by injection of angiotensin into the brain of the rat. J Physiol 210:457-474

Evered MD, Robinson MM (1983) Effects of captopril on salt appetite in sodium-replete rats and rats treated with desoxycorticosterone acetate (DOCA). J Pharmacol Exp Ther 225:416-421

Evered MD, Robinson MM, Richardson MA (1980) Captopril given intracerebroventricularly, subcutaneously or by gavage inhibits angiotensin-converting enzyme activity in the rat brain. Eur J Pharmacol 68:443-449

Evered MD, Robinson MM, Rose PA (1988) Effect of arterial pressure on drinking and urinary responses to angiotensin II. Am J Physiol 254:R69-R74

Fei DTW, Abraham SF, Coghlan JP, Denton DA, McKinley MJ, Scoggins BA, Weisinger RS (1982) Angiotensin II in CSF of sheep. Clin Exp Pharmacol Physiol 9:564-565

Ferguson AV, Renaud LP (1986) Systemic angiotensin acts at subfornical organ to facilitate activity of neurohypophysial neurons. Am J Physiol 251:R712-R717

Findlay AL, Epstein AN (1980) Increased sodium intake is somehow induced in rats by intravenous angiotensin II. Horm Behav 14:86-92

Fitts DA (1994) Angiotensin II receptors in SFO but not in OVLT mediate isoproterenol-induced thirst. Am J Physiol 267:R7-R15

Fitts DA, Tjepkes DS, Bright RO (1990) Salt appetite and lesions of the ventral part of the ventral median preoptic nucleus. Behav Neurosci 104:818-827

Fitts DA, Starbuck EM, Ruhf A (2000) Circumventricular organs and Ang II-induced salt appetite: blood pressure and connectivity. Am J Physiol Regul Integr Comp Physiol 279:R2277-R2286.

Fitzsimons JT (1969) The role of a renal thirst factor in drinking induced by extracellular stimuli. J Physiol (Lond) 201:349-368

Fitzsimons JT (1971) The effect on drinking of peptide precursors and of shorter chain peptide fragments of angiotensin II injected into the rat's diencephalon. J Physiol 214:295-303

Fitzsimons JT (1998) Angiotensin, thirst and sodium appetite. Physiol Rev 78:583-682

Fitzsimons JT, Kucharczyk J (1978) Drinking and haemodynamic changes induced in the dog by intracranial injection of components of the renin-angiotensin system. J Physiol 276:419-434

Fitzsimons JT, Simons B (1969) The effect on drinking in the rat of intravenous infusion of angiotensin, given alone or in combination with other stimuli of thirst. J Physiol 203:45-57

Fitzsimons JT, Kucharczyk J, Richards G (1978) Systemic angiotensin-induced drinking in the dog: a physiological phenomenon. J Physiol 276:435-448

Fluharty SJ, Epstein AN (1983) Sodium appetite elicited by intracerebroventricular infusion of angiotensin II in the rat: II. Synergistic interaction with systemic mineralocorticoids. Behav Neurosci 97:746-758

Fregly MJ (1980) Effect of the angiotensin converting enzyme inhibitor, captopril, on NaCl appetite of rats. J Pharmacol Exp Ther 215:407-412

Fregly MJ, Rowland NE (1991) Effect of a nonpeptide angiotensin II receptor antagonist, DuP 753, on angiotensin-related water intake in rats. Brain Res Bull 27:97-100

Galaverna O, De Luca LA Jr, Schulkin J, Yao SZ, Epstein AN (1992) Deficits in NaCl ingestion after damage to the central nucleus of the amygdala in the rat. Brain Res Bull 28:89-98

Ganten D, Hermann K, Bayer C, Unger T, Lang RE (1983) Angiotensin synthesis in the brain and increased turnover in hypertensive rats. Science 221:869-871

Geddes BJ, Parry LJ, Summerlee AJ (1994) Brain angiotensin-II partially mediates the effects of relaxin on vasopressin and oxytocin release in anesthetized rats. Endocrinology 134:1188-1192

Giles ME, Fernley RT, Nakamura Y, Moeller I, Aldred GP, Ferraro T, Penschow JD, McKinley MJ, Oldfield BJ (1999) Characterization of a specific antibody to the rat angiotensin II AT_1 receptor. J Histochem Cytochem 47:507-516

Gohlke P, Weiss S, Jansen A, Wienen W, Stangier J, Rascher W, Culman J, Unger T (2001) AT_1 receptor antagonist telmisartan administered peripherally inhibits central responses to angiotensin II in conscious rats. J Pharmacol Exp Ther 298:62-70

Gohlke P, Kox T, Jurgensen T, von Kugelgen S, Rascher W, Unger T, Culman T (2002) Peripherally applied candesartan inhibits central responses to angiotensin II in conscious rats. Naunyn-Schmiedeberg's Arch Pharmacol 365:477-483

Gutman MB, Ciriello J, Mogenson GJ (1988) Effects of plasma angiotensin II and hypernatremia on subfornical organ neurons. Am J Physiol 254:R746-R754

Hammer M, Olgaard K, Madsen S (1980) The inability of angiotensin II infusions to raise plasma vasopressin levels in haemodialysis patients. Acta Endocrinol (Copenh) 95:422-426

Herbert J, Forsling ML, Howes SR, Stacey PM, Shiers HM (1992) Regional expression of c-fos antigen in the basal forebrain following intraventricular infusions of angiotensin and its modulation by drinking either water or saline. Neuroscience 51:867-882

Hoffman WE, Phillips MI (1976) Evidence for sar1-ala8-angiotensin crossing the blood cerebrospinal fluid barrier to antagonize central effects of angiotensin II. Brain Res 109:541-552

Hoffman WE, Ganten U, Phillips MI, Schmid PG, Schelling P, Ganten D (1978) Inhibition of drinking in water-deprived rats by combined central angiotensin II and cholinergic receptor blockade. Am J Physiol 234:F41-F47

Hoffman WE, Weet JF, Phillips MI, Schmid PG (1979) Central effects of angiotensin II in water and saline loaded rats. Neuroendocrinology 28:289-296

Hogarty DC, Speakman EA, Puig V, Phillips MI (1992) The role of angiotensin, AT_1 and AT_2 receptors in the pressor, drinking and vasopressin responses to central angiotensin. Brain Res 586:289-294

Hogarty DC, Tran DN, Phillips MI (1994) Involvement of angiotensin receptor subtypes in osmotically induced release of vasopressin. Brain Res 637:126-132

Hsiao S, Epstein AN, Camardo JS (1977) The dipsogenic potency of peripheral angiotensin II. Horm Behav 8:129-140

Jensen LL, Harding JW, Wright JW (1992) Role of paraventricular nucleus in control of blood pressure and drinking in rats. Am J Physiol 262:F1068-F1075

Johnson AK, Epstein AN (1975) The cerebral ventricles as the avenue for the dipsogenic action of intracranial angiotensin. Brain Res 86:399-418

Johnson AK, Cunningham JT, Thunhorst RL (1996) Integrative role of the lamina terminalis in the regulation of cardiovascular and body fluid homeostasis. Clin Exp Pharmacol Physiol 23:183-191

Johnson AK, de Olmos J, Pastuskovas CV, Zardetto-Smith AM, Vivas L (1999) The extended amygdala and salt appetite. Ann N Y Acad Sci 877:258-280

Johnson AK, Mann JF, Rascher W, Johnson JK, Ganten D (1981) Plasma angiotensin II concentrations and experimentally induced thirst. Am J Physiol 240:R229-R234

Keil LC, Barbella YR, Dundore RL, Wurpel JN, Severs WB (1983) Vasopressin release induced by water deprivation: effects of centrally administered saralasin. Neuroendocrinology 37:401-405

Knepel W, Nutto D, Meyer DK (1982) Effect of transection of subfornical organ efferent projections on vasopressin release induced by angiotensin or isoprenaline in the rat. Brain Res 248:180-184

Kobayashi H, Uemura H, Wada M, Takei Y (1979) Ecological adaptation of angiotensin-induced thirst mechanism in tetrapods. Gen Comp Endocrinol 38:93-104

Lebrun CJ, Blume A, Herdegen T, Seifert K, Bravo R, Unger T (1995) Angiotensin II induces a complex activation of transcription factors in the rat brain: expression of Fos, Jun and Krox proteins. Neuroscience 65:93-99

Lee MC, Thrasher TN, Ramsay DJ (1981) Is angiotensin essential in drinking induced by water deprivation and caval ligation? Am J Physiol 240:R75-R80

Lehr D, Goldman HW, Casner P (1973) Renin-angiotensin role in thirst: paradoxical enhancement of drinking by angiotensin converting enzyme inhibitor. Science 182:1031-1034

Lenkei Z, Palkovits M, Corvol P, Llorens-Cortes C (1997) Expression of angiotensin type-1 (AT_1) and type-2 (AT_2) receptor mRNAs in the adult rat brain: a functional neuroanatomical review. Front Neuroendocrinol 18:383-439

Lind RW, Johnson AK (1983) A further characterization of the effects of AV3V lesions on ingestive behavior. Am J Physiol 245:R83-R90

Lind RW, Swanson LW, Ganten D (1985) Organization of angiotensin II immunoreactive cells and fibers in the rat central nervous system. An immunohistochemical study. Neuroendocrinology 40:2-24

Malayan SA, Reid IA (1976) Antidiuresis produced by injection of renin into the third cerebral ventricle of the dog. Endocrinology 98:329-335

Malayan SA, Keil LC, Ramsay DJ, Reid IA (1979) Mechanism of suppression of plasma renin activity by centrally administered angiotensin II. Endocrinology 104:672-675

Malvin RL, Mouw D, Vander AJ (1977) Angiotensin: physiological role in water-deprivation-induced thirst of rats. Science 197:171-173

Mangiapane ML, Thrasher TN, Keil LC, Simpson JB, Ganong WF (1983) Deficits in drinking and vasopressin secretion after lesions of the nucleus medianus. Neuroendocrinology 37:73-77

Mangiapane ML, Thrasher TN, Keil LC, Simpson JB, Ganong WF (1984) Role for the subfornical organ in vasopressin release. Brain Res Bull 13:43-47

Mann JF, Johnson AK, Ganten D (1980) Plasma angiotensin II: dipsogenic levels and angiotensin-generating capacity of renin. Am J Physiol 238:R372-R377

Massi M, Epstein AN (1990) Angiotensin/aldosterone synergy governs the salt appetite of the pigeon. Appetite 14:181-192

Mathai M, Evered MD, McKinley MJ (1997) Intracerebroventricular losartan inhibits postprandial drinking in sheep. Am J Physiol 272:R1055-R1059

Mathai ML, Evered MD, McKinley MJ (1998) Central losartan blocks natriuretic, vasopressin, and pressor responses to central hypertonic NaCl in sheep. Am J Physiol 275:R548-R554

May CN, McAllen RM, McKinley MJ (2000) Renal nerve inhibition by central NaCl and Ang II is abolished by lesions of the lamina terminalis. Am J Physiol 279:R1827-R1833

McKinley MJ, Denton DA, Park RG, Weisinger RS (1986) Ablation of subfornical organ does not prevent angiotensin-induced water drinking in sheep. Am J Physiol 250:R1052-R1059

McKinley MJ, Allen AM, Clevers J, Paxinos G, Mendelsohn FA (1987) Angiotensin receptor binding in human hypothalamus: autoradiographic localization. Brain Res 420:375-379

McKinley MJ, McAllen RM, Mendelsohn FAO, Allen AM, Chai S-Y, Oldfield BJ (1990) Circumventricular organs: neuroendocrine interfaces between the brain and the hemal milleu. Front Neuroendocrinol 11:91-127

McKinley MJ, Badoer E, Oldfield BJ (1992) Intravenous angiotensin II induces Fos-immunoreactivity in circumventricular organs of the lamina terminalis. Brain Res 594:295-300

McKinley MJ, Evered M, Mathai M, Coghlan JP (1994) Effects of central losartan on plasma renin and centrally mediated natriuresis. Kidney Int 46:1479-1482

McKinley MJ, Badoer E, Vivas L, Oldfield BJ (1995) Comparison of c-fos expression in the lamina terminalis of conscious rats after intravenous or intracerebroventricular angiotensin. Brain Res Bull 37:131-137

McKinley MJ, Colvill LM, Giles ME, Oldfield BJ (1997) Distribution of Fos-immunoreactivity in rat brain following a dipsogenic dose of captopril and effects of angiotensin receptor blockade. Brain Res 747:43-51

McKinley MJ, Burns P, Colvill L, Giles M, Sinnayah P, Sunn N, Oldfield BJ, Weisinger RS (2001) Neurons and neural pathways mediating the actions of circulating relaxin on the brain. In: Tregear GW, Ivell R, Bathgate RA, Wade JD (eds) Relaxin 2000. Kluwer, Dordrecht, pp 201-208

Mendelsohn FA, Quirion R, Saavedra JM, Aguilera G, Catt KJ (1984) Autoradiographic localization of angiotensin II receptors in rat brain. Proc Natl Acad Sci U S A 81:1575-1579

Mendelsohn FA, Allen AM, Clevers J, Denton DA, Tarjan E, McKinley MJ (1988) Localization of angiotensin II receptor binding in rabbit brain by in vitro autoradiography. J Comp Neurol 270:372-384

Meng H, Wielbo D, Gyurko R, Phillips MI (1994) Antisense oligonucleotide to AT_1 receptor mRNA inhibits central angiotensin induced thirst and vasopressin. Regul Pept 54:543-551

Miselis RR, Shapiro RE, Hand PJ (1979) Subfornical organ efferents to neural systems for control of body water. Science 205:1022-1025

Moe KE, Weiss ML, Epstein AN (1984) Sodium appetite during captopril blockade of endogenous angiotensin II formation. Am J Physiol 247:R356-R365

Morimoto S, Cassell MD, Sigmund CD (2002a) Neuron-specific expression of human angiotensinogen in brain causes increased salt appetite. Physiol Genomics 9:113-120
Morimoto S, Cassell MD, Sigmund CD (2002b) Glial- and neuronal-specific expression of the renin-angiotensin system in brain alters blood pressure, water intake, and salt preference. J Biol Chem 277:33235-33241
Morris M, Li P, Callahan MF, Oliverio MI, Coffman TM, Bosch SM, Diz DI (1999) Neuroendocrine effects of dehydration in mice lacking the angiotensin AT_{1a} receptor. Hypertension 33:482-486
Mouw D, Bonjour JP, Malvin RL, Vander A (1971) Central action of angiotensin in stimulating ADH release. Am J Physiol 220:239-242
Oldfield BJ, Ganten D, McKinley MJ (1989) An ultrastructural analysis of the distribution of angiotensin II in the rat brain. J Neuroendocrinol 1:123-128
Oldfield BJ, Badoer E, Hards DK, McKinley MJ (1994) Fos production in retrogradely labelled neurons of the lamina terminalis following intravenous infusion of either hypertonic saline or angiotensin II. Neuroscience 60:255-262
O'Neill TP, Brody MJ (1987) Role for the median preoptic nucleus in centrally evoked pressor responses. Am J Physiol 252:R1165-R1172
Phillips PA, Rolls BJ, Ledingham JG, Morton JJ, Forsling ML (1985) Angiotensin II-induced thirst and vasopressin release in man. Clin Sci (Lond) 68:669-674
Propper CR, Hillyard SD, Johnson WE (1995) Central angiotensin II induces thirst-related responses in an amphibian. Horm Behav 29:74-84
Qadri F, Waldmann T, Wolf A, Hohle S, Rascher W, Unger T (1998) Differential contribution of angiotensinergic and cholinergic receptors in the hypothalamic paraventricular nucleus to osmotically induced AVP release. J Pharmacol Exp Ther 285:1012-1018
Ramsay DJ, Keil LC, Sharpe MC, Shinsaki J (1978) Angiotensin II infusion increases vasopressin, ACTH and 11-hydroxycorticosteroid secretion. Am J Physiol 234:R66-R71
Ramsay DJ, Thrasher TN, Keil LC (1988) Neurohumoral influences on vasopressin. In: Cowley AW, Liard J-F, Ausiello DA (eds) Vasopressin: cellular and integrative functions. Raven Press, New York, pp 169-176
Reid IA, Brooks VL, Rudolph CD, Keil LC (1982) Analysis of the actions of angiotensin on the central nervous system of conscious dogs. Am J Physiol 243:R82-R91
Robinson MM, Evered MD (1987) Pressor action of intravenous angiotensin II reduces drinking response in rats. Am J Physiol 252:R754-R759
Rohmeiss P, Beyer C, Nagy E, Tschope C, Hohle S, Strauch M, Unger T (1995) NaCl injections in brain induce natriuresis and blood pressure responses sensitive to Ang II AT_1 receptors. Am J Physiol 269:F282-F288
Rowland NE, Rozelle A, Riley PJ, Fregly MJ (1992) Effect of nonpeptide angiotensin receptor antagonists on water intake and salt appetite in rats. Brain Res Bull 29:389-393
Rowland NE, Fregly MJ, Han L, Smith G (1996) Expression of Fos in rat brain in relation to sodium appetite: furosemide and cerebroventricular renin. Brain Res 728:90-96
Sakai RR, Nicolaidis S, Epstein AN (1986) Salt appetite is suppressed by interference with angiotensin II and aldosterone. Am J Physiol 251:R762-R768
Sakai RR, Chow SY, Epstein AN (1990) Peripheral angiotensin II is not the cause of sodium appetite in the rat. Appetite 15:161-170
Sakai RR, He PF, Yang XD, Ma LY, Guo YF, Reilly JJ, Moga CN, Fluharty SJ (1994) Intracerebroventricular administration of AT_1 receptor antisense oligonucleotides inhibits the behavioral actions of angiotensin II. J Neurochem 62:2053-2056
Schelling P, Hutchinson JS, Ganten U, Sponer G, Ganten D (1976) Impermeability of the blood-cerebrospinal fluid barrier for angiotensin II in rats. Clin Sci Mol Med Suppl 3:399s-402s
Schiffrin EL, Genest J (1982) Mechanism of captopril-induced drinking. Am J Physiol 242:R136-R140

Schinke M, Baltatu O, Bohm M, Peters J, Rascher W, Bricca G, Lippoldt A, Ganten D, Bader M (1999) Blood pressure reduction and diabetes insipidus in transgenic rats deficient in brain angiotensinogen. Proc Natl Acad Sci U S A 96:3975-3980

Schloorlemmer GHM, Johnson AK, Thunhorst RL (2001) Circulating angiotensin II mediates sodium appetite in adrenalectomised rats. Am J Physiol 281:R723-R729

Schmid HA, Simon E (1992) Effect of angiotensin II and atrial natriuretic factor on neurons in the subfornical organ of ducks and rats in vitro. Brain Res 588:324-328

Severs WB, Daniels-Severs A, Summy-Long J, Radio GJ (1971) Effects of centrally administered angiotensin II on salt and water excretion. Pharmacology 6:242-252

Shade RE, Share L (1975) Vasopressin release during nonhypotensive hemorrhage and angiotensin II infusion. Am J Physiol 228:149-154

Shade RE, Blair-West JR, Carey KD, Madden LJ, Weisinger RS, Denton DA (2002) Synergy between angiotensin and aldosterone in evoking sodium appetite in baboons. Am J Physiol Regul Integr Comp Physiol 283:R1070-R1078

Simpson JB, Routtenberg A (1973) Subfornical organ: site of drinking elicitation by angiotensin II. Science 181:1772-1775

Simpson JB, Epstein AN, Camardo JS Jr (1978) Localization of receptors for the dipsogenic action of angiotensin II in the subfornical organ of rat. J Comp Physiol Psychol 92:581-601

Sinnayah P, Kachab E, Haralambidis J, Coghlan JP, McKinley MJ (1997) Effects of angiotensinogen antisense oligonucleotides on fluid intake in response to different dipsogenic stimuli in the rat. Brain Res 50:43-50

Sinnayah P, Burns P, Wade JD, Weisinger RS, McKinley MJ (1999) Water drinking in rats resulting from intravenous relaxin and its modification by other dipsogenic factors. Endocrinology 140:5082-5086

Sladek CD, Joynt RJ (1980) Role of angiotensin in the osmotic control of vasopressin release by the organ-cultured rat hypothalamo-neurohypophyseal system. Endocrinology 106:173-178

Song K, Allen AM, Paxinos G, Mendelsohn FA (1992) Mapping of angiotensin II receptor subtype heterogeneity in rat brain. J Comp Neurol 316:467-484

Speth RC, Wamsley JK, Gehlert DR, Chernicky CL, Barnes KL, Ferrario CM (1985) Angiotensin II receptor localization in the canine CNS. Brain Res 326:137-143

Stornetta RL, Hawelu-Johnson CL, Guyenet PG, Lynch KR (1988) Astrocytes synthesize angiotensinogen in brain. Science 242:1444-1446

Stricker EM (1983) Thirst and sodium appetite after colloid treatment in rats: role of the renin-angiotensin-aldosterone system. Behav Neurosci 97:725-737

Summerlee AJS, Robertson GF (1995) Central administration of porcine relaxin stimulates drinking behaviour in rats: an effect mediated by central angiotensin II. Endocrine 3:377-381

Tanaka J, Kaba H, Saito H, Seto K (1985) Electrophysiological evidence that circulating angiotensin II sensitive neurons in the subfornical organ alter the activity of hypothalamic paraventricular neurohypophyseal neurons in the rat. Brain Res 342:361-365

Tanaka J, Saito H, Kaba H (1987) Subfornical organ and hypothalamic paraventricular nucleus connections with median preoptic nucleus neurons: an electrophysiological study in the rat. Exp Brain Res 68:579-585

Tanaka J, Hori K, Nomura M (2001) Dipsogenic response induced by angiotensinergic pathways from the lateral hypothalamic area to the subfornical organ in rats. Behav Brain Res 118:111-116

Thrasher TN, Simpson JB, Ramsay DJ (1982) Lesions of the subfornical organ block angiotensin-induced drinking in the dog. Neuroendocrinology 35:68–72

Thunhorst RL, Ehrlich KJ, Simpson JB (1990) Subfornical organ participates in salt appetite. Behav Neurosci 104:637-642

Thunhorst RL, Fitts DA (1994) Peripheral angiotensin causes salt appetite in rats. Am J Physiol 2267:R171-R177

Thunhorst RL, Xu Z, Cicha MZ, Zardetto-Smith AM, Johnson AK (1998) Fos expression in rat brain during depletion-induced thirst and salt appetite. Am J Physiol 274:R1807-R1814

Vivas L, Pastuskovas CV, Tonelli L (1995) Sodium depletion induces Fos immunoreactivity in circumventricular organs of the lamina terminalis. Brain Res 679:34-41

Weisinger RS, Denton DA, McKinley MJ, Muller AF, Tarjan E (1986) Angiotensin and Na appetite of sheep. Am J Physiol 251:R690-R699

Weisinger RS, Denton DA, Di Nicolantonio R, McKinley MJ, Muller AF, Tarjan E (1987) Role of angiotensin in sodium appetite of sodium-deplete sheep. Am J Physiol 253:R482-R488

Weisinger RS, Denton DA, Di Nicolantonio R, McKinley MJ (1988) The effect of captopril or enalaprilic acid on the Na appetite of Na-deplete rats. Clin Exp Pharmacol Physiol 15:55-65

Weisinger RS, Denton DA, R. DN, Hards DK, McKinley MJ (1990) Subfornical organ lesion decreases sodium appetite in the sodium-depleted rat. Brain Res 526:23-30

Weisinger RS, Blair-West JR, Burns P, Denton DA, McKinley MJ (1996) The role of angiotensin II in ingestive behaviour: a brief review of angiotensin II thirst and Na appetite. Regul Pept 661:73-81

Weisinger RS, Burns P, Colvill LM, Davern P, Giles ME, Oldfield BJ, McKinley MJ (2000) Fos immunoreactivity in the lamina terminalis of adrenalectomized rats and effects of angiotensin II type 1 receptor blockade or deoxycorticosterone. Neuroscience 98:167-180

Wright JW, Morseth SL, Abhold RH, Harding JW (1985) Pressor action and dipsogenicity induced by angiotensin II and III in rats. Am J Physiol 249:R514-R521

Wright JW, Quirk WS, Hanesworth JM, Harding JW (1988) Influence of aminopeptidase inhibitors on brain angiotensin metabolism and drinking in rats. Brain Res 441:215-220

Yamaguchi K, Hama H, Sakaguchi T, Negoro H, Kamoi K (1980) Effects of intraventricular injection of Sar1-Ala8-angiotensin II on plasma vasopressin level increased by angiotensin II and by water deprivation in conscious rats. Acta Endocrinol (Copenh) 93:407-412

Yamaguchi K, Sakaguchi T, Kamoi K (1982) Central role of angiotensin in the hyperosmolality- and hypovolaemia-induced vasopressin release in conscious rats. Acta Endocrinol (Copenh) 101:524-530

Zini S, Fournie-Zaluski MC, Chauvel E, Roques BP, Corvol P, Llorens-Cortes C (1996) Identification of metabolic pathways of brain angiotensin II and III using specific aminopeptidase inhibitors: predominant role of angiotensin III in the control of vasopressin release. Proc Natl Acad Sci U S A 93:11968-11973

Angiotensin Receptor Signaling in the Brain: Ionic Currents and Neuronal Activity

C. Sumners[1] · E. M. Richards[2]

[1] Department of Physiology and Functional Genomics, College of Medicine, University of Florida, 1600 SW Archer Road, P.O. Box 100274, Gainesville, FL 32610, USA
e-mail: csumners@phys.med.ufl.edu

[2] Department of Pharmacodynamics, College of Pharmacy, University of Florida, Gainesville, FL 32610, USA

Abstract The brain contains both major subtypes of angiotensin receptors, the angiotensin type 1 (AT_1) and angiotensin type 2 (AT_2) receptors. This chapter begins with a review of the distribution of these receptor subtypes in the brain and continues with a discussion of their physiological and behavioral functions, from which it is clear that the AT_1 and AT_2 receptors mediate very different central actions of angiotensin II (Ang II). However, the primary focus of this chapter is a discussion of the mechanisms through which Ang II, acting via AT_1 and AT_2 receptors, can produce rapid alterations in neuronal activity and ultimately functional changes. Hence, we review the studies that have demonstrated electrophysiological actions of Ang II in the brain, and then focus on the specific intracellular signaling molecules that link the angiotensin receptor subtypes to changes in the activity of neuronal membrane ionic currents and firing rate. From this discussion it is clear that the intracellular signaling mechanisms that couple AT_1 and AT_2 receptors to changes in neuronal activity are vastly different. When taking a broader view of Ang II actions in neurons, it is evident that the signaling molecules responsible for regulation of neuronal activity represent only a few of the intracellular pathways that are modulated by this peptide.

Keywords Angiotensin · Angiotensin type 1 receptor · Angiotensin type 2 receptor · Intracellular signaling · Neuronal activity · Brain

1 Introduction

One of the most striking physiological actions elicited by a naturally occurring substance is the drinking behavior produced by brain application of angiotensin II (Ang II) (reviewed by Fitzsimons 1998). This observation, along with the demonstration that Ang II increases blood pressure and hormone release via actions in the CNS, has led to decades of research investigating the brain renin-angiotensin system (De Gasparo et al. 2000). A major focus of this research has been the localization, whole body physiological functions and cellular actions of specific Ang II receptors in the CNS, an area that was greatly expanded with the discovery of the angiotensin type 1 (AT_1) and angiotensin type 2 (AT_2) receptor subtypes (Chiu et al. 1989; Whitebread et al. 1989). This chapter will begin with a brief overview of the expression and known physiological and behavioral functions of Ang II receptor subtypes in the CNS. However, the primary focus will be the intracellular processes by which Ang II elicits *rapid* changes in neuronal activity, which are thought to result in behavioral or physiological functions. Specifically, we discuss intracellular signaling mechanisms that link the AT_1 and AT_2 receptors to changes in the activity of neuronal membrane ionic currents and firing rate. Lastly, we will introduce the fact that many of the Ang II receptor-induced signaling mechanisms in neurons are not related to changes in neuronal activity; rather, they control longer-term actions of this peptide in neurons such as genomic or post-translational effects. While it will be apparent that much progress has been made, we have barely scratched the surface in understanding the central actions of Ang II at the cellular level in neurons, and decades of work still remain.

2 Angiotensin Receptors in the Brain

2.1 The Presubtype Era

It has been known since the 1960s that Ang II acts within the brain to elicit a pressor response and drinking behavior. (Booth 1968; Fitzsimons and Simons 1969; Buckley 1972). The first demonstrations of specific receptors for Ang II in the brain came a decade later, when receptor binding studies revealed high levels of Ang II-specific binding in large brain areas such as the hypothalamus, thalamus, septum and midbrain (Sirett et al. 1977; Baxter et al. 1980). This distribution was consistent with the pressor and dipsogenic actions of Ang II, since these brain regions contain cardiovascular and thirst control centers. Later studies provided a more detailed picture of Ang II receptor distribution in the brain. For example, receptor autoradiography was used to localize Ang II-specific binding sites to circumventricular organs such as the subfornical organ, area postrema and organum vasculosum of the lamina terminalis, which can be influenced by circulating Ang II. The same technique was used to demonstrate high levels of Ang II receptors within intrinsic brain nuclei, which are not accessed by blood-borne Ang II, such as the paraventricular nucleus, locus ceruleus and nucleus of the solitary tract (Van Houten et al. 1980; Harding et al. 1981; Mendelsohn et al. 1984). Low levels of Ang II receptor binding were also demonstrated in areas that are not directly involved in fluid balance or blood pressure regulation, such as the striatum and limbic structures. These early studies measured total (AT_1 + AT_2) Ang II receptors and demonstrated widespread distribution. More recent studies indicate that this distribution pattern is due to mostly separate localization of AT_1 and AT_2 receptor subtypes, as discussed in the next section.

2.2 Subtypes and Distribution

As described in the previous section, the presence of specific Ang II receptors within the brain was established during the 1970s and 1980s. Furthermore, a large number of physiological studies had demonstrated that these were functional receptors, based upon the ability of Ang II to modulate neuronal electrophysiological properties, fluid intake and blood pressure (reviewed by Phillips and Sumners 1998). The occurrence in mammalian tissues of two major subtypes of Ang II receptor was based on differential affinities for pharmacological agents (Chiu et al. 1989; Whitebread et al. 1989; Timmermans et al. 1993). The AT_1 receptor subtype has a high affinity for a series of non-peptide antagonists that include losartan, valsartan, candesartan and irbesartan, whereas the AT_2 receptor specifically binds PD123,319 or the peptide CGP42112A (De Gasparo et al. 2000). Subsequent cloning studies and analysis of the protein sequence predicted that both AT_1 and AT_2 receptors would be G-protein coupled. While similar in size (359 amino acids for AT_1 vs 363 amino acids for AT_2 receptors) these receptors are only 32%–34% identical with respect to amino acid sequence (Murphy et al. 1991; Sasaki el at. 1991; Kambayashi et al. 1993; Mukoyama et al. 1993). It is also clear that rodents contain two highly homologous subtypes of the AT_1 receptor, termed AT_{1A} and AT_{1B} receptors (Iwai et al. 1992; Kakar et al. 1992). Based upon the results of radioligand binding, quantitative autoradiography, immunostaining and in situ hybridization approaches, there appears to be discrete and mostly exclusive localization of AT_1 and AT_2 receptors in rat brain (Gehlert et al. 1990; Rowe et al. 1990; Obermuller et al. 1991; Tsutsumi and

Saavedra 1991; Song et al. 1992; Lenkei et al. 1997; Von Buhlen und Halback and Albrecht 1998; Hu et al. 2002). Numerous studies have indicated that central AT_1 receptors mediate the cardiovascular and fluid balance effects of Ang II in the brain (reviewed by Phillips and Sumners 199;, De Gasparo et al. 2000). The distribution of AT_1 receptors in adult rat brain is consistent with these functions. High concentrations of these sites are localized in nuclei such as the median preoptic nucleus, paraventricular nucleus, rostral ventrolateral medulla, NTS and within circumventricular organs such as the subfornical organ, organum vasculosum of the lamina terminalis and area postrema. These areas are devoid of AT_2 receptors. It should also be noted that adult rats have AT_1 receptors in areas not directly linked to fluid balance or blood pressure control, e.g., several limbic structures (Lenkei et al. 1997; Von Bohlen und Halbach and Albrecht et al. 1998).

In contrast, in adult rats the highest concentrations of AT_2 receptors are localized in areas that are involved in sensory and motor functions, such as the inferior olive, various thalamic nuclei, and the subthalamic nucleus. The presence of AT_2 receptors in the amygdala may suggest a role in fluid intake (Johnson and Thunhorst 1997). There is very little overlap between AT_1 and AT_2 receptors in brain areas, but two notable areas of co-localization are the amygdala and the locus ceruleus (Speth et al. 1991; Lenkei et al. 1997). One striking observation concerning the distribution of brain Ang II receptors is that the AT_2 receptors are widespread in neonatal CNS, only to decline to a more restricted pattern of expression in adults (Tsutsumi and Saavedra 1991). In contrast, AT_1 receptor expression appears to develop later in neonatal life and persists at higher levels than AT_2 receptors in adult brain (Nuyt et al. 2001).

A number of investigators have determined that the functional effects of Ang II in the brain are dependent upon its modulation of other neuronal systems. More specifically, it is clear that the AT_1 receptor-mediated effects of Ang II on cardiovascular regulation and fluid balance may involve direct modulation of catecholamergic, glutamate, GABA and substance P-containing neurons (Stadler et al. 1992; Dampney et al. 1996; Diz et al. 1998; Zhu et al. 1998; Tanaka et al. 2001; Hu et al. 2002). Consistent with these observations are data which indicate that AT_1 receptors are co-localized with catecholaminergic neurons in the medulla oblongata (Yang et al. 1997; Hirooka et al. 1996; Hu et al. 2002) and locus ceruleus (Speth et al. 1991), and with glutamate and GABA neurons in the rostral ventrolateral medulla (Hu et al. 2002). Based on the regional distribution, functional and co-localization studies, the impression might be that AT_1 and AT_2 receptors are restricted entirely to neurons. However, numerous in vitro studies have demonstrated the presence of specific functional AT_1 receptors in astrocyte glia cultured from rat brain (Sumners et al. 1991; Tallant and Higson 1997; Gebke et al. 1998; Muscella et al. 2000). The recent observations that AT_1 receptors are present on astrocytes in white matter tracts in adult rat cerebellum and periventricular region (Fogarty and Matute 2002) and in corpus callosum of postnatal mice (Bernstein et al. 1996) may indicate that glial cells have an important role in Ang II actions in the CNS in vivo.

2.3
Physiological and Behavioral Functions of Brain Ang II Receptors

The most well-known physiological actions of Ang II that are mediated by its brain receptors are regulatory effects on fluid balance and cardiovascular function. These actions include stimulatory effects of Ang II on water and sodium intake, vasopressin secretion, blood pressure, and sympathetic outflow, and modulation of baroreflex sensitivity (Fitzsimons 1998; Saavedra 1999; Averill and Diz 2002, McKinley et al. 2001; Zucker 2002). The majority of evidence indicates that these effects of Ang II are mediated by AT_1 receptors within specific hypothalamic and brainstem nuclei (Phillips and Sumners 1998; De Gasparo et al. 2002). AT_1 receptors in the subfornical organ, organum vasculosum of the lamina terminalis and median preoptic nucleus have a major role in the dipsogenic response, while AT_1 receptors in the paraventricular nucleus, rostral ventrolateral medulla and nucleus of the solitary tract are involved in the cardiovascular actions of this peptide. For example, injection of Ang II into the rostral ventrolateral medulla elicits an AT_1 receptor-mediated increase in sympathetic activity and blood pressure (Dampney et al. 2002). Studies from Davisson et al. (2000) using knockout mice indicate that the Ang II-induced drinking response requires AT_{1B} receptors while the pressor actions of centrally-injected Ang II is mediated by AT_{1A} receptors. While it is clear that AT_1 receptors have a primary role in the dipsogenic action of Ang II via the brain, a number of studies suggest that brain AT_2 receptors are important in mediating responses to thirst stimuli such as water deprivation and hypovolemia (Rowland and Fregly 1993; Hein et al. 1995; Lee et al. 1996). This implies that central AT_2 receptors may be a component of the final common neural pathway for drinking. Aside from the centrally mediated actions of Ang II on fluid homeostasis and cardiovascular control, a number of more recent studies indicate that this peptide has a role in the control of learning and memory and certain behaviors (for review see Gard 2002). For example, Ang II has a role in exploratory behavior (Ichiki et al. 1995), increases the acquisition of conditioned avoidance behavior (Braszko 2002) and inhibits sexual behavior in male rats (Breigeron et al. 2002). Thus, these studies are beginning to identify functions for the Ang II receptors that are localized outside the hypothalamus and brainstem regions.

In summary, Ang II has a number of important physiological effects that are controlled by specific receptor-mediated activation of neuronal pathways in the brain. The cellular bases of these physiological effects are modulatory actions on neuronal activity and action potentials (APs). In order to gain a complete picture of Ang II actions in the brain, it is important to understand the intracellular mechanisms by which Ang II alters neuronal activity. Thus, the remainder of this chapter will focus on the receptor-mediated actions of Ang II on neuronal activity, and the intracellular signaling mechanisms that underlie these effects.

3
Angiotensin II Receptors and Neuronal Activity

3.1
Electrophysiological Effects of Angiotensin II in the Brain

As discussed in the previous sections, the binding of Ang II to neuronal receptors located within specific nuclei of the hypothalamus and brainstem evokes changes in the activity of neuronal pathways and ultimately leads to physiological and behavioral alterations. These rapid changes in neuronal activity or firing rate occur through alterations in AP frequency and/or duration, which in turn depend on the activity of membrane ionic currents and their underlying channels. Thus, when considering the functional effects of Ang II in the brain, it is clearly important to understand the effects of Ang II on the electrophysiological properties of neurons. For this reason, a number of investigations have examined the actions of Ang II on neuronal activity and membrane ionic currents. These studies on the effects of Ang II have been performed in situ, in brain slices, in freshly isolated neurons, and in neurons in culture, and have mostly been performed in the brainstem and hypothalamus. Various electrophysiological recording techniques, including extracellular, voltage and current clamp recording procedures, have been utilized. In studies performed prior to the discovery of Ang II receptor subtypes, iontophoretic procedures were used to apply Ang II to brain neurons in anesthetized rats in situ, in the subfornical organ, paraventricular nucleus, septum and medulla. In each case, Ang II produced an increase in neuronal activity through an increase in firing rate, which was inhibited by non-subtype selective peptidergic Ang II receptor blockers (Felix and Schlegel 1978; Suga et al. 1979; Harding and Felix 1987; Chan et al. 1991). Iontophoretic application of Ang II onto the organum vasculosum of the lamina terminalis (contained within brain slices) resulted in increases in firing rate that were abolished by a nonselective peptidergic Ang II receptor antagonist (Knowles and Phillips 1980). Thus, these excitatory actions of Ang II in these brain regions are consistent with the dipsogenic and cardiovascular actions of this peptide.

The discovery of distinct Ang II receptor subtypes that display different molecular properties and physiological functions has precipitated many studies on the electrophysiological properties of these sites in the brain. For example, in situ recordings from anesthetized rats have revealed that iontophoretic injection of Ang II into the paraventricular nucleus (Ambuhl et al. 1992a) or amygdala (Albrecht et al. 2000) produces neuronal excitation, effects mediated by AT_1 receptors. Similar excitatory effects of Ang II, also mediated by AT_1 receptors, have been recorded from subfornical organ, paraventricular nucleus, median preoptic nucleus, and rostral ventrolateral medulla neurons contained in brain slices (Li and Ferguson 1993; Li and Guyenet 1995, 1996; Bai and Renaud 1998; Ono et al. 2001; Dewald et al. 2002). Once again these electrophysiological actions of Ang II via AT_1 receptors are consistent with its dipsogenic and cardiovascular actions at these sites.

A number of studies have begun to investigate the effects of Ang II on membrane ionic currents that underlie the observed neuronal excitation via AT_1 receptors. While it is clear that neurons contain multiple types of membrane ionic currents and channels, only certain of these are modulated by Ang II. For example, our whole cell voltage clamp recordings from neurons cultured from newborn rat hypothalamus and brainstem have demonstrated that Ang II elicits AT_1 receptor-mediated changes in particular K^+ and Ca^{2+} currents (Sumners et al. 2002). Specifically, Ang II produces an increase in total Ca^{2+} current (I_{Ca}) and a decrease in total K^+ current. These effects are due to respective increases in N-type Ca^{2+} current and decreases in both delayed rectifier K^+ current (I_{Kv}) and A-type K^+ current (I_A) (Sumners et al. 1996; Wang et al. 1997a; Evans et al. 2001). These changes in K^+ and Ca^{2+} currents are consistent with our observation that Ang II produces a positive chronotropic effect in the same cultured neurons (Wang et al. 1997b). The mechanisms by which the above changes in I_{Kv}, I_A and in N-type Ca^{2+} current lead to increases in firing rate include several effects, all of which increase the probability for AP generation. To date, the identified effects are increases in the number and amplitude of subthreshold oscillations in membrane potential, decreases in the amount of current needed to generate an AP, and increases in the amplitude and duration of early after depolarizations and the APs generated by EADs (Wang et al. 1997b).

Studies from adult rat brain are mostly consistent with these neonatal cell culture findings. For instance, inhibitory effects of Ang II on I_A have also been observed in neurons from the supraoptic nucleus (SON), subfornical organ and paraventricular nucleus magnocellular area contained in brain slices (Nagatomo et al. 1995; Ferguson and LI 1996; Li and Ferguson 1996). More recently, Washburn and Ferguson (2001) have described an AT_1 receptor-mediated stimulatory action of Ang II on N-type-, but not on L-type, Ca^{2+} current in subfornical organ neurons isolated from adult rat. One study has attempted to link the drinking response produced by central (i.c.v.) injection of Ang II in rats to activation of Ca^{2+} channels. In that study, Zhu and Herbert (1997) demonstrated that the Ang II dipsogenic response was reduced by pretreatment with an L-type Ca^{2+}-channel blocker but not by an N-type channel blocker. The reasons for the discrepancy between this data and the electrophysiological results of Washburn and Ferguson (2001) are likely to be multiple, but could include the fact that i.c.v. injected Ang II will spread to many brain nuclei, not simply the subfornical organ. Therefore, it is clear that progress has been made in defining the electrophysiological actions of Ang II, via AT_1 receptors, in the brain. Nonetheless, more work is needed to define the exact K^+ and Ca^{2+} currents that are modulated by Ang II in different brain regions and to define the underlying types of channel involved.

Investigations that have attempted to define the electrophysiological actions of Ang II in the brain, via AT_2 receptors have been relatively few in number compared with the AT_1 receptor studies. Nonetheless, it has been determined that AT_2 receptors within the brain are functional, at least from an electrophysiological standpoint. The studies of Ambuhl et al. (1992b) demonstrated that iontophoresis of Ang II onto the inferior olive of anesthetized rats produced neuronal excitation, mediated by AT_2 receptors. Using a similar approach, Albrecht et al. (2000) determined that Ang II elicits an AT_2 receptor-mediated increase in the firing rate of neurons in the amygdala. Analysis of the changes in membrane ionic currents that underlie these chronotropic effects has only been performed in neural cells in culture. Whole cell voltage clamp recordings from neonatal hypothalamus and brainstem neuronal cultures have revealed that Ang II elicits AT_2 receptor-mediated increases in both I_A and I_{Kv} (Kang et al. 1993, 1994). Current clamp analyses in similar cultures revealed that Ang II produced an increase in firing rate, due to a decrease in AP duration (Zhu et al. 2001). The increases in I_A and I_{Kv} may produce the positive chronotropic effect by increasing the rate of repolarization and shortening the AP refractory period. Another study, utilizing NG108–15 neuroblastoma cells, indicated that Ang II decreases T-type Ca^{2+} current via AT_2 receptors (Buisson et al. 1992).

Thus it is clear that Ang II, acting via AT_1 and AT_2 receptors, elicits discrete electrophysiological effects on central neurons. To date, the results indicate that activation of each receptor subtype produces an increase in neuronal activity, albeit via distinct effects on membrane ionic currents. In the following sections we will discuss the intracellular signaling mechanisms that are responsible for AT_1 and AT_2 receptor-mediated changes in neuronal membrane ionic currents and activity. The major focus of this discussion will be studies performed on neurons cultured from newborn rat hypothalamus and brainstem, since most research performed in this area has been done using this in vitro model. These cultures contain AT_1 and AT_2 receptors and a wide array of intracellular signaling molecules and ion channels. While neuronal cultures do not represent the intact adult rat brain, they do provide a more simplified system in which complex signal transduction mechanisms can be evaluated more easily.

3.2
AT_1 Receptor-Mediated Changes in Neuronal Activity: Intracellular Signaling Mechanisms

The modulation of membrane ionic currents and their underlying channels by G-protein-coupled receptors such as the AT_1 receptor can occur through two general mechanisms: direct (membrane-delimited) coupling of a G-protein subunit to the ion channel proteins or indirect modulation via generation of second messengers and activation of protein kinases or phosphatases (Levitan 1999). Studies from neuronal cultures indicate that the latter indirect mechanism is of primary importance in the AT_1 modulation of neuronal ionic currents and activity. Similar to its effects in rat brain (Seltzer et al. 1995), Ang II acts at AT_1 receptors in neuronal cultures to stimulate phosphoinositide (PI) hydrolysis (Sumners et al. 1991). This stimulation of PI hydrolysis results in generation of diacylglycerol (DAG) and inositol 1, 4, 5-triphosphate (IP_3), which is followed by increases in intracellular Ca^{2+} ($[Ca^{2+}]_{int}$), protein kinase (PKC) and calcium/calmodulin kinase II (CaMKII) activities (Sumners et al. 1996; Zhu et al. 1999). Both PKC and CaMKII have key roles in the AT_1 receptor-mediated changes in K^+ and Ca^{2+} currents in neurons. For example, the Ang II-induced stimulation of neuronal I_{Ca} is entirely abolished by inhibition of PKC (Sumners et al. 1996). In fact it appears that PKC acts to inhibit a negative influence of G subunits on in N-type Ca^{2+} current (Evans et al. 2001). It is also clear that the inhibitory effects of Ang II on neuronal I_A involve activation of PKC (Wang et al. 1997b). The inhibition of neuronal I_{Kv} occurs via a more complex series of intracellular signaling events. As with the modulation of I_{Ca} and I_A, the inhibition of I_{Kv} by Ang II involves G-protein-mediated activation of phospholipase C (PLC) and PI hydrolysis, since it is abolished by intracellular application of anti-$G_{q/11\alpha}$ antibodies or by the selective PLC inhibitor U73122 (Sumners et al. 1996; Pan et al. 2001). Furthermore, this Ang II effect is mimicked by intracellular application of a peptide corresponding to the third intracellular (G-protein interacting) loop of the AT_1 receptor (Zhu et al. 1997). Intracellular application of IP_3, DAG analogs or activated CaMKII (Sumners et al 1996; Zhu et al. 1999; Pan et al. 2001), as well as superfusion of PKC activating phorbol esters (Sumners et al. 1996), all cause decreases in I_{Kv}. Thus, unlike the effects of Ang II on I_{Ca} and I_A, these data suggest an involvement of both PKC and CaMKII in the reduction of I_{Kv} by Ang II. This was confirmed with the use of selective inhibitors for these kinases. For example, the PKC inhibitors calphostin C and PKC inhibitory peptide (PKCIP) partially inhibit this Ang II effect on I_{Kv} (Sumners et al. 1996; Zhu et al 1999). Similar attenuation is obtained with the use of a calmodulin antagonist (W-7) or by the CaMKII inhibitors KN-93 and CaMKII (281–302) (Sumners et al. 1996, Zhu et al. 1999). The fact that this Ang II response is abolished by chelation of $[Ca^{2+}]_{int}$ (Sumners et al. 1996) suggests that a Ca^{2+}-dependent PKC isozyme is involved. Our recent studies indicate that the Ca^{2+}-dependent PKC-α is of primary importance, but do not exclude the possibility that another Ca^{2+}-dependent PKC isozyme is involved (Pan et al. 2001; Sumners et al. 2002). Since changes in membrane ionic currents result in alteration in AP firing rate and neuronal activity, it is reasonable to suggest that AT_1 receptor-mediated chronotropic effect of Ang II in neurons also involves activation of PKC and CaMKII, and experiments have shown that this is indeed the case. Superfusion of neuronal cultures with Ang II produces an AT_1 receptor-dependent increase in firing rate that is attenuated by either calphostin C or by KN-93, and which is abolished by combined application of both inhibitors (Sun et al. 2002). The intracellular signaling pathways by which Ang II, acting via AT_1 receptors, modulates I_{Kv}, I_A, I_{Ca} and neuronal activity are summarized in Fig. 1.

These studies have raised a number of interesting questions. For example, why does the inhibitory effect of Ang II on I_{Kv} involve a dual regulation via PKC-α and CaMKII? One possible answer is that each kinase provides a unique regulatory influence on I_{Kv}, each of which leads to differential physiological functions. These influences may take the form of distinct actions by each kinase on the biophysical properties of the Kv2.2 channel protein, which has been

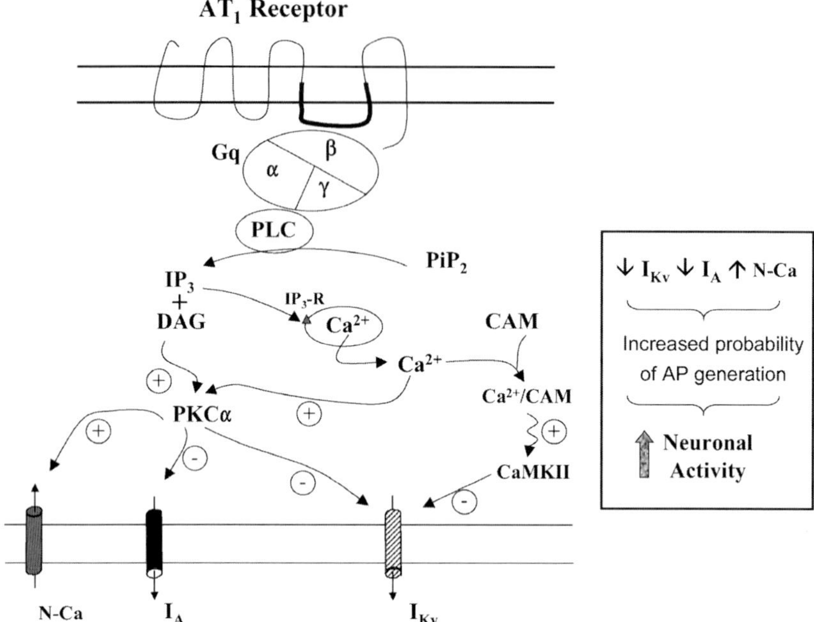

Fig. 1 Summary of the intracellular signaling pathways that couple AT_1 receptor to changes in membrane ionic currents and neuronal activity. *PLC*, phospholipase C; *PiP_2*, phosphatidylinositol 4, 5-bisphosphate; *IP_3*, inositol 1, 4, 5-trisphosphate; *DAG*, diacylglycerol; *IP_3-R*, IP_3 receptor; *PKCα*, protein kinase Cα; *CAM*, calmodulin; *CaMKII*, calcium/calmodulin-dependent kinase II; *N-Ca*, N-type Ca^{2+} channel. Signaling pathways were determined as described in the text, and the localization of PKCα and CaMKII with AT_1 receptors in responsive neurons was established through single cell RT-PCR. (Zhu et al. 1999)

shown to be involved in the reduction of neuronal I_{Kv} by Ang II (Gelband et al. 1999). A further implication from this data is that factors that regulate the activity of PKC-α or CaMKII independent of Ang II will influence the AT_1 receptor-mediated inhibition of neuronal I_{Kv}. Therefore, a factor that, for example, regulates PKC or CaMKII activity alone may blunt or amplify this action of Ang II, allowing for cross-talk and interaction with other neurotransmitter signaling pathways.

Another question raised by these studies is the nature of the molecular mechanisms by which Ang II, via PKC and CaMKII, regulates the activity of K^+ and Ca^{2+} channel proteins and thus K^+ and Ca^{2+} currents. Various mechanisms have been demonstrated for the regulation of channel proteins. These include, but are probably not limited to, phosphorylation/dephosphorylation of channel subunits (Chandy and Gutman 1995; Levitan 1999), conformational changes in channel proteins produced by phosphorylation of cytoskeletal elements (Jing et al. 1997; Peretz et al. 1996; Levin et al. 1996), or regulation mediated via interactions of G-protein subunits, kinases and channel proteins within lipid rafts

(Martens et al. 2000). None of these known mechanisms can be excluded with respect to Ang II-induced modulation of neuronal I_{Kv}, I_A or I_{Ca}.

A third question raised by these studies is whether the increases in neuronal activity produced by Ang II are directly linked to increased release of neurotransmitters such as norepinephrine or glutamate. This question is being approached in our laboratory through the combined use of electrophysiological and amperometric recording procedures.

In summary, the basic intracellular mechanisms through which Ang II produces an AT_1 receptor-mediated increase in neuronal activity have been defined in cell culture experiments. These mechanisms employ the activation of Ca^{2+}-dependent signaling molecules, providing a *rapid* means for increasing neuronal firing rate. A major question that remains is whether these signaling molecules are responsible for the AT_1 receptor-mediated action of Ang II in the rat brain in situ. Early indications are that this is indeed the case, since the dipsogenic effect of centrally administered Ang II appears to be dependent upon PKC and CaMKII activation (M.A. Fleegal and C. Sumners, unpublished data). Studies that produce selective inhibition of PKC and CaMKII in specific brain areas such as the subfornical organ and paraventricular nucleus will allow determination of whether these signaling molecules mediate the physiological actions of Ang II through its brain AT_1 receptors.

3.3
AT_2 Receptor-Mediated Changes in Neuronal Activity: Intracellular Signaling Mechanisms

From the previous section, it is apparent that AT_1 and AT_2 receptors differ in both molecular and pharmacological properties and mediate different physiological actions of Ang II. These differences extend to the cellular level in neurons, where Ang II elicits opposite effects on I_{Kv} and I_A via AT_1 and AT_2 receptors (Kang et al. 1993; Sumners et al. 1996). Based on this, it is reasonable to expect that the intracellular messengers that couple AT_1 and AT_2 receptors to cellular functions in neurons are different, and that has been shown to be the case. The stimulatory effect of Ang II via AT_2 receptors on I_{Kv} in neuronal cultures does not involve PI hydrolysis, generation of IP_3 or activation of Ca^{2+}-dependent PKC (Kang et al. 1994). This is consistent with the fact that AT_2 receptors do not influence PI hydrolysis nor PKC activity in neurons (Sumners et al. 1991, 1996) nor in other tissues or cells (reviewed by Gallinat et al. 2000; Nouet and Nahmias 2000; Stoll and Unger 2001). Various studies from neural cells in culture have shown that activation of AT_2 receptors modulates cyclic GMP and phosphodiesterase levels, and increases phosphotyrosine phosphatase activity (Sumners and Myers 1991; Bottari et al. 1992; Nahmias et al. 1995; Yamada et al. 1996). However, it was clear from our experiments that none of these messengers, nor cyclic AMP, are involved in the Ang II-induced increase in I_{Kv} (Kang et al. 1994). Rather, this Ang II effect involves a G-protein-coupled pathway since the increase in neuronal I_{Kv} produced by Ang II is abolished by pertussis toxin or by intracellular application of anti Giα antibodies (Kang et al. 1994). Further, this Ang II effect is mimicked by intracellular application of a peptide corresponding to the third intracellular (G-protein interacting) loop of the AT_2 receptor (Kang et al. 1995). This suggests that the AT_2 receptor-mediated increase in I_{Kv} in cultured neonatal neurons occurs via Gi, and is consistent with studies from rat fetus, which indicate that AT_2 receptors co-precipitate with Giα_2 and Giα_3 proteins (Zhang and Pratt 1996). It is also clear that the inhibition of T-type Ca^{2+} current in a neural cell line, the NG108–15 neuroblastoma cells, is G-protein-mediated (Buisson et al. 1995). However, the signaling pathway that mediates the Ang II stimulation of I_{Kv}, downstream of Gi, appears to be quite different from the calcium-dependent mechanisms that are responsible for the AT_1 receptor-mediated decrease in I_{Kv}. We had determined that activation of AT_2 receptors in neuronal cultures causes a rapid stimulation of phospholipase A_2 (PLA_2) activity, with subsequent generation of arachidonic acid (AA). These effects were abolished by pertussis toxin (Zhu et al. 1998). Considering that AA and metabolites of AA are known modulators of K^+ currents in neurons (Meves 1994; Kim et al. 1995), we examined the role of these factors in the AT_2 receptor-mediates stimulation of I_{Kv}. Our data indicated that this Ang II effect was mimicked by AA and by the 12-lipoxygenase (12-LO) metabolite of AA, 12S-HETE (Zhu et al. 1998). Furthermore, the stimulatory effects of Ang II, AA and 12S-HETE on neuronal I_{Kv} were attenuated but not abolished by selective inhibition of 12-LO, but not by inhibition of 5-lipoxygenase, cyclooxygenase or p450 monooxygenase enzymes (Zhu et al. 1998, 2000). These data indicate that a signaling pathway that includes Gi activation of PLA_2, generation of AA and subsequent metabolism of AA to 12-HETE is at least partially responsible for the Ang II stimulation of neuronal I_{Kv}. This pathway is probably also involved in the AT_2 receptor-mediated increase in neuronal I_A, since the recording procedures that are used to analyze I_{Kv} also measure I_A, and this transient K^+ current exhibited many of the same changes in response to the various treatments. However, there is an increased level of complexity to this signaling mechanism, since our earlier experiments in this area indicated that the stimulation of I_{Kv} and I_A by Ang II is completely abolished by selective inhibition of serine/threonine phosphatase type 2A (PP-2A) (Kang et al. 1994). Inhibition of PP-2A also abolished the stimulation of these K^+ currents by AA or by 12-LO, but not the Ang II-induced increase in AA generation (Zhu et al. 1998). Finally, it is clear that the chronotropic action of Ang II via AT_2 receptors involves activation of 12-LO and PP-2A (Zhu et al. 2001). Collectively, these data indicate that PP-2A is downstream of AA and 12S-HETE, and is perhaps a final common event in the signaling pathways through which Ang II modulates neuronal K^+ currents and firing rate.

This signaling pathway is summarized in Fig. 2 but, as with AT_1 receptor-mediated signaling events, there are still many unanswered questions. For example, if the PLA_2/AA/12S-HETE pathway is only partially responsible for the Ang II stimulation of I_{Kv} and I_A, which other mechanisms are involved? One possibility is membrane delimited coupling of Gi to the K^+ channel proteins involved in

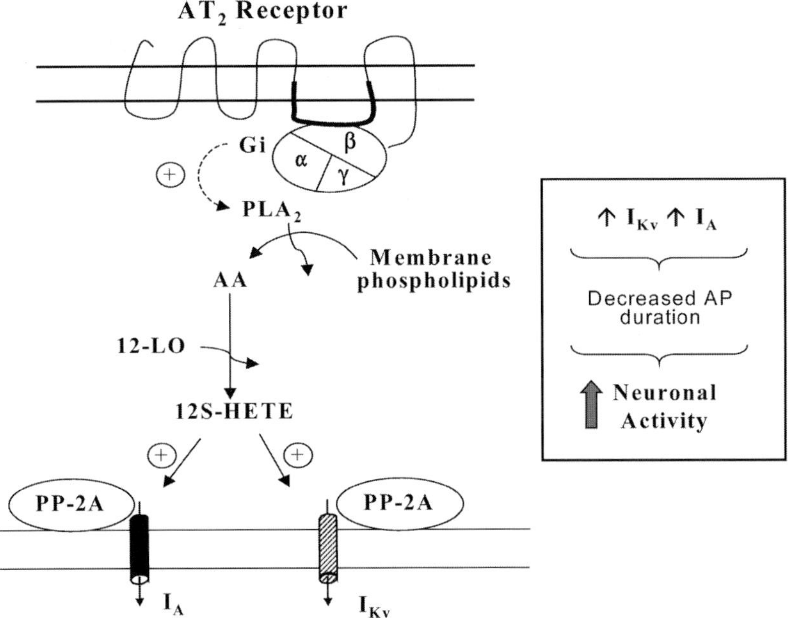

Fig. 2 Summary of the intracellular signaling pathways that couple AT_2 receptors to change in membrane ionic currents and neuronal activity. *PLA_2*, phospholipase A_2; *AA*, arachidonic acid; *12-LO*, 12-lipoxygenase; *12S-HETE*, 12-(s)-hydroxy-(5Z, 8Z, 10E, 14Z)-eicosatetraenoic acid; *PP-2A*, serine/threonine phosphatase 2A. Signaling pathways were determined as detailed in the text, and the localization of PLA_2 and 12-LO with AT_2 receptors in responsive neurons was established through single-cell RT-PCR. (Zhu et al. 2000)

this response, but there is no evidence to support this speculation as yet. In this scenario, PP-2A would still act as the final common event in channel regulation. A further question concerns the subtypes of K^+ channel protein that are involved in this Ang II action. Our preliminary pharmacological data indicate that the Kv2.2 channel may be involved in the regulation of I_{Kv} (M. Zhu and C. Sumners, unpublished data), similar to its role in the AT_1 receptor modulation of I_{Kv}. Once the identity of these channel proteins is clearly established, it may be possible to decipher the mechanisms by which 12S-HETE and PP-2A effect channel regulation. On another note, the multiplicity of signaling events that are involved in the AT_2 receptor modulation of neuronal activity again suggests that various factors may be able to regulate or temper this response, as postulated for the AT_1 receptor signaling pathways. Perhaps the most important questions relate to the functional significance of the observed AT_2 receptor-mediated changes in K^+ currents and neuronal activity. In particular, whether they translate to specific neuromodulatory effects (on catecholaminergic neurons, for example) and involve the same intracellular signaling events in situ. These questions will not be answered until the physiological functions of AT_2 receptors within the brain are better understood.

4
The Broader Perspective of Angiotensin II Receptor Signaling in Neurons

The main focus of this chapter has been the intracellular mechanisms that are responsible for the *rapid* changes in neuronal activity produced by Ang II. It is clear that the AT_1 and AT_2 receptor-mediated changes in neuronal firing occur via activation of very specific and unique signaling molecules. It is also apparent that each receptor subtype, upon stimulation by Ang II, activates more than one intracellular messenger pathway in order to alter neuronal firing. When taking a broader view of Ang II actions in neurons, it is evident that the signaling molecules responsible for regulation of neuronal activity represent only a few of the intracellular pathways that are modulated by this peptide. Rather, Ang II activation of AT_1 or AT_2 receptors in neurons results in the modulation of a wide variety of intracellular signaling molecules, many of which support the long-term actions of this peptide. A good example is the AT_1 receptor-mediated activation of Erk mitogen-activated protein (MAP) kinases (Huang et al. 1996; Yang et al. 1996) in neurons. This activation of Erk occurs via a Ras/Raf/MEK (MAP kinase kinase) pathway, and results in stimulation of c-fos and c-jun and synthesis of their respective protein products Fos and Jun, which combine to form the activator protein 1 (AP-1) transcription factor (Yang et al. 1996). Further examples are the Ang II-induced activation of Fos regulating kinase and Jun kinase, which elicit respective phosphorylation of Fos and Jun proteins, resulting in transactivation of the AP-1 complex (Huang et al. 1998). The time course of the stimulation and transactivation of Fos and Jun proteins is well beyond that required for the rapid electrophysiological changes (milliseconds to minutes) that produce alterations in firing rate. Rather, these signaling events are geared towards the AP-1-directed synthesis of tyrosine hydroxylase and dopamine β-hydroxylase in catecholaminergic neurons (Yang et al. 1996; Lu et al. 1996). The above are examples of the longer-term signaling pathways that couple to neuronal AT_1 receptors; a more extensive picture of these mechanisms is reviewed elsewhere (Richards et al. 1999; Sumners et al. 2002). Similar to the AT_1 receptors, neuronal AT_2 receptors also couple to a diverse array of intracellular signaling molecules, many of which are not involved in the rapid changes in neuronal activity. Prime examples are the activation/induction of various tyrosine and serine/threonine phosphatases (Huang et al. 1995; Bedecs et al. 1997; Horiuchi et al. 1997), which in turn inhibit the activity of Erk MAP kinases, a contributing event to the AT_2 receptor-mediated apoptosis of neural cells (Yamada et al. 1996). Finally, while it appears that the rapid and the longer-term actions of Ang II in neurons are served by mostly separate intracellular signaling events, it is likely that the earlier actions of Ang II can influence the later effects. For example, the changes in $[Ca^{2+}]_{int}$ that result from Ca^{2+} influx may lead to changes in the activity of enzymes that ultimately influence transcription factors.

5 Conclusions

It is evident that much progress has been made in understanding how Ang II affects ionic currents and neuronal activity at the cellular level. The use of brain tissue and cells derived from rodents has allowed the identification of specific AT_1 and AT_2 receptor-mediated electrophysiological actions in neurons, as well as the intracellular signaling events that control these actions. Despite this progress, many questions remain regarding the cellular and intracellular actions of Ang II in neurons, as set out in the above sections. Probably the most important questions, however, are as follows: (1) Do the discoveries made in rodent cells and tissues represent a physiological action of Ang II common to all species? (2) What is the relationship of these cellular events with physiological actions of Ang II. For example: Does Ang II, acting via its AT_1 receptors in a particular brain region, activate a cascade of intracellular signaling events that alters the activity of a specific neuronal pathway, and ultimately stimulate a physiological or behavioral change? Do the divergent signaling mechanisms that are activated by AT_1 receptors (PKC, CaMKII) interact with those activated by other transmitters such as vasopressin in the regulation of a specific physiological function? Or do they allow a particular neuron to serve multiple functions? The advent of genomic technologies that will allow the selective expression or inhibition of signaling molecules and channel proteins within neurons in specific brain regions may allow these questions to be addressed.

References

Albrecht D, Nitschke T, Von Bohlen Und Halbach O (2000) Various effects of angiotensin II on amygdaloid neuronal activity in normotensive control and hypertensive transgenic rats. FASEB J 7:925–931

Ambühl P, Felix D, Imboden H, Khosla MC, Ferrario CM (1992a) Effects of angiotensin analogues and angiotensin receptor antagonists on paraventricular neurones. Regul Pept 38:111–120

Ambühl P, Felix D, Imboden H, Khosla MC, Ferrario CM (1992b) Effects of angiotensin II and its selective antagonists on inferior olivary neurones. Regul Pept 41:19–26

Averill DB, Diz DI (2000) Angiotensin peptides and baroreflex control of sympathetic outflow: pathway and mechanisms of the medulla oblongata. Brain Res Bull 51:119–128

Bai D, Renaud LP (1998) ANG II AT1 receptors induce depolarization and inward current in rat median preoptic neurons in vitro. Am J Physiol 275:R632–R639

Baxter C, Horvath J, Duggin G, Tiller D (1979) Effect of age on angiotensin II receptors from rat brain. Clin Sci (Lond) 5:111s–1113s

Bedecs K, Elbaz N, Sutren M, Masson M, Susini C, Strosberg AD, Nahmias C (1997) Angiotensin II type 2 receptors mediate inhibition of mitogen-activated protein kinase cascade and functional activation of SHP-1 tyrosine phosphatase. Biochem J 325 (Pt 2):449–454

Bernstein M, Lyons SA, Moller T, Kettenmann H (1996) Receptor-mediated calcium signaling in glial cells from mouse corpus callosum slices. J Neurosci Res 46:152–163

Booth DA (1968) Mechanism of action of norepinephrine in eliciting an eating response on injection into the rat hypothalamus. J Pharmacol Exp Ther 160:336–348

Bottari SP, King IN, Reichlin S, Dahlstroem I, Lydon N, de Gasparo M (1992) The angiotensin AT2 receptor stimulates protein tyrosine phosphatase activity and mediates inhibition of particulate guanylate cyclase. Biochem Biophys Res Commun 183:206–211

Braszko JJ (2002) AT(2) but not AT(1) receptor antagonism abolishes angiotensin II increase of the acquisition of conditioned avoidance responses in rats. Behav Brain Res 131:79–86

Breigeiron MK, Morris M, Lucion AB, Sanvitto GL (2002) Effects of angiotensin II microinjected into medial amygdala on male sexual behavior in rat. Horm Behav 41:267–274

Buckley JP (1972) Actions of angiotensin on the central nervous system. Fed Proc 4:1332–1337

Buisson B, Bottari SP, de Gasparo M, Gallo-Payet N, Payet MD (1992) The angiotensin AT2 receptor modulates T-type calcium current in non-differentiated NG108–15 cells. FEBS Lett 309:161–164

Buisson B, Laflamme L, Bottari SP, de Gasparo M, Gallo Payet N, Payet MD (1995) A G protein is involved in the angiotensin AT2 receptor inhibition of the T-type calcium currents in non-differentiated NG108–15 cells. J Biol Chem 270:1670–1674

Chan RK, Chan YS, Wong TM (1991) Responses of cardiovascular neurons in the rostral ventrolateral medulla of the normotensive Wistar Kyoto and spontaneously hypertensive rats to iontophoretic application of angiotensin II. Brain Res 556:145–150

Chandy KB, Gutman GA (1995) Voltage-gated K^+ channel genes. In: North RA (ed) Ligand and voltage-gated ion channels. CRC, Boca Raton, pp 1–71

Chiu AT, Herblin WF, McCall DE, Ardecky RJ, Carini DJ, Duncia JV, Pease LJ, Wong PC, Wexler RR, Johnsin Al et al (1989) Identification of angiotensin II receptor subtypes. Biochem Biophys Res Commun 165:196–203

Dampney RA, Hirooka Y, Potts PD, Head GA (1996) Functions of angiotensin peptides in the rostral ventrolateral medulla. Clin Exp Pharmacol Physiol Suppl 3:S105–111

Dampney RA, Fontes MA, Hirooka Y, Horiuchi J, Potts PD, Tagawa T (2002) Role of angiotensin II receptors in the regulation of vasomotor neurons in the ventrolateral medulla. Clin Exp Pharmacol Physiol 5–6:467–472

Davisson RL, Oliverio MI, Coffman TM, Sigmund CD (2000) Divergent functions of angiotensin II receptor isoforms in the brain. J Clin Invest 106:103–106

De Gasparo M, Catt KJ, Inagami T, Wright JW, Unger T (2000) International union of pharmacology. XXIII. The angiotensin II receptors. Pharmacol Rev 52:415–472

Dewald M, Braun HA, Huber MT, Zwingmann D, Roth J, Voigt K (2002) Interactions of temperature and angiotensin II in paraventricular neurons of rats in vitro. Pflugers Arch 444:117–125

Diz DI, Westwood B, Bosch SM, Ganten D, Ferrario C (1998) NK1 receptor antagonist blocks angiotensin II responses in renin transgenic rat medulla oblongata. Hypertension 31:473–479

Evans J, Sumners C, Gelband C (2001) Relief of Gaa inhibition may underlie angiotensin II stimulation of neuronal K^+ current (abstract). Biophys J 79:548

Felix D, Schlegel W (1978) Angiotensin receptive neurones in the subfornical organ. Structure-activity relations. Brain Res 149:107–116

Ferguson AV, Li Z (1996) Whole cell patch recordings from forebrain slices demonstrate angiotensin II inhibits potassium currents in subfornical organ neurons. Regul Pept 66:55–58

Fitzsimons JT (1998) Angiotensin, thirst, and sodium appetite. Physiol Rev 78:583–686

Fitzsimons JT, Simons BJ (1969) The effect on drinking in the rat of intravenous infusion of angiotensin, given alone or in combination with other stimuli of thirst. J Physiol 203:45–57

Fleegal MA, Sumners C (2003) Drinking behavior elicited by central injection of angiotensin: roles for proteinkinase C and Ca^{2+}/calmodulin-dependent protein kinase II. Am J Physiol 258:R632–640

Fogarty DJ, Matute C (2002) Angiotensin receptor-like immunoreactivity in adult brain white matter astrocytes and oligodendrocytes. Glia 35:131–146

Gallinat S, Busche S, Raizada MK, Sumners C (2000) The angiotensin II type 2 receptor: an enigma with multiple variations. Am J Physiol Endocrinol Metab 278:E357–E374

Gard PR (2002) The role of angiotensin II in cognition and behavior. Eur J Pharmacol 438:1–14

Gebke E, Muller AR, Jurzak M, Gerstberger R (1998) Angiotensin II-induced calcium signalling in neurons and astrocytes of rat circumventricular organs. Neuroscience 85:509–520

Gehlert DR, Gackenheimer SL, Reel JK, Lin HS, Steinberg MI (1990) Non-peptide angiotensin II receptor antagonists discriminate subtypes of 125I-angiotensin II binding sites in the rat brain. Eur J Pharmacol 187:123–126

Gelband CH, Warth JD, Mason HS, Zhu M, Moore JM, Kenyon JL, Horowitz B, Sumners C (1999) Angiotensin II type 1 receptor-mediated inhibition of K+ channel subunit kv2.2 in brain stem and hypothalamic neurons. Circ Res 84:352–359

Harding JW, Felix D (1987) Angiotensin-sensitive neurons in the rat paraventricular nucleus: relative potencies of angiotensin II and angiotensin III. Brain Res 410:130–134

Harding JW, Stone LP, Wright JW (1981) The distribution of angiotensin II binding sites in rodent brain. Brain Res 205:265–274

Hein L, Barsh GS, Pratt RE, Dzau VJ, Kobilka BK (1995) Behavioral and cardiovascular effects of disrupting the angiotensin II type-2 receptor in mice. Nature 377:744–747

Hirooka Y, Head GA, Potts PD, Godwin SJ, Bendle RD, Dampney RA (1996) Medullary neurons activated by angiotensin II in the conscious rabbit. Hypertension 27:287–296

Horiuchi M, Hayashida W, Kambe T, Yamada T, Dzau VJ (1997) Angiotensin type 2 receptor dephosphorylates Bcl-2 by activating mitogen-activated protein kinase phosphatase-1 and induces apoptosis. J Biol Chem 272:19022–19026

Hu L, Zhu DN, Yu Z, Wang JQ, Sun ZJ, Yao T (2002) Expression of angiotensin II type 1 (AT(1)) receptor in the rostral ventrolateral medulla in rats. J Appl Physiol 92:2153–2161

Huang XC, Richards EM, Sumners C (1995) Angiotensin II type 2 receptor-mediated stimulation of protein phosphatase 2A in rat hypothalamic/brainstem neuronal cocultures. J Neurochem 65:2131–2137

Huang XC, Richards EM, Sumners C (1996) Mitogen-activated protein kinases in rat brain neuronal cultures are activated by angiotensin II type 1 receptors and inhibited by angiotensin II type 2 receptors. J Biol Chem 271:15635–15641

Huang XC, Deng T, Sumners C (1998) Angiotensin II stimulates activation of Fos-regulating kinase and c-Jun NH2-terminal kinase in neuronal cultures from rat brain. Endocrinology 139:245–251

Ichiki T, Labosky PA, Shiota C, Okuyama S, Imagawa Y, Fogo A, Niimura F, Ichikawa I, Hogan BL, Inagami T (1995) Effects on blood pressure and exploratory behaviour of mice lacking angiotensin II type-2 receptor. Nature 377:748–750

Iwai N, Inagami T (1992) Identification of two subtypes in the rat type I angiotensin II receptor. FEBS Lett 298:257–260

Jing J, Peretz T, Singer-Lahat D, Chikvashvili D, Thornhill WB, Lotan I (1997) Inactivation of a voltage-dependent K+ channel by beta subtype modulation by a phosphorylation-dependent interaction between distal C terminus of alpha subunit and cytoskeleton. J Biol Chem 272:14021–14024

Johnson AK, Thunhorst RL (1997) The neuroendocrinology of thirst and salt appetite: visceral sensory signals and mechanisms of central integration. Front Neuroendocrinol 18:292–353

Kakar SS, Sellers JC, Devor DC, Musgrove LC, Neill JD (1992) Angiotensin II type-1 receptor subtype cDNAs: differential tissue expression and hormonal regulation. Biochem Biophys Res Commun 183:1090–1096

Kambayashi Y, Bardhan S, Takahashi K, Tsuzuki S, Inui H, Hamakubo T, Inagami T (1993) Molecular cloning of a novel angiotensin II receptor isoform involved in phosphotyrosine phosphatase inhibition. J Biol Chem 268:24543–24546

Kang J, Sumners C, Posner P (1993) Angiotensin II type 2 receptor-modulated changes in potassium currents in cultured neurons. Am J Physiol 265:C607–C616

Kang J, Posner P, Sumners C (1994) Angiotensin II type 2 receptor stimulation of neuronal K+ currents involves an inhibitory GTP binding protein. Am J Physiol 267:C1389–C1397

Kang J, Richards EM, Posner P, Sumners C (1995) Modulation of the delayed rectifier K+ current in neurons by an angiotensin II type 2 receptor fragment. Am J Physiol 268:C278–C282

Kim D, Sladek CD, Aguado-Velasco C, Mathiasen JR (1995) Arachidonic acid activation of a new family of K+ channels in cultured rat neuronal cells. J Physiol 484 (Pt 3): 643–660

Knowles WD, Phillips MI (1980) Angiotensin II responsive cells in the organum vasculosum lamina terminals (OVLT) recorded in hypothalamic brain slices. Brain Res 197:256–259

Lee WJ, Kim KS, Yang EK, Lee JH, Lee EJ, Park JS, Kim HJ (1996) Effect of brain angiotensin II AT1, AT2 and cholinergic receptor antagonism on drinking in water-deprived rats. Regul Pept 66:41–46

Lenkei Z, Palkovits M, Corvol P, Llorens-Cortes C (1997) Expression of angiotensin type-1 (AT1) and type-2 (AT2) receptor mRNAs in the adult rat brain: a functional neuroanatomical review. Front Neuroendocrinol 18:383–439

Levin G, Chikvashvili D, Singer-Lahat D, Peretz T, Thornhill WB, Lotan I (1996) Phosphorylation of a K+ channel alpha subunit modulates inactivation conferred by a beta subunit. Involvement of cytoskeleton. J Biol Chem 271:29321–29328

Levitan IB (1999) Modulation of ion channels by protein phosphorylation. How the brain works. Adv Second Messenger Phosphoprotein Res 33:3–22

Li Z, Ferguson AV (1993) Angiotensin II responsiveness of rat paraventricular and subfornical organ neurons in vitro. Neuroscience 55:197–207

Li Z, Ferguson AV (1996) Electrophysiological properties of paraventricular magnocellular neurons in rat brain slices: modulation of IA by angiotensin II. Neuroscience 71:133–145

Li YW, Guyenet PG (1995) Neuronal excitation by angiotensin II in the rostral ventrolateral medulla of the rat in vitro. Am J Physiol. 268:R272–R277

Li YW, Guyenet PG (1996) Angiotensin II decreases a resting K+ conductance in rat bulbospinal neurons of the C1 area. Circ Res 78:274–282

Lu D, Yang H, Raizada MK (1996) Angiotensin II regulation of neuromodulation: downstream signaling mechanism from activation of mitogen-activated protein kinase. J Cell Biol 135:1609–1617

Martens JR, Navarro-Polance R, Coppock EA, Nishiyama A, Parshely L, Grobaski TD, Tamkun MM (2000) Differential targeting of Shaker-like potassium channels to lipid rafts. J Biol Chem 275:7443–7446

McKinley MJ, Allen AM, Mathai ML, May C, McAlle RM, Oldfield BJ, Weisinger RS (2001) Brain angiotensin and body fluid homeostasis. Jpn J Physiol 51:281–289

Mendelsohn FA, Quirion R, Saavedra JM, Aguilera G, Catt KJ (1984) Autoradiographic localization of angiotensin II receptors in rat brain. Proc Natl Acad Sci U S A 81:1575–1579

Meves H (1994) Modulation of ion channels by arachidonic acid. Prog Neurobiol 43:175–186

Mukoyama M, Nakajima M, Horiuchi M, Sasamura H, Pratt RE, Dzau VJ (1993) Expression cloning of type 2 angiotensin II receptor reveals a unique class of seven-transmembrane receptors. J Biol Chem 268:24539–24542

Murphy TJ, Alexander RW, Griendling KK, Runge MS, Bernstein KE (1991) Isolation of a cDNA encoding the vascular type-1 angiotensin II receptor. Nature 351:233–236

Muscella A, Aloisi F, Marsigliante S, Levi G (2000) Angiotensin II modulates the activity of Na+, K+-ATPase in cultured rat astrocytes via the AT1 receptor and protein kinase C-delta activation. J Neurochem 74:1325–1331

Nagatomo T, Inenaga K, Yamashita H (1995) Transient outward current in adult rat supraoptic neurones with slice patch-clamp technique: inhibition by angiotensin II. J Physiol 485 (Pt 1):87–96

Nahmias C, Cazaubon SM, Briend-Sutren MM, Lazard D, Villageois P, Strosberg AD (1995) Angiotensin II AT2 receptors are functionally coupled to protein tyrosine dephosphorylation in N1E-115 neuroblastoma cells. Biochem J 306 (Pt 1):87–92

Nouet S, Nahmias C (2000) Signal transduction from the angiotensin II AT2 receptor. Trends Endocrino Metab 11:1–6

Nuyt AM, Lenkei Z, Corvol P, Palkovits M, Llorens-Cortes C (2001) Ontogeny of angiotensin II type I receptor mRNAs in fetal and neonatal rat brain. J Comp Neurol 440:192–203

Obermuller N, Unger T, Culman J, Gohlke P, de Gasparo M, Bottari SP (1991) Distribution of angiotensin II receptor subtypes in rat brain nuclei. Nurosci Lett 132:11–15

Ono K, Honda E, Inenaga K (2001) Angiotensin II induces inward current in subfornical organ neurones of rats. J Neuroendocrinol 13:517–523

Pan SJ, Zhu M, Raizada MK, Sumners C, Gelband CH (2001) ANG II-mediated inhibition of neuronal delayed rectifier current: role of proteinkinase C-alpha. Am J Physiol Cell Physiol 281:C17–C23

Peretz T, Levin G, Moran O, Thornhill WB, Chikvashvili D, Lotan I (1996) Modulation by protein kinase C activation of rat brain delivery rectifier K+ channel expressed in Xenopus oocytes. FEBS Lett 381:71–76

Phillips MI, Sumners C (1998) Angiotensin II in central nervous system physiology. Regul Pept 78:1–11

Richards EM, Raizada MK, Gelband CH, Sumners C (1999) Angiotensin II type 1 receptor-modulated signaling pathways in neurons. Mol Neurobiol 19:25–41

Rowe BP, Grove KL, Saylor DL, Speth RC (1990) Angiotensin II receptor subtypes in the rat brain. Eur J Pharmacol 186:339–342

Rowland NE, Fregly MJ (1993) Brain angiotensin AT-2 receptor antagonism and water intake. Brain Res Bull 32:391–394

Saavedra JM (1999) Emerging features of brain angiotensin receptors. Regul Pept 85:31–45

Sasaki K, Yamano Y, Bardhan S, Iwai N, Murray JJ, Hasegawa M, Matsuda Y, Inagami T (1991) Cloning and expression of a complementary DNA encoding a bovine adrenal angiotensin II type-1 receptor. Nature 351:230–233

Seltzer AM, Zorad S, Saavedra JM (1995) Stimulation of angiotensin II AT1 receptors in rat median eminence increases phosphoinositide hydrolysis. Brain Res 705:24–30

Sirett NE, McLean AS, Bray JJ, Hubbard JI (1977) Distribution of angiotensin II receptors in rat brain. Brain Res 122:299–312

Song K, Allen AM, Paxinos G, Mendelsohn FA (1992) Mapping of angiotensin II receptor subtype heterogeneity in rat brain. J Comp Neurol 316:467–484

Speth RC, Grove KL, Rowe BP (1991) Angiotensin II and the locus coeruleus. Prog Brain Res 88:217–226

Stadler T, Veltmar A, Qadri F, Unger T (1992) Angiotensin II evokes noradrenaline release from the paraventricular nucleus in conscious rats. Brain Res 569:117–122

Stoll M, Unger T (2001) Angiotensin and its AT2 receptor: new insights into an old system. Regul Pept 99:175–182

Suga T, Suzuki M, Suzuki M (1979) Effects of angiotensin II on the medullary neurons and their sensitivity to acetylcholine and catecholamines. Jpn J Pharmacol 4:541–552

Sumners C, Myers LM (1991) Angiotensin II decreases cGMP levels in neuronal cultures from rat brain. Am J Physiol 260:C79–C87

Sumners C, Tang W, Zelezna B, Raizada MK (1991) Angiotensin II receptor subtypes are coupled with distinct signal-transduction mechanisms in neurons and astrocytes from the rat brain. Proc Natl Acad Sci U S A 88:7567–7571

Sumners C, Zhu M, Gelband CH, Posner P (1996) Angiotensin II type 1 receptor modulation of neuronal K^{+} and Ca^{2+} currents: intracellular mechanisms. Am J Physiol 271:C154–C163

Sumners C, Fleegal MA, Zhu M (2002) Angiotensin at1 receptor signalling in neurons. Clin Exp Pharacol Physiol 29:483–490

Sun C, Sumners C, Raizada MK (2002) Chronotropic action of angiotensin II in neurons via protein kinase C and CaMKII. Hypertension 39:562–566

Tallant EA, Higson JT (1997) Angiotensin II activates distinct signal transduction pathways in astrocytes isolated from neonatal rat brain. Glia 19:333–342

Tanaka J, Hayashi Y, Sakamaki K, Okumura T, Nomura M (2001) Activation of the subfornical organ enhances extracellular noradrenaline concentrations in the hypothalamic paraventricular nucleus in the rat. Brain Res Bull 54:421–425

Timmermans PB, Wong PC, Chiu AT, Herblin WF, Benfield P, Carini DJ, Lee RJ, Wexler RR, Saye JA, Smith RD (1993) Angiotensin II receptors and angiotensin II receptor antagonists. Pharmacaol Rev 45:205–251

Tsutsumi K, Saavedra JM (1991) Characterization and development of angiotensin II receptor subtypes (AT1 and AT2) in rat brain. Am J Physiol 261:R209–R216

Van Houten M, Schiffrin EL, Mann JF, Posner BI, Boucher R (1980) Radioautographic localization of specific binding sites for blood-borne angiotensin II in the rat brain. Brain Res 186:480–485

Von Bohlen und Halbach O, Albrecht D (1998) Mapping of angiotensin AT1 receptors in the rat limbic system. Regul Pept 78:51–56

Wang D, Gelband CH, Sumners C, Poser P (1997a) Mechanisms underlying the chronotropic effects of angiotensin II on cultured neurons from rat hypothalamus and brain stem. J Neurophysiol 78:1013–1020

Wang D, Sumners C, Poser P, Gelband CH (1997b) A-type K+ current in neurons cultured from neonatal rat hypothalamus and brain stem: modulation by angiotensin II. J Neurophysiol 78:1021–1029

Washburn DL, Ferguson AV (2001) Selective potentiation of N-type calcium channels by angiotensin II in rat subfornical organ neurones. J Physiol 536:667–675

Whitebread S, Mele M, Kamber B, de Gasparo M (1989) Preliminary biochemical characterization of two angiotensin II receptor subtypes. Biochem Biophys Res Commun 163:284–291

Yamada T, Horiuchi M, Dzau VJ (1996) Angiotensin II type 2 receptor mediates programmed cell death. Proc Natl Acad Sci U S A 93:156–160

Yang H, Lu D, Yu K, Raizada MK (1996) Regulation of neuromodulatory actions of angiotensin II in the brain neurons by the Ras-dependent mitogen-activated protein kinase pathway. J Neurosci 16:4047–4058

Yang SN, Lippoldt A, Jansson A, Phillips MI, Ganten D, Fuxe K (1997) Localization of angiotensin II AT1 receptor-like immunoreactivity in catecholaminergic neurons of the rat medulla oblongata. Neuroscience 81:503–515

Zhang J, Pratt RE (1996) The AT2 receptor selectively associates with Gialpha2 and Gialpha3 in the rat fetus. J Biol Chem 271:15026–15033

Zhu B, Herbert J (1997) Calcium channels mediate angiotensin II-induced drinking behavior and c-fos expression in the brain. Brain Res 778:206–214

Zhu DN, Moriguchi A, Mikami H, Higaki J, Ogihara T (1998) Central amino acids mediate cardiovascular response to angiotensin II in the rat. Brain Res Bull 45:189–197

Zhu M, Neubig RR, Wade SM, Posner P, Gelband CH, Sumners C (1997) Modulation of K+ and Ca2+ currents in cultured neurons angiotensin II type 1a receptor peptide. Am J Physiol 273:C1040–C1048

Zhu M, Gelband CH, Moore JM, Posner P, Sumners C (1998) Angiotensin II type 2 receptor stimulation of neuronal delayed-rectifier potassium current involves phospholipase A2 and arachidonic acid. J Neurosci 18:679–686

Zhu M, Gelband CH, Posner P, Sumners C (1999) Angiotensin II decreases neuronal delayed rectifier potassium current: role of calcium/calmodulin-dependent protein kinase. J Neurophysiol 82:1560–1568

Zhu M, Natarajan R, Nadler JL, Moore JM, Gelband CH, Sumners C (2000) Angiotensin II increases neuronal delayed rectifier K(+) current: role of 12-lipoxygenase metabolites of arachidonic acid. J Neurophysiol 84:2494–2501

Zhu M, Sumners C, Gelband CH, Posner P (2001) Chronotropic effects of angiotensin II via type 2 receptors brain neurons. J Neurophysiol 85:2177–2183

Zucker IH (2002) Brain angiotensin II: new insights into its role in sympathetic regulation. Circ Res 90:503–505

Angiotensin, Neuroplasticity and Stroke

A. Blume[1] · J. Culman[2]

[1] Institute of Zoology, University of Regensburg, Universitätsstraße 31, 93053 Regensburg, Germany
e-mail: annegret.blume@biologie.uni-regensburg.de

[2] Institute of Pharmacology, University Hospital of Schleswig-Holstein, Campus Kiel, Christian-Albrechts-University of Kiel, 24105 Kiel, Germany
e-mail: juraj.culman@pharmakologie.uni-kiel.de

Abstract The effector peptide of the renin-angiotensin system (RAS), angiotensin II (Ang II), can be directly generated in the brain independently of the peripheral RAS. Activation of angiotensin receptors in the brain induces the expression of inducible transcription factors involved in gene regulation. A number of animal studies and clinical trials have shown that the RAS is associated with the development of cardiovascular diseases and pathophysiological processes occurring in ischaemic stroke. Angiotensin-converting enzyme (ACE) inhibitors and selective AT_1 receptor antagonists protect hypertensive subjects against stroke by reduction of blood pressure and improve the recovery from ischaemic insult by normalisation of cerebral blood autoregulation and improvement of collateral blood flow to the ischaemic tissue. In animal models, inhibition of the brain RAS proved to be beneficial with respect to stroke incidence and outcome. Blockade of cerebrovascular or brain AT_1 receptors reduces the volume of ischaemic injury and improves the recovery from brain ischaemia. Remarkable progress has been made in elucidating the role of Ang II and its receptors in the control of pathophysiological events associated with stroke such as cerebral haemodynamics, apoptosis, inflammation, neuroregeneration and neuroprotection. In this chapter, we summarise the effects of Ang II in the brain on the expression of inducible transcription factors and possible target genes. The second part of the article reviews the effects of inhibition of ACE and angiotensin receptors in cerebral vessels and brain tissue on processes occurring during and after ischaemic injury and discusses the effects of AT_1 receptor blockade on cerebral haemodynamics, neuroprotection and neuroregeneration. This chapter may also provide a framework for the development of new therapeutic strategies in the prevention and treatment of ischaemic stroke.

Keywords Renin-angiotensin system · Brain · Angiotensin-converting enzyme · Angiotensin receptors · Inducible transcription factors · Neuroplasticity · AT_1 and AT_2 receptor antagonists · Cerebral haemodynamics · Neuroprotection · Stroke

1 Introduction

The hormonal renin-angiotensin system (RAS) with its effector peptide, angiotensin II (Ang II), has traditionally been linked to the regulation of salt and water homeostasis. In the last 2 decades evidence has accumulated that Ang II is formed in various tissues, including the brain. All components of the peripheral RAS have been found in the central nervous system and it is now firmly established that Ang II in the brain is synthesised independently of peripheral sources. In general, Ang II exerts its actions through activation of at least two receptor subtypes, referred to as the AT_1 and the AT_2 receptor. In the brain, angiotensin receptors are expressed in structures localised inside and outside the blood–brain barrier. AT_2 receptors are predominantly expressed in the foetal brain. After birth, the ratio of AT_1- to AT_2-receptor expression reverses, with the AT_1 receptor being the predominant receptor in the adult brain.

Most of the classic actions of Ang II in the brain, which include the central control of fluid and electrolyte homeostasis and the regulation of blood pressure, are mediated by AT_1 receptors. In other cell types such as vascular smooth muscle cells or fibroblasts Ang II acting on the AT_1 receptor promotes cell growth and proliferation. Since adult brain neurons cannot proliferate, effects mediated by AT_1 receptors in neurons are associated with rapid processes, involving neurotransmission or neuromodulation, or delayed processes, requiring synthesis of new proteins.

Over the last 10 years, numerous studies have indicated that Ang II, acting via AT_2 receptors, may modulate embryonic development, tissue regeneration and protection but also initiate processes leading to programmed cell death (apoptosis). There is also substantial evidence that the AT_2 receptor can offset or counteract the effects mediated by the AT_1 receptor in the brain, for example on water intake and vasopressin release.

This chapter focuses on the functions of Ang II in neuronal cells and the nervous system, especially on the effects of the peptide and its receptors on the expression of transcription factors. The second part of the article provides some aspects concerning the relevance of Ang II, angiotensin-converting enzyme (ACE), angiotensin receptors and their inhibition in cerebral vessels and brain tissue for processes occurring during and after ischaemic injury.

2 Angiotensin II and Neuroplasticity

Inducible transcription factors (ITF) are proteins which are able to regulate the expression of genes by eukaryotic RNA polymerase II after binding to specific nucleotide sequences in the non-coding promoter regions of the DNA. The base sequences can vary in one or two nucleotides between different promoters. These variations as well as nucleotide sequences next to the binding sites influence their affinity to the respective transcription factors and thus the regulatory activity of the proteins. Some transcription factors such as c-Fos or c-Jun bind to their consensus sequences (called AP-1 for activator protein 1, or CRE for CREB responsive element) only after dimerisation with a partner (Cohen et al. 1989). Other ITF such as the members of the Krox family do not dimerise (Chavrier et al. 1989). Interactions between members of different families of transcription factors are frequent, and the assortment of transcription factors bound to a promoter in response to a given stimulus decides whether transcription is enhanced, not altered or even inhibited. Therefore, the expression pattern of different ITFs which are induced by a given stimulus in a tissue or cell has a decisive influence on the group of target genes which are regulated by these factors (West et al. 2002).

Peripherally, Ang II acts as a direct vasoconstrictor, induces the release of aldosterone from the adrenal gland and enhances the retention of salt and water in the kidney. In addition, Ang II plays an important role in growth processes: the peptide promotes cell growth and regulates the formation of matrix proteins such as fibronectin, thrombospondin or collagen. The hormone also plays a role in pathological trophic responses which occur, for example, after injuries to blood vessels or in connection with cardiovascular or cerebral disease. The expression of ITFs such as c-Fos or c-Jun has been reported to be linked with these growth processes (Kawahara et al. 1988; Kim et al. 1995).

In the brain, all components of the renin-angiotensin system are present, generating the peptide hormone locally. The effects of Ang II in the brain include an elevation of blood pressure, a modulation of sympathetic nerve activity, natriuresis and the induction of drinking. Moreover, the peptide induces the release of pituitary hormones such as arginine vasopressin (AVP), oxytocin or adrenocorticotrophic hormone (ACTH) and interacts with other neurotransmitters such as norepinephrine in the brain (Culman et al. 2002).

Two Ang II receptor subtypes have been identified in the brain: the AT_1 receptor is responsible for most of the known effects of the peptide in the brain, especially for the central regulation of cardiovascular functions (Hohle et al. 1995). AT_1 receptors are located in high numbers in brain regions associated with the regulation of blood pressure and volume homeostasis, including circumventricular organs such as the subfornical organ (SFO) or the organum vasculosum of the lamina terminalis (OVLT), the median preoptic area (MnPO) and in hypothalamic nuclei such as the paraventricular (PVN) and supraoptic nuclei (SON). Additionally, AT_1 receptors are found, for example, in the suprachiasmatic nucleus and in the median eminence. In the hindbrain and brain stem, nuclei and regions such as the nucleus of the solitary tract (NTS), area postrema (AP), the rostral (RVLM) and caudal ventrolateral medulla (CVLM) and the intermediolateral column of the spinal cord express AT_1 receptors. The AT_2 receptor, in contrast, is quantitatively expressed in some thalamic and subthalamic nuclei, in the amygdala, in the inferior olive and the locus coeruleus (Lenkei et al. 1997; Tsutsumi and Saavedra 1991). This angiotensin receptor subtype is up-regulated following injury and ischaemic damage and takes part in such opposing effects as differentiation, neuronal regeneration and apoptosis depending on the cell type or tissue and the overall signal input in a given situation (Unger 1999).

A binding site for an angiotensin fragment, Ang IV (Ang 3–8) has been described and named AT4 receptor. Recently, this receptor has been identified as the enzyme insulin-regulated aminopeptidase (Albiston et al. 2001). In the brain, Ang IV is implicated in the facilitation of memory retrieval and retention (Braszko et al. 1988; Wright et al. 1993). The AT4 receptor is expressed in brain regions such as the hippocampus, cortex (cerebral and piriform), septum, arcuate nucleus and superior and inferior colliculi (Moeller et al. 1996; Roberts et al. 1995).

A number of studies have shown that Ang II induces the expression of ITF in specific brain nuclei, reaching the brain via the circumventricular organs which lack the blood–brain barrier after intravenous injection or directly after injection into the lateral brain ventricle. Two main responsive regions can be distinguished: one is a regulatory circuit which comprises some nuclei of the lamina terminalis and the hypothalamus, the other one comprises nuclei in the brain stem associated with cardiovascular regulation.

An injection of Ang II (100 pmol) into the lateral brain ventricle stimulates periventricular AT receptors to induce a pronounced and dense expression of AP-1 proteins and Krox proteins in distinct forebrain and hypothalamic nuclei: the SFO, MnPO, OVLT, PVN and SON (Herbert et al. 1992; Lebrun et al. 1995; McKinley et al. 1995; Rowland et al. 1994; Xu and Herbert 1994). The response is dose dependent and mediated via AT_1 receptors, since pretreatment with the AT_1 receptor antagonist losartan was able to abolish the expression completely (Lebrun et al. 1995). Intracerebroventricular (i.c.v.) injection of a higher dose of the peptide (250 nmol) or its constant infusion into the lateral ventricle (1 ng/min for 90 min) additionally induced the expression of c-Fos in the bed nucleus of the stria terminalis and the central amygdaloid nucleus (Herbert et al. 1992; McKinley et al. 1995). The Ang II-induced expression of ITF was not only restricted to specific brain nuclei, but also showed a temporally differentiated pattern, with some transcription factors appearing at an earlier time point and some later (Blume et al. 1998). Thus, each of the nuclei displays its own specific temporal and spatial expression pattern of ITF, indicating that the target genes which are regulated by the transcription factors are also specific in each region. The Ang II-induced release of arginine vasopressin from magnocellular SON and PVN neurons requires activation of adrenoceptors in these regions (Qadri et al. 1993; Veltmar et al. 1992). Activation of adrenergic receptors is also involved in the Ang II-induced expression of transcription factors in these areas. While the β-adrenoceptor antagonist propranolol did not influence Ang II-induced expression of c-Fos, c-Jun or Krox 24, pretreatment with the $\alpha 1$-adrenoceptor antagonist, prazosin, or the $\alpha 2$ adrenoceptor antagonist, yohimbine, inhibited the Ang II-induced ITF expression in the SON and PVN, but not in the SFO or MnPO (Blume et al. 2002).

Interestingly, in the SFO mainly the cells adjacent to the ventricle express ITF following stimulation of periventricular AT receptors. In contrast, following an intravenous injection of Ang II, cells in the centre of the SFO are stimulated to express ITF (McKinley et al. 1995). Intravenous Ang II reaches the brain via the circumventricular organs, which lack the blood–brain barrier. Besides the SFO, the pattern of ITF expression induced by i.v. Ang II closely resembles the pattern induced by i.c.v. Ang II: the OVLT, MnPO, PVN and SON exhibit a strong expression of ITF, while the central amygdaloid nucleus, the bed nucleus of the stria terminalis and hindbrain regions such as the AP and the NTS exhibit scattered immunopositive cells (McKinley et al. 1992, 1995; Rowland et al. 1994). I.c.v. pretreatment with the AT_1 receptor antagonist, losartan, blocks ITF expression in all the above-mentioned regions (Rowland et al. 1994). In contrast to i.c.v. Ang II, i.v. Ang II-induced expression of ITF in the PVN and SON is blocked by pretreatment with the AT_2 receptor antagonist PD 123 319 (Rowland et al. 1994). A lesion of the SFO abolishes ITF expression in the SON and PVN. The differences in the expression patterns of ITF after i.v. and i.c.v. Ang II might in part be due to the fact that Ang II activates the baroreceptor reflex when administered i.v. The baroreceptor reflex, in turn, activates noradrenergic fibres, originating in the A1 or A2 regions and terminating in the MnPO and hypothalamus. It seems possible that these fibres are responsible for the activation of ITF expression after peripheral stimulation with Ang II. Experiments with the vasoactive substance phenylephrine showed that a 2-h i.v. infusion of the substance induced Fos-like immunoreactivity in the SON and PVN, suggesting that baroreceptor mechanisms could play a role in ITF induction by i.v. Ang II, at least in these nuclei (McKinley et al. 1992).

Spontaneously hypertensive rats (SHRs) have a genetically determined upregulation of the renin angiotensin system, with increased expression of AT_1 receptors as well as elevated levels of Ang II. Treatment of these rats with Ang II, administered i.v. or i.c.v., resulted in an elevated expression of ITF in the SFO, MnPO, OVLT, SON and PVN (Blume et al. 1997; Rowland et al. 1995). This enhanced response is not due to the high blood pressure in these animals, since in

Wistar Kyoto rats, which displayed nephrogenic hypertension after aortic banding, levels of ITF expression were similar to those seen in normotensive Wistar or Wistar Kyoto rats. Possibly, elevated ITF levels in SHRs lead to the regulation of different target genes.

Ang IV and its potency to induce the expression of ITF has been studied by Roberts et al. (1995). They found that Ang IV (i.c.v.) induced an expression of c-Fos exclusively in brain regions which also express the AT4 receptor: the hippocampus (dentate gyrus, CA2, CA3) and the piriform cortex. Other brain regions which express AT4 receptors (e.g. the cerebral cortex, septum or arcuate nucleus of the hypothalamus) as well as regions which predominantly express AT_1 or AT_2 receptors showed no c-Fos in response to Ang IV (Roberts et al. 1995). The response to Ang IV was inhibited by a specific AT4 receptor antagonist, divalanal Ang IV. Thus, it seems that the two peptides, Ang II and Ang IV, possess distinct regulatory roles in the brain and do not interfere with each other (at least at the level of transcriptional control).

Endogenously released Ang II (Oldfield and McKinley 1994) resulted in an expression of c-Fos in exactly the same brain nuclei as Ang II administered i.v. or i.c.v.; these studies were a further step towards the characterisation of a regulatory network in the forebrain and the hypothalamus, which responds to Ang II. Peripheral Ang II, which reaches the brain via circumventricular organs, as well as central Ang II, which stimulates periventricular AT receptors, activate neurons in the SFO and OVLT, which respond (among other reactions described in detail in the chapter by F. Qadri, this volume) with the expression of ITF. The neuronal activation is then transmitted via angiotensinergic fibres to the MnPO and ultimately to the hypothalamic nuclei PVN and SON, resulting in an expression of ITF, and also accompanied by the release of arginine vasopressin into the circulation. That the expression of ITF in the hypothalamic nuclei is rather a secondary event and not a result of a direct stimulation of AT_1 receptors in these nuclei has been demonstrated by studies in which the lesion of the anteroventricular region of the third ventricle (which comprises the MnPO) as well as the lesion of the SFO resulted in an inhibition of the Ang II-induced ITF expression in the SON and PVN (Rowland et al. 1994; Xu and Herbert 1996). Moreover, low doses of Ang II (1 pmol, i.c.v.) result in an expression of ITF in the SFO and in the MnPO, while higher doses also evoke a response in hypothalamic nuclei, suggesting that the signal input that activates neurons in the SFO and MnPO has to reach a certain intensity before it is transmitted further in the network (Blume et al. 1998). The finding that adrenoceptors take part in Ang II-induced ITF expression in the SON and PVN, but not in the SFO or MnPO point in the same direction (Blume et al. 2002).

In peripheral tissues, Ang II-induced ITF expression is associated with trophic processes. Neurons, in contrast, are assumed to be postmitotic. Therefore, one would expect to find neuroplastic modifications in the brain rather than growth in the sense of proliferation. According to an accepted hypothesis, ITF are signal transducers, which convert the ligand receptor association into long-lasting changes in the protein expression pattern of a cell (Greenberg et al. 1986;

Morgan and Curran 1991). The morphological and physiological changes resulting from these modified protein expression patterns are summarised as neuroplasticity and include, for example, changes in the number of receptors, in the nature and frequency of cell–cell contacts or modifications of the Golgi apparatus (Takasu et al. 2002).

It is rather difficult to study the target genes which might be activated as a result of the Ang II-induced expression of ITF in vivo, because the brain nuclei where these processes occur are small and only the pooling of nuclei from about 60–80 animals would provide enough material for biochemical analyses. Another possibility is the extraction of RNA or protein from a particular nucleus and the surrounding tissue, based on the assumption that the changes in the individual nucleus are strong enough to manifest themselves in a larger tissue probe. This approach has recently been used to study the effects of Ang II-induced c-Fos expression in the nucleus of the solitary tract. The NTS as well as the adjacent AP are brain regions which express considerable amounts of AT_1 receptors, and the AP, with its lack of the blood–brain barrier, has been identified as a site of entry for circulating Ang II to promote sympathetic outflow (Ferrario et al. 1987). The NTS is a brain nucleus where primary baroreceptor afferents terminate, making the NTS an integration centre for the baroreceptor reflex (Diz et al. 2002). Endogenous or exogenously administered Ang II impairs the baroreceptor reflex; an inhibition of AT_1 receptors in the NTS results in an enhancement of baroreflex sensitivity (Campagnole-Santos et al. 1988; Casto and Phillips 1986; DiBona et al. 1995; Michelini and Bonagamba 1988; Murakami et al. 1996).

An involvement of c-Fos in these well-established responses has been shown recently, providing the first identification of a target gene of Ang II-induced c-Fos in vivo. In a series of publications, the authors showed that injection of Ang II into the NTS inhibited the baroreceptor reflex and induced the expression of c-Fos in this nucleus. Both, the c-Fos expression and the inhibition of the baroreceptor reflex were suppressed by pretreatment with an antisense oligomer directed at the initiation codon of the c-fos mRNA (Luoh and Chan 2001). A baroreceptor reflex activation induced by prolonged hypertension induced a transient decrease in AT_1 receptor mRNA in the dorsomedial medulla, a brain region which includes the NTS as well as the AP, the dorsal motor nucleus of the vagus and the hypoglossal nucleus (Wang et al. 2001). The AT_1 receptor has a consensus sequence for AP-1 proteins in its promoter region (Herzig et al. 1997). Pretreatment with the antisense oligomer against c-fos inhibited the restoration of the pressor response to Ang II as well as the re-expression of AT_1 receptors, suggesting that Ang II via activation of a transcription factor can counter-regulate the physiological down-regulation of its own AT_1 receptor and thus restore an important regulatory circuit of blood pressure control (Wang et al. 2001).

Besides the AT_1 receptor, other genes coding for RAS proteins also have AP-1 sequences in their promoters, for example, the rat AT_2 receptor gene (Ichiki and Inagami 1995; Murasawa et al. 1996). The acute phase response element (APRE) in the promoter of the angiotensinogen gene interacts with NF-κB, a transcription factor that also dimerises with c-Fos and c-Jun (Ron et al. 1990; Stein et al. 1993).

Another interesting candidate for a target gene may be the gene coding for vasopressin (AVP). AVP is released in response to Ang II from nerve terminals localised in the posterior pituitary. The promoter for the AVP precursor contains a CRE-like binding site (Mohr and Richter 1990). Therefore, it can be speculated that Ang II also initiates the resynthesis of AVP after its secretion from the pituitary. However, considering the large amounts of AVP present in the nerve terminals and high levels of mRNA coding for AVP in the cell bodies, it appears to be rather difficult to prove this hypothesis in in vivo experiments.

The gene for tyrosine hydroxylase (TH), the enzyme which catalyses the first step in catecholamine synthesis, can be induced by Ang II in brain neurons in culture (Yu et al. 1996). Since Ang II has been shown to interact with catecholamine systems in the brain (for reviews see Culman et al. 1995; Jenkins et al. 1995), it seems reasonable to hypothesise that Ang II is also involved in the TH gene regulation in vivo.

As already mentioned above, some effects of Ang II in the neuronal tissue such as alterations in gene expression leading to neuroplastic changes are believed to be mediated by ITF, a link between primary stimuli and long-term effects. Therefore, genes associated with neuroplasticity in the brain are possible targets of Ang II. Such targets may include genes coding for growth factors, like nerve growth factor (NGF) (Hengerer et al. 1990), genes encoding structural proteins like glial fibrillary acid protein (GFAP) (Masood et al. 1993), or the gene for the growth associated protein-43 (GAP-43), which plays a major role in axonal elongation and promotes changes in synapse morphology (Eggen et al. 1994).

Since AT receptor antagonists affect the expression of ITF after stroke, as will be discussed in the second part of this chapter, targets of Ang II in the brain may also play a role in the pathophysiology of stroke.

3
Angiotensin II and Stroke

Stroke is among the leading causes of death and neurological impairment worldwide. A stroke occurs when a blood vessel in the brain ruptures or is clogged by a thrombus. Thus, ischaemic stroke results from a transient or permanent occlusion of cerebral artery which leads to a reduction in cerebral blood flow. Neuronal cells require a constant delivery of oxygen, glucose and other nutrients to function properly. A local reduction or a complete arrest of blood supply prevents the delivery of oxygen and other substrates resulting in metabolic disturbances such as the deterioration of high-energy phosphates, membrane ion pump failure, efflux of intracellular potassium, and influx of sodium and calcium, chloride and water, leading finally to membrane depolarisation. The most important mechanisms occurring in stroke involve so-called excitotoxicity, inflammation and programmed cell death, apoptosis. The outcome of an ischaemic event mostly depends on the severity of the cerebral blood flow reduction and on the duration of the ischaemic insult (Dirnagl et al. 1999).

Ischaemic stroke, which is caused by an occlusion of a cerebral artery, accounts for almost 80% of all stroke cases. The most common type of the ischaemic stroke is the cerebral thrombosis, characterised by a thrombus which is attached to the wall of the artery, where it grows and blocks the flow of blood. The arteries in which the thrombus forms are usually damaged by atherosclerosis. The cerebral embolism represents the second form of the ischaemic stroke. This ischaemic event occurs when a thrombus (embolus) formed outside of the brain, typically in the heart, is carried by the blood stream to an artery leading to or in the brain and occludes it. Two other types of stroke are caused by bleeding (subarachnoid or intracerebral haemorrhage).

The mortality rate in ischaemic stroke victims is high. Within the first 3 months, 16%–25% of patients die of stroke. About 50% of stroke survivors are left with varying degrees of disability up to 1 year after their stroke and only 25% of patients are reported to fully recover. The neurological impairments and dysfunction in patients who have suffered from ischaemic stroke are influenced by many factors, including genetic factors, age, gender, vascular risk factors including previous strokes, predilection of cognitive impairment etc. Stroke incidence increases proportionally with the age; the incidence and prevalence are approximately equal for men and women, although women at all ages are more likely to die from a stroke than men. High blood pressure is the most important modifiable risk factor of stroke; about 50% of strokes may be attributed to hypertension. It has been estimated that lowering diastolic blood pressure by 5 mmHg can lower the risk of a first stroke by 35%–40% (Collins et al. 1990; MacMahon et al. 1990). Reduction of arterial blood pressure had beneficial effects in spontaneously hypertensive rats (SHRs) and reduced cerebral infarction after occlusion of the middle cerebral artery (Fujii et al. 1992). Hypertension is known to cause endothelial dysfunction, which predisposes to atherosclerosis. Hypertension may, therefore, lead to stroke by aggravating atherosclerosis in the aortic arch and cerebral arteries, by atherosclerosis in the small cerebral arteries and end arteries and by promoting heart diseases which may be complicated by stroke. While haemorrhagic stroke is proportionally related to the levels of blood pressure increases, ischaemic stroke is largely accounted for by atherosclerosis of the extracranial or intracranial arteries (Rossi et al. 1995). Therefore, atherosclerotic plaque stabilisation may be a new therapeutic strategy in the prevention of stroke (Gorelick 2002). Other important factors which contribute to stroke include diabetes mellitus, obesity and overweight, arterial fibrillation and hyperlipidaemia.

The deficiency in blood supply to a part of the brain results in neuronal loss and subsequent gliosis, which are typical characteristics of cerebral infarction. The degree of the neuronal damage depends on the extent of the cerebral blood flow (CBF) reduction and the duration of the ischaemia. When a cerebral artery is occluded and the CBF considerably decreases, cerebral electric activity fails, a deterioration of metabolic reactions and energy state occurs, which leads to the ischaemic cascade and neuronal death. A central ischaemic area, the ischaemic core, has been formed, the size of which depends mainly on the state of collateral arteries. The ischaemic area is surrounded by an ischaemic penumbra, a brain region with less CBF reduction in which perfusion was maintained by collateral circulation. The blood flow in this area lies between the two thresholds—the upper threshold of electrical failure (15–18 ml/100 g/min) and lower threshold of energy failure (10–12 ml/100 g/min). This area is characterised by electrical silence with only slightly elevated extracellular potassium concentrations. The collateral blood flow supplies sufficient oxygen and glucose to maintain the energy state at almost normal levels. This area is potentially recoverable if the perfusion is restored. The penumbra is unstable and dynamic in time, and usually it progresses towards infarction in a few hours, even though no further reduction in CBF occurred. Postischaemic inflammation and apoptosis mostly contribute to the lost of neurons and to the growth of structural lesion (Astrup et al. 1981; Dirnagl et al. 1999; Martinez-Vila and Sieira 2001).

4 Animal Models of Stroke

SHRs represent a well-established experimental tool in the hypertension research, an excellent model to study the mechanisms of the development of hypertension, responses of blood pressure to various pharmacological interventions and to investigate the development of the end-organ damage in hypertensive subjects. However, these animals do not develop stroke. A separate inbred strain of hypertensive animals, stroke-prone spontaneously hypertensive rats (SHR-SPs) develop stroke after feeding with a diet high in sodium and low in potassium and protein (Okamoto et al. 1973). These rats have contributed to identifying genetic factors in ischaemic stroke.

Experimental models of global and focal ischaemia have provided insight into the complex sequence of pathophysiological events which occur during and after stroke. Furthermore, these experimental models allow the assessment of the efficacy of potential therapeutic interventions. Two main global ischaemia models, bilateral carotid artery occlusion in the gerbil and four-vessel occlusion in the rat are used to study the outcome after global ischaemia in animals. The typical lesions of pyramidal cells in the hippocampus can be detected in both models. The most characteristic sensorimotor response in gerbils subjected to the bilateral carotid occlusion is the hyperlocomotion, which depends on the duration of the ischaemia. Global ischaemia induced by four-vessel occlusion in rats is rather characterised by a transient reduction in the locomotor activity during the first 24 h after ischaemia. At the later time points, when the rats are tested days or weeks after the ischaemic insult, little changes in the locomotor activity can be detected (Hunter et al. 1998). The global ischaemia models are believed to be pertinent in the relation to neurological deficits after a heart attack or cardiac arrest.

The most common case of stroke in humans is the occlusion of the middle carotid artery (MCA). The cortex, basal ganglia and the internal capsule represent the important brain regions supplied by the MCA and its small penetrating branches. According to some studies, focal infarctions of the MCA territory account for almost 80% of all ischaemic cerebral insults in humans. Permanent and transient occlusion of the MCA in mice and rats represent a well-established and the most frequently used animal models of focal cerebral ischaemia in humans. A permanent occlusion of the MCA can be achieved by a ligation of the artery or by electrocoagulation. Electrocoagulation of the main trunk of the MCA produces an area of damage involving the cortex and striatum (Chiamulera et al. 1993; for review see McAuley 1995). Bederson et al. (1986) have characterised precise anatomical sites of the MCA occlusion in the rat which produce uniform cerebral infarctions and developed a neurological grading system that can evaluate the degree of paresis after ischaemia. More detailed and complex neurological evaluation systems include quantification of a number of sensorimotor deficits (Garcia et al. 1995). Other modifications of the permanent MCA occlusion such as two-vessel occlusion and three-vessel occlusion models have been described to obtain a better-defined infarction area (McAuley 1995).

Models of transient occlusion of the MCA can explore the pathological events associated with re-establishing blood flow to the ischaemic territory. In stroke patients, the reperfusion can occur spontaneously through a resolution of an embolus or through clinical intervention. Modification of the same techniques, which are used for the permanent occlusion of the MCA, have been utilised to develop a model of transient focal ischaemia in the rat (McAuley 1995). Koizumi et al. (1986) described a method of an intraluminal thread model of MCA occlusion with subsequent reperfusion. This method and its later modification by Longa et al. (1989) became the most widely used models to study pathological events and therapeutic approaches in transient focal ischaemia. The effectiveness and reliability of both methods to produce acute ischaemia in the MCA territory have been critically re-evaluated and modified (Laing et al. 1993; Schmid-Elsaesser et al. 1998). The model of the intraluminal occlusion of the MCA with reperfusion is believed to be the most pertinent in relation to human ischaemic stroke. The ischaemic injury resembles that of embolic stroke patients and is associated with sensorimotor and cognitive dysfunction (Hunter et al. 1998). The transient occlusion of the MCA is the most frequently used model of focal cerebral ischaemia to test the efficacy of a large number of therapeutic interventions and neuroprotective compounds to prevent neuronal damage. Models of thromboembolic focal and multifocal ischaemia have been developed which try to resemble the events which occur in thromboembolic stroke. These models make it possible to examine the thrombolytic efficacy of compounds. The rat model of photothrombotic cortical ischaemia allows the study of the pathophysiological and morphological events in discrete cortical regions (Hunter et al. 1998; McAuley 1995).

5
Renin-Angiotensin System and Stroke

Evidence has accumulated that the RAS may play a central role in the development of cardiovascular diseases and stroke. Angiotensin-converting enzyme inhibitors (ACEI) and selective AT_1 receptor antagonists effectively lower blood pressure in hypertensive patients and prevent end-organ damage. Local tissue RASs, described in a number of organs, including the vascular wall, the heart and especially the brain, may considerably contribute to the pathogenesis of various processes linked to stroke such as atherosclerosis of the carotid artery and small brain arteries or regulation of cerebral blood flow during cerebral ischaemia. The brain RAS, which is regulated independently of the hormonal RAS, is believed to be involved in the initiation and regulation of events occurring during and after brain ischaemia. Recent pharmacological data and experiments carried out in transgenic mice overexpressing angiotensinogen or in AT_1 receptor knock-out mice indicate that the effector peptide of the RAS, angiotensin II (Ang II) may considerably contribute to pathophysiological processes in ischaemic brain tissue (Dai et al. 1999; Walther et al. 2002).

6
Genetics of the RAS and Ischaemic Stroke

Ischaemic stroke can be caused by a number of monogenic disorders, but the majority of cases of ischaemic stroke are multifactorial in aetiology. There is some evidence from epidemiological studies indicating genetic influences but the identification of individual causative mutations remains problematic. A candidate gene approach has been most frequently used in human studies. Association with polymorphisms in a variety of candidate genes have been investigated, including haemostatic genes, genes controlling homocysteine metabolism, the arterial nitric oxide synthase gene and others. The genetics of the ischaemic stroke has recently been reviewed (Hassan and Markus 2000). The angiotensin-converting enzyme (ACE) gene is one of the most extensively investigated candidate genes in ischaemic stroke. The enzyme generates the potent vasoactive peptide Ang II from Ang I. The *ACE* gene is known to have two polymorphic alleles I/D (insertion/deletion polymorphism). The *ACE DD* and *DI* genotypes are associated with high prevalence of hypertension, the *D* allele being linked to increased circulating levels of ACE. An association of the *ACE* gene with stroke with a relative risk of the order 1.5–2.5 has been reported in a number of studies. The *ACE I/D* genotype has been associated with ischaemic stroke in hypertensive patients, while and the *DD* genotype with lacunar stroke (Gorelick 2002). Szolnoki et al. (2001) have reported that the *ACE DD* genotype positively correlates with the occurrence of small-vessel infarctions rather than with large-vessel infarctions. Margaglione et al. (1996) evaluated the genotype of the *ACE* gene in 101 subjects with and 108 subjects without a history of ischaemic stroke and compared the two groups for major risk factors for ischaemic events. Deletion polymorphism of the *ACE* gene (*D/D* genotype) was shown to be more common in subjects with a history of stroke (relative risk 1.76; confidence interval, 1.02–3.05). A meta-analysis has evaluated seven studies on the risk of stroke in 1918 subjects vs 722 controls (Sharma 1998). The odds ratio for the *D* allele as an independent risk factor in ischaemic stroke was 1.31 (95% confidence interval, 1.06–1.61). The recessive *D* allele seems to be a modest but independent risk factor for ischaemic stroke. However, it should be noted that many of the studies were small and statistically underpowered, and various methodological difficulties seem to be responsible for the conflicting results regarding the *ACE* gene polymorphism and stroke. One variant of the angiotensinogen gene has been implicated in vascular diseases, but there were inconsistent results for its mutations M235T and T174M for ischaemic stroke (Sethi et al. 2001).

7
ACE Inhibitors and Stroke

ACE is predominantly localised on endothelial cells of vascular beds, including those of the brain. Enzymatic-and radioligand-binding assays demonstrated the presence of ACE in the carotid artery and cerebral microvessels in both normotensive rats and SHRs (Perich et al. 1992; Veniant et al. 1992).

Since hypertension is the most relevant risk factor for stroke, in hypertensive subjects, ACE inhibitors can primarily protect against brain stroke by reducing blood pressure. However, clinical studies and trials and experimental data indicate that additional mechanisms such as normalisation of cerebral blood flow autoregulation or increased arterial compliance may considerably contribute to the protective effects of ACE inhibitors against stroke (Chillon and Baumbach 2001; Saavedra et al. 2001; Torup et al. 1993; Veniant et al. 1992).

The Heart Outcomes and Prevention Evaluation (HOPE) Study was a double-blind, 2×2 factorial, randomised trial evaluating a 5-year treatment with ramipril and vitamin E for high-risk patients with a history of a cardiovascular disease (coronary artery disease, stroke or TIA, peripheral vascular disease, diabetes mellitus combined with another cardiovascular risk factor such as hypertension, elevated cholesterol, low high-density lipoprotein cholesterol, cigarette smoking). The primary outcome was myocardial infarction, stroke or death from cardiovascular events. The ACE inhibitor, ramipril, significantly reduced the risk of many cardiovascular events and complications in high-risk patients. A subgroup analysis revealed a reduction in the incidence of recurrent myocardial infarction (20%, $P<0.01$) and, especially, of the first stroke (32%, $P<0.001$) in the ramipril treatment group (Yusuf et al. 2000).

Perindopril Protection Against Recurrent Stroke Study (PROGRESS) was a double-blind, placebo-controlled, randomised trial designed to determine the effects of a blood-pressure lowering regimen in hypertensive and nonhypertensive patients with a history of any type of stroke except subarachnoid haemorrhage. The active group was treated with the ACE inhibitor perindopril alone or in combination with the diuretic indapamide. The primary outcome endpoint was recurrent stroke. Perindopril alone or in combination with the diuretic reduced blood pressure. Among participants treated with perindopril alone (in whom blood pressure was lowered by a mean of 5/3 mmHg), stroke risk was not discernibly different from that among subjects, who received single placebo. Combination therapy with perindopril and diuretic reduced blood pressure by 12/5 mmHg and stroke risk by 43%. This therapy regimen was associated with a lower risk of each of the main stroke subtypes: fatal or disabling stroke (46%), ischaemic stroke (36%) and cerebral haemorrhage (76%). The perindopril-treated group showed significant reductions in major vascular events and the first myocardial infarction. ACE inhibitor treatment also reduced dementia and cognitive decline among patients with stroke and stroke-related disability (PROGRESS collaborative group, 2001). In both studies, nonhypertensive subjects also benefited from the therapy with an ACE inhibitor.

Beneficial effects of ACE inhibitors on neurological outcome and recovery from cerebral ischaemia have been demonstrated in a number of studies carried out in hypertensive rats. Long-term treatment with the ACE inhibitor, cilazapril, prevented the occurrence of stroke in SHR-SP (Hajdu et al. 1991). In another study, cilazapril decreased neurological deficits and reduced infarction volume in SHR-SP subjected to focal cerebral ischaemia (Fujii et al. 1992). Similarly, chronic treatment with the ACE inhibitor, enalapril, considerably improved survival and prevented cerebrovascular lesions in salt-loaded SHR-SP independently of reductions in blood pressure (Stier et al. 1989). ACE inhibitors have also been demonstrated to exert beneficial effects on the metabolic and circulatory derangement in the ischaemic brain of SHRs (Sadoshima et al. 1993).

As stated above, ACE inhibitors can protect hypertensive subjects against stroke or improve recovery from ischaemic brain injury by reducing blood pressure, normalising cerebral blood flow autoregulation, improving collateral blood flow to the ischaemic tissue and by inhibiting Ang II formation in the brain.

CBF is controlled by autoregulatory mechanisms whereby CBF remains constant despite wide variations in systemic blood pressure. The control is exerted at the level of small resistance vessels. These vessels dilate in response to a drop in blood pressure or constrict, when the systemic blood pressure increases. Autoregulation of CBF has a lower and an upper blood pressure limit. When blood pressure drops below the lower limit of the cerebral autoregulation, small blood vessel dilation becomes inadequate and the CBF falls. When the blood pressure rises, the resistance vessels constrict until the upper limit of the cerebral autoregulation is reached. Above this upper limit, the vessels dilate and CBF increases. CBF in patients with uncomplicated essential hypertension is the same as in normotensive subjects. Chronic hypertension is associated with hypertrophy of the medial layer of cerebral arteries and arterioles. Vascular remodelling, comprising thickening of the media and an increase, in proportion of smooth muscle cells, in elastin and collagen, limits the ability of the resistance vessels to dilate. These changes protect the brain against high-perfusion pressure but, at the same time, impair dilation of these vessel at low pressure. In hypertensive subjects, the lower and upper limits of the CBF autoregulation are shifted towards higher perfusion pressures (Squire 1994; Strandgaard and Paulson 1992).

In animal models, ACE inhibitors had no effect on resting CBF but were shown to shift the lower and upper limits of the autoregulation towards lower blood pressure levels (Barry et al. 1984; Sadoshima et al. 1994; Squire 1994; Strandgaard and Paulson 1992). This effect would improve tolerance to hypotension while impairing that to hypertension. The mechanism may involve the inhibition of Ang II formation in cerebral vessels, resulting in the attenuation of constriction of large cerebral vessels. It has been proposed that reduced formation of Ang II by ACE inhibitors dilates the large cerebral arteries with compensatory constriction of smaller arteries and arterioles (Paulson et al. 1988). Although resting CBF is unaltered, smaller cerebral arteries possess a greater dilatory capacity, thus explaining the shift of the CBF autoregulation curve towards lower blood pressure. Recent findings indicate that this effect is, at least partly, mediated by bradykinin (Takada et al. 2001). However, Ang II acting on AT_1 receptors represents a growth factor, initiating proliferation, hypertrophy and growth in various tissues. Ang II was reported to increase the media width, media cross-sectional area and media/lumen ratio in rats by a non-pressor mechanism (Griffin et al. 1991). Long-term treatment with ACE inhibitors decreases the hypertrophy of the media of cerebral arteries and arterioles, increases their external diameter and normalises endothelial function (Veniant et al. 1992). Chronic administration of ACE inhibitors in hypertensive subjects and rats with genetic and experimental hypertension reverses the alteration in CBF autoregulation (Saavedra et al. 2001). ACE inhibitors have been demonstrated to effectively reduce blood pressure in patients without altering the resting CBF (Dyker et al. 1997; Frei and Muller-Brand 1986). These findings indicate that antihypertensive agents such as ACE inhibitors, which normalise the CBF autoregulation and protect the cerebral circulation in hypertensive patients, might be preferred for the treatment of hypertension in patients with advanced atherosclerosis.

Treatment with ACE inhibitors improved the recovery from cerebral ischaemia also in normotensive rats. Captopril injected intravenously 30 min prior to cerebral ischaemia improved neurological outcome on the 3rd day after experimental ischaemic stroke (Werner et al. 1991). Since in normotensive rats the CBF autoregulation is not impaired, the beneficial effects of ACE inhibitors may result from the inhibition of Ang II formation in the brain. A number of in vivo and ex vivo studies indicate that ACE in the brain can be inhibited after short- or long-term systemic treatment with ACE inhibitors; however, the results are rather equivocal (Unger et al. 1988). One of the most controversial topics concerning the central actions of ACE inhibitors is the question whether these compounds inhibit Ang II formation in brain structures inside the blood–brain barrier upon systemic administration (Gohlke et al. 1989). Moreover, some of the actions of ACE inhibitors may be unrelated to the inhibition of the brain RAS. Along with the reduction in the levels of Ang II, the neuroprotection conferred by ACE inhibitors can also be related to various effects, including an increase in bradykinin or other brain peptides, suppression of sympathetic activity and antioxidant effects. Therefore, questions as to which extent the inhibition of the Ang II formation in the ischaemic brain contributes to the beneficial effects of ACE inhibitors in stroke cannot be definitely answered as yet.

The use of ACE inhibitors in the acute phase of stroke as a therapeutic strategy to reduce neuronal injury and subsequent disability of patients is a matter of controversy. Immediately after cerebral occlusion, CBF is extremely low in the centre of the lesion and considerably reduced in the penumbra. In hypertensive subjects, cerebral vessels have a limited capacity to dilate. This can result in a profound reduction in collateral blood flow to the ischaemic core and to the penumbra, thereby leading to neurological deterioration and enlarged infarction area, especially when acute ischaemic stroke is accompanied by a drop in blood pressure.

However, blood pressure is usually increased in the acute phase of stroke and there is no agreed consensus on the treatment of hypertension following stroke. In the early poststroke phase, the CBF autoregulation is impaired and CBF becomes dependent on systemic blood pressure. Reduction of blood pressure may reduce blood flow to the ischaemic penumbra and neuronal death is likely to occur. Initiating antihypertensive therapy may, therefore, have deleterious effects. Alternatively, failure to reduce an abnormally high blood pressure can promote cerebral oedema and increase the size of the infarction. The loss of autoregulatory function further complicates the issue of blood pressure management in the acute phase of stroke (Bhalla et al. 2001; Fotherby and Panayiotou 1999; Squire 1994). Only a few studies report on effects of ACE inhibitors on CBF in patients suffering from stroke. Treatment of patients within 5 days after their first stroke with captopril had no effect on mean hemispheric blood flow (Waldemar et al. 1989). Similarly, the ACE inhibitor, perindopril, lowered blood pressure without affecting CBF in patients 2–7 days after symptoms of cerebral infarction. (Dyker et al. 1997). On the contrary, Naritomi et al. (1984) reported that ACE inhibition increased CBF in hypertensive patients with stroke.

8 AT_1 and AT_2 Receptor Antagonists and Stroke

The major actions of Ang II are mediated by two subtypes of G-protein-coupled angiotensin receptors: the AT_1 and the AT_2 receptor. The AT_1 receptor is widely distributed in adult tissues. All classical effects of Ang II such as vasoconstriction, aldosterone release, renal salt and water retention are mediated by this angiotensin receptor subtype. The AT_1 receptor can be selectively blocked by the family of so-called sartan compounds, the AT_2 receptor can be selectively inhibited by the nonpeptide ligands PD 123177 and PD 123319. The selective AT_1 receptor antagonists are increasingly used in antihypertensive treatment and the prevention of end-organ damage related to hypertension.

As already mentioned, some antihypertensive agents such as ACE inhibitors reduce arterial blood pressure without affecting basal cerebral blood flow. Several clinical trials were designed to evaluate whether AT_1 receptor antagonists can protect against brain ischaemia.

The Losartan Intervention For Endpoint reduction (LIFE) study was a double-blind, randomised, parallel-group trial in patients with essential hypertension. The aim of the study was to establish whether selective blockade of AT_1 receptors with losartan improves left ventricular hypertrophy beyond reducing blood pressure and reduces cardiovascular morbidity and death and to compare the effects of losartan with the β-adrenoceptor blocker, atenolol. Losartan was superior to atenolol in reducing the primary composite endpoints of cardiovascular death, myocardial infarction and stroke. Especially the outcome for stroke was highly significant in favour of losartan, showing a 24.9% relative risk reduction compared with atenolol (P=0.001) (Dahlof et al. 2002).

Acute Candesartan Cilexetil Evaluation in Stroke Survivors (ACCESS) was a double-blinded, randomised multicentre trial designed to evaluate the influence of an early moderate blood pressure reduction in patients with acute cerebral ischaemia in comparison to restrictive antihypertensive therapy. Inclusion criteria were initial blood pressure values of greater than 200/110 mmHg as a median value of two measurements in 30 min. Additionally, a motor paresis had to be present. Patients were randomised and treated double-blind for 7 days with placebo or candesartan cilexetil. Primary endpoints were patient morbidity, measured by neurological status, and mortality rates after 3 months. The combined endpoints of total mortality, cerebral complications and cardiovascular complications was reduced by 47.5% for patients treated with the AT_1 receptor antagonist candesartan (Schrader et al. 2003).

SCOPE was a multicentre, prospective, randomised, double-blind, parallel-group study which evaluated the prophylactic effects of candesartan cilexetil on major vascular events in elderly patients with mild hypertension. The primary outcome was cardiovascular death, nonfatal myocardial infarction and stroke; the secondary outcomes comprised the extent of cognitive function or dementia, total mortality and renal function. There was a statistically significant 28% reduction in the risk of nonfatal stroke in the candesartan-treated group. Other outcomes showed only non-statistically significant beneficial trends in the candesartan-treated group (Lithell et al. 2003).

A number of animal studies have also demonstrated protective effects of AT_1 receptor antagonists against stroke. Long-term treatment of SHR-SP with losartan or candesartan at doses which did not affect blood pressure reduced the incidence of stroke and cerebrovascular lesions (Inada et al. 1997; Stier et al. 1993). Losartan decreased the incidence and delayed the progression of renal damage and stroke and increased survival in salt-loaded Dahl salt-sensitive rats. In the losartan-treated rats, the development of hypertension was slightly delayed, but after 6 weeks of treatment, the systolic blood pressure reached values above 200 mmHg and was identical in both the vehicle-treated and losartan-treated groups of rats. Regression analyses revealed that the protective effect of losartan in increasing survival and reducing end-organ damage was beyond that expected for its blood pressure lowering effects (von Lutterotti et al. 1992).

Chronic systemic pretreatment of SHRs with candesartan has been shown to reduce neuronal injury resulting from focal brain ischaemia (Ito et al. 2002; Nishimura et al. 2000). Dai et al. (1999) reported that a long-term blockade of brain AT_1 receptors improved the recovery from focal brain ischaemia in normotensive rats.

In hypertensive subjects, AT_1 receptor antagonists can primarily prevent stroke by reducing blood pressure. Similarly to ACE inhibitors, AT_1 receptor antagonists can improve the recovery from ischaemic stroke by normalisation of cerebral blood flow autoregulation and by improvement in collateral blood flow to ischaemic tissue. Inhibition of AT_1 receptors in the brain may also contribute to the beneficial effects of AT_1 receptor antagonists (Culman et al. 2002).

Of all angiotensin receptor subtypes, the AT_1 receptor is predominantly expressed in brain arteries in the adult organism. Cerebral microvessels may express, along with ACE, other subtypes of angiotensin receptors, which have not yet been well characterised. Ang II can be formed locally and, together with circulating Ang II, contribute to CBF regulation (for a review see Saavedra et al. 2001).

Experimental evidence indicates that, after vascular occlusion, angiotensin II can restore the blood supply to ischaemic areas. Fernandez and co-workers (1986) observed that infusion of pressor doses of Ang II decreased the mortality rate in gerbils with unilateral carotid ligation through mechanisms unrelated to the hypertensive action of the peptide. The underlying mechanisms have not been clearly defined, but Ang II may exert these beneficial effects independently of the rise in blood pressure, through a fast opening of the reserve collateral circulation and/or through a much slower response comprising new vessel formation

In line with this assumption, mortality increased in progressively ligated gerbils treated with the ACE inhibitor enalapril (Achard et al. 2001; Kaliszewski et al. 1988). In another study, Dalmay et al. (2001) demonstrated that pretreatment with ACE inhibitors, but not with AT_1 receptor antagonists, significantly decreased survival of gerbils subjected to ischaemic stroke. The effects were not blood pressure dependent and obviously not mediated by AT_1 receptors. Moriuchi et al. (1998) have reported on the presence of an atypical angiotensin type 1-like receptor in the gerbil brain and in some peripheral tissues, which displays high affinity for Ang II but is not sensitive to displacement with selective ligands for the AT_1 and the AT_2 receptor such as losartan and CGP 42112 or PD 123319, respectively. This atypical angiotensin receptor might be responsible for the beneficial effects of Ang II in gerbils exposed to cerebral ischaemia.

The role of the individual angiotensin receptor subtypes in the regulation of CBF is unclear. Similarly to ACE inhibitors, the AT_1 receptor antagonist, candesartan, was reported to shift both the upper and lower limits of the CBF autoregulation curve towards lower blood pressure in SHR and in WKY normotensive rats, whereas baseline cerebral blood flow was unaffected (Vraamark et al. 1995). In contrast, Strömberg at al. (1993) reported that losartan shifted dose-dependently the upper limit of CBF towards higher pressures in Sprague-Dawley rats. Interestingly, the same authors found that the selective AT_2 receptor ligands, CGP 42112 and PD123319, also shifted the upper limit of the cerebral blood flow autoregulation towards higher pressures (Näveri et al. 1994; Stromberg et al. 1993). The conclusion of Näveri et al. (1994) was that both CGP 42112 and PD 123319 exerted their effects through stimulation of AT_2 receptors. This is a somewhat surprising finding because PD 123319 is an antagonist and CGP 42112 (at least at higher doses) an agonist for the AT_2 receptor (de Gasparo et al. 2000). These confusing findings exemplify that the role of angiotensin and its receptors in the regulation of the CBF is far from being understood. AT_2 receptor stimulation by PD 123319 was reported to increase survival in gerbils subjected to abrupt unilateral carotid ligation (Fernandez et al. 1994).

Recent findings indicate that, similarly to ACE inhibitors, AT_1 receptor antagonists can improve the recovery from ischaemic stroke by normalisation of cerebral blood flow autoregulation. Long-term pretreatment of SHRs with candesartan reduced infarction volume after temporal or permanent MCAO (Ito et al. 2002; Nishimura et al. 2000). Pretreatment of SHRs with candesartan for 2 weeks resulted in almost complete inhibition of AT_1 receptors in the cerebral arteries and in the brain areas involved in the central regulation of cerebral blood flow and the sympathetic system. In SHRs exposed to temporary MCAO with reperfusion, candesartan reduced the volume of injury predominantly in the affected cortical areas. The proposed mechanisms involve the shifting the autoregulatory curve to the left, in the direction of improved vasodilation, and the prevention of the decrease in blood flow in the marginal zone of ischaemia (Nishimura et al. 2000). The observed reduction in cerebral oedema immediately after MCAO and in the infarction volume resulted most probably from the normalisation of cerebrovascular autoregulation in the marginal ischaemic zone (Saavedra et al. 2001). Other effects observed in SHRs after long-term AT_1 receptor blockade such as an increase in the external diameter and reduction of the media thickness of the middle cerebral artery, resulting in an improved arterial compliance and increased capacity of cerebral arteries to dilate when CBF decreases, can also considerably contribute to the beneficial effects of AT_1 receptor antagonists in cerebral ischaemia (Ito et al. 2002).

However, AT_1 receptor antagonists may also improve recovery from stroke by mechanisms independent of the normalisation of cerebrovascular autoregulation or blood pressure reduction in hypertension. Our group has demonstrated that long-term blockade of brain AT_1 receptors by irbesartan improves neurological outcome of focal cerebral ischaemia and markedly reduces the expression of the AP-1 transcription factors c-Fos and c-Jun in the parietal and piriform cortices on the ligated side of the brain and completely abolishes the ischaemia-induced c-Fos expression in the hippocampus. In these experiments, the AT_1 receptor antagonists were infused intracerebroventricularly over a 5-day period before the induction of ischaemia at a dose which effectively inhibited brain but not vascular AT_1 receptors (Dai et al. 1999). Taken together, these findings demonstrate that AT_1 receptor antagonists may improve the recovery from stroke by restoration of blood flow after ischaemia and by blocking the biochemical and metabolic changes at the ischaemic cascade level, including the inhibition of ITF expression in the ischaemic brain tissue.

Since effective and long-lasting blockade of brain AT_1 receptors may protect against stroke and provide therapeutic benefits for patients suffering from brain ischaemia, we studied the access of systemically applied losartan, irbesartan, telmisartan and candesartan to brain AT_1 receptors. All studied antagonists penetrate the blood–brain barrier in a dose- and time-dependent manner to inhibit centrally mediated effects of Ang II. In our studies, candesartan produced the most effective and long-lasting inhibition of brain AT_1 receptors following systemic application (Culman et al. 1999; Gohlke et al. 2001, 2002a, 2002b).

We have also investigated the role of AT_2 receptors in the processes occurring during and after focal brain ischaemia. Although a long-term inhibition of brain AT_2 receptors did not affect the recovery from stroke, it prevented the beneficial effects of the AT_1 receptor blockade (Blume et al. 2000). Ischaemic lesions of the nervous system have been reported to increase the expression of AT_2 receptors in the brain (Makino et al. 1996; Zhu et al. 2000). When AT_1 receptors are inhibited, Ang II can increasingly interact with AT_2 receptors, as AT_1 receptor antagonists leave the AT_2 receptor unopposed and rather expose it to elevated Ang II levels. Activation of AT_2 receptors in brain tissue that has undergone ischaemic injury may, on one side, initiate neuroregenerative events or, on the other side, induce apoptosis when neurons are severely damaged (for a review see Unger 1999). Both these effects are important for the recovery from stroke.

9 Outlook

The findings that AT_1 receptor antagonists cross the blood–brain barrier, improve neurological outcome and prevent brain injury following brain ischaemia in experimental animals may provide a basis for new therapeutic strategies in the prevention and treatment of stroke. Similarly to AT_1 receptor antagonists, several classes of drugs such as excitatory amino acid antagonists, antioxidants inhibiting free radicals, drugs related to the NO system, various calcium antagonists and others have displayed neuroprotective effects in cerebral ischaemia and decreased neuronal damage in experimental animals. However, a number of clinical trials have produced negative results when the same drugs were tested in patients suffering from acute stroke. Many of these agents caused serious psychotomimetic or cardiovascular side effects. The studies had to be terminated and the clinical developments of these agents were discontinued (Martinez-Vila and Sieira 2001). On the other hand, AT_1 receptor antagonists and ACE inhibitors display a low incidence of side effects and are successfully used in antihypertensive treatment and the prevention of end-organ damage related to hypertension. The results of clinical trials have shown that inhibitors of RAS system reduce the incidence of stroke in high-risk patients with vascular diseases or diabetes mellitus combined with other risk factors. In the few clinical trials which have been conducted so far, AT_1 receptor antagonists displayed clinical efficacy in stroke-prone or stroke patients. The outcomes of these clinical studies raise the hope that this class of drugs may be a new therapeutic approach to reduce the magnitude and extent of brain tissue damage in stroke patients.

References

Achard J, Fournier A, Mazouz H, Caride VJ, Penar PL, Fernandez LA (2001) Protection against ischemia: a physiological function of the renin-angiotensin system. Biochem Pharmacol 62:261–271

Albiston AL, McDowall SG, Matsacos D, Sim P, Clune E, Mustafa T, Lee J, Mendelsohn FA, Simpson RJ, Connolly LM, Chai SY (2001) Evidence that the angiotensin IV (AT(4)) receptor is the enzyme insulin-regulated aminopeptidase. J Biol Chem 276:48623–48626

Astrup J, Siesjo BK, Symon L (1981) Thresholds in cerebral ischemia—the ischemic penumbra. Stroke 12:723–725

Barry DI, Paulson OB, Jarden JO, Juhler M, Graham DI, Strandgaard S (1984) Effects of captopril on cerebral blood flow in normotensive and hypertensive rats. Am J Med 76:79–85

Bederson JB, Pitts LH, Tsuji M, Nishimura MC, Davis RL, Bartkowski H (1986) Rat middle cerebral artery occlusion: evaluation of the model and development of a neurologic examination. Stroke 17:472–476

Bhalla A, Wolfe CD, Rudd AG (2001) Management of acute physiological parameters after stroke. QJM 94:167–172

Blume A, Lebrun CJ, Herdegen T, Bravo R, Linz W, Mollenhoff E, Unger T (1997) Increased brain transcription factor expression by angiotensin in genetic hypertension. Hypertension 29:592–598

Blume A, Seifert K, Lebrun CJ, Mollenhoff E, Gass P, Unger T, Herdegen T (1998) Differential time course of angiotensin-induced AP-1 and Krox proteins in the rat lamina terminalis and hypothalamus. Neurosci Lett 241:87–90

Blume A, Funk A, Gohlke P, Unger T, Culman J (2000) AT_2 receptor inhibition in the rat brain reverses the beneficial effects of AT_1 receptor blockade on neurological outcome after focal brain ischemia (abstract). Hypertension 36:656

Blume A, Neumann C, Dorenkamp M, Culman J, Unger T (2002) Involvement of adrenoceptors in the angiotensin II-induced expression of inducible transcription factors in the rat forebrain and hypothalamus. Neuropharmacology 42:281–288

Braszko JJ, Kupryszewski G, Witczuk B, Wisniewski K (1988) Angiotensin II-(3–8)-hexapeptide affects motor activity, performance of passive avoidance and a conditioned avoidance response in rats. Neuroscience 27:777–783

Campagnole-Santos MJ, Diz DI, Ferrario CM (1988) Baroreceptor reflex modulation by angiotensin II at the nucleus tractus solitarii. Hypertension 11:I167–I171

Casto R, Phillips MI (1986) Angiotensin II attenuates baroreflexes at nucleus tractus solitarius of rats. Am J Physiol 250:R193–R198

Chavrier P, Janssen-Timmen U, Mattei MG, Zerial M, Bravo R, Charnay P (1989) Structure, chromosome location, and expression of the mouse zinc finger gene Krox-20: multiple gene products and coregulation with the proto-oncogene c-fos. Mol Cell Biol 9:787–797

Chiamulera C, Terron A, Reggiani A, Cristofori P (1993) Qualitative and quantitative analysis of the progressive cerebral damage after middle cerebral artery occlusion in mice. Brain Res 606:251–258

Chillon JM, Baumbach GL (2001) Effects of an angiotensin-converting enzyme inhibitor and a beta-blocker on cerebral arteriolar dilatation in hypertensive rats. Hypertension 37:1388–1393

Cohen DR, Ferreira PC, Gentz R, Franza BR Jr, Curran T (1989) The product of a fos-related gene, fra-1, binds cooperatively to the AP-1 site with Jun: transcription factor AP-1 is comprised of multiple protein complexes. Genes Dev 3:173–184

Collins R, Peto R, MacMahon S, Hebert P, Fiebach NH, Eberlein KA, Godwin J, Qizilbash N, Taylor JO, Hennekens CH (1990) Blood pressure, stroke, and coronary heart disease. Part 2, Short-term reductions in blood pressure: overview of randomised drug trials in their epidemiological context. Lancet 335:827–838

Culman J, Hohle S, Qadri F, Edling O, Blume A, Lebrun C, Unger T (1995) Angiotensin as neuromodulator/neurotransmitter in central control of body fluid and electrolyte homeostasis. Clin Exp Hypertens 17:281–293

Culman J, von Heyer C, Piepenburg B, Rascher W, Unger T (1999) Effects of systemic treatment with irbesartan and losartan on central responses to angiotensin II in conscious, normotensive rats. Eur J Pharmacol 367:255–265

Culman J, Blume A, Gohlke P, Unger T (2002) The renin-angiotensin system in the brain: possible therapeutic implications for AT(1)-receptor blockers. J Hum Hypertens Suppl 16 3:S64–S70

Dahlof B, Devereux RB, Kjeldsen SE, Julius S, Beevers G, Faire U, Fyhrquist F, Ibsen H, Kristiansson K, Lederballe-Pedersen O, Lindholm LH, Nieminen MS, Omvik P, Oparil S, Wedel H (2002) Cardiovascular morbidity and mortality in the Losartan Intervention For Endpoint reduction in hypertension study (LIFE): a randomised trial against atenolol. Lancet 359:995–1003

Dai WJ, Funk A, Herdegen T, Unger T, Culman J (1999) Blockade of central angiotensin AT(1) receptors improves neurological outcome and reduces expression of AP-1 transcription factors after focal brain ischemia in rats. Stroke 30:2391–2398; discussion 2398–2399

Dalmay F, Mazouz H, Allard J, Pesteil F, Achard JM, Fournier A (2001) Non-AT(1)-receptor-mediated protective effect of angiotensin against acute ischaemic stroke in the gerbil. J Renin Angiotensin Aldosterone Syst 2:103–106

De Gasparo M, Catt KJ, Inagami T, Wright JW, Unger T (2000) International union of pharmacology. XXIII. The angiotensin II receptors. Pharmacol Rev 52:415–472

DiBona GF, Jones SY, Brooks VL (1995) ANG II receptor blockade and arterial baroreflex regulation of renal nerve activity in cardiac failure. Am J Physiol 269:R1189–R1196

Dirnagl U, Iadecola C, Moskowitz MA (1999) Pathobiology of ischaemic stroke: an integrated view. Trends Neurosci 22:391–397

Diz DI, Jessup JA, Westwood BM, Bosch SM, Vinsant S, Gallagher PE, Averill DB (2002) Angiotensin peptides as neurotransmitters/neuromodulators in the dorsomedial medulla. Clin Exp Pharmacol Physiol 29:473–482

Dyker AG, Grosset DG, Lees K (1997) Perindopril reduces blood pressure but not cerebral blood flow in patients with recent cerebral ischemic stroke. Stroke 28:580–583

Eggen BJ, Nielander HB, Rensen-de Leeuw MG, Schotman P, Gispen WH, Schrama LH (1994) Identification of two promoter regions in the rat B-50/GAP-43 gene. Brain Res Mol Brain Res 23:221–234

Fernandez LA, Spencer DD, Kaczmar T Jr (1986) Angiotensin II decreases mortality rate in gerbils with unilateral carotid ligation. Stroke 17:82–85

Fernandez LA, Caride VJ, Stromberg C, Naveri L, Wicke JD (1994) Angiotensin AT_2 receptor stimulation increases survival in gerbils with abrupt unilateral carotid ligation. J Cardiovasc Pharmacol 24:937–940

Ferrario CM, Barnes KL, Diz DI, Block CH, Averill DB (1987) Role of area postrema pressor mechanisms in the regulation of arterial pressure. Can J Physiol Pharmacol 65:1591–1597

Fotherby MD, Panayiotou B (1999) Antihypertensive therapy in the prevention of stroke: what, when and for whom? Drugs 58:663–674

Frei A, Muller-Brand J (1986) Cerebral blood flow and antihypertensive treatment with enalapril. J Hypertens 4:365–367

Fujii K, Weno BL, Baumbach GL, Heistad DD (1992) Effect of antihypertensive treatment on focal cerebral infarction. Hypertension 19:713–716

Garcia JH, Wagner S, Liu KF, Hu XJ (1995) Neurological deficit and extent of neuronal necrosis attributable to middle cerebral artery occlusion in rats. Statistical validation. Stroke 26:627–634; discussion 635

Gohlke P, Urbach H, Scholkens B, Unger T (1989) Inhibition of converting enzyme in the cerebrospinal fluid of rats after oral treatment with converting enzyme inhibitors. J Pharmacol Exp Ther 249:609–616

Gohlke P, Weiss S, Jansen A, Wienen W, Stangier J, Rascher W, Culman J, Unger T (2001) AT_1 receptor antagonist telmisartan administered peripherally inhibits central responses to angiotensin II in conscious rats. J Pharmacol Exp Ther 298:62–70

Gohlke P, Kox T, Jurgensen T, von Kugelgen S, Rascher W, Unger T, Culman J (2002a) Peripherally applied candesartan inhibits central responses to angiotensin II in conscious rats. Naunyn Schmiedebergs Arch Pharmacol 365:477–483

Gohlke P, Von Kugelgen S, Jurgensen T, Kox T, Rascher W, Culman J, Unger T (2002b) Effects of orally applied candesartan cilexetil on central responses to angiotensin II in conscious rats. J Hypertens 20:909–918

Gorelick PB (2002) Stroke prevention therapy beyond antithrombotics: unifying mechanisms in ischemic stroke pathogenesis and implications for therapy: an invited review. Stroke 33:862–875

Greenberg ME, Ziff EB, Greene LA (1986) Stimulation of neuronal acetylcholine receptors induces rapid gene transcription. Science 234:80–83

Griffin SA, Brown WC, MacPherson F, McGrath JC, Wilson VG, Korsgaard N, Mulvany MJ, Lever AF (1991) Angiotensin II causes vascular hypertrophy in part by a non-pressor mechanism. Hypertension 17:626–635

Hajdu MA, Heistad DD, Ghoneim S, Baumbach GL (1991) Effects of antihypertensive treatment on composition of cerebral arterioles. Hypertension 18:II15–II21

Hassan A, Markus HS (2000) Genetics and ischaemic stroke. Brain 123:1784–812

Hengerer B, Lindholm D, Heumann R, Ruther U, Wagner EF, Thoenen H (1990) Lesion-induced increase in nerve growth factor mRNA is mediated by c-fos. Proc Natl Acad Sci U S A 87:3899–3903

Herbert J, Forsling ML, Howes SR, Stacey PM, Shiers HM (1992) Regional expression of c-fos antigen in the basal forebrain following intraventricular infusions of angiotensin and its modulation by drinking either water or saline. Neuroscience 51:867–882

Herzig TC, Jobe SM, Aoki H, Molkentin JD, Cowley AW Jr, Izumo S, Markham BE (1997) Angiotensin II type1a receptor gene expression in the heart: AP-1 and GATA-4 participate in the response to pressure overload. Proc Natl Acad Sci U S A 94:7543–7548

Hohle S, Blume A, Lebrun C, Culman J, Unger T (1995) Angiotensin receptors in the brain. Pharmacol Toxicol 77:306–315

Hunter AJ, Mackay KB, Rogers DC (1998) To what extent have functional studies of ischaemia in animals been useful in the assessment of potential neuroprotective agents? Trends Pharmacol Sci 19:59–66

Ichiki T, Inagami T (1995) Transcriptional regulation of the mouse angiotensin II type 2 receptor gene. Hypertension 25:720–725

Inada Y, Wada T, Ojima M, Sanada T, Shibouta Y, Kanagawa R, Ishimura Y, Fujisawa Y, Nishikawa K (1997) Protective effects of candesartan cilexetil (TCV-116) against stroke, kidney dysfunction and cardiac hypertrophy in stroke-prone spontaneously hypertensive rats. Clin Exp Hypertens 19:1079–1099

Ito T, Yamakawa H, Bregonzio C, Terron JA, Falcon-Neri A, Saavedra JM (2002) Protection against ischemia and improvement of cerebral blood flow in genetically hypertensive rats by chronic pretreatment with an angiotensin II AT_1 antagonist. Stroke 33:2297–2303

Jenkins TA, Allen AM, Chai SY, Mendelsohn FA (1995) Interactions of angiotensin II with central catecholamines. Clin Exp Hypertens 17:267–280

Kaliszewski C, Fernandez LA, Wicke JD (1988) Differences in mortality rate between abrupt and progressive carotid ligation in the gerbil: role of endogenous angiotensin II. J Cereb Blood Flow Metab 8:149–154

Kawahara Y, Sunako M, Tsuda T, Fukuzaki H, Fukumoto Y, Takai Y (1988) Angiotensin II induces expression of the c-fos gene through protein kinase C activation and calcium ion mobilization in cultured vascular smooth muscle cells. Biochem Biophys Res Commun 150:52–59

Kim S, Kawamura M, Wanibuchi H, Ohta K, Hamaguchi A, Omura T, Yukimura T, Miura K, Iwao H (1995) Angiotensin II type 1 receptor blockade inhibits the expression of immediate-early genes and fibronectin in rat injured artery. Circulation 92:88–95

Koizumi J, Yoshida Y, Nakazawa T, Ooneda G (1986) Experimental studies of ischemic brain edema, 1: a new experimental model of cerebral embolism in rats in which recirculation can be induced in the ischemic area. Jpn J Stroke 8:1–8

Laing RJ, Jakubowski J, Laing RW (1993) Middle cerebral artery occlusion without craniectomy in rats. Which method works best? Stroke 24:294–297; discussion 297–298

Lebrun CJ, Blume A, Herdegen T, Seifert K, Bravo R, Unger T (1995) Angiotensin II induces a complex activation of transcription factors in the rat brain: expression of Fos, Jun and Krox proteins. Neuroscience 65:93–99

Lenkei Z, Palkovits M, Corvol P, Llorens-Cortes C (1997) Expression of angiotensin type-1 (AT_1) and type-2 (AT_2) receptor mRNAs in the adult rat brain: a functional neuroanatomical review. Front Neuroendocrinol 18:383–439

Lithell H, Hansson L, Skoog I, Elmfeldt D, Hofman A, Olofsson B, Trenkwalder P, Zanchetti A (2003) The Study on Cognition and Prognosis in the Elderly (SCOPE): principal results of a randomized double-blind intervention trial. J Hypertens 21:875–886

Longa EZ, Weinstein PR, Carlson S, Cummins R (1989) Reversible middle cerebral artery occlusion without craniectomy in rats. Stroke 20:84–91

Luoh SH, Chan SH (2001) Inhibition of baroreflex by angiotensin II via Fos expression in nucleus tractus solitarii of the rat. Hypertension 38:130–135

MacMahon S, Peto R, Cutler J, Collins R, Sorlie P, Neaton J, Abbott R, Godwin J, Dyer A, Stamler J (1990) Blood pressure, stroke, and coronary heart disease. Part 1, Prolonged differences in blood pressure: prospective observational studies corrected for the regression dilution bias. Lancet 335:765–774

Makino I, Shibata K, Ohgami Y, Fujiwara M, Furukawa T (1996) Transient upregulation of the AT_2 receptor mRNA level after global ischemia in the rat brain. Neuropeptides 30:596–601

Margaglione M, Celentano E, Grandone E, Vecchione G, Cappucci G, Giuliani N, Colaizzo D, Panico S, Mancini FP, Di Minno G (1996) Deletion polymorphism in the angiotensin-converting enzyme gene in patients with a history of ischemic stroke. Arterioscler Thromb Vasc Biol 16:304–309

Martinez-Vila E, Sieira PI (2001) Current status and perspectives of neuroprotection in ischemic stroke treatment. Cerebrovasc Dis 11:60–70

Masood K, Besnard F, Su Y, Brenner M (1993) Analysis of a segment of the human glial fibrillary acidic protein gene that directs astrocyte-specific transcription. J Neurochem 61:160–166

McAuley MA (1995) Rodent models of focal ischemia. Cerebrovasc Brain Metab Rev 7:153–180

McKinley MJ, Badoer E, Oldfield BJ (1992) Intravenous angiotensin II induces Fos-immunoreactivity in circumventricular organs of the lamina terminalis. Brain Res 594:295–300

McKinley MJ, Badoer E, Vivas L, Oldfield BJ (1995) Comparison of c-fos expression in the lamina terminalis of conscious rats after intravenous or intracerebroventricular angiotensin. Brain Res Bull 37:131–137

Michelini LC, Bonagamba LG (1988) Baroreceptor reflex modulation by vasopressin microinjected into the nucleus tractus solitarii of conscious rats. Hypertension 11:I75–I79

Moeller I, Paxinos G, Mendelsohn FA, Aldred GP, Casley D, Chai SY (1996) Distribution of AT4 receptors in the Macaca fascicularis brain. Brain Res 712:307–324

Mohr E, Richter D (1990) Sequence analysis of the promoter region of the rat vasopressin gene. FEBS Lett 260:305–308

Morgan JI, Curran T (1991) Stimulus-transcription coupling in the nervous system: involvement of the inducible proto-oncogenes fos and jun. Annu Rev Neurosci 14:421–451

Moriuchi R, Shibata S, Himeno A, Johren O, Hoe KL, Saavedra JM (1998) Molecular cloning and pharmacological characterization of an atypical gerbil angiotensin II type-1 receptor and its mRNA expression in brain and peripheral tissues. Brain Res Mol Brain Res 60:234–246

Murakami H, Liu JL, Zucker IH (1996) Blockade of AT_1 receptors enhances baroreflex control of heart rate in conscious rabbits with heart failure. Am J Physiol 271:R303–R309

Murasawa S, Matsubara H, Kijima K, Maruyama K, Ohkubo N, Mori Y, Iwasaka T, Inada M (1996) Down-regulation by cAMP of angiotensin II type 2 receptor gene expression in PC12 cells. Hypertens Res 19:271–279

Naritomi H, Shimizu T, Watanabe Y, Murata S, Sawada T (1994) Effects of angiotensin converting enzyme inhibitor alacepril on cerebral blood flow in hypertensive stroke patients. Curr Ther Res 55:1446–1454

Näveri L, Stromberg C, Saavedra JM (1994) Angiotensin II AT_2 receptor stimulation extends the upper limit of cerebral blood flow autoregulation: agonist effects of CGP 42112 and PD 123319. J Cereb Blood Flow Metab 14:38–44

Nishimura Y, Ito T, Saavedra JM (2000) Angiotensin II AT(1) blockade normalizes cerebrovascular autoregulation and reduces cerebral ischemia in spontaneously hypertensive rats. Stroke 31:2478–2486

Okamoto K, Yamori Y, Nagaoka A (1974) Establishment of the stroke-prone spontaneously hypertensive rat (SHR). Circ Res 33/34 [Suppl 1]:I143–I153

Oldfield BJ, McKinley MJ (1994) Distribution of Fos in rat brain resulting from endogenously-generated angiotensin II. Kidney Int 46:1567–1569

Paulson OB, Waldemar G, Andersen AR, Barry DI, Pedersen EV, Schmidt JF, Vorstrup S (1988) Role of angiotensin in autoregulation of cerebral blood flow. Circulation 77:I55–I58

Perich R, Jackson B, Paxton D, Johnston CI (1992) Characterization of angiotensin converting enzyme in isolated cerebral microvessels from spontaneously hypertensive and normotensive rats. J Hypertens 10:149–153

PROGRESS Collaborative Group (2001) Randomised trial of a perindopril-based blood-pressure-lowering regimen among 6105 individuals with previous stroke or transient ischaemic attack. Lancet 358:1033–1041

Qadri F, Culman J, Veltmar A, Maas K, Rascher W, Unger T (1993) Angiotensin II-induced vasopressin release is mediated through alpha-1 adrenoceptors and angiotensin II AT_1 receptors in the supraoptic nucleus. J Pharmacol Exp Ther 267:567–574

Roberts KA, Krebs LT, Kramar EA, Shaffer MJ, Harding JW, Wright JW (1995) Autoradiographic identification of brain angiotensin IV binding sites and differential c-Fos expression following intracerebroventricular injection of angiotensin II and IV in rats. Brain Res 682:13–21

Ron D, Brasier AR, Wright KA, Tate JE, Habener JF (1990) An inducible 50-kilodalton NF kappa B-like protein and a constitutive protein both bind the acute-phase response element of the angiotensinogen gene. Mol Cell Biol 10:1023–1032

Rossi G, Rossi A, Sacchetto A, Pavan E, Pessina AC (1995) Hypertensive cerebrovascular disease and the renin-angiotensin system. Stroke 26:1700–1706

Rowland NE, Li BH, Rozelle AK, Smith GC (1994) Comparison of fos-like immunoreactivity induced in rat brain by central injection of angiotensin II and carbachol. Am J Physiol 267:R792–R798

Rowland NE, Li BH, Fregly MJ, Smith GC (1995) Fos induced in brain of spontaneously hypertensive rats by angiotensin II and co-localization with AT-1 receptors. Brain Res 675:127–134

Saavedra JM, Ito T, Nishimura Y (2001) The role of angiotensin AT(1) receptors in the regulation of cerebral blood flow and brain ischemia. J Renin Angiotensin Aldosterone Syst 2 [Suppl 1]:S102–S109

Sadoshima S, Fujii K, Ooboshi H, Ibayashi S, Fujishima M (1993) Angiotensin converting enzyme inhibitors attenuate ischemic brain metabolism in hypertensive rats. Stroke 24:1561–1566; discussion 1566–1567

Sadoshima S, Nagao T, Ibayashi S, Fujishima M (1994) Inhibition of angiotensin-converting enzyme modulates the autoregulation of regional cerebral blood flow in hypertensive rats. Hypertension 23:781–785

Schmid-Elsaesser R, Zausinger S, Hungerhuber E, Baethmann A, Reulen HJ (1998) A critical reevaluation of the intraluminal thread model of focal cerebral ischemia: evidence of inadvertent premature reperfusion and subarachnoid hemorrhage in rats by laser-Doppler flowmetry. Stroke 29:2162–2170

Schrader J, Luders S, Kulschewski A, Berger J, Zidek W, Treib J, Einhaupl K, Diener HC, Dominiak P, on behalf of the ACCESS Study Group (2003) The ACCESS study: evaluation of acute candesartan cilexetil therapy in stroke survivors. Stroke. 34:1699–1703

Sethi AA, Tybjaerg-Hansen A, Gronholdt ML, Steffensen R, Schnohr P, Nordestgaard BG (2001) Angiotensinogen mutations and risk for ischemic heart disease, myocardial infarction, and ischemic cerebrovascular disease. Six case-control studies from the Copenhagen City Heart Study. Ann Intern Med 134:941–954

Sharma P (1998) Meta-analysis of the ACE gene in ischaemic stroke. J Neurol Neurosurg Psychiatry 64:227–230

Squire IB (1994) Actions of angiotensin II on cerebral blood flow autoregulation in health and disease. J Hypertens 12:1203–1208

Stein B, Baldwin AS Jr, Ballard DW, Greene WC, Angel P, Herrlich P (1993) Cross-coupling of the NF-kappa B p65 and Fos/Jun transcription factors produces potentiated biological function. EMBO J 12:3879–3891

Stier CT Jr, Benter IF, Ahmad S, Zuo HL, Selig N, Roethel S, Levine S, Itskovitz HD (1989) Enalapril prevents stroke and kidney dysfunction in salt-loaded stroke-prone spontaneously hypertensive rats. Hypertension 13:115–121

Stier CT Jr, Adler LA, Levine S, Chander PN (1993) Stroke prevention by losartan in stroke-prone spontaneously hypertensive rats. J Hypertens Suppl 11:S37–S42

Strandgaard S, Paulson OB (1992) Regulation of cerebral blood flow in health and disease. J Cardiovasc Pharmacol 19:S89–S93

Stromberg C, Naveri L, Saavedra JM (1993) Nonpeptide angiotensin AT_1 and AT_2 receptor ligands modulate the upper limit of cerebral blood flow autoregulation in rats. J Cereb Blood Flow Metab 13:298–303

Szolnoki Z, Somogyvari F, Kondacs A, Szabo M, Fodor L (2001) Evaluation of the roles of the Leiden V mutation and ACE I/D polymorphism in subtypes of ischaemic stroke. J Neurol 248:756–761

Takada J, Ibayashi S, Nagao T, Ooboshi H, Kitazono T, Fujishima M (2001) Bradykinin mediates the acute effect of an angiotensin-converting enzyme inhibitor on cerebral autoregulation in rats. Stroke 32:1216–1219

Takasu MA, Dalva MB, Zigmond RE, Greenberg ME (2002) Modulation of NMDA receptor-dependent calcium influx and gene expression through EphB receptors. Science 295:491–495

Torup M, Waldemar G, Paulson OB (1993) Ceranapril and cerebral blood flow autoregulation. J Hypertens 11:399–405

Tsutsumi K, Saavedra JM (1991) Quantitative autoradiography reveals different angiotensin II receptor subtypes in selected rat brain nuclei. J Neurochem 56:348–351

Unger T (1999) The angiotensin type 2 receptor: variations on an enigmatic theme. J Hypertens 17:1775–1786

Unger T, Badoer E, Ganten D, Lang RE, Rettig R (1988) Brain angiotensin: pathways and pharmacology. Circulation 77:I40–I54

Veltmar A, Culman J, Qadri F, Rascher W, Unger T (1992) Involvement of adrenergic and angiotensinergic receptors in the paraventricular nucleus in the angiotensin II-induced vasopressin release. J Pharmacol Exp Ther 263:1253–1260

Veniant M, Clozel JP, Kuhn H, Clozel M (1992) Protective effect of cilazapril on the cerebral circulation. J Cardiovasc Pharmacol 19:S94–S99

Von Lutterotti N, Camargo MJ, Campbell WG Jr, Mueller FB, Timmermans PB, Sealey JE, Laragh JH (1992) Angiotensin II receptor antagonist delays renal damage and stroke in salt-loaded Dahl salt-sensitive rats. J Hypertens 10:949–957

Vraamark T, Waldemar G, Strandgaard S, Paulson OB (1995) Angiotensin II receptor antagonist CV-11974 and cerebral blood flow autoregulation. J Hypertens 13:755–761

Waldemar G, Vorstrup S, Andersen AR, Pedersen H, Paulson OB (1989) Angiotensin-converting enzyme inhibition and regional cerebral blood flow in acute stroke. J Cardiovasc Pharmacol 14:722–729

Walther T, Olah L, Harms C, Maul B, Bader M, Hortnagl H, Schultheiss HP, Mies G (2002) Ischemic injury in experimental stroke depends on angiotensin II. FASEB J 16:169–176

Wang LL, Chan SH, Chan JY (2001) Fos protein is required for the re-expression of angiotensin II type 1 receptors in the nucleus tractus solitarii after baroreceptor activation in the rat. Neuroscience 103:143–151

Werner C, Hoffman WE, Kochs E, Rabito SF, Miletich DJ (1991) Captopril improves neurologic outcome from incomplete cerebral ischemia in rats. Stroke 22:910–914

West AE, Griffith EC, Greenberg ME (2002) Regulation of transcription factors by neuronal activity. Nat Rev Neurosci 3:921–931

Wright JW, Miller-Wing AV, Shaffer MJ, Higginson C, Wright DE, Hanesworth JM, Harding JW (1993) Angiotensin II(3–8) (ANG IV) hippocampal binding: potential role in the facilitation of memory. Brain Res Bull 32:497–502

Xu Z, Herbert J (1994) Regional suppression by water intake of c-fos expression induced by intraventricular infusions of angiotensin II. Brain Res 659:157–168

Xu Z, Herbert J (1996) Effects of unilateral or bilateral lesions within the anteroventral third ventricular region on c-fos expression induced by dehydration or angiotensin II in the supraoptic and paraventricular nuclei of the hypothalamus. Brain Res 713:36–43

Yu K, Lu D, Rowland NE, Raizada MK (1996) Angiotensin II regulation of tyrosine hydroxylase gene expression in the neuronal cultures of normotensive and spontaneously hypertensive rats. Endocrinology 137:3566–3576

Yusuf S, Sleight P, Pogue J, Bosch J, Davies R, Dagenais G (2000) Effects of an angiotensin-converting-enzyme inhibitor, ramipril, on cardiovascular events in high-risk patients. The Heart Outcomes Prevention Evaluation Study Investigators. N Engl J Med 342:145–153

Zhu YZ, Chimon GN, Zhu YC, Lu Q, Li B, Hu HZ, Yap EH, Lee HS, Wong PT (2000) Expression of angiotensin II AT_2 receptor in the acute phase of stroke in rats. Neuroreport 11:1191–1194

Part 4
Tissues
Heart

Role of Angiotensin II in Cardiac Remodeling

J. Díez

Facultad de Medicina C/Irunlarrea 1, 31080 Pamplona, Spain
e-mail: jadimar@unav.es

Abstract In response to mechanical and/or metabolic stress, the heart undergoes structural remodeling involving cardiomyocyte hypertrophy and myocardial fibrosis. A number of factors, including angiotensin II, are involved in the process of cardiac remodeling. Angiotensin II has endocrine, autocrine and paracrine properties that influence the behavior of cardiac cells and extracellular matrix via the type 1 receptor. For instance, angiotensin II has been shown to induce collagen types I and III deposition in a number of cardiac conditions characterized by the presence of myocardial fibrosis. Angiotensin II-dependent fibrosis can be the result of both stimulation of synthesis and secretion of fibrillar collagen precursors by cardiac fibroblasts and inhibition of collagen fiber degradation by matrix metalloproteinases. Emerging evidence is providing support that myocardial fibrosis is associated with a progressive decline in cardiac function, namely diastolic function overtime. Thus, to reduce the risk of heart failure that accompanies chronic cardiomyopathies, fibrosis must be targeted for pharmacological intervention. Available experimental and clinical data suggest that agents that inhibit angiotensin-converting enzyme or block type 1 receptor may provide such a cardioprotective effect.

Keywords Angiotensin · Cardiac remodeling collagen · Heart failure · Myocardial fibrosis

1 Introduction

In response to mechanical and/or metabolic stress, the heart undergoes structural remodeling involving cardiomyocyte hypertrophy and myocardial fibrosis. Cardiomyocyte hypertrophy involves an increase in contractile and embryonic protein content, which occurs when transcription of genes that encode for these cardiac proteins are activated (Hunter et al. 1999). Myocardial fibrosis is the result of exaggerated deposition of collagen types I and III fibers, a consequence of the predominance of synthesis over degradation of collagen types I and III molecules (Weber 1997). Cardiac remodeling is accompanied by progressive decline cardiac function over time, which underlies the pathogenesis of heart failure in patients with chronic cardiac conditions (Cohn et al. 2000).

It is now accepted that a number of systemically and locally expressed factors have key roles in the process of cardiac remodeling (Swynghedauw 1999). One of these factors is angiotensin II (ANG II). Whereas the role of ANG II in cardiomyocyte hypertrophy is well established and has been recently reviewed (Lijnen and Petrov 1999; Yamazaki et al. 1999), emerging evidence suggests that this peptide also induces myocardial fibrosis. Thus, this chapter will review the available information supporting the fibrotic role of ANG II in the heart, specifically the hypertensive left ventricle, as well as the potential cellular and molecular mechanisms involved.

2 General Aspects of Myocardial Fibrosis

2.1 Molecular Aspects

Collagen types I and III are the major fibrillar collagens produced by fibroblasts in the adult heart. They exhibit the characteristic triple helical conformation of 3 polypeptide α chains. Fibrillar collagen in the heart provides structural scaffolding for cardiomyocytes and coronary vessels, and imparts physical properties, such as stiffness and resistance to deformation, to cardiac tissue (Weber 1989). In addition, fibrillar collagen may also act as a link between contractile elements of adjacent cardiomyocytes, and as a conduit for information necessary for cell function.

Table 1 Factors modulating collagen types I and III turnover in the myocardium

Factors
Factors that facilitate the predominance of synthesis over degradation
Vasoactive substances
Angiotensin II, endothelin-1, catecholamines
Growth factors
Transforming growth factor-β_1, platelet derived growth factor, basic fibroblast growth factor, insulin-like growth factor-1
Hormones
Aldosterone, deoxycorticosterone
Cytokines
Interleukin-1
Adhesion molecules
Osteopontin
Factors that facilitate the predominance of degradation over synthesis
Vasoactive substances
Bradykinin, prostaglandins, nitric oxide, natriuretic peptides
Growth factors
Hepatocyte growth factor
Hormones
Glucocorticoids
Cytokines
Tumoral necrosis factor-α, interferon-γ
Endogenous peptides
N-acetyl-seryl-aspartyl-lysyl-proline (Ac-SDKP)

As in other organs, collagen turnover in normal adult heart results from the equilibrium between synthesis and degradation of collagen types I and III molecules (Burlew and Weber 2000). Synthesis of collagen molecules follows the normal pattern of protein synthesis, but differs in that the newly formed α chains undergo a number of post-translational modifications. Extracellular degradation of collagen fibers is mediated by collagenase and other members of the matrix metalloproteinase (MMP) family of zinc-containing endoproteinases. The active form of collagenase can be inhibited by interaction with specific, naturally occurring tissue inhibitors of MMPs (TIMPs).

A number of factors have been described that may alter the balance between synthesis and degradation of fibrillar collagen (Table 1) (Burlew and Weber 2000). Predominance of synthesis over degradation leads to increased interstitial and perivascular deposition of collagen types I and III, and consequently to fibrosis that accompanies cardiac diseases such as hypertensive heart disease (Fig. 1), ischemic cardiomyopathy, diabetic cardiomyopathy, and hypertrophic cardiomyopathy.

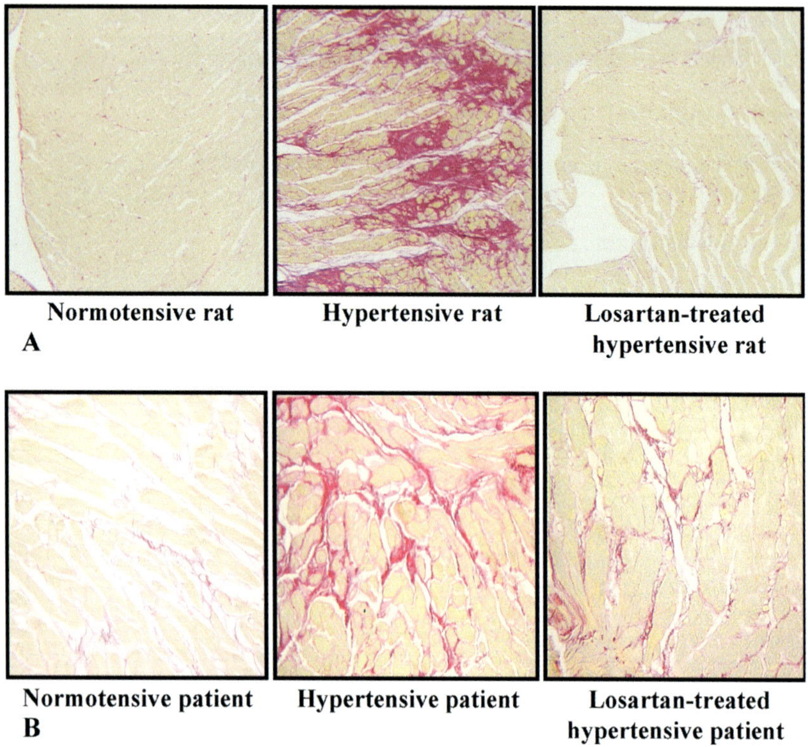

Fig. 1A, B Histological sections of the left ventricle from rats (**A**) and interventricular septal specimen biopsies from humans (**B**). Sections were stained with picrosirius red, and the interstitial collagen is identified in *red*. ×20

2.2 Pathophysiological Aspects

Several arguments support the concept that myocardial fibrosis is particularly important in the development of heart failure associated with cardiac remodeling, namely, in hypertensive heart disease (Diez et al. 2001). Firstly, interstitial fibrosis contributes to ventricular wall stiffness and consequently impairs cardiac compliance, contributing to impaired diastolic function. Secondly, since neither the collagen network nor the fibroblasts contribute to systolic contraction, increased collagen deposition and fibroblast volume means that systolic work is being performed by a smaller proportion of the cardiac mass, contributing to systolic dysfunction. Thirdly, perivascular fibrosis leads to an increased distance over which oxygen must diffuse and therefore potentially lowers the PaO_2 of working cardiomyocytes. Finally, electrical coupling of cardiomyocytes may be impaired by accumulation of collagen proteins and fibroblasts, since such accumulation causes morphological separation of cardiomyocytes.

3
Role of ANG II in Myocardial Fibrosis

Observations that myocardial fibrosis is present not only in the left ventricle but also in the right, and in the interventricular septum of subjects with arterial hypertension, suggests that apart from hypertension, some systemically and/or locally produced humoral factor may also contribute to hypertensive myocardial fibrosis. Various lines of evidence support a role for ANG II as one potential candidate.

3.1
In Vivo Evidence

3.1.1
Animal Models

Endogenous elevations in circulating ANG II that accompany unilateral renal artery stenosis (Sun et al. 1997) or infusion of exogenous ANG II (Jalil et al. 1991) are associated with increased blood pressure and fibrosis. The appearance of such fibrous tissue formation is preceded by increased expression of angiotensin II type 1 (AT_1) receptors, transforming growth factor-β_1 (TGF-β_1), and mRNA for collagen types I and III (Everett et al. 1994). In addition, fibrosis development involves fibroblast proliferation and differentiation into myofibroblasts (Campbell et al. 1995). Two observations suggest that the ability of ANG II to induce cardiac fibrosis in these models is independent of its hypertensive action. Firstly, fibrosis in the renal artery stenosis model develops both in low-pressure right and left atria and right ventricle, and in high-pressure left ventricle (Brilla et al. 1990). Secondly, cardiac fibrosis in the ANG II infusion model can be prevented by either angiotensin converting enzyme (ACE) inhibitors or AT_1 receptor antagonists, but not hydralazine or prazosin, in spite of similar antihypertensive efficacy of these compounds (Kim et al. 1995; Crawford et al. 1994). The critical role of ANG II in hypertension-associated cardiac fibrosis is further supported by the observation that experimental infrarenal aortic binding, which does not induce ANG II, causes blood pressure elevation and cardiomyocyte hypertrophy but not cardiac fibrosis (Brilla et al. 1990).

The hypertensive Ren2 rat provides a well-established model of ANG II-dependent cardiac hypertrophy (Lee et al. 1996). Several studies have revealed that interstitial and perivascular fibrosis, along with extensive collagen types I and III deposition, are present in Ren2 rats (Bishop et al. 2000; Pinto et al. 2000; Rothermund et al. 2002). Increased cardiac renin and ANG II levels have also been described in this transgenic rat model (Lee et al. 1996). In addition, cardiac lesions in Ren2 rats are very sensitive to ACE inhibition and AT_1 receptor antagonism (Teisman et al. 1998). As a result, the development of hypertrophy and fibrosis in the heart of these animals has been attributed, at least partially, to local activation of the cardiac renin-angiotensin system.

3.1.2 Pharmacological Studies

Experimental Findings. Pharmacological interventions with ACE inhibitors or AT_1 receptor antagonists have underscored the potential importance of ANG II in the mediation of cardiac fibrosis in pathological conditions such as primary hypertension. In rats with spontaneous hypertension (SHRs) and left ventricular hypertrophy, myocardial fibrosis is regressed by treatment with the ACE inhibitor lisinopril (Brilla et al. 1991). This effect occurred independently of the drug's antihypertensive effect (Brilla et al. 1991). On the other hand, it has been found that chronic AT_1 receptor antagonism with losartan resulted in reversal of fibrosis, inhibition of post-transcriptional synthesis of procollagen type I, inhibition of TIMP-1 expression and stimulation of collagenase activity in the left ventricle of adult SHRs (Varo et al. 1999, 2000) (Fig. 1A). Analysis of individual data showed that the intensity of these myocardial changes was independent of the antihypertensive efficacy of the drug (Varo et al. 1999, 2000).

Clinical Findings. The fibrogenic role of ANG II in humans has been investigated in three recent prospective trials of limited size, of patients with biopsy-proven myocardial fibrosis, essential hypertension and left ventricular hypertrophy. Schwartzkopff et al. (2000) studied 14 patients before and after 1 year of treatment with the ACE inhibitor perindopril. Structural analysis revealed diminution of perivascular and interstitial fibrosis following treatment. The observed regression of fibrosis upon ACE inhibitor treatment was observed in the non-pressure-overloaded right ventricle, indicating that the antifibrotic effect was not solely due to a reduction in left ventricular pressure. Brilla et al. (2000) randomized 35 previously treated patients with controlled blood pressure to receive either the ACE inhibitor lisinopril or the diuretic hydrochlorothiazide for 6 months. Only patients randomized to lisinopril had a significant reduction in myocardial fibrosis. Blood pressure reduction was similar in patients treated with either lisinopril or hydrochlorothiazide. Lopez et al. (2001) studied 37 treated patients with uncontrolled blood pressure. After randomization, 21 patients were assigned to the AT_1 receptor antagonist losartan and 16 to the calcium channel blocker amlodipine for 12 months. Whereas myocardial fibrosis decreased significantly in losartan-treated patients, this parameter remained unchanged in amlodipine-treated patients (Fig. 1B). Similar reductions in blood pressure from both losartan-treated and amlodipine-treated patients were noted. Collectively, these observations support the concept that in addition to pressure overload, ANG II induces myocardial fibrosis in essential hypertension.

3.2 Cellular and Molecular Mechanisms

Increasingly accumulated evidence strongly indicates that ANG II exerts multiple profibrotic effects in the heart, including induction of fibroblast hyperplasia, activation of collagen biosynthetic pathways and inhibition of collagen degrada-

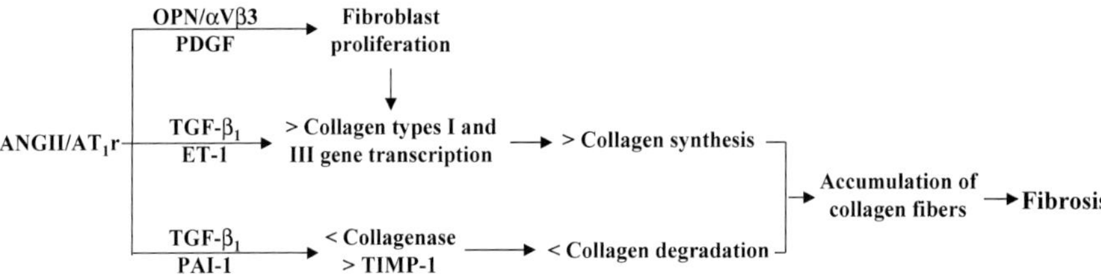

Fig. 2 Summary of angiotensin II (*ANG II*) effects on fibrillar collagen metabolism acting via the angiotensin II type 1 receptor (*AT_1r*) in cardiac fibroblasts. Cofactors of these effects are *OPN*, osteopontin; *$\alpha V\beta 3$*, $\alpha V\beta 3$ integrin ; *TGF-β_1*, transforming growth factor-beta1; *ET-1*, endothelin-1; *PAI-1*, plasminogen activator inhibitor-1. (*TIMP-1*, tissue inhibitor of matrix metalloproteinase-1)

tive pathways (Fig. 2). In addition, available data indicate that these effects may result from either the direct action of ANG II or a synergistic cooperation between this peptide and other profibrotic factors.

3.2.1 Stimulation of Fibroblast Proliferation

In vitro studies of rat and human cardiac fibroblasts have shown that ANG II stimulates cell proliferation via the AT_1 receptor (Sadoshima and Izumo 1993). Results in the literature indicate that the proliferative response of fibroblasts to ANG II might well be mediated by stimulation of synthesis of growth or inflammatory substances such as platelet-derived growth factor (PDGF) and cytokines, by integrin activation due to adhesion proteins or by a combination of these mechanisms (Bouzegrhane and Thibault 2002; Schnee and Hsueh 2000).

For instance, ANG II strongly up-regulates the expression of osteopontin and its ligand, $\alpha V\beta 3$ integrin, in rat and human cardiac fibroblasts (Ashizawa et al. 1996; Graf et al. 2000). Interestingly, elevated left ventricular osteopontin expression has been reported in the Ren2 rat model, which is characterized by high myocardial ANG II concentrations (Rothermund et al. 2002). Monoclonal antibodies directed against either osteopontin or $\alpha V\beta 3$ integrin completely blocked the mitogenic effects of ANG II on cultured rat cardiac fibroblasts (Ashizawa et al. 1996), thus suggesting that osteopontin mediates ANG II-induced fibroblast proliferation via an integrin-dependent pathway.

3.2.2 Stimulation of Collagen Synthesis

Although different signaling pathways of the AT_1 receptor may subserve direct ANG II-induced collagen synthesis in cardiac fibroblasts (Dostal et al. 1996), recent data suggest that the MAP/ER kinase pathway appears to play a major role (Tharaux et al. 2000). The end result of signaling mechanisms is activation of transcription factors, which bind to various *cis*-acting elements in the regulatory sequences of $\alpha 1$ and $\alpha 2$ collagen type I, and $\alpha 1$ collagen type III genes (Tharaux et al. 2000; Ghiggeri et al. 2000). This, in turn, will couple with gene expression and the synthesis of collagen types I and III precursor molecules (Crabos et al. 1994; Villarreal et al. 1993).

On the other hand, a number of studies provide strong evidence that in cardiac fibroblasts, ANG II indirectly regulates collagen synthesis via specific growth factors (Dostal 2001). Principal candidates include TGF-β_1 and endothelin-1.

In fact, ANG II has been shown to induce collagen type I gene expression via activation of TGF-β_1 signaling pathways (e.g., connective tissue growth factor, and Smad proteins), and these effects were blocked by the AT_1 receptor antagonist losartan (Hao et al. 2000). ANG II has also been shown to increase TGF-β_1 expression in cultured cardiac fibroblasts, via stimulation of the AT_1 receptor (Sadoshima and Izumo 1993). Recent data suggest that a Krüppel-like zinc-finger transcription factor 5 (KLF5; also known as BTEB2 and IKLF) is critically involved in ANG II-induced TGF-β_1 expression, collagen synthesis and development of cardiac fibrosis (Shindo et al. 2002). Besides up-regulation of cardiac gene TGF-β_1 expression, ANG II has been reported to convert latent TGF-β_1 to the active protein in the in vivo heart (Tomita et al. 1998).

Endothelin-1 is synthesized and released by cardiac fibroblasts in response to the interaction between ANG II and the AT_1 receptor (Gray et al. 1998), and has been shown to stimulate the synthesis of collagen types I and III in these cells (Guarda et al. 1993). In several rat models of arterial hypertension, blockade of endothelin receptors is associated with a decrease in left ventricular collagen accumulation (Ammarguellat et al. 2001; Yamamoto et al. 2002a).

Some in vivo evidence suggests that ANG II also influences post-translational processing of cardiac fibrillar collagen. It has been shown that ANG II infusion is associated with stimulation of prolyl 4-hydroxylase (an enzyme that mediates hydroxylation of procollagen α–chains in the endoplasmic reticulum of cardiac fibroblasts) in the rat left ventricle (Leipala et al. 1988) In addition, it has been reported that immunoreactive prolyl 4-hydroxylase concentrations decrease significantly in ventricles of postmyocardial infarcted rats treated with the AT_1 receptor antagonist losartan (Ju et al. 1997).

3.2.3 Inhibition of Collagen Degradation

In addition to collagen synthesis, ANG II stimulation of the AT_1 receptor has been shown to regulate collagen degradation by attenuating interstitial collagenase activity in adult rat (Sadoshima and Izumo 1993) and human (Funck et al. 1997) cardiac fibroblasts, and by enhancing TIMP-1 production in rat heart endothelial cells (Chua et al. 1996).

A number of factors may mediate the inhibitory effects of ANG II on cardiac collagen degradation (e.g., TGF-β_1 and plasminogen activator inhibitor-1, PAI-1). Cell culture studies on human fibroblasts show that exposure of these cells to TGF-β_1 in the presence of other growth factors (e.g., epidermal growth factor and basic fibroblastic growth factor) resulted in down-regulation of collagenase and up-regulation of TIMP-1 (Edwards et al. 1987; Chua et al. 1991). Similar findings have been reported in the fibrotic myocardium of TGF-β_1 transgenic mice (Seeland et al. 2002).

Activation of the AT_1 receptor in human cardiac fibroblasts has been shown to promote stimulation of PAI-1 expression (Kawano et al. 2000). This stimulatory effect has been confirmed in the left ventricle of ANG II-induced hypertensive rats (Kobayashi et al. 2002). PAI-1 inhibits the activation of collagenase and other matrix metalloproteinases, and thereby collagen degradation (Yamamoto and Saito 1998; Dollery et al. 1995).

3.3 Emerging Aspects

3.3.1 Communication Between Cardiomyocytes and Fibroblasts

Cross-talk between fibroblasts and cardiomyocytes has been demonstrated to be an important factor in ANG II-stimulated cardiomyocyte growth by ANG II (Harada et al. 1997; Sil and Sen 1997). Similarly, emerging evidence suggests that cardiomyocyte-originated factors may modulate the response of fibroblasts to ANG II.

Matsusaka et al. (1999) have reported that in AT_1 receptor chimeric mice, in response to ANG II infusion, cardiac fibroblasts proliferate through both local and systemic actions of the peptide. Importantly, the former is determined by the AT_1 receptor of neighboring cardiomyocytes, indicating that nonidentified growth factors may be produced by activated cardiomyocytes, which stimulate fibroblasts in a paracrine fashion.

Molkentin et al. (1998) generated transgenic mice overexpressing activated forms of calcineurin or NF-AT3, the calcium-dependent signal transducing molecule that is activated by various stimuli, including ANG II. These mice develop cardiomyocyte hypertrophy and cardiac fibrosis. Since the transgene expression is controlled by the α–myosin heavy chain gene promoter and hence thought to be limited to cardiomyocytes, the above results suggest that activated cardiomyocytes can generate a paracrine factor that stimulates fibroblast activity leading to fibrosis.

Pathak et al. (2001) have reported that expression of collagen types I and III mRNAs, in cardiac fibroblasts coincubated with ANG II, is significantly increased when fibroblasts were also cocultured with cardiomyocytes. In addition, when the supernatants from cardiomyocyte cultures were added to the fibroblasts, similar up-regulation of collagen was observed. These findings suggest that a cardiomyocyte-derived soluble cofactor is an important mediator of ANG II-induced collagen synthesis in cardiac fibroblasts.

Osteopontin has been identified in cultured rat cardiomyocytes (Singh et al. 1995). Furthermore, cardiomyocytes appear to be the primary source of osteopontin in the fibrotic myocardium of rats and humans with pressure overload-induced heart failure (Graf et al. 1997; Singh et al. 1999). Thus, it remains to be determined whether either cardiomyocyte- or fibroblast-derived osteopontin contributes effectively to the profibrotic effects of ANG II.

3.3.2 Role of the Angiotensin II Type 2 (AT_2) Receptor

The role of AT_2 receptor signaling in cardiac collagen metabolism is still debatable because of conflicting data. Some studies suggest that AT_1 and AT_2 receptors induce phenotypically opposite responses. In experiments performed in fibroblasts isolated from myopathic hamster hearts, Ohkubo et al. (1997) provided evidence that the AT_2 receptor is a negative regulator of cell growth and collagen synthesis. In support of this, Tsutsumi et al. (1998) found that the AT_2 receptor is up-regulated in failing human hearts with interstitial fibrosis, and that AT_2 receptors present in fibroblasts exert an inhibitory effect on ANG II-induced mitogen signals (e.g., MAPK activation). Liu et al. (1997) reported that the antifibrotic effect of AT_1 receptor antagonists is mediated in part by activation of AT_2 receptor in a model of myocardial infarction-induced heart failure in rats. One possibility is that activation of the AT_2 receptor inhibits collagen synthesis by stimulating kinin production (Kim et al. 1999; Liu et al. 1997).

Conversely, other evidence suggests that simultaneous activation of both AT_1 and AT_2 receptors may stimulate a signaling pathway that results in up-regulation of collagen in parallel, rather than in opposite, directions. Brilla et al. (1994) reported that ANG II stimulated collagen synthesis via both AT_1 and AT_2 receptors in cultured adult rat cardiac fibroblasts, and that ANG II-induced inhibition of collagenase activity was specifically mediated by the AT_2 receptor. In support of these findings, Ichihara et al. (2001) reported that ANG II-induced cardiac fibrosis was attenuated in mice lacking the AT_2 receptor gene.

3.3.3 Contribution of ANG II-Induced Cardiomyocyte Death

Myocardial fibrosis may develop in the absence (e.g., reactive fibrosis) or the presence (replacement fibrosis) of cardiomyocyte loss. As proposed by Anversa et al. (1996), death of individual cardiomyocytes may be more common than generally believed, and this phenomenon may stimulate discrete healing processes contributing to interstitial fibrosis. This proposal is further supported by the finding that fibrosis is associated with cardiomyocyte loss in the left ventricle of hypertensive patients (Olivetti et al. 1994). Furthermore, Bing et al. (1997) reported that failing hearts from SHRs present colocalization of collagen $\alpha 1$ type I gene expression to areas of focal cardiomyocyte degeneration.

Various experimental and clinical observations would suggest that ANG II may induce cardiomyocyte cell death and this, in turn, contributes to myocardial fibrosis. Previous studies in rats showed that ANG II infusion induces focal necrosis of cardiomyocytes, with subsequent fibroblast activation and collagen deposition (Kabour et al. 1995; Ratajska et al. 1994). Cardiomyocyte injury was unrelated to the hypertensive or enhanced adrenergic effects of ANG II and was prevented by ACE inhibition and AT_1 receptor antagonism.

On the other hand, it has been reported that the binding of ANG II to the AT_1 receptor induces apoptosis of rat cardiomyocytes (Cigola et al. 1997; Kajstura et al. 1997). In addition, an association has been found between myocardial ACE activity, cardiomyocyte apoptosis and interstitial fibrosis in SHRs (Diez et al. 1997) and patients with essential hypertension (Gonzalez et al. 2002). Furthermore, chronic treatment with the AT_1 receptor antagonist losartan reduced both cardiomyocyte apoptosis and fibrosis in SHRs (Fortuño et al. 1998) and hypertensive patients (Gonzalez et al. 2002), these effects being independent of the hypotensive effect of the compound.

3.3.4 Interactions with Aldosterone

Aldosterone has been shown to stimulate collagen synthesis via the mineralocorticoid receptor in isolated cardiac fibroblasts (Brilla et al. 1994). Since ANG II increases aldosterone secretion and chronic aldosterone excess is associated with marked interstitial and perivascular fibrosis that can be prevented with spironolactone treatment (Brilla et al. 1993), an increase in this mineralocorticoid, thus, may be a mechanism for ANG II-induced cardiac fibrosis in some forms of arterial hypertension.

Alternatively, an increase in the density of AT_1 receptors has been observed in the heart of aldosterone salt-treated rats (Sun and Weber 1993). In addition, the AT_1 receptor antagonist losartan prevents fibrosis and up-regulation of collagen types I and III mRNAs in the heart of aldosterone salt-treated rats (Robert et al. 1999). Taken together, these findings support the hypothesis that one mechanism by which aldosterone induces cardiac fibrosis involves ANG II acting via AT_1 receptors. Since aldosterone production is activated in the hypertrophied left ventricle of SHRs (Takeda et al. 2000) and hypertensive patients (Yamamoto et al. 2002b), it is thus possible that aldosterone contributes to ANG II-mediated myocardial fibrosis in primary hypertension.

4 Conclusions and Perspectives

Structural homogeneity of cardiac tissue is governed by mechanical and humoral factors that regulate cell growth, apoptosis, phenotype, and extracellular matrix turnover. ANG II has endocrine, autocrine and paracrine properties that influence the behavior of cardiac cells and matrix via AT_1 receptor binding. Thus, various paradigms have been suggested, including ANG II-mediated up-regulation of collagen types I and III formation and deposition in cardiac conditions such as hypertensive heart disease. To reduce the risk of heart failure that accompanies hypertensive heart disease, its adverse structural remodeling (e.g., myocardial hypertrophy and fibrosis) must be targeted for pharmacological intervention. Thus, cardioprotective agents must reverse not only the exaggerated growth of cardiac cells, but also regress existing abnormalities in fibrillar collagen. Available experimental and clinical data suggest that agents interfering with either ACE or AT_1 receptor may provide such a cardioprotective effect.

References

Ammarguellat F, Larouche II, Schiffrin EL (2001) Myocardial fibrosis in DOCA-salt hypertensive rats: effect of endothelin (ETA) receptor antagonism. Circulation 103:319–324

Anversa P, Kajstura J, Olivetti G (1996) Myocyte death in heart failure. Curr Opin Cardiol 11:245–251

Ashizawa N, Graf K, Do YS, Nunohiro T, Giachelli CM, Meehan WP, Tuan TL, Hsueh WA (1996) Osteopontin is produced by rat cardiac fibroblasts and mediates AII-induced DNA synthesis and collagen gel contraction. J Clin Invest 98:2218–2227

Bing OHL, Ngo HQ, Humphries DE, Robinson KG, Lucey EC, Carver W, Brooks WW, Conrad CH, Hayes JA, Goldstein RH (1997) Localization of α_1(I) collagen mRNA in myocardium from the spontaneously hypertensive rat during the transition from compensatory hypertrophy to failure. J Mol Cell Cardiol 29:2335–2344

Bishop JE, Lindhal G (1999) Regulation of cardiovascular collagen synthesis by mechanical load. Cardiovasc Res 42:27–44

Bishop JE, Kiernan LA, Montgomery HE, Gohlke P, McEwan JR (2000) Raised blood pressure, not renin-angiotensin systems, causes cardiac fibrosis in TGR m(Ren2)27 rats. Cardiovasc Res 47:57–67

Bouzegrhane F, Thibault G (2002) Is angiotensin II a proliferative factor of cardiac fibroblasts? Cardiovasc Res 53:304–312

Brilla CG, Janicki JS, Weber KT (1991) Impaired diastolic function and coronary reserve in genetic hypertension: role of interstitial fibrosis and medial thickening of intramyocardial coronary arteries. Circ Res 69:107–115

Brilla CG, Matsubara LS, Weber KT (1993) Anti-aldosterone treatment and the prevention of myocardial fibrosis in primary and secondary aldosteronism. J Mol Cell Cardiol 25:563–575

Brilla CG, Zhou G, Matsubara L, Weber KT (1994) Collagen metabolism in cultured adult rat cardiac fibroblasts: response to angiotensin II and aldosterone. J Mol Cell Cardiol 26:809–820

Brilla CG, Funck RC, Rupp RH (2000) Lisinopril-mediated regression of myocardial fibrosis in patients with hypertensive heart disease. Circulation 102:1388–1393

Burlew BS, Weber KT (2000) Connective tissue and the heart. Functional significance and regulatory mechanisms. Cardiol Clin 18:435–442

Campbell JE, Janicki JS, Weber KT (1995) Temporal differences in fibroblast proliferation and phenotype expression in response to chronic administration of angiotensin II or aldosterone. J Mol Cell Cardiol 27:1545–1560

Chua CC, Chua BH, Zhao ZY, Krebs C, Diglio C, Perrin E (1991) Effect of growth factors on collagen metabolism in cultured human heart fibroblasts. Connect Tissue Res 26:271–281

Chua CH, Hamdy RC, Chua BH (1996) Angiotensin II induces TIMP-1 production in rat heart endothelial cells. Biochem Biophys Acta 1311:175–180

Cigola E, Kajstura J, LI B, Meggs LG, Anversa P (1997) Angiotensin II activates programmed myocyte cell death in vitro. Exp Cell Res 231:363–371

Cohn JN, Ferrari R, Sharpe N, on behalf of An International Forum on Cardiac Remodeling (2000) Cardiac remodeling-concepts and clinical implications: a consensus paper from an international forum on cardiac remodeling. J Am Coll Cardiol 35:569–582

Crabos M, Roth M, Hahn AW, Erne P (1994) Characterization of angiotensin II receptors in cultured adult rat cardiac fibroblasts. Coupling to signaling systems and gene expression. J Clin Invest 93:2372–2378

Crawford DC, Chobanian AV, Brecher P (1994) Angiotensin II induces fibronectin expression associated with cardiac fibrosis in the rat. Circ Res 74:727–739

Diez J, Panizo A, Hernandez M, Vega F, Sola I, Fortuño MA, Pardo J (1997) Cardiomyocyte apoptosis and cardiac angiotensin-converting enzyme in spontaneously hypertensive rats. Hypertension 30:1029–1034

Diez J, Lopez B, Gonzalez B, Querejeta R (2001) Clinical aspects of hypertensive myocardial fibrosis. Curr Opin Cardiol 16:328–335

Dollery CM, McEwan JR, Henney AM (1995) Matrix metalloproteinases and cardiovascular diseases. Circ Res 77:863–868

Dostal DE (2001) Regulation of cardiac collagen. Angiotensin and cross-talk with local growth factors. Hypertension 37:841–844

Dostal DE, Booz GW, Baker KM (1996) Angiotensin II signalling pathways in cardiac fibroblasts: conventional versus novel mechanisms in mediating cardiac growth and function. Mol Cell Biochem 157:15–21

Edwards DR, Murphy G, Reynolds JJ, Whitham SE, Docherty AJ, Angel P, Heath JK (1987) Transforming growth factor beta modulates the expression of collagenase and metalloproteinase inhibitor. EMBO J 6:1899–1904

Everett AD, Tufro-McReddie A, Fisher A, Gomez RA (1994) Angiotensin receptor regulates cardiac hypertrophy and transforming growth factor-β1 expression. Hypertension 23:587–592

Fortuño MA, Ravassa S, Etayo JC, Diez J (1998) Overexpression of Bax-α protein and enhanced apoptosis in the left ventricle of spontaneously hypertensive rats: effects of AT_1 blockade with losartan. Hypertension 32:280–286

Funck RC, Wilke A, Rupp H, Brilla CG (1997) Regulation and role of myocardial matrix remodeling in hypertensive heart disease. Adv Exp Mol Biol 432:35–44

Ghiggeri GM, Oleggini R, Musante L, Garidi G, Gusmano R, Ravazzolo R (2000) A DNA element in the α1 type III collagen promoter mediates a stimulatory response by angiotensin II. Kidney Int 58:537–548

Gonzalez A, Lopez B, Ravassa S, Querejeta R, Larman M, Diez J, Fortuño MA (2002) Stimulation of cardiac apoptosis in essential hypertension. Potential role of angiotensin II. Hypertension 39:75–80

Graf K, Do YS, Ashizawa N, Meehan WP, Glachelli CM, Marboe CC, Fleck E, Hsueh WA (1997) Myocardial osteopontin expression is associated with left ventricular hypertrophy. Circulation 96:3063–3071

Graf K, Neuss M, Stawowy P, Hsueh WA, Pleck E, Law RE (2000) Angiotensin II and alpha(V)beta(3) integrin expression in rat neonatal cardiac fibroblasts. Hypertension 35:978–984

Gray MO, Long CS, Kalinyak JE, Li HT, Karliner JS (1998) Angiotensin II stimulates cardiac myocyte hypertrophy via paracrine release of TGF-beta 1 and endothelin-1 from fibroblasts. Cardiovasc Res 40:352–363

Guarda E, Katwa LC, Myers PR, Tyagi SC, Weber KT (1993) Effects of endothelins on collagen turnover in cardiac fibroblasts. Cardiovasc Res 27:2130–2134

Hao J, Wang B, Jones SC, Jassal DS, Dixon IM (2000) Interaction between angiotensin II and Smad proteins in fibroblasts in failing heart and in vitro. Am J Physiol 279: H3020–H3030

Harada M, Itoh H, Nakagawa O, Ogawa Y, Miyamoto Y, Kuwahara K, Ogawa E, Iguki T, Yamashita J, Masuda I, Yoshimasa T, Tanaka I, Saito Y, Nakao K (1997) Significance of ventricular myocytes and nonmyocytes during cardiocyte hypertrophy. Evidence for endothelin-1 as a paracrine hypertrophic factor from cardiac nonmyocytes. Circulation 96:3737–3744

Hunter JJ, Grace A, Chien KR (1999) Molecular and cellular biology of cardiac hypertrophy and failure. In: Chien KR (ed) Molecular basis of cardiovascular disease. WB Saunders, Philadelphia, p 211

Ichihara S, Senbonmatsu T, Price E Jr, Ichiki T, Gaffney FA, Inagami T (2001) Angiotensin II type 2 receptor is essential for left ventricular hypertrophy and cardiac fibrosis in chronic angiotensin II-induced hypertension. Circulation 104:346–351

Jalil JE, Janicki JS, Pick R, Weber KT (1991) Coronary vascular remodeling and myocardial fibrosis in the rat with renovascular hypertension: response to captopril. Am J Hypertens 4:51–55

Ju H, Zhao S, Jassal DS, Dixon IM (1997) Effect of AT1 receptor blockade on cardiac collagen remodeling after myocardial infarction. Cardiovasc Res 35:223–232

Kabour A, Henegar JR, Devineni VR, Janicki JS (1995) Prevention of angiotensin II induced myocyte necrosis and coronary vascular damage by lisinopril and losartan in the rat. Cardiovasc Res 29:543–548

Kajstura J, Cigola E, Malhotra A, LI P, Cheng W, Meggs LG, Amversa P (1997) Angiotensin II induces apoptosis of adult ventricular myocytes in vitro. J Mol Cell Cardiol 29:859–870

Kawano H, Do YS, Kawano Y, Starnes V, Barr M, Law RE, Hsueh WA (2000) Angiotensin II has multiple profibrotic effects in human cardiac fibroblasts. Circulation 101:1130–1137

Kim NN, Villegas S, Summerour SR, Villarreal FJ (1999) Regulation of cardiac fibroblast extracellular matrix production by bradykinin and nitric oxide. J Mol Cell Cardiol 31:457–466

Kim S, Ohta K, Hamaguchi A, Yukimura T, Miura K, Iwao H (1995) Angiotensin II-induced cardiac phenotypic modulation and remodeling in vivo in rats. Hypertension 25:1252–1259

Kobayashi N, Nakano S, Mita S, Kobayashi T, Honda T, Tsubokou Y, Matsuoka H (2002) Involvement of Rho-kinase pathway for angiotensin II-induced plasminogen activator inhibitor-1 gene expression and cardiovascular remodeling in hypertensive rats. J Pharmacol Exp Ther 301:459–466

Lee MA, Bohm M, Paul M, Bader M, Ganten U, Ganten D (1996) Physiological characterization of the hypertensive transgenic rat TGR(mREN2)27. Am J Physiol 270:E919–E929

Leipala JA, Takala TE, Ruskoaho H, Myllyla R, Kainulainen H, Hassinen IE, Anttinen H, Vihko V (1988) Transmural distribution of biochemical markers of total protein and collagen synthesis, myocardial contraction speed and capillary density in the rat left ventricle in angiotensin II-induced hypertension. Acta Physiol Scand 133:325–333

Lijnen P, Petrov V (1999) Renin-angiotensin system, hypertrophy and gene expression in cardiac myocytes. J Mol Cell Cardiol 31:949–970

Liu YH, Yang XP, Sharov VG, Nass O, Sabbah HN, Peterson E, Carretero OA (1997) Effects of angiotensin-converting enzyme inhibitors and angiotensin II type 1 receptor antagonists in rats with heart failure: role of kinins and angiotensin II type 2 receptors. J Clin Invest 99:1926–1935

Lopez B, Querejeta R, Varo N, Gonzalez A, Larman M, Martinez Ubago JL, Diez J (2001) Usefulness of serum carboxy-terminal propeptide of procollagen type I in assessment of the cardioreparative ability of antihypertensive treatment in hypertensive patients. Circulation 104:286–291

Matsusaka T, Katori H, Inagami T, Fogo A, Ichikawa I (1999) Communication between myocytes and fibroblasts in cardiac remodeling in angiotensin chimeric mice. J Clin Invest 103:1451–1458

Molkentin JD, Lu JR, Antos CL, Markham B, Richardson J, Robbins J, Grant SR, Olson EN (1998) A calcineurin-dependent transcriptional pathway for cardiac hypertrophy. Cell 93:215–228

Ohkubo N, Matsubara H, Nozawa Y, Mori Y, Murasawa S, Kijima K, Maruyama K, Masaki H, Tsutumi Y, Shibazaki Y, Iwasaka T, Inada M (1997) Angiotensin type 2 receptors are reexpressed by cardiac fibroblasts from failing myopathic hamster hearts and inhibit cell growth and fibrillar collagen metabolism. Circulation 96:3954–3962

Olivetti G, Melissari M, Balbi T, Quaini F, Cigola E, Sonnenblick EH, Anversa P (1994) Myocyte cellular hypertrophy is responsible for ventricular remodeling in the hypertrophied heart of middle aged individuals in the absence of cardiac failure. Cardiovasc Res 28:1199–1208

Pathak M, Sarkar S, Vellaichamy E, Sen S (2001) Role of myocytes in myocardial collagen production. Hypertension 37:833–840

Pinto YM, Pinto-Sietsma SJ, Philipp T, Engler S, Kossmehl P, Hocher B, Marqueardt H, Sethmann S, Lauster R, Merker HJ, Paul M (2000) Reduction in left ventricular messenger RNA for transforming growth factor beta (1) attenuates left ventricular fibrosis and improves survival without lowering blood pressure in the hypertensive TGR(mRen2)27 rat. Hypertension 36:747–754

Ratajska A, Campbell SE, Sun Y, Weber KT (1994) Angiotensin II associated cardiac myocyte necrosis: role of adrenal catecholamines. 28:684–690

Robert V, Heymes C, Silvestre J-S, Sabri A, Swynghedauw B, Delcayre C (1999) Angiotensin AT_1 receptor subtype as a cardiac target of aldosterone. Role in aldosterone-salt-induced fibrosis. Hypertension 33:981–986

Rothermund L, Kreutz R, Kossmehl P, Fredersdorf S, Shakibaei M, Schulze-Tamzil G, Paul M, Grimm D (2002) Early onset of chondroitin sulfate and osteopontin expression in angiotensin II-dependent left ventricular hypertrophy. Am J Hypertens 15:644–652

Sadoshima J, Izumo S (1993) Molecular characterization of angiotensin II-induced hypertrophy of cardiac myocytes and hyperplasia of cardiac fibroblasts. Circ Res 73:413–423

Schnee JM, Hsueh WA (2000) Angiotensin II, adhesion, and cardiac fibrosis. Cardiovasc Res 46:264–268

Schwartzkopff B, Brehm M, Mundehenke M, Strauer BE (2000) Repair of coronary arterioles after treatment with perindopril in hypertensive heart disease. Hypertension 36:220–225

Seeland U, Haeuseler C, Hinrichs R, Rosenkranz S, Pfitzner T, Scharffetter-Kochanek K, Bohm M (2002) Myocardial fibrosis in transforming growth factor-β_1 (TGF-β_1) transgenic mice is associated with inhibition of interstitial collagenase. Eur J Clin Invest 32:295–303

Shindo T, Manabe I, Fukushima Y, Tobe K, Aizawa K, Miyamoto S, Kawai-Kowase K, Moriyama N, Imai Y, Kawakami H, Nishimatsu H, Ishikawa T, Suzuki T, Morita H, Maemura K, Sata M, HIirata Y, Komukai M, Kagechica H, Kadowaki T, Kurabayashi M, Nagai R (2002) Krüppel-like zinc-finger transcription factor KLF5/BTEB2 is a target for angiotensin II signaling and an essential regulator of cardiovascular remodeling. Nat Med 8:856–863

Sil P, Sen S (1997) Angiotensin II and myocyte growth. Role of fibroblasts. Hypertension 30:209–216

Singh K, Ballingand J, Fischer T, Smit T, Kelly R (1995) Glucocorticoids increase osteopontin expression in cardiac myocytes and microvascular endothelial cells. J Biol Chem 270:28471–28478

Singh K, Sirokman G, Communal C, Robinson KG, Conrad CH, Brooks WW, Bing OHL, Colucci WS (1999) Myocardial osteopontin expression coincides with the development of heart failure. Hypertension 33:663–670

Sun Y, Weber KT (1993) Ang II and aldosterone receptor binding in rat heart and kidney: response to chronic angiotensin II or aldosterone. J Lab Clin Med 122:404–411

Sun Y, Ramires FJA, Weber KT (1997) Fibrosis of atria and great vessels in response to angiotensin II or aldosterone infusion. Cardiovasc Res 35:138–147

Swynghedauw B (1999) Molecular mechanisms of myocardial remodeling. Physiol Rev 79:215–262

Takeda Y, Yoneda T, Demure M, Miyamori I, Mabuchi H (2000) Cardiac aldosterone production in genetically hypertensive rats. Hypertension 36:495–500

Teisman AC, Pinto YM, Buikema H, Flesch M, Bohm M, Paul M, Van Gilst WH (1998) Dissociation of blood pressure reduction from end-organ damage in TGR(mREN2)27 transgenic hypertensive rats. J Hypertens 16:1759–1765

Tharaux PL, Chatziantoniou C, Fakhouri F, Dussaule JC (2000) Angiotensin II activates collagen I gene through a mechanism involving the MAP/ER kinase pathway. Hypertension 36:330–336

Tomita H, Egahsira K, Ohara Y, Takemoto M, Koyanagi M, Katoh M, Yamamoto H, Tamaki K, Shimokawa H, Takeshita A (1998) Early induction of transforming growth factor-β via angiotensin II type 1 receptors contributes to cardiac fibrosis induced by long-term blockade of nitric oxide synthesis in rats. Hypertension 32:273–279

Tsutsumi Y, Matsubara H, Ohkubo N, Mori Y, Nozawa Y, Murusawa S, Kijima K, Maruyama K, Masaki H, Moriguchi Y, Shibasaki Y, Kamihata H, Inada M, Iwasaka T (1998) Angiotensin II type 2 receptor is upregulated in human heart with interstitial fibrosis, and cardiac fibroblasts are the major cell type for its expression. Circ Res 83:1035–1046

Varo N, Etayo JC, Zalba G, Beaumont J, Iraburu MJ, Montiel C, Gil MJ, Monreal I, Diez J (1999) Losartan inhibits the postranscriptional synthesis of collagen type I and reverses left ventricular fibrosis in spontaneously hypertensive rats. J Hypertens 17:101–114

Varo N, Iraburu M, Varela M, Lopez B, Etayo JC, Diez J (2000) Chronic AT_1 blockade stimulates extracellular collagen type I degradation and reverses myocardial fibrosis in spontaneously hypertensive rats. Hypertension 35:1197–1202

Villarreal FJ, Kim NN, Ungab GD, Printz MP, Dillmann WH (1993) Identification of functional angiotensin II receptors on cardiac fibroblasts. Circulation 88:2849–2861

Weber KT (1989) Cardiac interstitium in health and disease: the fibrillar collagen network. J Am Coll Cardiol 13:1637–1652

Weber KT (1997) Extracellular matrix remodeling in heart failure. A role for de novo angiotensin II generation. Circulation 96:4065–4082

Yamamoto K, Saito H (1998) A pathological role of increased expression of plasminogen activator inhibitor-1 in human or animal disorders. Int J Hematol 68:371–385

Yamamoto K, Masuyama T, Sakata Y, Nishikawa N, Mano T, Hori M (2002a) Prevention of diastolic heart failure by endothelin type A receptor antagonist through inhibition of ventricular structural remodeling in hypertensive heart. J Hypertens 20:753–761

Yamamoto N, Yasue H, Mizuno Y, Yoshimura M, Fujii H, Nakayama M, Harada E, Nakamura S, Ito T, Ogawa H (2002b) Aldosterone is produced from ventricles in patients with essential hypertension. Hypertension 39:958–962

Yamazaki T, Komuro I, Yazaki Y (1999) Role of the renin-angiotensin system in cardiac hypertrophy. Am J Cardiol 83:53H–57H

Pathophysiology of Cardiac AT_1 and AT_2 Receptors

J. Fielitz · V. Regitz-Zagrosek

Klinik für Innere Medizin—Kardiologie, Deutsches Herzzentrum Berlin,
Augustenburger Platz 1, 13353 Berlin, Germany
e-mail: vrz@dhzb.de

Abstract Cardiac angiotensin II type-1 (AT_1) and angiotensin II type-2 (AT_2) play a crucial role in mediating the myocardial effects of angiotensin II (Ang II). AT_1 mediates most of the known Ang II effects, whereas the role of AT_2 is still controversial. AT_2 is the predominant subtype in the human heart and its activation is proposed to counteract AT_1-mediated effects. Expression and regulation of AT_1 and AT_2 is disease dependent and locally inhomogeneous. ATR regulation occurs at different levels and involves rapid desensitization, receptor internalization, attenuation of transcription and post-transcriptional mRNA destabilization. Also, alternative splicing of AT_1 and AT_2 mRNA may represent a mechanism of ATR regulation. AT_1 splice patterns differ between controls and failing hearts, possibly leading to differences in AT_1 mRNA translation. AT_1 content is reduced in LVH, heart failure and in transplanted human hearts. However, data on AT_2 regulation are conflicting. Signal transduction following AT_1 stimulation is mediated by activation of PKC, the MAP kinase pathway and immediate early genes. In addition, the JAK-STAT pathway, tyrosine phosphorylation and NF-κB seem to be involved. In contrast, AT_2 receptor stimulates protein tyrosine phosphatase activity and leads to inhibition of phosphotyrosine phosphatase 1B in fibroblasts. Regulation of the AT_2 receptor in human heart occurs also by elements in intron 1. Factors such as nitric oxide, hypercholesterolemia, statins and estrogens are closely related to the expression of the cardiac ATR. Nitric oxide, statins and estrogen can attenuate Ang II-mediated adverse effects and may in part have protective effects in cardiovascular disease. Cardiac RAS and collagen synthesis are activated in human aortic valve stenosis and regurgitation. However, Ang II seems not to be a necessary factor for LVH in most animal models. Genetic data also suggest that AT_2 has a role in human LVH. The +1675 G/A-polymorphism of the X-chromosomal located AT_2 receptor modulates LV structure in white, young male subjects with normal or mildly elevated blood pressure and is associated with LVH in these subjects. Further understanding of the disease-specific regulation and function of AT_1 and AT_2 may lead to a differential pharmacotherapy in LVH and early stages of heart failure.

Keywords Angiotensin · Receptor · Myocardium · Hypertrophy · Genetics

Abbreviations

ATR	Angiotensin II receptor
AT_1	Angiotensin II receptor type 1
AT_2	Angiotensin II receptor type 2
RAS	Renin-angiotensin system
LVH	Left ventricular hypertrophy

1 Introduction

Cardiac angiotensin II type-1 (AT_1) and angiotensin II type-2 (AT_2) play a crucial role in mediating the myocardial effects of angiotensin II (Ang II). In this chapter, we will discuss AT_1 and AT_2 subtype-specific receptor expression, regulation in cardiac diseases, cellular localization, signal transduction, and myocardial effects. A number of phenomena are similar in rodent and in human hearts, but in a number of aspects, the human heart seems to differ from animal hearts. Specific consideration will be given to the human heart.

2
Receptor Expression and Regulation in Cardiac Disease

Expression of angiotensin II receptors (ATRs) in rodent and human hearts has known for more than 10 years (Grady et al. 1991; Sechi et al. 1993). Both subtypes were found in the myocardium (Timmermans et al. 1992; Sechi et al. 1992). Cloning of the AT_1 and AT_2 receptor allowed more specific expression analysis of AT_1 and AT_2 (Kambayashi et al. 1993; Koike et al. 1995). Papers that described regulation with cardiac hypertrophy in animal models followed (Lopez et al. 1994). Shortly thereafter in 1995, we and others described the disease-dependent regulation of AT_1 and AT_2 in the human heart (Regitz-Zagrosek et al. 1995; Rogg et al. 1996). To assess the chamber localization, subtype distribution, and regulation of human myocardial ATRs in heart failure, we determined the binding of Ang II and the subtype-specific antagonists Dup 753 (AT_1-specific) and PD 123319 (AT_2-specific) in atria from patients with normal or moderately impaired cardiac function and in atria and ventricles from explanted end-stage failing hearts. The number of Ang II binding sites in sarcolemmal fractions was significantly reduced in explanted end-stage failing hearts in comparison with control subjects and moderate heart failure. The dissociation constants were comparable in nonfailing and failing hearts. In nonfailing human hearts, 69% of binding sites were classified as AT_2; whereas 33% were classified as AT_1. In explanted hearts, comparable ratios of 66% AT_2 sites and 34% AT_1 sites were found. AT_1 mRNA content was reduced to about one-third of the level in control subjects in end-stage failing hearts (Fig. 1). We concluded that ATRs in human myocardium are present in relatively low numbers, and AT_2 is the predominant subtype. A significant loss of ATRs occurs in end-stage heart failure. The loss of ATRs affects both subtypes to a comparable degree. The data suggested that the decrease in receptor density was due to a decrease in steady-state mRNA abundance.

Further studies also confirmed that both normal and hypertrophied human myocardium predominantly contains the AT_2 receptor subtype and that these receptors are down-regulated in hypertrophied tissues (Nozawa et al. 1996). Later studies also aimed at correlation with cardiac function (Zisman et al. 1998). In order to asses ATR regulation in patients with moderately decreased cardiac function, AT_1 mRNA content was measured in right ventricular endomyocardial biopsies from patients with heart failure due to dilated cardiomyopathy (DCM). In the biopsies from patients with heart failure, a 68% decrease in AT_1 mRNA content was found in comparison with controls. A close correlation existed between AT_1 mRNA content and left and right ventricular ejection fraction (AT_1/LVEF: r=0.43, p<0.05; AT_1/RVEF: r=0.59, p<0.01) (Fig. 2). The loss of ventricular AT_1 mRNA in human heart failure corresponded to the loss of AT_1 protein described previously, suggesting that down-regulation occurs at transcriptional level (Regitz-Zagrosek et al. 1997).

Interestingly, down-regulation of ATRs also occurred in transplanted human hearts (Regitz-Zagrosek et al. 1996a). Right ventricular biopsies from patients

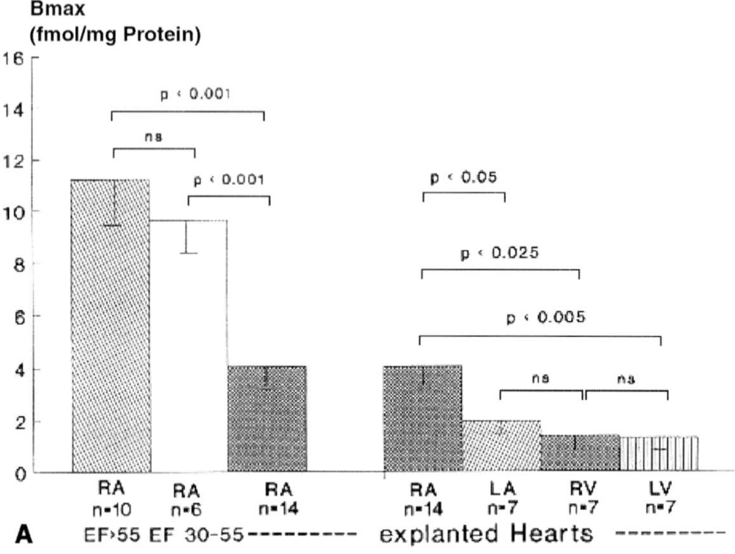

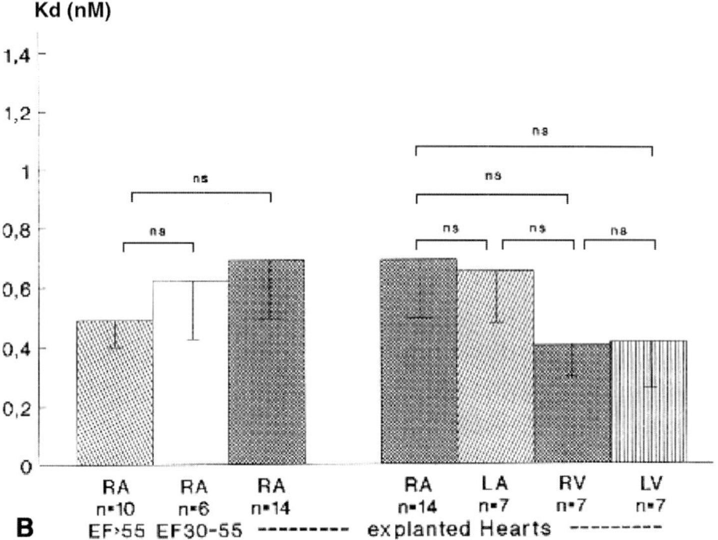

Fig. 1. A The number of binding sites (B_{max}) in right atrial (*RA*) sarcolemmal membrane fractions from explanted end-stage failing hearts (EFH) was significantly reduced in comparison with control subjects [left ventricular ejection fraction (*EF*), >55%] or moderate heart failure (EF, 30%–55%). B_{max} in the left atrium (*LA*) and in the right and left ventricles (*RV* and *LV*) of explanted end-stage failing hearts was

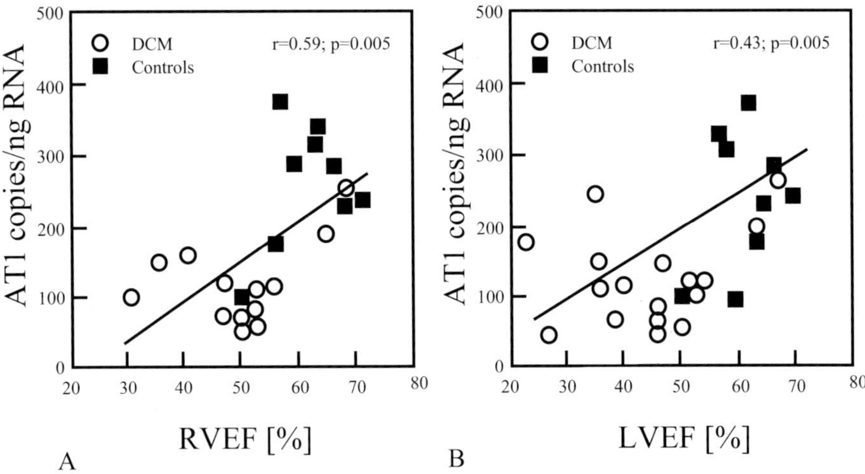

Fig. 2A, B Correlation between AT_1 mRNA expression and right and left ventricular ejection fraction (*RVEV, LVEF*). Patients with DCM are depicted as *circles* and controls are depicted as *squares*. The number of AT_1 copies is plotted against RVEF (**A**) and LVEF (**B**) and the equation of the regression line is indicated in the figure. Correlation coefficients are $r=0.59$, $p<0.01$ for AT_1/RVEF and $r=0.43$, $p<0.05$ for AT_1/LVEF. There is no significant correlation between LV or RVEF and AT_1 within the DCM group (LVEF/AT_1, $r=-0.30$, $p=0.3$, n.s.; RVEF, $r=+0.40$, $p=0.19$, n.s.)

after heart transplantation with normal cardiac function were compared with patients with normal cardiac function undergoing diagnostic catheterization to exclude myocarditis, and in whom no evidence of cardiac disease was found. AT_1 mRNA content was reduced by 46% in right ventricular biopsies after heart transplantation compared with controls (Fig. 3). No association between the extent of AT_1 down-regulation and clinical or hemodynamic variables was detect-

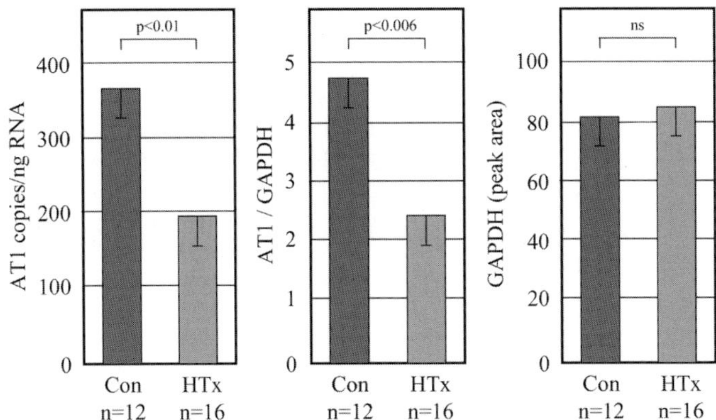

Fig. 3 Number of AT_1 mRNA copies/ng RNA, AT_1/GAPDH mRNA ratios, and GAPDH mRNA contents in transplanted and in control hearts. The number of AT_1 copies/ng RNA and the AT_1/GAPDH ratios were significantly reduced in the transplanted hearts

ed. Thus, in the human heart ventricular AT_1 is down-regulated after heart transplantation and the decrease in AT_1 mRNA is not associated with altered systolic function. This has implications for the understanding of AT_1 function, suggesting that it does not contribute to systolic function at rest (Regitz-Zagrosek et al. 1996a).

Whereas down-regulation of AT_1 at mRNA and protein levels was confirmed by all authors, data on AT_2 regulation were conflicting, including up-, down- and no regulation.

We determined subtype numbers and distribution of AT_2 by ligand-binding studies in ventricular myocardium from patients with end-stage heart failure due to coronary artery disease and DCM. We found that about 50%–80% of ATRs belonged to the AT_2 subtype in the right and left ventricles from patients with end-stage heart failure, confirming that AT_2 is the dominant ATR subtype in the human heart (Regitz-Zagrosek et al. 1996b; Regitz-Zagrosek et al. 1998). A down-regulation of AT_2 receptors was shown in hypertrophied human myocardium (Nozawa et al. 1996). Others found no regulation of AT_2 gene expression in human heart failure (Asano et al. 1997; Haywood et al. 1997). However, according to Tsutsumi et al., an AT_2 receptor up-regulation has also been reported in failing hearts caused by myocardial infarction and DCM (Tsutsumi et al. 1998). Fibroblasts present in the interstitial regions were the major cell type responsible for AT_2 expression (Tsutsumi et al. 1998). The expression level of AT_2 receptor seems therefore to be determined by the extent of interstitial fibrosis associated with heart failure and seems to vary more in different models than AT_1 expression.

In ischemic heart disease, ATR regulation is locally inhomogeneous. Radioligand-binding studies, in agreement with previous investigations, showed that AT_2 receptors represent 70%–77% of the sites in noninfarcted myocardium (Wharton et al. 1998). Endocardial, interstitial, perivascular and infarcted regions in the ventricles of patients with end-stage ischemic heart disease exhibited a significantly greater density of high-affinity AT_2 binding compared with adjacent noninfarcted myocardium. Regions displaying the relative increase in AT_2 binding sites corresponded to areas of fibroblast proliferation and collagen deposition, shown by picrosirius red staining. AT_1 binding sites were localized to nerves, occurred at relatively low density in coronary vessels and represented only 23%–29% of myocardial Ang II binding sites. The border zone between infarcted and noninfarcted myocardium characteristically contained numerous microvessels, exhibiting perivascular AT_2 receptors and endothelial angiotensin-converting enzyme (ACE) activity.

Complementary information on the regulation of ATRs was obtained in animal models, including pressure overload, volume overload and myocardial ischemia.

On the cellular level, it was shown that mechanical stress plays a pivotal role in the development of cardiac hypertrophy during hemodynamic overload, and Ang II secreted from stretched myocytes mediates mechanical stretch-induced left ventricular hypertrophy (LVH). Both AT_1 receptor and AT_2 receptor mRNA levels were up-regulated by stretching of neonatal rat myocytes with similar time courses. This could facilitate our understanding of ATR up-regulation in load-induced LVH (Kijima et al. 1996).

To determine the distribution of cardiac AT_1 and AT_2 in LVH, Ang II, alone or in the presence of the AT_1 inhibitor losartan or an AT_2 inhibitor, was infused into isolated rat hearts with LVH from aortic banding. Control left ventricles contained predominantly the AT_1 subtype, whereas LVH ventricles contained primarily the putative AT_2 subtype (Lopez et al. 1994). This suggested that ATR subtype redistribution occurs in LVH with AT_1 subtype down-regulation.

To investigate the regulation of the AT_1 in cardiac volume overload, AT_1 mRNA content was measured in the atria, left and right ventricle of rats with an aortocaval shunt, produced by infrarenal aortocaval puncture 4 weeks earlier. Heart weight was increased in the shunt animals, with the greatest increase in the atria. The AT_1 mRNA content was significantly increased in the atria, but no change was found in the right or left ventricle. The findings document that atrial hypertrophy in cardiac volume overload goes in parallel with a significant increase in atrial AT_1 mRNA content (Bauer et al. 1997).

Regional changes in ATR density also occur after experimental acute myocardial infarction (MI). It has been shown that AT_1 and AT_2 are expressed on a relatively small number of cardiomyocytes in vivo and that an observed up-regulation of receptors after infarction involves other cells than cardiomyocytes (Busche et al. 2000). It has also been shown that the time-dependent increase in AT_1 and AT_2 is associated with early remodeling and is independent from AT_1 blockade (Zhu et al. 2000). Acute MI was induced in rats by left coronary artery ligation (Lefroy et al. 1996). The number of Ang II binding sites was unchanged at 18 h, but was increased at 7 days in the infarcted region of the left ventricle compared with the noninfarcted and with the LV myocardium of sham-operated control animals. The increased Ang II binding density was still present, but diminished, at 8 months after acute MI and was identified as AT_1 binding sites. The regional increase in AT_1 receptor density in the infarcted region of myocardium was associated with fibroblast infiltration and collagen deposition. The infarct scar and the cardiac fibroblasts within it expressed high levels of ATR and therefore were believed to represent potential targets for the actions of Ang II after acute MI.

3 Regulation of Myocardial ATRs by Growth Factors

To determine which mechanisms regulate ATRs at the molecular and cellular level, the effect of growth factors was studied. We found a protein kinase C (PKC)-dependent regulation of the human AT_1 promoter in vascular smooth muscle cells (VSMCs) mediated by the transcription factors activator protein 1 (AP1) and Ets, which also plays a key role in cell growth, death, and differentiation (Regitz-Zagrosek 1996b). AT_1 mRNA expression in cultured bovine VSMCs increased twofold after 8 h of PKC activation with phorbol-acetate (PMA), whereas stimulation with forskolin did not alter the AT_1 mRNA level. The expression of AT_1 promoter/exon 1 luciferase constructs transfected into VSMCs increased 2.4-fold with PMA stimulation. In-gel kinase assays demonstrated rapid phosphorylation of mitogen-activating protein kinases (MAPK) ERK1 and ERK2 by PMA. Electrophoretic gel mobility shift assays showed sequence-specific binding of nuclear proteins from PMA-activated VSMCs, identified as AP-1 complex in competition assays, to a radiolabeled AT_1-promoter fragment. Site-specific mutagenesis destroying the AP-1 site, the adjacent polyoma enhancer activator 3 element, or both sites simultaneously, indicated that both sites together are necessary and sufficient to control basal and PMA-induced activation of the human AT_1 promoter. The capability of the phorbol ester PMA to activate the human AT_1 promoter via an AP-1 element suggests a prominent role for PKC/MAPK and Ets proteins in AT_1 regulation.

4
Effects of Nitric Oxide on ATRs

Another interesting modulator of cardiac ATRs is nitric oxide (NO) (Katoh et al. 1998). The chronic administration of the inhibitor of NO synthesis, L-NAME to normal rats increased the arterial blood pressure and the number of AT_1 and AT_2 receptors, during the 1st week of L-NAME administration. However, ATR expression returned to control levels after 4 weeks of treatment. Inflammatory changes (monocyte infiltration and myofibroblast formation) in perivascular areas surrounding coronary vessels and myocardial interstitial spaces were observed during the 1st week. Immunohistochemistry revealed that myofibroblasts expressed AT_1 receptor. AT_1 receptor blockade or cotreatment with L-arginine, but not cotreatment with hydralazine, prevented the L-NAME-induced increase in ATRs and inflammatory changes. This suggested that rat cardiac ATRs are up-regulated at an early phase of chronic inhibition of NO synthesis, which may contribute to cardiovascular inflammatory changes in an early phase and to remodeling at the later phase after inhibition of NO synthesis (Katoh et al. 1998).

Furthermore, continuous infusion of Ang II in subpressor and pressor doses leads to an increase of NO synthases 1 (NOS1, 160-kDa) and 3 (NOS3, 135-kDa) to approximately 200% in the rat myocardium. Immunohistochemistry in this study showed that NOS1 and NOS3 are differentially expressed in myocardial cells. Ang II infusion enhanced the tissue immunoreactivity of both isoforms in their specific locations but did not change the distribution throughout the myocardium. Consequently, long-term increases in circulating Ang II levels could up-regulate the NOS1 and NOS3 protein expression and therefore the function of the local myocardial NO system (Tambascia et al. 2001).

In summary, Ang II and NO seem to closely modulate their function with NO attenuating Ang II-mediated adverse effects.

5 Regulation of ATRs by Statins and Estrogens

Statins are drugs for the treatment of hypercholesterolemia and have been reported to interact with elevated blood pressure (Glorioso et al. 1999). A reduction in blood pressure associated with the use of statins has been reported in patients with untreated hypertension and in patients treated with antihypertensive drugs, particularly ACEI and calcium channel blockers (Glorioso et al. 1999; Borghi et al. 2000). The mechanism responsible for the hypotensive effect seems to be partially related to the interaction with ATRs.

Hypercholesterolemia leads to a significant increase in Ang II-induced blood pressure elevation in men. Moreover, an up-regulation of AT_1 receptor expression was found in hypercholesterolemic individuals compared with normocholesterolemic men. Hypercholesterolemia could therefore enhance biological effects of Ang II in men. Cholesterol-lowering treatment with statins reversed the elevated blood pressure response to Ang II infusion ($p<0.05$) and down-regulated AT_1 receptor density ($p<0.05$) (Nickenig et al. 1999). The capacity of statins to improve blood pressure control may contribute to its efficacy in the prevention of cardiovascular events.

In rats transgenic for human renin and angiotensinogen, the statin cerivastatin ameliorates Ang II-induced hypertension, cardiac hypertrophy, fibrosis, and remodeling independently of cholesterol reduction (Dechend et al. 2001). Cerivastatin and fluvastatin suppressed the AT_1 receptor promoter activity but did not affect AT_1 receptor mRNA stability, suggesting that the suppression occurs at the transcriptional level (Ichiki et al. 2001). Down-regulation of AT_1 receptor may be a part of the pluripotent cholesterol-independent effects of statins on the cardiovascular system.

In addition, a major impact of estrogens on AT_1 receptor expression was found. In ovariectomized rats, estrogen-deficiency leads to an AT_1 receptor increase and estradiol substitution provoked a time-dependent down-regulation of AT_1 receptor mRNA and protein (Nickenig et al. 1998; Krishnamurthi et al. 1999). Estrogen-induced down-regulation of AT_1 receptor mRNA and protein is mediated through activation of estrogen receptors. A possible mechanism for estrogen-induced AT_1 mRNA down-regulation is an estrogen-mediated reduction in AT_1 receptor mRNA half-life, whereas transcription rate is not altered. (Nickenig et al. 2000). Furthermore, cytosolic RNA binding proteins that recognize the 5′ leader sequence of the AT_1 receptor mRNA are modulated by estrogen in an inverse manner to AT_1 receptor regulation. In vitro assays suggested that this inhibits AT_1 receptor translation (Krishnamurthi et al. 1999). These findings may in part account for the blood pressure lowering effects of estrogen.

6
Mechanisms of ATR Regulation

Regulatory mechanisms of AT_1 involve rapid desensitization, receptor internalization, attenuation of transcription and post-transcriptional mRNA destabilization (Nickenig and Murphy 1994; Hein et al. 1997; Wang et al. 1997). Down-regulation of AT_1 surface receptor expression is frequently induced by growth factors such as epidermal growth factor, basic fibroblast growth factor (bFGF) and platelet-derived growth factor BB. This down-regulation is based to a major degree on a significant shortening of ATR mRNA half-life and to a minor degree on a decrease in transcription rate (Nickenig and Murphy 1994; Wang et al. 1997). Rapid desensitization of AT_1 probably depends upon phosphorylation due to PKC, as it can be induced by activation of PKC with a short burst of PMA stimulation (Brock et al. 1985; Abdellatif et al. 1991). Another mechanism of ATR regulation is internalization of AT_1 receptors together with Ang II as a receptor–ligand complex. Endocytosis seems, however, not to be a regulatory mechanism for AT_2 upon agonist stimulation, showing evidence for subtype-specific receptor sorting and internalization (Hein et al. 1997).

AT_2 is regulated independently from AT_1. Little is known about the precise mechanisms in the human heart. The growth-dependent expression of AT_2 receptor in RT3 cells, a mouse fibroblast cell line that expresses only AT_2 receptors, is regulated by transcription factors interferon regulatory factor-1 and -2 (Horiuchi et al. 1995). The regulatory effects of Ang II and several growth factors, including insulin-like growth factor 1 (IGF-1), bFGF and transforming growth factor beta1 (TGFβ1) on the AT_2 receptor were also studied in RT3 cells. The findings demonstrated that the main mechanism by which Ang II regulates the AT_2 receptor is by increasing the rate of AT_2 mRNA translation, whereas the stimulatory or inhibitory effects of the growth factors (IGF-1, bFGF and TGFβ1) on AT_2 expression were exerted at the transcriptional level (Li et al. 1999). Regulation of AT_2 gene in PC12W cells also occurs by the PKC-calcium pathway (Kijima et al. 1996).

To study the regulation of the AT 2 receptor in human heart, we cloned the complete mRNA sequence and thereafter the promoter sequence from a human genomic library. A functional AT_2 promoter, with greater than 90% homology with the mouse promoter and 35% homology with the human AT_1 promoter containing numerous cis-acting sequences for basal (TFIID) and inducible (AP-1, PEA-3, CBF) transcription factors in the first 1,000 bp was identified (Regitz-Zagrosek et al. 1996b). To investigate in detail the mechanisms of human AT_2 receptor gene regulation, we than functionally characterized the promoter and downstream regions of the gene. 5′ terminal deletion mutants from −1417/+100 to −46/+100 elicited significant but low functional activity in luciferase reporter gene assays with PC12W cells. Inclusion in the promoter constructs of intron 1 and the transcribed region of the human AT_2 gene up to the translation start enhanced luciferase activity up to 12-fold, whereas fusion of the promoter to the spliced 5′ untranslated region of human AT_2 cDNA did not, which indicated an enhancement caused by intronic sequence elements. Mapping of intron 1 re-

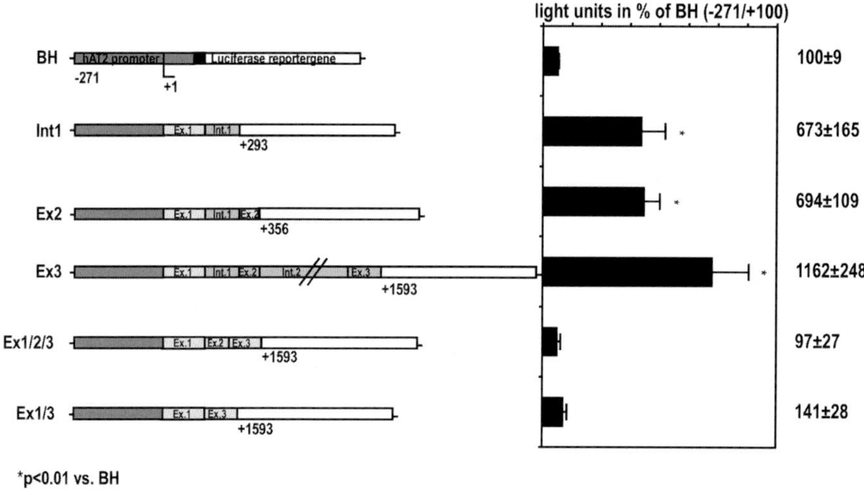

Fig. 4 Transcription of the human AT_2 gene in reporter gene assays. AT_2 core promoter fragments and fragments including intron 1 and 2 were cloned in front of a luciferase reporter gene in the PGL2 basic vector and transfected into cells. Activity of gene transcription was significantly increased by including intron 1 and 2 into the transcripts. Omission of intron 1 and 2 from the constructs sign reduced luciferase activity

vealed that a 12-bp sequence in the center of the intron was required for the increase in promoter activity. Mutation of the 12-bp region led to altered protein binding and markedly decreased luciferase activity. Cloned into a promoterless luciferase vector, a 123-bp intron 1 fragment was able to direct reporter gene expression to the same activity as occurred in conjunction with the 5– flanking region (Fig. 4). These results indicate that sequence elements in intron 1 are necessary for efficient transcription of human AT_2. In reporter gene assays, intron 1 might by itself function as a promoter and initiate transcription from an alternative start point (Warnecke et al. 1999b).

7
AT_1/2 Splice Variants in the Human Heart

Regulation of ATR may also be mediated by the effects of splice variants, which are described for AT_1 and AT_2. Alternatively spliced human AT_1 mRNAs are translated at different efficiencies and encode two receptor isoforms (Curnow et al. 1995). AT_2 splice variants in the human heart were described by Wharton et al. (Wharton et al. 1998). He found that specific myocardial AT_2 receptor mRNA transcripts occurred in human hearts and exhibited alternative splicing of untranslated 5′ exons.

A more detailed analysis and functional characterization of alternatively spliced AT_1 and AT_2 receptor transcripts in the human heart was conducted by

Regitz-Zagrosek's group (Warnecke et al. 1999a). They searched for AT_1 and AT_2 mRNA splice patterns specific to chamber localization or to cardiac function and analyzed their effect on protein expression in transfection experiments. The exon composition of the AT_1 and AT_2 mRNA transcripts in normal and diseased human hearts was analyzed using a reverse transcription polymerase chain reaction followed by HPLC quantitation of the amplicons. Atrial and ventricular samples and endomyocardial biopsies from patients with normal and severely impaired cardiac function were compared. AT_1 transcripts with the exon composition 1/2/5 and 1/5 represented 92%–98% of all AT_1 mRNAs. For more detail, in right atria from controls Ex1/2/5 was 48% of all AT_1 splice variants compared to heart failure probes, where it was 41% in the right atrium, 38% in RV and 36% in LV (Fig. 5). Right atrial Ex1–5 in controls was 44% of all AT_1 mRNAs compared to heart failure probes, where it was 50% in the right

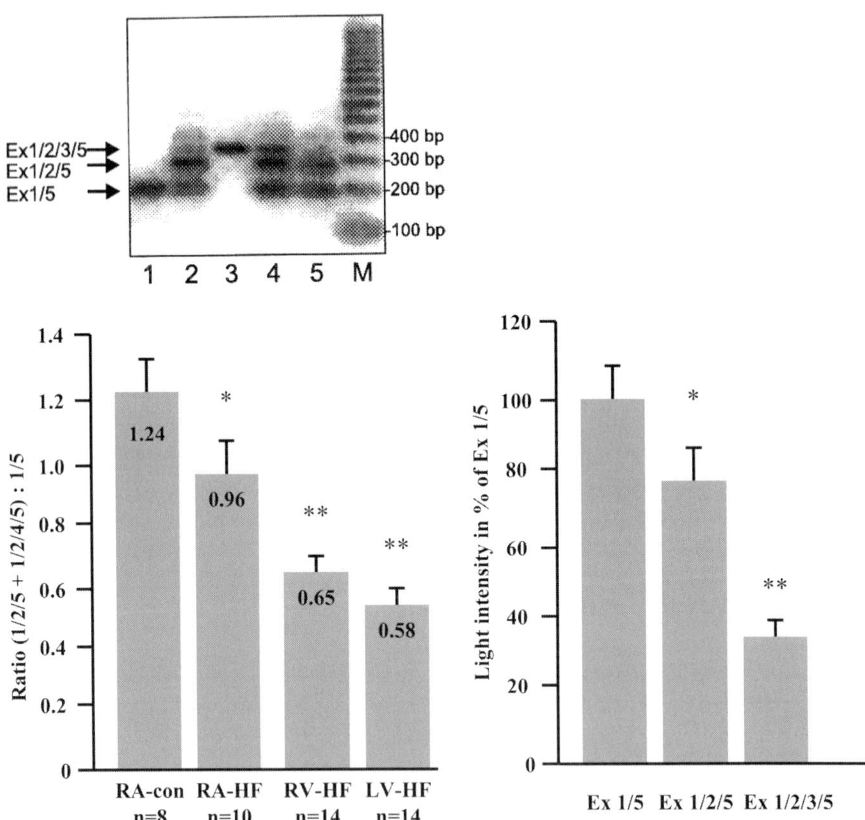

Fig. 5. A Splice variants of the AT_1 mRNA in human hearts. AT_1 mRNA transcripts with different exon composition are shown. Only transcripts 1/5, 1/2/5 and 1/2/3/5 were present in the human heart. **B** Distribution of different AT_1 transcripts in the human atria and ventricles in healthy and diseased hearts. **C** Protein synthesis resulting from the different transcripts of the human *AT_1* gene which are expressed in the myocardium

atrium, 60% in RV and 62% in LV. Transcript 1/2/3/5 represented 8%–9% in the atria and 2% in ventricles in failing and nonfailing myocardium. Since exon 2 reduces translational efficiency in vitro, the ratios of transcripts with (1/2/5+1/2/3/5) and without (1/5) exon 2 were compared. These were highest in normal atria, second in atria from failing hearts and lowest in failing left ventricles. Endomyocardial biopsy specimens confirmed significant differences between controls and heart failure. Of the two identified AT_2 transcripts, mRNA 1/2/3 was the most abundant in the human heart (92%), without many disease-dependent changes.

The various splice isoforms affected protein expression at least in reporter gene assays. The presence of the untranslated exons 2 and 3 of AT_1 inhibited translational activity as did exon 2 in the AT_2 constructs. Thus, alternative splicing may be a mechanism of ATR regulation in vivo. In the human heart, AT_1 splice patterns differ distinctly between atria and ventricles and to a lesser degree between controls and failing hearts. This may lead to differences in AT_1 mRNA translation into protein in the various cardiac areas and under different pathophysiological conditions.

8
Cellular Localization of ATR in the Heart

The AT_1 receptor was in general found evenly distributed over the myocardium at low concentrations. Higher local concentrations of AT_1 were observed on nervous tissue. However, high densities of AT_2 receptor were frequently associated with fibrous tissue (Brink et al. 1996). The association of AT_1 and AT_2 with fibrous tissues was studied in an animal model of heart failure using Bio14.6 cardiomyopathic hamsters (Ohkubo et al. 1997). AT_2 sites were increased by 153% during heart failure. AT_1 numbers were increased by 72% in the hypertrophy stage and then decreased to the control level during heart failure. Autoradiography and immunocytochemistry revealed that both the AT_2 receptor and the AT_1 receptor were localized at higher densities in fibroblasts present in fibrous regions. Surrounding myocardium predominantly expressed AT_1, but the level of expression was less than that in fibrous regions. Cardiac fibroblasts isolated from cardiomyopathic hamster hearts during heart failure expressed AT_2. Ang II stimulated net collagenous protein production by 48% and pretreatment with an AT_2 antagonist, PD123319, evoked a further elevation (83%). Thus, AT_2 was re-expressed by cardiac fibroblasts present in fibrous regions in failing hearts and the increased AT_2 was claimed to exert an anti-AT_1 action on the progression of interstitial fibrosis during cardiac remodeling (Ohkubo et al. 1997).

A second study confirmed the association of AT_2 receptor with interstitial fibrosis in the human heart (Tsutsumi et al. 1998). Tsutsumi et al. analyzed AT_2 expression in LV samples from patients with failing hearts caused by idiopathic dilated cardiomyopathy (DCM). The expression of AT_2 was markedly increased in DCM hearts. AT_2 sites were highly localized in the interstitial region in either nonfailing or failing heart, whereas AT_1 was evenly distributed over the myocardium at lower densities. AT_2 receptor expression was up-regulated in failing hearts, and fibroblasts present in the interstitial regions were the major cell type responsible for its expression.

9 Signal Transduction and Intracellular Effects

AT_1 stimulation is mediated by the generation of phospholipid-derived second messengers, activation of PKC, the MAP kinase pathway and of immediate early genes. In addition, phosphorylation and dephosphorylation of tyrosine kinases have been associated with AT_1- and AT_2-mediated signal transduction. Ang II activates the MAP kinase pathway via G protein coupling in different ways. For example, Ang II may activate c-Jun NH2-terminal kinase in cultured cardiac myocytes through an increase in intracellular Ca2+ and activation of PKC, and the activated c-Jun NH2-terminal kinase may regulate gene expression by activating AP-1 during Ang II-induced cardiac hypertrophy (Kudoh et al. 1997).

In addition to coupling to conventional G-protein signal transduction pathways, the AT_1 receptor also increases the tyrosine phosphorylation of several intracellular substrates, including the STAT (signal transducers and activators of transcription) family of transcription factors, in rat cardiac fibroblasts and myocytes. Ang II stimulates the tyrosine phosphorylation and nuclear translocation of STAT 1 (STAT 91) and STAT 3 (STAT 92). Ang II induces the formation of a complex of STAT proteins termed SIF (sis-inducing factor), which binds the DNA sequence SIE (sis-inducing element), present in the promotor element of many genes. Thus, the JAK-STAT pathway has an important role in Ang II-mediated effects on gene transcription, cardiac and vascular cellular growth and development, and inflammatory responses (Dostal et al. 1997).

Ang II has also been reported to induce the tyrosine phosphorylation of the gamma-isoform of phospholipase C, pp120 and pp125FAK. Furthermore, Ang II seems to modulate the activity of the soluble cytoplasmic tyrosine kinase pp60c-src, and this tyrosine kinase has been implicated in the phosphorylation of some of the above proteins (Schieffer et al. 1996).

Ang II-dependent induction of nuclear factor-kappaB (NF-κB) is mainly mediated through AT_1. AT_1 and AT_2 receptors share some signaling pathways (oxygen radicals and ceramide); however, tyrosine kinases only participate in AT_1/NF-κB responses (Ruiz-Ortega et al. 2000). Ang II stimulates VCAM-1 mRNA and protein expression via the AT_1 receptor in an NF-κB-dependent way in endothelial cells. (Pueyo et al. 2000). NF-κB regulates genes involved in cardiac pathophysiology and its activation could be a part of the negative effects of Ang II in heart failure.

10
Effects on Gene Transcription

AT_1 receptor stimulation also interferes with calcineurin-dependent pathways in rats. Treatment of Dahl salt-sensitive (DS) rats with a nonantihypertensive dose of the selective AT_1 receptor blocker candesartan or the calcineurin inhibitor FK506 attenuated both calcineurin activity and its mRNA expression in the heart, as well as the development of cardiac hypertrophy and fibrosis. Treatment with candesartan, but not FK506, prevented the up-regulation of ACE and *TGF-β1* gene expression. Both candesartan and FK506 prevented the load-induced induction of fetal-type cardiac genes. Calcineurin may therefore be a part in AT_1 receptor-mediated Ang II signaling in vivo (Nagata et al. 2002).

Mechanical stretch activates angiotensinogen and fibronectin gene expression mainly via AT_1 receptor in cardiomyocytes and VSMCs. An AT_1 receptor antagonist significantly decreased these stretch-mediated increases in angiotensinogen and fibronectin mRNA levels, whereas the AT_2 receptor antagonist (PD 123319) did not affect the induction (Tamura et al. 2000).

11
AT_2

Signal transduction of AT_2 was difficult to study since few primary cells express the receptor in culture and on these it is rapidly down-regulated. Few data are available from endothelial cells. Most studies on AT_2 signaling were conducted in special cell lines such as PC12W, RT3, or in neuroblastoma cells or transfected cells. The up-regulation of a zinc finger homeodomain enhancer protein was found in PCW12 cells and confirmed in coronary endothelial cells (Stoll et al. 2002). In PC12W cells, the AT_2 receptor is reported to stimulate protein tyrosine phosphatase activity and to mediate inhibition of particulate guanylate cyclase (Bottari et al. 1992). AT_2 receptors also mediate the inhibition of the MAP kinase cascade and functional activation of SHP-1 tyrosine phosphatase in N1E-115 neuroblastoma cells (Bedecs et al. 1997). AT_2 receptor activation leads to protein tyrosine dephosphorylation in N1E-115 cells and support a possible role for AT_2 receptors in the negative regulation of cell proliferation (Nahmias et al. 1995).

Recently, an interaction between the AT_2 and the ErbB3 receptor, a member of the epidermal growth factor receptor family, was found in the yeast 2-hybrid system, which suggests new possibilities for AT_2 function (Knowle et al. 2000).

A number of studies suggest significant cross talk between AT_2 and other receptors. Cross-talk between ATR and the tyrosine kinases and phosphatases has been reported (Inagami 1999; Inagami et al. 1999). The study by Elbaz et al. demonstrates negative intracellular cross-talk between AT_2 and insulin receptors. AT_2 receptor stimulation leads to inhibition of insulin-induced extracellular signal-regulated protein kinase (ERK) activity and cell proliferation in transfected Chinese hamster ovary cells (Elbaz et al. 2000).

In adult rat ventricular myocytes, Fischer et al. examined the role of Ang II signaling, via AT_1 and AT_2 receptors, on the activation of the ERKs and on the expression of the MAP kinase phosphatase (MKP-1). Ang II increased MKP-1 mRNA levels within 15 min via AT_2-receptor activation. They speculated on constitutive as well as inducible suppression of ERKs and c-jun NH2-terminal protein kinases by MKP 1 in adult cardiac myocytes (Fischer et al. 1998). These mechanisms do not, however, apply to cardiac endothelial cells or cardiac fibroblasts (see the next section).

To overcome the down-regulation of ATR in primary cardiac cells, we performed adenovirus-mediated overexpression and stimulation of the human AT_2 receptor in porcine cardiac fibroblasts. Transduction of cardiac fibroblasts with the adenoviral AT_2 receptor expression vector led to a six- to tenfold higher AT_2 expression than endogenous AT_1 receptor expression. The overexpressed AT_2 receptor had the same apparent molecular mass as the endogenous AT_2 receptor in rat PC12W cells. Unexpectedly, the fibroblast proliferation rate was not significantly lower in AT_2 receptor expression than in antisense-transduced (TA2) cells upon stimulation with Ang II or Ang II plus the AT_1 receptor blocker irbesartan. Collagen 1 mRNA revealed no differences between AT_2 receptor-transduced fibroblasts and antisense controls when stimulated with Ang II (1 μM, 24 h) plus Irbesartan and 10 ng/ml TGF-β1 (Fig. 6). Ang II stimulation of the endogenous AT_1 receptor increased ERK 1 and ERK 2 activities. This response was reduced by irbesartan, but PD123319 had no effect. The time course and magnitude of Ang II-stimulated ERK1 and ERK2 activation was identical in AT_2 receptor-transduced and control cells. Also, neither simultaneous nor Ang II prestimulation induced gene expression of the MAP kinase phosphatase 1 or modulated PMA-stimulated ERK1 and ERK2 activation in AT_2 receptor-transduced fibroblasts. Similar results were obtained in AT_2 receptor-transduced human umbilical vein endothelial cells, and in PC12W cells. Using a tyrosine phosphatase assay, we observed an inhibition of phosphotyrosine phosphatase activity by 30% after 5 min Ang II stimulation of AT_2 receptor-expressing porcine fibroblasts. Immunoprecipitation-tyrosine phosphatase assays revealed that inhibition of phosphotyrosine phosphatase 1B, which regulates insulin signaling, contributed to this effect. In conclusion, stimulation of the overexpressed human AT_2 receptor in porcine cardiac fibroblasts inhibited tyrosine phosphatase activity but had no significant effect on fibroblast functions related to cardiac fibrosis. It is conceivable that possible antifibrotic AT_2 receptor effects are species specific and/or require the interaction between fibroblasts and cardiomyocytes, probably via paracrine factors or mechanical load (Warnecke et al. 2001).

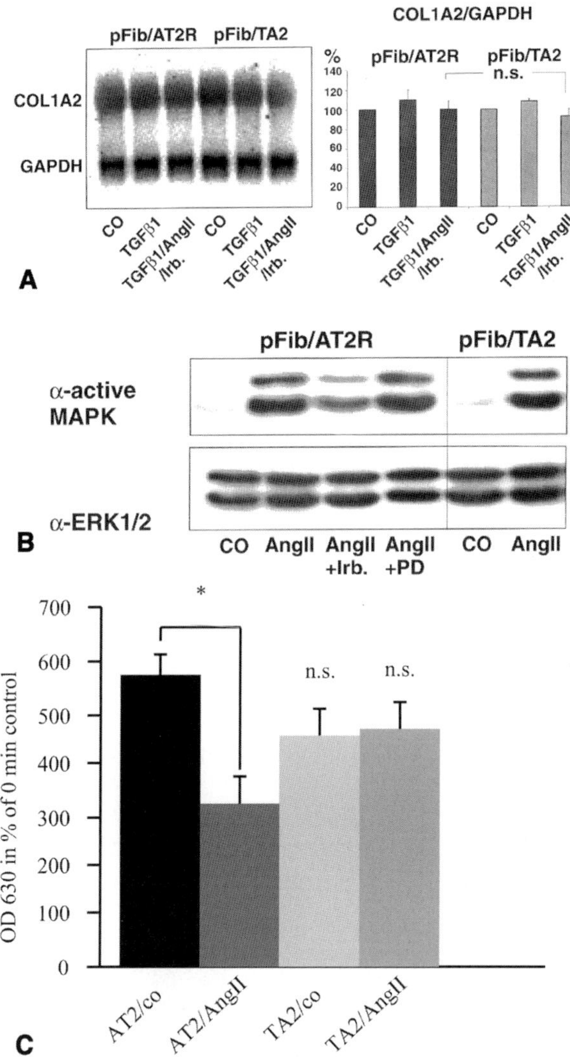

Fig. 6. A Collagen synthesis in porcine cardiac fibroblasts transduced with AT_2 receptor (*pFib/AT$_2$R*) or antisense (*pFib/TA2*) receptor constructs. Basal and TGFβ1 and Ang II stimulated collagen I alpha2 (*COL1A2*) mRNA levels in sense AT_2- and antisense TA2-transduced fibroblasts were determined by Northern blot analysis. **B** MAP kinase activities of AT_2 receptor overexpressing porcine fibroblasts compared to antisense controls. MAP kinase activity in fibroblasts was determined by immunoblot analysis using an antibody specific for the dually phosphorylated and thus activated forms of ERK 1 and ERK2. Identical blots were stained with a phosphorylation independent antibody against ERK1/2. **C** Protein tyrosine phosphatase (*PTPase*) inhibition by the overexpressed AT_2 receptor. AT_2- and TA2-transduced fibroblasts were stimulated with 1 µM Ang II in the presence of an AT_1 antagonist and PTPase activity was measured

12
Effects of Ang II on Isolated Cardiovascular Cells

It has been shown by several groups that Ang II induces fibroblast proliferation and collagen synthesis in neonatal rat cells (Sun et al. 1997). Since Ang II has been linked to fibrosis in the human heart but the precise mechanisms of its action are still unknown, we investigated whether Ang II directly affects the collagen mRNA content in the human myocardium and in isolated human cardiac fibroblasts or the growth factors TGFβ-1 and osteopontin are involved as mediators in this process. In a first set of experiments, the direct effect of Ang II on collagen I, TGFβ-1 and osteopontin mRNA content in fresh samples of human atrial myocardium was determined after short time stimulation. After 4 h, Ang II-induced increased expression of both TGFβ-1 and osteopontin mRNA in the atrial samples, whereas the expression of collagen I mRNA was unchanged (Fig. 7). Stimulation with TGFβ-1 led to an increase in collagen I and III mRNA. In a second protocol, to assess the effects of longer stimulation periods, the effects of angiotensin II and its potential mediator TGFβ-1 on collagen I, III and fibronectin mRNA expression and on proliferation of cultured human cardiac fibroblasts were determined. Ang II caused a dose-dependent stimulation of proliferation but did not affect collagen I, II or fibronectin mRNA content after 24 h. In contrast, TGFβ-1 significantly increased collagen I and III mRNA expression. Thus, in contrast to the neonatal rat cells, Ang II does not directly increase collagen or fibronectin mRNA in fibroblasts isolated from adult human hearts. However, it does increase TGFβ-1 and osteopontin mRNA expression. Since TGFβ-1 induces collagen I and III mRNA in atrial samples and in isolated

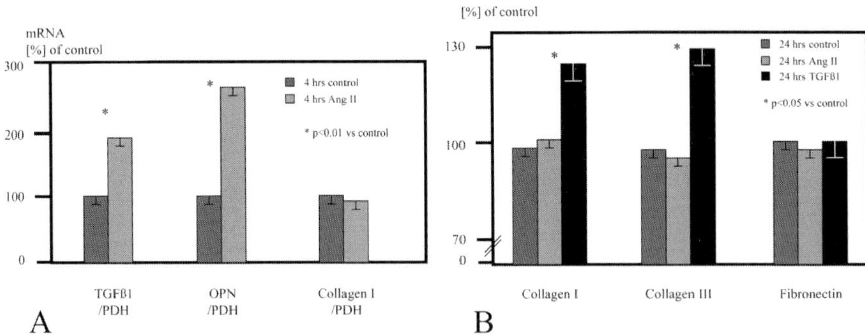

Fig. 7. A TGFβ-1, osteopontin and collagen I mRNA expression in atrial samples after 4 h of Ang II stimulation. Significant increases were observed for TGFβ-1/PDH and osteopontin/PDH (*OPN*) but not for collagen I/PDH mRNA. Mean values from ten hearts, with duplicate samples. Levels of significance for differences between groups are based on the ANOVA procedure. **B** Induction by TGFβ-1 of Col I, Col III and FN mRNA in cultured human cardiac fibroblasts. Regulation of Col I, III and FN mRNA expression by Ang II and TGFβ-1. After 24 h, Col I and III mRNA expression was significantly increased by TGFβ-1, whereas Ang II had no effect. The results given are the mean values from seven hearts. All stimulation data were related to the respective negative controls after 24 h and expressed as percentages of the controls. *Col I*, collagen I; *Col III*, collagen III; *FN*, fibronectin

cardiac fibroblasts, it may represent a necessary mediator of the Ang II effects in the human heart (Kupfahl et al. 2000).

13 Coronary Endothelial Cells

Clinical data suggest a link between the activation of the renin-angiotensin-aldosterone system (RAS) and cardiovascular ischemic events. This may be related to the interaction of ATR with the function of the endothelium. Growth regulation and differential regulation of gene expression by AT_1 and AT 2 was found in endothelial cells (Fischer et al. 2000). Leukocyte accumulation in the vessel wall is a hallmark of early atherosclerosis and plaque progression. E-selectin, vascular cell adhesion molecule-1 (VCAM-1) and intercellular adhesion molecule-1 (ICAM-1) are adhesion molecules participating in mediating interactions between leukocytes and endothelial cells and have been found to be expressed in atherosclerotic plaques. Graefe et al. investigated whether Ang II influences the endothelial expression of E-selectin. In coronary endothelial cells derived from explanted human hearts, Ang II induced a concentration-dependent increase in E-selectin expression. In addition, Ang II induced E-selectin-dependent leukocyte adhesion under flow conditions. The AT_1-receptor antagonist DUP 753 significantly reduced E-selectin-dependent adhesion, whereas the AT_2-receptor antagonist PD 123177 had no inhibitory effect. Therefore, it is suggested that AT_1 receptors up-regulate E-selectin expression and leukocyte adhesion on coronary endothelial cells (Grafe et al. 1997).

14 Cardiomyocytes

There is not much doubt that Ang II induces hypertrophy in isolated myocytes. The role of AT_2 is discussed in a much more controversial and yet unsolved manner. Molecular characterization of the stretch-induced adaption of cultured cardiac cells was used as an in vitro model of load-induced cardiac hypertrophy and confirmed a significant role of Ang II via AT_1 (Sadoshima and Izumo 1993).

Cardiac hypertrophy often occurs in response to both hemodynamic and neurohumoral factors. To study whether activation of the RAS by itself may induce a cardiac growth response, the acute effects of Ang II on cardiac protein synthesis were studied in isolated rat hearts (Inagami 1999). New protein synthesis in isolated buffer-perfused adult rat hearts was measured. Ang II stimulated protein synthesis in left and right ventricles roughly three- to fourfold, but failed to induce c-fos and c-jun mRNA. Thus Ang II induced protein synthesis but the exact pathway remained elusive (Schunkert et al. 1995).

Booz et al. compared the ability of Ang II to induce hypertrophy of neonatal rat ventricular myocytes with that of endothelin-1 in isolated rat myocytes. They concluded that Ang II-induced myocyte growth is tempered because of low AT_1 expression and an antigrowth effect of AT_2. These findings, if confirmed in adult hearts, have potential clinical significance in that regression of hypertension-induced cardiac hypertrophy by AT_1 antagonists may be in part due to an unopposed antigrowth effect of Ang II mediated via AT_2 (Booz and Baker 1996).

Based on the capacity of AT_1 to induce hypertrophy, Thienelt et al. tested the hypothesis that AT_1 receptor blockade will inhibit the acute induction of proto-oncogenes and protein synthesis by the elevation of systolic wall stress in isolated beating adult rat hearts. However, elevation of systolic load was associated with a twofold increase in c-fos and c-myc mRNA levels that was not blocked by losartan. The rate of phenylalanine incorporation into cardiac proteins was increased 2.7-fold in hearts subjected to increased systolic load compared with control hearts. However, AT_1 receptor blockade with losartan did not prevent the stimulation of protein synthesis. Thus, the acute growth responses induced by systolic pressure overload in adult rat hearts do not depend on AT_1 receptor activation (Thienelt et al. 1997).

In agreement with the above-mentioned results, long-term AT_1 receptor blockade did not regress LVH in rats with persistent systolic pressure overload due to ascending aortic stenosis. However, both AT_1-receptor blockade and calcium antagonization with amlodipine improved in vivo left ventricular end-diastolic pressure together with the normalization of left ventricular ACE mRNA levels (Weinberg et al. 1997).

Taken together, the data may suggest that Ang II can contribute to myocardial hypertrophy, but AT_1 activation is not a necessary factor in most animal models.

The role of AT_2 in LVH is even more difficult to define. Up-regulation of AT_2 receptors has been described in hypertrophied hearts. Activation of AT_2 receptors is proposed to counteract growth effects of AT_1 receptor in response to Ang II. Thus, in hypertrophied hearts, the AT_2 receptor may mediate inhibitory effects on the new cardiac protein synthesis in response to acute Ang II stimulation. If adult normal and hypertrophied rat hearts were perfused with Ang II 10^{-8} mol/l plus prazosin 10^{-7} mol/l or Ang II plus an AT_2 blocker, Ang II increased the rate of phenylalanine incorporation by 74±27%, whereas treatment with PD123319 did not increase protein synthesis compared with Ang II alone (32±11% vs Ang II alone, p=n.s.). In hypertrophied hearts, Ang II alone increased the rate of phenylalanine incorporation only by 23±13% and treatment with PD123319 (n=7) induced a 76±21% increase in new LV protein synthesis compared with Ang II alone ($p<0.05$). AT_2 receptor blockade in Ang II-stimulated hypertrophied hearts was associated with enhanced membrane PKC translocation and reduced left ventricular cGMP content. These data support the hypothesis that in adult hypertrophied rat hearts, inhibition of cardiac AT_2 receptors, which are up-regulated in chronic LV hypertrophy, amplifies the immediate LV growth response to Ang II. This appears to be related to augmented Ang II-stimulated PKC activation and suppression of cGMP signaling (Bartunek et al. 1999).

These data suggest that Ang II activation of the cardiac AT_2 receptor subtype inhibits the effects of Ang II on the immediate growth response in the adult heart (Lorell 1999).

However, there are also conflicting data about the role of AT_2 in LVH in mice. Ichihara et al. showed that AT_2 is important for the development of LVH and cardiac fibrosis induced by Ang II infusion in AT_2 knock-out mice. Wild type but not AT_2 knock-out mice developed concentric LVH, prominent fibrosis, and impaired diastolic relaxation after Ang II infusion over a period of 3 weeks. Thus, loss of AT_2 abolished LVH and cardiac fibrosis in mice with Ang II-induced hypertension in this model (Ichihara et al. 2001).

Human data are more difficult to obtain. We sought to determine whether the cardiac RAAS is activated in human aortic valve disease depending on left ventricular function, and whether it is linked to a concomitant regulation of the extracellular matrix components. In LV biopsies from patients with aortic valve stenosis and aortic valve regurgitation and from control subjects, we quantitated mRNAs for angiotensin-converting enzyme (ACE), chymase, TGF-β1, collagen I, collagen III and fibronectin by reverse-transcription PCR.

Protein, ACE and TGF-β1 mRNA were significantly increased in patients with aortic stenosis and aortic regurgitation (1.5- to 2.1-fold) and correlated with each other. The increase occurred also in patients with normal systolic function. Collagen I and III and fibronectin mRNAs were both up-regulated approximately twofold in patients with aortic stenosis and aortic regurgitation. In aortic stenosis, collagen and fibronectin mRNA expression levels were positively correlated with left ventricular end-diastolic pressure and inversely with left ventricular ejection fraction (LVEF). Thus, In human hearts, pressure and volume overload increases cardiac ACE and TGF-β1 in the early stages. This activation of the cardiac RAS may contribute to the observed increase in collagen I and III and fibronectin mRNA expression. The increase in extracellular matrix already exists in patients with a normal LVEF, and it increases with functional impairment (Fielitz et al. 2001).

Thus, AT_1 receptors probably mediate the growth and fibrosis-promoting effects of angiotensin and RAS activation in human aortic stenosis. The role of AT_2 could not yet be defined in a human setting. There are, however, genetic data that suggest that AT_2 has a role in human LVH.

15 Effects of AT_1 and AT_2 Polymorphisms on LV Function

The physiological effects of polymorphisms of ATR are still poorly understood. Long-term effects of genetic variants can be studied in cross-sectional linkage studies. We examined the short-term effects of genetic polymorphisms of the AT_1- and AT_2-receptor subtypes in humans by means of Ang II infusion. We selected two polymorphisms with potential functional relevance, –2228 G/A in the AT_1 promotor region, and 1675 G/A in the *AT_2* gene (Zhang et al. 2000;

Erdmann et al. 2000). The AT_2 polymorphism is located in intron 1, which has a regulatory function (Nickenig and Murphy 1994).

In 120 male, white, young (26±3 years) subjects with known AT_1 *−2228 G/A* and AT_2 *+1675 G/A* genotypes, with normal or mildly elevated blood pressure, changes in mean arterial blood pressure, aldosterone levels, glomerular filtration rate, and renal plasma flow were measured in response to Ang II infusion. Infusion of Ang II resulted in an increase in mean arterial pressure, serum aldosterone levels and glomerular filtration rate, and in a decrease in renal plasma flow (all $p<0.001$). However, at similar baseline mean arterial pressure, aldosterone levels, and renal hemodynamics, the response to Ang II did not significantly differ across the AT_1- and AT_2-receptor genotypes, with the sample size of our study being adequate to detect relevant differences across the genotypes with a power greater than 90% for all parameters. Thus, the response to Ang II infusion does not differ across the AT_1- and AT_2-receptor genotypes examined in our study. However, long-term effects of variants of *ATR* genes cannot be ruled out with this approach (Delles et al. 2000).

To analyze whether these gene variants are associated with the long-term effects of Ang II on LVH, we analyzed, in the same 120 white, young male subjects with normal or mildly elevated blood pressure, LV structure by two-dimensional guided M-mode echocardiography and determined the +1675 G/A polymorphism. Hypertensive subjects with the *A* allele had a greater left ventricular posterior and relative wall thickness as well as left ventricular mass index than those with the *G* allele. Confounding factors, i.e., body mass index and surface area, plasma Ang II, sodium excretion, systolic and diastolic ambulatory blood pressure, were similar between the two genotypes. In normotensive subjects, relative wall thickness and left ventricular mass index were nearly identical across the two genotypes, with similar confounding variables. These data indicate that the X-chromosomal located +1675 G/A polymorphism of the AT_2 receptor gene is associated with left ventricular structure in young male humans and with early structural changes of the heart due to arterial hypertension (Delles et al. 2000; Schmieder et al. 2001) (Fig. 8).

Similar results were also obtained in the GLAECO and GLAEOLD study cohorts. In this study, we genotyped 1,968 patients from both population samples for the 1675 G/A AT_2 variant. In the GLAEOLD study, AT_2 *1675 A* allele carriers were more common in males with LVH than in males without LVH (60% vs 46%; $p<0.05$). The effect could not, however, be confirmed in the GLAECO study, which may have been due to efferent criteria for LVH in both studies. The *A* allele was also more common in females with MI or episodes of coronary ischemia than in females without evidence of ischemia (86% vs 73%) in the GLAEOLD study.

Thus, even though the precise mechanism has not been fully understood, there is good evidence that the AT_2 receptor modifies LV structure in humans.

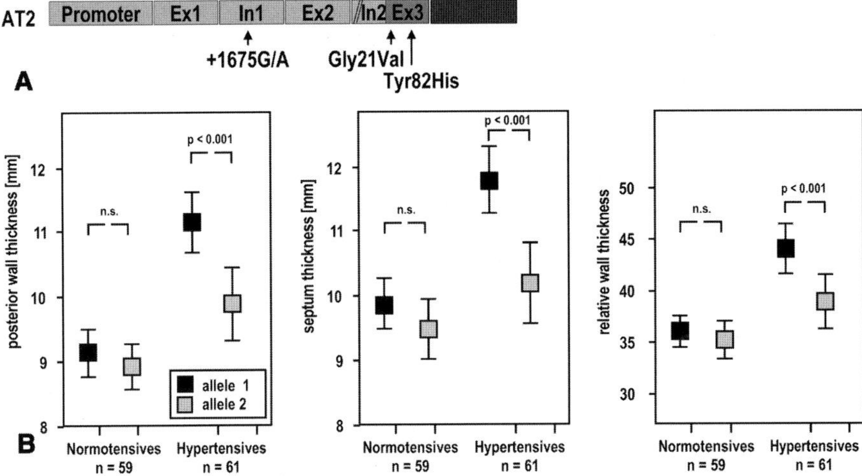

Fig. 8. A Polymorphisms in the *AT* gene. The AT_2 gene has two noncoding exons 1 and 2 and one coding exon 3. Several polymorphisms are known. +1675 G/A is located in intron 1. Gly 21Val was recently claimed to be associated with mental retardation but indeed represents a rare polymorphism. **B** Echocardiographic measures of LV wall thickness in male normotensive and hypertensive carriers of the *G* and *A* allele. Posterior wall, septum and relative wall thickness were determined

16 Effects of ATR in Myocardial Ischemia

The role of ATR in myocardial ischemia was even more under discussion than its role in cardiac hypertrophy. A few direct effects have been described that interfere with ischemic nonvascular events. Modulation of the expression and activity of the sodium-proton exchanger (NHE-1) and the sodium-bicarbonate symporter may interfere with ischemic myocyte damage and occurs by AT_1 and AT_2 in different manners (Sandmann et al. 2001). The regulation of calpain I and II and its inhibitor calpastatin may interfere with the activation of proteolytic processes as a consequence of ischemia (Sandmann et al. 2001). However, the positive effects of ACE inhibitors and ATR antagonists after myocardial infarction are probably due to a major part to their positive effects on myocardial remodeling and only to a lesser degree on their effects on myocardial ischemia. This is supported by the experimental finding that early, but not very early treatment was most efficient in maintaining LV function after rat myocardial infarction, and that delayed installation of treatment was also effective (Xia et al. 2001). This is in agreement with clinical studies, but their discussion is beyond the scope of this chapter.

References

Abdellatif MM, Neubauer CF, Lederer WJ et al (1991) Angiotensin-induced desensitization of the phosphoinositide pathway in cardiac cells occurs at the level of the receptor. Circ Res 69:800–809

Asano K, Dutcher DL, Port JD et al (1997) Selective downregulation of the angiotensin II AT_1-receptor subtype in failing human ventricular myocardium. Circulation 95:1193–1200

Bartunek J, Weinberg EO, Tajima M et al (1999) Angiotensin II type 2 receptor blockade amplifies the early signals of cardiac growth response to angiotensin II in hypertrophied hearts. Circulation 99:22–25

Bauer P, Regitz-Zagrosek V, Kalisch H et al (1997) Myocardial angiotensin receptor type 1 gene expression in a rat model of cardiac volume overload. Basic Res Cardiol 92:139–146

Bedecs K, Elbaz N, Sutran M et al (1997) Angiotensin II type 2 receptors mediate inhibition of mitogen-activated protein kinase cascade and functional activation of SHP-1 tyrosine phosphatase. Biochem J 325:449–454

Booz GW, Baker KM (1996) Role of type 1 and type 2 angiotensin receptors in angiotensin II-induced cardiomyocyte hypertrophy. Hypertension 28:635–640

Borghi C, Prandin MG, Costa FV et al (2000) Use of statins and blood pressure control in treated hypertensive patients with hypercholesterolemia. J Cardiovasc Pharmacol 35:549–555

Bottari SP, King IN, Reichlin S et al (1992) The angiotensin AT_2 receptor stimulates protein tyrosine phosphatase activity and mediates inhibition of particulate guanylate cyclase. Biochem Biophys Res Commun 183:206–211

Brink M, Erne P, de Gasparo M et al (1996) Localization of the angiotensin II receptor subtypes in the human atrium. J Mol Cell Cardiol 28:1789–1799

Brock TA, Rittenhouse SE, Powers CW et al (1985) Phorbol ester and 1-oleoyl-2-acetylglycerol inhibit angiotensin activation of phospholipase C in cultured vascular smooth muscle cells. J Biol Chem 260:14158–14162

Busche S, Gallinat S, Bohle RM et al (2000) Expression of angiotensin AT(1) and AT(2) receptors in adult rat cardiomyocytes after myocardial infarction. A single-cell reverse transcriptase-polymerase chain reaction study. Am J Pathol 157:605–611

Curnow K, Pascoe L, Davies E et al (1995) Alternatively spliced human type 1 angiotensin II receptor mRNAs are translated at different efficiencies and encode two receptor isoforms. Mol Endocrinol 9:1250–1262

Dechend R et al (2001) Amelioration of angiotensin II-induced cardiac injury by a 3-hydroxy-3-methylglutaryl coenzyme a reductase inhibitor. Circulation 104:576–581

Delles C, Erdmann J, Jacobi J et al (2000) Lack of association between polymorphisms of angiotensin II receptor genes and response to short-term angiotensin II infusion. J Hypertens 18:1573–1578

Dostal DE, Hunt RA, Kule CE et al (1997) Molecular mechanisms of angiotensin II in modulating cardiac function: intracardiac effects and signal transduction pathways [in process citation]. J Mol Cell Cardiol 29:2893–2902

Elbaz N, Bedecs K, Masson M et al (2000) Functional trans-inactivation of insulin receptor kinase by growth-inhibitory angiotensin II AT_2 receptor [in process citation]. Mol Endocrinol 14:795–804

Erdmann J, Guse M, Kallisch H et al (2000) Novel intronic polymorphism (+1675G/A) in the human angiotensin II subtype 2 receptor gene. Hum Mutat 15:487

Fielitz J, Hein S, Mitrovic V et al (2001) Activation of the cardiac renin-angiotensin system and increased myocardial collagen expression in human aortic valve disease. J Am Coll Cardiol 37:1443–1449

Fischer JW, Stoll M, Hahn AW et al (2001) Differential regulation of thrombospondin-1 and fibronectin by angiotensin II receptor subtypes in cultured endothelial cells. Cardiovasc Res 51:784–791

Fischer TA, Sinhg K, O'Hare DS et al (1998) Role of AT_1 and AT_2 receptors in regulation of MAPKs and MKP-1 by ANG II in adult cardiac myocytes. Am J Physiol 275:H906–H916

Glorioso N, Troffa C, Filigheddu F et al (1999) Effect of the HMG-CoA reductase inhibitors on blood pressure in patients with essential hypertension and primary hypercholesterolemia. Hypertension 34:1281–1286

Grady EF, Sechi LA, Griffin EAet al (1991) Expression of AT_2 receptors in the developing rat fetus. J Clin Invest 88:921–933

Grafe M, Auch-Schwelk W, Zakrzewicz A et al (1997) Angiotensin II-induced leukocyte adhesion on human coronary endothelial cells is mediated by E-selectin. Circ Res 81:804–811

Haywood GA, Gullestad L, Katsuya T et al (1997) AT_1 and AT_2 angiotensin receptor gene expression in human heart failure. Circulation 95:1201–1206

Hein L Meinel L, Pratt RE et al (1997) Intracellular trafficking of angiotensin II and its AT_1 and AT_2 receptors: evidence for selective sorting of receptor and ligand. Mol Endocrinol 11:1266–1277

Horiuchi M et al (1995) The growth-dependent expression of angiotensin II type 2 receptor is regulated by transcription factors interferon regulatory factor-1 and -2. 270:20225–20230

Ichihara S, Senbonmatsu C, Price E Jr et al (2001) Angiotensin II type 2 receptor is essential for left ventricular hypertrophy and cardiac fibrosis in chronic angiotensin II-induced hypertension. Circulation 104:346–351

Ichiki T, Takeda K, Tokounu T et al (2001) Downregulation of angiotensin II type 1 receptor by hydrophobic 3-hydroxy-3-methylglutaryl coenzyme A reductase inhibitors in vascular smooth muscle cells. Arterioscler Thromb Vasc Biol 21:1896–1901

Inagami T (1999) Molecular biology and signaling of angiotensin receptors: an overview. J Am Soc Nephrol 10:S2–S7

Inagami T, Eguchhi S, Numaguchi K et al (1999) Cross-talk between angiotensin II receptors and the tyrosine kinases and phosphatases. J Am Soc Nephrol 10:S57–S61

Kambayashi Y, Bardhan S, Takahashi K et al (1993) Molecular cloning of a novel angiotensin II receptor isoform involved in phosphotyrosine phosphatase inhibition. J Biol Chem 268:24543–24546

Katoh M, Egashira K, Usui M et al (1998) Cardiac angiotensin II receptors are upregulated by long-term inhibition of nitric oxide synthesis in rats. Circ Res 83:743–751

Kijima K, Matsubara H, Murasawa S et al (1996) Mechanical stretch induces enhanced expression of angiotensin II receptor subtypes in neonatal rat cardiac myocytes. Circ Res 79:887–897

Knowle D, Ahmed S, Pulakat L (2000) Identification of an interaction between the angiotensin II receptor sub-type AT_2 and the ErbB3 receptor, a member of the epidermal growth factor receptor family. Regul Pept 87:73–82

Koike G, Winer ES, Horiuchi M et al (1995) Cloning, characterization, and genetic mapping of the rat type 2 angiotensin II receptor gene. Hypertension 26:998–1002

Krishnamurthi K, Verbalis JG, Zheng W et al (1999) Estrogen regulates angiotensin AT_1 receptor expression via cytosolic proteins that bind to the 5′ leader sequence of the receptor mRNA. Endocrinology 140:5435–5438

Kudoh S, Komuro I, Misuno T et al (1997) Angiotensin II stimulates c-Jun NH2-terminal kinase in cultured cardiac myocytes of neonatal rats. Circ Res 80:139–146

Kupfahl C, Pink D, Friedrich K et al (2000) Angiotensin II directly increases transforming growth factor beta1 and osteopontin and indirectly affects collagen mRNA expression in the human heart. Cardiovasc Res 46:463–475

Lefroy DC, Wharton J, Crake T et al (1996) Regional changes in angiotensin II receptor density after experimental myocardial infarction. J Mol Cell Cardiol 28:429–440

Li JY, Avallet O, Berthelon MC et al (1999) Transcriptional and translational regulation of angiotensin II type 2 receptor by angiotensin II and growth factors. Endocrinology 140:4988–4994

Lopez JJ, Laurel BH, Ingelfinger JR et al (1994) Distribution and function of cardiac angiotensin AT_1- and AT_2-receptor subtypes in hypertrophied rat hearts. Am J Physiol 267:H844–H852

Lorell BH (1999) Role of angiotensin AT_1, and AT_2 receptors in cardiac hypertrophy and disease. Am J Cardiol 83:48H–52H

Nagata K, Somura F, Obata K et al (2002) AT_1 receptor blockade reduces cardiac calcineurin activity in hypertensive rats. Hypertension 40:168–174

Nahmias C, Cazaubom SM, Briend-Sutren MM et al (1995) Angiotensin II AT_2 receptors are functionally coupled to protein tyrosine dephosphorylation in N1E-115 neuroblastoma cells. Biochem J 306:87–92

Nickenig G, Murphy TJ (1994) Down-regulation by growth factors of vascular smooth muscle angiotensin receptor gene expression. 46:653–659

Nickenig G, Baumer AT, Grohe C et al (1998) Estrogen modulates AT_1 receptor gene expression in vitro and in vivo. Circulation 97:2197–2201

Nickenig G et al (1999) Statin-sensitive dysregulated AT_1 receptor function and density in hypercholesterolemic men. Circulation 100:2131–2134

Nickenig G, Strehlow K, Wassmann S et al (2000) Differential effects of estrogen and progesterone on AT(1) receptor gene expression in vascular smooth muscle cells. Circulation 102:1828–1833

Nozawa Y, Miyake H, Haruno A et al (1996) Down-regulation of angiotensin II receptors in hypertrophied human myocardium. Clin Exp Pharmacol Physiol 23:514–518

Ohkubo N, Matsubara H, Nozawa Y et al (1997) Angiotensin type 2 receptors are reexpressed by cardiac fibroblasts from failing myopathic hamster hearts and inhibit cell growth and fibrillar collagen metabolism. Circulation 96:3954–3962

Pueyo ME, Gonzalez W, Nicoletti A et al (2000) Angiotensin II stimulates endothelial vascular cell adhesion molecule-1 via nuclear factor-kappaB activation induced by intracellular oxidative stress. Arterioscler Thromb Vasc Biol 20:645–651

Regitz-Zagrosek V, Friedel N, Heymann A et al (1995) Regulation, chamber localization, and subtype distribution of angiotensin II receptors in human hearts. Circulation 91:1461–1471

Regitz-Zagrosek V, Fielitz J, Hummel M et al (1996a) Decreased expression of ventricular angiotensin receptor type 1 mRNA after human heart transplantation. J Mol Med 74:777–782

Regitz-Zagrosek V, Neuss M, Warnecke C et al (1996b) Subtype 2 and atypical angiotensin receptors in the human heart. Basic Res Cardiol 91 [Suppl 2]:73–77

Regitz-Zagrosek V, Fielitz J, Dreyesse R et al (1997) Angiotensin receptor type 1 mRNA in human right ventricular endomyocardial biopsies: downregulation in heart failure. Cardiovasc Res 35:99–105

Regitz-Zagrosek V, Fielitz J, Fleck E (1998) Myocardial angiotensin receptors in human hearts. Basic Res Cardiol 93 [Suppl 2]:37–42

Rogg H, de Gasparo M, Graedel E et al (1996) Angiotensin II-receptor subtypes in human atria and evidence for alterations in patients with cardiac dysfunction [see comments]. Eur Heart J17:1112–1120

Ruiz-Ortega M, Lorenzo O, Ruperez M et al (2000) Angiotensin II activates nuclear transcription factor kappaB through AT(1) and AT(2) in vascular smooth muscle cells: molecular mechanisms. Circ Res 86:1266–1272

Sadoshima J, Izumo S (1993) Molecular characterization of angiotensin II-induced hypertrophy of cardiac myocytes and hyperplasia of cardiac fibroblasts. Critical role of the AT_1 receptor subtype. Circ Res 73:413–423

Sandmann S, Yu M, Kaschina E et al (2001) Differential effects of angiotensin AT_1 and AT_2 receptors on the expression, translation and function of the Na+-H+ exchanger and Na+-HCO3- symporter in the rat heart after myocardial infarction. J Am Coll Cardiol 37:2154–2165

Schieffer B, Bernstein KE, Marrero MB (1996) The role of tyrosine phosphorylation in angiotensin II mediated intracellular signaling and cell growth. J Mol Med 74:85–91

Schmieder RE, Erdmann J, Delles C et al (2001) Effect of the angiotensin II type 2-receptor gene (+1675 G/A) on left ventricular structure in humans. J Am Coll Cardiol 37:175–182

Schunkert H, Sadoshima J, Corneilus T et al (1995) Angiotensin II-induced growth responses in isolated adult rat hearts. Evidence for load-independent induction of cardiac protein synthesis by angiotensin II. 76:489–497

Sechi LA, Griffin EA, Grady EF et al (1992) Characterization of angiotensin II receptor subtypes in rat heart. 71:1482–1489

Sechi LA, Sechi G, Di Carli S et al (1993) [Angiotensin receptors in the rat myocardium during pre- and postnatal development]. Cardiologica 38:471–476

Stoll M, Hahn AW, Jonas U et al (2002) Identification of a zinc finger homeodomain enhancer protein after AT(2) receptor stimulation by differential mRNA display. Arterioscler Thromb Vasc Biol 22:231–237

Sun Y et al (1997) Fibrous tissue and angiotensin II. J Mol Cell Cardiol 29:2001–2012

Tambascia RC, Fonesca PM, Corat PD et al (2001) Expression and distribution of NOS1 and NOS3 in the myocardium of angiotensin II-infused rats. Hypertension 37:1423–1428

Tamura K et al (2000) Expression of renin-angiotensin system and extracellular matrix genes in cardiovascular cells and its regulation through AT_1 receptor. Mol Cell Biochem 212:203–209

Thienelt CD, Weinberg EO, Bartunek J et al (1997) Load-induced growth responses in isolated adult rat hearts. Role of the AT_1 receptor. J Am Coll Cardiol 95:2677–2683

Timmermans PB, Chiu AT, Herblin WF et al (1992) Angiotensin II receptor subtypes. Am J Hypertens 5:406–410

Tsutsumi Y, Matsubara H, Ohkubo N et al (1998) Angiotensin II type 2 receptor is upregulated in human heart with interstitial fibrosis, and cardiac fibroblasts are the major cell type for its expression. Circ Res 83:1035–1046

Wang X, Nickenig G, Murphy TJ (1997) The vascular smooth muscle type I angiotensin II receptor mRNA is destabilized by cyclic AMP-elevating agents. Mol Pharmacol 52:781–787

Warnecke C, Surder D, Curth R et al (1999a) Analysis and functional characterization of alternatively spliced angiotensin II type 1 and 2 receptor transcripts in the human heart. J Mol Med 77:718–727

Warnecke C, Willich T, Holzmeister J et al (1999b) Efficient transcription of the human angiotensin II type 2 receptor gene requires intronic sequence elements. Biochem J 340:17–24

Warnecke C, Kaup D, Marienfeld U et al (2001) Adenovirus-mediated overexpression and stimulation of the human angiotensin II type 2 receptor in porcine cardiac fibroblasts does not modulate proliferation, collagen I mRNA expression and ERK1/ERK2 activity, but inhibits protein tyrosine phosphatases. J Mol Med 79:510–521

Weinberg EO, Lee MA, Weigner M et al (1997) Angiotensin AT_1 receptor inhibition. Effects on hypertrophic remodeling and ACE expression in rats with pressure-overload hypertrophy due to ascending aortic stenosis. Circulation 95:1592–1600

Wharton J, Morgan K, Rutherford RA et al (1998) Differential distribution of angiotensin AT_2 receptors in the normal and failing human heart. J Pharmacol Exp Ther 284:323–336

Xia QG, Chung O, Spitznagel H et al (2001) Significance of timing of angiotensin AT_1 receptor blockade in rats with myocardial infarction-induced heart failure. Cardiovasc Res 49:110–117

Zhang X et al (2000) Evaluation of three polymorphisms in the promoter region of the angiotensin II type I receptor gene. J Hypertens 18:267–272

Zhu YZ et al (2000) Effects of losartan on haemodynamic parameters and angiotensin receptor mRNA levels in rat heart after myocardial infarction. J Renin Angiotensin Aldosterone Syst 1:257–262

Zisman LS, Asano K, Dutcher DL et al (1998) Differential regulation of cardiac angiotensin converting enzyme binding sites and AT_1 receptor density in the failing human heart. Circulation 98:1735–1741

Activation of the Renin-Angiotensin System After Myocardial Infarction

W. M. Aartsen · J. F. M. Smits · M. J. A. P. Daemen

Department of Pharmacology and Toxicology,
Cardiovascular Research Institute Maastricht (CARIM),
Universiteit Maastricht, P.O. Box 616, 6200 MD Maastricht, The Netherlands
e-mail: j.smits@farmaco.unimaas.nl

Abstract After myocardial infarction (MI), the heart adapts to the loss of contractile units by cardiac remodelling, a phenomenon which includes processes such as wound healing, ventricular dilatation and ventricular hypertrophy. The renin-angiotensin system (RAS) is a neurohormonal system that has been suggested to play an important role during cardiac remodelling after MI. Immediately after MI, the expression and activity of many RAS components are elevated, not only in the circulation, but also in the heart. Increased circulating angiotensin II will lead to increased vascular resistance, which supports arterial pressure in a situation of reduced cardiac output. Locally elevated angiotensin II is supposed to influence the processes of fibrosis and hypertrophy, because of its growth-promoting capacity. RAS activation is appropriate during the early cardiac adaptations after MI; however, prolonged activation could eventually worsen cardiac performance. High vascular resistance increases cardiac stroke work and augments oxygen consumption. Although new vessels emerge in the noninfarcted myocardium, the capillary to myocyte-fibre ratio is decreased, as is oxygen availability. Hence, disproportionate hypertrophy decreases the capillary density even further and excessive fibrosis induces cardiac stiffness. By inhibiting the RAS activity, either via ACE inhibition or AT_1 receptor antagonism, post-MI left ventricular dysfunction can be reduced, leading to a substantially higher rate of survival. These beneficial effects have been studied extensively but have, so far, not elucidated a uniform responsible mechanism. The exact roles of the AT_2 receptor and the bradykinin type 2 receptor are still uncertain. The present review discusses the proposed contributions of the RAS in the process of post-MI cardiac remodelling.

Keywords Myocardial infarction · Cardiac remodelling · RAS expression · RAS inhibition · Mechanisms

1
Myocardial Infarction

Acute myocardial infarction (MI) can develop when a part of the coronary circulation is suddenly occluded. The cardiac segment normally fed by the now occluded artery becomes ischaemic and prolonged oxygen deprivation leads to cell death. Dead cardiomyocytes are the trigger of the wound healing response, involving proliferation, differentiation and apoptosis of many different cell types. This cellular activity is mainly initiated and regulated by growth factors and leads to replacement of the demised cells by scar tissue (Cleutjens et al. 1999). In the early inflammatory phase, macrophages, fibroblast-like cells and endothelial cells invade the infarcted area. Macrophages are responsible for the removal of dead cells and cell debris. Fibroblast-like cells and endothelial cells proliferate and create new networks of collagen and small vessels (granulation tissue). In the following fibrogenic phase, the scar matures into a very collagen-rich segment from which the cells have disappeared (Struijker-Boudier et al. 1995; Sun and Weber 2000).

After MI, optimal cardiac performance can no longer be achieved due to the loss of contractile units and the altered structure. Alterations in the architecture are accompanied by changes in the forces that the heart endures (Anversa et al. 1991). The heart adapts to alterations in stress (pressure load) and strain (volume load) through ventricular dilation and hypertrophy. The increased ventricular volume compensates for the reduced ejection fraction (Ertl et al. 1993). The infarct area (Pfeffer and Braunwald 1990), as well as the surviving myocardium enlarge. To permit cell rearrangement, the connective tissue matrix has to be disrupted and reorganised (Creemers et al. 2000). Moreover, the cardiomyocytes in the noninfarcted area are triggered to increase the number of contractile units, resulting in a hypertrophied left ventricular wall. Before cells start growing, the mechanical alteration must be translated into a biochemical growth signal (Mackenna et al. 2000). This biochemical signal differs according to the type of haemodynamic load. Although the exact signalling pathways are still a mystery, several hypotheses have been proposed. Via their seven-transmembrane receptors, angiotensin II, catecholamines and endothe-

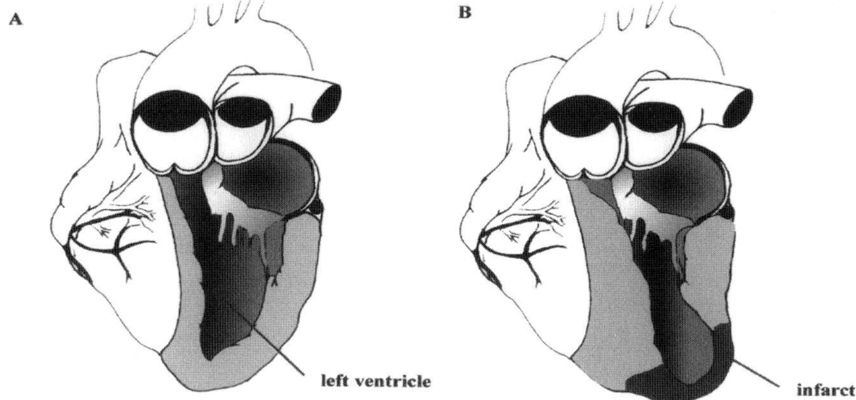

Fig. 1A, B Schematic drawing of the heart. **A** In the normal individual. **B** After myocardial infarction

lin can activate protein kinase C (PKC). Growth factors such as insulin-like growth factor (IGF) and transforming growth factor (TGF-β) bind their tyrosine kinase receptors. Both types of receptor activation lead, via the complex network of second messengers (e.g. MAPK, JAK/STAT), to increased DNA, mRNA and protein synthesis and cellular dedifferentiation (Swynghedauw 1991; Katz 1995). Neovascularisation and fibrosis are observed in the noninfarcted myocardium in reaction to hypertrophy and dilation (Gaasch 1994). Although new vessels develop, the capillary to myocyte fibre ratio is decreased, as is oxygen availability. The enhanced interstitial fibrosis of the myocardium causes stiffness. In time, the left ventricle becomes more hypertrophied and stiff, which will finally lead to a severely impaired contractile function, resulting in heart failure. Thus, myocardial infarction induces a primary alteration in the function of the infarcted region as well as time-dependent secondary changes in the noninfarcted myocardium (Fig. 1).

2 Expression of the Renin-Angiotensin System After MI

2.1 Circulating RAS Activation After MI

Several neurohormonal systems are activated in the early phase following MI. These systems help to initiate the process of wound healing and to translate the mechanical forces into growth responses. Together they create a compensatory response, which preserves cardiac performance. However, prolonged activation of these systems may lead the heart from the compensatory state into heart failure (Sigurdsson et al. 1993). One of the activated neurohormonal systems is the renin-angiotensin system (RAS). After MI, components of both circulating (or plasma) and local (or tissue) RAS are up-regulated. Within 24 h after hospitalisation, plasma renin activity measured in patients with MI is 2.7-fold higher compared to patients without MI (Blumenfeld et al. 2000). Plasma angiotensin II levels are also elevated in the early days after acute MI (Michorowski and Ceremuzynski 1983; Sigurdsson et al. 1993). In patients with uncomplicated infarction, plasma renin and angiotensin levels return to normal levels during the 1st week after infarction (McAlpine et al. 1988). These data are confirmed by experimentally induced myocardial infarction, which causes an immediate elevation of renin activity and angiotensin II levels in rat and dog plasma (Ertl et al. 1985; Lindpainter et al. 1993). During the period of compensation, the concentrations of RAS components seem to be normal in both humans and rats (Dzau et al. 1981; Ceiler et al. 1998). This reduction of RAS activation is delayed when infarctions are large and RAS activation is sustained if MI is complicated by heart failure (Dargie et al. 1987; McAlpine et al. 1988; Cleland et al. 1996). Activation of the RAS after MI strongly relates to the extent of left ventricular dysfunction (Vaughan et al. 1990). Moreover, persistent activation of the RAS after MI is a strong indicator for the development of heart failure in time (McAlpine et al. 1988; Francis et al. 1990). During heart failure, plasma renin and angiotensin II concentrations increase in humans (Dzau et al. 1981) and rats (Gaasch 1994). However, activation during heart failure was not confirmed in all experimental models (Hirsch et al. 1991; Kelly et al. 1997; Ceiler et al. 1998) (Fig. 2).

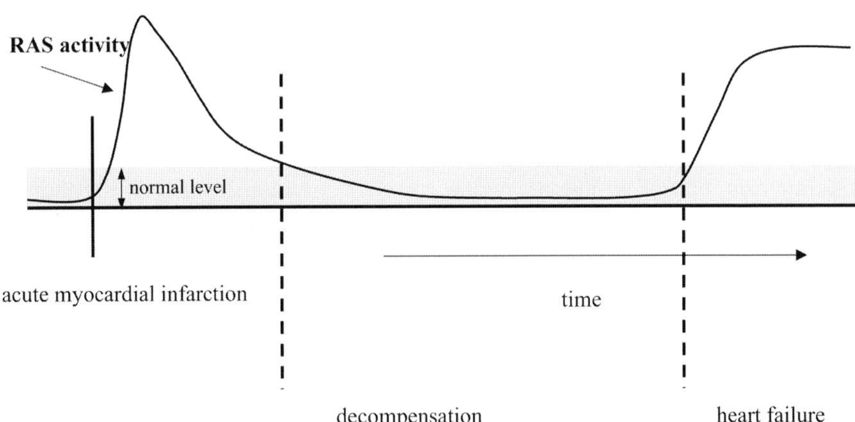

Fig. 2 Plasma RAS activation at the different stages of cardiac remodelling after myocardial infarction. (Data from Weber 1997)

2.2 Local RAS Activation After MI

The RAS is expressed locally in a number of tissues including the heart (Danser 1996). The basal expression of RAS components is elevated early after MI. In the rat heart, angiotensinogen mRNA expression is detectable as early as 5 days after experimentally induced MI (Passier et al. 1996). This expression is localised in the noninfarcted left ventricle (Lindpainter et al. 1993). Also, renin expression is induced early after MI (Hirsch et al. 1999). Between 2 and 7 days after MI, renin mRNA is expressed in the border zone of infarcted rat hearts (Passier et al. 1996) and in macrophages in the infarct area (Sun et al. 2001). Angiotensin-converting enzyme (ACE) is found in rat macrophages invading the necrotic area within 3–7 days after MI. Within the same week, left ventricular ACE activity is approximately threefold higher and especially localised in the border zone around the infarct area (Passier et al. 1995; Hokimoto et al. 1996). At the onset of scar tissue formation, ACE expression on the vascular endothelial cells is elevated and myofibroblasts become ACE- and renin-positive (Falkenhahn et al. 1995; Sun and Weber 2000; Sun et al. 2001).

One of the regulating factors of RAS expression and angiotensin II production around the infarct may be local wall stress. In vitro, mechanically stretched myocytes produce angiotensin II and show an autocrine hypertrophic response (Sadoshima et al. 1993). After ventricular dysfunction and cardiac overloading, the ability of myocytes to produce angiotensin II is even potentiated (Zhang et al. 1995). In vivo, cardiac overloading enhances ACE mRNA and protein expression in the surviving myocardium remote from the infarct region (Hirsch et al. 1991). Moreover, cardiac angiotensin II formation correlates strongly with the end-systolic stress measured in the left ventricle (Serneri et al. 2001).

The produced angiotensin II affects the regions expressing the angiotensin II receptors. In the rat, AT_1 (mainly AT_{1a}) and AT_2 receptor populations are elevated in the infarct region (Lefroy et al. 1996; Nio et al. 1995). Receptor density is increased from 3 days to 8 weeks after MI and mainly in the regions of fibroblast infiltration and collagen deposition. Myofibroblasts are predominantly responsible for the elevated infarct receptor density with a majority of AT_1 receptors (Sun and Weber 1996). In humans, the expression of the AT_1 receptor is increased after MI and this expression positively correlates with the extent of fibrosis (Ohtani et al. 1997). In contrast to the induced expression of AT_1 receptor found directly after MI, the AT_1 receptor population on the failing human heart decreases (Haywood et al. 1997; Serneri et al. 2001). Cardiac fibroblasts also express AT_2 receptors (Matsubara et al. 1994; Wharton et al. 1998). While ventricular function is impaired over time, the AT_2 receptor expression seems to remain constant (Regitz-Zagrosek et al. 1998; Tsutsumi et al. 1998) or seems to increase (Ohkubo et al. 1997).

Expression of angiotensin receptors after MI is less abundant on cardiomyocytes compared to fibroblasts (Lefroy et al. 1996). Viable rat cardiomyocytes elevate their angiotensin receptor population within 2–3 days after MI (Reiss et

Table 1 Activation of RAS after myocardial infarction

	Expression (mRNA)	Time point	Cell type	Species
Angiotensinogen	↑	5 days	Fibroblasts, myocytes	Rat
Renin	↑	2–7 days	Fibroblasts, myocytes (border zone), macrophages, myofibroblasts (infarct)	Rat
ACE	↑	3–7 days	Macrophages, fibroblasts, myocytes, endothelial and smooth muscle cells	Rat, human
AT_1 receptor	↑	3 days	Fibroblasts (infarct), myocytes (?)	Mouse, rat, rabbit
AT_2 receptor	↑	7 days	Fibroblasts (infarct), myocytes (?)	Rat, human

Data were taken from Busche et al. 2000; Campbell et al. 1997; Falkenhahn et al. 1995; Fujii et al. 1995; Hirsch et al. 1999; Hokimoto et al.,1996; Kijima et al. 1996; Lefroy et al. 1996; Lindpainter et al. 1993; Matsubara et al. 1994; Nio et al. 1995; Ohtani et al. 1997; Passier et al.1995, 1996; Sadoshima et al. 1993; Sun et al. 1996a,b, 2000, 2001; Suzuki et al. 1993; Wharton et al. 1998; Zhang et al. 1995.

al. 1993). The predominant receptor subtype expressed by myocytes is AT_1 (Zhang et al. 1995). AT_2 receptor mRNA is observed in cardiomyocytes, but its expression seems not to be enhanced within 24 h after MI (Busche et al. 2000). In contrast, mechanical stretch applied to neonatal rat myocytes in vitro resulted in the up-regulation of both receptor subtypes (Sadoshima et al. 1993; Kijima et al. 1996). Although a lot of information is available on the angiotensin II receptor expression after MI and during cardiac failure, there is no consensus on the exact expression pattern. It appears that the cardiac expression pattern of angiotensin II receptors after MI depends on (a) species, (b) cell type, (c) time point (early after MI, fibrogenic phase or heart failure) and (d) localisation within the heart (noninfarcted zone, border zone or infarct zone) (Gallagher et al. 1998a; Staufenberger et al. 2001). However, it is clear that the total angiotensin binding capacity increases after MI, together with the elevated possibility to generate angiotensin II (Reiss et al. 1993) (Table 1).

3 The Role of RAS Activation After MI

3.1 The Role of the Circulating RAS

Circulating RAS activation immediately after MI is thought to be a reflex response to the decreasing stroke volume. It supports arterial pressure by increasing the vascular resistance and supports cardiac output by increasing cardiac contractility and heart rate. Although the RAS activation may be appropriate when the heart is normal, under the circumstances of MI, the physiological effects may become deleterious over time: resistance, contractility and heart rate increase myocardial oxygen consumption in a situation of subnormal oxygen supply (Sigurdsson et al. 1993)

3.2
The Role of the Local RAS

After MI, all RAS components are expressed in the heart. The functional advantage of locally active RAS lies especially in the reaction rate, which is accelerated by a few orders of magnitude when comparing paracrine to endocrine activation. Thus, a very efficient system is present at the site of cardiac infarction and might influence processes such as wound healing, cardiac hypertrophy and fibrosis, which affect the cardiac structure and function. The AT_1 receptor is known to mediate most of the currently described effects of angiotensin II such as vasoconstriction, aldosterone secretion and vasopressin secretion. This description of the authentic angiotensin actions only slightly lifts the veil. Angiotensin II influences an enormous number of cellular processes, directly and indirectly. Under appropriate conditions angiotensin II can act as a mitogen. Activation of AT_1 receptors induces expression of genes, including proto-oncogenes, many growth factors [PDGF (Cook et al. 2001), TGF-β (Weber 1997) and IGF-I (Delafontaine et al. 1996)] and growth factor receptors (Sadoshima and Izumo 1993; Huckle and Earp 1994), which leads to growth of the cells that express the AT_1 receptor. Stimulation of the AT_2 receptors might lead to cellular alterations counteracting the AT_1-induced effects, for instance, the induction of apoptosis (Horiuchi et al. 1997). Moreover, AT_2 receptor activation decreases ACE activity, which indirectly attenuates the AT_1 receptor actions (Hunley et al. 2000). More recently, mice overexpressing the AT_2 receptor exhibited better cardiac function after MI (Yang Z et al. 2002). However, much remains to be elucidated on the exact function of the AT_2 receptor, especially after MI.

3.2.1
Effects of Angiotensin II on Fibroblasts

Stimulation of neonatal and adult rat fibroblasts with angiotensin II results in DNA synthesis, mRNA expression and protein synthesis of fibronectin and collagen I/III, via the AT_1 receptor expressed on the cell-surface (Schrob et al. 1993; Villarreal et al. 1993). Human fibroblasts stimulated with angiotensin II respond with an up-regulation in laminin and fibronectin mRNA expression, but collagen mRNA expression is not up-regulated (Kawano et al. 2000). The opposite occurs during stimulation of the AT_2 receptor; down-regulation of fibronectin and collagen synthesis as well as inhibition of the mitogenic response induced by stimulation of the AT_1 receptor (Ohkubo et al. 1997; Tsutsumi et al. 1998). Angiotensin II influences fibroblast proliferation and the production of extracellular matrix proteins (Weber 1997). However, angiotensin II is mitogenic for rat fibroblasts isolated from sham-operated and the noninfarcted myocardium, whereas it induces hypertrophy in fibroblasts isolated from the infarct area (Staufenberger et al. 2001). These latter data indicate that the surrounding cells, like myocytes, are affecting the response of fibroblasts towards angiotensin II.

3.2.2
Effects of Angiotensin II on Cardiomyocytes

Stimulation of AT_1 receptors on neonatal rat myocytes leads to activation of growth-related genes, which is followed by cell growth (Huckle and Earp 1994; Kim et al. 1995). This growth response is inhibited by activation of the AT_2 receptor (Booz and Baker 1996). Not only the myocyte's size can be affected by angiotensin II, but also its function. Contractility is closely related to ion currents (Morgan 1991). Recently it has been described that cardiomyocytes isolated from hypertrophied heart have depressed contractile reserve (Ito et al. 2000). Angiotensin II can alter ion currents via both the AT_1 and AT_2 receptors (Zhang et al. 1995). Contractility is supposed to be increased via stimulation of the AT_1 receptor due to an increase in cellular calcium. Furthermore, the calcium sensitivity of the myofilaments is enhanced in the presence of angiotensin II (Spinale et al. 1997; Meissner et al. 1998). The exact effects of changes in ion currents after stimulation of the AT_2 receptors have not been elucidated yet.

Thus, cardiomyocytes are equipped with an autocrine RAS, which may support myocyte growth and contractile function. This cardiomyocyte growth response to angiotensin II is transformed in the presence of fibroblasts (Booz et al. 1999). Conditioned medium of angiotensin II-stimulated fibroblasts contains trophic components for myocytes (Kim et al. 1995) and induces mRNA expression of angiotensinogen in these myocytes (Sil and Sen 1997). After MI, when many fibroblasts are activated and invade the myocardium, local RAS might influence cardiac remodelling in an autocrine and paracrine manner.

4
RAS Inhibition After MI

4.1
Effects of RAS Inhibition After MI

Positive influences of ACE inhibition on the survival rate and ventricular remodelling after myocardial infarction have been extensively studied in humans (McAlpine et al. 1987; Pfeffer et al. 1992; AIRE Study Investigators 1993) and animal models (Mulder et al. 1997; Ryoke et al. 1999). Major ACE inhibitor trials (e.g. SAVE and CONSENSUS II) have been reviewed (Huckell et al. 1997a, 1997b). The main conclusions are (a) a reduction of the post-MI left ventricular dysfunction and heart failure and (b) a substantial mortality and morbidity benefit to post-MI patients. Animal studies on this subject have been very helpful to elucidate the mechanisms underlying the beneficial effects of ACE inhibition after MI. Prolonged captopril treatment of rats with MI demonstrated that increased survival rate associated with by smaller infarcts and smaller end-diastolic volumes (Pfeffer et al. 1987). Cardiac enlargement and fibrosis subsequent to myocardial infarction can be reduced by ACE inhibition both in rats and patients (Pfeffer et al. 1982; Krimpen van et al. 1991). AT_1 receptor blockade after

MI prevented or attenuated cardiac loading, left ventricular expansion and hypertrophy (Capasso et al. 1994; van Kats et al. 2000). Moreover, fibrosis of the noninfarcted myocardium was reduced and capillary density slightly increased compared to nontreated groups (Smits et al. 1992; Holtz 1998). A study in single AT_{1a} knockout mice showed that after MI, wild-type animals exhibited more LV enlargement, fibrosis, ventricular dysfunction and mortality (Harada et al. 1999).

4.2 Mechanisms Involved in the Response to RAS Inhibition After MI

Although the mechanisms activated during ACE inhibition and AT_1 receptor blockade are distinct, the effects of ACE inhibitors and AT_1 receptor antagonist are highly comparable. Yet no consensus has been reached on the exact mechanisms involved in their beneficial effects. It is still questioned whether the reduced AT_1 activation or the bradykinin receptor type 2 activation (BK_2 receptor) is crucial for the beneficial effects of ACE inhibition (Martorana et al. 1990). It is not exactly clear what the role of AT_2 receptor activation might be during AT_1 receptor antagonism. This receptor might be involved in both structural and functional remodelling early after MI. There is a balance between two receptors in normal hearts; however, this balance is shifted towards the AT_2 receptor after MI. Stimulation of the AT_2 receptor leads to coronary vasodilation (Schuijt et al. 2001) and AT_2 receptor overexpression after MI improves the systolic function of the infarcted left ventricle (Yang, et al. 2002). The opposite is found with AT_2 receptor antagonism, that is a reduced DNA synthesis, endothelial cell proliferation and stroke volume in infarcted rat hearts at 14 days after MI (Smits et al. 1995 Kuizinga et al. 1998).

Remarkable observations have been made in the comparison between ACE inhibition and AT_1 receptor blockade. It is surprising that cardiac ACE activity is not only impaired during ACE inhibition, but also during AT_1 receptor blockade. Antagonism of the BK_2 receptor attenuated positive effects on hypertrophy of both ACE inhibition and AT_1 receptor blockade (Liu et al. 1997; Tsutsumi et al. 1999). Similar results were found in BK_2 receptor-deficient mice (Yang et al. 2001). ACE inhibitors and AT_1 antagonists have also been combined, which yielded synergistic effects. Results show less inflammatory cellular infiltration and less collagen deposition after MI in rats treated with both fosinopril and valsartan (Yu et al. 2001).

Reduction of cardiac loading might be one explanation for the prevention of ventricular enlargement and fibrosis during both ACE inhibition and AT_1 receptor antagonism. This would suggest that any type of effective antihypertensive treatment should result in a regression of hypertrophy that is proportional to the degree of blood pressure reduction. However, this correlation is not supported by experimental evidence. Classic vasodilators are very effective in reducing blood pressure but the reduction of cardiac hypertrophy is not in all cases related to the reduction in pressure (Raya et al. 1989; Smits et al. 1991). Beneficial effects on cardiac contractility are even found with low-dose ACE inhibition without reduction of blood pressure (Zhu et al. 1997), suggesting a direct positive effect of ACE inhibition on the cardiac function. Several explanations for these actions have been hypothesised (Unger and Golhke 1990).

1. Cardiomyocytes isolated from hypertrophied heart have depressed contractile reserve (Ito et al. 2000). Myocardial infarction has a negative effect on the contractility of rat trabeculae due to a disturbed calcium-handling. AT_1 receptor antagonism improves the calcium homeostasis, which might lead to a better contractile function (Daniëls et al. 2001).
2. The coronary flow is enhanced during ACE inhibition, most likely by reduction of coronary resistance (Gilst van et al. 1988; Nikolaidis et al. 2002). This enhanced coronary flow was also seen in the absence of changes in the systemic vascular resistance or plasma renin activity (Grinstead and Young 1991). These data point in the direction of the cardiac RAS and its involvement in cardiac remodelling after MI. Inhibition of the cardiac RAS might not only improve coronary flow, but also reduce cardiac hypertrophy and fibrosis, all in favour of the cardiac function.
3. Another hypothesis is the potentiation of local bradykinin action (Duncan et al. 1997). Kinins have both short-term and long-term cardioprotective effects after MI. Short-term protection is due to reduction of the ischaemia-reperfusion injury and increase of coronary flow (Yoshida et al. 2000). Long-term effects involve reductions in ventricular hypertrophy (Wollert et al. 1997), fibrosis (Gallagher et al. 1998b) and progression to heart failure (Kokkonen et al. 2000). During ACE inhibition the bradykinin degeneration is reduced, which might be beneficial to the cardiac function.
4. Angiotensin II is known to facilitate noradrenaline release from the sympathetic nerve-endings (Brasch et al. 1993). Inhibition of angiotensin II production decreases the noradrenaline overflow (Wenting et al. 1983). Although sympathetic activation directly after MI is part of the adaptive reflex (Graham et al. 2002), prolonged activation might harm the myocardium and ACE inhibitors might be beneficial through indirect inhibition of the sympathetic activity (Persson et al. 2002). However, inhibition of the sympathetic activity is not the only mechanism involved because β-adrenoreceptor blockade is effective in reducing the cardiac loading, but does not prevent cardiac hypertrophy as much as ACE inhibition. Moreover, synergetic effects have been demonstrated during combined therapy with ramipril and metoprolol (Theres et al. 2000). Still more information is needed on the pathophysiological role of RAS after MI to understand the therapeutic potential of interference with this system.

References

Acute Infarction Ramipril (AIRE) Study Investigators (1993) Effect of ramipril on mortality and morbidity of survivors of acute myocardial infarction with clinical evidence of heart failure. Lancet 342:821–828

Anversa P, Olivetti G, Capasso KM (1991) Cellular basis of ventricular remodeling after myocardial infarction. Am J Cardiol 68:7D–16D

Blumenfeld JD, Sealey JE, Alderman MH, Cohen H, Lappin R, Catanzaro DF, Laragh JH (2000) Plasma renin activity in emergency department and its independent association with acute myocardial infarction. Am J Hypertens 13:855–863

Booz GW, Baker KM (1996) Role of type 1 and type 2 angiotensin receptors in angiotensin II-induced cardiomyocyte hypertrophy. Hypertension 28:635–640

Booz GW, Dostal DE, Baker KM (1999) Paracrine actions of cardiac fibroblasts on cardiomyocytes: implications for the cardiac renin-angiotensin system. Am J Cardiol 83:44H–47H

Brasch H, Sieroslawski L, Dominiak P (1993) Angiotensin II increases norepinephrine release from atria by acting on angiotensin subtype 1 receptors. Hypertension 22:699–704

Busche S, Gallinat S, Bohle RM, Reinecke A, Seebeck J, Franke F, Fink L, Zhu M, Sumners C, Unger T (2000) Expression of angiotensin AT1 and AT2 receptors in adult rat cardiomyocytes after myocardial infarction. Am J Pathol 157:605–611

Capasso JM, Li P, Meggs LG, Herman HV, Anversa P (1994) Efficacy of angiotensin-converting enzyme inhibition and AT1 receptor blockade on cardiac pump performance after myocardial infarction in rats. J Cardiovasc Pharmacol 23:584–593

Ceiler DL, Schiffers PMH, Nelissen-Vrancken HJM, Smits JFM (1998) Time-related adaptations in plasma neurohormone levels and hemodynamics after myocardial infarction in the rat. J Card Fail 4:131–138

Cleland JGF, Cowburn PJ, Morgan K (1996) Neuroendocrine activation after myocardial infarction: causes and consequences. Heart 76 [Suppl 3]:53–59

Cleutjens JPM, Blankesteijn WM, Daemen MJAP, Smits JFM (1999) The infarcted myocardium: simply dead tissue, or lively target for therapeutic interventions. Cardiovasc Res 44:232–241

Cook JL, Zhang Z, Re RN (2001) In vitro evidence for an intracellular site of angiotensin action. Circ Res 89:1138–1146

Creemers E, Cleutjens J, Smits J, Heymans S, Moons L, Collen D, Daemen M, Carmeliet P (2000) Disruption of the plasminogen gene in mice abolishes wound healing after myocardial infarction. Am J Pathol 156:1865–1873

Daniëls MCG, Keller RS, De Tombe PP (2001) Losartan prevents contractile dysfunction in rat myocardium after left ventricular myocardial infarction. Am J Physiol 281:H2150–H2158

Danser AHJ (1996) Local renin-angiotensin systems. Mol Cell Biochem 157:211–216

Dargie HJ, McAlpine HM, Morton JJ (1987) Neuroendocrine activation in acute myocardial infarction. J Cardiovasc Pharmacol 9 [Suppl 2]:S21–S24

Delafontaine P, Brink M, Du J (1996) Angiotensin II modulation of insulin-like growth factor I expression in the cardiovascular system. Trends Cardiovasc Med 6:187–193

Duncan AM, Burrell LM, Kladis A, Campbell DJ (1997) Angiotensin and bradykinin peptides in rats with myocardial infarction. J Card Fail 3:41–52

Dzau VJ, Colucci WS, Hollenberg NK, Willimas GH (1981) Relation of renin-angiotensin-aldosterone system to clinical state in congestive heart failure. Circulation 63:645–651

Ertl G, Meesman M, Kochsiek K (1985) On the mechanism of renin release during experimental myocardial ischemia. Eur J Clin Invest 15:375–381

Ertl G, Gaudron P, Hu K (1993) Ventricular remodeling after myocardial infarction. Experimental and clinical studies. Basic Res Cardiol 88 [Suppl 1]:125–137

Falkenhahn M, Franke K, Bohle RM, Zhu YC, Stauss HM, Bachmann S, Danilov S, Unger T (1995) Cellular distribution of angiotensin converting enzyme after myocardial infarction. Hypertension 25:219–226

Francis GS, Benedict C, Johnstone DE, Kirlin PC, Nicklas J, Liang CS, Kubo SH, Rudin-Toretsky E, SOLVD Investigators (1990) Comparison of neuroendocrine activation in patients with left ventricular dysfunction with and without congestive heart failure. Circulation 82:1724–1729

Gaasch WH (1994) Diagnosis and treatment of heart failure based on left ventricular systolic or diastolic dysfunction. JAMA 271:1276–1280

Gallagher AM, Bahnson TD, Yu H, Kim NN, Printz MP (1998a) Species variability in angiotensin receptor expression by cultures cardiac fibroblasts and the infarcted heart. Am J Physiol 274: H801–H809

Gallagher AM, Yu H, Printz MP (1998b) Bradykinin-induced reductions of collagen gene expression involve prostacyclin. Hypertension 32:84–88

Graham LN, Smith PA, Stoker JB, Mackintosh AF, Mary DASG (2002) Time course of sympathetic neural hyperactivity after uncomplicated acute myocardial infarction. Circulation 106:793–797

Grinstead WC, Young JB (1991) The myocardial renin-angiotensin system: existence, importance and clinical implications. Am Heart J 123:1039–1045

Harada K, Sugaya T, Murakami K, Yazaki Y, Komuro I (1999) Angiotensin II type 1A receptor knockout mice display less left ventricular remodeling and improved survival after myocardial infarction. Circulation 100:2093–2099

Haywood GA, Gullestad L, Katsuya T, Hutchinson HG, Pratt RE, Horiuchi M, Fowler MB (1997) AT1 and AT2 angiotensin receptor gene expression in human heart failure. Circulation 95:1201–1206

Hirsch AT, Talsness CE, Schunkert H, Paul M, Dzau VJ (1991) Tissue-specific activation of cardiac angiotensin converting enzyme in experimental heart failure. Circ Res 69:475–482

Hirsch AT, Opsahl JA, Lunzer MM, Katz SA (1999) Active renin and angiotensinogen in cardiac interstitial fluid after myocardial infarction. Am J Physiol 276:H1818–H1826

Hokimoto S, Yasue H, Fujimoto K, Yamamoto H, Nakao K, Kaikita K, Sakata R, Miyamoto E (1996) Expression of angiotensin-converting enzyme in remaining viable myocytes of human ventricles after myocardial infarction. Circulation 94:1513–1518

Holtz J (1998) Role of ACE inhibition or AT1 blockade in the remodeling following myocardial infarction. Basic Res Cardiol 93 [Suppl 2]:92–100

Horiuchi M, Hayashida W, Kambe T, Yamada T, Dzau VJ (1997) Angiotensin type 2 receptor dephosphorylates bcl-2 by activating mitogen-activated protein kinase phosphotase-1 and induces apoptosis. J Biol Chem 272:19022–19026

Huckell VF, Bernstein V, Cairns JA, Crowell R, Dagenais GR, Higginson LA, Isserow S, Laramee P, Liu P, McCans JL, Orchard RC, Prewitt R, Quinn BP, Samson M, Turazza F, Warnica JW, Wielgosz A (1997a) Angiotensin-converting enzyme inhibition in myocardial infarction—Part 1: clinical data. Can J Cardiol 13:161–169

Huckell VF, Bernstein V, Crowell R, Dagenais GR, Higginson LA, Isserow S, Laramee P, Liu P, McCans JL, Orchard RC, Prewitt R, Quinn BP, Samson M, Turazza F, Warnica JW, Wielgosz A (1997b) Angiotensin-converting enzyme inhibition in myocardial infarction—Part 2: clinical issues and controversies. Can J Cardiol 13:173–182

Huckle WR, Earp HS (1994) Regulation of cell proliferation and growth by angiotensin II. Prog Growth Factor Res 5:177–194

Hunley TE, Tamura M, Stoneking BJ, Nishimura H, Ichiki T, Inagami T, Kon V (2000) The angiotensin type II receptor tonically inhibits angiotensin-converting enzyme in AT2 null mutant mice. Kidney Int 57:570–577

Ito K, Yan X, Tajima M, Su Z, Barry WH, Lorell BH (2000) Contractile reserve and intracellular calcium regulation in mouse myocytes from normal and hypertrophied failing hearts. Circ Res 87:588–595

Katz AM (1995) The cardiomyopathy of overload: an unnatural growth response. Eur Heart J 16:110–114

Kawano H, Do YS, Kawano Y, Starnes V, Barr M, Law RE, Hsueh WA (2000) angiotensin II has multiple profibrotic effects in human cardiac fibroblasts. Circulation 101:1130–1137

Kelly MP, Kahr O, Aalkjaer C, Cumin F, Samani NJ (1997) Tissue expression of components of the renin-angiotensin system in experimental post-infarction heart failure in rats: effects of heart failure and angiotensin-converting enzyme inhibitor treatment. Clin Sci 92:455–465

Kijima K, Matsubara H, Murasawa S, Maruyama K, Mori Y, Ohkubo N, Komuro Y, Iwasaka T, Inada M (1996) Mechanical stretch induces enhanced expression of angiotensin II receptor subtypes in neonatal rat cardiac myocytes. Circ Res 79:887–897

Kim NN, Villarreal FJ, Printz MP, Lee AA, Dillmann WH (1995) Trophic effects of angiotensin II on neonatal rat cardiac myocytes are mediated by cardiac fibroblasts. Am J Physiol 269: E426–E437

Kokkonen JO, Lindstedt KA, Kuoppala A, Kovanen PT (2000) Kinin-degrading pathways in the human heart. Trends Cardiovasc Med 10:42–45

Kuizinga MC, Smits JFM, Arends JW, Daemen MJAP (1998) AT2 receptor blockade reduces cardiac interstitial cell function after rat myocardial infarction. J Mol Cell Cardiol 30:425–434

Lee S, Kramer CM, Mankad S, Yoo SE, SandbergmK (2001) Combined angiotensin converting enzyme inhibition and angiotensin AT1 receptor blockade up-regulates myocardial AT2 receptors in remodeled myocardium post-infarction. Cardiovasc Res 51:131–139

Lefroy DC, Wharton J, Crake T, Knock GA, Rutherford RAD, Suzuki T, Morgan K, Polak JM, Poole-Wilson PA (1996) Regional changes in angiotensin II receptor density after experimental myocardial infarction. J Mol Cell Cardiol 28:429–440

Lindpainter K, Lu W, Niedermajer N, Schieffer B, Just H, Ganten D, Drexler H (1993) Selective activation of cardiac angiotensinogen gene expression in post-infarction ventricular remodeling in the rat. J Mol Cell Cardiol 25:133–143

Liu Y-H, Yang X-P, Sharov VG, Nass O, Sabbah NH, Peterson E, Carretero OA (1997) Effects of angiotensin-converting enzyme inhibitors and angiotensin II type 1 receptor antagonists in rats with heart failure. J Clin Invest 99:1926–1935

Mackenna D, Summerour SR, Vilarreal FJ (2000) Role of mechanical factors in modulating cardiac fibroblast function and extracellular matrix synthesis. Cardiovasc Res 46:257–263

Martorana PA, Kettenbach B, Breipohl G, Linz W, Schölkens BA (1990) Reduction of infarct size by local angiotensin-converting enzyme inhibition is abolished by a bradykinin antagonist. Eur J Pharmacol 182:395–396

Matsubara H, Kanasaki M, Murasawa S, Tsukaguchi Y, Nio Y, Inada M (1994) Differential gene expression and regulation of angiotensin II receptor subtypes in rat cardiac fibroblasts and cardiomyocytes in culture. J Clin Invest 93:1592–1601

McAlpine H, Morton JJ, Leckie B, Dargie HJ (1987) Haemodynamic effects of captopril in acute left ventricular failure complicating myocardial infarction. J Cardiovasc Pharmacol 9 [Suppl 2]: S25–S30

McAlpine HM, Morton JJ, Leckie B, Rumley A, Gillen G, Dargie HJ (1988) Neuroendocrine activation after myocardial infarction. Br Heart J 60:117–124

Meissner A, Min JY, Simon R (1998) Effects of angiotensin II on inotropy and intracellular Ca2+ handling in normal and hypertrophied rat myocardium. J Mol Cell Cardiol 30:2507–2518

Michorowski B, Ceremuzynski L (1983) The renin-angiotensin-aldosterone system and the clinical course of acute myocardial infarction. Eur Heart J 4:259–264

Morgan JP (1991) Abnormal intracellular modulation of calcium as a major cause of cardiac dysfunction. N Engl J Med 325:625–632

Mulder P, Devaux B, Richard V, Henry JP, Wimart MC, Thibout E, Macé B, Thuillez C (1997) Early versus delayed angiotensin-converting enzyme inhibition in experimental chronic heart failure. Circulation 95:1314–1319

Nikolaidis LA, Doverspike A, Huerbin R, Hentosz T, Shannon RP (2002) Angiotensin-converting enzyme inhibitors improve coronary flow reserve in dilated cardiomyopathy by a bradykinin-mediated, nitric oxide-dependent mechanism. Circulation 105:2785–2790

Nio Y, Matsubara H, Murasawa S, Kanasaki M, Inada M (1995) Regulation of gene transcription of angiotensin II receptor subtypes in myocardial infarction. J Clin Invest 95:46–54

Ohkubo N, Matsubara H, Nozawa Y, Mori Y, Murasawa S, Kijima K, Maruyama K, Masaki H, Tsutumi Y, Shibazaki Y, Iwasaka T, Inada M (1997) Angiotensin type 2 receptors are reexpressed by cardiac fibroblasts from failing myopathic hamster hearts and inhibit cell growth and fibrillar collagen metabolism. Circulation 96:3954–3962

Ohtani S, Fujiwara H, Hasegawa K, Doyama K, Inada T, Tanaka M, Fujiwara T, Sasayama S (1997) Up-regulated expression of angiotensin type 1 receptor gene in human pathologic hearts. J Card Fail 3:303–310

Passier RCJJ, Smits JFM, Verluyten MJA, Studer R, Drexler H, Daemen MJAP (1995) Activation of angiotensin-converting enzyme expression in infarct zone following myocardial infarction. Am J Physiol 269:H1268–H1276

Passier RCJJ, Smits JFM, Verluyten MJA, Daemen MJAP (1996) Expression and localization of renin and angiotensinogen in rat heart after myocardial infarction. Am J Physiol 271:H1040–H1048

Persson H, Andréasson K, Kahan T, Eriksson SV, Tidgren B, Hjemdahl P, Hall C, Erhardt L (2002) Neurohormonal activation in heart failure after acute myocardial infarction treated with beta-receptor antagonists. Eur J Heart Fail 4:73–82

Pfeffer MA, Braunwald E (1990) Ventricular remodeling after myocardial infarction. Experimental observations and clinical implications. Circulation 81:1161–1172

Pfeffer JM, Pfeffer MA, Mirsky I, Braunwald E (1982) Regression of left ventricular hypertrophy and prevention of left ventricular dysfunction by captopril in spontaneously hypertensive rat. Proc Natl Acad Sci U S A 79:3310–3314

Pfeffer JM, Pfeffer MA, Braunwald E (1987) Hemodynamic benefits and prolonged survival with long-term captopril therapy in rats with myocardial infarction and heart failure. Circulation 75 [Suppl I]:I149–I155

Pfeffer MA, Braunwald E, Moyé LA, Batsa L, Brown EJ, Cuddy TE, Davis BR, Geltman EM, Goldman S, Flaker GC, Klein M, Lamas GA, Packer M, Rouleau J, Rouleau JL, Rutherford J, Wertheimer JH, Hawkins CM (1992) Effect of captopril on mortality and morbidity in patients with left ventricular dysfunction after myocardial infarction. N Engl J Med 327:669–677

Raya TE, Gay RG, Aguirre M, Goldman S (1989) Importance of venodilatation in prevention of left ventricular dilatation after chronic large myocardial infarction in rats: a comparison of captopril and hydralazine. Circ Res 64:330–337

Regitz-Zagrosek V, Fielitz J, Fleck E (1998) Myocardial angiotensin receptors in human hearts. Basic Res Cardiol 93:37–42

Reiss K, Capasso JM, Huang HE, Meggs LG, Li P, Anversa P (1993) ANG II receptors, c-myc and c-jun in myocytes after myocardial infarction and ventricular failure. Am J Physiol 264:H760–H769

Ryoke T, Gu Y, Mao L, Hongo M, Clark RG, Peterson KL, Ross J (1999) Progressive cardiac dysfunction and fibrosis in the cardiomyopathic hamster and effects of growth hormone and angiotensin converting enzyme inhibition. Circulation 100:1734–1743

Sadoshima J, Izumo S (1993) Molecular characterization of angiotensin II induced hypertrophy in cardiac myocytes and hyperplasia of cardiac fibroblasts. Critical role of the AT1 receptor subtype. Circ Res 73:413–423

Sadoshima JI, Xu Y, Slayter HS, Izumo S (1993) Autocrine release of angiotensin II mediates stretch-induced hypertrophy of cardiac myocytes in vitro. Cell 75:977–984

Schrob W, Booz GW, Dostal DE, Conrad KM, Chang KC, Baker KM (1993) Angiotensin II is mitogen in neonatal rat cardiac fibroblasts. Circ Res 72:1245–1254

Schuijt MP, Badshew M, van Veghel R, Vries de R, Saxena PR, Schoemaker RG, Danser AHJ (2001) AT2 receptor-mediated vasodilation in the heart: effect of myocardial infarction. Am J Physiol 281:H2590–H2596

Serneri GGN, Boddi M, Cecioni I, Vanni S, Coppo M, Pap ML, Bandinelli B, Bertolozzi I, Polidori G, Toscano T, Maccherinin M, Modesti PA (2001) Cardiac angiotensin II formation in clinical course of heart failure and its relationship with left ventricular function. Circ Res 88:961–968

Sigurdsson A, Held P, Swedberg K (1993) Short- and long-term neurohormonal activation following acute myocardial infarction. Am Heart J 126:1068–1076

Sil P, Sen S (1997) Angiotensin II and myocyte growth. Hypertension 30:209–216

Smits JFM, Cleutjens JPM, Krimpen CV, Schoemaker RG, Daemen MJAP (1991) Cardiac remodeling in hypertension and following myocardial infarction: effects of arteriolar vasodilators. Basic Res Cardiol 86:133–139

Smits JFM, van Krimpen C, Schoemaker RG, Cleutjens JPM, Daemen MJAP (1992) Angiotensin II receptor blockade after myocardial infarction in rats: effects on hemodynamics, myocardial DNA synthesis, and interstitial collagen content. J Cardiovasc Pharmacol 20:772–778

Smits JFM, Passier RCJJ, Nelissen-Vrancken HJMG, Cleutjens JPM, Kuizinga MC, Daemen MJAP (1995) Does ACE inhibition limit structural changes in the heart following myocardial infarction? Eur Heart J 16:N46–N51

Spinale FG, Holzgrefe HH, Mukherjee R, Webb ML, Hird RB, Cavallo MJ, Powell JR, Koster WH (1997) Angiotensin-converting enzyme inhibition and angiotensin II subtype-1 receptor blockade during the progression of left ventricular dysfunction: differential effects on myocyte contractile processes. J Pharmacol Exp Ther 283:1082–1094

Staufenberger S, Jacobs M, Brandstätter K, Hafner M, Regitz-Zagrosek V, Ertl G, Schrob W (2001) Angiotensin II type 1 receptor regulation and differential trophic effects on rat cardiac myofibroblasts after acute myocardial infarction. J Cell Physiol 187:326–335

Struijker-Boudier HAJ, Smits JFM, De Mey JGR (1995) Pharmacology of cardiac and vascular remodeling. Ann Rev Pharmacol Toxicol 35:509–539

Sun Y, Weber KT (1996) Cells expressing angiotensin II receptors in fibrous tissue of rat heart. Cardiovasc Res 31:518–525

Sun Y, Weber KT (2000) Infarct scar: a dynamic tissue. Cardiovasc Res 46:250–256

Sun Y, Zhang JQ, Weber KT (2001) Renin expression at sites of repair in the infarcted rat heart. J Mol Cell Cardiol 33:995–1003

Swynghedauw B (1991) Remodeling of the heart in chronic pressure overload. Basic Res Cardiol 86:99–105

Theres HP, Wagner KD, Romberg D, Feig C, Strube S, Leiterer KP, Günter J, Stangl K, Baumann G, Schimke I (2000) Combined treatment with ramipril and metoprolol prevents changes in the creatine kinase isoenzyme system and improves hemodynamic function in rat hearts after myocardial infarction. Cardiovasc Drugs Ther 14:597–606

Tsutsumi Y, Matsubara H, Ohkubo N, Mori Y, Nozawa Y, Murasawa S, Kijima K, Maruyama K, Masaki H, Moriguchi Y, Shibasaki Y, Kamihata H, Inada M, Iwasaka T (1998) Angiotensin II type 2 receptor is upregulated in human heart with interstitial fibrosis, and cardiac fibroblasts are major cell type for its expression. Circ Res 83:1035–1046

Tsutsumi Y, Matsubara H, Masaki H, Kurihara H, Murasawa S, Takai S, Miyazaki M, Nozawa Y, Ozono R, Nakagawa K, Miwa Y, Kawada N, Mori Y, Shibasaki Y, Tanaka Y,

Fujiyama S, Koyama Y, Fujiyama A, Takahashi H, Iwasaka T (1999) Angiotensin II type 2 receptor overexpression activates the vascular kinin system and causes vasodilation. J Clin Invest 104:925–935

Unger T, Golhke P (1990) Tissue renin-angiotensin systems in the heart and vasculature: possible involvement in the cardiovascular actions of converting enzyme inhibitors. Am J Cardiol 65:3I–10I

Van Gilst WH, Scholtens E, de Graeff PA, de Langen CDJ, Wesseling H (1988) Differential influences of angiotensin converting-enzyme inhibitors on the coronary circulation. Circulation 77 [Suppl I]:I24–I29

Van Kats JP, Duncker DJ, Haitsma DB, Schuijt MP, Niebuur R, Stubenitsky R, Boomsma F, Schalekamp MADH, Verdouw PD, Danser AHJ (2000) Angiotensin converting enzyme inhibition and angiotensin II type 1 receptor blockade prevent cardiac remodeling in pigs after myocardial infarction. Role of tissue angiotensin II. Circulation 102:1556–1563

Van Krimpen C, Smits JFM, Cleutjens JPM, Debets JJM, Schoemaker RG, Struyker-Boudier HAJ, Daemen MJAP (1991) DNA synthesis in the non-infarcted cardiac interstitium after left coronary artery ligation in the rat: effects of captopril. J Mol Cell Cardiol 23:1245–1253

Vaughan DE, Lamas GA, Pfeffer MA (1990) Role of left ventricular dysfunction in selective neurohumoral activation in recovery phase of anterior wall acute myocardial infarction. Am J Cardiol 66:529–532

Villarreal FJ, Kim NN, Ungab GD, Printz MP, Dilmann WH (1993) Identification of functional angiotensin II receptors on rat cardiac fibroblasts. Circulation 88:2849–2861

Weber KT (1997) Extracellular matrix remodeling in heart failure. A role for de novo angiotensin II generation. Circulation 96:4065–4082

Wenting GJ, Man in 't veld AJ, Woittiez AJ, Boomsma F, Laird-Meeter K, Simoons ML, Hugenholtz PG, Schalekamp MADH (1983) Effects of captopril in acute and chronic heart failure, correlations with plasma levels of noradrenaline, renin and aldosterone. Br Heart J 49:65–76

Wharton J, Morgan K, Rutherford RAD, Catravas JD, Chester A, Whitehead BF, Leval de MR, Yacoub MH, Polak JM (1998) Differential distribution of angiotensin AT2 receptors in the normal and failing human heart. J Pharmacol Exp Ther 284:323–336

Wollert KC, Studer R, Doerfer K, Schieffer E, Holubarsch C, Just H, Drexler H (1997) Differential effects of kinins on cardiomyocyte hypertrophy and interstitial collagen matrix in the surviving myocardium after myocardial infarction in the rat. Circulation 95:1910–1917

Yang XP, Liu YH, Mehta D, Cavasin MA, Shesely E, Xu J, Liu F, Carretero OA (2001) Diminished cardioprotective response to inhibition of angiotensin-converting enzyme and angiotensin II type 1 receptor in B2 kinin receptor gene knockout mice. Circ Res 88:1072–1079

Yang Z, Bove CM, French BA, Epstein FH, Berr SS, DiMariaJM, Gibson JJ, Carey RM, Kramer CM (2002) Angiotensin II type 2 receptor overexpression preserves left ventricular function after myocardial infarction. Circulation 106:106–111

Yoshida H, Zhang JJ, Chao L, Chao J (2000) Kallikrein gene delivery attenuates myocardial infarction and apoptosis after myocardial ischemia and reperfusion. Hypertension 35:25–31

Zhang X, Dostal DE, Reiss K, Cheng W, Kajstura J, Li P, Huang H, Sonnenblick EH, Meggs LG, Baker KM, Anversa P (1995) Identification and activation of autocrine renin-angiotensin system in adult ventricular myocytes. Am J Physiol 269:H1791–H1802

Zhu YC, Zhu YZ, Gohlke P, Strauss HM, Unger T (1997) Effects of angiotensin-converting enzyme inhibition and angiotensin II AT1 receptor antagonism on cardiac parameters in left ventricular hypertrophy. Am J Cardiol 80:110A–117A

Part 4
Tissues
Kidney and Adrenal Gland

Angiotensin II, the Kidney and Hypertension

O. Grisk · R. Rettig

Department of Physiology, University of Greifswald,
Greifswalder Strasse 11C, 17495 Karlsburg, Germany
e-mail: grisko@uni-greifswald.de

Abstract Angiotensin II produced locally and within the systemic circulation participates in the regulation of renal development, renal hemodynamics and renal epithelial transport. Proper renal morphogenesis critically depends on angiotensin type I receptor-mediated actions of angiotensin II. Angiotensin II causes mainly vasoconstriction of the renal vasculature. Under certain conditions, angiotensin II may also have vasodilatory actions in the renal circulation. Vascular actions of angiotensin II involve differential signal transduction mechanisms depending on the renal vessel segment. Numerous transport processes involved in water, electrolyte and acid base homeostasis are influenced by angiotensin II along the entire nephron. Increased renal angiotensin II levels contribute to pathological conditions such as renal inflammation and arterial hypertension.

Keywords Angiotensin II · Kidney · Hemodynamics · Epithelial transport · Sodium chloride · Arterial pressure · Inflammation · Oxygen radicals

1 Introduction

Angiotensin II is involved in the regulation of renal development and function. Via its actions on the kidney, it contributes to the maintenance of fluid and electrolyte homeostasis as well as long-term arterial pressure control (Guyton 1991). Juxtaglomerular cells within the kidney release renin into the circulation to generate angiotensin I from plasma angiotensinogen. In addition, the kidney possesses all components of the renin-angiotensin system and is able to generate angiotensin II locally (Braam et al. 1993; Rohrwasser et al. 1999).

Angiotensinogen is expressed in the proximal tubule (Braam et al. 1993; Rohrwasser et al. 1999) and in glomerular podocytes (Mentzel et al. 1997). Besides occurring in juxtaglomerular cells, renin synthesis and secretion has been demonstrated in principal cells of the connecting tubules (Rohrwasser et al. 1999). Angiotensin I-converting enzyme (ACE) is present in the renal cortex as well as in the outer and inner stripe of the outer medulla (Correa et al. 1995). Both angiotensin receptor types (AT_1 and AT_2) are present in the kidney. AT_1 receptors can be demonstrated immunohistochemically throughout the kidney (Harrison-Bernard et al. 1997; Miyata et al. 1999). Immunohistochemically detected sites of AT_1 receptor protein expression include the renal vasculature, glomerular mesangial cells, the proximal and distal tubule as well as the collecting duct (Harrison-Bernard et al. 1997; Miyata et al. 1999). Mice and rats have two AT_1 receptor subtypes (AT_{1A} and AT_{1B}), whereas most other mammals including humans have only a single type of AT_1 receptor. On the mRNA level, both AT_1 receptor isoforms (AT_{1A} and AT_{1B}) have been demonstrated to be present along the entire rat nephron and in the renal vasculature (Bouby et al. 1997; Miyata et al. 1999). AT_{1A} receptor mRNA has been reported to be more abundant than AT_{1B} receptor mRNA along the entire rat nephron, except for the glomerulus where more AT_{1B} than AT_{1A} receptor mRNA has been found (Bouby et al. 1997). Although less abundant than the AT_1 receptor, AT_2 receptor protein and mRNA are also expressed in the adult kidney, including glomeruli, epithelial cells and vasculature (Miyata et al. 1999; Ozono et al. 1997).

Based on physiological experiments and quantitative analyses of factors involved in long-term arterial pressure regulation, strong evidence has been provided that the kidney contributes importantly to the pathogenesis and maintenance of arterial hypertension (Harrison-Bernard et al. 1997). Cross-transplantation studies in several experimental forms of genetic hypertension demonstrated that arterial pressure can be normalized with a kidney graft from histocompatible normotensive donors (Grisk and Rettig 2001). The importance of angiotensin II for the pathogenesis of hypertension is illustrated by the fact that brief ACE inhibition in young spontaneously hypertensive rats (SHR) produces a long-term arterial pressure reduction (Harrap et al. 1990) and that kidney specific overexpression of components of the renin-angiotensin system in transgenic mice is associated with hypertension (Sigmund 2001).

The present chapter summarizes recent data on the role of angiotensin II in renal development and function as well as its role in the pathophysiology of experimental hypertension.

2 Development

Angiotensin II receptor types show a characteristic renal expression pattern during ontogeny and mediate distinct effects of angiotensin II on renal development and growth. In fetal rat kidneys at gestational day 14, AT_2 receptor mRNA is more abundant than AT_1 receptor mRNA (Norwood et al. 1997). AT_2 receptors are mainly found in metanephric mesenchyme and in developing epithelia such as primitive tubules and S-shaped bodies (Norwood et al. 1997; Ozono et al. 1997). Conversely, AT_1 receptors as well as angiotensinogen are not found in metanephric mesenchyme and pretubular aggregates but are abundant in the ureteric bud and more mature regions of the developing kidney (Norwood et al. 1997; Prieto et al. 2001). During late gestation, renal AT_2 receptor expression declines and AT_1 receptor expression increases, with AT_1 receptors being far more abundant than AT_2 receptors in mature kidneys (Norwood et al. 1997; Robillard et al. 1995; Shanmugam et al. 1995).

Evidence for a role of angiotensin II in renal development comes from data obtained in knockout mice (Esther et al. 1996; Niimura et al. 1995; Nishimura et al. 1999; Tsuchida et al. 1998) and from experiments with pharmacological blockade of the renin-angiotensin system in neonatal rats. Mice lacking either angiotensinogen (Niimura et al. 1995), angiotensin I-converting enzyme (Esther et al. 1996) or both AT_1 receptor subtypes (AT_{1A} and AT_{1B}) (Tsuchida et al. 1998) show similar abnormal phenotypes, including arterial hypotension, growth retardation and abnormal renal architecture. The kidneys of these mice display widening of the calyx and atrophy of the papilla as well as the medulla, resulting in morphological changes similar to hydronephrosis with urinary tract obstruction. Interestingly, there is hypertrophy of the renal vasculature. Functional defects include a reduced urine concentrating ability. These severe changes in renal morphology and function are not seen in mice lacking only one AT_1 receptor subtype (AT_{1A} or AT_{1B}) (Cervenka et al. 1999a; Chen et al. 1997). In knockout mice lacking the AT_2 receptor, renal development is largely normal, although some animals show kidney and urinary tract anomalies that resemble congenital malformation syndromes sometimes found in humans (Nishimura et al. 1999). When angiotensinogen knockout mice were crossed with transgenic animals expressing the rat angiotensinogen gene in brain and liver but not in the kidney, the resulting progeny had no renal malformations (Kang et al. 2002). These findings indicate that the presence and actions of angiotensin II, but not the local production of the peptide within the kidney are required for normal renal development.

In rodents, nephrogenesis continues during the first 2 postnatal weeks. Thus, newborn rats are suitable animal models for investigating renal developmental phenomena. Rats treated with an ACE inhibitor or an AT_1 receptor antagonist during the neonatal period show a dilated renal pelvis, an atrophied renal papilla, tubular distensions, thickening of intrarenal arteries, and disturbed arrangement of peritubular capillaries and vasa recta (Guron et al. 2000; McCausland et al. 1997; Tufro-McReddie et al. 1995). These morphological abnormalities are remarkably similar to those seen in mice with targeted inactivation of the genes for angiotensinogen, ACE, and both AT_1 receptor isoforms, respectively (Esther et al. 1996; Niimura et al. 1995; Nishimura et al. 1999; Tsuchida et al. 1998). Glomerular morphology has been found not to be seriously affected by these treatment regimens by some authors (Guron et al. 1997; McCausland et al. 1997), while others found decreased glomerular number and increased glomerular volume in rats treated with losartan during the first 12 postnatal days (Woods and Rasch 1998). Neonatal AT_2 receptor blockade has no detrimental effects on rat renal morphology (McCausland et al. 1997; Tufro-McReddie et al. 1995). On the functional level, neonatal ACE inhibitor treatment in rats causes reduced renal concentration ability and reductions in renal blood flow and glomerular filtration rate (Guron et al. 1998).

The mechanisms leading to permanent renal defects in response to targeted inactivation of renin-angiotensin system genes or pharmacological blockade of the renin-angiotensin system in rodents are largely unknown. They may involve a lack of AT_2 receptor-mediated apoptosis of undifferentiated mesenchymal cells during renal development (Maric et al. 1998; Nishimura et al. 1999). The absence of the growth- and proliferation-stimulating effects of angiotensin II on renal vascular smooth muscle cells and epithelial cells, which are mediated mainly via AT_1 receptors, may also be of importance (Aizawa et al. 2001; Guron and Friberg 2000; Wang et al. 1997). In addition, interactions between angiotensin II and other growth factors involved in nephrogenesis are currently discussed (Guron and Friberg 2000).

Factors involved in renal development have been implicated in the pathogenesis of arterial hypertension. Thus, angiotensinogen and ACE knockout mice are hypotensive (Esther et al. 1996; Tanimoto et al. 1994),whereas arterial pressure is elevated and more sensitive to DOCA-salt treatment in AT_2 receptor knockout mice compared to wild type mice (Gross et al. 2000). Serious renal lesions induced by neonatal ACE inhibitor treatment in rats are not accompanied by arterial pressure elevations (Guron and Friberg 2000). On the other hand, elevations in arterial pressure have been found in adult Sprague-Dawley rats (Woods and Rasch 1998) after neonatal losartan treatment.

Experimental paradigms to influence fetal programming of organ development revealed that prenatal disturbances in renal development may contribute to the development of arterial hypertension (Langley-Evans et al. 1999; Vehaskari et al. 2001; Woods et al. 2001). Both moderate global dietary restriction and dietary protein restriction in pregnant normotensive rats induce permanent arterial pressure elevations in the offspring (Langley-Evans et al. 1999; Ozaki et al. 2001; Vehaskari et al. 2001; Woods et al. 2001). Maternal protein restriction causes a decrease in glomerular number (Langley-Evans et al. 1999; Vehaskari et al. 2001; Woods et al. 2001) but no major change in glomerular filtration rate (Langley-Evans et al. 1999; Woods et al. 2001). Tissue renin mRNA expression, tissue renin concentrations and tissue angiotensin II levels are reduced in kidneys of neonatal offspring from mothers exposed to dietary protein restriction during pregnancy (Woods et al. 2001). Plasma renin activity is lower in young offspring from protein-restricted mothers while plasma aldosterone concentration is higher than in control animals (Vehaskari et al. 2001).

Neonatal and young SHR show characteristic alterations of the intrarenal components of the renin-angiotensin system. Thus, in 1-week-old SHR, angiotensin II receptor binding is elevated compared to normotensive Wistar-Kyoto rats (WKY) in all kidney regions, while radioligand binding to ACE is diminished (Correa et al. 1995). In 4- and 6-week-old SHR, basal and isoproterenol-stimulated renin release have been reported to be increased compared to normotensive WKY rats, while the ability of angiotensin II to inhibit renin release is compromised (Henrich and Levi 1991). Furthermore, angiotensin II receptor binding in renal brush border preparations and proximal tubular AT_1 receptor mRNA expression is increased in 4-week-old SHR compared to age-matched WKY (Cheng et al. 1998; Henrich and Levi 1991). Most of these strain differences disappear as the rats mature and hypertension develops in SHR (Cheng et al. 1998; Correa et al. 1995; Henrich and Levi 1991).

3
Hemodynamics

Under resting conditions, the kidneys receive approximately 20% of cardiac output. In order to maintain normal renal excretory function, renal blood flow (RBF) and glomerular filtration rate (GFR) must be kept within relatively close boundaries. This is achieved by a combination of interacting regulatory and autoregulatory mechanisms, including vascular contraction in response to increased intraarterial pressure (myogenic response) and the tubuloglomerular feedback (TGF) mechanism.

Angiotensin II has been demonstrated to contribute to the autoregulation of RBF and GFR (Sorenson et al. 2000; Wang et al. 1997). Thus, it has been shown that acute AT_1 receptor blockade reduces the ability of the kidney to autoregulate GFR in response to stepwise reductions in renal perfusion pressure in anesthetized rats (Wang et al. 1997). Similarly, acute ACE inhibition with captopril reduced the ability of the kidney to autoregulate RBF in response to stepwise reductions in renal perfusion pressure (Sorenson et al. 2000). When plasma angiotensin II concentration was maintained at a constant high level by simultaneous infusions of captopril and angiotensin II, the lower limit of RBF autoregulation was shifted to lower renal perfusion pressures, i.e., RBF autoregulation was improved (Sorenson et al. 2000).

The main renal hemodynamic effects of angiotensin II are to raise renal vascular resistance and to lower renal blood flow by its action on vascular smooth muscle AT_1 receptors (Arendshorst et al. 1999; Toke et al. 1995). Angiotensin II also causes AT_2 receptor-mediated endothelium-dependent vasodilatation in the renal preglomerular vasculature, which involves cytochrome P-450 metabolites of arachidonic acid (Arima et al. 1997). In vivo studies performed in rats demonstrated that the vasoconstrictor action of angiotensin II is mainly confined to the cortical circulation resulting in a decrease in cortical blood flow (Badzynska et al. 2002; Gross et al. 1998; Walker et al. 1999). In contrast, endogenous angiotensin II has only minor effects on renal medullary blood flow, which accounts for approximately 5% of total RBF (Gross et al. 1998). Intravenous infusions of angiotensin II result in increases in renal medullary blood flow, which are inhibited by AT_1 receptor blockade and depend on nitric oxide formation (Badzynska et al. 2002; Walker et al. 1999). Others have shown a biphasic response of medullary blood flow to intravenous angiotensin II consisting of a transient constriction followed by sustained dilatation (Sarkis et al. 2003). Both dilatation and constriction are inhibited by AT_1 but not AT_2 receptor blockade (Sarkis et al. 2003).

Renal cortical vascular resistance is regulated mainly by preglomerular resistance vessels. Glomerular capillary pressure is set by renal arterial pressure and the lumen diameter of preglomerular resistance vessels as well as efferent arterioles. Angiotensin II has been shown to constrict both renal vascular segments with similar potency (Arendshorst et al. 1999; Helou and Marchetti 1997). However, the signal transduction mechanisms in response to angiotensin II appear to be different in afferent and efferent arterioles (Fig. 1). Whereas resting membrane potentials are similar in both vessel types (around −40 mV) angiotensin II-induced vasoconstriction is associated with significant membrane depolarization only in afferent but not in efferent arterioles (Loutzenhiser et al. 1997). Furthermore, angiotensin II-induced increases in cytosolic free Ca^{2+} concentrations are greater in afferent than in efferent arterioles (Helou and Marchetti 1997). This difference is particularly pronounced in the outer renal cortex (Helou and Marchetti 1997). In afferent arterioles, the angiotensin II-induced increase in intracellular Ca^{2+} concentration is mainly due to the activation of voltage-operated Ca^{2+} channels (Helou and Marchetti 1997; Iversen and Arendshorst 1998; Loutzenhiser and Loutzenhiser 2000). The L-type Ca^{2+} channel blocker nifedipine blocks the angiotensin II-induced Ca^{2+} influx in afferent but not in efferent arterioles (Loutzenhiser and Loutzenhiser 2000; Takenaka et al. 1997). It has been suggested that the angiotensin II-induced constriction of efferent arterioles may be mainly due to Ca^{2+} release from intracellular stores (Imig et al. 2000). This hypothesis is supported by the finding that depletion of intracellular Ca^{2+} stores by pharmacological blockade of the sarcoplasmatic reticulum Ca^{2+} ATPase reduces the angiotensin II-induced vasoconstriction in these vessels (Imig et al. 2000). By releasing Ca^{2+} from intracellular stores in efferent arteri-

Fig. 1 Major steps involved in AT_1 receptor-mediated vascular smooth muscle contraction in afferent and efferent arterioles. In afferent arterioles, angiotensin II induces generation of inositol trisphosphate (*IP_3*) via G protein-mediated activation of phospholipase C (*PLC*). IP_3 induces Ca^{2+} release from the sarcoplasmic reticulum (*SR*) via its action on IP_3 receptors (*IP_3R*). Extracellular Ca^{2+} enters the afferent arteriolar smooth muscle cell via receptor-operated (*ROC*) and voltage-operated calcium channels (*VOC*). Activation of an outward Cl^- current is involved in membrane depolarization after angiotensin II bind-

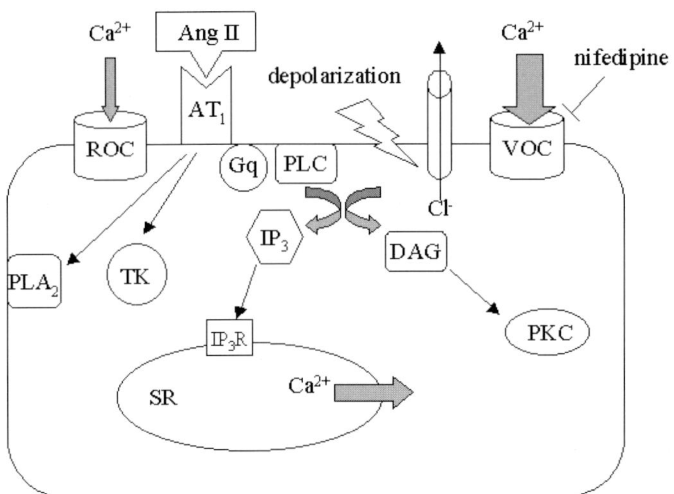

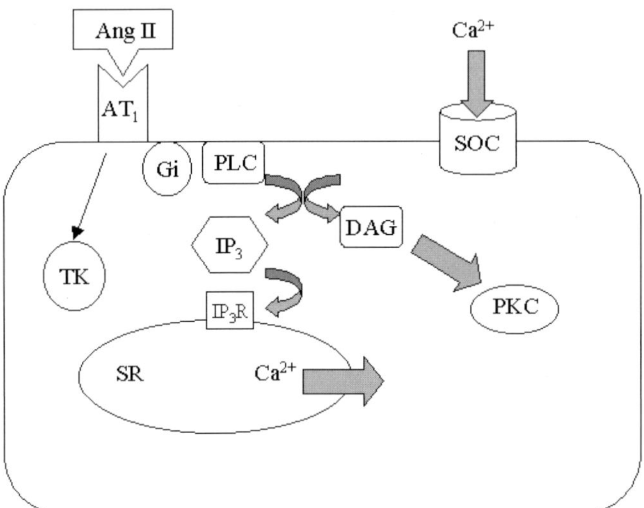

ing to its AT_1 receptor. Approximately 90% of the Ca^{2+} influx from the extracellular compartment can be inhibited by the L-type Ca^{2+} channel blocker nifedipine. In addition, activation of protein kinase C (*PKC*) by diacylglycerol (*DAG*), tyrosine kinase activity (*TK*) and phospholipase A_2 are involved in the activation of the contractile apparatus in the afferent arteriole. In efferent arteriolar smooth muscle cells, angiotensin II does not consistently cause depolarization. PLC is activated via Gi proteins causing Ca^{2+} release from the sarcoplasmic reticulum. Depletion of intracellular Ca^{2+} stores activates store-operated Ca^{2+} channels (*SOC*) within the cell membrane, causing Ca^{2+} influx from the extracellular space. This Ca^{2+} influx is not sensitive to nifedipine. Activation of PKC and TK are part of the signal transduction pathway causing activation of the contractile apparatus

oles, angiotensin II may also activate nifedipine-insensitive store-operated Ca^{2+} influx from the extracellular space (Loutzenhiser and Loutzenhiser 2000).

Activation of phospholipase C (PLC) is centrally involved in the angiotensin II-induced signal transduction pathways of both afferent and efferent arterioles (Arendshorst et al. 1999; Takenaka et al. 1997). In afferent arterioles, PLC mainly acts through the production of inositol trisphosphate (IP_3). The increased IP_3 facilitates Ca^{2+} release from intracellular stores, activates chloride channels and opens voltage-dependent Ca^{2+} channels (Nagahama et al. 2000). PLC-mediated production of diacylglycerol (DAG) and subsequent activation of protein kinase C (PKC) appears to be less important in preglomerular vessels. In contrast, PKC plays an obligatory role in angiotensin II-induced constriction of efferent arterioles where it activates Ca^{2+} channels in the cell membrane and modulates IP_3-induced Ca^{2+} release from intracellular stores (Nagahama et al. 2000).

In addition to the mechanisms mentioned above, recent evidence suggests that angiotensin II may affect several other intracellular signaling pathways in afferent and efferent arterioles (Andresen et al. 2001; Carmines et al. 2001; Croft et al. 2000; Imig and Deichmann 1997; Inoue et al. 1999). Thus it has been shown that tyrosine kinase-mediated phosphorylation contributes to the angiotensin II-induced constriction of both afferent and efferent arterioles (Carmines et al. 2001). Furthermore, it has been demonstrated that angiotensin II activates phospholipase A_2 and that inhibition of phospholipase A_2 blunts angiotensin II-induced vasoconstriction in afferent arterioles, thus linking the renovascular actions of the peptide to the metabolism of arachidonic acid (Imig and Deichmann 1997). Arachidonic acid metabolites generated by the lipoxygenase pathway augment and metabolites generated by the epoxygenase pathway attenuate preglomerular vasoconstriction in response to angiotensin II (Imig and Deichmann 1997). Studies in freshly isolated preglomerular rat vascular smooth muscle cells (VSMCs) show that angiotensin II via AT_2 receptors leads to increased production of the cytochrome P-450-derived arachidonic acid metabolite 20-hydroxyeicosatetraenoic acid (20-HETE), which acts as a vasoconstrictor (Croft et al. 2000). Experiments in cultured preglomerular VSMCs demonstrate that angiotensin II via AT_1 receptors enhances β-adrenoreceptor-induced cAMP formation (Inoue et al. 1999) and the activity of phospholipase D (Andresen et al. 2001), which may link angiotensin II signaling to adrenergic mechanisms and mitogen-activated protein kinases.

Renal medullary blood flow is autoregulated to a lesser degree than cortical blood flow (Cowley 1997; Mattson 2003). The poor autoregulation of renal medullary blood flow has been implicated in long-term arterial pressure regulation as well as pressure natriuresis and diuresis (Cowley 1997). Although in vivo studies performed in rats and mice do not provide evidence for a major role of angiotensin II as a medullary vasoconstrictor (Badzynska et al. 2002; Gross et al. 1998; Qi et al. 2002; Walker et al. 1999), in vitro experiments have shown that angiotensin II induces vasoconstriction in outer medullary descending vasa recta (Pallone 1994). This constriction is associated with depolarization of pericytes and involves a depolarizing Cl^- current, which opens voltage-dependent

Ca^{2+} channels (Zhang et al. 2001). The differential effects of angiotensin II on renal medullary vasculature in vitro and in vivo may be explained by modulation through paracrine factors released from surrounding epithelia upon angiotensin II administration in vivo. Among these factors are nitric oxide and metabolites of cyclooxygenase-2 (Pallone et al. 2003; Qi et al. 2002).

GFR depends on glomerular capillary pressure and the ultrafiltration coefficient (K_f). The latter depends on the filtration surface area and the hydraulic conductivity of the filtration barrier, both of which are affected by glomerular mesangial cells and epithelial cells (podocytes) (Douglas and Hopfer 1994; Pavenstädt 2000; Stockand and Sansom 1998). Glomerular mesangial cells surround capillary loops of the glomerulus (Stockand and Sansom 1998). They are able to contract in response to angiotensin II via an AT_1 receptor-dependent mechanism which involves an increase in cytosolic Ca^{2+} activity (Douglas and Hopfer 1994; Schlatter et al. 1995; Stockand and Sansom 1998). Angiotensin II-induced contraction of mesangial cells may influence the pressure gradient between afferent and efferent arterioles and reduces K_f (Douglas and Hopfer 1994; Stockand and Sansom 1998). Glomerular epithelial cells have been implicated in changes of glomerular filtration barrier function (Pavenstädt 2000). Both types of angiotensin II receptors (AT_1 and AT_2) have been identified in these cells (Sharma et al. 1998). Angiotensin II elicits a depolarization, which can be antagonized by AT_1 receptor blockade (Gloy et al. 1997). Furthermore, angiotensin II causes intracellular accumulation of cAMP, which can be prevented only by simultaneous blockade of AT_1 and AT_2 receptors (Sharma et al. 1998), suggesting that glomerular epithelial cells respond to angiotensin II in a manner distinct from that of mesangial cells or proximal tubular epithelial cells.

TGF couples afferent arteriolar tone and single nephron GFR to tubular reabsorption. When the amount of electrolytes that pass the macula densa increases, the afferent arteriole of the same nephron constricts and the single-nephron GFR is reduced. While the mechanisms underlying the TGF response are currently not completely understood, adenosine appears to play a major role as a mediator (Thomson et al. 2000). Various lines of evidence suggest that angiotensin II can substantially modify TGF. Thus, AT_{1A} receptor knockout mice do not show a TGF response and the response is reduced in heterozygous $AT_{1A}{}^{+/-}$ mice compared to homozygous $AT_{1A}{}^{+/+}$ controls (Schnermann et al. 1997). Similar data have been observed with $ACE^{-/-}$ knockout mice (Schnermann 1999). Pharmacological interventions with the renin-angiotensin system in rats yielded controversial results. Thus, acute ACE inhibitor treatment as well as acute and chronic AT_1 receptor blockade blunted the maximum TGF response in Sprague-Dawley rats. In contrast, the maximum TGF response was not affected by either acute AT_1 receptor blockade in WKY (Brännstrom et al. 1996) or chronic ACE inhibitor treatment in Sprague-Dawley rats (Turkstra et al. 2000). Thus, the exact role of angiotensin II in the regulation of the TGF response remains to be elucidated.

Angiotensin II-induced changes in renal hemodynamics and abnormalities in renal vascular reactions to angiotensin II play a major role in the development of several forms of arterial hypertension. Both systemic and intrarenal infusions of angiotensin II without spillover into the systemic circulation cause arterial hypertension in rats (Stevenson et al. 2000; Wang et al. 2000). Rats made hypertensive by systemic angiotensin II infusion show a decreased ability to autoregulate GFR and a slight impairment in autoregulating RBF (Wang et al. 2000). The slope of the pressure–natriuresis relationship is greatly reduced, while the autoregulation of medullary blood flow seems not to be impaired in this form of hypertension (Wang et al. 2000). Intrarenal infusion of angiotensin II is associated with a shift of the renal pressure–flow relationship to elevated perfusion pressures, suggesting a reduction of preglomerular lumen diameter (Stevenson et al. 2000). In two-kidney, one-clip renovascular hypertensive rats, RBF and GFR are reduced in the nonclipped kidney (Wang et al. 1997). The autoregulation of RBF is well preserved, whereas the ability to autoregulate GFR is reduced in these kidneys (Wang et al. 1997). AT_1 receptor blockade does not affect the ability to autoregulate RBF but it completely abolishes the ability to autoregulate GFR in the nonclipped kidney of two-kidney, one-clip hypertensive rats (Wang et al. 1997).

Renal vascular responsiveness to angiotensin II is increased in several forms of experimental hypertension, including SHR (Akishita et al. 1999; Haddad and Garcia 1996; Vyas et al. 2000), Lyon hypertensive rats (Sarkis et al. 2003), renovascular hypertensive rats (Cruz and Escalante 1999), and *ren-2* transgenic rats (Jacinto et al. 1999). In contrast to the angiotensin II-induced responses, renal vascular responsiveness to norepinephrine is normal in SHR and in *ren-2* transgenic rats (Haddad and Garcia 1996; Jacinto et al. 1999), suggesting that the enhanced responsiveness of the renal vasculature to angiotensin II in these models is not due to unspecific hypertension-induced vascular remodeling.

Several recent studies (Dukacz et al. 1999; Dukacz and Kline 1999; Endo et al. 1998; Haddad and Garcia 1996; Jackson et al.1999; Lu et al. 1994; Vyas et al. 2000) have provided evidence for an important role of angiotensin II-dependent hemodynamic mechanisms in the pathophysiology of genetic hypertension in SHR. As mentioned earlier, there is increased AT_1 receptor density in afferent arterioles, which may be involved in the enhanced responsiveness of these vessels to angiotensin II in SHR (Haddad and Garcia 1996). Furthermore, there is enhanced G_i protein-dependent signaling (Jackson et al.1999) and a down-regulation of G_s protein expression in response to angiotensin II in SHR renal resistance vessels (Vyas et al. 2000). The dose response curve of angiotensin II-induced activation of phospholipase D in SHR renal microvascular smooth muscle cells is shifted to the left (Andresen et al. 2001). AT_2 receptor-mediated dilatation of afferent arterioles has been demonstrated in normotensive WKY rats but not in SHR (Endo et al. 1998), indicating that AT_2 receptor-dependent mechanisms may contribute to the hypertensiogenic actions of angiotensin II on the kidney. Further evidence for a role of angiotensin II-dependent renal hemodynamic mechanisms comes from pharmacological experiments that interfere with the renin-angiotensin system. Thus, renal medullary ACE inhibition increases medullary blood flow and renal sodium excretion and lowers arterial pressure in SHR (Lu et al. 1994). Furthermore, systemic short-term ACE inhibition and short-term AT_1 receptor antagonism as well as long-term systemic ACE inhibition increase medullary blood flow responsiveness to changes in renal perfusion pressure, which may improve pressure natriuresis in SHR (Dukacz et al. 1999; Dukacz and Kline 1999).

In addition to renal hemodynamic mechanisms, impaired glomerular function and reduced glomerular filtration surface area have been implicated in the development of hypertension in SHR (Brenner et al. 1988). Four-to-five-week-old SHR show a threefold increase in glomerular AT_1 receptor binding compared to age-matched WKY rats (Haws et al. 1994). Although less pronounced, this strain difference in glomerular AT_1 receptor binding is also present in adult animals (Haws et al. 1994). Glomeruli from SHR of the stroke-prone substrain show increased levels of angiotensinogen, ACE and AT_{1A} as well as AT_{1B} receptor mRNA (Obata et al. 2000). Chronic ACE inhibitor treatment reduced the mRNA levels of the components of the renin-angiotensin system to those seen in normotensive rats (Obata et al. 2000). The effect of ACE inhibitor treatment on renin-angiotensin system gene expression were associated with reductions in urinary albumin excretion and glomerulosclerosis (Obata et al. 2000). On the other hand, chronic ACE inhibitor treatment did not affect glomerular capillary growth and filtration surface area in juvenile SHR (Black et al. 2001).

A further mechanism by which angiotensin II may contribute to hypertension via its renal actions may relate to the TGF response. Increased reactivity of the TGF mechanism may contribute to the compromised electrolyte and water excretion in hypertension. The reactivity of the TGF mechanism is increased in SHR and Milan hypertensive rats (Arendshorst et al. 1999) but not in hypertensive fawn-hooded rats (Verseput et al. 1998). It has been shown with several experimental modifications that blockade of AT_1 receptors reduces the maximum range and sensitivity of the TGF response in SHR (Arendshorst et al. 1999; Brännstrom et al. 1996).

4 Epithelial Transport

Angiotensin II contributes to renal tubular sodium reabsorption by its hemodynamic effects and via the release of aldosterone from the adrenal cortex. In addition, angiotensin II affects tubular epithelial electrolyte transport by direct actions on tubular epithelial cells via AT_1 receptors (Douglas and Hopfer 1994; Hiranyachattada and Harris 1996; Quan and Baum 1996) located at both the luminal and basolateral membranes (Cogan 1990; Douglas and Hopfer 1994). In contrast to AT_1 receptors in renal vasculature and glomeruli, AT_1 receptors in the proximal tubulus may be positively regulated, i.e., AT_1 receptor mRNA levels and AT_1 receptor binding increase with rising angiotensin II concentrations (Cheng et al. 1995). On the other hand, tubular AT_1 receptors are down-regulated in several forms of experimental hypertension with elevated angiotensin II levels (Wang et al. 1999).

In the proximal tubule—where about two-thirds of the filtered volume and electrolytes are reabsorbed—angiotensin II exerts a biphasic effect on epithelial sodium transport with low concentrations of the peptide increasing and high concentrations decreasing tubular sodium reabsorption (Harris and Young 1977; Hiranyachattada and Harris 1996). Thus, it has been known for more than 2 decades that peritubular angiotensin II concentrations in the picomolar range, which correspond to physiological plasma angiotensin II concentrations, stimulate sodium reabsorption, whereas peritubular angiotensin II concentrations in the nanomolar range are inhibitory (Harris and Young 1977). It should be noted, however, that angiotensin II levels in renal tissue, and in particular in renal tubular fluid, are much higher than plasma levels (Braam et al. 1993; Cervenka et al. 1999b; Imig et al. 1999; Navar et al. 2002; Quan and Baum 1996). While the dose-response curve for the effects of angiotensin II on proximal tubular sodium transport is shifted to the right when intraluminal rather than peritubular administration of the peptide is considered, physiologically occurring angiotensin II concentrations in the proximal tubular fluid may be high enough to elicit inhibitory effects. Thus, it has been reported that AT_1 receptor inhibition by losartan given intratubularly increases proximal tubular fluid absorption in anesthetized Sprague-Dawley rats (Hiranyachattada and Harris 1996), while renal interstitial infusion of losartan increases fractional sodium reabsorption (Peng and Knox 1995). The effects of angiotensin II on sodium reabsorption are most prominent in the early proximal tubule (S1 segment) and it has been estimated that about 15% of the total renal tubular sodium reabsorption may be under the direct control of the peptide (Cogan 1990).

While there is general agreement that the actions of angiotensin II on proximal tubular sodium transport are mainly, if not exclusively mediated via AT_1 type receptors, reports on the angiotensin II-induced signaling mechanisms in this section of the nephron have been variable (Fig. 2). Enhanced angiotensin II-induced proximal tubular sodium reabsorption may involve stimulation of the Na^+-H^+ exchanger in the apical membrane (Geibel et al. 1990; Houillier et al. 1996; Reilly et al. 1995), the Na^+-K^+-ATPase in the basolateral membrane (Aperia et al. 1994; Bharatula et al. 1998; Houillier et al. 1996), or the Na^+/HCO_3^- cotransporter (Geibel et al. 1990; Reilly et al. 1995), which is also located in the basolateral membrane. Angiotensin II-induced increase and decrease in Na^+-H^+ exchanger activity and in Na^+-K^+-ATPase activity have been reported with low doses of the peptide stimulating the membrane electrolyte transporters, whereas high doses tend to be inhibitory (Aperia et al. 1994; Bharatula et al. 1998; Houillier et al. 1996).

Proximal tubular cells share with glomerular mesangial cells the feature that angiotensin II profoundly inhibits adenylate cyclase, whereas activation of PLC, which is the major angiotensin II signal transduction pathway in most other target tissues of the peptide, is less pronounced (Cogan 1990; Douglas and Hopfer 1994; Schelling et al. 1994; Thekkumkara and Linas 2002). Which signaling enzymes couple angiotensin II to sodium transport in the proximal tubule is currently not clear (Rangel et al. 2002; Thekkumkara et al. 1998). Thus, inhibition

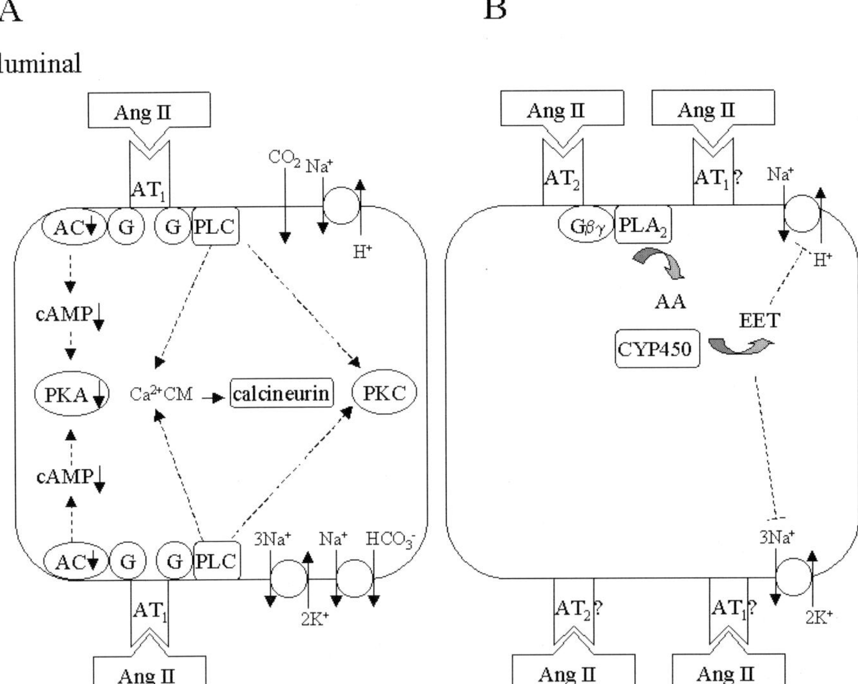

Fig. 2A, B Mechanisms involved in angiotensin II-mediated biphasic effects on proximal tubular Na^+ reabsorption. **A** Angiotensin II increases proximal tubular Na^+ reabsorption at picomolar to nanomolar concentrations. This is due to stimulation of apical Na^+-H^+-exchanger activity and basolateral Na^+-K^+-ATPase activity via AT_1 receptors. Signal transduction involves inhibition of adenylate cyclase (*AC*), activation of phospholipase C (*PLC*) and protein kinase C (*PKC*) as well as stimulation of protein phosphatase 2B (calcineurin). **B** Angiotensin II decreases proximal tubular Na^+ reabsorption at nanomolar to micromolar concentrations. Inhibition of apical tubular acidification via the Na^+-H^+ exchanger appears to be mediated via AT_2 receptors. The role of the AT_1 receptor is less clear. In several studies, the inhibition of proximal tubular sodium reabsorption could be blocked by losartan. High concentrations of angiotensin II activate phospholipase A_2 (*PLA_2*), which causes arachidonic acid (*AA*) release from membrane lipids. Epoxyeicosatetraenoic-acids (*EETs*) formed from AA by cytochrome P450 (*CYP450*) epoxygenase activity are causally linked to inhibition of apical Na^+-H^+ exchanger activity and basolateral Na^+-K^+-ATPase activity

of adenylate cyclase reduces protein kinase A activity, which—if it was involved in transcellular sodium transport—should in turn lead to increased Na^+-H^+ exchanger activity. However, blockade of PKA failed to stimulate proximal tubular Na^+-H^+ exchanger activity (Houillier et al. 1996), indicating that this enzyme may not be involved in the regulation of tubular sodium transport by angiotensin II. The situation is similarly complex with PKC, which is activated by the PLC pathway. Thus, PKC has been reported to stimulate or inhibit Na^+-H^+ exchanger activity (Houillier et al. 1996; Tse et al. 1993) and to activate or inactivate Na^+-K^+-ATPase (Bertorello 1992; Middleton et al. 1993). The stimulation of proximal tubular Na^+-K^+-ATPase activity by angiotensin II has been reported to be associated with the activation of Ca^{2+}/calmodulin-dependent protein phosphatase 2B (calcineurin) (Aperia et al. 1994).

While both the low-dose angiotensin II-induced stimulation and the high-dose angiotensin II-induced inhibition of proximal tubular sodium transport are potently inhibited by the AT_1 receptor-specific blocker losartan (Bharatula et al. 1998; Hiranyachattada and Harris 1996), recent evidence suggests that AT_2 receptors may also play a role as mediators of angiotensin II-induced inhibition of sodium transport (Haithcock et al. 1999). The signal transduction mechanisms related to angiotensin II-induced inhibition of tubular sodium reabsorption may involve cytochrome P450-dependent metabolites of arachidonic acid (Douglas and Hopfer 1994; Haithcock et al. 1999; Harwalkar et al. 1998; Houillier et al. 1996).

Further angiotensin II effects on proximal tubular transport include the activation of H^+-ATPase activity (Wagner et al. 1998) and the stimulation of proximal tubular ammonia secretion during acidosis via an AT_1 receptor-dependent mechanism (Nagami 2002). The AT_1 receptor antagonist losartan inhibits urate uptake through an AT_1 receptor-independent mechanism by interfering with an anion exchanger in the proximal tubule (Edwards et al. 1996).

In addition to having profound effects on proximal tubular fluid and electrolyte reabsorption, angiotensin II may be an important regulator of the function of the thick ascending limb of Henle (TAL). In this nephron segment, the Na^+-K^+-$2Cl^-$ cotransporter is the major apical carrier responsible for luminal uptake and thus transcellular absorption of sodium. The Na^+-K^+-$2Cl^-$ cotransporter in the TAL is inhibited by picomolar concentrations of angiotensin II, whereas it is activated by nanomolar concentrations (Amlal et al. 1998). Both the inhibitory and the stimulatory effects of angiotensin II on Na^+-K^+-$2Cl^-$ cotransporter activity are mediated through AT_1 receptors (Amlal et al. 1998). The inhibitory effect appears to be mediated by cytochrome P-450-dependent 20-HETE formation, whereas the stimulatory effect seems to be due to PKC activation (Amlal et al. 1998). In addition to regulating Na^+-K^+-$2Cl^-$ cotransporter activity, angiotensin II affects apical K^+ channel activity in the TAL (Lu et al. 1996). Again, the effect is biphasic, with low concentrations of angiotensin II inhibiting and high concentrations stimulating K^+ channel activity. As with the effects of angiotensin II on Na^+-K^+-$2Cl^-$ cotransporter activity, the basal inhibitory effect of the peptide on apical K^+ channel activity appears to be mediated by cytochrome P-450-dependent 20-HETE formation, whereas the stimulatory effect may be mediated via NO (Lu et al. 1996).

Further actions of angiotensin II on TAL tubular cells include a biphasic effect on ^{86}Rb uptake (Ferreri et al. 1998) with low doses of the peptide being stimulatory and high doses being inhibitory as well as an inhibitory effect on bicarbonate reabsorption (Good et al. 1999). The latter effect does not show a concentration-dependent biphasic behavior, which is in contrast to the effects of angiotensin II on many other transport mechanisms in the tubular system, including the effects of the peptide on bicarbonate reabsorption in the proximal tubulus (Good et al. 1999).

Further downstream the tubular system angiotensin II directly stimulates fluid as well as sodium and bicarbonate reabsorption in the distal tubule (Berreto-Chaves and Mello-Aires 1996; Levine et al. 1996; Wang and Giebisch 1996). The distal tubule is a heterogeneous structure and different transport mechanisms are present in early and late segments. In vivo microperfusion experiments revealed that angiotensin II in the picomolar range stimulates fluid, sodium and bicarbonate reabsorption in the early distal tubule (Berreto-Chaves and Mello-Aires 1996; Wang and Giebisch 1996). The effects of angiotensin II on fluid and electrolyte reabsorption have been reported to be abolished by the luminal application of an AT_1 receptor antagonist and of the specific Na^+-H^+-exchange inhibitor 5-(*N*,*N*-hexamethylene)amiloride (Berreto-Chaves and Mello-Aires 1996; Wang and Giebisch 1996). In the late distal tubule, angiotensin II also stimulates fluid and sodium reabsorption via specific AT_1 receptor-mediated mechanisms (Berreto-Chaves and Mello-Aires 1996; Levine et al. 1996; Wang and Giebisch 1996) and may (Berreto-Chaves and Mello-Aires 1996; Levine et al. 1996) or may not (Wang and Giebisch 1996) activate bicarbonate reabsorption.

The role of angiotensin II as a regulator of tubular transport systems also comprises effects on the collecting duct (Peti-Peterdi et al. 2001; Tojo et al. 1995; Weiner et al. 1995), which is lined with principal cells as well as type A and B intercalated cells. Type A intercalated cells secrete H^+ and type B intercalated cells secrete bicarbonate into the lumen; both cell types are involved in the regulation of acid base homeostasis. In the rat microdissected cortical collecting duct, angiotensin II has been shown to inhibit H^+-ATPase but not H^+-K^+-ATPase activity in a dose-dependent manner (Tojo et al. 1995). Furthermore, in microperfused rabbit, outer but not inner cortical collecting duct angiotensin II has been shown to enhance B cell apical bicarbonate secretion through basolateral AT_1 receptor stimulation (Weiner et al. 1995). In addition, recent evidence suggests that angiotensin II may directly stimulate epithelial sodium channel activity in the cortical collecting duct via AT_1 receptors (Peti-Peterdi et al. 2001).

Via its actions on renal tubular epithelial cells, angiotensin II may contribute to the pathogenesis of arterial hypertension (Cogan 1990; Navar et al. 2002). While elevated intrarenal angiotensin II concentrations have been found in several forms of experimental hypertension including two-kidney, one-clip hypertensive rats, *ren2* transgenic rats, and angiotensin II-infused rats (Navar et al. 2002), their exact contributions to increased tubular sodium and fluid reabsorption as well as to the development of hypertension in these models remain to be established. Thus, the addition of the AT_1 receptor blocker losartan to the intratubular fluid has been reported to increase transepithelial fluid intake in the proximal tubulus of Sprague-Dawley rats (Hiranyachattada and Harris 1996). On the other hand, renal interstitial infusion of losartan has been found to increase fractional sodium excretion in the same rat strain (Peng and Knox 1995). The effects of AT_1 receptor blockade on renal sodium excretion have also been investigated in the nonclipped kidney of two-kidney, one-clip hypertensive rats

(Cervenka et al. 1999b). Although these kidneys undergo marked depletion of renin content and renin mRNA, intrarenal angiotensin II levels are not suppressed and are much higher than can be explained on the basis of plasma levels. Direct infusion of the AT_1 receptor antagonist candesartan into the renal artery of nonclipped kidneys increased fractional sodium excretion independently of its hemodynamic effects (Cervenka et al. 1999b). These data suggest that locally formed intrarenal angiotensin II in the nonclipped kidney of two-kidney, one-clip hypertensive rats may contribute to arterial hypertension in this model by enhancing tubular sodium reabsorption.

The effects of specific AT_1 receptor blockade on tubular sodium reabsorption occur despite significant down-regulation of tubular AT_1 receptors (AT_{1A} subtype), which has been reported in two-kidney, one-clip hypertensive rats and other experimental models of hypertension with high angiotensin II levels, including renal wrap hypertension and chronic intravenous angiotensin II infusion (Wang et al. 1999). Interestingly, the AT_{1A} receptor was down-regulated in both the nonischemic and the ischemic kidney of two-kidney, one-clip and renal wrap hypertensive rats, whereas the AT_2 receptor was down-regulated only in the ischemic kidney (Wang et al. 1999), opening up the possibility that differential regulation of angiotensin receptor subtypes may play a role in the pathophysiology of renal hypertension.

SHR show multiple abnormalities in tubular transport mechanisms (Aldred et al. 2000; Parenti et al. 2000; Thomas et al. 1990), which may be subject to regulation by angiotensin II (Cheng et al. 1998; Parenti et al. 2000; Thomas et al. 1990) and which may contribute to hypertension in this strain (Wood et al. 1993). Thus, selective intrarenal infusion of the AT_1 receptor blocker valsartan via the suprarenal artery reduced arterial pressure in SHR at doses that left blood pressure unaltered when infused systemically (Wood et al. 1993). While this observation strongly suggests that intrarenal AT_1 receptor-mediated effects contribute to the pathophysiology of hypertension in SHR, the underlying mechanisms are currently unclear.

At 4 weeks of age, i.e., during the developmental phase of hypertension, SHR show increased AT_1 receptor mRNA levels and increased AT_1 receptor-specific immunoreactive staining in proximal tubular cells compared to age-matched WKY rats (Cheng et al. 1998). In adult SHR, these differences were no longer detectable (Cheng et al. 1998). Furthermore, 5-week-old SHR showed increased proximal tubular Na^+-H^+ exchanger and H^+-ATPase activities compared with normotensive Donryu rats (Aldred et al. 2000). Although angiotensin II activates mitogen-activated protein kinases and the latter have been shown to regulate Na^+-H^+ exchanger activity, recent evidence suggests that this regulatory pathway is unlikely to account for the increased activity of the Na^+-H^+ exchanger in the proximal tubular epithelium of SHR (Parenti et al. 2000). Nevertheless, the above findings support the hypothesis that angiotensin II may act on proximal tubular AT_1 receptors to stimulate apical Na^+-H^+ exchanger activity, thereby causing increased tubular sodium reabsorption in young SHR during the developmental phase of hypertension. On the other hand, 5- and 12-week-old SHR had similar basal proximal tubular fluid reabsorption rates as age-matched WKY rats (Thomas et al. 1990). Angiotensin II added to the peritubular fluid in picomolar concentrations stimulated fluid reabsorption. In 5-week-old SHR, the dose-response curve for this effect was shifted to the right (Thomas et al. 1990), indicating that angiotensin II stimulates proximal tubular fluid reabsorption to a lesser extent in young SHR than in WKY.

5 Inflammation

Hypertension causes renal damage, which contributes to a further arterial pressure rise and may ultimately lead to renal failure. The progression of renal damage is associated with inflammatory reactions within the kidney. These inflammatory reactions are mediated to a certain extent by angiotensin II (Zatz and Fujihara 2002). In addition to pressure-induced damage to the renal vasculature and glomeruli, there are two major pathways by which angiotensin II can enhance renal inflammatory responses. First, mononuclear cells and macrophages possess all major components of the renin-angiotensin system (Okamura et al. 1999) and stimulation of this system may activate regulatory and effector mechanisms of the immune system such as cytokines and oxygen radicals. Second, angiotensin II may induce the expression of transcription factors, adhesion molecules and cytokines in various renal cell types, which in turn may cause or facilitate leukocyte and mononuclear infiltration into the kidney (Luft et al. 1999; Zatz and Fujihara 2002).

The link between angiotensin II and immunological processes within the kidney is illustrated by several observations (Kang et al. 2002; Luft et al. 1999; Mervaala et al. 2000; Ritz et al. 2000; Rodriguez-Iturbe et al. 2001a; Ruiz-Ortega et al. 2001). ACE inhibitors and angiotensin II receptor blockers have anti-inflammatory effects within the damaged kidney (Luft et al. 1999; Ritz et al. 2000; Ruiz-Ortega et al. 2001). Immunosuppressive treatment can lower arterial pressure and reduce renal inflammation in double transgenic rats with human renin and angiotensinogen transgenes (Mervaala et al. 2000) and in angiotensin II-infused rats (Rodriguez-Iturbe et al. 2001a). Angiotensinogen-deficient mice expressing a rat angiotensinogen transgene in the liver and the brain but not in the kidney showed less renal damage than animals with intact renal angiotensinogen gene expression, despite similar blood pressures (Kang et al. 2002).

When infused intravenously, angiotensin II up-regulated the transcription factors activator protein-1 (AP-1) and nuclear factor κB (NF-κB) in glomerular mesangial and epithelial cells as well as in tubular epithelial and mononuclear infiltrating cells (Ruiz-Ortega et al. 2001). It also induced the expression of AP-1 and NF-κB in cultured tubular epithelial cells (Ruiz-Ortega et al. 2001). Both types of angiotensin II receptors were involved in these responses. Thus, AT_1 receptor blockade diminished NF-κB and abolished AP-1 activity in glomerular and tubular cells, whereas AT_2 receptor blockade diminished mononuclear cell infiltration and NF-κB activity without any effect on AP-1 (Ruiz-Ortega et al.

2001). Angiotensin II has also been demonstrated to induce the expression of the chemotactic cytokine RANTES in rat glomeruli in vivo and in cultured rat glomerular epithelial cells (Wolf et al. 1997b). The effect appears to be mediated via AT_2 receptors (Wolf et al. 1997b).

Increased cytokine expression has also been observed in hypertensive animal models with high angiotensin II levels (Hilgers et al. 2001; Mervaala et al. 2001). Thus, the expression of macrophage chemotactic protein-1 (MCP-1) has been reported to be elevated in the nonclipped kidney of two-kidney, one-clip hypertensive rats (Hilgers et al. 2001). Renal MCP-1 expression was substantially reduced by low-dose losartan treatment, which did not affect systemic blood pressure (Hilgers et al. 2001). Furthermore, interleukin-6 expression is increased in kidneys of double transgenic rats harboring the genes for human renin and angiotensinogen (Mervaala et al. 2001).

6 Nitric Oxide and Oxygen Radicals

Nitric oxide (NO) and oxygen radicals modulate vascular tone as well as epithelial transport and are involved in inflammatory processes. NO and oxygen radicals are closely related with regard to their generation and may have mutually antagonistic and synergistic actions (Rodriguez-Iturbe et al. 2001a; Schnackenberg 2002).

NO and cGMP antagonize many physiological effects of angiotensin II such as regulation of blood pressure and sodium transport in the kidney. It has been shown that angiotensin II activates the NO/cGMP signaling pathway in several tissues, including renal interstitium (Siragy and Carey 1997), isolated renal resistance arteries (Thorup et al. 1999), and isolated renal proximal tubules (Zhang and Mayeux 1998). Which angiotensin II receptor type is involved remains controversial, as the effect has been reported to be inhibited by either specific AT_1 (Thorup et al. 1999; Zhang and Mayeux 1998) or AT_2 receptor antagonists (Gohlke et al. 1998; Siragy and Carey 1997). As revealed by renal microdialysis experiments in conscious sodium-depleted rats, high levels of endogenously formed angiotensin II may exert a tonic stimulatory effect on renal interstitial fluid cGMP levels (Siragy and Carey 1997). In contrast to the results mentioned above, angiotensin II failed to stimulate NO production in cultured mouse proximal epithelial cells, but inhibited the stimulatory effect of the combined administration of lipopolysaccharide and γ-interferon on NO synthesis through an AT_1 receptor-dependent mechanism (Wolf et al. 1997a).

The functional consequences of angiotensin II-induced stimulation of cGMP synthesis may involve inhibition of Na^+-K^+-ATPase activity. cGMP has been shown to inhibit renal Na^+-K^+-ATPase activity (McKee et al. 1994). Angiotensin II stimulates renal Na^+-K^+-ATPase activity at low concentrations, but the effect is lost at high concentrations. In isolated rat proximal tubulus inhibition of NO synthase or of guanylyl cyclase both unmasked the stimulatory effect of angiotensin II on Na^+-K^+-ATPase activity when high concentrations of the peptide were administered (Zhang and Mayeux 2001). These data indicate that the failure of high concentrations of angiotensin II to stimulate Na^+-K^+-ATPase activity may be due to the simultaneous activation of the NO/cGMP signaling pathway, which serves as a negative regulatory component.

In addition to the stimulation of NO by angiotensin II, there are several known functional interactions between the two compounds. Thus, inhibition of NO synthesis leads to glomerular arteriolar vasoconstriction, decreased glomerular ultrafiltration, reduced single nephron GFR and decreased tubular fluid reabsorption (Baylis and Qiu 1996; De Nicola et al. 1992). All of these effects have been reported to be substantially attenuated or even prevented by specific AT_1 receptor blockade (Baylis and Qiu 1996; De Nicola et al. 1992). In contrast, the renal vasoconstrictor actions of angiotensin II have been reported to be enhanced by NO synthase inhibition (Baylis et al. 1994; Sanchez-Mendoza et al. 1998).

Reactive oxygen species such as superoxide anions, hydroxyl radicals and hydrogen peroxide are formed during redox reactions in mitochondria, cytosol and in close proximity to the plasma membrane. The vascular synthesis of reactive oxygen species is increased by angiotensin II via AT_1 receptor-mediated activation of NAPDH oxidase (Schnackenberg 2002; Wang et al. 2001) and superoxide dismutase (Fukai et al. 1999). In the kidney, superoxide anions act as vasoconstrictors (Schnackenberg 2002). The renal vasoconstrictor effects and the hypertensive effects of oxygen radicals (especially superoxide anions) may in part be due to antagonism of the NO system, i.e., through formation of peroxynitrite ($ONOO^-$) (Schnackenberg 2002). Because of their close functional relationship, reactive oxygen species and NO are discussed together with regard to their role in renal hypertensive actions of angiotensin II.

Peroxynitrite is formed by NO and molecular oxygen or superoxide and reacts with tyrosine to form nitrotyrosine. The latter can be detected by specific immunohistochemical methods. To the extent that nitrotyrosine formation is due to enhanced NO synthesis rather than unspecific oxidative stress, immunohistochemically detected nitrotyrosine can be used as a footprint for locally released NO. With this method, enhanced signals have been found in the extraglomerular mesangium of the ischemic kidney as well as in cortical arteries and arterioles of the nonischemic kidney of two-kidney, one-clip hypertensive rats (Bosse and Bachmann 1997). On the other hand, experiments in rats with aortic coarctation between both renal arteries provided evidence that NO formation and functional antagonism of the NO system to the vasoconstrictor effects of angiotensin II are greatly reduced in the kidney lying above the coarctation (Sanchez-Mendoza et al. 1998). The exact role of the renal NO system in these experimental models of hypertension remains to be elucidated.

Experimental hypertension caused by chronic angiotensin II infusion is mediated at least in part by an increased formation of oxygen radicals and a reduced availability of NO (Nishiyama et al. 2001; Ortiz et al. 2001). It has been shown that vitamin E and the superoxide dismutase mimetic 4-hydroxy-2,2,6,6-tetramethyl-1-piperidinoxyl (tempol) attenuate angiotensin II-induced hypertension (Nishiyama et al. 2001; Ortiz et al. 2001). The antioxidant treatment was associated with increased urinary nitrite/nitrate excretion (Ortiz et al. 2001) and NO synthase inhibition greatly blunted the blood pressure-lowering effect of tempol (Nishiyama et al. 2001), indicating that the beneficial effects of the antioxidants was at least partly due to increased availability of NO. An important role for NO in angiotensin II-induced hypertension is further supported by the finding that the ability of neuronal nitric oxide synthase-derived NO to counteract the afferent arteriolar response to increased distal tubular flow (i.e., the TGF response) is reduced in this experimental model of hypertension (Ichihara et al. 1999).

Hypertensive double transgenic rats carrying the human renin and angiotensinogen genes have elevated tissue angiotensin II levels (Luft et al. 1999) and the pathogenesis of this form of hypertension may be similar to that of chronic low-dose angiotensin II-infusion hypertension. Hypertension in double transgenic rats is associated with impaired endothelium-dependent vasodilatation of renal arterial rings and decreased urinary nitrite/nitrate excretion. Furthermore, the activity of renal xanthine oxidoreductase, a hypoxia-inducible enzyme capable of generating reactive oxygen species, is elevated (Mervaala et al. 2001). All of these alterations are relieved by AT_1 receptor blockade (Mervaala et al. 2001). These data suggest that angiotensin II causes endothelial dysfunction associated with increased oxidative stress and an impaired availability of NO.

Interactions between angiotensin II and the renal formation of reactive oxygen species and NO may also be involved in the development of genetic hypertension in SHR (Dukacz et al. 2001; Feng et al. 2001; Schnackenberg et al. 1998; Szentivanyi et al. 1999; Vaziri et al. 2000; Zou et al. 1998). Thus, the oxygen scavenger tempol reduces arterial pressure and renal vascular resistance in SHR (Feng et al. 2001; Schnackenberg et al. 1998;), whereas selective impairment of the NO system in the renal medulla enhances the hypertensive action of angiotensin II (Szentivanyi et al. 1999). While the vasoconstrictor activity of angiotensin II in the renal medulla is normally buffered by the NO system (Zou et al. 1998), there appears to be an impairment of the NO system in the kidney of SHR. Thus, intravenous infusions of angiotensin II have been shown to dose-dependently reduce renal medullary blood flow in SHR but not in WKY rats (see above) (Dukacz et al. 2001). Infusion of the NO synthase inhibitor L-NAME into the renal medulla unmasked angiotensin II sensitivity in WKY rats, while the NO precursor L-arginine given into the renal medulla abolished the responses to angiotensin II in SHR (Dukacz et al. 2001). The sensitivity of medullary blood flow to angiotensin II in SHR was also reduced by tempol (Feng et al. 2001). Together these results indicate that medullary blood flow in SHR is sensitive to angiotensin II and suggest that this effect may be due to an impaired counter-regulatory effect of NO.

The TGF response to increased distal tubular flow is enhanced in SHR and can be normalized by AT_1 receptor inhibition (Arendshorst et al. 1999). The mechanisms underlying this effect are currently not well understood but may involve a beneficial effect of AT_1 receptor inhibition on oxygen radical formation. NO buffers and inhibition of NO synthase in the macula densa enhances the TGF response. There is evidence that the production of oxygen radicals in the juxtaglomerular apparatus opposes the buffering effect of NO on the TGF response (Wilcox and Welch 2000) and that this effect is exaggerated in SHR (Welch et al. 2000). Perfusion of the efferent arteriole with tempol has been shown to increase the enhancing effect of NO synthase inhibition on the TGF response in SHR but not in WKY rats (Welch and Wilcox 2001), implying that the dampening effects of NO on the TGF response had been prevented in SHR because of excessive oxygen radical production. This effect of tempol was not seen in SHR treated with an AT_1 receptor inhibitor in which the effect of NO synthase inhibition on the TGF response was normal, presumably due to the suppression of oxygen radical formation by AT_1 receptor inhibition (Welch and Wilcox 2001). Together these data suggest that an impaired functional antagonism between angiotensin II and the NO system due to increased oxygen radical formation may be a renal mechanism that contributes to primary hypertension in SHR.

7 Summary and Conclusions

Angiotensin II has profound actions on renal development and function. The renal effects of angiotensin II are mediated by both AT_1 and AT_2 receptors and are involved in the pathophysiology of several forms of hypertension. Renal interstitial and tubular fluid angiotensin II concentrations are much higher than can be explained on the basis of plasma levels, indicating that the peptide is generated locally. During renal development, AT_1 and AT_2 receptor-mediated mechanisms contribute to an appropriate balance between cellular proliferation and growth as well as apoptotic cell death and are therefore part of the mechanisms ensuring proper nephrogenesis.

The main renal hemodynamic effects of angiotensin II are to raise renal vascular resistance and to lower renal blood flow via AT_1 receptor-dependent mechanisms. Furthermore, angiotensin II enhances the tubuloglomerular feedback response, which in turn reduces renal blood flow and glomerular filtration rate. The hemodynamic responses to angiotensin II are increased in several forms of experimental hypertension.

In addition to its renal hemodynamic actions, angiotensin II profoundly affects tubular epithelial transport. Many of the effects of angiotensin II on tubular epithelial transport systems show a biphasic dose-response relationship with low concentrations of angiotensin II increasing and high concentrations of the peptide inhibiting tubular reabsorption.

Beyond the effects of angiotensin II on renal vasculature and the tubular system, recent evidence suggests that angiotensin II may interfere with inflammatory processes and oxidative stress mechanisms in the kidney, thereby contributing significantly to hypertensive renal end-organ damage.

References

Aizawa T, Ishizaka N, Kurokawa K et al (2001) Different effects of angiotensin II and catecholamine on renal cell apoptosis and proliferation in rats. Kidney Int 59:645–653

Akishita M, Yamada H, Dzau VJ, Horiuchi M (1999) Increased vasoconstrictor response of the mouse lacking angiotensin II type 2 receptor. Biochem Biophys Res Commun 261:345–349

Aldred KL, Harris PJ, Eitle E (2000) Increased proximal tubule NHE-3 and H^+-ATPase activities in spontaneously hypertensive rats. J Hypertens 18:623–628

Amlal H, LeGoff C, Vernimmen C, Soleimani M, Paillard M, Bichara M (1998) ANG II controls Na^+-K^+NH4^+-$2Cl^-$ cotransport via 20-HETE and PKC in medullary thick ascending limb. Am J Physiol 274:C1047–C1056

Andresen BT, Jackson EK, Romero GG (2001) Angiotensin II signaling to phospholipase D in renal microvascular smooth muscle cells in SHR. Hypertension 37:635–639

Aperia A, Holtback U, Syren ML, Svensson LB, Fryckstedt J, Greengard P (1994) Activation/deactivation of renal Na^+ K^+-ATPase: a final common pathway for regulation of natriuresis. FASEB J 8:436–439

Arendshorst WJ, Brännstrom K, Ruan X (1999) Actions of angiotensin II on the renal microvasculature. J Am Soc Nephrol 10 [Suppl 11]:S149–S161

Arima S, Endo Y, Yaoita H et al (1997) Possible role of P-450 metabolite of arachidonic acid in vasodilator mechanism of angiotensin II type 2 receptor in the isolated microperfused rabbit afferent arteriole. J Clin Invest 100:2816–2823

Badzynska B, Grzelec-Mojzesowicz M, Dobrowolski L, Sadowski J (2002) Differential effect of angiotensin II on blood circulation in the renal medulla and cortex of anaesthetised rats. J Physiol 538:159–166

Barreto-Chaves ML, Mello-Aires M (1996) Effect of luminal angiotensin II and ANP on early and late cortical distal tubule HCO_3^- reabsorption. Am J Physiol 271:F977–F984

Baylis C, Qiu C (1996) Importance of nitric oxide in the control of renal hemodynamics. Kidney Int 49:1727–1731

Baylis C, Harvey J, Engels K (1994) Acute nitric oxide blockade amplifies the renal vasoconstrictor actions of angiotensin II. J Am Soc Nephrol 5:211–214

Bertorello AM (1992) Diacylglycerol activation of protein kinase C results in a dual effect on Na^+ K^+-ATPase activity from intact renal proximal tubule cells. J Cell Sci 101:343–347

Bharatula M, Hussain T, Lokhandwala MF (1998) Angiotensin II AT_1 receptor/signaling mechanisms in the biphasic effect of the peptide on proximal tubular Na^+ K^+-ATPase. Clin Exp Hypertens 20:465–480

Black MJ, Briscoe TA, Dunstan HJ, Bertram JF, Johnston CI (2001) Effect of angiotensin-converting enzyme inhibition on renal filtration surface area in hypertensive rats. Kidney Int 60:1837–1843

Bosse HM, Bachmann S (1997) Immunohistochemically detected protein nitration indicates sites of renal nitric oxide release in Goldblatt hypertension. Hypertension 30:948–952

Bouby N, Hus-Citharel A, Marchetti J, Bankir L, Corvol P, Llorens-Cortes C (1997) Expression of type 1 angiotensin II receptor subtypes and angiotensin II-induced calcium mobilization along the rat nephron. J Am Soc Nephrol 8:1658–1667

Braam B, Mitchell KD, Fox J, Navar LG (1993) Proximal tubular secretion of angiotensin II in rats. Am J Physiol 264:F891–F898

Brännstrom K, Morsing P, Arendshorst WJ (1996) Exaggerated tubuloglomerular feedback activity in genetic hypertension is mediated by ANG II and AT_1 receptors. Am J Physiol 270:F749–F755

Brenner BM, Garcia DL, Anderson S (1988) Glomeruli and blood pressure. Less of one, more the other? Am J Hypertens 1:335–347

Carmines PK, Fallet RW, Che Q, Fujiwara K (2001) Tyrosine kinase involvement in renal arteriolar constrictor responses to angiotensin II. Hypertension 37:569–573

Cervenka L, Mitchell KD, Oliverio MI, Coffman TM, Navar LG (1999a) Renal function in the AT_{1A} receptor knockout mouse during normal and volume-expanded conditions. Kidney Int 56:1855–1862

Cervenka L, Wang CT, Mitchell KD, Navar LG (1999b) Proximal tubular angiotensin II levels and renal functional responses to AT_1 receptor blockade in nonclipped kidneys of Goldblatt hypertensive rats. Hypertension 33:102–107

Chen X, Li W, Yoshida H et al (1997) Targeting deletion of angiotensin type 1B receptor gene in the mouse. Am J Physiol 272:F299–F304

Cheng HF, Becker BN, Burns KD, Harris RC (1995) Angiotensin II upregulates type 1 angiotensin II receptors in renal proximal tubule. J Clin Invest 95:2012–2019

Cheng HF, Wang JL, Vinson GP, Harris RC (1998) Young SHR express increased type 1 angiotensin II receptors in renal proximal tubule. Am J Physiol 274:F10–F17

Cogan MG (1990) Angiotensin II: a powerful controller of sodium transport in the early proximal tubule. Hypertension 15:451–458

Correa FM, Viswanathan M, Ciuffo GM, Tsutsumi K, Saavedra JM (1995) Kidney angiotensin II receptors and converting enzyme in neonatal and adult Wistar-Kyoto and spontaneously hypertensive rats. Peptides 16:19–24

Cowley AW (1997) Role of the renal medulla in volume and arterial pressure regulation. Am J Physiol 273:R1–R15

Croft KD, McGiff JC, Sanchez-Mendoza A, Carroll MA (2000) Angiotensin II releases 20-HETE from rat renal microvessels. Am J Physiol 279:F544–F551

Cruz BV, Escalante B (1999) Renal vascular interaction of angiotensin II and prostaglandins in renovascular hypertension. J Cardiovasc Pharmacol 34:21–27

De Nicola L, Blantz RC, Gabbai FB (1992) Nitric oxide and angiotensin II. Glomerular and tubular interaction in the rat. J Clin Invest 89:1248–1256

Douglas JG, Hopfer U (1994) Novel aspect of angiotensin receptors and signal transduction in the kidney. Annu Rev Physiol 56:649–669

Dukacz SA, Kline RL (1999) Differing effects of enalapril and losartan on renal medullary blood flow and renal interstitial hydrostatic pressure in spontaneously hypertensive rats. J Hypertens 17:1345–1352

Dukacz SA, Adams MA, Kline RL (1999) Short- and long-term enalapril affect renal medullary hemodynamics in the spontaneously hypertensive rat. Am J Physiol 276:R10–R16

Dukacz SA, Feng MG, Yang LF, Lee RM, Kline RL (2001) Abnormal renal medullary response to angiotensin II in SHR is corrected by long-term enalapril treatment. Am J Physiol 280:R1076–R1084

Edwards RM, Trizna W, Stack EJ, Weinstock J (1996) Interaction of nonpeptide angiotensin II receptor antagonists with the urate transporter in rat renal brush-border membranes. J Pharmacol Exp Ther 276:125–129

Endo Y, Arima S, Yaoita H, Tsunoda K, Omata K, Ito S (1998) Vasodilation mediated by angiotensin II type 2 receptor is impaired in afferent arterioles of young spontaneously hypertensive rats. J Vasc Res 35:421–427

Esther C-RJ, Howard TE, Marino EM, Goddard JM, Capecchi MR, Bernstein KE (1996) Mice lacking angiotensin-converting enzyme have low blood pressure, renal pathology, and reduced male fertility. Lab Invest 74:953–965

Feng MG, Dukacz SA, Kline RL (2001) Selective effect of tempol on renal medullary hemodynamics in spontaneously hypertensive rats. Am J Physiol 281:R1420–R1425

Ferreri NR, Escalante BA, Zhao Y, An SJ, McGiff JC (1998) Angiotensin II induces TNF production by the thick ascending limb: functional implications. Am J Physiol 274:F148–F155

Fukai T, Siegfried MR, Ushio-Fukai M, Griendling KK, Harrison DG (1999) Modulation of extracellular superoxide dismutase expression by angiotensin II and hypertension. Circ Res 85:23–28

Geibel J, Giebisch G, Boron WF (1990) Angiotensin II stimulates both Na^+-H^+ exchange and Na^+/HCO_3^- cotransport in the rabbit proximal tubule. Proc Natl Acad Sci USA 87:7917–7920

Gloy J, Henger A, Fischer KG et al (1997) Angiotensin II depolarizes podocytes in the intact glomerulus of the rat. J Clin Invest 99:2772–2781

Gohlke P, Pees C, Unger T (1998) AT_2 receptor stimulation increases aortic cyclic GMP in SHRSP by a kinin-dependent mechanism. Hypertension 31:349–355

Good DW, George T, Wang DH (1999) Angiotensin II inhibits HCO_3^- absorption via a cytochrome P-450-dependent pathway in MTAL. Am J Physiol 276:F726–F736

Grisk O, Rettig R (2001) Renal transplantation studies in genetic hypertension. News Physiol Sci 16:262–265

Gross V, Kurth TM, Skelton MM, Mattson DL, Cowley AW (1998) Effects of daily sodium intake and ANG II on cortical and medullary renal blood flow in conscious rats. Am J Physiol 274:R1317–R1323

Gross V, Milia AF, Plehm R, Inagami T, Luft FC (2000) Long-term blood pressure telemetry in AT_2 receptor-disrupted mice. J Hypertens 18:955–961

Guron G, Friberg P (2000) An intact renin-angiotensin system is a prerequisite for normal renal development. J Hypertens 18:123–137

Guron G, Adams MA, Sundelin B, Friberg P (1997) Neonatal angiotensin-converting enzyme inhibition in the rat induces persistent abnormalities in renal function and histology. Hypertension 29:91–97

Guyton AC (1991) Blood pressure control—special role of the kidneys and body fluids. Science 252:1813–1816

Haddad G, Garcia R (1996) Characterization and hemodynamic implications of renal vascular angiotensin II receptors in SHR. J Mol Cell Cardiol 28:351–361

Haithcock D, Jiao H, Cui XL, Hopfer U, Douglas JG (1999) Renal proximal tubular AT_2 receptor: signaling and transport. J Am Soc Nephrol 10 [Suppl 11]:S69–S74

Harrap SB, Van der Merwe WM, Griffin SA, Macpherson F, Lever AF (1990) Brief angiotensin converting enzyme inhibitor treatment in young spontaneously hypertensive rats reduces blood pressure long-term. Hypertension 16:603–614

Harris PJ, Young JA (1977) Dose-dependent stimulation and inhibition of proximal tubular sodium reabsorption by angiotensin II in the rat kidney. Pflugers Arch 367:295–297

Harrison-Bernard LM, Navar LG, Ho MM, Vinson GP, el Dahr SS (1997) Immunohistochemical localization of ANG II AT_1 receptor in adult rat kidney using a monoclonal antibody. Am J Physiol 273:F170–F177

Harwalkar S, Chang CH, Dulin NO, Douglas JG (1998) Role of phospholipase A2 isozymes in agonist-mediated signaling in proximal tubular epithelium. Hypertension 31:809–814

Haws RM, Shaul PW, Arant B-SJ, Atiyeh BA, Seikaly MG (1994) Glomerular losartan (DuP 753)-sensitive angiotensin II receptor density is increased in young spontaneously hypertensive rats. Pediatr Res 35:671–676

Helou CM, Marchetti J (1997) Morphological heterogeneity of renal glomerular arterioles and distinct $[Ca^{2+}]i$ responses to ANG II. Am J Physiol 273:F84–F96

Henrich WL, Levi M (1991) Ontogeny of renal renin release in spontaneously hypertensive rat and Wistar-Kyoto rat. Am J Physiol 260:F530–F535

Hilgers KF, Hartner A, Porst M, Veelken R, Mann JF (2001) Angiotensin II type 1 receptor blockade prevents lethal malignant hypertension: relation to kidney inflammation. Circulation 104:1436–1440

Hiranyachattada S, Harris PJ (1996) Modulation by locally produced luminal angiotensin II of proximal tubular sodium reabsorption via an AT_1 receptor. Br J Pharmacol 119:617–618

Houillier P, Chambrey R, Achard JM, Froissart M, Poggioli J, Paillard M (1996) Signaling pathways in the biphasic effect of angiotensin II on apical Na/H antiport activity in proximal tubule. Kidney Int 50:1496–1505

Ichihara A, Imig JD, Navar LG (1999) Neuronal nitric oxide synthase-dependent afferent arteriolar function in angiotensin II-induced hypertension. Hypertension 33:462–466

Imig JD, Deichmann PC (1997) Afferent arteriolar responses to ANG II involve activation of PLA_2 and modulation by lipoxygenase and P-450 pathways. Am J Physiol 273:F274–F282

Imig JD, Navar GL, Zou LX et al (1999) Renal endosomes contain angiotensin peptides, converting enzyme, and AT_{1A} receptors. Am J Physiol 277:F303–F311

Imig JD, Cook AK, Inscho EW (2000) Postglomerular vasoconstriction to angiotensin II and norepinephrine depends on intracellular calcium release. Gen Pharmacol 34:409–415

Inoue T, Mi Z, Gillespie DG, Dubey RK, Jackson EK (1999) Angiotensin receptor subtype 1 mediates angiotensin II enhancement of isoproterenol-induced cyclic AMP production in preglomerular microvascular smooth muscle cells. J Pharmacol Exp Ther 288:1229–1234

Iversen BM, Arendshorst WJ (1998) ANG II and vasopressin stimulate calcium entry in dispersed smooth muscle cells of preglomerular arterioles. Am J Physiol 274:F498–F508

Jacinto SM, Mullins JJ, Mitchell KD (1999) Enhanced renal vascular responsiveness to angiotensin II in hypertensive ren-2 transgenic rats. Am J Physiol 276:F315–F322

Jackson EK, Herzer WA, Vyas SJ, Kost CK (1999) Angiotensin II-induced renal vasoconstriction in genetic hypertension. J Pharmacol Exp Ther 291:329–334

Kang N, Walther T, Tian XL et al (2002) Reduced hypertension-induced end-organ damage in mice lacking cardiac and renal angiotensinogen synthesis. J Mol Med 80:359–366

Langley-Evans SC, Sherman RC, Welham SJ, Nwagwu MO, Gardner DS, Jackson AA (1999) Intrauterine programming of hypertension: the role of the renin-angiotensin system. Biochem Soc Trans 27:88–93

Levine DZ, Iacovitti M, Buckman S, Burns KD (1996) Role of angiotensin II in dietary modulation of rat late distal tubule bicarbonate flux in vivo. J Clin Invest 97:120–125

Loutzenhiser K, Loutzenhiser R (2000) Angiotensin II-induced Ca^{2+} influx in renal afferent and efferent arterioles: differing roles of voltage-gated and store-operated Ca^{2+} entry. Circ Res 87:551–557

Loutzenhiser R, Chilton L, Trottier G (1997) Membrane potential measurements in renal afferent and efferent arterioles: actions of angiotensin II. Am J Physiol 273:F307–F314

Lu M, Zhu Y, Balazy M, Reddy KM, Falck JR, Wang W (1996) Effect of angiotensin II on the apical K^+ channel in the thick ascending limb of the rat kidney. J Gen Physiol 108:537–547

Lu S, Mattson DL, Cowley AW (1994) Renal medullary captopril delivery lowers blood pressure in spontaneously hypertensive rats. Hypertension 23:337–345

Luft FC, Mervaala E, Muller DN et al (1999) Hypertension-induced end-organ damage: a new transgenic approach to an old problem. Hypertension 33:212–218

Maric C, Aldred GP, Harris PJ, Alcorn D (1998) Angiotensin II inhibits growth of cultured embryonic renomedullary interstitial cells through the AT_2 receptor. Kidney Int 53:92–99

Mattson DL (2003) Importance of the renal medullary circulation in the control of sodium excretion and blood pressure. Am J Physiol 284:R13–R27

McCausland JE, Bertram JF, Ryan GB, Alcorn D (1997) Glomerular number and size following chronic angiotensin II blockade in the postnatal rat. Exp Nephrol 5:201–209
McKee M, Scavone C, Nathanson JA (1994) Nitric oxide, cGMP, and hormone regulation of active sodium transport. Proc Natl Acad Sci USA 91:12056–12060
Mentzel S, Van Son JP, De Jong AS et al (1997) Mouse glomerular epithelial cells in culture with features of podocytes in vivo express aminopeptidase A and angiotensinogen but not other components of the renin-angiotensin system. J Am Soc Nephrol 8:706–719
Mervaala E, Muller DN, Park JK et al (2000) Cyclosporin A protects against angiotensin II-induced end-organ damage in double transgenic rats harboring human renin and angiotensinogen genes. Hypertension 35:360–366
Mervaala EM, Cheng ZJ, Tikkanen I et al (2001) Endothelial dysfunction and xanthine oxidoreductase activity in rats with human renin and angiotensinogen genes. Hypertension 37:414–418
Middleton JP, Khan WA, Collinsworth G, Hannun YA, Medford RM (1993) Heterogeneity of protein kinase C-mediated rapid regulation of Na/K-ATPase in kidney epithelial cells. J Biol Chem 268:15958–15964
Miyata N, Park F, Li XF, Cowley A-WJ (1999) Distribution of angiotensin AT_1 and AT_2 receptor subtypes in the rat kidney. Am J Physiol 277:F437–F446
Nagahama T, Hayashi K, Ozawa Y, Takenaka T, Saruta T (2000) Role of protein kinase C in angiotensin II-induced constriction of renal microvessels. Kidney Int 57:215–223
Nagami GT (2002) Enhanced ammonia secretion by proximal tubules from mice receiving NH_4Cl: role of angiotensin II. Am J Physiol 282:F472–F477
Navar LG, Harrison-Bernard LM, Nishiyama A, Kobori H (2002) Regulation of intrarenal angiotensin II in hypertension. Hypertension 39:316–322
Niimura F, Labosky PA, Kakuchi J et al (1995) Gene targeting in mice reveals a requirement for angiotensin in the development and maintenance of kidney morphology and growth factor regulation. J Clin Invest 96:2947–2954
Nishimura H, Yerkes E, Hohenfellner K et al (1999) Role of the angiotensin type 2 receptor gene in congenital anomalies of the kidney and urinary tract, CAKUT, of mice and men. Mol Cell 3:1–10
Nishiyama A, Fukui T, Fujisama Y, Rahman M et al (2001) Systemic and regional hemodynamic responses to tempol in angiotensin II-infused hypertensive rats. Hypertension 37:77–83
Norwood VF, Craig MR, Harris JM, Gomez RA (1997) Differential expression of angiotensin II receptors during early renal morphogenesis. Am J Physiol 272:R662–R668
Obata J, Nakamura T, Takano H et al (2000) Increased gene expression of components of the renin-angiotensin system in glomeruli of genetically hypertensive rats. J Hypertens 18:1247–1255
Okamura A, Rakugi H, Ohishi M et al (1999) Upregulation of renin-angiotensin system during differentiation of monocytes to macrophages. J Hypertens 17:537–545
Ortiz MC, Manriquez MC, Romero JC, Juncos LA (2001) Antioxidants block angiotensin II-induced increases in blood pressure and endothelin. Hypertension 38:655–659
Ozaki T, Nishina H, Hanson MA, Poston L (2001) Dietary restriction in pregnant rats causes gender-related hypertension and vascular dysfunction in offspring. J Physiol 530:141–152
Ozono R, Wang ZQ, Moore AF, Inagami T, Siragy HM, Carey RM (1997) Expression of the subtype 2 angiotensin (AT_2) receptor protein in rat kidney. Hypertension 30:1238–1246
Pallone TL (1994) Vasoconstriction of outer medullary vasa recta by angiotensin II is modulated by prostaglandin E. Am J Physiol 266:F850–F857
Pallone TL, Zhang Z, Rhinehart K (2003) Physiology of the renal medullary microcirculation. Am J Physiol 284:F253–F266

Parenti A, Cui XL, Hopfer U, Ziche M, Douglas JG (2000) Activation of MAPKs in proximal tubule cells from spontaneously hypertensive and control Wistar-Kyoto rats. Hypertension 35:1160–1166

Pavenstädt H (2000) Roles of the podocyte in glomerular function. Am J Physiol 278:F173–F179

Peng Y, Knox FG (1995) Comparison of systemic and direct intrarenal angiotensin II blockade on sodium excretion in rats. Am J Physiol 269:F40–F46

Peti-Peterdi J, Warnock DG, Bell PD (2001) Angiotensin II directly stimulates ENaC activity in the cortical collecting duct via AT(1) receptors. J Am Soc Nephrol 13:1131–1135

Prieto M, Dipp S, Meleg-Smith S, el Dahr SS (2001) Ureteric bud derivatives express angiotensinogen and AT_1 receptors. Physiol Genomics 6:29–37

Qi Z, Hao CM, Langenbach RI et al (2002) Opposite effects of cyclooxygenase-1 and -2 activity on the pressor response to angiotensin II. J Clin Invest 110:61–69

Quan A, Baum M (1996) Endogenous production of angiotensin II modulates rat proximal tubule transport. J Clin Invest 97:2878–2882

Rangel LB, Caruso-Neves C, Lara LS, Lopes AG (2002) Angiotensin II stimulates renal proximal tubule Na^+-K^+-ATPase activity through the activation of protein kinase C. Biochim Biophys Acta 1564:310–316

Reilly AM, Harris PJ, Williams DA (1995) Biphasic effect of angiotensin II on intracellular sodium concentration in rat proximal tubules. Am J Physiol 269:F374–F380

Ritz E, Schomig M, Wagner J (2000) Counteracting progression of renal disease: a look into the future. Kidney Int Suppl 75, S71–S76

Robillard JE, Page WV, Mathews MS, Schutte BC, Nuyt AM, Segar JL (1995) Differential gene expression and regulation of renal angiotensin II receptor subtypes (AT_1 and AT_2) during fetal life in sheep. Pediatr Res 38:896–904

Rodriguez-Iturbe B, Pons H, Herrera-Acosta J, Johnson RJ (2001a) Role of immunocompetent cells in nonimmune renal diseases. Kidney Int 59:1626–1640

Rodriguez-Iturbe B, Pons H, Quiroz Y et al (2001b) Mycophenolate mofetil prevents salt-sensitive hypertension resulting from angiotensin II exposure. Kidney Int 59:2222–2232

Rohrwasser A, Morgan T, Dillon HF et al (1999) Elements of a paracrine tubular renin-angiotensin system along the entire nephron. Hypertension 34:1265–1274

Ruiz-Ortega M, Lorenzo O, Ruperez M, Blanco J, Egido J (2001) Systemic infusion of angiotensin II into normal rats activates nuclear factor-kappaB and AP-1 in the kidney: role of AT_1 and AT_2 receptors. Am J Pathol 158:1743–1756

Sanchez-Mendoza A, Hong E, Escalante B (1998) The role of nitric oxide in angiotensin II-induced renal vasoconstriction in renovascular hypertension. J Hypertens 16:697–703

Sarkis A, Liu KL, Lo M, Benzoni D (2003) Angiotensin II and renal medullary blood flow in Lyon rats. Am J Physiol 284:F365–F372

Schelling JR, Singh H, Marzec R, Linas SL (1994) Angiotensin II-dependent proximal tubule sodium transport is mediated by cAMP modulation of phospholipase C. Am J Physiol 267:C1239–C1245

Schlatter E, Ankorina I, Haxelmans S, Kleta R (1995) Effects of diadenosine polyphosphates, ATP and angiotensin II on cytosolic Ca^{2+} activity and contraction of rat mesangial cells. Pflugers Arch 430:721–728

Schnackenberg CG (2002) Physiological and pathophysiological roles of oxygen radicals in the renal microvasculature. Am J Physiol 282:R335–R342

Schnackenberg CG, Welch WJ, Wilcox CS (1998) Normalization of blood pressure and renal vascular resistance in SHR with a membrane-permeable superoxide dismutase mimetic: role of nitric oxide. Hypertension 32:59–64

Schnermann J (1999) Micropuncture analysis of tubuloglomerular feedback regulation in transgenic mice. J Am Soc Nephrol 10:2614–2619

Schnermann JB, Traynor T, Yang T et al (1997) Absence of tubuloglomerular feedback responses in AT_{1A} receptor-deficient mice. Am J Physiol 273:F315–F320

Shanmugam S, Lenkei ZG, Gasc JM, Corvol PL, Llorens-Cortes CM (1995) Ontogeny of angiotensin II type 2 (AT_2) receptor mRNA in the rat. Kidney Int 47:1095–1100

Sharma M, Sharma R, Greene AS, McCarthy ET, Savin VJ (1998) Documentation of angiotensin II receptors in glomerular epithelial cells. Am J Physiol 274:F623–F627

Sigmund CD (2001) Genetic manipulation of the renin-angiotensin system in the kidney. Acta Physiol Scand 173:67–73

Siragy HM, Carey RM (1997) The subtype 2 (AT_2) angiotensin receptor mediates renal production of nitric oxide in conscious rats. J Clin Invest 100:264–269

Sorensen CM, Leyssac PP, Skott O, Holstein-Rathlou NH (2000) Role of the renin-angiotensin system in regulation and autoregulation of renal blood flow. Am J Physiol 279:R1017–R1024

Stevenson KM, Edgley AJ, Bergstrom G, Worthy K, Kett MM, Anderson WP (2000) Angiotensin II infused intrarenally causes preglomerular vascular changes and hypertension. Hypertension 36:839–844

Stockand JD, Sansom SC (1998) Glomerular mesangial cells: electrophysiology and regulation of contraction. Physiol Rev 78:723–744

Szentivanyi MJ, Maeda CY, Cowley A-WJ (1999) Local renal medullary L-NAME infusion enhances the effect of long-term angiotensin II treatment. Hypertension 33:440–445

Takenaka T, Suzuki H, Fujiwara K et al (1997) Cellular mechanisms mediating rat renal microvascular constriction by angiotensin II. J Clin Invest 100:2107–2114

Tanimoto K, Sugiyama F, Goto Y et al (1994) Angiotensinogen-deficient mice with hypotension. J Biol Chem 269:31334–31337

Thekkumkara T, Linas SL (2002) Role of internalization in AT_{1A} receptor function in proximal tubule epithelium. Am J Physiol 282:F623–F629

Thekkumkara TJ, Cookson R, Linas SL (1998) Angiotensin (AT_{1A}) receptor-mediated increases in transcellular sodium transport in proximal tubule cells. Am J Physiol 274:F897–F905

Thomas D, Harris PJ, Morgan TO (1990) Altered responsiveness of proximal tubule fluid reabsorption of peritubular angiotensin II in spontaneously hypertensive rats. J Hypertens 8:407–410

Thomson S, Bao D, Deng A, Vallon V (2000) Adenosine formed by 5′-nucleotidase mediates tubuloglomerular feedback. J Clin Invest 106:289–298

Thorup C, Kornfeld M, Goligorsky MS, Moore LC (1999) AT_1 receptor inhibition blunts angiotensin II-stimulated nitric oxide release in renal arteries. J Am Soc Nephrol 10:S220–S224

Tojo A, Tisher CC, Madsen KM (1995) Angiotensin II regulates H^+-ATPase activity in rat cortical collecting duct. Am J Physiol 267:F1045–F1051

Toke A, Meyer TW (2001) Hemodynamic effects of angiotensin II in the kidney. Contrib Nephrol 135:34–46

Tse CM, Levine SA, Yun CH et al (1993) Functional characteristics of a cloned epithelial Na^+/H^+ exchanger (NHE3): resistance to amiloride and inhibition by protein kinase C. Proc Natl Acad Sci USA 90:9110–9114

Tsuchida S, Matsusaka T, Chen X et al (1998) Murine double nullizygotes of the angiotensin type 1A and 1B receptor genes duplicate severe abnormal phenotypes of angiotensinogen nullizygotes. J Clin Invest 101:755–760

Tufro-McReddie A, Romano LM, Harris JM, Ferder L, Gomez RA (1995) Angiotensin II regulates nephrogenesis and renal vascular development. Am J Physiol 269:F110–F115

Turkstra E, Braam B, Koomans HA (2000) Normal TGF responsiveness during chronic treatment with angiotensin-converting enzyme inhibition: role of AT_1 receptors. Hypertension 36:818–823

Vaziri ND, Ni Z, Oveisi F, Trnavsky-Hobbs DL (2000) Effect of antioxidant therapy on blood pressure and NO synthase expression in hypertensive rats. Hypertension 36:957–964

Vehaskari VM, Aviles DH, Manning J (2001) Prenatal programming of adult hypertension in the rat. Kidney Int 59:238–245

Verseput GH, Braam B, Provoost AP, Koomans HA (1998) Tubuloglomerular feedback and prolonged ACE-inhibitor treatment in the hypertensive fawn-hooded rat. Nephrol Dial Transplant 13:893–899

Vyas SJ, Blaschak CM, Chinoy MR, Jackson EK (2000) Angiotensin II-induced changes in G-protein expression and resistance of renal microvessels in young genetically hypertensive rats. Mol Cell Biochem 212:121–129

Wagner CA, Giebisch G, Lang F, Geibel JP (1998) Angiotensin II stimulates vesicular H^+-ATPase in rat proximal tubular cells. Proc Natl Acad Sci USA 95:9665–9668

Walker LL, Rajaratne AA, Blair-West JR, Harris PJ (1999) The effects of angiotensin II on blood perfusion in the rat renal papilla. J Physiol 519:273–278

Wang CT, Chin SY, Navar LG (2000) Impairment of pressure-natriuresis and renal autoregulation in ANG II-infused hypertensive rats. Am J Physiol 279:F319–F325

Wang DH, Du Y, Zhao H, Granger JP, Speth RC, Dipette DJ (1997) Regulation of angiotensin type 1 receptor and its gene expression: role in renal growth. J Am Soc Nephrol 8:193–198

Wang HD, Xu S, Johns DG et al (2001) Role of NADPH oxidase in the vascular hypertrophic and oxidative stress response to angiotensin II in mice. Circ Res 88:947–953

Wang T, Giebisch G (1996) Effects of angiotensin II on electrolyte transport in the early and late distal tubule in rat kidney. Am J Physiol 271:F143–F149

Wang X, Aukland K, Iversen BM (1997) Acute effects of angiotensin II receptor antagonist on autoregulation of zonal glomerular filtration rate in renovascular hypertensive rats. Kidney Blood Press Res 20:225–232

Wang ZQ, Millatt LJ, Heiderstadt NT, Siragy HM, Johns RA, Carey RM (1999) Differential regulation of renal angiotensin subtype AT_{1A} and AT_2 receptor protein in rats with angiotensin-dependent hypertension. Hypertension 33:96–101

Weiner D, New AR, Milton AE, Tisher CC (1995) Regulation of luminal alkalinization and acidification in the cortical collecting duct by angiotensin II. Am J Physiol 269:F730–F738

Welch WJ, Tojo A, Wilcox CS (2000) Roles of NO and oxygen radicals in tubuloglomerular feedback in SHR. Am J Physiol 278:F769–F776

Welch WJ, Wilcox CS (2001) AT_1 receptor antagonist combats oxidative stress and restores nitric oxide signaling in the SHR. Kidney Int 59:1257–1263

Wilcox CS, Welch WJ (2000) Interaction between nitric oxide and oxygen radicals in regulation of tubuloglomerular feedback. Acta Physiol Scand 168:119–124

Wolf G, Ziyadeh FN, Schroeder R, Stahl RA (1997a) Angiotensin II inhibits inducible nitric oxide synthase in tubular MCT cells by a posttranscriptional mechanism. J Am Soc Nephrol 8:551–557

Wolf G, Ziyadeh FN, Thaiss F, Tomaszewski J, Caron RJ et al (1997b) Angiotensin II stimulates expression of the chemokine RANTES in rat glomerular endothelial cells. J Clin Invest 100:1047–1058

Wood JM, Schnell CR, Levens NR (1993) Kidney is an important target for the antihypertensive action of an angiotensin II receptor antagonist in spontaneously hypertensive rats. Hypertension 21:1056–1061

Woods LL, Rasch R (1998) Perinatal ANG II programs adult blood pressure, glomerular number, and renal function in rats. Am J Physiol 275:R1593–R1599

Woods LL, Ingelfinger JR, Nyengaard JR, Rasch R (2001) Maternal protein restriction suppresses the newborn renin-angiotensin system and programs adult hypertension in rats. Pediatr Res 49:460–467

Zatz R, Fujihara CK (2002) Mechanisms of progressive renal disease: role of angiotensin II, cyclooxygenase products and nitric oxide. J Hypertens 20:S37–S44
Zhang C, Mayeux PR (1998) Angiotensin II signaling activities the NO-cGMP pathway in rat proximal tubules. Life Sci 63:PL75–PL80
Zhang C, Mayeux PR (2001) NO/cGMP signaling modulates regulation of Na^+-K^+-ATPase activity by angiotensin II in rat proximal tubules. Am J Physiol 280:F474–F479
Zhang Z, Huang JM, Turner MR, Rhinehart KL, Pallone TL (2001) Role of chloride in constriction of descending vasa recta by angiotensin II. Am J Physiol 280:R1878–R1886
Zou AP, Wu F, Cowley A-WJ (1998) Protective effect of angiotensin II-induced increase in nitric oxide in the renal medullary circulation. Hypertension 31:271–276

Angiotensin and Aldosterone Biosynthesis

A. M. Capponi[1] · M. F. Rossier[2]

[1] Division of Endocrinology and Diabetology, University Hospital, 24 rue Micheli-du-Crest, 1211 Geneva 14, Switzerland
e-mail: alessandro.capponi@medecine.unige.ch

[2] Laboratory of Clinical Chemistry, Department of Pathology, University Hospital, 1211 Geneva 14, Switzerland

Abstract Angiotensin II (Ang II) is, with extracellular potassium, the principal regulator of the biosynthesis of aldosterone, the main mineralocorticoid in man. Because various cardiovascular pathological states are often associated with inappropriate levels of circulating aldosterone, it is of importance to understand the intracellular molecular mechanisms leading to aldosterone output in the adrenal zona glomerulosa cell. The present chapter reviews the current knowledge on the events that are triggered by angiotensin II, from the interaction with membrane receptors to the final output of aldosterone. The initial signaling processes involve complex changes in intracellular phospholipid and calcium homeostasis resulting from effects of Ang II on various effector enzymes (phospholipases) and calcium channels, as well as on intracellular calcium stores. Concomitantly, a series of kinase pathways (protein kinase C, mitogen-activated protein kinase, tyrosine kinases, etc.) are activated. These intracellular signals then mediate a variety of localized responses along the entire cascade of events leading from uptake of cholesterol, the precursor of all steroids, to cholesterol supply to the mitochondrial enzymatic machinery that will process cholesterol to aldosterone. Indeed, Ang II increases HDL-cholesterol import at the cell surface, cholesterol ester hydrolysis in lipid droplets, scavenger receptor class B type I(HDL receptor), steroidogenic acute regulatory (StAR) protein or aldosterone synthase gene expression in the nucleus, or StAR-mediated cholesterol importation into mitochondria. In addition, recent work has shown that aldosterone can also be synthesized and act in nonclassic steroidogenic tissues such as brain, vessels and, most importantly, the heart, where it may have deleterious effects in some pathological situations such as heart failure or myocardial infarction.

Keywords Cholesterol · Mineralocorticoid · Calcium · Adrenal · Heart

1 Introduction

Although angiotensin II (Ang II) had been identified more than 20 years earlier (Braun-Menéndez et al. 1939; Page and Helmer 1940) and the structure of aldosterone had been known for several years (Simpson et al. 1954), the functional link between these two hormones only became apparent in the early 1960s, when it was shown that intravenous infusions of Ang II in normal humans resulted in increased secretion or urinary excretion rates of aldosterone (Laragh et al. 1960; Biron et al. 1961). This crucial finding was soon confirmed in the dog (Ganong et al. 1962) and later in other animal species (Müller 1988).

The renin-angiotensin-aldosterone system functions as the main modulator of salt and water homeostasis in the body. Whereas Ang II directly stimulates sodium reabsorption in the kidney proximal tubule, the fine tuning of sodium balance requires adrenal modulation of the biosynthesis of aldosterone, which enhances distal tubular sodium reabsorption and therefore exactly adjusts natriuresis to the dietary sodium load. This system is particularly useful in situations of low sodium availability, to prevent sodium wasting and dehydration. However, in high-salt-consuming societies, its action is often inadequate and can lead to salt-sensitive hypertension.

Ang II is thus recognized as the main regulator of aldosterone secretion together with extracellular potassium (Vallotton et al. 1995). In contrast, adrenocorticotropic hormone (ACTH), although efficiently activating aldosterone biosynthesis in vitro, only exerts a transient stimulatory effect in vivo. ACTH can therefore be considered more as a modulator of glomerulosa cell steroidogenesis. This is supported, for example, by the fact that circulating levels of aldosterone are much more influenced by position (orthostatic vs supine) and therefore by the activation of the renin-angiotensin system, than by the circadian rhythm, as it is the case for cortisol.

Hyperaldosteronism is associated with high blood pressure and hypokalemia. Because these problems can be efficiently solved by controlling aldosterone secretion, it is particularly important to determine the etiology of the aldosterone excess. For this purpose, renin activity is generally measured in order to assess whether aldosteronism is primary and due, for example, to the presence of an adrenal adenoma, or secondary and resulting from the activation of renin in a situation of volume contraction. In the latter case, Ang II is directly responsible for the high rate of aldosterone secretion. Like other steroids, aldosterone, because of its hydrophobicity, cannot be stored within the steroidogenic adrenal glomerulosa cell and aldosterone secretion faithfully reflects the rate of its biosynthesis. Considerable progress has been achieved during the last 25 years in the understanding of the intracellular mechanisms elicited by Ang II in controlling aldosterone biosynthesis, and several reviews have been recently published on this topic (Foster et al. 1997; Lumbers 1999; Mulrow 1999; Weir and Dzau 1999). In the present chapter, we will focus our attention on the cellular signaling induced by Ang II in adrenal glomerulosa cells as well as on the regulation by this hormone of the various steps of steroidogenesis. We will also briefly mention the main factors that modulate the action of Ang II on aldosterone secretion.

In addition to their role in the control of salt and water homeostasis, both Ang II and aldosterone, when chronically produced, have been involved in the remodeling of blood vessels, heart and kidney. Although the circulating hormones may contribute to this process, the locally produced hormones probably play a major role. A section of this review will deal with this novel avenue in the field of Ang II-controlled aldosterone biosynthesis in nonclassic tissues.

2 Angiotensin II-Induced Intracellular Signaling in the Activation of Aldosterone Biosynthesis

2.1 Receptor Internalization and G Protein Activation

As for other peptide hormones, the first step in Ang II action is its binding to a membrane receptor expressed at the surface of the target cell. We will first discuss the receptor subtype linked to stimulation of aldosterone secretion, the modulation of this receptor and the signal transduction through G proteins.

2.1.1 Angiotensin Receptor Subtypes

Most of the classic functions attributed to Ang II (vasoconstriction, salt and water retention, growth stimulation) are mediated by activation of the AT_1 receptor subtype (Bernstein and Berk 1993; Timmermans et al. 1993). This receptor is a member of the seven transmembrane domain receptor superfamily and is coupled to various G proteins. Its effectors include phospholipase C, D and A_2, adenylyl cyclase and calcium channels. Two distinct subtypes of this receptor, AT_{1A} and AT_{1B}, have been cloned and sequenced, and in contrast to most of the other actions of Ang II, the control of aldosterone biosynthesis and secretion appears to be essentially mediated by the AT_{1B} isoform (Gigante et al. 1997).

The AT_2 receptor is not involved in the control of aldosterone secretion, given that aldosterone production induced by Ang II in rat, bovine and rabbit adrenal glomerulosa cells is prevented by the AT_1 selective antagonists losartan (DuP753) and CV-11974, but not by the AT_2 antagonists PD123177 or PD123319 (Hajnoczky et al. 1992a; Tanabe et al. 1998; Wada et al. 1994). However, the AT_2 receptor could play a role in the adrenal glomerulosa cell in the production of an ouabain-like substance (Shah et al. 1998). Interestingly, in the mid-gestation ovine fetus, Ang II can stimulate aldosterone production only after AT_2 receptor inhibition, suggesting the presence of a strong antagonizing signal generated by this receptor at this period of development (Moritz et al. 1999). Screening of human aldosterone-producing adenomas or carcinomas did not reveal any activating mutation in the coding region of AT_1 receptor gene, as it is the case for TSH or LH receptors, which could have been responsible for some form of hyperaldosteronism (Davies et al. 1997; Sachse et al. 1997).

2.1.2 Receptor Internalization and Modulation

Controlling the level of the AT_1 receptor expression at the cell surface is a key mechanism for modulating aldosterone production. In this regard, the effect of sodium depletion or high potassium diet on aldosterone secretion in vivo could be partly due to AT_1 receptor gene up-regulation (Inagami 1995). On the other hand, a rapid desensitization of the system can be achieved by retrieval of the receptor from the cell surface. Indeed, early observations suggest that the reduction of aldosterone response to Ang II after 6 h of cell incubation with the hormone is linked to a decrease in the capacity of the cells to form inositol phosphates in response to a subsequent stimulation with Ang II (Enyedi and Spat 1987). Binding studies in bovine glomerulosa cells revealed a rapid shift of the AT_1 receptor from a high-affinity to a low-affinity state in response to Ang II, probably due to receptor dissociation from G proteins (Boulay et al. 1994), while a loss of total Ang II binding sites was observed upon pretreatment with higher concentrations of Ang II, an effect that was maximal after 14 h and sensitive to temperature, a parameter that has been shown to control receptor recycling (Richard et al. 1997a, 1997b). A time-dependent decrease in AT_1 receptor messenger RNA levels was also observed after similar pretreatment of the cells for 24 h. Using a monoclonal antibody against the N-terminal part of the rat AT_1 receptor, it has been directly shown that receptor recycling to the plasma membrane is constitutive or regulated by unknown factors and that retention of the receptor at the surface is sufficient to stimulate aldosterone output (Vinson et al. 1994b). Indeed, while AT_1 receptor internalization has been proposed to be required for PKC activation (Vinson et al. 1995), preventing this process stimulates inositol trisphosphate release and steroidogenesis (Kapas et al. 1994b).

2.1.3 G Protein Coupling

Seven transmembrane domain receptors depend upon specific G protein activation to transmit their signal into the cell, and the AT_1 receptor is no exception. Coupling this receptor to a cyclase inhibitory protein (G_i) has been recognized for a long time, at least in rat glomerulosa cells, where Ang II inhibits ACTH-stimulated cAMP production. However, the lack of effect of pertussis toxin—which blocks both G_i and G_o—on other Ang II-induced responses such as PLC activation, the rise of cytosolic calcium concentration and aldosterone secretion, rapidly suggested receptor coupling to additional G proteins of the $G_{q/11}$ family, as classically observed for other hormones linked to similar intracellular messenger pathways (Catt et al. 1997, 1988; Hausdorff et al. 1987).

Interestingly, cytoskeleton-disrupting agents such as cytochalasin B and colchicine have been shown to decrease the Ang II-induced inositol phosphate response in rat glomerulosa cells, suggesting that both microfilament and microtubule integrity is essential for PLC activation (Cote et al. 1997). This was supported by the finding that $G_{q/11}$ alpha subunit is associated with the cytoskeleton and rapidly redistributed, together with these cytoskeletal elements, toward the plasma membrane upon Ang II stimulation.

In addition to cAMP modulation, the pertussis toxin-sensitive G proteins activated by Ang II in glomerulosa cells have been more recently proposed to be involved in the rapid (within minutes) regulation of voltage-operated calcium channels. Indeed, while G_i apparently mediates a positive modulatory effect of Ang II on the activity of T-type calcium channels (Lu et al. 1996), a G_o protein could be responsible for L channel inhibition by the same hormone (Maturana et al. 1999b). This complex modulation explains why pertussis toxin changes Ca^{2+} influx without affecting Ins(1,4,5)P_3 production (Kojima et al. 1986a).

2.2 The Calcium Messenger System

Calcium mobilization is the principal signaling mechanism recruited by Ang II to exert its effect on its various target cells, including adrenal glomerulosa cells

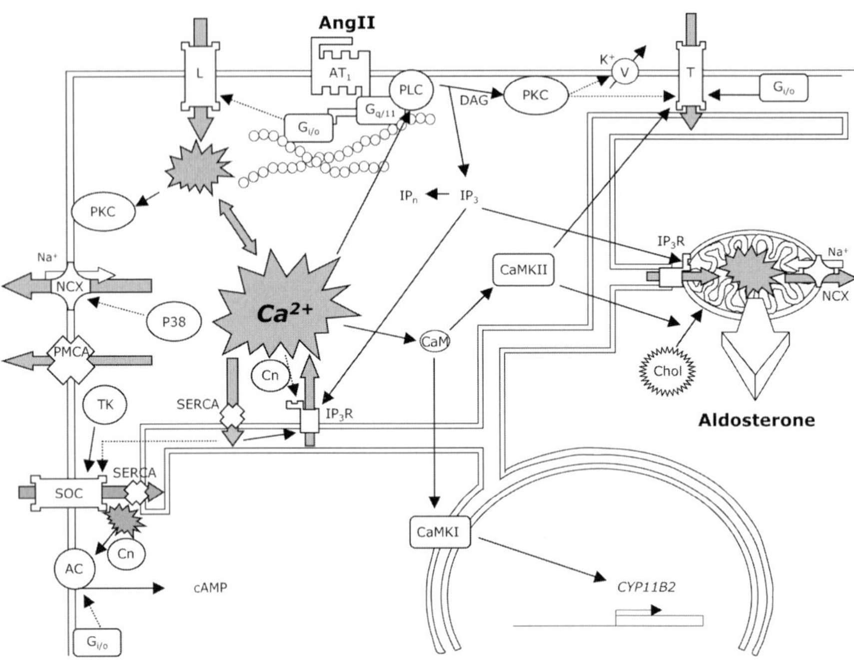

Fig. 1 Angiotensin II-induced calcium signaling in adrenal glomerulosa cells. Model of the various pathways of the calcium messenger system in adrenal glomerulosa cells involved in angiotensin II-induced aldosterone biosynthesis. *Solid arrows* indicate stimulatory and *dotted arrows* inhibitory actions of an agent on its target, while *gray arrows* indicate calcium fluxes. See text for additional explanations. *Ang II*, angiotensin II; *AT_1*, Ang II receptor subtype 1; *PLC*, phosphoinositide-specific phospholipase C; *$G_{q/11}$*, heterotrimeric G protein of the q/11 family; *$G_{i/o}$*, pertussis toxin-sensitive heterotrimeric G protein of the i/o family; *L*, L-type calcium channel; *T*, T-type calcium channel; *SOC*, store-operated calcium channel; *V*, membrane potential; *DAG*, diacylglycerol; *PKC*, protein kinase C; *IP_3*, inositol 1,4,5-trisphosphate; *IP_n*, IP_3 metabolites; *IP_3R*, IP_3 receptor; *CaM*, calmodulin; *CaMKI*, *CaMKII*, CaM-dependent protein kinase type I, type II, respectively; *Chol*, cholesterol-rich lipid droplet; *NCX*, sodium/calcium exchanger; *P38*, p38 mitogen-activated protein kinase (MAPK); *PMCA*, plasma membrane calcium ATPase; *TK*, tyrosine kinase; *SERCA*, sarcoendoplasmic reticulum calcium ATPase; *Cn*, calcineurin; *AC*, adenylyl cyclase; *CYP11B2*, gene coding for aldosterone synthase

(Barrett et al. 1989; Ganguly and Davis 1994; Spät 1988; Spät et al. 1991). We will now describe the various pathways of Ca^{2+} as well as the Ca^{2+} effectors directly involved in the stimulation of aldosterone secretion by Ang II. A general scheme of the Ca^{2+} signaling in adrenal glomerulosa cells in relationship to Ang II-induced aldosterone synthesis is proposed in Fig. 1.

2.2.1 Calcium Influx

The requirement for extracellular Ca^{2+} and for its influx into glomerulosa cells in the steroidogenic response to Ang II has been recognized for more than 20 years (Braley et al. 1984; Capponi et al. 1984; Fakunding and Catt 1982; Fakunding et al. 1979; Foster et al. 1981; Kojima et al. 1985c). Various pathways for this Ca^{2+} influx, each with its specific regulation, have been characterized.

Voltage-Operated Channels. The first demonstrations of the involvement of voltage-operated Ca^{2+} channels (VOC) in the steroidogenic response to Ang II were essentially pharmacological. Various drugs such as verapamil, dihydropyridines or nickel have been extensively used to inhibit the response to Ang II in vitro and in vivo (Fakunding and Catt 1980; Millaret al. 1982; Struthers et al. 1982; Finkel et al. 1984; Scriabine et al. 1984; Schiebinger et al. 1986; Fitzpatrick and McKenna 1992). The presence of binding sites for dihydropyridines on adrenal capsular membranes has also been demonstrated (Finkel et al. 1984; Aguilera and Catt 1986). The expression in aldosterone-producing cells from various species of two different types of calcium channels, a low-threshold T-type and a high-threshold L-type channel, was shown first using an electrophysiological approach (Cohen et al. 1988; Matsunaga et al. 1987a, 1987b; Payet et al. 1994; Rossier et al. 1998a) and was later confirmed using molecular techniques (Horváth et al. 1998; Lesouhaitier et al. 2001; Schrier et al. 2001). However, the relatively high pharmacological concentrations of calcium antagonists employed in early studies for blocking aldosterone secretion were unable to discriminate between the two pathways. Later, using more selective drugs such as omega-Aga IIIA toxin (Barrett et al. 1995), tetrandrine (Rossier et al. 1993), peripheral-type benzodiazepines (Python et al. 1993), calciseptine (Rossier et al. 1996b), or mibefradil (Lotshaw 2001a; Rossier et al. 1998b), it has been clearly demonstrated that T-type but not L-type channels control aldosterone production (see below in this section). This specificity of function explains in part the relatively poor efficiency in vivo of the classic calcium antagonists such as dihydropyridines or diltiazem currently used as antihypertensive drugs, to reduce circulating aldosterone levels (Conlin et al. 1998; Freed et al. 1991; Leonetti et al. 1987; Rossier and Capponi 2000).

If Ang II can activate VOCs, it is primarily because the hormone is able to depolarize the glomerulosa cell membrane. A modulation by the hormone of the electrophysiological properties of glomerulosa cells has been known for many years (Quinn et al. 1987). Application of Ang II to isolated cells induces a biphasic response, a brief hyperpolarizing phase is followed by a sustained depolarization. This last effect is achieved by inhibition of potassium conductances that normally maintain a very negative voltage in resting cells (Brauneis et al. 1991; Lotshaw 1997b; Quinn et al. 1987). Whereas the cell depolarization induced by extracellular potassium is stable and reversible, that elicited by Ang II appears much more complex, consisting of a slow sustained component superimposed with small amplitude fluctuations that are occasionally observed, suggesting that Ang II affects various conductances (Lotshaw 2001b).

It is, however, noteworthy that, in contrast to what has been observed in Y1 cells (Hescheler et al. 1988), no action potentials are evoked by Ang II in glomerulosa cells, probably because of the much more negative resting potential maintained in resting cells.

Although the hormone-induced depolarization is always due to inhibition of K^+ permeability (Brauneis et al. 1991) and similar K^+ currents are expressed in human, rat and bovine glomerulosa cells (Payet et al. 1994), differences among species have been seen in the characteristics of the K^+ currents involved (Kanazirska et al. 1992; Payet et al. 1995; Vassilev et al. 1992). The effect of Ang II on membrane potential is reversible and entirely blocked by AT_1 antagonists (Lotshaw 1997a). Investigation of the modulation by Ang II of $^{86}Rb^+$ or $^{43}K^+$ efflux has shown that Ang II probably exerts its effect on K^+ conductance via PKC activation (Lobo and Marusic 1986, 1988; Lobo et al. 1990). Mimicking the Ang II effect on the K^+ permeability through addition of sn-1,2 dioctanoylglycerol led to an increase in aldosterone production, probably resulting from the activation of VOCs upon cell depolarization.

Several groups then confirmed the functional link between K^+ conductance modulation and aldosterone secretion by demonstrating that Ang II-induced steroidogenesis is markedly affected by valinomycin, a K^+ ionophore (Shepherd et al. 1992), or by various K^+ channel modulators (Hadjokas and Goodfriend 1991; Lotshaw 1997b). The fundamental role of glomerulosa cell depolarization has been proposed to be the basis of the synergistic interaction between the hormone and extracellular K^+ in the control of aldosterone production (Chen et al. 1999).

Whereas cell depolarization is the initial event required for the activation of VOCs, a subsequent modulation of the channel activity can be achieved through G proteins (Schulz et al. 1990), through activation of protein kinases and phosphatases (Trautwein and Hescheler 1990) and by controlling the channel expression levels. If the first two mechanisms occur rapidly (within minutes), the third one takes hours or even days. Calcium channel modulation by Ang II in steroidogenic and nonsteroidogenic tissues has been recently reviewed (Rossier and Capponi 2000a). Rapid modulation of both L and T channels in glomerulosa cells by Ang II has been convincingly demonstrated and will be discussed now, but it must be remembered that a modulation of channel expression could also occur indirectly via an aldosterone-induced mineralocorticoid receptor stimulation (Lesouhaitier et al. 2001).

Ang II has been shown to modulate T channel activity either positively or negatively (Rossier and Capponi 2000a). Indeed, a shift in response to Ang II of the channel activation curve toward more negative potential values has been attributed to activation of a G_i protein in calf glomerulosa cells (Cohen et al. 1988; Lu et al. 1996; McCarthy et al. 1993), while the same hormone displayed an opposite effect in adult bovine cells through the activation of PKC (Rossier et al. 1995). Moreover, T channels could also be positively controlled by Ang II-stimulated calmodulin-dependent kinase II (CaMKII) (Fern et al. 1995; Lu et al. 1994).

Given the essential role played by T channels in aldosterone secretion, a complex modulation of their activity probably allows the integration of several signals generated by Ang II and other modulators to finely tune steroidogenesis in glomerulosa cells.

A negative L channel modulation has also been described upon stimulation by Ang II. This effect of the hormone appears to be mediated by a pertussis toxin-sensitive G_o protein (Maturana et al. 1999b; Rossier et al. 1998a) and does not involve PKC or tyrosine kinases (Maturana et al. 1999a). This action of Ang II could explain the reduction by the hormone of the K^+-induced cytosolic Ca^{2+} signal (Balla et al. 1991; Capponi et al. 1987; Aptel et al. 1996; Maturana et al. 1999b).

Store-Operated Channels. The lower sensitivity to dihydropyridines of the steroidogenic response to Ang II as compared to the response to potassium (Kojima et al. 1984b), as well as the different pharmacological profiles of the $^{45}Ca^{2+}$ influx induced by each agonist (Kramer 1993; Spät et al. 1989), suggested early that the hormone can mobilize calcium from additional sources to exert its effect on aldosterone production. A voltage-independent influx of calcium, called capacitative influx (Putney and McKay 1999), is also induced in adrenal glomerulosa cells by Ang II (Ambroz and Catt 1992). This influx involves particular channels, the store-operated calcium channels (SOCs), that are activated, probably through conformational coupling (Irvine 1990; Putney 1999), upon intracellular store emptying by inositol 1,4,5-trisphosphate (Ins(1,4,5)P_3-mediated) or by SERCA pump-inhibiting agents such as thapsigargin.

A portion of the steroidogenic response to Ang II is clearly dependent on this capacitative influx (Burnay et al. 1994; Rohacs et al. 1994). Thapsigargin, which mobilizes Ca^{2+} from a pool largely coincident with the pool sensitive to Ang II (Ely et al. 1991), exerts in itself only a weak activation of aldosterone production, but markedly potentiates the response to VOC activation by potassium (Hajnoczky et al. 1991).

Tyrosine kinase inhibitors reduce both thapsigargin and Ang II-evoked Ca^{2+} capacitative influx and steroidogenesis, suggesting that these kinases are required for a maximal stimulation of the influx (Aptel et al. 1999). Even though SOC isoforms are expressed in most of the cell types and their characterization at the molecular levels, as well as their tissue distribution are actively investigated at the present time, no information concerning their specific isoform expression in adrenal glomerulosa cells is available at this time.

Functional Specificity of Calcium Channels and Signal Confinement. The presence in glomerulosa cells of multiple types of calcium channels that are simultaneously activated upon Ang II challenge could a priori appear as a redundancy of the Ca^{2+} entry pathways, but specific functions have been attributed for each channel (Rossier 1997). Indeed, specific roles for T- and L-type channels with respect to aldosterone biosynthesis were first proposed by Barrett et al. (Barrett et al. 1995), who observed that selective inhibition of L channels with omega-Aga IIIA toxin led to a potentiation instead of an inhibition of aldosterone production. Similarly, it has been shown that dihydropyridines such as nifedipine exert an inhibitory action on aldosterone only when employed at concentrations equivalent to those required for blocking T channel activity. In contrast, lower concentrations of drugs showing L channel selectivity were stimulatory in terms of steroidogenesis (Rossier et al. 1996b). Because the cytosolic Ca^{2+} response to K^+ was exclusively dependent on L channel activity, it was proposed that the steroidogenic Ca^{2+} entering through T channels is conveyed to its relevant organelle, the mitochondrion, through a sort of intracellular Ca^{2+} pipeline (Rossier 1997).

The role of Ca^{2+} influx through L-type channels is still unclear. Because Ca^{2+} entering the cell through these channels is clearly detected in the cytosol, it could be responsible for the stimulation of various Ca^{2+}-sensitive enzymes, such as kinases, phosphatases or proteases, that are present in this compartment of the cell during periods of stimulation (hours to days), where genomic induction of steroidogenic enzymes could be required. Moreover, it is noteworthy that selective inhibition of these channels influences the peak of cytosolic Ca^{2+} resulting from the Ins1,4,5P_3-mediated Ca^{2+} release from the stores (Spät et al. 1996), as well as the frequency of Ang II-induced Ca^{2+} oscillations in single glomerulosa cells (Rossier and Capponi 2002; Rössig et al. 1996). These results therefore suggest that a part of the calcium coming from L channels could be used to refill intracellular pools after Ca^{2+} release into the cytosol.

Whereas the primary function attributed to the capacitative Ca^{2+} influx through the SOC is to refill intracellular stores, additional specific functions have been proposed. Indeed, in bovine glomerulosa cells, Ang II potentiates, in a Ca^{2+}-dependent manner, ACTH-induced cAMP production (Baukal et al. 1994) and it has been demonstrated that this effect is mediated by SOCs but not VOCs (Burnay et al. 1998).

These few examples demonstrate that, as in other cell types and due to its poor diffusion in the cytosol, Ca^{2+} signaling in adrenal glomerulosa cells is highly compartmentalized and specific functions, linked to the presence of specific proteins in the channel microenvironment, can therefore be attributed to the various types of Ca^{2+} channels expressed in these cells (Rossier 1997).

2.2.2
Calcium Release

Angiotensin II has been shown early to stimulate $^{45}Ca^{2+}$ efflux from adrenal glomerulosa cells (Elliott et al. 1985; Foster and Rasmussen 1983; Kojima et al. 1985a), indicating that the hormone is able to mobilize Ca^{2+} from intracellular stores, as well as activate an influx into the cell (Catt et al. 1987). Since the discovery by Michael Berridge of the role played by Ins(1,4,5)P_3 in this process (Berridge and Irvine 1984), the metabolism of this intracellular messenger in response to Ang II has been extensively studied in adrenal glomerulosa cells.

Generation and Metabolism of Ins(1,4,5)P_3. Angiotensin accelerated phosphatidyl inositol turnover in bovine glomerulosa cells in proportion to hormone dose (Elliott et al. 1982). Because several drugs inhibiting the aldosterone response to the hormone did not affect phosphatidyl inositol metabolism, it was proposed that this event should occur relatively early in the response to Ang II. The hydrolysis of phosphatidyl inositol 4,5-bisphosphate, induced by Ang II-activated phosphoinositide-specific phospholipase C (PLC), leads to the formation of Ins(1,4,5)P_3 and diacylglycerol rich in arachidonic acid (Balla et al. 1986; Capponi et al. 1986b; Farese et al. 1984; Foster et al. 1990; Hunyady et al. 1990; Kojima et al.1984a; Rossier et al. 1988; Whitley et al. 1987), is pertussis toxin-insensitive (Baukal et al. 1988; Kojima et al. 1986a) and is well correlated with the steroidogenic response to the hormone (Boulay et al. 1990; Rossier et al. 1988). The availability of the inositol phospholipid pool could be regulated by a wortmannin-sensitive phosphatidylinositol 4-kinase in glomerulosa cells (Nakanishi et al. 1994; Nakanishi et al. 1995).

The Ca^{2+} requirement for this process has been investigated by several groups. While PLC activation appeared Ca^{2+}-dependent in some instances (Foster et al. 1990), opposite results have been also reported (Enyedi et al. 1985; Farese et al. 1984; Kojima et al.1984a). In fact, a biphasic production of Ins(1,4,5)P_3 has been shown to occur during stimulation with Ang II and the first, rapid peak, which is Ca^{2+}-independent, would allow an initial release of Ca^{2+} from the stores that would serve as a priming signal before a later Ca^{2+}-dependent elevation of Ins(1,4,5)P_3 (Balla et al. 1989; Balla et al. 1994; Woodcock et al. 1988).

Interestingly, while the phosphoinositide turnover elicited by Ang II, vasopressin and endothelin is very similar and probably involves the same pool of lipids, only Ang II can maintain a large and sustained aldosterone production, suggesting that additional steroidogenic mechanisms are activated by Ang II, which are absent when cells are stimulated with the two other agonists (Enyedi et al. 1988; Woodcock et al. 1990a, 1990b).

Analysis by HPLC of the inositol phosphates produced upon Ang II stimulation then revealed the presence of various $InsP_3$ isomers, as well as of higher phosphorylated forms of inositol (tetrakis- and pentakisphosphate) (Balla et al. 1986; Balla et al. 1987; Baukal et al. 1988; Rossier et al. 1988). These metabolites of Ins(1,4,5)P_3 increased in glomerulosa cells with distinct kinetics, suggesting their sequential interconversion (Balla et al. 1987; Rossier et al. 1986). Moreover, some of the enzymes involved in Ins(1,4,5)P_3 metabolism are regulated by Ca^{2+} itself, therefore providing a mechanism for a rapid feedback control on messenger levels (Balla et al. 1988; Rossier et al. 1986).

Inositol 1,4,5-Trisphosphate-Induced Release of Calcium from Intracellular Stores. Permeabilized glomerulosa cells have been used to demonstrate the ability of Ins(1,4,5)P_3 to release calcium from intracellular stores (Kojima et al. 1984a; Rossier et al. 1987), while binding assays revealed the presence of specific receptors for Ins(1,4,5)P_3 in these cells (Baukal et al. 1985; Guillemette et al. 1987; Poitras et al. 1993). Receptor characterization showed similar properties as those of receptors purified from other sources (Guillemette et al. 1990). The discrepancy between the apparent affinity of the receptor for its ligand and Ca^{2+} release efficiency of Ins(1,4,5)P_3 was explained by the complex regulation of receptor activation by cytosolic and intraluminal Ca^{2+} and by various other modulators (Guillemette and Segui 1988; Ribeiro-do-Valle et al. 1994). The three subtypes of Ins(1,4,5)P_3 receptors are expressed at different levels in rat (Enyedi et al.1994) and bovine (Poitras et al. 2000) glomerulosa cells, type 1 being predominantly present in each species.

The precise subcellular localization of the Ins(1,4,5)P_3 receptors has not been defined yet, but the Ins(1,4,5)P_3 binding sites have been shown to be distinct from the endoplasmic reticulum membranes, at least when the distribution of classic biochemical markers are compared (Rossier et al. 1989), suggesting the presence in glomerulosa cells of specialized organelles such as calciosomes for storing intracellular Ca^{2+} (Rossier and Putney 1991).

Whereas the role of Ins(1,4,5)P_3 in Ca^{2+} release from intracellular stores in glomerulosa cells has been clearly demonstrated, the function of the numerous other Ins(1,4,5)P_3 metabolites such as inositol tetrakisphosphates or inositol trisphosphate isomers, remains to be defined.

Cytosolic Calcium Oscillations. Measurement of cytosolic Ca^{2+} fluctuations at the single cell level revealed the oscillatory nature of the response to Ang II (Quinn et al. 1988; Johnson et al. 1989; Rössig et al. 1996). In spite of the strong dependence on extracellular calcium availability for the maintenance of Ca^{2+} oscillations, this process appears to primarily involve the regulation of Ca^{2+} release by Ins(1,4,5)P_3 (Rossier and Capponi 2002). Indeed, glomerulosa cells have no action potentials that could be responsible for transient openings of the voltage-operated channels embedded in the plasma membrane, potassium is unable to elicit Ca^{2+} oscillations, and, most importantly, thapsigargin-induced inhibition of SERCA pumps or prevention of Ins(1,4,5)P_3 receptor modulation by Ca^{2+}-dependent enzymes both abolish the oscillatory pattern of the Ca^{2+} signal. As previously mentioned, inhibition of voltage-operated channels, as well as changing the Ca^{2+} concentration in the medium, directly affects the frequency of the Ca^{2+} peaks, but not their amplitude. Globally, these data strongly suggest that Ca^{2+} influx is required for replenishing the stores and that the rate of this process controls the delay to the next pulse of release.

The role of calcium oscillations in the control of steroidogenesis is still poorly understood, but their transient nature and their possible frequency modulation are considered to allow a differential activation of the various Ca^{2+}-sensitive effectors in the cell without inducing the toxic effects related to a sustained intracellular elevation of the cation. Since in glomerulosa cells, a part of this calcium is transferred into the mitochondria, where it can exert its steroidogenic action (see Sect. 2.2.3.2), one may speculate that mitochondrial calcium could display similar oscillatory patterns, preventing a sustained rise in the organelle that could result in cell apoptosis.

2.2.3
Calcium Effectors

Calcium-Sensitive Protein Kinases and Phosphatases. Adrenal glomerulosa cells have been shown to contain the Ca^{2+}-activated, phospholipid-dependent protein kinase C (PKC) (Kojima et al. 1984a). While PKC is effectively activated by Ca^{2+}, and therefore can be considered as a target for this messenger, Ang II primarily affects kinase activity through the production of diacylglycerol (Nishizuka 1984). Indeed, although direct activation of the enzyme with high Ca^{2+} concentrations has been reported in some systems, DAG is generally required for sensitizing PKC to more physiological Ca^{2+} fluctuations. For example, PKC translocation from the cytosol to the membranes in bovine adrenal glomerulosa cells, an indirect index of PKC activation, was achieved with Ang II, but not with potassium, in spite of the stronger Ca^{2+} response elicited by the latter (Lang and Vallotton 1987). Because of the complex activation of PKC by Ang II, its modulation and function in steroidogenesis will be discussed in detail later.

Calmodulin, a classic effector of cellular calcium, is present in glomerulosa cells (Koletsky et al. 1983), and a number of calmodulin inhibitors have been shown to inhibit aldosterone secretion induced by Ang II (Balla and Spät 1982; Ganguly et al. 1992; Ganguly and Waldron 1994; Wilson et al. 1984) or by Ca^{2+} itself (Capponi et al. 1988). Interestingly, in the guinea pig adrenal cortex, a higher activity of calmodulin-dependent protein kinase (CaMK) has been reported in the zona glomerulosa as compared to the inner cortex (Kubo and Strott 1988) and substrates were identified in rat cells (Kigoshi et al. 1988). More recently, the role of the type II CaMK (CaMKII) in the aldosterone response to Ang II has been evaluated with specific inhibitors (Clyne et al. 1995; Ganguly et al. 1995; Pezzi et al. 1996). In the human adrenocortical cell line H295R, the specific inhibition by KN93 of the acute aldosterone response to Ang II and K^+, but not of the response to addition of 22R-hydroxy-cholesterol (used as a Ca^{2+}-independent substrate for steroidogenesis), suggests that CaMKII may be involved in the control of cholesterol mobilization to the mitochondria.

As previously mentioned, the positive modulation of T-type Ca^{2+} channels by CaMKII in glomerulosa cells or heterologous systems (Barrett et al. 2000; Lu et al. 1994; Wolfe et al. 2002) reveals a short positive feedback loop in calcium signaling.

Calmodulin-dependent kinases are apparently also involved in a more sustained stimulation of glomerulosa cells. Indeed, the Ca^{2+}-regulated expression of aldosterone synthase is calmodulin-dependent (Pezzi et al. 1997) and requires the activation of the CaMKI isoform, as shown by its overexpression in H295R cells, or the expression of constitutively active forms of the enzyme (Condon et al. 2002).

Much less is known about the role of the calmodulin-dependent protein phosphatase, calcineurin, on glomerulosa cell function. Nevertheless, this enzyme has been proposed to be involved in the potentiation of agonist-stimulated cAMP formation by Ang II in bovine glomerulosa cells (Baukal et al. 1994) and to negatively modulate the activity of the Ins(1,4,5)P_3 receptor, probably by reversing the activation of the receptor by PKC-mediated phosphorylation (Poirier et al. 2001). Interestingly, maintaining a high phosphorylation state of the Ins(1,4,5)P_3 receptor by blocking calcineurin with FK506 blocks Ang II-induced cytosolic Ca^{2+} oscillations that are replaced by a plateau level of high Ca^{2+}.

Mitochondrial Calcium Homeostasis. The rate-limiting step of steroid synthesis is the conversion of cholesterol to pregnenolone, an early step of steroidogenesis that occurs within the mitochondria and can be directly stimulated by increasing cytosolic Ca^{2+} levels within physiological ranges (Python et al. 1995). The demonstration that preventing Ca^{2+} influx into the mitochondrial matrix through the ruthenium-sensitive uniporter inhibits aldosterone production induced in permeabilized bovine glomerulosa cells by rising ambient calcium strongly supported a crucial role for intramitochondrial calcium in the control of steroidogenesis (Capponi et al. 1988). The same permeabilized cell model was also used to show that calcium homeostasis in mitochondria is directly dependent on the cytosolic Ca^{2+} and Na^+ concentrations, mitochondrial Ca^{2+} influx increasing in parallel with cytosolic Ca^{2+} and mitochondrial efflux increasing in parallel with cytosolic Na^+ (Rossier et al. 1987). Importantly, above a given cytosolic Ca^{2+} concentration ($[Ca^{2+}]_c$), defined as the mitochondrial set-point, influx largely exceed efflux and large amounts of Ca^{2+} accumulated within the organelle. This is reflected by a large amplification of the mitochondrial Ca^{2+} signal as compared to the cytosolic response, particularly when elicited upon Ang II-induced Ca^{2+} release from intracellular stores (Brandenburger et al. 1996). Moreover, blocking Ca^{2+} extrusion from the mitochondria with a specific inhibitor of the mitochondrial Na^+/Ca^{2+} exchanger has been shown to result in increased mitochondrial Ca^{2+} levels and increased steroid production (Brandenburger et al. 1996).

The higher mitochondrial Ca^{2+} response to Ang II than to extracellular potassium, as well as the presence of many contacts between the endoplasmic reticulum and the outer mitochondrial membrane in glomerulosa cells (Brandenburger et al. 1999) led to the hypothesis that, as in other cell types (Rizzuto et al. 1993), Ca^{2+} is released by Ins(1,4,5)P_3 in proximity of the mitochondria and could even be transferred into the organelle in a quasi-synaptic manner.

Whereas it is clear that a rise in Ca^{2+} in the mitochondrial matrix is sufficient for increasing cholesterol supply to the P450 side chain cleavage (P450scc) enzyme (Cherradi et al. 1996, 1997b) and therefore for stimulating steroidogenesis, its molecular mode of action in the organelle is still poorly understood (Rossier et al. 1996a). High concentrations of the cation are able to activate the permeability transition pore, but this extreme situation leads to cell apoptosis and is probably not the mechanism of activation of steroidogenesis. A swelling of mitochondria is generally associated with a sustained load of Ca^{2+} into the organelle, possibly as a consequence of modulation of ionic conductances in the inner membrane. This swelling has been postulated to increase the number of contact sites between the inner and the outer mitochondrial membranes, therefore favoring the transfer of cholesterol to the P450scc enzyme.

Side chain cleavage of cholesterol requires NADPH, which is formed by reduction of NADP at the expense of NADH. The formation of reduced pyridine nucleotides has been examined in rat glomerulosa cells and was shown to be correlated to, and dependent on Ca^{2+} signaling upon stimulation with Ang II (Pralong et al. 1994; Rohacs et al. 1997a, 1997b). Although increased NADPH production in the mitochondria probably contributes to the control of steroidogenesis by Ang II, it is unlikely that this process is strictly rate limiting, because addition of hydroxylated cholesterol derivatives, which freely cross the mitochondrial intermembrane space, results in pregnenolone formation in a calcium-independent manner (Python et al. 1995).

2.2.4 Calcium Extrusion

After Ca^{2+} mobilization from intracellular Ca^{2+} stores or from the extracellular medium and activation of the various Ca^{2+}-sensitive targets in the cell, Ca^{2+} signaling is terminated by its extrusion out of the cytosol. This process is achieved either by pumping Ca^{2+} back into the stores, essentially through a thapsigargin-sensitive SERCA pump, or outside the cell. The first mechanism is apparently required during the decreasing phase of Ca^{2+} oscillations, because thapsigargin abolishes these fluctuations (Rossier and Capponi 2002; Rössig et al. 1996). Calcium extrusion out of the cell has been first shown by measuring $^{45}Ca^{2+}$ efflux upon Ang II stimulation (Elliott et al. 1985; Foster and Rasmussen 1983; Kojima et al.1985a), but its apparent activation by the hormone essentially reflected Ang II action on Ca^{2+} release from the stores and the activation of plasma membrane Ca^{2+} pumping activity by the resulting rise in cytosolic Ca^{2+} concentration. In fact, it has been recently demonstrated that Ca^{2+} extrusion after Ins(1,4,5)P_3-mediated Ca^{2+} release principally occurs through the plasma membrane Na^+/Ca^{2+} exchanger, and that this process is inhibited (rather than activated) by Ang II in a P38 MAPK-dependent manner (Startchik et al. 2002). Preventing this control of Ca^{2+} extrusion by Ang II resulted in a parallel decrease in the Ca^{2+} signal and in aldosterone production. It was therefore suggested that Ang II, in addition to activating Ca^{2+} entry into the cytosol of glomerulosa cells, also reduces its extrusion in order to prevent the establishment of futile cycles of Ca^{2+} across the plasma membrane and thus to optimize the intracellular Ca^{2+} signal.

2.3 Kinase Activation and Lipid Metabolism

2.3.1 Protein Kinase C

Activation. Diacylglycerol (DAG) is produced parallel to Ins(1,4,5)P_3 upon activation of PLC by Ang II (Hunyady et al. 1990; Kojima et al. 1986b), and therefore the combination of diacylglycerol and cytosolic Ca^{2+} elevation leads to the activation of PKC (Rasmussen et al. 1995). A biphasic response to Ang II was observed in bovine glomerulosa cells, with a second rise in DAG being less dependent on extracellular calcium than Ins(1,4,5)P_3 production (Hunyady et al. 1990). This observation therefore suggests that upon sustained stimulation, a major part of DAG is derived from sources other than phosphoinositides (Bollag et al. 1990). It has been proposed that the different fatty acid composition of the DAG produced from various sources could result in differential activation of PKC. For example, sn-1,2 dioctanoylglycerol (DiC8) mimics the effect of Ang II on aldosterone production and potassium permeability in bovine glomerulosa cells (Lobo et al. 1990), but not sn-1,2 oleoylacetylglycerol (OAG) or phorbol esters (Lobo and Marusic 1991), suggesting that DiC8 and OAG could affect distinct PKC isoenzymes or even PKC-independent pathways. Indeed, the family of PKCs consists of various isoenzymes with similar but not identical activation characteristics and substrate specificity. Both alpha and epsilon PKC isoforms have been detected in rat glomerulosa cells (Natarajan et al.1994), while zeta is also present in some species such as hamster and in the human H295R cell line (LeHoux et al. 2001). The classic and new isoforms (alpha and epsilon) are activated by DAG, whereas the atypical zeta isoform has been shown to be regulated by the PI3 kinase-derived PtdIns(3,4,5)P_3. The lipoxygenase products, 12- or 15-HETE also caused a marked increase in PKC activity, predominantly by activation of the epsilon isoform.

Activation of the enzyme can be assessed by measuring specific substrate phosphorylation or enzyme translocation from the cytosol to the membrane fraction. The latter effect has been shown to occur upon stimulation of glomerulosa cells with Ang II (but not K^+) and to be Ca^{2+}-dependent (Lang and Vallotton 1987; Farago et al. 1988). One report also suggests that Ang II-stimulated PKC translocation requires AT_1 receptor internalization, but Ins(1,4,5)P_3 formation (and aldosterone production) is insensitive to receptor retention in the plasma membrane (Kapas et al. 1994).

Substrates and Function. The temporal pattern of protein phosphorylation has been analyzed in bovine glomerulosa cells upon stimulation with Ang II or PMA, in order to identify specific PKC substrates, and compared to the kinetics of aldosterone secretion (Barrett et al. 1986). Interestingly, PMA activation of protein phosphorylation required a concomitant elevation of Ca^{2+} induced by an ionophore, even though in other studies, PMA alone was sufficient to induce PKC translocation to the membrane. At least 12 cytosolic proteins have been then identified as PKC substrates in rat glomerulosa cells (Kigoshi et al. 1988). More recently, a synthetic peptide that is a specific substrate for PKC has been shown to be phosphorylated within 30 min of stimulation with Ang II. Phosphorylation rapidly ceased upon Ca^{2+} entry blockade with dihydropyridines or upon inhibition of PKC, demonstrating a critical role of Ca^{2+} entry in the regulation of PKC activity by Ang II (Kojima et al. 1994).

Although both activation of PKC (with phorbol esters) and of Ca^{2+}-dependent mechanisms (elicited with a calcium ionophore) were proposed early on to be both required at specific times for an optimal stimulation of aldosterone production (Kojima et al. 1984a), the role of PKC in glomerulosa cell steroidogenesis has been particularly controversial. Indeed, PKC activation with phorbol esters induced an inhibition of aldosterone production elicited by various agonists, including Ang II, in rat and bovine glomerulosa cells, while pretreatment of the enzyme with inhibitors enhanced the steroidogenic response, demonstrating that PKC activation in these cells exerts a negative feed back control on aldosterone (Aptel et al. 1996; Capponi et al. 1989; Hajnoczky et al. 1992b; Kojima et al. 1986b; Lang et al. 1991). This was supported by the observation that pharmacological stimulation of PKC inhibits Ang II-induced activation of PLC (Kojima et al. 1986b) and T-type Ca^{2+} channel activity (Rossier et al. 1995), two PKC actions leading to reduction of important sources of steroidogenic calcium. PKC stimulation also reduces the activity of the aldosterone synthase promoter (LeHoux et al. 2001), suggesting acute as well as long-term negative effects of PKC on aldosterone steroidogenesis. In addition, down-regulation of PKC activity by extended exposure of rat glomerulosa cells to phorbol esters did not impair Ang II-stimulated aldosterone secretion (Nakano et al. 1990), or even enhanced the stimulatory effect of the hormone (Hajnoczky et al. 1992b), showing that PKC activation is not an absolute requisite for Ang II cell signaling.

Nevertheless, a positive action of PKC has also been observed (Bodart et al. 1995; Ganguly and Waldron 1994; Kapas et al. 1995), apparently on the late steps of aldosterone biosynthesis (Vinson et al. 1989).

Conflicting conclusions on the role of PKCs in the control of aldosterone production by Ang II probably result from the complexity of the signaling involving these enzymes or from the ability of the commonly used pharmacological agents to mimic or prevent Ang II stimulation of PKCs, for example in terms of kinetics. Finally, it should be remembered that the various PKC isoforms expressed in glomerulosa cells (Natarajan et al. 1994) and their specific substrates may modulate a variety of cellular functions that are not yet completely clarified (Ganguly and Davis 1994).

2.3.2 Mitogen-Activated Protein Kinases

Mitogen-activated protein kinases (MAPKs) are a family of ubiquitous and highly conserved serine-threonine protein kinases activated by diverse stimuli, including Ang II (Widmann et al. 1999). Mammalian MAPKs have been classified into three groups, which differ in their activation pathway: the MAPK extracellular signal-regulated kinase (ERK or p42/44 MAPK), the MAPK c-Jun N-terminal kinase (JNK) and the p38 MAPK. Angiotensin II, through its AT_1 receptor, has been shown to activate some of these kinases in various cell types and to exert its control on cell proliferation via this mechanism. We will now discuss specifically the effect of these kinases on aldosterone production (see Fig. 2).

Activation. Angiotensin II stimulates ERK in bovine and rat glomerulosa cells (McNeill et al. 1998; Tian et al. 1998). In the rat adrenal, the enzyme is essentially expressed in the zona glomerulosa and the medulla. Upon stimulation with

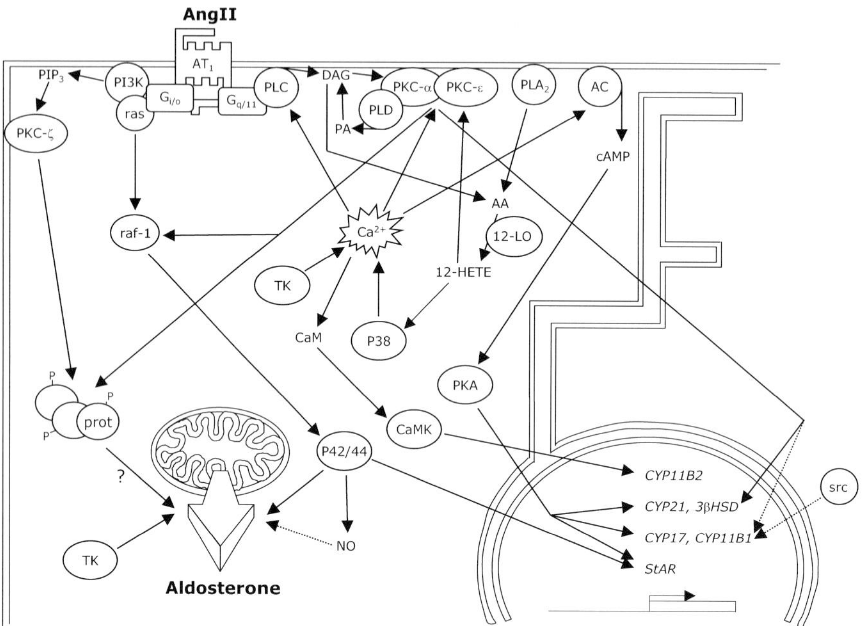

Fig. 2 Angiotensin II-induced signaling through kinases and lipases. Putative involvement of kinases and lipases in angiotensin II-induced aldosterone biosynthesis in adrenal glomerulosa cells. The abbreviations used are the same as in Fig. 1. *PKC-α, -ε, -ζ*, protein kinase C of the α-, ε-, and ζ-isoform; *PLD*, phospholipase D; *PA*, phosphatidic acid; *PLA$_2$*, phospholipase A_2; *AA*, arachidonic acid; *12-LO*, 12-lipoxygenase; *12-HETE*, 12-hydroxyeicosatetraenoic acid; *PKA*, cAMP-dependent protein kinase; *src*, Src kinase; *StAR*, steroidogenic acute regulatory protein; *CYP17*, *CYP11B1*, *CYP21*, genes coding for cytochrome P450 enzymes 17-hydroxylase, 11-hydroxylase, and 21-hydroxylase; *3βHSD*, gene coding for 3β-hydroxysteroid dehydrogenase; *NO*, nitric oxide; *ras*, small ras G protein; *PI3K*, phosphatidylinositol 3-kinase; *PIP$_3$*, phosphatidylinositol 3,4,5-trisphosphate; *raf-1*, raf-1 kinase; *prot*, various phosphorylated proteins; *CaMK*, CaM-dependent protein kinase; *P42/44*, p42/44 mitogen-activated protein kinase (*ERK*)

Ang II, the activity in glomerulosa cells is redistributed from the cytosol to the nucleus. Although neither Ca^{2+} influx nor Ca^{2+} release was sufficient for activating MAPK in bovine cells, Ca^{2+} appeared to play a permissive role in the response to Ang II (Tian et al. 1998). In the same study, it was demonstrated that ras and raf-1 kinase, two classic activators of the MAPK pathway, were stimulated by Ang II with time courses that correlated with that of MAPK activation, but pharmacological activation of PKC was sufficient to activate both MAPK and raf-1. It therefore appears that Ang II stimulates multiple pathways to MAPK activation in bovine glomerulosa cells via PKC and ras/raf-1 kinase. The PKC-independent pathway leading to raf-1 kinase activation was then shown to involve a pertussis toxin-sensitive G protein, phosphatidylinositol 3-kinase and ras, in addition to being negatively modulated by Ca^{2+} (Smith et al. 1999). Interestingly, MAPK basal activity was modulated in rat glomerulosa cells by low- or high-sodium diets in parallel to AT_1 receptors (McNeill and Vinson 2000).

In contrast, the activation of the p38 MAPK by Ang II in glomerulosa cells is poorly documented. Nevertheless, it has been recently reported that Ang II leads to a rapid activation of p38 MAPK in bovine glomerulosa cells and that this response is partially prevented by caffeine (Startchik et al. 2002). A similar effect of Ang II was previously observed in H295R cells, in which kinase activation by the hormone was proposed to be mediated by the lipoxygenase product, 12-HETE (Gu et al. 2003). In the same study, no activation of JNK by Ang II was observed.

Steroidogenic Function. Activation of MAPK is primarily linked to cell proliferation and growth-promoting effects of Ang II on bovine adrenocortical cells, previously attributed to 12-lipoxygenase and PKC (Natarajan et al. 1992) are probably mediated by MAPKs. In contrast, a direct function of these kinases on aldosterone biosynthesis is still unclear. While p42/44 MAPK inhibition with the PD098059 compound was reported not to affect aldosterone production induced by Ang II in rat glomerulosa cells (Cote et al. 1998) or in H295R cells (Yang et al. 1999), the same treatment reduced StAR expression and aldosterone production in Ang II-stimulated bovine cells (Osman et al. 2002). Moreover, transfection of newborn rat adrenal glands with a plasmid containing the EJ-Ha-ras oncogene not only led to a proliferative cell line, but these cells produced high levels of aldosterone when supplied with corticosterone and deoxycorticosterone in the presence of Ang II (Maume et al. 1991). On the other hand, MAPK is known to stimulate nitric oxide (NO) production in many cell types, an agent that inhibits steroidogenesis by binding to the heme group of P450 enzymes, particularly the P450 side chain cleavage and aldosterone synthase enzymes (Peterson et al. 2001). Further studies are clearly required to determine whether activation of p42/44 MAPK by Ang II is a mean to control acute and/or chronic aldosterone production.

In contrast, p38 MAPK inhibition has been clearly shown to reduce aldosterone secretion evoked by Ang II (Gu et al. 2003) and it has been recently suggested that the permissive kinase action could be exerted through the optimization of the Ca^{2+} signal evoked by the hormone (Startchik et al. 2002).

2.3.3 Protein Kinase A

Classically, Ang II has been considered as negatively linked to cAMP production and protein kinase A (PKA) activation in adrenal glomerulosa cells. Indeed, inhibition of the adenylyl cyclase activity through a Gi protein has been reported in rat adrenal glomerulosa cells (Hausdorff et al. 1987; Woodcock and Johnston 1984). However, as previously mentioned, in bovine cells, Ang II was found to significantly increase cAMP production in response to ACTH (Baukal et al. 1994; Burnay et al. 1994), and, at least in these cells, cAMP can therefore be considered as a contributor to Ang II-induced steroidogenesis. The role of cAMP and PKA activation in the steroidogenic response to ACTH has been extensively studied. For example, activation of this signaling pathway with forskolin or cAMP analogs stimulates the expression of the steroidogenic acute regulatory (StAR) protein and various steroidogenic enzymes. In H295R cells, the role of PKA on the various enzymes has been compared to that of PKC and CaMK (Bird et al. 1998a, 1998b). While PKA induces the expression of all the enzymes tested except aldosterone synthase (CYPB11B2), the latter was specifically activated by CaMK, and PKC enhanced the expression of the 3βHSD and CYP21 enzyme while repressing CYP17 and CYP11B1. This differential regulation of enzyme subsets by each kinase probably reflects mechanisms involved in adrenal cortex development and zonation. Indeed, CaMK and PKC, essentially expressed and activated in the zona glomerulosa, lead to mineralocorticoid synthesis, while PKA, primarily activated by ACTH throughout the cortex, may have a more general action on steroidogenesis and favor glucocorticoid production in cells displaying low PKC and CaMK activity.

2.3.4 Tyrosine Kinases

Stimulation of the AT_1 receptor may also trigger a variety of different tyrosine kinases (TK) and stimulate the early gene response controlling cell growth (Bernstein and Berk 1993; Kapas et al. 1995; Marrero et al. 1995; Smith and Timmermans 1994). In bovine and rat glomerulosa cells, various TK inhibitors have been reported to reduce Ang II-induced aldosterone secretion (Aptel et al. 1999; Bodart et al. 1995; Kapas et al. 1995). In one report, however, tyrphostin-23, commonly used as an inhibitor of TKs, was shown to increase steroidogenesis, but through a direct inhibition of the cAMP phosphodiesterase and therefore an activation of the PKA-dependent pathway (Andreis et al. 2000). The negative effect of TK inhibitors on steroidogenesis has been partially attributed to the permissive role played by some TKs in the activation of the capacitative Ca^{2+} influx by Ang II (Aptel et al. 1999; Bodart et al.1995). However, a comparison of the effect of various pharmacological drugs on the conversion of 11-deoxycorticosterone to aldosterone and the use of hydroxylated cholesterol as a substrate for steroidogenesis revealed that specific TKs could also be involved in the regulation of the late steps of steroidogenesis (Aptel et al. 1999). In particular, the src TK appears to be responsible for the inhibition of CYP17 mRNA expression and therefore increases aldosterone formation by repressing alternative steroidogenic pathways (Sirianni et al. 2001).

Clearly, Ang II appears to activate several types of TKs in glomerulosa cells that will exert a global positive control at various steps of steroidogenesis, but further investigations are required to precisely identify which kinases are involved and to determine their specific action.

2.3.5
Phospholipases and Lipoxygenases

In addition to the activation, through a G_q protein, of a phosphoinositide-specific PLC responsible for the formation of Ins(1,4,5)P_3 and DAG, Ang II also exerts its effects in glomerulosa cells through stimulation of additional phospholipases such as PLA2 and PLD, and it has long been recognized that DAG produced during sustained cell stimulation may be derived from other membrane lipids, such as phosphatidylcholine. The products of these lipases can then be further metabolized to produce additional cellular messengers involved in the regulation of multiple pathways.

Whereas DAG can be produced directly by the action of PLC on phospholipids, this compound can also be generated by the combined activities of PLD and phosphatidic acid phosphohydrolase. Activation of PLD by Ang II has been shown in bovine glomerulosa cells, as well as the ability of exogenous enzyme to elicit DAG production and potentiate the steroidogenic response to calcium channel agonists (Bollag et al. 1990). In contrast to the response to carbachol, activation of the enzyme by Ang II was sustained in these cells (Jung et al. 1998) and appeared to involve PKC, but not Ca^{2+} (Bollag et al. 2002). This possible mechanism has to be considered in relationship with the previously mentioned biphasic production of DAG in response to Ang II. Finally, interfering with PLD-generated signals inhibited Ang II-induced aldosterone secretion but not that elicited by addition of hydroxycholesterol derivatives, which bypass the signaling pathways controlling early steps of steroidogenesis.

Phospholipase A2 (PLA2) may produce arachidonic acid (AA), which, by itself or through its many metabolites (Campbell et al. 1991), can modulate aldosterone production. Indeed, PLA2 inhibitors reduce aldosterone secretion, while exogenous PLA2 or addition of AA transiently activates steroidogenesis, apparently through modulation of the Ca^{2+} signaling (Enyedi et al. 1981; Kojima et al. 1985b). In the same studies, it was shown that cyclooxygenase inhibitors had no effect, while lipoxygenase inhibitors markedly reduced Ang II-stimulated aldosterone secretion.

Glomerulosa cells produce 12-hydroxyeicosatetraenoic acid (12-HETE), the 12-lipoxygenase product, in response to Ang II, but not upon stimulation with other agonists such as K^+ or ACTH (Gu et al. 1994; Nadler et al. 1987; Natarajan et al. 1990). Although also present in glomerulosa cells, 15-HETE does not affect aldosterone formation and is not increased by Ang II (Natarajan et al. 1988a). Arachidonic acid used for this reaction has been shown to be essentially produced from DAG by the action of DAG lipase rather than by hydrolysis of phosphoinosites by PLA2 (Hunyady et al. 1985; Natarajan et al. 1990; Natarajan et al. 1988b).

The action of lipoxygenase products has been partially elucidated and these compounds appear to modulate both PKC activity and the cellular calcium messenger system. Indeed, treatment of rat glomerulosa cells with 12- or 15-HETE caused a marked increase of PKC activity in the membrane fraction, predominantly of the epsilon isoform (Natarajan et al. 1994). Adding 12-HETE (but not 15-HETE) to these cells has also been recently shown to induce a concentration-dependent rise in cytosolic Ca^{2+} levels that was sustained for several minutes and also observed in Ca^{2+}-free medium, suggesting mobilization from intracellular stores (Stern et al. 1993). Moreover, a lipoxygenase blocker significantly reduced Ang II-induced Ca^{2+} signaling in parallel to aldosterone production. This effect of 12-HETE on Ca^{2+} could be explained by a modulation of Ca^{2+} extrusion through the Na^+/Ca^{2+} exchanger mediated by p38 MAPK, as recently proposed (Startchik et al. 2002).

3 Cholesterol Metabolism and Steroidogenesis

Aldosterone is the final product of a cascade involving:

1. Cholesterol ester supply to intracellular storage organelles
2. Mobilization and hydrolysis of cholesterol esters
3. Free cholesterol supply to the inner mitochondrial membrane
4. Mitochondrial metabolism of cholesterol by tissue-specific enzymes

Each of these steps constitutes a potential regulatory impact site for activators of mineralocorticoid biosynthesis. We will next summarize the current state of knowledge on the regulation of the various steps leading from cholesterol to aldosterone.

3.1 Cholesterol Supply for Aldosterone Biosynthesis

Three main mechanisms can contribute to cholesterol homeostasis in mammalian somatic cells:

1. Intracellular de novo cholesterol synthesis from acetyl coenzyme A
2. Low-density lipoprotein (LDL)-mediated delivery of cholesterol esters
3. Selective uptake of high density lipoprotein (HDL) cholesterol esters

Depending on whether plasma cholesterol is carried mainly in LDL or HDL form, animal species are believed to derive their cholesterol for steroidogenesis mainly through endocytosis via the LDL receptor (human) or through a nonendocytic pathway involving HDL (rodents).

3.1.1
Endogenous Cholesterol Synthesis

The rate-limiting enzyme in de novo cholesterol synthesis is hydroxymethylglutaryl coenzyme A (HMG CoA) reductase, the enzyme that converts β-hydroxy-β-methyglutaryl-CoA to mevalonic acid. In the adrenal cortex of various species, the expression and activity of this enzyme is powerfully regulated by ACTH (Balasubramaniam et al. 1977; Brody and Black 1991; Lehoux et al. 1989; Mason and Rainey 1987; Rainey et al. 1992). In the guinea pig, the aldosterone response to ACTH appears to depend essentially upon activation of de novo cholesterol synthesis (Brody and Black 1991). Interestingly, however, the activity of Acyl-CoA:cholesterol acyltransferase (ACAT), the enzyme that esterifies cholesterol to fatty acids for storage in lipid droplets, is insensitive to ACTH in the outer zone of the guinea pig adrenal cortex (Brody and Black 1988), suggesting an almost maximal activity of the enzyme in the basal state. Whether Ang II or K^+, which are even more important key players in the acute stimulation of aldosterone production than ACTH, also modulate the activity of enzymes involved in de novo cholesterol biosynthesis, is unknown.

3.1.2
LDL Cholesterol and Aldosterone Synthesis

LDL receptor activity (Kovanen et al. 1979), mRNA (Russell et al. 1983) and protein (Kroon et al. 1984) were demonstrated in the bovine adrenal cortex almost 2 decades ago. ACTH increases the levels of LDL receptor mRNA in the rat adrenal but not in the hamster (Lehoux and Lefebvre 1991), and Ang II stimulates specific LDL binding in zona glomerulosa-rich primary cultures of bovine adrenal cortical cells (Leitersdorf et al. 1985).

In spite of this early recognition of the presence and regulation of LDL receptors in the adrenal cortex and, more specifically, in zona glomerulosa cells, controversial results have been produced with respect to the contribution of LDL cholesterol esters to aldosterone biosynthesis. Indeed, LDL was shown to increase both K^+- (Campbell 1982) and ACTH-induced (Nagy et al. 1984) aldosterone production in the rat. In contrast, in bovine adrenocortical cells, conflicting data showed that LDL either had no effect on basal and stimulated aldosterone synthesis, thus not appearing to be the preferred source of cholesterol (Simpson et al. 1989), or increased the response to Ang II and ACTH (Leitersdorf et al. 1985). In the latter work, Ang II treatment enhanced specific LDL binding and uptake, although there was preferential uptake of the cholesterol ester moiety, a process that would exclude LDL receptor-mediated endocytosis (Brown and Goldstein 1986). In other studies, it was found that exogenous LDL cholesterol is the main source of precursor for steroid biosynthesis in both human adult and fetal adrenals and bovine adrenals (Higashijima et al. 1987). Finally, in the guinea pig, LDL enhanced steroid production, with the exception of aldosterone, to a greater extent in glomerulosa than in fasciculata cells. However, the aldosterone response to ACTH treatment was enhanced in the presence of LDL but the stimulation of LDL utilization was smaller compared to that observed in fasciculata cells (Black 1987).

Therefore, although a potential involvement of LDL as a regulated source of cholesterol precursor for aldosterone biosynthesis in glomerulosa cells was reported several years ago, marked species differences and contradictory results have hampered the building of a coherent understanding of this process.

3.1.3 Selective Uptake of HDL Cholesterol Esters

The recent discovery and identification of a specific, well-defined receptor for HDL, SR-BI (Acton et al. 1996; Rigotti et al. 1997; Williams et al. 1999), which promotes uptake of HDL cholesterol esters through a selective pathway that is distinct from LDL receptor-mediated endocytosis of LDL (Brown and Goldstein 1986), has cast new light on the mechanisms of fuel supply to steroidogenic cells and could solve some of the above contradictions. SR-BI is expressed mainly in the liver and steroidogenic tissues, and it is particularly abundant in the adrenal cortex (Landschulz et al. 1996). The delivery of HDL cholesterol esters involves binding of the α-helical repeats of apolipoprotein A-I (apoA-I), the protein moiety of HDL, to SR-BI. SR-BI is a 57-kDa membrane protein (apparent molecular weight on gel, 82–86 kDa) possessing two transmembrane domains that flank a large extracellular loop (Williams et al. 1999). The interaction of apoA-I with this extracellular loop is thought to lead to the formation of a lipophilic channel through which cholesterol esters move down their concentration gradient to the plasma membrane (Williams et al. 1999).

Thus, only the lipid content of HDL enters the cell through SR-BI, whereas the apoA-I-containing moiety is not internalized by endocytosis, as it is the case for LDL. This feature made it possible to develop a test for visualizing and imaging the fate of HDL cholesterol esters. This elegant technique, which was first applied to ovarian granulosa cells (Reaven et al. 1996), involves reconstituting apoE-free HDL—to avoid potential recognition by LDL receptors—with a fluorescent cholesterol ester derivative, BODIPY-CE (4,4-difluoro-5,7-dimethyl-4-bora-3*a*,4*a*-diaza-S-indacene-3-dodecanoate cholesterol ester). Interestingly, this molecule cannot be hydrolyzed by neutral cholesterol ester hydrolases and therefore labels specifically cholesterol esters that are internalized through the selective uptake pathway and accumulate within lipid droplets (Reaven et al. 1996).

It was recognized soon after the discovery of SR-BI that the selective uptake of HDL cholesterol esters plays a major role in adrenal steroidogenesis, maintaining up to 90% of the supply of cholesterol destined for steroid biosynthesis in rodents. Thus, in apoA-I-deficient mice, both basal and stress-stimulated corticosteroid production are dramatically reduced (Plump et al. 1996). SR-BI is the major pathway for the delivery of HDL cholesterol to the adrenal steroidogenic machinery, as shown by the fact that an antibody directed against a segment of the putative extracellular domain of mouse SR-BI prevents selective HDL cholesterol ester uptake in mouse adrenocortical cells (Temel et al. 1997). HDL cholesterol esters are taken up preferentially over other major HDL phospholipids in these same cells, via a nonaqueous pathway (Rodrigueza et al. 1999). Moreover, adrenal SR-BI expression is regulated by ACTH in vivo in mice (Rigotti et al. 1996; Sun et al. 1999), in murine adrenocortical cell lines (Rigotti et al. 1996; Wang et al. 996), and by cAMP derivatives in human adrenocortical carcinoma NCI-H295 cells (Martin et al. 1999) In bovine zona glomerulosa and in human adrenocortical carcinoma NCI-H295R cells, Ang II selectively promotes cholesterol ester uptake from HDL through an increase in the expression of SR-BI (Cherradi et al. 2001; Capponi 2002).

3.2 Mobilization and Hydrolysis of Cholesterol Esters

Adrenal glomerulosa cells, like most steroid hormone-producing cells, store fuel for steroidogenesis in the form of cholesterol esters within cytosolic lipid droplets (Toth et al. 1997). These contain a large core of neutral lipid covered by a phospholipid monolayer, which is in turn coated by a proteinaceous layer including, among others, lipid droplet-specific proteins called perilipins A, B and C (Greenberg et al. 1993; Lu et al. 2001). Perilipin expression is limited to steroidogenic cells and adipocytes; in the latter cell type, perilipin A has been shown to translocate (Clifford et al. 2000) and to modulate protein kinase A-mediated lipolysis (Souza et al. 2002). In adrenal cortical and Leydig cells, perilipins are associated with cholesteryl ester droplets (Servetnick et al. 1995). Activation of PKA causes detachment of perilipin from the lipid droplet surface (Fong et al. 2002). Whereas perilipin ablation results in leanness and aberrant adipocyte lipolysis (Martinez-Botas et al. 2000; Tansey et al. 2001), little is known on the effects of perilipin knock-out on adrenal steroidogenesis.

Hormonal stimulation of steroid-producing cells results in prompt mobilization of cholesterol esters from intracellular lipid droplets (Boyd et al. 1983; Vahouny et al. 1984). These cholesterol esters are hydrolyzed to free cholesterol by the cholesterol ester hydrolase enzyme (CEH), whose physicochemical properties and regulation have been thoroughly investigated in the adrenal cortex (Brody and Black 1988; Kraemer et al. 2002; Lee et al. 1997; Mikami et al. 1984; Nishikawa et al. 1981). Neutral cholesterol esterase activity purified from the adrenal has been shown to be identical to HSL purified from adipose tissue (Cook et al. 1982). In addition, immunoreactive hormone-sensitive lipase (HSL) (Kraemer et al. 1993) and HSL mRNA (Kraemerr et al. 1991) can be detected in adrenal tissue. Most importantly, HSL deficiency in mice is associated with dramatic changes in adrenal neutral CEH activity (Kraemer et al. 2002) and with profound morphological changes—a marked accumulation of lipid droplets—in the adrenal cortex (Li et al. 2002).

The regulation of CEH activity through cAMP-dependent mechanisms has been extensively studied (Beckett and Boyd 1977; Brody and Black 1988; Klemcke 1992; Mikami et al. 1984; Nishikawa et al. 1981; Nishikawa et al. 1988). In the adrenal cortex, ACTH induces marked reductions of cholesterol ester levels with a concomitant increase in CEH activity (Vahouny et al. 1984). Like HSL (Belfrage et al. 1980; Cordle et al. 1986; Stralfors and Belfrage 1983), its homologue in adipose tissue, CEH is activated in vitro following phosphorylation by cAMP-dependent protein kinase (PKA) (Colbran et al. 1986; Cook et al. , 1982 1983; Sonnenborn et al. 1982). The Ang II challenge also leads to an increase in the CEH phosphorylation state and activity in bovine zona glomerulosa cells, an effect that occurs via activation of the p42/p44 MAPK pathway (Cherradi et al. 2003).

3.3 Intramitochondrial Cholesterol Transfer

Once free cholesterol has been generated from cholesterol esters by CEH, it must be transferred to the mitochondria in order to be metabolized into steroids. This process has remained until now the least understood. Clearly, the cytoskeleton must play an important role in this transfer (Feuilloley and Vaudry 1996). In addition, steroidogenic cell-specific proteins such as sterol-carrier protein-2 (SCP-2) (Baum et al. 1997; Puglielli et al. 1995; Yanase et al. 1996) or steroidogenesis activatory polypeptide (SAP) (Pedersen and Brownie 1983) appear to contribute to some extent to cholesterol supply to the mitochondria.

When it has reached the outer mitochondrial membrane, cholesterol has to overcome an almost insurmountable obstacle, the aqueous intermembrane space, before it can reach the inner mitochondrial membrane, where the enzymes of the steroidogenic cascade, such as the P450 side-chain cleavage (P450scc) enzyme, are located. Over the last 7–8 years, a major breakthrough has been achieved with the discovery of the steroidogenic acute regulatory (StAR) protein (Clark et al. 1994) and of its crucial role in mediating intramitochondrial transfer. Furthermore, a critical importance of the StAR protein in developmental processes of the adrenal cortex has also been evidenced in StAR knockout mice (Caron et al. 1997; Hasegawa et al. 2000). Numerous exhaustive reviews have been published on the StAR protein (Christenson and Strauss 2000; Clark and Stocco 1996; Stocco 1997, 2001; Stocco and Clark 1996a, 1996b). Research has focused lately on the mode of action of the StAR protein and on the regulation of its expression under trophic hormone stimulation. We will summarize hereafter the latest most significant findings in these two areas.

3.3.1
Mode of Action of the StAR Protein

The StAR protein precursor contains an N-terminal mitochondrial targeting sequence and it was initially thought that intramitochondrial cholesterol transfer to the P450scc enzyme was facilitated during StAR precursor import into the mitochondria via contact sites (Clark and Stocco 1996; Stocco and Clark 1996a) It was later found, however, that the C-terminal region of the protein is crucial for its biological activity (Arakane et al. 1996), and this observation was corroborated by the fact that all the known mutations in lipoid congenital adrenal hyperplasia, a dramatic disease state in which patients are practically unable to synthesize any steroids, are localized within the C-terminal domain of the StAR protein (Bose et al. 1996).

It is now known that StAR belongs to a family of proteins showing homology to the StAR protein and sharing a 200- to 210-amino acid sequence termed the StAR-related lipid transfer (START) domain (Romanowski et al. 2002; Soccio et al. 2002; Tsujishita and Hurley 2000). Both the StAR protein (Petrescu et al. 2001) and the START domain (Tsujishita and Hurley 2000) have been shown to bind cholesterol. StAR appears to adopt a molten globule conformation during its import into mitochondria (Bose et al. 1999; Song et al. 2001) and to interact with the outer mitochondrial membrane (Arakane et al. 1998; Bose et al. 2002). The precise molecular mechanism of StAR-mediated mitochondrial import of cholesterol is not yet fully understood. It would appear that an interaction is required with some factor(s) at the surface of the mitochondria. One of the potential partners of the StAR protein at the outer mitochondrial membrane that appears to be of critical importance could be the peripheral-type benzodiazepine receptor (Amri et al. 1999; West et al. 2001).

3.3.2
Regulation of StAR Protein Expression

Because of the vital importance of the StAR protein, it is not surprising that considerable attention has been and still is dedicated to the understanding of the mechanisms controlling its expression. Soon after the cloning of the protein, the sequence of the human (Sugawara et al. 1996, 1997b) and the mouse (Caron et al. 1997) promoter region of the StAR gene have been characterized. Various consensus sequences of known response elements have been identified and the corresponding transcription factors shown to contribute to the regulation of StAR gene expression. We briefly describe hereafter the principal transcription factors participating in this control. A more detailed review has recently been published by Reinhart et al. (1999b).

CREBP/CREM. Most trophic agents acting on steroidogenic tissues (gonads, adrenal cortex) recruit the cAMP-PKA signaling system. Canonical half-sites for CRE (cAMP response element) in the StAR promoter have recently been shown to bind CREB (cAMP response-element binding protein) and CREM (CREB/CRE modulator) in MA-10 mouse Leydig cells (Manna et al. 2002) and to be functionally involved in the acute regulation of StAR gene expression.

C/EBP. The CCAAT/enhancer binding proteins (C/EBPα, C/EBPβ, etc.) are a family of basic region/leucine zipper transcription factors regulating the differentiation and the function of numerous cells types (Lekstrom-Himes and Xanthopoulos 1998). Two consensus C/EBPβ response elements have been identified in the StAR promoter (Reinhart et al. 1999a) and C/EBPβ is required for StAR gene transcription in both rat ovarian cells (Silverman et al. 1999) and MA-10 testicular Leydig cells (Reinhart et al.1999a).

GATA. The GATA family of zinc finger transcription factors are also important modulators of differentiation during vertebrate development (Weiss and Orkin 1995). In reproductive steroidogenic tissues, GATA-1/4/6 appear to participate in the control of StAR gene transcription, presumably through cooperation with other transcription factors (Reinhart et al. 1999a; Silverman et al. 1999; Tremblay et al. 2002). In the mouse and human adrenal, GATA-4 and GATA-6 appear to show differential expression patterns throughout development and adulthood (Kiiveri et al. 2002).

SF-1. In all species, several putative binding sites for steroidogenic factor-1 (SF-1), also called Ad4BP, are present in the StAR gene promoter (Caron et al. 1997; Clark and Combs 1999; Rust et al. 1998; Sandhoff et al. 1998; Sugawara et al. 1997a, 1997b). The number and localization of these binding sites vary from one species to the other. SF-1 is a nuclear transcription factor that was first identified in adrenal cortical cells (Morohashi et al. 1993). The orphan nuclear receptor SF-1 plays a critical role in adrenal and gonadal differentiation, development, and function (Parker and Schimmer 1997). Furthermore, SF-1 has also been shown to regulate the expression of genes encoding cytochrome P450 hydroxylases, and to efficiently transactivate the StAR gene in transient transfection assays in various cell types (Rust et al. 1998; Sandhoff et al. 1998; Sugawara et al. 1997b; Wooton-Kee and Clark 2000). Although the extent of SF-1 involvement in the regulation of StAR gene expression may present species- and cell type-dependent differences (Silverman et al. 1999), it appears that activation of the cAMP signaling pathway leads to increased phosphorylation (Gyles et al. 2001) and/or expression (Aesoy et al. 2002) of SF-1 protein. Liver receptor homologue-1 (LRH or NR5A2), which is closely related to SF-1 and is involved in bile acid biosynthesis, appears to also be expressed in human adrenal cells and to enhance reporter activity driven by the StAR promoter in co-transfection experiments (Sirianni et al. 2002).

DAX-1. The StAR gene promoter also bears a binding site for another orphan member of the nuclear receptor superfamily, DAX-1 (*d*osage-sensitive sex reversal, *a*drenal hypoplasia congenita critical region on the *X*-chromosome, gene *1*)

(Yu et al. 1998; Lalli and Sassone-Corsi 1999). DAX-1 has been shown to act as a powerful repressor of StAR gene expression. Indeed, overexpression of DAX-1 in Y-1 mouse adrenal tumor cells inhibits steroid synthesis and DAX-1 represses both basal and cAMP-induced StAR promoter activity by binding to DNA hairpin secondary structures on the StAR gene promoter or to the SF-1 protein itself (Zazopoulos et al. 1997). Furthermore, overexpression of DAX-1 in Y-1 adrenocortical cells impairs basal and cAMP-stimulated steroid production (Tamai et al. 1996). Conversely, cAMP down-regulates DAX-1 expression in cultured rat Sertoli cells (Tamai et al. 1996). Finally, SF-1 and DAX-1 are colocalized in various endocrine and steroidogenic tissues, suggesting that these two nuclear proteins may be linked in function.

SREBP. Lipoproteins regulate the expression of the StAR protein and StAR promoter activity (Reyland et al. 2000). Because of the importance of cholesterol as a precursor for steroidogenesis, the potential involvement of sterol regulatory element-binding proteins (SREBPs), a family of transcription factors involved in cholesterol metabolism (Brown and Goldstein 1997) and in modulating StAR gene expression has been investigated. Indeed, a near-consensus binding site for SREBP-1a has been identified in a proximal portion of the StAR promoter (–115 to –30) that is the target for the action of multiple transcription factors, and it appears that the StAR promoter is conditionally responsive to high SREBP-1a (Christenson et al. 2001). SREBP-1a could be involved in selective combinations with other transcriptional cofactors (Shea-Eaton et al. 2001).

YY1 and c-FOS. The multifunctional transcription factor Yin Yang 1 (YY1) has been shown to repress StAR gene expression, presumably by binding to SREBP-1a and preventing its binding to DNA (Nackley et al. 2002). Similarly, three putative activator protein-1 (AP-1) binding elements are present in the rat StAR promoter, and treatment of rats with PGF2α, which led to increases of c-Fos, resulted in a repression of StAR mRNA in rat ovaries (Shea-Eaton et al. 2002). Furthermore, c-Fos was shown to bind to AP-1 sites in the StAR promoter (Shea-Eaton et al. 2002), thus suggesting a c-Fos-mediated repression of StAR gene transcription, at least in ovarian cells.

Modulation of Transcription Factor Expression by Ang II. In the adrenal zona glomerulosa cell, the expression of the StAR protein is rapidly increased by factors that activate mineralocorticoid biosynthesis. Indeed, Ang II and ACTH have been shown to stimulate StAR mRNA and StAR protein expression and to concomitantly increase aldosterone production in bovine zona glomerulosa cells (Cherradi et al. 1997, 1998). Moreover, in human H295R adrenocortical carcinoma cells, which bear only very few ACTH receptors, challenge with cAMP analogs, used to mimic adenylyl cyclase activation, Ang II or K+, leads to StAR mRNA and protein expression, via a SF-1-dependent mechanism (Clark and Combs 1999; Clark et al. 1995). Whereas the initial signal transduction mechanisms mediating the steroidogenic action of these activators of aldosterone biosynthesis are well characterized (Capponi and Rossier 1996), the events occurring downstream of Ca^{2+} or cAMP signal generation and leading to the induction of StAR protein expression are still poorly understood. Nevertheless, it appears that Ang II powerfully represses the expression of DAX-1 in bovine glomerulosa cells, an effect that is accompanied by increased StAR protein expression and aldosterone production (Osman et al. 2002).

3.4 Regulation by Ang II of Aldosterone Synthase Expression

The capacity of the zona glomerulosa of the adrenal cortex to produce aldosterone depends essentially on the level of expression of aldosterone synthase (*CYP11B2*). The transcriptional regulation of aldosterone synthase is controlled by Ang II and K^+ (Denner et al. 1996; Rainey 1999; Shibata et al. 1991; Yagci and Muller 1996) and appears to be mediated by calmodulin-dependent kinase I (Condon et al. 2002). Whereas cholesterol transfer into the mitochondria, as described above, corresponds to an acute regulatory phase of aldosterone biosynthesis, increased expression of aldosterone synthase under Ang II control occurs during a chronic stimulatory phase

4 Modulation of Angiotensin II-Induced Aldosterone Secretion

If aldosterone production by adrenal glomerulosa cells is primarily controlled by Ang II (and potassium), its secretion is modulated, positively or negatively, by many paracrine or endocrine factors. We will now briefly describe the role of the local, intra-adrenal renin-angiotensin system and the effect of the principal modulators of Ang II action: potassium, adrenocorticotropic hormone (ACTH) and atrial natriuretic peptide (ANP). We will also indicate how these factors are believed to affect the cellular steroidogenic mechanisms activated by Ang II.

It has been assumed for a long time that the Ang II responsible for the activation of steroidogenesis in glomerulosa cells was coming from the periphery, as a result of the processing of circulating angiotensinogen by renin and converting enzyme. However, local synthesis of renin and intra-adrenal generation of Ang II has long been recognized (Ganten et al. 1983; Mulrow 1988), as recently reviewed (Mulrow 1999). The highest renin activity is localized in the zona glomerulosa where the enzyme apparently participates in the regulation of aldosterone production. Indeed, changes in electrolyte balance can change adrenal renin production; a low-sodium (or high-potassium) diet increasing renin and aldosterone in glomerulosa cells, while a high-sodium diet diminishes both (Mulrow et al. 1988). This adrenal renin-angiotensin system is highly stimulated upon nephrectomy, a situation in which plasma renin is reduced to undetectable levels, and could contribute to the resulting elevation of aldosterone production (Baba et al. 1986). Inhibition of the adrenal renin system is additional evidence of its physiological involvement. For example, in vitro studies revealed that ACE inhibitors prevent the aldosterone response of isolated glomerulosa cells exposed to angiotensin I, to potassium or ACTH (Horiba et al. 1990; Oda et al. 1991; Yamaguchi et al. 1990), and even to Ang II itself (Vinson et al. 1996). Similarly, the AT_1 receptor antagonist, losartan, inhibited basal, ACTH- and K^+-stimulated aldosterone production from bovine and rat glomerulosa cells (Chiou et al. 1994; Gupta et al. 1995), suggesting that locally produced Ang II could participate in the steroidogenic action of potassium and ACTH.

4.1 Potassium

Potassium is in itself a powerful agonist of aldosterone secretion and the sensitivity of adrenal glomerulosa cells to low variations of extracellular potassium concentrations, within the physiological range, makes the cation, besides Ang II, the main physiological regulator of aldosterone. However, the state of potassium balance can also alter the response of aldosterone to Ang II. Indeed, K^+ strongly potentiates steroidogenesis when combined with submaximal concentrations of Ang II, which has been recognized both in vitro and in vivo (Linde et al. 1981; Vallotton et al. 1995). Although the exact mechanism responsible for this potentiation is not entirely elucidated and could involve the expression of enzymes responsible for the late steps of aldosterone biosynthesis (Muller et al. 1989; Tremblay et al. 1992), it is probably primarily linked to the ability of each agonist to differentially mobilize the Ca^{2+} messenger system in glomerulosa cells. This is suggested by the persistence of the synergistic effect of K^+ on aldosterone production when Ang II is replaced with thapsigargin, a drug inducing Ca^{2+} release from intracellular stores and stimulating the capacitative Ca^{2+} influx, a response elicited by Ang II but not K^+ (Burnay et al. 1994; Hajnoczky et al. 1991). It has been also proposed that synergy could result from the concomitant effect of K^+ on membrane potential and of Ang II on voltage-operated calcium channels, such that together they promote enhanced steady-state Ca^{2+} influx (Chen et al. 1999)

Potassium is not only able to potentiate the aldosterone response to Ang II, but it is also permissive for the hormone action. Indeed, K^+-depleted rats fail to respond to sodium depletion by increasing aldosterone, even though the renin-angiotensin system is adequately stimulated (Boyd et al. 1973). Similarly, in circumstances under which the renin-angiotensin is stimulated, secondary hyperaldosteronism is induced and results in both sodium retention and potassium loss. This situation can lead to severe hypokalemia, particularly when combined with volume depletion, resulting for example from abuse of laxatives and diuretics, or in the case of Bartter's syndrome (Vallotton et al. 1995). Under these conditions, whereas plasma renin activity is maximally stimulated, aldosterone concentration is only modestly elevated. Upon administration of K^+ to these patients, one can observe a dramatic rise in aldosterone secretion, which is explained by the removal of what is often called hypokalemic brake. This phenomenon illustrates the control exerted by potassium on the steroidogenic response to Ang II and demonstrates the risk of missing a diagnosis of hyperaldosteronism if mineralocorticoids are assayed prior to normalization of plasma potassium levels.

4.2 Adrenocorticotropic Hormone

Through its general trophic action on the adrenal cortex and its stimulation, via the cAMP pathway, of the expression of various steroidogenic enzymes, ACTH exerts a positive action on Ang II-stimulated aldosterone response and can therefore be considered as an important modulator. Indeed, a low response to Ang II has been associated with low concentrations of ACTH (Foster et al. 1997). ACTH appears to be necessary in a permissive way to maintain the early steps of steroidogenesis of glomerulosa cells (Boyd et al. 1972).

Nevertheless, several antagonist interactions between Ang II and ACTH, at the level of the cellular messengers, have been also reported. In rat adrenal glomerulosa cells, ACTH inhibits Ang II-stimulated Ins(1,4,5)P_3 formation, in a cAMP-independent manner (Woodcock 1989). Moreover, an antagonist peptide of the ACTH receptor has been shown to increase aldosterone formation in these cells through a mechanism involving the activation of the Ang II receptor (Malendowicz et al. 1998). Conversely, as previously mentioned, in rat cells (but not in bovine cells), Ang II reduces ACTH-elicited cAMP production through a Gi protein (Woodcock and Johnston 1984). Angiotensin II also inhibits the induction of CYP17 by ACTH in a concentration-dependent manner, as shown in bovine glomerulosa cells (Galtier et al. 1996).

4.3 Atrial Natriuretic Peptide

Atrial natriuretic peptide (ANP) is a potent negative regulator of aldosterone synthesis that can alter the sensitivity of the adrenal glomerulosa cells to Ang II. This effect has been demonstrated both in vivo and in vitro and could play a significant physiological role in sensitizing the adrenal upon sodium depletion, a situation where ANP is reduced (Atarashi et al. 1985; Chartier et al. 1984; Franco-Saenz et al. 1989). The mechanism of this inhibition (affecting both the responses to Ang II and to ACTH) is not completely elucidated (Ganguly 1992). ANP increases cGMP levels in adrenal glomerulosa cells (Matsuoka et al. 1985) and activates a guanylyl cyclase activity in the adrenal cell particulate fraction (Tremblay et al. 1986), but the role of cGMP in the inhibition of aldosterone secretion remains ambiguous. Indeed, various permeable analogs of cGMP, which mimic ANP action in other cell types, were unable to inhibit, and even slightly increased, basal or stimulated aldosterone production from adrenal glomerulosa cells (Barrett and Isales 1988; Elliott and Goodfriend 1986; Ganguly et al. 1989; Matsuoka et al. 1987).

In spite of early reports suggesting that ANP-induced inhibition of aldosterone does not involve the Ca^{2+} signal induced by Ang II (Apfeldorf et al. 1988;

Capponi et al. 1986a; Ganguly et al. 1989; Takagi et al. 1988, Cherradi et al. 1998), other results showed that ANP could act by interfering with the appropriate changes in Ca^{2+} fluxes elicited by the agonist (Barrett et al. 1991; Chartier and Schiffrin 1987; McCarthy et al. 1990). Interestingly, ANP has been reported to modulate L- and T-type channels in opposite ways, increasing currents through L-type while decreasing those occurring through T-type channels. This opposite regulation could explain the inhibition of steroidogenesis, linked to the activity of T channels, without reduction of the cytosolic Ca^{2+} concentration, which depends upon L channel activity.

4.4 Other Factors

A series of other hormones and neurotransmitters have also been shown to modulate the effect of Ang II on aldosterone synthesis in the adrenal glomerulosa cell. Thus, for example, endothelin can potentiate the aldosterone response to Ang II (Cozza et al. 1992; Cozza and Gomez-Sanchez 1993; Nussdorfer et al. 1999). Insulin also appears to affect the steroidogenic effect of Ang II in rat glomerulosa cells (Petrasek et al. 1992). Moreover, in human adrenal cells, short-term insulin treatment results in inhibition of Ang II-induced aldosterone synthesis, possibly via inhibition of the 12-lipoxygenase pathway, whereas chronic treatment leads to a potentiation of Ang II action due in part to the up-regulation of cytochrome P-450scc enzyme levels (Natarajan et al. 1995).

The presence of nerve endings in the adrenal cortex has been demonstrated morphologically (Vinson et al. 1994a), and direct effects on aldosterone production in vitro have been reported for some neurotransmitters such as pituitary adenylyl cyclase-activating polypeptide in frog adrenal cells (Yon et al. 1994), vasopressin in rat adrenal glomerulosa cells (Mazzocchi et al. 1993) and in human adrenal glands (Guillon et al. 1995), substance P (Mazzocchi et al. 1995), or serotonin in man (Lefebvre et al. 1995)

5 Extra-Adrenal Aldosterone Production

5.1 The Heart as a Source of Aldosterone

Recent in vitro and in vivo studies have reported that extra-adrenal tissues such as vascular endothelial (Takeda et al. 1996) and smooth muscle cells (Hatakeyama et al. 1994) express aldosterone synthase under the control of Ang II and are able to produce aldosterone. Moreover, similar findings have been reported in the brain (Gomez-Sanchez et al. 1997) and in particular in the heart (Silvestre, 1998, 1999b; Young and Funder 2000). An increase in cardiac aldosterone production as well as in aldosterone synthase mRNA expression has been observed in genetically hypertensive rats (Takeda et al. 2000a) and following experimental myocardial infarction (MI) in the rat, and this increase is mediated by Ang II acting via the cardiac AT_1 receptor subtype (Silvestre et al. 1998, 1999a). Indeed, there is evidence for Ang II generation in the heart: neonatal rat cardiomyocytes are known to produce Ang II (Sadoshima et al. 1993). Moreover, in canine ventricular myocytes, the renin-angiotensin system is upregulated with heart failure (Barlucchi et al. 2001). There is also increased renin mRNA expression in the border zone of the infarcted left ventricle, with a possible role for intracardiac Ang II in infarct healing (Passier et al. 1996). Lastly, cardiac Ang II concentration is increased two- to fourfold following myocardial infarction, this increase being more marked in the infarcted zone than in noninfarcted tissue (Hirsch et al. 1991; Silvestre et al. 1999a; Sun et al. 2001).

In addition, in the rat submitted to high-sodium intake, cardiac aldosterone production and activity of aldosterone synthase are increased (Takeda et al. 2000b). In humans, recent studies have shown that plasma aldosterone levels are higher in the interventricular vein and coronary sinus than in aortic root in patients with failing ventricles (Mizuno et al. 2001) and with essential hypertension (Yamamoto et al. 2002), and that some steroidogenic genes can be detected in the heart (Kayes-Wandover and White 2000; Young et al. 2001), including aldosterone synthase (Yoshimura et al. 2002), suggesting a possibility for endogenous aldosterone synthesis, at least in the failing heart. However, whether the aldosterone that is produced in cardiac tissue results from active cholesterol transformation as in the adrenal cortex or is merely the result of the extraction from blood of aldosterone and/or of some of its precursors (Hayashi et al. 2001) such as pregnenolone, progesterone or corticosterone remains to be determined. In this respect, it is noteworthy that a significant StAR mRNA expression has been detected in neonatal rat cardiomyocytes in culture and that this expression can be modulated by Ang II. Furthermore, after myocardial infarction in the rat, StAR RNA levels are more than doubled in the noninfarcted area of the left ventricle, a phenomenon that can be prevented by losartan or spironolactone (Casal et al. 2003).

5.2 The Heart as a Target of Aldosterone

In addition to being a potential source of aldosterone, the heart also appears to act as a target for the mineralocorticoid. Indeed, mineralocorticoid receptors are present in cardiac myocytes and fibroblasts (Young et al. 1994). Aldosterone exerts rapid positive inotropic effects in the isolated rat heart (Harada et al. 2001), as well as longer-term effects. Aldosterone treatment, for example, induces left ventricular hypertrophy (Young et al. 1995), fibrosis, and an increase in ventricular collagen I and III mRNAs within 2 weeks (Robert et al. 1999). This delay may however be shortened since a bolus injection of deoxycorticosterone induces a collagen III increase in only 2 days, likely due to the high hormone concentration acutely reached under those conditions (Fujisawa et al. 2001; Robert et al. 1999). Alterations of cardiac function are observed in transgenic mice models of cardiac mineralocorticoid receptor down-expression (Beggah et al. 2002) or overexpression (Le Menuet et al. 2001). Moreover, the expression of some genes of the renin-angiotensin system is also modulated by aldosterone in the heart, for example, genes for the AT_1 receptor (Robert et al. 1999) and angiotensin-converting enzyme (Harada et al. 2001). All these effects can be prevented by simultaneous administration of spironolactone, an aldosterone antagonist. Finally, strong indirect evidence for an active involvement of aldosterone in contributing to heart failure was recently provided by the RALES trial, in which a 30% improvement in mortality risk has been observed in patients with severe heart failure who have received spironolactone in addition to their classic treatment (Pitt et al. 1999). Newer aldosterone antagonists devoid of side effects such as eplerenone (McMahon 2001) also prevent the development of cardiac damage through a mechanism linked to selective aldosterone antagonism in the heart (Martinez et al. 2002).

On the other hand, Rocha et al. (2000) have recently shown that the deleterious effects of aldosterone treatment in the rat can be suppressed by adrenalectomy and that myocardial necrosis and renal arteriopathy are restored by exogenous aldosterone. These results would tend to rule out a role of cardiac aldosterone in cardiac dysfunction. Nevertheless, it appears clear that aldosterone plays a critical role in the vascular inflammatory phenotype induced by Ang II in the heart (Rocha et al. 2002a, 2002b).

6 Conclusion

The biosynthesis of aldosterone from its precursor, cholesterol, is an extremely complex process involving numerous steps that are localized and confined to specific compartments of the steroidogenic cell. When stimulating aldosterone production through a pleiotropic signaling system, Ang II is able to modulate the functioning of a large number of these steps, be it HDL-cholesterol import at the cell surface, cholesterol ester hydrolysis in lipid droplets, StAR protein or aldosterone synthase expression, among others, in the nucleus or StAR-mediated cholesterol importation into mitochondria. These concerted actions, which are integrated in time and intracellular location, result in the fine tuning of aldosterone production for an appropriate maintenance of both acute and long-term salt and water homeostasis.

References

Acton S, Rigotti A, Landschulz KT, Xu S, Hobbs HH, Krieger M (1996) Identification of scavenger receptor SR-B1 as a high density lipoprotein receptor. Science 271:518–520

Aesoy R, Mellgren G, Morohashi K, Lund J (2002) Activation of cAMP-dependent protein kinase increases the protein level of steroidogenic factor-1. Endocrinology 143:295–303

Aguilera G, Catt KJ (1986) Participation of voltage-dependent calcium channels in the regulation of adrenal glomerulosa function by angiotensin II and potassium. Endocrinology 118:112–128

Ambroz C, Catt KJ (1992) Angiotensin II receptor-mediated calcium influx in bovine adrenal glomerulosa cells. Endocrinology 131:408–414

Amri H, Li H, Culty M, Gaillard JL, Teper G, Papadopoulos V (1999) The peripheral-type benzodiazepine receptor and adrenal steroidogenesis. Curr Opin Endocrinol Diabetes 6:179–184

Andreis PG, Neri G, Tortorella C, Gottardo L, Nussdorfer GG (2000) Tyrphostin-23 enhances steroid-hormone secretion from dispersed human and rat adrenocortical cells. Endocr Res 26:319–332

Apfeldorf WJ, Isales CM, Barrett PQ (1988) Atrial natriuretic peptide inhibits the stimulation of aldosterone secretion but not the transient increase in intracellular free calcium concentration induced by angiotensin II addition. Endocrinology 122:1460–1465

Aptel HBC, Johnson EIM, Vallotton MB, Rossier MF, Capponi AM (1996) Demonstration of an angiotensin II-induced negative feedback effect on aldosterone synthesis in isolated rat adrenal zona glomerulosa cells. Mol Cell Endocrinol 119:105–111

Aptel HBC, Burnay MM, Rossier MF, Capponi AM (1999) The role of tyrosine kinases in capacitative calcium influx-mediated aldosterone production in bovine adrenal zona glomerulosa cells. J Endocrinol 163:131–138

Arakane F, Sugawara T, Nishino H, Liu ZM, Holt JA, Pain D, Stocco DM, Miller WL, Strauss JF (1996) Steroidogenic acute regulatory protein (StAR) retains activity in the absence of its mitochondrial import sequence: implications for the mechanism of StAR action. Proc Natl Acad Sci USA 93:13731–13736

Arakane F, Kallen CB, Watari H, Foster JA, Sepuri NV, Pain D, Stayrook SE, Lewis M, Gerton GL, Strauss JF (1998) The mechanism of action of steroidogenic acute regulatory protein (StAR)—StAR acts on the outside of mitochondria to stimulate steroidogenesis. J Biol Chem 273:16339–16345

Atarashi K, Mulrow PJ, Franco-Saenz R (1985) Effect of atrial peptides on aldosterone production. J Clin Invest 76:1807–1811

Baba K, Doi Y, Franco-Saenz R, Mulrow PJ (1986) Mechanisms by which nephrectomy stimulates adrenal renin. Hypertension 8:997–1002

Balasubramaniam S, Goldstein JL, Faust JR, Brunschede GY, Brown MS (1977) Lipoprotein-mediated regulation of 3-hydroxy-3-methylglutaryl coenzyme A reductase activity and cholesteryl ester metabolism in the adrenal gland of the rat. J Biol Chem 252:1771–1779

Balla T, Spat A (1982) The effect of various calmodulin inhibitors on the response of adrenal glomerulosa cells to angiotensin II and cyclic AMP. Biochem Pharmacol 31:3705–3707

Balla T, Baukal AJ, Guillemette G, Morgan RO, Catt KJ (1986) Angiotensin-stimulated production of inositol trisphosphate isomers and rapid metabolism through inositol 4-monophosphate in adrenal glomerulosa cells. Proc Natl Acad Sci USA 83:9323–9327

Balla T, Guillemette G, Baukal AJ, Catt KJ (1987) Metabolism of inositol 1,3,4-trisphosphate to a new tetrakisphosphate isomer in angiotensin-stimulated adrenal glomerulosa cells. J Biol Chem 262:9952–9955

Balla T, Baukal AJ, Guillemette G, Catt KJ (1988) Multiple pathways of inositol polyphosphate metabolism in angiotensin-stimulated adrenal glomerulosa cells. J Biol Chem 263:4083–4091

Balla T, Hausdorff WP, Baukal AJ, Catt KJ (1989) Inositol polyphosphate production and regulation of cytosolic calcium during the biphasic activation of adrenal glomerulosa cells by angiotensin II. Arch Biochem Biophys 270:398–403

Balla T, Hollo Z, Varnai P, Spät A (1991) Angiotensin II inhibits potassium-induced calcium signal generation in rat adrenal glomerulosa cells. Biochem J 273:399–404

Balla T, Nakanishi S, Catt KJ (1994) Cation sensitivity of inositol 1,4,5-trisphosphate production and metabolism in agonist-stimulated adrenal glomerulosa cells. J Biol Chem 269:16101–16107

Barlucchi L, Leri A, Dostal DE, Fiordaliso F, Tada H, Hintze TH, Kajstura J, Nadal-Ginard B, Anversa P (2001) Canine ventricular myocytes possess a renin-angiotensin system that is upregulated with heart failure. Circ Res 88:298–304

Barrett PQ, Isales CM (1988) The role of cyclic nucleotides in atrial natriuretic peptide-mediated inhibition of aldosterone secretion. Endocrinology 122:799–808

Barrett PQ, Kojima I, Kojima K, Zawalich K, Isales CM, Rasmussen H (1986) Temporal patterns of protein phosphorylation after angiotensin II, A23187 and/or 12-O-tetradecanoylphorbol 13-acetate in adrenal glomerulosa cells. Biochem J 238:893–903

Barrett PQ, Bollag WB, Isales CM, McCarthy RT, Rasmussen H (1989) Role of calcium in angiotensin II-mediated aldosterone secretion. Endocr Rev 10:496–518

Barrett PQ, Ertel EA, Smith MM, Nee JJ, Cohen CJ (1995) Voltage-gated calcium currents have two opposing effects on the secretion of aldosterone. Am J Physiol 268:C985–C992

Barrett PQ, Isales CM, Bollag WB, McCarthy RT (1991) Calcium channels and aldosterone secretion: modulation by potassium and atrial natriuretic peptide. Am J Physiol 261:F706–F719

Barrett PQ, Lu HK, Colbran R, Czernik A, Pancrazio JJ (2000) Stimulation of unitary T-type Ca(2+) channel currents by calmodulin-dependent protein kinase II [In Process Citation]. Am J Physiol Cell Physiol 279:C1694–C1703

Baukal AJ, Guillemette G, Rubin R, Spät A, Catt KJ (1985) Binding sites for inositol trisphosphate in the bovine adrenal cortex. Biochem Biophys Res Comm 133:532–538

Baukal AJ, Balla T, Hunyady L, Hausdorff WP, Guillemette G, Catt KJ (1988) Angiotensin II and guanine nucleotides stimulate formation of inositol 1,4,5-trisphosphate and its metabolites in permeabilized adrenal glomerulosa cells. J Biol Chem 263:6087–6092

Baukal AJ, Hunyady L, Catt KJ, Balla T (1994) Evidence for participation of calcineurin in potentiation of agonist-stimulated cyclic AMP formation by the calcium-mobilizing hormone, angiotensin II. J Biol Chem 269:24546–24549

Baum CL, Reschly EJ, Gayen AK, Groh ME, Schadick K (1997) Sterol carrier protein-2 overexpression enhances sterol cycling and inhibits cholesterol ester synthesis and high density lipoprotein cholesterol secretion. J Biol Chem 272:6490–6498

Beckett GJ, Boyd GS (1977) Purification and control of cholesterol ester hydrolase and evidence for the activation of the enzyme by a phosphorylation. Eur J Biochem 72:223–233

Beggah AT, Escoubet B, Puttini S, Cailmail S, Delage V, Ouvrard-Pascaud A, Bocchi B, Peuchmaur M, Delcayre C, Farman N, Jaisser F (2002) Reversible cardiac fibrosis and heart failure induced by conditional expression of an antisense mRNA of the mineralocorticoid receptor in cardiomyocytes. Proc Natl Acad Sci USA 99:7160–7165

Belfrage P, Fredrikson G, Nilsson NO, Stralfors P (1980) Regulation of adipose tissue lipolysis: phosphorylation of hormones sensitive lipase in intact rat adipocytes. FEBS Lett 111:120–124

Bernstein KE, Berk BC (1993) The biology of angiotensin II receptors. Am J Kidney Dis 22:745–754

Berridge MJ, Irvine RF (1984) Inositol trisphosphate, a novel second messenger in cellular signal transduction. Nature 312:315–321

Bird IM, Mason JI, Rainey WE (1998a) Battle of kinases: integration of adrenal responses to cyclic AMP, diacylglycerol, and calcium at the level of steroidogenic cytochromes P450 and 3beta HSD expression in H295R. Endocr Res 24:345–354

Bird IM, Mason JI, Rainey WE (1998b) Protein kinase A, protein kinase C, and calcium-regulated expression of 21-hydroxylase cytochrome P450 in H295R human adrenocortical cells. J Clin Endocrinol Metab 83:1592–1597

Biron P, Koiw E, Nowaczynski W, Brouillet J, Genest J (1961) The effects of intravenous infusions of valine-5-angiotensin II and other pressor agents on urinary electrolytes and corticosteroids, including aldosterone. J Clin Invest 40:338–347

Black VH (1987) Lipoprotein requirements for secretion of ultraviolet-absorbing corticosteroids by guinea pig adrenocortical cells in vitro: inner versus outer cortices; zona glomerulosa versus zona fasciculata. Endocrinology 120:640–650

Bodart V, Ong H, De Lean A (1995) A role for protein tyrosine kinase in the steroidogenic pathway of angiotensin II in bovine zona glomerulosa cells. J Steroid Biochem Mol Biol 54:55–62

Bollag WB, Barrett PQ, Isales CM, Liscovitch M, Rasmussen H (1990) A potential role for phospholipase D in the angiotensin II-induced stimulation of aldosterone secretion from bovine adrenal glomerulosa cells. Endocrinology 127:1436–1443

Bollag WB, Jung E, Calle RA (2002) Mechanism of angiotensin II-induced phospholipase D activation in bovine adrenal glomerulosa cells. Mol Cell Endocrinol 192:7–16

Bose HS, Sugawara T, Strauss JF, Miller WL (1996) The pathophysiology and genetics of congenital lipoid adrenal hyperplasia. N Engl J Med 335:1870–1878

Bose HS, Whittal RM, Baldwin MA, Miller WL (1999) The active form of the steroidogenic acute regulatory protein, StAR, appears to be a molten globule. Proc Natl Acad Sci USA 96:7250–7255

Bose HS, Lingappa VR, Miller WL (2002) Rapid regulation of steroidogenesis by mitochondrial protein import. 417:87–91

Boulay G, Gallo-Payet N, Guillemette G (1990) Implication of phospholipase C in the steroidogenic action of angiotensin II. Eur J Pharmacol 189:267–275

Boulay G, Chretien L, Richard DE, Guillemette G (1994) Short-term desensitization of the angiotensin II receptor of bovine adrenal glomerulosa cells corresponds to a shift from a high to a low affinity state. Endocrinology 135:2130–2136

Boyd GS, McNamara B, Suckling KE, Tocher DR (1983) Cholesterol metabolism in the adrenal cortex. J Steroid Biochem 19:1017–1027

Boyd JE, Page RB, Mulrow PJ (1972) The effect of hypophysectomy on the conversion of corticosterone to aldosterone in the sodium-depleted rat. Endocrinology 90:827–829

Boyd J, Mulrow PJ, Palmore WP, Silvo P (1973) Importance of potassium in the regulation of aldosterone production. Circ Res 32 [Suppl 1]:39–45

Braley L, Menachery A, Brown E, Williams G (1984) The effects of extracellular K+ and angiotensin II on cytosolic Ca++ and steroidogenesis in adrenal glomerulosa cells. Biochem Biophys Res Commun 123:810–815

Brandenburger Y, Arrighi J-F, Rossier MF, Maturana AD, Vallotton MB, Capponi AM (1999) Measurement of perimitochondrial calcium concentration in bovine adrenal glomerulosa cells with aequorin targeted to the outer mitochondrial membrane. Biochem J 341:745–753

Brandenburger Y, Kennedy ED, Python CP, Rossier MF, Vallotton MB, Wollheim CB, Capponi AM (1996) Possible role for mitochondrial calcium in angiotensin II- and potassium-stimulated steroidogenesis in bovine adrenal glomerulosa cells. Endocrinology 137:5544–5551

Braun-Menéndez, Fasciolo JC, Leloir LF, Muñoz JM (1939) La sustancia hipertensora de la sangre del riñon isquemiado. Rev Soc Arg Biol 15:420–425

Brauneis U, Vassilev PM, Quinn SJ, Williams GH, Tillotson DL (1991) Angiotensin II blocks potassium currents in zona glomerulosa cells from rat, bovine and human adrenals. Am J Physiol 260:E772–E779

Brody RI, Black VH (1988) Acyl-coenzyme A:cholesterol acyltransferase and cholesterol ester hydrolase in the outer and inner cortices of the guinea pig adrenal: effects of adrenocorticotropin and dexamethasone. Endocrinology 122:1722–1731

Brody RI, Black VH (1991) Differential ACTH response of immunodetectable HMG CoA reductase and cytochromes P450(17 alpha) and P450(21) in guinea pig adrenal outer zone cell types, zona glomerulosa and zona fasciculata. Endocr Res 17:195–208

Brown MS, Goldstein JL (1986) A receptor-mediated pathway for cholesterol homeostasis. Science 232:34–47

Brown MS, Goldstein JL (1997) The SREBP pathway: regulation of cholesterol metabolism by proteolysis of a membrane-bound transcription factor. Cell 89:331–340

Burnay MM, Python CP, Vallotton MB, Capponi AM, Rossier MF (1994) Role of the capacitative calcium influx in the activation of steroidogenesis by angiotensin II in adrenal glomerulosa cells. Endocrinology 135:751–758

Burnay MM, Vallotton MB, Capponi AM, Rossier MF (1998) Angiotensin II potentiates the adrenocorticotrophic hormone-induced cyclic AMP formation in bovine adrenal glomerulosa cells through a capacitative calcium influx. Biochem J 330:21–27

Campbell DJ (1982) Effect of rat plasma lipoproteins on aldosterone production by rat zona glomerulosa cells in vitro. J Steroid Biochem 17:709–711

Campbell WB, Brady MT, Rosolowsky LJ, Falck JR (1991) Metabolism of arachidonic acid by rat adrenal glomerulosa cells: synthesis of hydroxyeicosatetraenoic acids and epoxyeicosatrienoic acids. Endocrinology 128:2183–2194

Capponi AM (2002) Regulation of cholesterol supply for mineralocorticoid biosynthesis. Trends Endocrinol Metab 13:118–121

Capponi AM, Rossier MF (1996) Regulation of aldosterone secretion. Curr Opin Endocrinol Diabetes 3:248–257

Capponi AM, Lew PD, Jornot L, Vallotton MB (1984) Correlation between cytosolic free calcium and aldosterone production in bovine adrenal glomerulosa cells. J Biol Chem 259:8863–8869

Capponi AM, Lew PD, Wuthrich R, Vallotton MB (1986a) Effects of atrial natriuretic peptide on the stimulation by angiotensin II of various target cells. J Hypertension 4 [Suppl 2]:S61–S65

Capponi AM, Rossier MF, Lang U, Lew PD, Vallotton MB (1986b) Comparison of the signal transduction mechanisms for angiotensin II in adrenal zona glomerulosa and vascular smooth muscle cells. J Hypertension 4 [Suppl 6]:S419–S420

Capponi AM, Lew PD, Vallotton MB (1987) Quantitative analysis of the cytosolic free calcium dependency of aldosterone production in bovine adrenal glomerulosa cells: different requirements for angiotensin II and potassium. Biochem J 247:335–340

Capponi AM, Rossier MF, Davies E, Vallotton MB (1988) Calcium stimulates steroidogenesis in permeabilized bovine adrenal cortical cells. J Biol Chem 263:16113–16117

Capponi AM, Johnson EIM, Rossier MF, Lang U, Vallotton MB (1989) The calcium messenger system in angiotensin II-induced aldosterone production. In: Mantero F, Takeda R, Scoggins BA, Biglieri EG, Funder JW (eds) The adrenal gland and hypertension : from cloning to clinic. Serono Symposia Publications, Vol 57, Raven Press, New York, pp 45–52

Caron KM, Ikeda Y, Soo SC, Stocco DM, Parker KL, Clark BJ (1997) Characterization of the promoter region of the mouse gene encoding the steroidogenic acute regulatory protein. Mol Endocrinol 11:138–147

Caron KM, Soo SC, Wetsel WC, Stocco DM, Clark BJ, Parker KL (1997) Targeted disruption of the mouse gene encoding steroidogenic acute regulatory protein provides insights into congenital lipoid adrenal hyperplasia. Proc Natl Acad Sci USA 94:11540–11545

Casal AJ, Silvestre JS, Delcayre C, Capponi AM (2003) Expression and modulation of steroidogenic acute regulatory protein messenger ribonucleic acid in rat cardiocytes and after myocardial infarction. Endocrinology 144:1861–1868

Catt KJ, Carson MC, Hausdorff WP, Leach-Harper CM, Baukal AJ, Guillemette G, Balla T, Aguilera G (1987) Angiotensin II receptors and mechanisms of action in adrenal glomerulosa cells. J Steroid Biochem 27:915–927

Catt KJ, Balla T, Baukal AJ, Hausdorff WP, Aguilera G (1988) Control of glomerulosa cell function by angiotensin II: transduction by G-proteins and inositol polyphosphates. Clin Exp Pharmacol Physiol 15:501–515

Chartier L, Schiffrin EL (1987) Role of calcium in effects of atrial natriuretic peptide on aldosterone production in adrenal glomerulosa cells. Am J Physiol 252:E485–E491

Chartier L, Schiffrin E, Thibault G, Garcia R (1984) Atrial natriuretic factor inhibits the stimulation of aldosterone secretion by angiotensin II, ACTH and potassium in vitro and angiotensin II-induced steroidogenesis in vivo. Endocrinology 115:2026–2028

Chen XL, Bayliss DA, Fern RJ, Barrett PQ (1999) A role for T-type calcium channels in synergistic control of aldosterone production by angiotensin II and potassium. Am J Physiol 276:F674–F683

Cherradi N, Rossier MF, Vallotton MB, Capponi AM (1996) Calcium stimulates intramitochondrial cholesterol transfer in bovine adrenal glomerulosa cells. J Biol Chem 271:25971–25975

Cherradi N, Rossier MF, Vallotton MB, Timberg R, Friedberg I, Orly J, Wang XJ, Stocco DM, Capponi AM (1997) Submitochondrial distribution of three key steroidogenic proteins (steroidogenic acute regulatory protein, P450 side-chain cleavage and 3β-hydroxysteroid dehydrogenase isomerase enzymes) upon stimulation by intracellular calcium in adrenal glomerulosa cells. J Biol Chem 272:7899–7907

Cherradi N, Brandenburger Y, Rossier MF, Vallotton MB, Stocco DM, Capponi AM (1998) Atrial natriuretic peptide inhibits calcium-induced steroidogenic acute regulatory protein gene transcription in adrenal glomerulosa cells. Mol Endocrinol 12:962–972

Cherradi N, Bideau M, Arnaudeau S, Demaurex N, James RW, Azhar S, Capponi AM (2001) Angiotensin II promotes selective uptake of high density lipoprotein cholesterol esters in bovine adrenal glomerulosa cells and human adrenocortical carcinoma cells through induction of scavenger receptor class b type 1. Endocrinology 142:4540–4549

Cherradi N, Pardo B, Greenberg AS, Kraemer FB, Capponi AM (2003) Angiotensin II activates cholesterol ester hydrolase in bovine adrenal glomerulosa cells through phosphorylation mediated by p42/p44 MAP kinase. Endocrinology 144:4905–4915

Chiou CY, Kifor I, Moore TJ, Williams GH (1994) The effect of losartan on potassium-stimulated aldosterone secretion in vitro. Endocrinology 134:2371–2375

Christenson LK, Strauss JF III (2000) Steroidogenic acute regulatory protein (StAR) and the intramitochondrial translocation of cholesterol. Biochim Biophys Acta 1529:175–187

Christenson LK, Osborne TF, Mcallister JM, Strauss JF III (2001) Conditional response of the human steroidogenic acute regulatory protein gene promoter to sterol regulatory element binding protein-1a. Endocrinology 142:28–36

Clark BJ, Combs R (1999) Angiotensin II and cyclic adenosine 3',5'-monophosphate induce human steroidogenic acute regulatory protein transcription through a common steroidogenic factor-1 element. Endocrinology 140:4390–4398

Clark BJ, Stocco DM (1996) StAR—a tissue specific acute mediator of steroidogenesis. Trends Endocrinol Metab 7:227–233

Clark BJ, Wells J, King SR, Stocco DM (1994) The purification, cloning and expression of a novel luteinizing hormone-induced mitochondrial protein in MA-10 mouse Leydig tumor cells. Characterization of the steroidogenic acute regulatory protein (StAR). J Biol Chem 269:28314–28322

Clark BJ, Pezzi V, Stocco DM, Rainey WE (1995) The steroidogenic acute regulatory protein is induced by angiotensin II and K^+ in H295R adrenocortical cells. Mol Cell Endocrinol 115:215–219

Clifford GM, Londos C, Kraemer FB, Yeaman SJ (2000) Translocation of hormone-sensitive lipase and perilipin upon lipolytic stimulation of rat adipocytes. J Biol Chem 275:5011–5015

Clyne CD, Nguyen A, Rainey WE (1995) The effects of KN62, a Ca2+/calmodulin-dependent protein kinase II inhibitor, on adrenocortical cell aldosterone production. Endocr Res 21:259–265

Cohen CJ, McCarthy RT, Barrett PQ, Rasmussen H (1988) Calcium channels in adrenal glomerulosa cells: potassium and angiotensin II increase T-type calcium current. Proc Natl Acad Sci USA 85:2412–2416

Colbran RJ, Garton AJ, Cordle SR, Yeaman SJ (1986) Regulation of cholesterol ester hydrolase by cyclic AMP-dependent protein kinase. FEBS Lett 201:257–261

Condon JC, Pezzi V, Drummond BM, Yin S, Rainey WE (2002) Calmodulin-dependent kinase I regulates adrenal cell expression of aldosterone synthase. Endocrinology 143:3651–3657

Conlin PR, Seely EW, Hollenberg NK, Williams GH (1998) Dissociation of vascular and adrenal responsiveness to angiotensin II following calcium channel blockade. Endocr Res 24:127–139

Cook KG, Yeaman SJ, Stralfors P, Fredrikson G, Belfrage P (1982) Direct evidence that cholesterol ester hydrolase from adrenal cortex is the same enzyme as hormone-sensitive lipase from adipose tissue. Eur J Biochem 125:245–249

Cook KG, Colbran RJ, Snee J, Yeaman SJ (1983) Cytosolic cholesterol ester hydrolase from bovine corpus luteum. Its purification, identification and relationship to hormone-sensitive lipase. Biochim Biophys Acta 752:46–53

Cordle SR, Colbran RJ, Yeaman SJ (1986) Hormone-sensitive lipase from bovine adipose tissue. Biochim Biophys Acta 887:51–57

Cote M, Payet MD, Dufour MN, Guillon G, Gallo-Payet N (1997) Association of the G protein alpha(q)/alpha11-subunit with cytoskeleton in adrenal glomerulosa cells: role in receptor-effector coupling. Endocrinology 138:3299–3307

Cote M, Muyldermans J, Chouinard L, Gallo-Payet N (1998) Involvement of tyrosine phosphorylation and MAPK activation in the mechanism of action of ACTH, angiotensin II and vasopressin. Endocr Res 24:415–419

Cozza EN, Chiou S, Gomez-Sanchez CE (1992) Endothelin-1 potentiation of angiotensin II stimulation of aldosterone production. Am J Physiol 262:R85–R89

Cozza EN, Gomezsanchez CE (1993) Mechanisms of ET-1 potentiation of angiotensin-II stimulation of aldosterone production. Am J Physiol 265:E179–E183

Davies E, Bonnardeaux A, Plouin PF, Corvol P, Clauser E (1997) Somatic mutations of the angiotensin II (AT1) receptor gene are not present in aldosterone-producing adenoma. J Clin Endocrinol Metab 82:611–615

Denner K, Rainey WE, Pezzi V, Bird IM, Bernhardt R, Mathis JM (1996) Differential regulation of 11 beta-hydroxylase and aldosterone synthase in human adrenocortical H295R cells. Mol Cell Endocrinol 121:87–91

Elliott ME, Goodfriend TL (1986) Inhibition of aldosterone synthesis by atrial natriuretic factor. Fed Proc 45:2376–2381

Elliott ME, Alexander RC, Goodfriend TL (1982) Aspects of angiotensin action in the adrenal. Key roles for calcium and phosphatidyl inositol. Hypertension 4:52–58

Elliott ME, Siegel FL, Hadjokas NE, Goodfriend TL (1985) Angiotensin effects on calcium and steroidogenesis in adrenal glomerulosa cells. Endocrinology 116:1051–1059

Ely JA, Ambroz C, Baukal AJ, Christensen SB, Balla T, Catt KJ (1991) Relationship between agonist- and thapsigargin-sensitive calcium pools in adrenal glomerulosa cells: thapsigargin-induced calcium mobilization and entry. J Biol Chem 266:18635–18641

Enyedi P, Spat A, Antoni FA (1981) Role of prostaglandins in the control of the function of adrenal glomerulosa cells. J Endocrinol 91:427–437

Enyedi P, Buki B, Muscsi I, Spat A (1985) Polyphosphoinositide metabolism in adrenal glomerulosa cells. Mol Cell Endocrinol 41:105–112

Enyedi P, Spat A (1987) The mechanism of angiotensin-induced desensitization of adrenal glomerulosa cells. Mol Cell Endocrinol 51:83–86

Enyedi P, Balla T, Antoni FA, Spat A (1988) Effect of angiotensin II and arginine vasopressin on aldosterone production and phosphoinositide turnover in rat adrenal glomerulosa cells: a comparative study. J Mol Endocrinol 1:117–124
Enyedi P, Szabadkai G, Horváth A, Szilágyi L, Gráf L, Spät A (1994) Inositol 1,4,5-trisphosphate receptor subtypes in adrenal glomerulosa cells. Endocrinology 134:2354–2359
Fakunding JL, Catt KJ (1980) Dependence of aldosterone stimulation in adrenal glomerulosa cells on calcium uptake: effects of lanthanum and verapamil. Endocrinology 107:1345–1353
Fakunding JL, Catt KJ (1982) Calcium-dependent regulation of aldosterone production in isolated adrenal glomerulosa cells: effects of the ionophore A-23187. Endocrinology 110:2006–2010
Fakunding JL, Chow R, Catt KJ (1979) The role of calcium in the stimulation of aldosterone production by adrenocorticotropin, angiotensin II, and potassium in isolated glomerulosa cells. Endocrinology 105:327–333
Farago A, Seprodi J, Spat A (1988) Subcellular distribution of protein kinase C in rat adrenal glomerulosa cells. Biochem Biophys Res Commun 156:628–633
Farese RV, Larson RE, Davis JS (1984) Rapid effect of angiotensin II on polyphosphoinositide metabolism in the rat adrenal glomerulosa. Endocrinology 114:302–304
Fern RJ, Hahm MS, Lu H-K, Liu L-P, Gorelick FS, Barrett PQ (1995) Calcium/calmodulin-dependent protein kinase II activation and regulation of adrenal glomerulosa calcium signaling. Am J Physiol 269:F751–F760
Feuilloley M, Vaudry H (1996) Role of the cytoskeleton in adrenocortical cells. Endocr Rev 17:269–288
Finkel MS, Aguilera G, Catt KJ, Keiser HR (1984) [3H]nitrendipine binding to adrenal capsular membranes. Life Sci 35:905–910
Fitzpatrick SC, McKenna TJ (1992) Evidence for a tonic inhibitory role of nifedipine-sensitive calcium channels in aldosterone biosynthesis. J Steroid Biochem Mol Biol 42:575–580
Fong TH, Yang CC, Greenberg AS, Wang SM (2002) Immunocytochemical studies on lipid droplet-surface proteins in adrenal cells. J Cell Biochem 86:432–439
Foster R, Rasmussen H (1983) Angiotensin-mediated calcium efflux from adrenal glomerulosa cells. Am J Physiol 245:E281–E287
Foster R, Lobo MV, Rasmussen H, Marusic ET (1981) Calcium: its role in the mechanism of action of angiotensin II and potassium in aldosterone production. Endocrinology 109:2196–2201
Foster RH, Davis JS, Farese RV (1990) External calcium is required for activation of phospholipase C by angiotensin II in adrenal glomerulosa cells. Mol Cell Biochem 95:157–166
Foster RH, MacFarlane CH, Bustamante MO (1997) Recent progress in understanding aldosterone secretion. Gen Pharmacol 28:647–651
Franco-Saenz R, Atarashi K, Takagi M, Takagi M, Mulrow PJ (1989) Effect of atrial natriuretic factor on renin and aldosterone. J Cardiovasc Pharmacol 13 [Suppl 6]:S31–S35
Freed MI, Rastegar A, Bia MJ (1991) Effects of calcium channel blockers on potassium homeostasis. Yale J Biol Med 64:177–186
Fujisawa G, Dilley R, Fullerton MJ, Funder JW (2001) Experimental cardiac fibrosis: differential time course of responses to mineralocorticoid-salt administration. Endocrinol 142:3625–3631
Galtier A, Liakos P, Keramidas M, Feige JJ, Chambaz EM, Defaye G (1996) ACTH angiotensin II and TGF beta participate in the regulation of steroidogenesis in bovine adrenal glomerulosa cells. Endocr Res 22:607–612
Ganguly A (1992) Atrial natriuretic peptide-induced inhibition of aldosterone secretion: a quest for mediator(s). Am J Physiol 263:E181–E194
Ganguly A, Davis JS (1994) Role of calcium and other mediators in aldosterone secretion from the adrenal glomerulosa cells. Pharmacol Rev 46:417–447

Ganguly A, Waldron C (1994) Comparative effects of a highly specific protein kinase C inhibitor, calphostin C and calmodulin inhibitors on angiotensin-stimulated aldosterone secretion. J Steroid Biochem Mol Biol 50:253–260

Ganguly A, Chiou S, West LA, Davis JS (1989) Atrial natriuretic factor inhibits angiotensin-induced aldosterone secretion: not through cGMP or interference with phospholipase C. Biochem Biophys Res Commun 159:148–154

Ganguly A, Chiou S, Fineberg NS, Davis JS (1992) Greater importance of Ca(2+)-calmodulin in maintenance of ang II- and K(+)-mediated aldosterone secretion: lesser role of protein kinase C. Biochem Biophys Res Commun 182:254–261

Ganguly A, Li L, Haxton M (1995) Inhibition of angiotensin II- and potassium-mediated aldosterone secretion by KN-62 suggests involvement of Ca(2+)-calmodulin dependent protein kinase II in aldosterone secretion. Biochem Biophys Res Commun 209:916–920

Ganong WF, Mulrow PJ, Boryczka A, Cera G (1962) Evidence for a direct effect of angiotensin II on adrenal cortex of the dog. Proc Soc Exp Biol Med 109:381–384

Ganten D, Hermann K, Unger T, Lang RE (1983) The tissue renin-angiotensin systems: focus on brain angiotensin, adrenal gland and arterial wall. Clin Exp Hypertens A 5:1099–1118

Gigante B, Rubattu S, Russo R, Porcellini A, Enea I, De Paolis P, Savoia C, Natale A, Piras O, Volpe M (1997) Opposite feedback control of renin and aldosterone biosynthesis in the adrenal cortex by angiotensin II AT1-subtype receptors. Hypertension 30:563–568

Gomez-Sanchez CE, Zhou MY, Cozza EN, Morita H, Foecking MF, Gomez-Sanchez EP (1997) Aldosterone biosynthesis in the rat brain. Endocrinology 138:3369–3373

Greenberg AS, Egan JJ, Weck SA, Moos MC, Londos C, Kimmel AR (1993) Isolation of cDNAs for perilipins A and B: sequence and expression of lipid droplet-associated proteins of adipocytes. Proc Natl Acad Sci USA 90:12035–12039

Gu JL, Natarajan R, Ben-Ezra J, Valente G, Scott S, Yoshimoto T, Yamamoto S, Rossi JJ, Nadler J (1994) Evidence that a leukocyte type of 12-lipoxygenase is expressed and regulated by angiotensin II in human adrenal glomerulosa cells. Endocrinology 134:70–77

Gu J, Wen Y, Mison A, Nadler JL (2003) 12-Lipoxygenase pathway increases aldosterone production, 3',5'-cyclic adenosine monophosphate response element-binding protein phosphorylation, and p38 mitogen-activated protein kinase activation in H295R human adrenocortical cells. Endocrinology 144:534–543

Guillemette G, Segui JA (1988) Effects of pH, reducing and alkylating reagents on the binding and Ca2+ release activities of inositol 1,4,5-triphosphate in the bovine adrenal cortex. Mol Endocrinol 2:1249–1255

Guillemette G, Balla T, Baukal AJ, Spät A, Catt KJ (1987) Intracellular receptors for inositol 1,4,5-trisphosphate in angiotensin II target tissues. J Biol Chem 262:1010–1015

Guillemette G, Favreau I, Boulay G, Potier M (1990) Solubilization and partial characterization of inositol 1,4,5-trisphosphate receptor of bovine adrenal cortex reveal similarities with the receptor of rat cerebellum. Mol Pharmacol 38:841–847

Guillon G, Trueba M, Joubert D, Grazzini E, Chouinard L, Côté M, Payet MD, Manzoni O, Barberis C, Roert M, Gallo-Payet N (1995) Vasopressin stimulates steroid secretion in human adrenal glands: comparison with angiotensin-II effects. Endocrinology 136:1285–1295

Gupta P, Franco-Saenz R, Mulrow PJ (1995) Locally generated angiotensin II in the adrenal gland regulates basal, corticotropin-, and potassium-stimulated aldosterone secretion. Hypertension 25:443–448

Gyles SL, Burns CJ, Whitehouse BJ, Sugden D, Marsh PJ, Persaud SJ, Jones PM (2001) ERKs regulate cyclic AMP-induced steroid synthesis through transcription of the steroidogenic acute regulatory (StAR) gene. J Biol Chem 276:34888–34895

Hadjokas NE, Goodfriend TL (1991) Inhibition of aldosterone production and angiotensin action by drugs affecting potassium channels. Pharmacology 43:141–150

Hajnoczky G, Varnai P, Hollo Z, Christensen SB, Balla T, Enyedi P, Spät A (1991) Thapsigargin-induced increase in cytoplasmic calcium concentration and aldosterone production in rat adrenal glomerulosa cells: interaction with potassium and angiotensin II. Endocrinology 128:2639–2644

Hajnoczky G, Csordas G, Bago A, Chiu AT, Spät A (1992a) Angiotensin II exerts its effect on aldosterone production and potassium permeability through receptor subtype AT1 in rat adrenal glomerulosa cells. Biochem Pharmacol 43:1009–1012

Hajnoczky G, Varnai P, Buday L, Farago A, Spät A (1992b) The role of protein kinase-C in control of aldosterone production by rat adrenal glomerulosa cells: activation of protein kinase-C by stimulation with potassium. Endocrinology 130:2230–2236

Harada E, Yoshimura M, Yasue H, Nagakawa O, Nagakawa M, Harada M, Mizuno Y, Nakayama M, Shimasaki Y, Ito T, Nakamura S, Kuwahara K, Saito Y, Nakao K, Ogawa H (2001) Aldosterone induces angiotensin-converting-enzyme gene expression in cultured neonatal rat cardiocytes. Circulation 104:137–139

Hasegawa T, Zhao L, Caron KM, Majdic G, Suzuki T, Shizawa S, Sasano H, Parker KL (2000) Developmental roles of the steroidogenic acute regulatory protein (StAR) as revealed by StAR knockout mice. Mol Endocrinol 14:1462–1471

Hatakeyama H, Miyamori I, Fujita T, Takeda Y, Takeda R, Yamamoto H (1994) Vascular aldosterone. Biosynthesis and a link to angiotensin II-induced hypertrophy of vascular smooth muscle cells. J Biol Chem 269:24316–24320

Hausdorff WP, Sekura RD, Aguilera G, Catt KJ (1987) Control of aldosterone production by angiotensin II is mediated by two guanine nucleotide regulatory proteins. Endocrinology 120:1668–1678

Hayashi M, Tsutamoto T, Wada A, Maeda K, Mabuchi N, Tsutsui T, Matsui T, Fujii M, Matsumoto T, Yamamoto T, Horie H, Ohnishi M, Kinoshita M (2001) Relationship between transcardiac extraction of aldosterone and left ventricular remodeling in patients with first acute myocardial infarction: extracting aldosterone through the heart promotes ventricular remodeling after acute myocardial infarction. J Am Coll Cardiol 38:1375–1382

Hescheler J, Rosenthal W, Hinsch K-D, Wulfern M, Trautwein W, Schulz G (1988) Angiotensin II-induced stimulation of voltage-dependent calcium currents in an adrenal cortical cell line. EMBO J 7:619–624

Higashijima M, Nawata H, Kato H, Ibayashi H (1987) Studies on lipoprotein and adrenal steroidogenesis: I. Roles of low density lipoprotein- and high density lipoprotein-cholesterol in steroid production in cultured human adrenocortical cells. Endocrinologia Japonica 34:635–645

Hirsch AT, Talsness CE, Schunkert H, Paul M, Dzau VJ (1991) Tissue-specific activation of cardiac angiotensin converting enzyme in experimental heart failure. Circ Res 69:475–482

Horiba N, Nomura K, Shizume K (1990) Exogenous and locally synthesized angiotensin II and glomerulosa cell functions. Hypertension 15:190–197

Horváth A, Szabadkai G, Varnai P, Aranyi T, Wollheim CB, Enyedi P (1998) Voltage-dependent calcium channels in adrenal glomerulosa cells and in insulin producing cells. Cell Calcium 23:33–42

Hunyady L, Balla T, Enyedi P, Spat A (1985) The effect of angiotensin II on arachidonate metabolism in adrenal glomerulosa cells. Biochem Pharmacol 34:3439–3444

Hunyady L, Baukal AJ, Bor M, Ely JA, Catt KJ (1990) Regulation of 1,2-diacylglycerol production by angiotensin II in bovine adrenal glomerulosa cells. Endocrinology 126:1001–1008

Inagami T (1995) Recent progress in molecular and cell biological studies of angiotensin receptors. Curr Opin Nephrol Hypertens 4:47–54

Irvine RF (1990) "Quantal" calcium release and the control of calcium entry by inositol phosphates: a possible mechanism. FEBS Lett 263:5–9

Johnson EIM, Capponi AM, Vallotton MB (1989) Cytosolic free calcium oscillates in single bovine adrenal glomerulosa cells in response to angiotensin II stimulation. J Endocrinol 122:391–402

Jung E, Betancourt-Calle S, Mann-Blakeney R, Foushee T, Isales CM, Bollag WB (1998) Sustained phospholipase D activation in response to angiotensin II but not carbachol in bovine adrenal glomerulosa cells. Biochem J 330:445–451

Kanazirska MV, Vassilev PM, Quinn SJ, Tillotson DL, Williams GH (1992) Single potassium channels in adrenal zona glomerulosa cells: II. Inhibition by angiotensin II. Am J Physiol 263:E760–E765

Kapas S, Hinson JP, Puddefoot JR, Ho MM, Vinson GP (1994) Internalization of the type I angiotensin II receptor (AT1) is required for protein kinase C activation but not for inositol trisphosphate release in the angiotensin II stimulated rat adrenal zona glomerulosa cell. Biochem Biophys Res Commun 204:1292–1298

Kapas S, Purbrick A, Hinson JP (1995) Role of tyrosine kinase and protein kinase C in the steroidogenic actions of angiotensin II, alpha-melanocyte-stimulating hormone and corticotrophin in the rat adrenal cortex. Biochem J 305:433–438

Kayes-Wandover KM, White PC (2000) Steroidogenic enzyme gene expression in the human heart. J Clin Endocrinol Metab 85:2519–2525

Kigoshi T, Uchida K, Morimoto S (1988) Existence of endogenous substrate proteins for calcium/calmodulin-dependent and calcium/phospholipid-dependent protein kinases in rat adrenal glomerulosa cells. J Steroid Biochem 29:277–283

Kiiveri S, Liu J, Westerholm-Ormio M, Narita N, Wilson DB, Voutilainen R, Heikinheimo M (2002) Differential expression of GATA-4 and GATA-6 in fetal and adult mouse and human adrenal tissue. Endocrinology 143:3136–3143

Klemcke HG (1992) Hormonal and stressor-associated changes in porcine adrenocortical cholesterol ester hydrolase activity. J Steroid Biochem Mol Biol 43:725–739

Kojima I, Kojima K, Kreutter D, Rasmussen H (1984a) The temporal integration of the aldosterone secretory response to angiotensin occurs via two intracellular pathways. J Biol Chem 259:14448–14457

Kojima K, Kojima I, Rasmussen H (1984b) Dihydropyridine calcium agonist and antagonist effects on aldosterone secretion. Am J Physiol 247:E645–E650

Kojima I, Kojima K, Rasmussen H (1985a) Effects of ANG II and K+ on Ca efflux and aldosterone production in adrenal glomerulosa cells. Am J Physiol 248:E36–E43

Kojima I, Kojima K, Rasmussen H (1985b) Possible role of phospholipase A2 action and arachidonic acid metabolism in angiotensin II-mediated aldosterone secretion. Endocrinology 117:1057–1066

Kojima I, Kojima K, Rasmussen H (1985c) Role of calcium fluxes in the sustained phase of angiotensin II-mediated aldosterone secretion from adrenal glomerulosa cells. J Biol Chem 260:9177–9184

Kojima I, Shibata H, Ogata E (1986a) Pertussis toxin blocks angiotensin II-induced calcium influx but not inositol trisphosphate production in adrenal glomerulosa cells. FEBS Lett 204:347–351

Kojima I, Shibata H, Ogata E (1986b) Phorbol ester inhibits angiotensin II-induced activation of phospholipase C in adrenal glomerulosa cells. Biochem J 237:253–258

Kojima I, Kawamura N, Shibata H (1994) Rate of calcium entry determines the rapid changes in protein kinase C activity in angiotensin II-stimulated adrenal glomerulosa cells. Biochem J 297:523–528

Koletsky RJ, Brown EM, Williams GH (1983) Calmodulin-like activity and calcium-dependent phosphodiesterase in purified cells of the rat zona glomerulosa and zona fasciculata. Endocrinology 113:485–490

Kovanen PT, Basu SK, Goldstein JL, Brown MS (1979) Low density lipoprotein receptors in bovine adrenal cortex. II. Low density lipoprotein binding to membranes prepared from fresh tissue. Endocrinology 104:610–616

Kraemer FB, Tavangar K, Hoffman AR (1991) Developmental regulation of hormone-sensitive lipase mRNA in the rat: changes in steroidogenic tissues. J Lipid Res 32:1303–1310

Kraemer FB, Patel S, Saedi MS, Sztalryd C (1993) Detection of hormone-sensitive lipase in various tissues. I. Expression of an HSL/bacterial fusion protein and of anti-HSL antibodies. J Lipid Res 34:663–671

Kraemer FB, Shen WJ, Natu V, Patel S, Osuga J, Ishibashi S, Azhar S (2002) Adrenal neutral cholesteryl ester hydrolase: identification, subcellular distribution, and sex differences. Endocrinology 143:801–806

Kramer RE (1993) Effects of diltiazem on calcium metabolism in cultured bovine glomerulosa cells: relationships to the actions of angiotensin II and potassium. J Pharmacol Exp Ther 266:374–384

Kroon PA, Thompson MG, Chao YS (1984) A comparison of the low-density-lipoprotein receptor from bovine adrenal cortex, rabbit and rat liver and adrenal glands by lipoprotein blotting. Biochem J 223:329–335

Kubo M, Strott CA (1988) Calcium-dependent protein kinase activity and protein phosphorylation in zones of the adrenal cortex. J Steroid Biochem 29:407–413

Lalli E, Sassone-Corsi P (1999) DAX-1 and the adrenal cortex. Curr Opin Endocrinol Diabetes 6:185–190

Landschulz KT, Pathak RK, Rigotti A, Krieger M, Hobbs HH (1996) Regulation of scavenger receptor, class B, type I, a high density lipoprotein receptor, in liver and steroidogenic tissues of the rat. J Clin Invest 98:984–995

Lang U, Vallotton MB (1987) Angiotensin II but not potassium induces subcellular redistribution of protein kinase C in bovine adrenal glomerulosa cells. J Biol Chem 262:8047–8050

Lang U, Daniel C, Chardonnens D, Capponi AM, Vallotton MB (1991) Modulatory effects of protein kinase C activation in aortic smooth muscle and adrenal glomerulosa cells during angiotensin II stimulation. In: Reid E, Cook GMW, Luzion JP (eds)Cell signalling: experimental strategies. Royal Society of Chemistry, Cambridge, UK, pp 115–126

Laragh JH, Angers M, Kelly WG, Liebermann S (1960) Hypotensive agents and pressor substances. The effect of epinephrine, norepinephrine, angiotensin II and others on the secretory rate of aldosterone in man. JAMA 174:234–240

Le Menuet D, Isnard R, Bichara M, Viengchareun S, Muffat-Joly M, Walker F, Zennaro MC, Lombes M (2001) Alteration of cardiac and renal functions in transgenic mice overexpressing human mineralocorticoid receptor. J Biol Chem 276:38911–38920

Lee TG, Lee YH, Kim JH, Kim HS, Suh PG, Ryu SH (1997) Immunological identification of cholesterol ester hydrolase in the steroidogenic tissues, adrenal glands and testis. Biochim Biophys Acta 1346:103–108

Lefebvre H, Contesse V, Delarue C, Legrand A, Kuhn JM, Vaudry H, Wolf LM (1995) The serotonin-4 receptor agonist cisapride and angiotensin II exert additive effects on aldosterone secretion in normal man. J Clin Endocrinol Metab 80:504–507

LeHoux JG, Lefebvre A (1991) Short-term effects of ACTH on the low-density lipoprotein receptor mRNA level in rat and hamster adrenals. J Mol Endocrinol 6:223–230

LeHoux JG, Lefebvre A, De Medicis E, Belisle S, Bellabarba D (1989) The enhancing effect of adrenocorticotropin on adrenal 3-hydroxy-3- methylglutaryl coenzyme A reductase messenger ribonucleic acid level is inhibited by aminoglutethimide but not by cycloheximide. Endocrinology 125:158–164

LeHoux JG, Dupuis G, Lefebvre A (2001) Control of CYP11B2 gene expression through differential regulation of its promoter by atypical and conventional protein kinase C isoforms. J Biol Chem 276:8021–8028

Leitersdorf E, Stein O, Stein Y (1985) Angiotensin II stimulates receptor-mediated uptake of LDL by bovine adrenal cortical cells in primary culture. Biochim Biophys Acta 835:183–190

Lekstrom-Himes J, Xanthopoulos KG (1998) Biological role of the CCAAT/enhancer-binding protein family of transcription factors. J Biol Chem 273:28545–28548

Leonetti G, Terzoli L, Zanchetti A (1987) Calcium antagonists and responsiveness of the adrenal glands to aldosterone-releasing stimuli in hypertensive patients. J Hypertens Suppl 5:S119–S122

Lesouhaitier O, Chiappe A, Rossier MF (2001) Aldosterone increases T-type calcium currents in human adrenocarcinoma (H295R) cells by inducing channel expression. Endocrinology 142:4320–4330

Li H, Brochu M, Wang SP, Rochdi L, Cote M, Mitchell G, Gallo-Payet N (2002) Hormone-sensitive lipase deficiency in mice causes lipid storage in the adrenal cortex and impaired corticosterone response to corticotropin stimulation. Endocrinology 143:3333–3340

Linde R, Winn S, Latta D, Hollifield J (1981) Graded dose effects of angiotensin II on aldosterone production in man during various levels of potassium intake. Metabolism 30:549–553

Lobo MV, Marusic ET (1986) Effect of angiotensin II, ATP, and ionophore A23187 on potassium efflux in adrenal glomerulosa cells. Am J Physiol 250:E125–E130

Lobo MV, Marusic ET (1988) Angiotensin II causes a dual effect on potassium permeability in adrenal glomerulosa cells. Am J Physiol 254:E144–E149

Lobo MV, Marusic ET (1991) Contrasting effects of sn-1,2-dioctanoyl glycerol as compared to other protein kinase C activators in adrenal glomerulosa cells. J Steroid Biochem Mol Biol 39:323–327

Lobo MV, Mendoza RR, Marusic ET (1990) sn-1,2 dioctanoylglycerol mimics the effects of angiotensin II on aldosterone production and potassium permeability in isolated bovine glomerulosa cells. J Steroid Biochem 35:29–33

Lotshaw DP (1997a) Characterization of angiotensin II-regulated potassium conductance in rat adrenal glomerulosa cells. J Membrane Biol 156:261–277

Lotshaw DP (1997b) Effects of potassium channel blockers on potassium channels, membrane potential, and aldosterone secretion in rat adrenal zona glomerulosa cells. Endocrinology 138:4167–4175

Lotshaw DP (2001) Role of membrane depolarization and T-type Ca2+ channels in angiotensin II and K+ stimulated aldosterone secretion. Mol Cell Endocrinol 175:157–171

Lu H-K, Fern RJ, Nee JJ, Barrett PQ (1994) Calcium-dependent activation of T-type calcium channels by calmodulin-dependent protein kinase II. Am J Physiol 267:F183–F189

Lu H-K, Fern RJ, Luthin D, Linden J, Liu L-P, Cohen CJ, Barrett PQ (1996) Angiotensin II stimulates T-type calcium channel currents via activation of a G protein, Gi. Am J Physiol 271:C1340–C1349

Lu X, Gruia-Gray J, Copeland NG, Gilbert DJ, Jenkins NA, Londos C, Kimmel AR (2001) The murine perilipin gene: the lipid droplet-associated perilipins derive from tissue-specific, mRNA splice variants and define a gene family of ancient origin. Mamm Genome 12:741–749

Lumbers ER (1999) Angiotensin and aldosterone. Regul Pept 80:91–100

Malendowicz LK, Rebuffat P, Nussdorfer GG, Nowak KW (1998) Corticotropin-inhibiting peptide enhances aldosterone secretion by dispersed rat zona glomerulosa cells. J Steroid Biochem Mol Biol 67:149–152

Manna PR, Dyson MT, Eubank DW, Clark BJ, Lalli E, Sassone-Corsi P, Zeleznik AJ, Stocco DM (2002) Regulation of steroidogenesis and the steroidogenic acute regulatory protein by a member of the cAMP response-element binding protein family. Mol Endocrinol 16:184–199

Marrero MB, Schieffer B, Paxton WG, Heerdt L, Berk BC, Delafontaine P, Bernstein KE (1995) Direct stimulation of Jak/STAT pathway by the angiotensin II AT1 receptor. Nature 375:247–250

Martin G, Pilon A, Albert C, Vallé M, Hum DW, Fruchart JC, Najib J, Clavey V, Staels B (1999) Comparison of expression and regulation of the high-density lipoprotein receptor SR-B1 and the low-density lipoprotein receptor in human adrenocortical carcinoma NCI-H295 cells. Eur J Biochem 261:481–491

Martinez DV, Rocha R, Matsumura M, Oestreicher E, Ochoa-Maya M, Roubsanthisuk W, Williams GH, Adler GK (2002) Cardiac damage prevention by eplerenone: comparison with low sodium diet or potassium loading. Hypertension 39:614–618

Martinez-Botas J, Anderson JB, Tessier D, Lapillonne A, Chang BH, Quast MJ, Gorenstein D, Chen KH, Chan L (2000) Absence of perilipin results in leanness and reverses obesity in Lepr(db/db) mice. Nat Genet 26:474–479

Mason JI, Rainey WE (1987) Steroidogenesis in the human fetal adrenal: a role for cholesterol synthesized de novo. J Clin Endocrinol Metab 64:140–147

Matsunaga H, Maruyama Y, Kojima I, Hoshi T (1987a) Transient calcium channel current characterized by a low threshold voltage in zona glomerulosa cells of rat adrenal cortex. Pflugers Arch 408:351–355

Matsunaga H, Yamashita N, Maruyama Y, Kojima I, Kurokawa K (1987b) Evidence for two distinct voltage-gated calcium channel currents in bovine adrenal glomerulosa cells. Biochem Biophys Res Commun 149:1049–1054

Matsuoka H, Ishii M, Sugimoto T, Hirata Y, Sugimoto T, Kangawa K, Matsuo H (1985) Inhibition of aldosterone production by alpha-human atrial natriuretic polypeptide is associated with an increase in cGMP production. Biochem Biophys Res Commun 127:1052–1056

Matsuoka H, Ishii M, Hirata Y, Atarashi K, Sugimoto T, Kangawa K, Matsuo H (1987) Evidence for lack of a role of cGMP in effect of alpha-hANP on aldosterone inhibition. Am J Physiol 252:E643–E647

Maturana AD, Burnay MM, Capponi AM, Vallotton MB, Rossier MF (1999a) Angiotensin II type 1 receptor activation modulates L- and T-type calcium channel activity through distinct mechanisms in bovine adrenal glomerulosa cells. J Recept Signal Transduc Res 19:509–520

Maturana AD, Casal AJ, Demaurex N, Vallotton MB, Capponi AM, Rossier MF (1999b) Angiotensin II negatively modulates L-type calcium channels through a pertussis toxin-sensitive G protein in adrenal glomerulosa cells. J Biol Chem 274:19943–19948

Maume G, Filali-Ansary A, Giannini E, Hathout Y, Fischbach M, Maume BF (1991) Aldosterone biosynthesis induced by ACTH and angiotensin II in newborn rat adrenocortical cells transfected by c-EJ-Ha-ras oncogene. Biochem Biophys Res Commun 175:596–603

Mazzocchi G, Markowska A, Malendowicz LK, Musajo F, Meneghelli V, Nussdorfer GG (1993) Evidence that endogenous arginine-vasopressin (AVP) is involved in the maintenance of the growth and steroidogenic capacity of rat adrenal zona glomerulosa. J Steroid Biochem Mol Biol 45:251–256

Mazzocchi G, Macchi C, Malendowicz LK, Nussdorfer GG (1995) Evidence that endogenous substance-P (SP) is involved in the maintenance of the growth and steroidogenic capacity of rat adrenal zona glomerulosa. Neuropeptides 29:53–58

McCarthy RT, Isales CM, Bollag WB, Rasmussen H, Barrett PQ (1990) Atrial natriuretic peptide differentially modulates T- and L-type calcium channels. Am J Physiol 258:F473–F478

McCarthy RT, Isales CM, Rasmussen H (1993) T-type calcium channels in adrenal glomerulosa cells: GTP-dependent modulation by angiotensin II. Proc Nat Acad Sci USA 90:3260–3264

McMahon EG (2001) Recent studies with eplerenone, a novel selective aldosterone receptor antagonist. Curr Opin Pharmacol 1:190–196

McNeill H, Vinson GP (2000) Regulation of MAPK activity in response to dietary sodium in the rat adrenal gland. Endocr Res 26:879–883

McNeill H, Puddefoot JR, Vinson GP (1998) MAP kinase in the rat adrenal gland. Endocr Res 24:373–380

Mikami K, Nishikawa T, Saito Y, Tamura Y, Matsuoka N, Kumagai A, Yoshida S (1984) Regulation of cholesterol metabolism in rat adrenal glands: effect of adrenocorticotropin, cholesterol, and corticosteroids on acyl-coenzyme A synthetase and cholesterol ester hydrolase. Endocrinology 114:136–140

Millar JA, Struthers AD, Beastall GH, Reid JL (1982) Effect of nifedipine on blood pressure and adrenocortical responses to trophic stimuli in humans. J Cardiovasc Pharmacol 4 [Suppl 3]:S330-S334

Mizuno Y, Yoshimura M, Yasue H, Sakamoto T, Ogawa H, Kugiyama K, Harada E, Nakayama M, Nakamura S, Ito T, Shimasaki Y, Saito Y, Nakao K (2001) Aldosterone production is activated in failing ventricle in humans. Circulation 103:72–77

Moritz KM, Boon WC, Wintour EM (1999) Aldosterone secretion by the mid-gestation ovine fetus: role of the AT2 receptor. Mol Cell Endocrinol 157:153–160

Morohashi K, Zanger UM, Honda S, Hara M, Waterman MR, Omura T (1993) Activation of CYP11A and CYP11B gene promoters by the steroidogenic cell-specific transcription factor, Ad4BP. Mol Endocrinol 7:1196–1204

Müller J (1988) Regulation of aldosterone biosynthesis. Monographs on endocrinology. Springer Verlag, Berlin Heidelberg New York

Müller J, Lauber M, Schmid C (1989) Potassium-induced aldosterone biosynthesis in cultured rat zona glomerulosa cells. Am J Physiol 256:E475–E482

Mulrow PJ (1988) Production of renin by adrenal glomerulosa cells. Trans Am Clin Climatol Assoc 100:126–131

Mulrow PJ (1999) Angiotensin II and aldosterone regulation. Regul Pept 80:27–32

Mulrow PJ, Kusano E, Baba K, Shier D, Doi Y, Franco-Saenz R, Stoner G, Rapp J (1988) Adrenal renin: a possible local hormonal regulator of aldosterone production. Cardiovasc Drugs Ther 2:463–471

Nackley AC, Shea-Eaton W, Lopez D, Mclean MP (2002) Repression of the steroidogenic acute regulatory gene by the multifunctional transcription factor Yin Yang 1. Endocrinology 143:1085–1096

Nadler JL, Natarajan R, Stern N (1987) Specific action of the lipoxygenase pathway in mediating angiotensin II-induced aldosterone synthesis in isolated adrenal glomerulosa cells. J Clin Invest 80:1763–1769

Nagy K, Koroknai L, Spat A (1984) Effect of lipoproteins on aldosterone production by isolated glomerulosa cells. J Steroid Biochem 20:789–791

Nakanishi S, Catt KJ, Balla T (1994) Inhibition of agonist-stimulated inositol 1,4,5-trisphosphate production and calcium signaling by the myosin light chain kinase inhibitor, wortmannin. J Biol Chem 269:6528–6535

Nakanishi S, Catt KJ, Balla T (1995) A wortmannin-sensitive phosphatidylinositol 4-kinase that regulates hormone-sensitive pools of inositolphospholipids. Proc Natl Acad Sci USA 92:5317–5321

Nakano S, Carvallo P, Rocco S, Aguilera G (1990) Role of protein kinase C on the steroidogenic effect of angiotensin II in the rat adrenal glomerulosa cell. Endocrinology 126:125–133

Natarajan R, Stern N, Hsueh W, Do Y, Nadler J (1988a) Role of the lipoxygenase pathway in angiotensin II-mediated aldosterone biosynthesis in human adrenal glomerulosa cells. J Clin Endocrinol Metab 67:584–591

Natarajan R, Stern N, Nadler J (1988b) Diacylglycerol provides arachidonic acid for lipoxygenase products that mediate angiotensin II-induced aldosterone synthesis. Biochem Biophys Res Commun 156:717–724

Natarajan R, Dunn WD, Stern N, Nadler J (1990) Key role of diacylglycerol-mediated 12-lipoxygenase product formation in angiotensin II-induced aldosterone synthesis. Mol Cell Endocrinol 72:73–80

Natarajan R, Gonzales N, Hornsby PJ, Nadler J (1992) Mechanism of angiotensin II-induced proliferation in bovine adrenocortical cells. Endocrinology 131:1174–1180

Natarajan R, Lanting L, Xu L, Nadler J (1994) Role of specific isoforms of protein kinase C in angiotensin II and lipoxygenase action in rat adrenal glomerulosa cells. Mol Cell Endocrinol 101:59–66

Natarajan R, Kathuria S, Lanting L, Gonzales N, Nadler JL (1995) Differential short- and long-term effects of insulin on Ang II action in human adrenal glomerulosa cells. Am J Physiol 268:E100-E106

Nishikawa T, Mikami K, Saito Y, Tamura Y, Kumagai A (1981) Studies on cholesterol esterase in the rat adrenal. Endocrinology 108:932–936

Nishikawa T, Mikami K, Saito Y, Tamura K, Yoshida S (1988) Functional differences in cholesterol ester hydrolase and acyl-coenzyme-A/cholesterol acyltransferase between the outer and inner zones of the guinea pig adrenal cortex. Endocrinology 122:877–883

Nishizuka Y (1984) Turnover of inositol phospholipids and signal transduction. Science 225:1365–1370

Nussdorfer GG, Rossi GP, Malendowicz LK, Mazzocchi G (1999) Autocrine-paracrine endothelin system in the physiology and pathology of steroid-secreting tissues (review). Pharmacol Rev 51:403–437

Oda H, Lotshaw DP, Franco-Saenz R, Mulrow PJ (1991) Local generation of angiotensin II as a mechanism of aldosterone secretion in rat adrenal capsules. Proc Soc Exp Biol Med 196:175–177

Osman H, Murigande C, Nadakal A, Capponi AM (2002) Repression of DAX-1 and induction of SF-1 expression: two mechanisms contributing to the activation of aldosterone biosynthesis in adrenal glomerulosa cells. J Biol Chem 277:41259–41267

Page IH, Helmer OM (1940) A crystalline pressor substance (angiotonin) resulting from the reaction between renin and renin-activator. J Exp Med 71:29–42

Parker KL, Schimmer BP (1997) Steroidogenic factor 1: a key determinant of endocrine development and function. Endocr Rev 18:361–377

Passier RC, Smits JF, Verluyten MJ, Daemen MJ (1996) Expression and localization of renin and angiotensinogen in rat heart after myocardial infarction. Am J Physiol 271:H1040–H1048

Payet MD, Durroux T, Bilodeau L, Guillon G, Gallo-Payet N (1994) Characterization of potassium and calcium ionic currents in glomerulosa cells from human adrenal glands. Endocrinology 134:2589–2598

Payet MD, Bilodeau L, Drolet P, Ibarrondo J, Guillon G, Gallo-Payet N (1995) Modulation of a calcium-activated potassium channel by angiotensin II in rat adrenal glomerulosa cells: involvement of a G protein. Mol Endocrinol 9:935–947

Pedersen RC, Brownie AC (1983) Cholesterol side-chain cleavage in the rat adrenal cortex: isolation of a cycloheximide-sensitive activator peptide. Proc Natl Acad Sci USA 80:1882–1886

Peterson JK, Moran F, Conley AJ, Bird IM (2001) Zonal expression of endothelial nitric oxide synthase in sheep and rhesus adrenal cortex. Endocrinology 142:5351–5363

Petrasek D, Jensen G, Tuck M, Stern N (1992) In vitro effects of insulin on aldosterone production in rat zona glomerulosa cells. Life Sci 50:1781–1787

Petrescu AD, Gallegos AM, Okamura Y, Strauss JF III, Schroeder F (2001) Steroidogenic acute regulatory protein binds cholesterol and modulates mitochondrial membrane sterol domain dynamics. J Biol Chem 276:36970–36982

Pezzi V, Clark BJ, Ando S, Stocco DM, Rainey WE (1996) Role of calmodulin-dependent protein kinase II in the acute stimulation of aldosterone production. J Steroid Biochem Mol Biol 58:417–424

Pezzi V, Clyne CD, Ando S, Mathis JM, Rainey WE (1997) Calcium-regulated expression of aldosterone synthase is mediated by calmodulin and calmodulin-dependent protein kinases. Endocrinology 138:835–838

Pitt B, Zannad F, Remme WJ, Cody R, Castaigne A, Perez A, Paliensky J, Wittes J (1999) The effect of spironolactone on morbidity and mortality in patients with severe heart failure. N Engl J Med 341:709–717

Plump AS, Erickson SK, Weng W, Partin JS, Breslow JL, Williams DL (1996) Apolipoprotein A-I is required for cholesteryl ester accumulation in steroidogenic cells and for normal adrenal steroid production. J Clin Invest 97:2660–2671

Poirier SN, Poitras M, Chorvatova A, Payet MD, Guillemette G (2001) FK506 blocks intracellular Ca2+ oscillations in bovine adrenal glomerulosa cells. Biochemistry 40:6486–6492

Poitras M, Bernier S, Servant M, Richard DE, Boulay G, Guillemette G (1993) The high affinity state of inositol 1,4,5-trisphosphate receptor is a functional state. J Biol Chem 268:24078–24082

Poitras M, Poirier SN, Laflamme K, Simoneau M, Escher E, Guillemette G (2000) Different populations of inositol 1,4,5-trisphosphate receptors expressed in the bovine adrenal cortex. Receptors Channels 7:41–52

Pralong WF, Spat A, Wollheim CB (1994) Dynamic pacing of cell metabolism by intracellular Ca2+ transients. J Biol Chem 269:27310–27314

Puglielli L, Rigotti A, Greco AV, Santos MJ, Nervi F (1995) Sterol carrier protein-2 is involved in cholesterol transfer from the endoplasmic reticulum to the plasma membrane in human fibroblasts. J Biol Chem 270:18723–18726

Putney JW Jr (1999) "Kissin' cousins": intimate plasma membrane-endoplasmic reticulum interactions underlie capacitative calcium entry. Cell 99:5–8

Putney JW Jr, McKay RR (1999) Capacitative calcium entry channels. Bioessays 21:38–46

Python CP, Rossier MF, Vallotton MB, Capponi AM (1993) Peripheral-type benzodiazepines inhibit calcium channels and aldosterone production in adrenal glomerulosa cells. Endocrinology 132:1489–1496

Python CP, Laban OP, Rossier MF, Vallotton MB, Capponi AM (1995) The site of action of calcium in the activation of steroidogenesis: studies in calcium-clamped bovine adrenal zona glomerulosa cells. Biochem J 305:569–576

Quinn SJ, Cornwall MC, Williams GH (1987) Electrophysiological responses to angiotensin II of isolated rat adrenal glomerulosa cells. Endocrinology 120:1581–1589

Quinn SJ, Williams GH, Tillotson DL (1988) Calcium oscillations in single adrenal glomerulosa cells stimulated by angiotensin II. Proc Nat Acad Sci USA 85:5754–5758

Rainey WE (1999) Adrenal zonation: clues from 11 beta-hydroxylase and aldosterone synthase. Mol Cell Endocrinol 151:151–160

Rainey WE, Rodgers RJ, Mason JI (1992) The role of bovine lipoproteins in the regulation of steroidogenesis and HMG-CoA reductase in bovine adrenocortical cells. Steroids 57:167–173

Rasmussen H, Isales CM, Calle R, Throckmorton D, Anderson M, Gasalla-Herraiz J, McCarthy RT (1995) Diacylglycerol production, calcium influx, and protein kinase C activation in sustained cellular responses. Endocrine Rev 16:649–681

Reaven E, Tsai L, Azhar S (1996) Intracellular events in the "selective" transport of lipoprotein-derived cholesteryl esters. J Biol Chem 271:16208–16217

Reinhart AJ, Williams SC, Clark BJ, Stocco DM (1999a) SF-1 (steroidogenic factor-1) and C/EBP beta (CCAAT/enhancer binding protein-beta) cooperate to regulate the murine StAR (steroidogenic acute regulatory) promoter. Mol Endocrinol 13:729–741

Reinhart AJ, Williams SC, Stocco DM (1999b) Transcriptional regulation of the StAR gene. Mol Cell Endocrinol 151:161–169

Reyland ME, Evans RM, White EK (2000) Lipoproteins regulate expression of the steroidogenic acute regulatory protein (StAR) in mouse adrenocortical cells. J Biol Chem 275:36637–36644

Ribeiro-do-Valle RM, Poitras M, Boulay G, Guillemette G (1994) The important discrepancy between the apparent affinity observed in Ca2+ mobilization studies and the Kd measured in binding studies is a consequence of the quantal process by which inositol 1,4,5-trisphosphate releases Ca2+ from bovine adrenal cortex microsomes. Cell Calcium 15:79–88

Richard DE, Chretien L, Caron M, Guillemette G (1997a) Stimulation of the angiotensin II type I receptor on bovine adrenal glomerulosa cells activates a temperature-sensitive internalization-recycling pathway. Mol Cell Endocrinol 129:209–218

Richard DE, Laporte SA, Bernier SG, Leduc R, Guillemette G (1997b) Desensitization of AT1 receptor-mediated cellular responses requires long term receptor down-regulation in bovine adrenal glomerulosa cells. Endocrinology 138:3828–3835

Rigotti A, Edelmann ER, Seifert P, Iqbal SN, Demattos RB, Temel RE, Krieger M, Williams DL (1996) Regulation by adrenocorticotropic hormone of the in vivo expression of scavenger receptor class B type I (SR-BI), a high density lipoprotein receptor, in steroidogenic cells of the murine adrenal gland. J Biol Chem 271:33545–33549

Rigotti A, Trigatti B, Babitt J, Penman M, Xu S, Krieger M (1997) Scavenger receptor BI—a cell surface receptor for high density lipoprotein. Curr Opin Lipidol 8:181–188

Rizzuto R, Brini M, Murgia M, Pozzan T (1993) Microdomains with high calcium close to inositol trisphosphate-sensitive channels that are sensed by neighboring mitochondria. Science 262:744–747

Robert V, Heymes C, Silvestre JS, Sabri A, Swynghedauw B, Delcayre C (1999) Angiotensin AT1 receptor subtype as a cardiac target of aldosterone: role in aldosterone-salt-induced fibrosis. Hypertension 33:981–986

Rocha R, Stier CT Jr, Kifor I, Ochoa-Maya MR, Rennke HG, Williams GH, Adler GK (2000) Aldosterone: a mediator of myocardial necrosis and renal arteriopathy. Endocrinology 141:3871–3878

Rocha R, Martin-Berger CL, Yang P, Scherrer R, Delyani J, McMahon E (2002a) Selective aldosterone blockade prevents angiotensin ii/salt-induced vascular inflammation in the rat heart. Endocrinology 143:4828–4836

Rocha R, Rudolph AE, Frierdich GE, Nachowiak DA, Kekec BK, Blomme EA, McMahon EG, Delyani JA (2002b) Aldosterone induces a vascular inflammatory phenotype in the rat heart. Am J Physiol Heart Circ Physiol 283:H1802–H1810

Rodrigueza WV, Thuahnai ST, Temel RE, Lund-Katz S, Phillips MC, Williams DL (1999) Mechanism of scavenger receptor class B type I-mediated selective uptake of cholesteryl esters from high density lipoprotein to adrenal cells. J Biol Chem 274:20344–20350

Rohacs T, Bago A, Deak F, Hunyady L, Spät A (1994) Capacitative calcium influx in adrenal glomerulosa cells: possible role in angiotensin II response. Am J Physiol 267:C1246–C1252

Rohacs T, Nagy G, Spät A (1997a) Cytoplasmic calcium signaling and reduction of mitochondrial pyridine nucleotides in adrenal glomerulosa cells in response to potassium, angiotensin II and vasopressin. Biochem J 322:785–792

Rohacs T, Tory K, Dobos A, Spät A (1997b) Intracellular calcium release is more efficient than calcium influx in stimulating mitochondrial NAD(P)H formation in adrenal glomerulosa cells. Biochem J 328:525–528

Romanowski MJ, Soccio RE, Breslow JL, Burley SK (2002) Crystal structure of the Mus musculus cholesterol-regulated START protein 4 (StarD4) containing a StAR-related lipid transfer domain. Proc Natl Acad Sci USA 99:6949–6954

Rossier MF (1997) Confinement of intracellular calcium signaling in secretory and steroidogenic cells. Eur J Endocrinol 137:317–325

Rossier MF, Capponi AM (2000a) Angiotensin II and calcium channels. Vitam Horm 60:229–284

Rossier MF, Capponi AM (2000b) Antagonistes calciques et inhibition de la sécrétion d'aldostérone. Métab Horm Nutr IV:101–106

Rossier MF, Capponi AM (2002) Cytosolic calcium oscillations in signal transduction pathways. Mini Rev Med Chem 2:353–360

Rossier MF, Dentand IA, Lew PD, Capponi AM, Vallotton MB (1986) Interconversion of inositol 1,4,5-trisphosphate to inositol 1,3, 4,5-tetrakisphosphate and 1,3,4-trisphosphate in permeabilized adrenal glomerulosa cells is calcium-sensitive and ATP-dependent. Biochem Biophys Res Commun 139:259–265

Rossier MF, Krause K-H, Lew PD, Capponi AM, Vallotton MB (1987) Control of cytosolic free calcium by intracellular organelles in bovine adrenal glomerulosa cells: effects of sodium and inositol 1,4,5-trisphosphate. J Biol Chem 262:4053–4058

Rossier MF, Capponi AM, Vallotton MB (1988) Inositol trisphosphate isomers in angiotensin II-stimulated adrenal glomerulosa cells. Mol Cell Endocrinol 57:163–168

Rossier MF, Capponi AM, Vallotton MB (1989) The inositol 1,4,5-trisphosphate-binding site in adrenal cortical cells is distinct from the endoplasmic reticulum. J Biol Chem 264:14078–14084

Rossier MF, Putney JW Jr (1991) The identity of the calcium-storing, inositol 1,4,5-trisphosphate-sensitive organelle in non-muscle cells: calciosomes, endoplasmic reticulum ... or both? Trends Neurosci 14:310–314

Rossier MF, Python CP, Capponi AM, Schlegel W, Kwan CY, Vallotton MB (1993) Blocking T-type calcium channels with tetrandrine inhibits steroidogenesis in bovine adrenal glomerulosa cells. Endocrinology 132:1035–1043

Rossier MF, Aptel HBC, Python CP, Burnay MM, Vallotton MB, Capponi AM (1995) Inhibition of low threshold calcium channels by angiotensin II in adrenal glomerulosa cells through activation of protein kinase C. J Biol Chem 270:15137–15142

Rossier MF, Burnay MM, Brandenburger Y, Cherradi N, Vallotton MB, Capponi AM (1996a) Sources and sites of action of calcium in the regulation of aldosterone biosynthesis. Endocrine Res 22:579–588

Rossier MF, Burnay MM, Vallotton MB, Capponi AM (1996b) Distinct functions of T-type and L-type calcium channels during activation of bovine adrenal glomerulosa cells. Endocrinology 137:4817–4826

Rossier MF, Burnay MM, Maturana AD, Capponi AM (1998a) Duality of the voltage-dependent calcium influx in adrenal glomerulosa cells. Endocrine Res 24:443–447

Rossier MF, Ertel EA, Vallotton MB, Capponi AM (1998b) Inhibitory action of mibefradil on calcium signaling and aldosterone synthesis in bovine adrenal glomerulosa cells. J Pharmacol Exp Ther 287:824–831

Rössig L, Zólyomi A, Catt KJ, Balla T (1996) Regulation of angiotensin II-stimulated calcium oscillations by calcium influx mechanisms in adrenal glomerulosa cells. J Biol Chem 271:22063–22069

Russell DW, Yamamoto T, Schneider WJ, Slaughter CJ, Brown MS, Goldstein JL (1983) cDNA cloning of the bovine low density lipoprotein receptor: feedback regulation of a receptor mRNA. Proc Natl Acad Sci USA 80:7501–7505

Rust W, Stedronsky K, Tillmann G, Morley S, Walther N, Ivell R (1998) The role of SF-1/Ad4BP in the control of the bovine gene for the steroidogenic acute regulatory (StAR) protein. J Molecular Endocrinol 21:189–200

Sachse R, Shao XJ, Rico A, Finckh U, Rolfs A, Reincke M, Hensen J (1997) Absence of angiotensin II type 1 receptor gene mutations in human adrenal tumors. Eur J Endocrinol 137:262–266

Sadoshima J, Xu Y, Slayter HS, Izumo S (1993) Autocrine release of angiotensin II mediates stretch-induced hypertrophy of cardiac myocytes in vitro. Cell 75:977–984

Sandhoff TW, Hales DB, Hales KH, Mclean MP (1998) Transcriptional regulation of the rat steroidogenic acute regulatory protein gene by steroidogenic factor 1. Endocrinology 139:4820–4831

Schiebinger RJ, Braley LM, Menachery A, Williams GH (1986) Unique calcium dependencies of the activating mechanism of the early and late aldosterone biosynthetic pathways in the rat. J Endocrinol 110:315–325

Schiffrin EL, Gutkowska J, Lis M, Genest J (1982) Relative roles of sodium and calcium ions in the steroidogenic response of isolated rate adrenal glomerulosa cells. Hypertension 4:36–42

Schrier AD, Wang H, Talley EM, Perez-Reyes E, Barrett PQ (2001) The alpha1 H T-type calcium channel is the predominant subtype expressed in bovine and rat zona glomerulosa. Am J Physiol 280:C265–C272

Schulz G, Rosenthal W, Hescheler J, Trautwein W (1990) Role of G proteins in calcium channel modulation. Annu Rev Physiol 52:275–292

Scriabine A, Anderson CL, Janis RA, Kojima K, Rasmussen H, Lee S, Michal U (1984) Some recent pharmacological findings with nitrendipine. J Cardiovasc Pharmacol 6 [Suppl 7]:S937–S943

Servetnick DA, Brasaemle DL, Gruia-Gray J, Kimmel AR, Wolff J, Londos C (1995) Perilipins are associated with cholesteryl ester droplets in steroidogenic adrenal cortical and Leydig cells. J Biol Chem 270:16970–16973

Shah JR, Laredo J, Hamilton BP, Hamlyn JM (1998) Different signaling pathways mediate stimulated secretions of endogenous ouabain and aldosterone from bovine adrenocortical cells. Hypertension 31:463–468

Shea-Eaton WK, Trinidad MJ, Lopez D, Nackley A, Mclean MP (2001) Sterol regulatory element binding protein-1a regulation of the steroidogenic acute regulatory protein gene. Endocrinology 142:1525–1533

Shea-Eaton W, Sandhoff TW, Lopez D, Hales DB, Mclean MP (2002) Transcriptional repression of the rat steroidogenic acute regulatory (StAR) protein gene by the AP-1 family member c-Fos. Mol Cell Endocrinol 188:161–170

Shepherd RM, Fraser R, Kenyon CJ (1992) Membrane permeability to potassium and the control of aldosterone synthesis: effects of valinomycin and cromakalim in bovine adrenocortical cells. J Mol Endocrinol 9:165–173

Shibata H, Ogishima T, Mitani F, Suzuki H, Murakami M, Saruta T, Ishimura Y (1991) Regulation of aldosterone synthase cytochrome P-450 in rat adrenals by angiotensin II and potassium. Endocrinology 128:2534–2539

Silverman E, Eimerl S, Orly J (1999) CCAAT enhancer-binding protein beta and GATA-4 binding regions within the promoter of the steroidogenic acute regulatory protein (StAR) gene are required for transcription in rat ovarian cells. J Biol Chem 274:17987–17996

Silvestre JS, Robert V, Heymes C, Aupetit-Faisant B, Mouas C, Moalic JM, Swynghedauw B, Delcayre C (1998) Myocardial production of aldosterone and corticosterone in the rat. Physiological regulation. J Biol Chem 273:4883–4891

Silvestre JS, Heymes C, Oubénaïssa A, Robert V, Aupetit-Faisant B, Carayon A, Swynghedauw B, Delcayre C (1999a) Activation of cardiac aldosterone production in rat myocardial infarction: effect of angiotensin II blockade and role in cardiac fibrosis. Circulation 99:2694–2701

Silvestre JS, Robert V, Delcayre C (1999b) The cardiac endocrine aldosterone system. Curr Opin Endocrinol Diabetes 6:204–209

Simpson HD, Shepherd R, Shepherd J, Fraser R, Lever AF, Kenyon CJ (1989) Effects of cholesterol and lipoproteins on aldosterone secretion by bovine zona glomerulosa cells. J Endocrinol 121:125–131

Simpson SA, Tait JF, Wettstein A, Neher R, von Euw J, Schindler O, Reîchstein T (1954) Konstitution des Aldosterons, des neuen Mieralocorticoids. Experientia 10:132–133

Sirianni R, Sirianni R, Carr BR, Pezzi V, Rainey WE (2001) A role for src tyrosine kinase in regulating adrenal aldosterone production. J Mol Endocrinol 26:207–215

Sirianni R, Seely JB, Attia G, Stocco DM, Carr BR, Pezzi V, Rainey WE (2002) Liver receptor homologue-1 is expressed in human steroidogenic tissues and activates transcription of genes encoding steroidogenic enzymes. J Endocrinol 174:R13 R17

Smith RD, Timmermans PB (1994) Human angiotensin receptor subtypes. Curr Opin Nephrol Hypertens 3:112–122

Smith RD, Baukal AJ, Dent P, Catt KJ (1999) Raf-1 kinase activation by angiotensin II in adrenal glomerulosa cells: roles of Gi, phosphatidylinositol 3-kinase, and Ca2+ influx. Endocrinology 140:1385–1391

Soccio RE, Adams RM, Romanowski MJ, Sehayek E, Burley SK, Breslow JL (2002) The cholesterol-regulated StarD4 gene encodes a StAR-related lipid transfer protein with two closely related homologues, StarD5 and StarD6. Proc Natl Acad Sci USA 99:6943–6948

Song M, Shao H, Mujeeb A, James TL, Miller WL (2001) Molten-globule structure and membrane binding of the N-terminal protease-resistant domain (63–193) of the steroidogenic acute regulatory protein (StAR). Biochem J 356:151–158

Sonnenborn U, Eiteljörge G, Trzeciak WH, Kunau WH (1982) Identical catalytic subunit in both molecular forms of hormone-sensitive cholesterol esterase from bovine adrenal cortex. FEBS Letters 145:271–276

Souza SC, Muliro KV, Liscum L, Lien P, Yamamoto MT, Schaffer JE, Dallal GE, Wang X, Kraemer FB, Obin M, Greenberg AS (2002) Modulation of hormone-sensitive lipase and protein kinase A-mediated lipolysis by perilipin A in an adenoviral reconstituted system. J Biol Chem 277:8267–8272

Spät A (1988) Stimulus-secretion coupling in angiotensin-stimulated adrenal glomerulosa cells. J Steroid Biochem 29:443–453

Spät A, Balla I, Balla T, Cragoe EJ Jr, Hajnoczky G, Hunyady L (1989) Angiotensin II and potassium activate different calcium entry mechanisms in rat adrenal glomerulosa cells. J Endocrinol 122:361–370

Spät A, Enyedi P, Hajnoczky G, Hunyady L (1991) Generation and role of calcium signal in adrenal glomerulosa cells. Exp Physiol 76:859–885

Spät A, Rohacs T, Horvath A, Szabadkai G, Enyedi P (1996) The role of voltage-dependent calcium channels in angiotensin-stimulated glomerulosa cells. Endocr Res 22:569–576

Startchik I, Morabito D, Lang U, Rossier MF (2002) Control of calcium homeostasis by angiotensin II in adrenal glomerulosa cells through activation of p38 MAPK. J Biol Chem 277:24265–24273

Stern N, Yanagawa N, Saito F, Hori M, Natarajan R, Nadler J, Tuck M (1993) Potential role of 12 hydroxyeicosatetraenoic acid in angiotensin II-induced calcium signal in rat glomerulosa cells. Endocrinology 133:843–847

Stocco DM (1997) The steroidogenic acute regulatory (StAR) protein two years later—an update. Endocrine 6:99–109

Stocco DM (2001) StAR protein and the regulation of steroid hormone biosynthesis. Annu Rev Physiol 63:193–213

Stocco DM, Clark BJ (1996a) Regulation of the acute production of steroids in steroidogenic cells. Endocr Rev 17:221–244

Stocco DM, Clark BJ (1996b) Role of the steroidogenic acute regulatory protein (StAR) in steroidogenesis. Biochem Pharmacol 51:197–205

Stralfors P, Belfrage P (1983) Phosphorylation of hormone-sensitive lipase by cyclic AMP-dependent protein kinase. J Biol Chem 258:15146–15152

Sugawara T, Holt JA, Kiriakidou M, Strauss JF (1996) Steroidogenic factor 1-dependent promoter activity of the human steroidogenic acute regulatory protein (StAR) gene. Biochemistry 35:9052–9059

Sugawara T, Kiriakidou M, Mcallister JM, Holt JA, Arakane F, Strauss JF (1997a) Regulation of expression of the steroidogenic acute regulatory protein (StAR) gene: a central role for steroidogenic factor 1. Steroids 62:5–9

Sugawara T, Kiriakidou M, Mcallister JM, Kallen CB, Strauss JF (1997b) Multiple steroidogenic factor 1 binding elements in the human steroidogenic acute regulatory protein gene 5′-flanking region are required for maximal promoter activity and cyclic AMP responsiveness. Biochemistry 36:7249–7255

Sun Y, Wang N, Tall AR (1999) Regulation of adrenal scavenger receptor-Bl expression by ACTH and cellular cholesterol pools. J Lipid Res 40:1799–1805

Sun Y, Zhang J, Zhang JQ, Weber KT (2001) Renin expression at sites of repair in the infarcted rat heart. J Mol Cell Cardiol 33:995–1003

Takagi M, Franco-Saenz R, Shier D, Mulrow PJ (1988) Effect of atrial natriuretic factor on calcium fluxes in adrenal glomerulosa cells. Hypertension 11:433–439

Takeda Y, Miyamori I, Yoneda T, Hatakeyama H, Inaba S, Furukawa K, Mabuchi H, Takeda R (1996) Regulation of aldosterone synthase in human vascular endothelial cells by angiotensin II and adrenocorticotropin. J Clin Endocrinol Metab 81:2797–2800

Takeda Y, Yoneda T, Demura M, Miyamori I, Mabuchi H (2000a) Cardiac aldosterone production in genetically hypertensive rats. Hypertension 36:495–500

Takeda Y, Yoneda T, Demura M, Miyamori I, Mabuchi H (2000b) Sodium-induced cardiac aldosterone synthesis causes cardiac hypertrophy. Endocrinology 141:1901–1904

Tamai KT, Monaco L, Alastalo TP, Lalli E, Parvinen M, Sassone-Corsi P (1996) Hormonal and developmental regulation of DAX-1 expression in sertoli cells. Mol Endocrinol 10:1561–1569

Tanabe A, Naruse M, Arai K, Naruse K, Yoshimoto T, Seki T, Imaki T, Kobayashi M, Miyazaki H, Demura H (1998) Angiotensin II stimulates both aldosterone secretion and DNA synthesis via type 1 but not type 2 receptors in bovine adrenocortical cells. J Endocrinol Invest 21:668–672

Tansey JT, Sztalryd C, Gruia-Gray J, Roush DL, Zee JV, Gavrilova O, Reitman ML, Deng CX, Li C, Kimmel AR, Londos C (2001) Perilipin ablation results in a lean mouse with aberrant adipocyte lipolysis, enhanced leptin production, and resistance to diet-induced obesity. Proc Natl Acad Sci USA 98:6494–6499

Temel RE, Trigatti B, Demattos RB, Azhar S, Krieger M, Williams DL (1997) Scavenger receptor class B, type I (SR-BI) is the major route for the delivery of high density lipoprotein cholesterol to the steroidogenic pathway in cultured mouse adrenocortical cells. Proc Natl Acad Sci USA 94:13600–13605

Tian Y, Smith RD, Balla T, Catt KJ (1998) Angiotensin II activates mitogen-activated protein kinase via protein kinase C and Ras/Raf-1 kinase in bovine adrenal glomerulosa cells. Endocrinology 139:1801–1809

Timmermans PB, Wong PC, Chiu AT, Herblin WF, Benfield P, Carini DJ, Lee RJ, Wexler RR, Saye JA, Smith RD (1993) Angiotensin II receptors and angiotensin II receptor antagonists. Pharmacol Rev 45:205–251

Toth IE, Szabo D, Bruckner GG (1997) Lipoproteins, lipid droplets, lysosomes, and adrenocortical steroid hormone synthesis: morphological studies. Microsc Res Tech 36:480–492

Trautwein W, Hescheler J (1990) Regulation of cardiac L-type calcium current by phosphorylation and G proteins. Annu Rev Physiol 52:257–274

Tremblay A, Parker KL, LeHoux JG (1992) Dietary potassium supplementation and sodium restriction stimulate aldosterone synthase but not 11 beta-hydroxylase P-450 messenger ribonucleic acid accumulation in rat adrenals and require angiotensin II production. Endocrinology 130:3152–3158

Tremblay J, Gerzer R, Pang SC, Cantin M, Genest J, Hamet P (1986) ANF stimulation of detergent-dispersed particulate guanylate cyclase from bovine adrenal cortex. FEBS Lett 194:210–214

Tremblay JJ, Hamel F, Viger RS (2002) Protein kinase A-dependent cooperation between GATA and CCAAT/enhancer-binding protein transcription factors regulates steroidogenic acute regulatory protein promoter activity. Endocrinology 143:3935–3945

Tsujishita Y, Hurley JH (2000) Structure and lipid transport mechanism of a StAR-related domain. Nat Struct Biol 7:408–414

Vahouny GV, Chanderbhan R, Noland BJ, Scallen TJ (1984) Cholesterol ester hydrolase and sterol carrier proteins. Endocr Rese 10:473–505

Vallotton MB, Rossier MF, Capponi AM (1995) Potassium-angiotensin interplay in the regulation of aldosterone biosynthesis. Clin Endocrinol 42:111–119

Vassilev PM, Kanazirska MV, Quinn SJ, Tillotson DL, Williams GH (1992) Potassium channels in adrenal zona glomerulosa cells: I. Characterization of distinct channel types. Am J Physiol 263:E752–E759

Vinson GP, Laird SM, Whitehouse BJ, Hinson JP (1989) Specific effects of agonists of the calcium messenger system on secretion of 'late-pathway' steroid products by intact tissue and dispersed cells of the rat adrenal zona glomerulosa. J Mol Endocrinol 2:157–165

Vinson GP, Hinson JP, Toth IE (1994a) The neuroendocrinology of the adrenal cortex. J Neuroendocrinol 6:235–246

Vinson GP, Ho MM, Puddefoot JR, Teja R, Barker S (1994b) Internalisation of the type I angiotensin II receptor (AT1) and angiotensin II function in the rat adrenal zona glomerulosa cell. J Endocrinol 141:R5–R9

Vinson GP, Ho MM, Puddefoot JR, Teja R, Barker S, Kapas S, Hinson JP (1995) Internalisation of the type I angiotensin II receptor (AT1) and angiotensin II function in the rat adrenal zona glomerulosa cell. Endocr Res 21:211–217

Vinson GP, Teja R, Ho MM, Hinson JP, Puddefoot JR (1996) Role of the tissue renin-angiotensin system in the response of the rat adrenal to exogenous angiotensin II. Endocr Res 22:589–593

Wada T, Inada Y, Sanada T, Ojima M, Shibouta Y, Noda M, Nishikawa K (1994) Effect of an angiotensin II receptor antagonist, CV-11974, and its prodrug, TCV-116, on production of aldosterone. Eur J Pharmacol 253:27–34

Wang N, Weng W, Breslow JL, Tall AR (1996) Scavenger receptor BI (SR-BI) is up-regulated in adrenal gland in apolipoprotein A-I and hepatic lipase knock-out mice as a response to depletion of cholesterol stores. In vivo evidence that SR-BI is a functional high density lipoprotein receptor under feedback control. J Biol Chem 271:21001–21004

Weir MR, Dzau VJ (1999) The renin-angiotensin-aldosterone system: a specific target for hypertension management. Am J Hypertens 12:205S–213S

Weiss MJ, Orkin SH (1995) GATA transcription factors: key regulators of hematopoiesis. Exp Hematol 23:99–107

West LA, Horvat RD, Roess DA, Barisas BG, Juengel JL, Niswender GD (2001) Steroidogenic acute regulatory protein and peripheral-type benzodiazepine receptor associate at the mitochondrial membrane. Endocrinology 142:502–505

Whitley GS, Hyatt PJ, Tait JF (1987) Angiotensin II-induced inositol phosphate production in isolated rat zona glomerulosa and fasciculata/reticularis cells. Steroids 49:271–286

Widmann C, Gibson S, Jarpe MB, Johnson GL (1999) Mitogen-activated protein kinase: conservation of a three-kinase module from yeast to human. Physiol Rev 79:143–180

Williams DL, Connelly MA, Temel RE, Swarnakar S, Phillips MC, de la Llera-Moya M, Rothblat GH (1999) Scavenger receptor BI and cholesterol trafficking. Curr Opin Lipidol 10:329–339

Wilson JX, Aguilera G, Catt KJ (1984) Inhibitory actions of calmodulin antagonists on steroidogenesis in zona glomerulosa cells. Endocrinology 115:1357–1363

Wolfe JT, Wang H, Perez-Reyes E, Barrett PQ (2002) Stimulation of recombinant Ca(v)3.2, T-type, Ca(2+) channel currents by CaMKIIgamma(C). J Physiol 538:343–355

Woodcock EA (1989) Adrenocorticotropic hormone inhibits angiotensin II-stimulated inositol phosphate accumulation in rat adrenal glomerulosa cells. Mol Cell Endocrinol 63:247–253

Woodcock EA, Johnston CI (1984) Inhibition of adenylate cyclase in rat adrenal glomerulosa cells by angiotensin II. Endocrinology 115:337–341

Woodcock EA, Smith AI, Schmauk-White LB (1988) Angiotensin II-stimulated phosphatidylinositol turnover in rat adrenal glomerulosa cells has a complex dependence on calcium. Endocrinology 122:1053–1059

Woodcock EA, Little PJ, Tanner JK (1990a) Inositol phosphate release and steroidogenesis in rat adrenal glomerulosa cells. Comparison of the effects of endothelin, angiotensin II and vasopressin. Biochem J 271:791–796

Woodcock EA, Tanner JK, Caroccia LM, Little PJ (1990b) Mechanisms involved in the stimulation of aldosterone production by angiotensin II, vasopressin and endothelin. Clin Exp Pharmacol Physiol 17:263–267

Wooton-Kee CR, Clark BJ (2000) Steroidogenic factor-1 influences protein-deoxyribonucleic acid interactions within the cyclic adenosine 3,5-monophosphate-responsive regions of the murine steroidogenic acute regulatory protein gene. Endocrinology 141:1345–1355

Yagci A, Muller J (1996) Induction of steroidogenic enzymes by potassium in cultured rat zona glomerulosa cells depends on calcium influx and intact protein synthesis. Endocrinology 137:4331–4338

Yamaguchi T, Naito Z, Stoner GD, Franco-Saenz R, Mulrow PJ (1990) Role of the adrenal renin-angiotensin system on adrenocorticotropic hormone- and potassium-stimulated aldosterone production by rat adrenal glomerulosa cells in monolayer culture. Hypertension 16:635–641

Yamamoto N, Yasue H, Mizuno Y, Yoshimura M, Fujii H, Nakayama M, Harada E, Nakamura S, Ito T, Ogawa H (2002) Aldosterone is produced from ventricles in patients with essential hypertension. Hypertension 39:958–962

Yanase T, Hara T, Sakai Y, Takayanagi R, Nawata H (1996) Expression of sterol carrier protein 2 (SCP2) in human adrenocortical tissue. Eur J Endocrinol 134:501–507

Yon L, Chartrel N, Feuilloley M, Demarchis S, Fournier A, Derijk E, Pelletier G, Roubos E, Vaudry H (1994) Pituitary adenylate cyclase-activating polypeptide stimulates both adrenocortical cells and chromaffin cells in the frog adrenal gland. Endocrinology 135:2749–2758

Yoshimura M, Nakamura S, Ito T, Nakayama M, Harada E, Mizuno Y, Sakamoto T, Yamamuro M, Saito Y, Nakao K, Yasue H, Ogawa H (2002) Expression of aldosterone synthase gene in failing human heart: quantitative analysis using modified real-time polymerase chain reaction. J Clin Endocrinol Metab 87:3936–3940

Young MJ, Funder JW (2000) Aldosterone and the heart. Trends Endocrinol Metab 11:224–226

Young M, Fullerton M, Dilley R, Funder JW (1994) Mineralocorticoids, hypertension and cardiac fibrosis. J Clin Invest 93:2578–2583

Young M, Head G, Funder J (1995) Determinants of cardiac fibrosis in experimental hypermineralocorticoid states. Am J Physiol 269:E657–E662

Young MJ, Clyne CD, Cole TJ, Funder JW (2001) Cardiac steroidogenesis in the normal and failing heart. J Clin Endocrinol Metab 86:5121–5126

Yu RN, Achermann JC, Ito M, Jameson JL (1998) The role of DAX-1 in reproduction. Trends Endocrinol Metab 5:169–175

Zazopoulos E, Lalli E, Stocco D, Sassone-Corsi P (1997) DNA binding and transcriptional repression by DAX-1 blocks steroidogenesis. Nature 390:311–315

Aldosterone: Clinical Aspects

T. L. Goodfriend

University of Wisconsin Medical School,
William S. Middleton Memorial Veterans Hospital, Room C-3127,
2500 Overlook Terrace, Madison, WI 53705, USA
e-mail: tgoodfri@facstaff.wisc.edu

Abstract Aldosterone is a potent mineralocorticoid synthesized by the outer zone of the adrenal cortex, the zona glomerulosa. Production of aldosterone is stimulated by angiotensin and potassium, but may also be affected by over 20 other endogenous and exogenous substances. Because of its sensitivity to angiotensin, secretion of aldosterone is altered by diseases that affect the kidneys and perturb their release of renin. Examples include renal artery stenosis, which increases renin secretion, and diabetes, which reduces it. Other diseases that reduce perfusion of the kidneys also alter aldosterone secretion. Examples include heart failure and cirrhosis of the liver, both of which stimulate renin release and increase aldosterone secretion. Disorders of the adrenal cortex can affect aldosterone secretion. Adenomas of the zona glomerulosa overproduce aldosterone independent of angiotensin. Hypertrophy of the zona glomerulosa causes overproduction of aldosterone because the cells are hypersensitive to angiotensin. The role of aldosterone in common cardiovascular diseases such as heart failure, low-renin essential hypertension, and the hypertension that accompanies obesity, is uncertain. The growing use of plasma aldosterone and renin measurements and the development of new aldosterone antagonists should clarify the importance of this steroid in a variety of disease states.

Keywords Aldosterone · Hypertension · Adrenal cortex · Potassium · Blood pressure · Angiotensin · Renin · Adrenal zona glomerulosa

Abbreviations

ACTH	Adrenocorticotropic hormone
APA	Aldosterone-producing adenoma
IHA	Idiopathic hyperaldosteronism
LREH	Low-renin essential hypertension
PAI-1	Plasminogen-activator inhibitor-1
RAAS	Renin-angiotensin-aldosterone system

1 Introduction

Aldosterone is one of the two most potent humoral effectors of the renin-angiotensin-aldosterone system (RAAS). Although angiotensin and aldosterone can both stimulate tubular reabsorption of sodium from the urine, aldosterone is largely responsible for the sodium retention and extracellular fluid volume expansion that accompanies activation of the RAAS. This action helps animals and humans survive threats to vascular homeostasis. On the other hand, aldosterone mediates some of the pathological manifestations of many human disease states in which sodium retention is dangerous. In addition to its well-known effects on sodium balance, aldosterone can influence synthesis of macromolecules in the cardiovascular system, suggesting a role for the steroid in hypertrophy and fibrosis. Lessening the influence of aldosterone by reducing its production or antagonizing its action has become an important component of therapeutics.

2 Biosynthesis and Regulation

Aldosterone is produced by the outermost cellular layer of the adrenal cortex, the zona glomerulosa. Steroidogenesis in these cells follows the same biosynthetic pathway found in other zones of the adrenal cortex and theca cells of the ovaries. The precursor is cholesterol, which is processed in mitochondria and cytosol. The rate-determining step in steroidogenesis is transport of cholesterol

Table 1 Potential stimuli and inhibitors of aldosterone secretion

Stimuli	Inhibitors
Known relevance to humans	
Angiotensin	
Potassium	
Probable relevance to humans	
ACTH (acute)	ACTH (chronic)
	Atrial natriuretic peptide
	Dopamine
Unknown relevance to humans	
Serotonin	Somatostatin
Catecholamines	Adrenomedullin
Bradykinin	Interleukin 1B
Vasoactive intestinal peptide	TNF-α
Substance P	Substance P
Neuropeptide Y	Neuropeptide Y
Fatty acid epoxides	Oleic and other nonesterified fatty acids
Prostaglandin E	Hyperosmolality
Enkephalins	Hypoxia
Interleukin 6	

from cytosol to the inner matrix of mitochondria where it undergoes side-chain cleavage, producing pregnenolone (Quinn and Williams 1988). Zona glomerulosa cells contain an enzyme complex, unique to that zone, that oxidizes the #18-carbon of corticosterone to form the aldehyde that gives aldosterone its name. The aldehyde at position 18 can bond with the hydroxyl group at position 11, protecting aldosterone from 11-hydroxysteroid dehydrogenases that inactivate other 11-hydroxy steroids such as hydrocortisone (Edwards and Hayman 1991). Zona glomerulosa cells lack the 17-hydroxylase that characterizes other steroidogenic cells such as those producing cortisol and sex steroids.

Aldosterone production is sensitive to a wide range of potential regulators, listed in Table 1 (Quinn and Williams 1988; Ehrhard-Bornstein et al. 1998). The length of this list implies that regulation of aldosterone secretion is complex, especially when compared to regulation of hydrocortisone or estrogen. The dominant stimuli of aldosterone production in intact animals and humans are angiotensin II and potassium (K^+). The best-known inhibitor is atrial natriuretic peptide. Adrenocorticotropic hormone (ACTH) stimulates aldosterone production transiently, but prolonged ACTH exposure inhibits zona glomerulosa cells. The effect of dopamine to inhibit aldosterone secretion is best demonstrated by the stimulatory effect of dopamine antagonists; direct administration of dopamine is largely ineffective (Carey 1982). Most of the compounds in Table 1 have been examined only in vitro, and their relevance to normal or abnormal regulation is unknown. As described below, situations arise in clinical medicine where aldosterone production is elevated or depressed out of proportion to plasma renin activity or potassium concentration, and it is in these situations that the less well-recognized regulators may play a role.

Aldosterone regulation is especially complex because the various stimuli and inhibitors interact with each other. For example, angiotensin's actions are powerfully affected by the level of potassium in extracellular fluid, and vice versa (Young et al. 1984). That synergy probably derives from the separate signal transduction mechanisms activated by potassium and angiotensin, which both stimulate the common rate-determining step, cholesterol side-chain cleavage. Another characteristic feature of aldosterone regulation is the ability of some dietary and environmental elements to alter the adrenal cell's signal transduction and steroidogenic machinery and its sensitivity to humoral regulators. For example, dietary sodium restriction or potassium supplementation increases zona glomerulosa cell content of angiotensin receptors and steroidogenic enzymes. Under these dietary influences, angiotensin's effects are amplified.

Feedback regulation of the renin-angiotensin-aldosterone system (RAAS) is mediated more by physiologic variables than by the effector hormones themselves. By contrast with the hypothalamic-pituitary-adrenal axis whose effector, cortisol, *directly* inhibits release of hypothalamic and pituitary tropins, aldosterone does not *directly* affect production of renin or angiotensin II. Instead, feedback is *indirect*, mediated by blood pressure, perfusion of the kidneys, and extracellular potassium concentration. For example, when aldosterone secretion is stimulated by a shrinking blood volume and consequent renin release, sodium retention induced by aldosterone restores blood volume and renal perfusion, turning off renin secretion. In a second example, when a dietary bolus of potassium elevates serum potassium and aldosterone secretion, the steroid favors potassium excretion by the kidney, extracellular potassium is lowered toward normal, and the rate of aldosterone secretion returns to baseline. Atrial natriuretic peptide buffers this system at the high end, reducing aldosterone secretion when intravascular volume is expanded excessively (Goodfriend et al. 1984). There is no known direct feedback control by aldosterone on the other regulators listed in Table 1. Thus, ACTH, dopamine, melanocyte-stimulating hormone, and serotonin may affect aldosterone secretion, but aldosterone has no known effects on release of those hormones.

Despite the many complexities in aldosterone regulation, the central role of angiotensin as a stimulus is unchallenged and crucial to understanding many clinical situations involving sodium, potassium, blood volume and blood pressure.

3 Clinical Situations Where Aldosterone Levels Are Increased by Angiotensin

3.1 Renal Diseases

Under ordinary circumstances, juxtaglomerular cells of the kidney release renin when sodium depletion, extracellular volume contraction, or circulatory impairment threatens perfusion of the kidneys and other vital organs. Aldosterone secretion is stimulated by the increment in angiotensin, and aldosterone and angiotensin effect sodium retention to restore renal perfusion. This is a situation where renin release seems appropriate to maintain total body circulation as well as the health of the kidneys. However, when juxtaglomerular cells are afflicted by ischemia or inflammation resulting from local diseases such as renal artery stenosis or glomerulonephritis, they release renin and stimulate aldosterone production in amounts greater than appropriate for total body fluid status (Bakris and Gavras 1993). In these situations, sodium retention and fluid volumes exceed amounts needed for perfusion of vital organs, and the result can be hypertension and edema. This is most dramatic in malignant hypertension, characterized in many patients by a vicious cycle of renal damage, renin release, aldosterone secretion, extremely high arterial pressure, and further renal damage from the pressure (McAllister el al. 1971).

In the conditions mentioned above, aldosterone is increased in parallel with renin, suggesting that its secretion is, it would seem, appropriate for homeostasis. Still, the increased circulating levels of aldosterone may be intrinsically harmful. This idea is supported by experiments in which aldosterone antagonists reduce histologic manifestations of renal damage in animal models of renovascular disease (Greene et al. 1996). The mechanism by which aldosterone damages parenchymal cells of the kidney is not known, but the toxicity is independent of tubular sodium reabsorption and systemic hypertension. Since it is prevented by agents such as spironolactone, aldosterone's cytotoxicity must be mediated by mineralocorticoid receptors, but the postreceptor events that damage tissue have not been identified. In the conditions where aldosterone levels are elevated in response to increased renin, both the physiologic and pathologic consequences are probably the result of combined actions of peptides of the angiotensin series and aldosterone. One example of this synergy is the production of plasminogen-activator inhibitor (PAI-1), stimulated maximally by the combined effects of the hexapeptide angiotensin IV and aldosterone (Brown et al. 2000).

In the nephrotic syndrome, the juxtaglomerular cells are normal. They release large amounts of renin because the loss of albumin in the urine shrinks extracellular volume and decreases perfusion of the kidney. The adrenals respond by secreting large amounts of aldosterone. It is not known whether those large amounts of angiotensin and aldosterone damage the kidney further.

Deviations from proportionality between renin and aldosterone can be seen when the renal disease affects serum potassium. In Bartter's and Gitelman's syndromes, for example, renal tubular defects cause elevated renin release and excessive urinary excretion of potassium (Gordon et al. 1993). High renin and angiotensin levels stimulate aldosterone secretion, which further accentuates potassium loss. However, when serum potassium reaches very low levels, aldosterone secretion is impaired and is less than one might predict from the very high plasma renin activity.

Renal disease may directly influence the extent of aldosterone's influence on potassium. If the disease that triggers renin release impairs renal excretion of potassium, high levels of aldosterone may not be able to deplete body stores or lower potassium concentration in serum. Similarly, if the renal disease reduces glomerular filtration of sodium, reducing the amounts that reach the distal tubule, high levels of aldosterone may not cause hypokalemia.

3.2 Extrarenal Diseases and Dietary Ingredients That Affect Aldosterone

Increased plasma levels of aldosterone accompany elevated levels of renin when renal perfusion is compromised by clinical conditions originating outside of the kidneys. Principal among these is circulatory failure, frequently caused by heart disease. Activation of the RAAS in circulatory failure is caused by decreased renal perfusion amplified by heightened sympathetic nervous system activity (Cody et al. 1986).

Although the increment in aldosterone is proportional to the increment in renin and angiotensin, and although these increases appropriately reflect circulatory compromise, aldosterone and angiotensin may aggravate the underlying heart disease in several ways. By retaining sodium and increasing intravascular volume, aldosterone may increase cardiac output and ventricular work. By constricting peripheral arterioles, angiotensin may increase cardiac afterload and oxygen demand while impeding coronary blood flow. By stimulating fibrosis in vessel walls and hypertrophy in the myocardium, aldosterone and angiotensin may contribute to deleterious remodeling of cardiovascular structures (Weber 2001). By stimulating production of an inhibitor of plasminogen activation, aldosterone and angiotensin may impair clot lysis and accelerate atherosclerosis (Brown et al. 2000).

For the reasons listed above, activation of the RAAS and production of increased amounts of aldosterone are now viewed as counterproductive responses to heart failure, even though they may seem to be appropriate responses to decreased renal perfusion. This is another example of damage caused by the combined effects of angiotensins and aldosterone. Blockade of the RAAS, including direct antagonism of aldosterone, is a cornerstone of current therapy for heart failure (Pitt et al. 1999).

Liver disease frequently causes perturbations of the RAAS. The best example is cirrhosis with ascites. The fibrotic liver and fluid-filled peritoneal cavity impede venous return to the heart, so perfusion of the kidneys is reduced, stimulating renin release. If albumin production is also impaired, intravascular volume is compromised, and renin release stimulated even more. Increased aldosterone production ensues, causing sodium retention and potassium excretion. Aldosterone levels are further increased if hepatic metabolism of steroids is impaired (Burmeister et al. 1986).

Cirrhosis can cause another vicious cycle in which excessive secretion of aldosterone provides the extra sodium that supports intraperitoneal fluid accumulation, which, in turn, further impedes venous return and stimulates still more renin release. Advanced liver disease may provide one brake to this vicious cycle by hampering synthesis of renin substrate. Cirrhosis with ascites is commonly treated with cautious administration of aldosterone antagonists. Treatment with antagonists of angiotensin production or action are usually not used, because angiotensin's pressor properties help maintain renal function and systemic blood pressure, and loss of these properties would be life-threatening.

In animals and humans living under primitive conditions, the most common stimulation of the RAAS occurs following hemorrhage, diarrhea, or salt and water deprivation. Here, elevated production of aldosterone fits into the concerted response that protects intravascular volume. If the predominant threat is dehydration, so that extracellular fluid becomes hypertonic, aldosterone secretion may be lower than plasma renin levels would predict. Hypertonicity inhibits aldosterone secretion by a direct effect on signal transduction in zona glomerulosa cells (Schneider et al. 1985). This may prevent excessive retention of sodium when serum sodium levels are abnormally high.

4 Clinical Conditions in Which Aldosterone Is Increased Independent of Angiotensin

4.1 Adrenal Tumors

Adenomas comprised of neoplastic zona glomerulosa cells produce aldosterone independent of angiotensin II. They are called aldosteronomas or aldosterone-producing adenomas (APA) (Conn et al. 1964). Because they are neoplastic and independent of the renin/angiotensin system, their production of aldosterone is far in excess of the amounts needed to maintain adequate intravascular fluid volume and normal extracellular potassium concentration. The clinical manifestations of aldosteronomas include hypertension and usually (but not always) hypokalemia. The clinical picture is called Conn's syndrome. Although the large amounts of aldosterone produced by these tumors result in expanded extracellular fluid volume and abnormally large accumulations of total body sodium, they do not cause peripheral edema. That this situation is confusing reflects our ignorance of the mechanism of edema rather than our ignorance of the effects of aldosterone. Apparently, aldosterone-induced sodium retention alone is not sufficient to cause edema, and edema can result from microvascular changes in the absence of sodium retention, as seen in patients receiving dihydropyridine calcium channel blocking drugs, for example. There is more to edema than salt and water.

Aldosteronomas produce inappropriately large amounts of aldosterone, and the resulting increase in arterial pressure and blood volume provides ample perfusion of the kidneys. As a result, renin release is suppressed to very low levels. The diagnostic hallmark of APAs is the combination of elevated plasma (or urine) aldosterone and depressed plasma renin activity.

Although most of the major clinical manifestations of APAs can be explained by the renal tubular actions of excessive aldosterone, there is some evidence that the degree of cardiac hypertrophy in these patients exceeds that predicted by the high arterial pressure (Schlaich et al. 2000). Since renin and angiotensin levels are suppressed in APA, the most likely mediator of that ventricular hypertrophy is aldosterone.

Aldosteronomas are not completely autonomous. The amount of aldosterone produced by these tumors depends to some extent on potassium. This is probably a reflection of the need for intracellular potassium to support cell functions such as synthesis of critical proteins involved in steroidogenesis. Because aldosteronomas cause potassium depletion, their output of aldosterone may be unimpressive until total body potassium is replete (Cain et al. 1972). In addition to potassium, ACTH exerts a stimulatory effect on steroid production by APAs. ACTH levels in humans usually fall during the morning hours, while renin activity increases with ambulation. Reflecting the primacy of ACTH over angiotensin in stimulating aldosteronomas, patients with Conn's syndrome produce less and less aldosterone as they stand and move around in the morning, while normal subjects produce more and more (Schambelan et al. 1976). The best treatment of APAs is surgical removal.

Carcinomas of the adrenal that produce aldosterone have been described. These are virtually unresponsive to classic regulators and are dangerously malignant. Fortunately, they are rare (Neville and Symington 1966).

4.2 Adrenal Hyperplasia

The adrenal zona glomerulosa can produce abnormally large amounts of aldosterone, out of proportion to the amount of circulating renin and angiotensin, by becoming hypertrophic, but not adenomatous. This clinical situation is termed idiopathic hyperaldosteronism (IHA) (Brown JJ et al. 1979). Patients with IHA have increased total body sodium, increased extracellular fluid, decreased total body potassium and hypertension. They frequently are hypokalemic, but not always. The adrenal cortex in IHA does not contain a single distinct neoplastic adenoma, but it usually contains one or more nodules. Although the nodules appear to be formed as a result of the same stimuli that produce hyperplasia and increased aldosterone production in these glands, it is not clear that the nodules themselves produce aldosterone (Nagae et al. 1991). The clinical manifestations of IHA can all be explained by the renal actions of excessive aldosterone.

By contrast with aldosterone-producing adenomas, adrenals in cases of idiopathic hyperaldosteronism respond to angiotensin. In fact, they overrespond. This has been graphically demonstrated by investigators who derived curves depicting the levels of aldosterone in response to infused angiotensin II (Brown RD et al. 1979). The slope of this curve was steeper in patients with IHA than in normal subjects or patients with adenomas. The basis of hyperresponsiveness to angiotensin II in IHA is unknown, but it may involve unusual amounts of one or more of the nonclassic secretagogues and inhibitors listed in Table 1. This mysterious humoral environment apparently can induce both growth and hyperresponsiveness in the adrenal cortex, but not necessarily in the same cells; the nodules may show the effects of growth promotion, and other cells produce excess aldosterone.

4.3 Low-Renin Essential Hypertension

Approximately 25% of patients with essential hypertension have a combination of low circulating renin activity and normal levels of plasma aldosterone (Schalekamp et al. 1974). These people are hyperresponsive to angiotensin compared to normal subjects, and they form a continuum with patients who have hyperplastic adrenals (IHA) (Brown et al. 1979b). It is not certain that aldosterone is responsible for the elevated blood pressure in low-renin essential hypertension (LREH). Plasma levels of the steroid are within the normal range, not as high as in AHA or IHA. One explanation for this situation is hyperresponsiveness of both the adrenal and blood vessels to angiotensin. It is most likely that LREH is not a single entity. Some patients may have inapparent adrenal hypertrophy or adenomas, some may simply have the depressed renin production that accompanies aging, and some may be producing an as-yet-unidentified pressor that weakly stimulates aldosterone production.

5 Miscellaneous Clinical Conditions with High Aldosterone

5.1 Obesity

A correlation has been demonstrated between the amount of visceral or upper-body fat and plasma levels of aldosterone (Goodfriend et al. 1998, 1999). Aldosterone levels in obese subjects decline when they lose weight (Tuck et al. 1981). These observations imply that visceral fat stimulates secretion of aldosterone. The special role of visceral fat, as opposed to other fat deposits, may derive from its location within the portal venous circulation. Fatty acids released from visceral adipocytes would flow directly to the liver. The liver can oxidize fatty acids, and it may be that some of those oxidized derivatives stimulate adrenal steroidogenesis. Stimulation of aldosterone production by linoleic acid epoxides has been demonstrated in vitro, but the relevance of these oxidized fatty acids to clinical conditions remains uncertain (Goodfriend et al. 2002).

5.2 Essential Hypertension

The role of aldosterone in essential hypertension has been studied for decades but is still uncertain. It is conventional wisdom among clinicians that some hypertensive patients who do not have evidence of adrenal pathology respond dramatically to treatment with aldosterone antagonists (Jeunemaitre et al. 1988). This observation alone provides strong evidence for the role of aldosterone in sustaining elevated blood pressure in these patients. The growing use of the plasma aldosterone/renin ratio to search for aldosteronomas and idiopathic hyperaldosteronism and the availability of new aldosterone antagonists should help define patients in whom aldosterone is an important pressor despite the absence of classic adrenal syndromes (McKenna et al. 1991; Weinberger et al. 2002). It will then be important to clarify the mechanism by which aldosterone became important to those patients. The possibilities include nonclassic aldosterone regulators (Table 1), mediators that sensitize the adrenal to secretagogues, desensitize it to inhibitors, or sensitize the kidney to aldosterone itself.

Some patients with essential hypertension have plasma aldosterone levels that fail to suppress with infusion of saline solutions (Conlin et al. 1993). This would suggest that these nonmodulators have adrenal glands that are stimulated by nonclassic secretagogues, not the renin-angiotensin system. However, nonmodulators can be converted to modulators by treatment with angiotensin-converting enzyme inhibitors, suggesting that their pathology resides in unusual angiotensin formation, not unusual aldosterone secretion.

6 Miscellaneous Clinical Conditions with Low Aldosterone

6.1 Hyporeninemic Hypoaldosteronism

Some renal diseases interfere with release of active renin. In some cases, pro-renin is released, but not activated (Luetscher et al. 1985). In these conditions, aldosterone levels are reduced in proportion to the low renin activity. The most common disease associated with these changes is diabetes mellitus (de Châtel et al. 1977). Impaired release of active renin may precede other renal manifestations of diabetes. Other renal diseases that may cause impaired release of active renin include various inflammatory conditions and lead poisoning. The clinical picture reflects the reduced electrolyte influence of aldosterone, causing hyperkalemia and systemic acidosis—the opposite of the hypokalemic alkalosis seen in patients with aldosteronomas. Another name for hyporeninemic hypoaldosteronism is renal tubular acidosis type IV or RTA-Type IV. It is one of the most common causes of hyperkalemia in hospitalized patients with otherwise normal renal function. The persistence of low aldosterone secretion in the face of increased potassium emphasizes the synergy between angiotensin and potassium in regulating aldosterone production.

One congenital defect in steroidogenesis can cause parallel suppression of renin release and reduced aldosterone secretion: 17-hydroxylase deficiency (Biglieri et al. 1966). This enzyme is not normally present in the zona glomerulosa, but its absence from the zona fasciculata impacts indirectly on aldosterone production. In patients with 17-hydroxylase deficiency, feedback inhibition of ACTH by glucocorticoids is relaxed. The resultant excess ACTH stimulates synthesis of steroids, some of which have mineralocorticoid activity such as deoxycorticosterone and corticosterone. Sodium retention caused by these steroids suppresses renin release and reduces secretion of aldosterone. Although mineralocorticoid activity is not deficient in these patients, glucocorticoid and sex steroid functions need to be replaced by exogenous hormone therapy. When glucocorticoids are administered, ACTH release is suppressed, steroid secretory patterns return to low levels, and mineralocorticoid regulation reverts to the RAAS again.

6.2 Hyper-reninemic Hypoaldosteronism

Impaired secretion of aldosterone can lower circulating levels of the hormone despite normal or high plasma renin activity. Excluding inhibition of angiotensin-converting enzyme activity, low aldosterone in the face of normal or high angiotensin II must be caused by a problem in the adrenal zona glomerulosa. Although the anatomic site of the impairment can be described, the biochemical lesion may be difficult to identify. For example, seriously ill patients can display low plasma aldosterone and high renin activity levels (Findling et al. 1987). Some of these patients have damaged adrenals, but most of them have normal glucocorticoid levels and normal cortisol responses to ACTH. Possible explanations of selective failure of the zona glomerulosa range from the inhibitory effects of prolonged ACTH to inhibition by high levels of unesterified fatty acids and low levels of oxygen (Goodfriend et al. 1993). Some of the drugs commonly used in seriously ill patients can inhibit aldosteronogenesis. These include heparin, cyclosporine, dopamine, and the preservative chlorbutanol found in some parenteral preparations (Sequeira and McKenna 1986).

Congenital defects in steroidogenesis result in low aldosterone levels despite high plasma renin activity (Melby and Azar 2001). The enzymes whose deficiency would directly impair aldosterone biosynthesis are 11-hydroxylase, 21-hydroxylase and corticosterone methyl oxidase (18-hydroxylase and oxidase.) However, 11-hydroxylase deficiency induces excess ACTH release and production of other mineralocorticoids such as deoxycorticosterone, so the lack of aldosterone is not clinically apparent.

7 Conclusion

Aldosterone is the effector of the RAAS that bears greatest responsibility for maintaining intravascular fluid volume and normal potassium levels. It also influences synthesis of macromolecules in cardiovascular structures. Except for some neoplastic or destructive diseases of the adrenal, plasma levels of aldosterone reflect the production of angiotensin by the RAAS. Use of aldosterone receptor antagonists will reveal the precise role of aldosterone, as distinct from angiotensin, in human physiology and pathology.

References

Bakris GL, Gavras H (1993) Renin in acute and chronic renal failure: implications for treatment. In: Robertson JIS, Nicholls MG (eds) The renin-angiotensin system, vol 2. Gower, London, pp 56.1–56.14

Biglieri EG, Herron MA, Brust N (1966) 17-Hydroxylation deficiency in man. J Clin Invest 45:1946–1954

Brown JJ, Lever AF, Robertson JI, et al (1979) Are idiopathic hyperaldosteronism and low-renin hypertension variants of essential hypertension? Ann Clin Biochem 16:380–388

Brown NJ, Kim K-S, Chen Y-Q, Blevins LS, Nadeau JH, Meranze SG, Vaughan DE (2000) Synergistic effect of adrenal steroids and angiotensin II on plasminogen activator inhibitor-1 production. J Clin Endocrinol Metab 85:336–344

Brown RD, Wisgerhof M, Carpenter PC, Brown G, Jiang N-S, Kao P, Hegstad R (1979) Adrenal sensitivity to angiotensin II and undiscovered aldosterone stimulating factors in hypertension. J Steroid Biochem 11:1043–1050

Burmeister P, Scholmerich J, Diener W, Gerok W (1986) Renin, aldosterone and arginine vasopressin in patients with liver cirrhosis: The influence of ascites retransfusion. Eur J Clin Invest 16:117–123

Cain JP, Tuck ML, Williams GH et al (1972) The regulation of aldosterone secretion in primary aldosteronism. Am J Med 53:627–637

Carey RM (1982) Acute dopaminergic inhibition of aldosterone secretion is independent of angiotensin II and adrenocorticotropin. J Clin Endocrinol Metab 54:463–469

Cody RJ, Covit AB, Schaer GL, Laragh JH, Sealey JE, Feldschuh J (1986) Sodium and water balance in chronic congestive heart failure. J Clin Invest 77:1441–1452

Conlin PR, Moore TJ, Williams GH, Hollenberg NK (1993) Rapid modulation of renal and adrenal responsiveness to angiotensin II. Hypertension 22:832–838

Conn JW, Knopf RF, Nesbit RM (1964) Clinical characteristics of primary aldosteronism from an analysis of 145 cases. Am J Surg 107:159–172

De Châtel R, Weidmann P, Flammer J, Ziegler WH, Beretta-Piccoli C, Vetter W, Reubi FC (1977) Sodium, renin, aldosterone, catecholamines and blood pressure in diabetes mellitus. Kidney Int 12:412–421

Edwards CRW, Hayman A (1991) Enzyme protection of the mineralocorticoid receptor: evidence in favour of the hemi-acetal structure of aldosterone. In: Bonvalet JP et al. (eds) Aldosterone: fundamental aspects. John Libbey Eurotext, London, pp 67–76

Ehrhart-Bornstein M, Hinson J, Bornstein S, Scherbaum W, Vinson G (1998) Intraadrenal interactions in the regulation of adrenocortical steroidogenesis. Endocr Rev 19:101–143

Findling JW, Waters VO, Raff H (1987) The dissociation of renin and aldosterone during critical illness. J Clin Endocrinol Metab 64:592–595

Goodfriend TL, Ball DL, Elliott ME, Chabhi A, Duong T, Raff H, Schneider EG, Brown RD, Weinberger MH (1993) Fatty acids may regulate aldosterone secretion and mediate some of insulin's effects on blood pressure. Prostaglandins Leukot Essent Fatty Acids 48:43–50

Goodfriend TL, Elliott ME, Atlas SA (1984) Actions of synthetic atrial natriuretic factor on bovine adrenal glomerulosa. Life Sci 35:1675–1682

Goodfriend TL, Egan BM, Kelley DE (1998) Aldosterone in obesity. Endocr Res 24:789–796

Goodfriend TL, Kelley DE, Goodpaster BH, Winters SJ (1999) Visceral obesity and insulin resistance are associated with plasma aldosterone levels in women. Obes Res 7:355–362

Goodfriend TL, Ball DL, Gardner HW (2002) An oxidized derivative of linoleic acid affects aldosterone secretion by adrenal cells in vitro. Prostaglandins Leukot Essent Fatty Acids 67:163–167

Gordon RD, Klemm SA, Tunny TJ (1993) Renin in Bartter's syndrome. In: Robertson JIS, Nicholls MG (eds) The renin-angiotensin system, vol 2. Gower, London, pp 64.1–64.15

Greene EL, Kren S. Hostetter TH (1996) Role of aldosterone in the remnant kidney model in the rat. J Clin Invest 98:1063–1068

Jeunemaitre X, Kreft-Jais C, Chatellier G, Julien J, Degoulet P, Plouin P-F, Menard J, Corvol P (1988) Long-term experience of spironolactone in essential hypertension. Kidney Int 34:S14–S17

Luetscher JA, Kraemer FB, Wilson DM, Schwartz HC, Bryer-Ash M (1975) Increased plasma inactive renin in diabetes mellitus. N Engl J Med 312:1412–1417

McAllister RG, Van Way CW, Dayani K, Anderson WJ, Temple E, Michelakis AM, Coppage WS, Oates JA (1971) Malignant hypertension: effect of therapy on renin and aldosterone. Circ Res 28, 29 [Suppl 2]:160–173

McKenna TJ, Sequeira SJ, Heffernan A et al (1991) Diagnosis under random conditions of all disorders of the renin- angiotensin-aldosterone axis, including primary hyperaldosteronism. J Clin Endocrinol Metab 73:952–957

Melby JC, Azar ST (2001) Hypoaldosteronism and mineralocorticoid resistance. In: DeGroot LJ, Jameson JL (eds) Endocrinology, 4th edn, vol 3. WB Saunders, Philadelphia, pp 1845–1850

Nagae A, Murakami E, Hiwada K et al (1991) Primary aldosteronism with cortisol overproduction from bilateral multiple adrenal adenomas. Jpn J Med Sci Biol 30:26–31

Neville AM, Symington T (1966) Pathology of primary aldosteronism. Cancer 19:1854–1868

Pitt B, Zannad F, Remme WJ et al. (1999) The effect of spironolactone on morbidity and mortality in patients with severe heart failure. N Engl J Med 341:709–717

Quinn SJ, Williams GH (1988) Regulation of aldosterone secretion. Annu Rev Physiol 50:409–426

Schalekamp MA, Lebel M, Beevers DG et al (1974) Body-fluid volume in low-renin hypertension. Lancet 2:310–311

Schambelan M, Brust NL, Chang BC et al (1976) Circadian rhythm and effect of posture on plasma aldosterone concentration in primary aldosteronism. J Clin Endocrinol 43:115–131

Schlaich MP, Schobel HP, Hilgers K, Schmieder RE (2000) Impact of aldosterone on left ventricular structure and function in young normotensive and mildly hypertensive subjects. Am J Cardiol 85:1199–1206

Schneider EG, Radke KJ, Ulderich D, Taylor R Jr (1985) Effect of osmolality on aldosterone secretion. Endocrinology 116:1621–1626

Sequeira SJ, McKenna TJ (1986) Chlorbutanol, a new inhibitor of aldosterone biosynthesis identified during examination of heparin effect on aldosterone production. J Clin Endocrinol Metab 63:780–784

Tuck ML, Sowers J, Dornfeld L, Kledzik G, Maxwell M (1981) The effect of weight reduction on blood pressure, plasma renin activity, and plasma aldosterone levels in obese patients. N Engl J Med 304:930–933

Weber KT (2001) Aldosterone in congestive heart failure. N Engl J Med 345:1689–1697

Weinberger MH, Roniker B, Krause SL et al (2002) Eplerenone, a selective aldosterone blocker, in mild-to-moderate hypertension. Am J Hypertens 15:709–716

Young DB, Smith MJ Jr, Jackson TE, Scott RE (1984) Multiplicative interaction between angiotensin II and K concentrations in stimulation of aldosterone. Am J Physiol 247:E328–E335

Part 5
Inhibition of the Renin-Angiotensin System
ACE Inhibitors

Interactions Between the Renin-Angiotensin and the Kallikrein-Kinin System

P. Wohlfart · G. Wiemer

Disease Group Cardiovascular, Aventis Pharma Industriepark Höchst, Building H 825, 65926 Frankfurt, Germany
e-mail: paulus.wohlfart@aventis.com

Abstract Angiotensin-I-converting enzyme (ACE) has turned out to be a central target molecule in the successful treatment of many important cardiovascular diseases. The value of ACE inhibitors as therapeutic options has grown with every year over the past 2 decades, but our knowledge regarding the underlying mechanisms also increased. Today it seems that multiple interactions into the renin-angiotensin (RAS) and the kallikrein-kinin (KKS) system exist. This chapter summarizes what is currently known about these interactions between RAS and KKS, with a specific focus on two target organs: the endothelium and the heart.

Keywords Angiotensin-converting-enzyme · Angiotensin-converting enzyme inhibitor · ACE2 · Ramipril · Renin · Kininase · Kallikrein · Kininogen · Kinin · Bradykinin · Kallidin · Angiotensins · Angiotensin I · Angiotensin II · B_2 kinin receptor · B_1 kinin receptor · Nitric oxide

1
Introduction

Small molecular weight inhibitors of the angiotensin-I-converting enzyme (ACE) bewitched the scientific community, because of their clear efficacy in important cardiovascular indications such as hypertension, congestive heart failure and prevention of atherosclerotic complications. The more scientists worked with these compounds in preclinical and clinical studies, the more it became clear that the mechanism of action needed to be subdivided into several signal transduction pathways, some of them only vaguely defined. The initial view was simplistic in that these inhibitors mainly interfere with the renin-angiotensin system (RAS) by blocking angiotensin-II (ANG-II) formation. Several studies demonstrated that interactions with the kallikrein-kinin system (KKS) contribute, via accumulation of bradykinin, to protective effects. Today, even this two-pronged mode of action does not sufficiently explain all facets of ACE inhibitors. Additional components seem to exist such as resensitization of bradykinin-subtype B_2-receptors (B_2 kinin receptors). It has become more and more apparent that blocking ACE modulates both RAS and KKS at numerous interaction points. The aim of this chapter is to review our current knowledge on these interactions, especially with a focus on what we have learned from ACE inhibition.

2
Early Clues for Interactions—Late Confirmation in Humans

As is known today, the two different proteolytic reactions catalyzed by ACE, ANG-II formation and bradykinin degradation (see Fig. 1), have synergistic effects in experimental and clinical models. In the first days of ACE inhibitors however, kinin-related effects were often not clearly delineated despite early evidence delivered by several groups (for a detailed review see Linz et al. 1999). In preclinical models, the contribution of KKS was supported very early, when Clappison and co-workers studied the effects of captopril on renal hemodynamics in anesthetized dogs (Clappison et al. 1981). Although ACE inhibition led to increases in circulating angiotensin-I (ANG-I), plasma renin activity, and urinary bradykinin, no increase in circulating bradykinin was observed. Furthermore, despite infusion of a competitive ANG-II receptor antagonist, ACE inhibition was still associated with an increase in renal blood flow and mean arterial pressure. The inability to detect plasma levels of bradykinin in many studies misled researchers to exclude the kallikrein-kinin system as one possible route of action. As was later known, difficulties in the measurement of plasma kinins accounted for false-negative values of circulating bradykinin while under ACE inhibition. In 1994, Nussberger and co-workers first demonstrated an increase in plasma bradykinin under ACE inhibition in normal human subjects, based on improved extraction protocols and new high-affinity antisera (Pellacani et al.

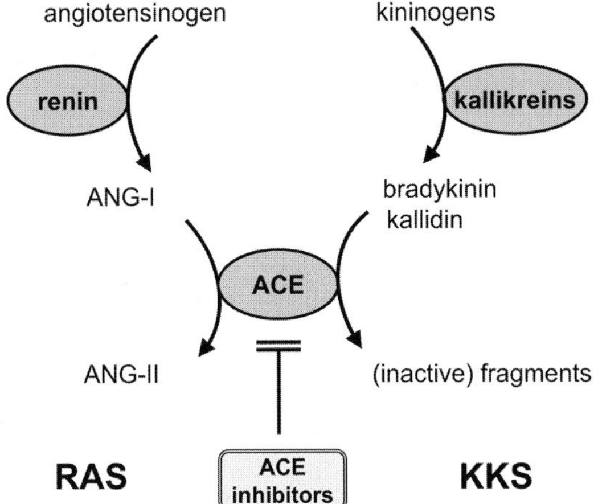

Fig. 1 Interaction between the renin angiotensin system (*RAS*) and the kallikrein-kinin system (*KKS*). When angiotensin I-converting enzyme (*ACE*) is inhibited angiotensin-II formation (*ANG-II*) is reduced, whereas bradykinin and its N-terminally elongated analog kallidin accumulate. Both kinins activate the B_2 kinin receptor

1994). Since then, the two-pronged mode of action of RAS inhibition and KKS activation has become well established.

3
Systemic Versus Local Renin-Angiotensin and Kallikrein-Kinin Systems

Although acute effects of ACE inhibition can be linked to transiently decreased levels of circulating ANG-II, subchronic and chronic ACE inhibition lead to a dose-dependent rise in levels of circulating active renin and blood ANG-I, and no further reductions in plasma ANG-II levels (Moser et al. 1990). This compensatory rise in plasma renin led to scrutinization of local RAS and KKS and their respective contributions to the long-term effects of ACE inhibition. The arrival of modern molecular biology and functional genomics provided answers to this question. To date, our accrued molecular knowledge on the expression of precursors, enzymes, receptors and effector molecules within the RAS and KKS have risen dramatically (see Table 1). Many of these components are specifically if not selectively expressed in cardiovascular tissues. Today the question is not whether local systems exist, but in which orchestrations these components of local systems contribute. To summarize all knowledge of locally expressed RAS and KKS in different organs and diseases and the way they interact would be far beyond the scope of this chapter. We would like to focus on the endothelium and the heart, not arbitrarily, but rather because of the important roles ACE inhibitors play in both organs.

Table 1 Components of the renin angiotensin and the kallikrein kinin system

Precursor	Protease	Peptides[a]	Receptors		Downstream signal transduction
Angiotensinogen	Renin	Specific conversion of angiotensinogen to ANG-I	Renin	Renin receptor	Binding of renin to its receptor enhances enzymatic activity of angiotensin-independent signal transduction by intracellular kinases
	ACE	Converts ANG-I to ANG-II degrades bradykinin	ANG-II	AT_1 receptor subtype	Intracellular signaling via calcium and kinases mediates nongenomic and genomic effects
	Chymase	Converts ANG-I to ANG-II		AT_2 receptor subtype	Activation of phosphatases Activation of tissue kallikrein and B_2 kinin receptors Nitric oxide and prostacyclin
	ACE-2	Converts ANG-I to Ang-(1–9) Converts ANG-II to Ang-(1–7) Nonangiotensin peptides[b]	?		
	NEP 24.11[c] and peptidase 24.15	ANG-I to Ang-(1–7)	Ang-(1–7)	Ang-(1–7) receptor[d]	Activation of kallikreins? B_2 kinin receptor-mediated release of nitric oxide
Kininogen	Tissue kallikrein	Forms bradykinin (BK)	BK	B_2 kinin receptor	Nitric oxide and prostacyclin
	Kininase-I	Forms des-Arg^9-BK	des-Arg^9-BK	B_1 kinin receptor	Nitric oxide among other messengers?

[a] Not all angiotensin and bradykinin peptides are listed.

[b] For example, apelin-13 and dynorphin A 1–13.

[c] Neutral endopeptidase 24.11.

[d] Receptor not yet cloned.

4 Endothelial Renin-Angiotensin and Kallikrein-Kinin Systems

Bakhle et al. (1965) discovered that ACE largely resided in pulmonary tissues, without specifying the cell type that contained ACE. Approximately half of all endothelial cells within the human body are located within the pulmonary vasculature. Ryan and co-workers were among the first to demonstrate the major contribution of the endothelium to lung ACE activity (Ryan et al. 1975; Ryan 1986). Four years before, Yang et al. (1971) showed that ACE is identical to kininase II. Thus, basic information had been collected very early to argue for the endothelium as an important interface between RAS and KKS. Many researchers may have underestimated the rich expression of ACE in endothelial cells, and assumed that the endothelium covers only passively blood vessels, with ACE localized on the luminal side and serving only into the blood compartment. This paradigm shifted when it was recognized that the endothelium plays a dominant role in cardiovascular homeostasis. The trigger for this shift may have been the identification of nitric oxide as an endothelial-derived relaxation factor. In the period following, several second messengers released and converted by endothelial cells were characterized, e.g., prostacyclin or endothelin.

Since 1990 the relative contribution of the KKS in response to ACE inhibitors was further investigated in cultured endothelial cells (Schini et al.1990; Wiemer et al. 1991; Brotherton 1996; Wiemer et al. 1996; Wohlfart et al. 1997). Activation of endothelial B_2 kinin receptors, under ACE inhibition, leads via different signaling pathways to the release of prostacyclin, nitric oxide and endothelial-derived hyperpolarizing factor. Because ACE-inhibitors act in an autocrine manner, accumulation of locally formed kinins was postulated as the activating mechanism upstream of the B_2 kinin receptor. Indeed, in venous effluents from isolated rat hearts perfused with the ACE inhibitor ramipril, increased bradykinin concentrations were detected (Baumgarten et al. 1993). How endothelial cells synthesize kinins is less well characterized. We demonstrated that cultured human endothelial cells express significant amounts of tissue kallikrein, which likely serves in the local generation of kinins (Dedio et al. 2001). Menetton and co-workers provided elegant evidence for the importance of tissue kallikrein in cardiac and vascular physiology by generating mice lacking tissue kallikrein (Menetton et al. 2001). Among several observed cardiovascular abnormalities, these mice displayed an impaired flow-induced vasodilatation. Thus, tissue kallikrein seems to be the main kinin-generating enzyme *in vivo.*

Within the endothelium, many studies have revealed local expression of several RAS components. Early enzymatic and immunological measurements indicated that cultured endothelial cells express renin (Lilly et al. 1985). Due to controversies concerning the source of vascular renin, Xiao et al. (2000) studied renin mRNA and protein expression in cultured bovine aortic endothelial cells. Using monoclonal antibodies against human renin, immunocytochemical analyses revealed positive immune reactivity in the cytoplasm of the endothelium. In-situ hybridization showed renin mRNA in cytoplasm. In addition, ANG-II was quantified in conditioned medium. These findings demonstrate that endothelial cells contain all the essential components of the RAS, namely renin, angiotensinogen and ACE. It is worth mentioning that the vasculature consists of more than the endothelium. However, although RAS components have been described in nonendothelial vascular cell types such as smooth muscle cells, the interaction between RAS and KKS in these vascular cells remains poorly defined.

Sraer and co-workers further expanded the frontiers of vascular biology when they cloned a human gene, which when transfected into cells expressed not only renin- and pro-renin-specific binding sites, but also renin-dependent intracellular signal transduction (Nguyen et al. 2002). Binding of renin to this receptor seems to accelerate the conversion of angiotensinogen to ANG-I. By confocal microscopy, the receptor was localized to the mesangium of glomeruli, and to the subendothelium of coronary and kidney arteries associated with smooth muscle cells and colocalized with renin. Because this receptor may also bind renin derived from the circulation, it may form an important balancing point between local and systemic RAS. Due to the intrinsic activity of this receptor, it seems possible that renin may exert actions independent of angiotensins.

5 Resensitization of B_2 Kinin Receptors by ACE Inhibitors

Endothelial bradykinin accumulation served to explain the mode of action of ACE inhibitors on endothelial cells. Surprisingly, an additional mechanism was identified leading to a similar downstream activation of kinin receptor cascades. Erdös and co-workers studied the effect of enalaprilat on the binding of radioactive labeled bradykinin to vehicle cells stably transfected with human B_2 kinin receptor, alone or in combination with ACE (Minshall et al. 1997). The ACE inhibitor enhanced the binding of bradykinin, but only in the presence of both B_2 kinin receptor and ACE. High-affinity binding sites were protected against receptor internalization. All these effects were not found in cell lines devoid of ACE expression, and therefore can be explained by a direct effect of ACE inhibition on B_2 kinin receptor desensitization. Importantly, some of these mechanisms were confirmed in native human endothelial cells with a different ACE inhibitor (Benzing et al. 1999). Pre-treatment with ramiprilat significantly attenuated the recovery of B_2 kinin receptors in caveolin-rich membranes, which form within endothelial cells an important sub-cellular compartment enriched in signaling molecules. Moreover, ramiprilat resulted in enhanced recovery of B2 receptor signaling in endothelial cells prestimulated with bradykinin. Further details of bradykinin receptor desensitization were investigated in vehicle cell lines (Marcic and Erdös 2000; Marcic et al. 2000). However, several important questions remain unanswered: for example, how does this interaction between the two membrane proteins occur, specifically under ACE inhibition.

Whereas the results from these cellular experiments were further supported by studies in intact coronary arteries (Danser et al. 2000; Tom et al. 2001), kinin accumulation in a micro milieu containing both membrane proteins was mentioned as an alternative explanation for the potentiation (Tom et al. 2002). It will be very difficult, if not impossible, to prove any influence of ACE inhibitors on kinin accumulation in micro milieus. Currently, the cellular results indicate B_2 kinin receptor resensitizing mechanisms as a more likely class effect of ACE inhibitors.

6 Effects of ACE Inhibition on B_1 Kinin Receptors

Along with the focus on ACE and B_2 kinin receptor interactions and on bradykinin accumulation, it should be taken into account that treatment with ACE inhibitors not only decreases ANG-II and increases bradykinin levels, but may also favor the generation of the B_1 kinin receptor agonist, des-Arg^9-bradykinin, via the enzyme kininase-I (for a review see Marceau et al. 1998). As B_1 and B_2 kinin receptors are activated by different kinin peptides, it may be very interesting to work on the interaction between ACE, ACE inhibitors and the B_1 kinin receptor system in more detail. This kinin receptor subtype is involved in inflammation and hyperalgesia; its role in cardiovascular disease is less well defined. In a recent study, Bascands and co-workers demonstrated the induction of vascular and renal B_1 kinin receptors in normotensive rats and mice under chronic treatment with an ACE inhibitor (Marin-Castano et al. 2002). ACE inhibitors may also exert direct effects on the B_1 kinin receptor (Ignjatovic et al. 2002). In this study with cultured cells, nanomolar concentrations of ACE inhibitors directly activated human B_1 kinin receptor, in the absence of ACE and a B_1 kinin receptor agonist. These inhibitors may bind to a Zn^{2+}-binding motif, which is present in the B_1 kinin receptor and in ACE but not in the B_2 kinin receptor. The only limitation of this study was the use of cultured cells. Direct actions of ACE inhibitors on B_1 kinin receptor signaling should be confirmed in a whole vessel, perhaps after a cytokine challenge, to up-regulate this receptor subtype. Regardless of this insufficient confirmation in experimental animal models, a contributive role for B_1 kinin receptor to the beneficial clinical effects of ACE inhibitors, cannot be excluded. Figure 2 summarizes the mechanisms how ACE inhibitors utilize a local endothelial KKS, via different transduction pathways,

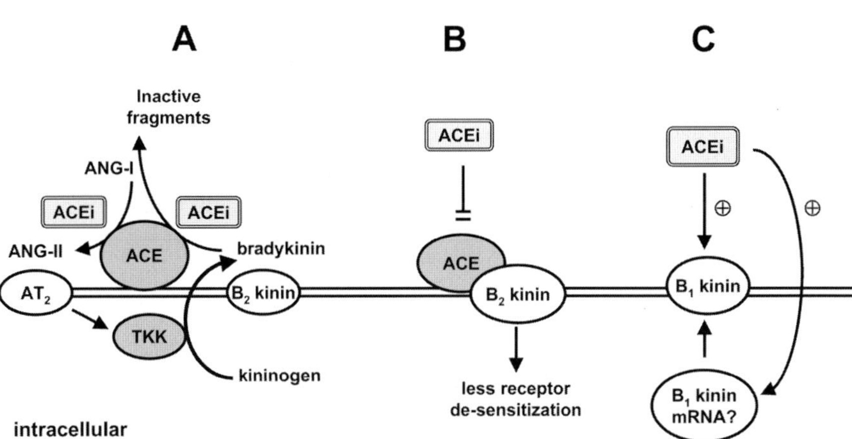

Fig. 2A–C Interference sites of ACE inhibitors and the RAS and KKS as described in detail in the text. **A** By blocking bradykinin degradation, these peptides accumulate. ACE inhibition seems not to couple directly to bradykinin anabolism in contrast to AT_2-receptors. **B** On endothelial cell surfaces, increased receptor densities have been observed under ACE inhibition. Shown here is a direct stabilization of B_2 kinin receptors in a ternary complex with ACE and ACE inhibitor, which may be one way to explain the effects of ACE inhibitors on B_2 kinin receptor sequestration. **C** ACE inhibition may also interfere with the B_1 kinin receptor signaling. A direct activation of this bradykinin receptor subtype has been described only in cultured cells, whereas transcriptional B_1 kinin receptor up-regulation has been found in vivo in rat and mice

namely kinin accumulation, post-transcriptional B_2 kinin receptor resensitization and B_1 kinin receptor activation.

7 Direct Outside-in Signaling by ACE Inhibitors

ACE is a so-called ectoprotein with a large extracellular domain containing both catalytic sites, a single transmembrane helix and a short intracellular tail. The entirely extracellular domain is released, in a regulated manner, in a process called ectodomain shedding, by the actions of metalloproteinases called ACE secretase (Connolly et al. 1995). For decades, the rather short transmembrane and cytoplasmic tail of ACE were regarded solely as an anchor for the extracellularly directed proteolytic activity. This view was recently questioned, when Fleming and colleagues reported that casein kinase 2 (CK-2) phosphorylated ACE at its short cytoplasmic tail (Kohlstedt et al. 2002a). Incubation of cultured endothelial cells with ANG-I and either bradykinin or an ACE inhibitor (ramiprilat or perindoprilat), enhanced this intracellular phosphorylation (Kohlstedt et al. 2002b). Furthermore, a CK-2 inhibitor inhibited both ectoprotein shedding and phosphorylation of the cytoplasmic tail. Specific intracellular proteins such as BiP, a chaperone, several PKC isoforms and an unknown protein with a mass of 225 kDa may associate with ACE and transduce intracellular phosphorylation into translocalization and ectoprotein shedding (Santhamma and Sen 2000; Kohlstedt et al. 2002b). This circular mechanism of outside-in followed by inside-out signaling certainly needs to be investigated further. It may open new sights onto the actions of ACE inhibitors, independent of ANG-II and bradykinin. It seems that by influencing ACE-ectoprotein shedding, and hence the ratio of tissue bound versus circulating ACE, ACE inhibitors interfere with the RAS in a totally unexpected way, different to other RAS modulators.

8 Angiotensin II Receptors in Endothelial Cells

Both known ANG-II receptor subtypes are expressed in rat (Stoll et al. 1995) and human coronary artery endothelial cells (Li et al. 1999, 2000). AT_1 receptor activation promotes growth and apoptosis of endothelial cells, whereas the AT_2 subtype has anti-proliferative effects. For a detailed review, the reader is referred to other chapters in this handbook. Here, we would like to focus on the role of the AT_2 receptor as a point of intersection between RAS and KKS. Because of low expression of this receptor subtype in normal tissues, it took a rather long time to identify and elaborate on its role in cardiac and vascular functions. Early work on cultured endothelial cells suggested beneficial counterbalancing effects of AT_2 stimulation through activation of the KKS and increased release of endothelial-derived nitric oxide (Wiemer et al. 1993; Korth et al. 1995; Seyedi et al. 1995). Tissue kallikrein appears to be a mediator in this signal transduction process, as suggested by the blockade of ANG-II-induced relaxation in human umbilical veins by a specific tissue kallikrein inhibitor (Dedio et al. 2001). However, the signaling pathways downstream of the receptor are still poorly characterized, with only some pieces of the puzzle known, for example, activation of serine/threonine phosphatases. The KKS-activating mechanism of AT_2 stimulation was confirmed in vivo under chronic AT_1 receptor blockade in stroke-prone spontaneously hypertensive rats (Gohlke et al. 1996; Linz et al. 2000). Specific blockade of the AT_1 receptor redirects endogenously released ANG-II to the AT_2. Stimulation of AT_2 with a subsequent increase in kinin synthesis, activation of kinin receptors, and release of endothelial-derived NO best explained why aortic tissue content of cGMP as surrogate marker for endothelial NO was increased under chronic AT_1 receptor blockade.

Additional evidence was delivered in transgenic and knockout mice. Overexpression of the AT_2 receptor by a factor of 5, in vascular smooth muscle did not affect aortic AT_1 receptor expression (Tsutsumi et al. 1999). However, chronic infusion of ANG-II completely abolished the AT_1-mediated pressor effect seen in wild-type animals. Aortic explants from transgenic mice showed significant increases in cGMP. Blocking B_2 kinin receptors or NO synthases, as well as removal of the endothelium, abolished these AT2-mediated effects. In B_2 kinin receptor knockout mice, subchronically infused with low dose ANG-II, mean arterial blood pressure was higher on day 12 than in control animals (Cervenka et al. 2001). In contrast, no difference between transgenic and knockout animals could be observed when norepinephrine, instead of ANG-II, was used to increase blood pressure. These results from transgenic and knockout mice confirmed that the KKS selectively buffers the activities of PAS.

Like ACE inhibition, AT_2 stimulation also involves local kinins as paracrine and autocrine mediators. However, kinin synthesis is activated by AT_2 receptor stimulation, in contrast to ACE inhibitors, which blocks the degradation of permanently released kinin. Can we expect higher efficacy by activating vascular KKS via direct AT_2 receptor stimulation? In spite of theoretical speculations, our experimental knowledge remains restricted by the lack of potent AT_2 agonists. Nevertheless, it would be worth comparing the vascular effects of an ACE inhibitor with those of a selective AT_1 receptor antagonist as another means of redirecting ANG-II to AT_2. Coadministration of a specific bradykinin receptor antagonist may further elucidate the involvement of the KKS in response to specific treatments. Unfortunately, such a study has not been performed so far.

To date, ANG-II is not the only active angiotensin peptide generated within the RAS. At least two other biologically active fragments are formed by further cleavage of ANG-II, namely ANG-(2–8) (ANG-III) and ANG-(3–8) (ANG-IV). Although binding sites of these peptides have been described in cellular studies, specific receptors for these fragments have not yet been cloned. At this time, little is known of the possible interactions between RAS and KKS as mediated by these peptides. In contrast, another angiotensin peptide ANG-(1–7) has definitely been demonstrated to be involved in the interaction between RAS and KKS. Briefly, this peptide is directly formed from ANG-I by neutral endopeptidase and metalloendopeptidase. At least two different mechanisms explaining how this peptide exerts its action are being discussed: ANG-(1–7) is converted by ACE to ANG-(1–5) and therefore may serve, at higher concentrations, as a competitive ACE inhibitor. Alternatively, hitherto undiscovered receptors may mediate B_2 kinin receptor activation with subsequent NO release from endothelial cells. The reader is referred to the chapter by Ferrario et al. in this handbook in which ANG-(1–7) is discussed in greater detail.

9
The Heart as a Target Organ

With the focused interest on vascular KKS and RAS in the past decade, one wonders whether all the discovered mechanisms sufficiently explain all the beneficial effects of ACE inhibition in heart disease patients, especially those with heart failure. The heart muscle comprises mostly cardiomyocytes. In heart failure, a certain number of cardiomyocytes die off. A remodeling process starts in which cardiac fibroblasts replace the lost cardiomyocytes as scar tissue. Although cardiac muscles are well-vascularized and thus are sufficiently supplied with nutrients and oxygen, the relative contribution of endothelial versus cardiomyocyte RAS and KKS to ACE inhibition remains in question.

Roscher and co-workers provided experimental evidence for functionally coupled B_2 kinin receptors in primary cultures of beating neonatal rat cardiomyocytes (Kasel et al. 1996). Radio-labeled bradykinin was much less degraded in cell cultures incubated with captopril or ramipril, indicating a role for ACE in bradykinin degradation at the level of cardiomyocytes. Yamyama et al. (2000) detected kininogen and kallikrein mRNA expressions in neonatal rat cardiomyocytes and whole ventricle. All these results provide a basis for ACE inhibition, at the level of cardiomyocytes, by kinin receptor-mediated mechanisms.

ACE inhibitors block left ventricular hypertrophy in vivo. A component of this effect has been attributed to cardiac accumulation of bradykinin. In a series of elegant cell culture experiments, Ritchie and co-workers demonstrated that bradykinin mediates this effect via endothelial-derived nitric oxide (Ritchie et al. 1998; Rosenkranz et al. 2002). When incubating rat neonatal and adult cardiomyocytes, bradykinin was found to be a hypertrophic agonist. However, in cocultures of rat adult cardiomyocytes with bovine endothelial cells, bradykinin did not induce hypertrophy. Furthermore, bradykinin abolished ANG-II-induced hypertrophy via a nitric oxide and cyclic GMP-dependent pathway. These results favor more a paracrine effect of ACE inhibitors on cardiomyocyte remodeling via endothelial cells.

Many components of the RAS were identified in the heart at both mRNA and protein levels and have been shown to be activated in experimental overload models (Suzuki et al. 1993). Cardiac myocytes and fibroblasts are able to produce ANG-I and ANG-II, thus providing direct evidence that cardiac tissue has its own functional RAS (Dostal et al. 1992). Furthermore, not only does ANG-II indirectly promote left ventricular hypertrophy and remodeling via an increase in peripheral resistance and pressure overload, but also directly via AT_1-mediated signal transduction in cardiomyocytes and cardiac fibroblasts. These include short-term intracellular responses such as calcium mobilization and alterations of ionic conductances. Various longer-term effects are also triggered by AT_1 receptor activation, including release of growth factors such as transforming growth factor $\beta 1$ and insulin-like growth factor, stimulation of fetal phenotype reprogramming, and release of local hormones, e.g., endothelin and norepinephrine (for a review, see the chapters by Conchon and Clauser, Vauquelin and Vanderheyden, Wassman and Nickening, Ferrario et al., and Regitz-Zagrosek, this volume). Cardiomyocytes possess AT_2 receptors, but no physiological role has been ascribed to date; potentially, they may be counteracting AT_1-mediated effects (Yamazaki et al. 1998).

Research on cardiac RAS was furthered encouraged by the cloning of ACE2, a novel ACE-related carboxy peptidase (Donoghue et al. 2000). In contrast to ACE, ACE2 hydrolyzes solely the carboxy-terminal amino acid of ANG-I, to generate angiotensin-(1–9) instead of ANG-II. Angiotensin-(1–9) is further converted to shorter angiotensin peptides by ACE and other convertases. Despite these initial findings, which peptide is the main substrate of ACE2 remains to be investigated. Using purified ACE2 against a panel of 126 biological peptides, eleven of the peptides were hydrolyzed by ACE2, of which three peptides exhibited high catalytic efficiency: ANG-II, apelin-13, and dynorphin. Additionally, ACE2 enzymatic activity is clearly higher with ANG-II than ANG-I as substrate (Vickers et al. 2002).

Nonredundancy between ACE and ACE2 is further supported by the fact that ACE inhibitors do not inhibit ACE2 activity, and both proteins display different tissue expression profiles. ACE2 transcripts are specifically expressed in heart, kidney and testis of 23 human tissues tested. Immunohistochemistry demonstrated ACE2 protein expression predominantly in the endothelium of coronary and intrarenal vessels and in renal tubular epithelium. Targeted disruption of ACE2 in mice results in a severe cardiac contractile defect, increased ANG-II levels, and up-regulation of hypoxia-induced genes in the heart, demonstrating an essential role for ACE2 in heart function (Crackower et al. 2002). Despite some sequence and domain homologies, it seems that ACE2 and ACE are biochemically, physiologically and pharmacologically distinct.

Among several potential ANG-II-forming pathways, chymases have been identified as the protease responsible for the formation of more than 80% of ANG-II in the human heart (Urata et al. 1990). Although this finding has evoked great interest in ANG-II formation pathways as alternative targets to ACE, the functional importance and pathophysiological relevance of these pathways, and especially of chymase, remains to be clarified. Research on this topic is certainly complicated in that, like ACE, ACE2 and chymase also use non-angiotensin peptides as substrates. Therefore, it will be difficult to assign effects in more complex genetic and pharmacological models solely to the RAS. The only selective protease identified in decades of research on PAS is still renin.

10
Conclusions

Angiotensin I-converting enzyme is an important central node in interactions between the RAS and KKS. Inhibitors of this protease may work via mechanisms other than those of ANG-II reduction and bradykinin accumulation such as B_2 kinin receptor resensitization and ACE-mediated outside-in signaling. ACE is not the only intersection between both systems; the vascular AT_2 receptor forms another important nodal point. Today, we know that local RAS and KKS exist within the endothelium and heart muscle. Molecular biology has enabled us to recognize new emerging players in the orchestra of RAS and KKS components such as renin receptor, ACE2 and chymases.

Acknowledgements. We would like to thank Dr. Jürgen Dedio for carefully reviewing this manuscript and for helpful discussions.

References

Baumgarten CR, Linz W, Kunkel G, Schölkens BA, Wiemer G (1993) Ramiprilat increases bradykinin outflow from isolated hearts of rat. Br J Pharmacol 108:293–955

Benzing T, Fleming I, Blaukat A, Müller-Esterl W, Busse R (1999) Angiotensin-converting enzyme inhibitor ramiprilat interferes with the sequestration of the B_2 kinin receptor within the plasma membrane of native endothelial cells. Circulation 99:2034–2040

Brotherton AFA (1996) Induction of prostacyclin biosynthesis is closely associated with guanosine 3',5'-cyclic monophosphate accumulation in cultured human endothelium. J Clin Invest 78:1253–1260

Cervenka L, Maly J, Karasova L, Simova M, Vitko S, Hellerova S, Heller J, El-Dahr SS (2001) Angiotensin II-induced hypertension in bradykinin B_2 receptor knockout mice. Hypertension 37:967–973

Clappison BH, Anderson WP, Johnston CI (1981) Renal hemodynamics and renal kinins after angiotensin-converting enzyme inhibition. Kidney Int 20:615–620

Connolly C, Oppong SY, Turner AJ, Hooper NM (1995) Purification and characterization of the angiotensin converting enzyme secretase. Biochem Soc Trans 23:551S

Crackower MA, Sarao R, Oudit GY, Yagil C, Kozieradzki I, Scanga SE, Oliveira-dos-Santos AJ, da Costa J, Zhang L, Pei Y, Scholey J, Ferrario CM, Manoukian AS, Chappell MC, Backx PH, Yagil Y, Penninger JM (2002) Angiotensin-converting enzyme 2 is an essential regulator of heart function. Nature 417:822–828

Danser AHJ, Tom B, de Vries R, Saxena PR (2000) L-NAME resistant bradykinin-induced relaxation in porcine coronary arteries is NO-dependent: effect of ACE inhibition. Br J Pharmacol 131:195–202

Dedio J, Wiemer G, Rütten H, Dendorfer A, Schölkens BA, Müller-Esterl W, Wohlfart P (2001) Tissue kallikrein KLK1 is expressed de novo in endothelial cells and mediates relaxation of human umbilical veins. Biol Chem 382:1483–1490

Donoghue M, Hsieh F, Baronas E, Godbout K, Gosselin M, Stagliano N, Donovan M, Woolf B, Robison K, Jeyaseelan R, Breitbart RE, Acton S (2000) A novel angiotensin-converting enzyme-related carboxypeptidase (ACE2) converts angiotensin I to angiotensin 1–9. Circ Res 2000 87:1–9

Dostal DE, Rothblum KN, Conrad KM, Cooper GR, Baker KM (1992) Detection of angiotensin I and II in cultured rat cardiac myocytes and fibroblasts. Am J Physiol 263:C851–C863

Gohlke P, Linz W, Schölkens BA, Wiemer G, Unger T (1996) Cardiac and vascular effects of long-term losartan treatment in stroke-prone spontaneously hypertensive rats. Hypertension 28:397–402

Ignjatovic T, Tan F, Brovkovych V, Skidgel RA, Erdös EG (2002) Novel mode of action of angiotensin I converting enzyme inhibitors: direct activation of bradykinin B1 receptor. J Biol Chem 277:16847–16852

Kasel AM, Faussner A, Pfeifer A, Muller U, Werdan K, Roscher AA (1996) B_2 bradykinin receptors in cultured neonatal rat cardiomyocytes mediate a negative chronotropic and negative inotropic response. Diabetes 45 [Suppl 1]:S44–S50

Kohlstedt K, Shoghi F, Müller-Esterl W, Busse R, Fleming I (2002a). CK2 phosphorylates the angiotensin-converting enzyme and regulates its retention in the endothelial cell plasma membrane. Circ Res 91:749–756

Kohlstedt K, Müller-Esterl W, Busse R, Fleming I (2002b) The angiotensin converting enzyme (ACE) is a signal transduction molecule. Circulation 106 [Suppl]:1112

Korth P, Fink E, Linz W, Schölkens BA, Wohlfart P, Wiemer G (1995) Angiotensin II receptor subtype-stimulated formation of endothelial cyclic GMP and prostacyclin is accompanied by an enhanced release of endogenous kinins. Pharm Pharmacol Lett 5:124–127

Li D, Yang B, Philips MI, Mehta JL (1999) Pro-apoptotic effects of ANG II in human coronary artery endothelial cells: role of AT1 receptor and PKC activation. Am J Physiol 1999 276:H786–H792

Li D, Saldeen T, Romeo, F, Mehta J (2000) Oxidized LDL upregulates angiotensin II type 1 receptor expression in cultured human coronary artery endothelial cells. Circulation 102:1970–1976

Lilly LS, Pratt RE, Alexander RW, Larson DM, Ellison KE, Gimbrone MA Jr, Dzau VJ (1985) Renin expression by vascular endothelial cells in culture. Circ Res 57:312–318

Linz W, Wohlfart P, Schölkens BA, Malinski T, Wiemer G (1999) Interactions among ACE, kinins and NO. Cardiovasc Res 43:549–561

Marin-Castano ME, Schanstra JP, Neau E, Praddaude F, Pecher C, Ader JL, Girolami JP, Bascands JL (2002) Induction of functional bradykinin B_1-receptors in normotensive rats and mice under chronic angiotensin-converting enzyme inhibitor treatment. Circulation 105:627–632

Marceau F, Hess JF, Bacharov DR (1998) The B_1-receptors for kinins. Pharmacol Rev 50:357–386

Marcic BM, Erdös EG (2000) Protein kinase C and phosphatase inhibitors block the ability of angiotensin I-converting enzyme inhibitors to resensitize the receptor to bradykinin without altering the primary effects of bradykinin. J Pharmacol Exp Ther 294:605–612

Marcic B, Deddish PA, Skidgel RA, Erdös EG, Minshall RD, Tan F (2000) Replacement of the transmembrane anchor in angiotensin I-converting enzyme (ACE) with a glycosylphosphatidylinositol tail affects activation of the B_2 bradykinin receptor by ACE inhibitors. J Biol Chem 275:16110–16118

Meneton P, Bloch-Faure M, Hagege AA, Ruetten H, Huang W, Bergaya S, Ceiler D, Gehring D, Martins I, Salmon G, Boulanger CM, Nussberger J, Crozatier B, Gasc JM, Heudes D, Bruneval P, Doetschman T, Menard J, Alhenc-Gelas F (2001) Cardiovascular abnormalities with normal blood pressure in tissue kallikrein-deficient mice. Proc Natl Acad Sci U S A 27:2634–2639

Minshall RD, Tan F, Nakamura F, Rabito SF, Becker RP, Marcic B, Erdös EG (1997) Potentiation of the actions of bradykinin by angiotensin I-converting enzyme inhibitors. The role of expressed human bradykinin B_2 receptors and angiotensin I-converting enzyme in CHO cells. Circ Res 81:848–856

Mooser V, Nussberger J, Juillerat L, Burnier M, Waeber B, Bidiville J, Pauly N, Brunner HR (1990) Reactive hyperreninemia is a major determinant of plasma angiotensin II during ACE inhibition. J Cardiovasc Pharmacol 15:276–282
Nguyen G, Delarue F, Burckle C, Bouzhir L, Giller T, Sraer JD (2002) Pivotal role of the renin/prorenin receptor in angiotensin II production and cellular responses to renin. J Clin Invest 109:1417–1427
Pellacani A, Brunner HR, Nussberger J (1994) Plasma kinins increase after angiotensin-converting enzyme inhibition in human subjects. Clin Sci (Lond) 87:567–574
Ritchie RH, Marsh JD, Lancaster WD, Diglio CA, Schiebinger RJ (1998) Bradykinin blocks angiotensin II-induced hypertrophy in the presence of endothelial cells. Hypertension 31:39–44
Rosenkranz AC, Hood SG, Woods RL, Dusting GJ, Ritchie RH (2002) Acute antihypertrophic actions of bradykinin in the rat heart: importance of cyclic GMP. Hypertension 40:498–503
Ryan JW, Ryan US, Schultz DR, Whitaker C, Chung A (1975) Subcellular localization of pulmonary angiotensin-converting enzyme (kininase II). Biochem J 146:497–499
Ryan US (1986) Metabolic activity of endothelium: modulations of structure and function. Annu Rev Physiol 48:263–277
Santhamma K, Sen I (2000). Specific cellular proteins associate with angiotensin-converting enzyme and regulate its intracellular transport and cleavage-secretion. J Biol Chem 275:23253–23258
Seyedi N, Xu X, Nasjletti A, Hintze T (1995) Coronary kinin generation mediates nitric oxide release after angiotensin receptor stimulation. Hypertension 26:164–170
Schini VB, Boulanger C, Regoli D, Vanhoutte PM (1990) Bradykinin stimulates the production of cyclic GMP via activation of B_2 kinin receptors in cultured porcine aortic endothelial cells. J Pharmacol Exp Ther 252:581–585
Stoll M, Steckelings UM, Paul M, Bottari SP, Metzger R, Unger T (1995) The angiotensin AT_2-receptor mediates inhibition of cell proliferation in coronary endothelial cells. J Clin Invest 95:651–657
Suzuki J, Masubara H, Urakami M, Inada M (1993) Rat angiotensin II (type 1A) receptor mRNA regulation and subtype expression in myocardial growth and hypertrophy. Circ Res 73:439–447
Tom B, de Vries R, Saxena PR, Danser AHJ (2001) Bradykinin potentiation by angiotensin-(1–7) and angiotensin-converting enzyme (ACE) inhibitors correlates with ACE C- and N-domain blockade. Hypertension 28:95–99
Tom B, Dendorfer A, de Vries R, Saxena PR, Danser AHJ (2002) Bradykinin potentiation by ACE inhibitors: a matter of metabolism. Br J Pharmacol 137:276–284
Tsutsumi Y, Matsubara H, Masaki H, Kurihara H, Murasawa S, Takai S, Miyazaki M, Nozawa Y, Ozono R, Nakagawa K, Miwa T, Kawada N, Mori Y, Shibasaki Y, Tanaka Y, Fujiyama S, Koyama Y, Fujiyama A, Takahashi H, Iwasaka T (1999) Angiotensin II type 2 receptor overexpression activates the vascular kinin system and causes vasodilation. J Clin Invest 104:925–935
Urata H, Kinoshita A, Misono KS, Bumpus FM, Husain A (1990) Identification of a highly specific chymase as the major angiotensin II-forming enzyme in the human heart. J Biol Chem 265:22348–22357
Vickers C, Hales P, Kaushik V, Dick L, Gavin J, Tang J, Godbout K, Parsons T, Baronas E, Hsieh F, Acton S, Patane M, Nichols A, Tummino P (2002) Hydrolysis of biological peptides by human angiotensin-converting enzyme-related carboxypeptidase. J Biol Chem 277:14838–14843
Wohlfart P, Dedio J, Wirth K, Schölkens BA, Wiemer G (1997) Different B_1 kinin receptor expression and pharmacology in endothelial cells of different origins and species. J Pharmacol Exp Ther 280:1109–1116

Wiemer G, Schölkens BA, Becker RHA, Busse R (1991) Ramiprilat enhances endothelial autoacoid formation by inhibiting breakdown of endothelium-derived bradykinin. Hypertension 18:558–563

Wiemer G, Schölkens BA, Busse R, Wagner A, Heitsch H, Linz W (1993) The functional role of angiotensin II-subtype AT_2-receptors in endothelial cells and isolated ischemic rat hearts. Pharm Pharmacol Lett 3:24–27

Wiemer G, Pierchala B, Mesaros S, Schölkens BA, Malinski T (1996) Direct measurement of nitric oxide release from cultured endothelial cells stimulated by bradykinin or ramiprilat. Endothelium 4:119–125

Xiao F, Puddefoot JR, Vinson GP (2000) The expression of renin and the formation of angiotensin II in bovine aortic endothelial cells. J Endocrinol 164:207–214

Yamazaki T, Komuro I, Yazaki Y (1998) Signalling pathways for cardiac hypertrophy. Cell Signal 10:693–698

Yayama K, Nagaoka M, Takano M, Okamoto H (2000) Expression of kininogen, kallikrein and kinin receptor genes by rat cardiomyocytes. Biochim Biophys Acta 1495:69–77

ACE Inhibitors: Pharmacology

P. Gohlke[1] · B. A. Schölkens[2]

[1] Institute of Pharmacology, University Hospital Schleswig-Holstein, Campus Kiel, Hospitalstrasse 4, 24105 Kiel, Germany
e-mail: peter.gohlke@pharmakologie.uni-kiel.de
[2] Aventis Deutschland GmbH, Frankfurt, Germany

Abstract More than 15 inhibitors of the angiotensin-converting enzyme (ACE) are now clinically available worldwide. ACE inhibitors can be divided into three chemical classes according to their zinc ligand. They mainly differ in their elimination half-life, potency, lipophilicity and the route of elimination. ACE inhibitors act by blocking the systemic and local generation of angiotensin II (Ang II) from angiotensin I (Ang I) and by inhibiting the degradation of kinins. Experimental studies have shown that both actions of the ACE inhibitors are important for their antihypertensive and organ-protective actions. ACE inhibitors effectively lower blood pressure and prevent or reverse the hypertension-induced cardiac and vascular structural changes. In addition, ACE inhibitors can improve cardiac function and prevent cardiac remodelling in animals with experimentally induced heart failure and are beneficial when administered at various time points before and after myocardial infarction. ACE inhibitors can effectively prevent or regress endothelial dysfunction, induced, for example, by high blood pressure, and exert antiatherosclerotic effects. Inhibition of the local formation of proinflammatory Ang II has been shown to be a promising therapeutic approach to stabilize plaque and prevent its rupturing. Several studies have also revealed nephroprotective effects of ACE inhibitors. Finally, ACE inhibition improve insulin resistance and insulin sensitivity and reduce albuminuria and microalbuminuria in experimental models of diabetes mellitus. In summary, ACE inhibitors not only lower blood pressure but also positively influence a number of cardiovascular risk factors such as cardiac and vascular hypertrophy, endothelial dysfunction, atherosclerosis or insulin resistance. These effects help to explain their clinical benefit in heart failure, postmyocardial infarction, chronic renal insufficiency and diabetes mellitus.

Keywords ACE inhibitor · Pharmacology · Hypertension · Congestive heart failure · Myocardial infarction · Hypertrophy · Endothelial dysfunction · Atherosclerosis · Nephroprotection · Diabetes mellitus

1 Introduction

The first specific angiotensin converting enzyme (ACE) inhibitors were small peptides, isolated from snake venom (Ferreira et al. 1970; Ondetti et al. 1971). The most potent nonapeptide, teprotide, turned out to be an effective drug for the treatment of hypertension but had to be applied by the intravenous route (Gavras et al. 1974). The development of orally available, nonpeptide ACE inhibitors was achieved by Cushman et al. (1977) and Ondetti et al. (1977). ACE inhibitors were first introduced in 1981 for the treatment of hypertension. Since then several additional indications have been identified as a result of clinical studies showing a reduction in morbidity and mortality in congestive heart failure, postmyocardial infarction, chronic renal insufficiency, atherosclerotic diseases and diabetes mellitus following ACE inhibitor treatment. ACE, the target enzyme of ACE inhibitors, is widely distributed in different tissue and cell types. Most prominently, it is localized at the luminal surface of the vascular endothelium but has also been found in body fluids (plasma, cerebrospinal fluid), in epithelial cells of the kidney (tubular brush border), gastrointestinal tract, testis and also in epithelial and neuronal structures in the brain (Erdös 1975; Erdös and Skidgel 1987). ACE is involved in the renin angiotensin system, where it converts angiotensin I (Ang I) to angiotensin II (Ang II), but it also hydrolyses other biologically active peptides such as bradykinin, substance P, enkephalins and luteinizing hormone releasing hormone (LHRH) (Unger et al. 1990; Salvetti 1990; Erdös 1975). Due to this broad substrate specificity and distribution pattern ACE has been implicated in a number of physiological and pathophysiological processes such as neuropeptide metabolism, immunity, reproduction or digestion (Ehlers and Riordan 1989).

2 Pharmacology

2.1 Classification of ACE Inhibitors

ACE inhibitors are slow- and tight-binding competitive inhibitors (K_i values 10^{-10}-10^{-11}) of ACE. They interact with the constituents of the enzyme's active site, i.e. the positively charged group, the hydrophobic pocket, the Zn^{2+} ion and auxiliary binding sites. The inhibitory activities of ACE inhibitors are largely determined by the strength of binding of the zinc ligand and by the number of auxiliary binding sites within the active centres of ACE (Bünning 1987; Ondetti 1988; Wyvratt 1988). Differences in binding characteristics may largely help to explain the differences with respect to potency and duration of action between several ACE inhibitors.

ACE inhibitors can be classified in three general chemical classes according to their zinc ligand. (Kostis 1989; Ondetti 1988; Unger et al. 1990; Wyvratt 1988) (Fig. 1). Captopril, the first clinically available ACE inhibitor, belongs to the group of sulfhydryl (SH)-containing ACE inhibitors. The SH group of captopril and its analogues make up a strong zinc ligand, but undergo a complex pattern of reversible modifications through interactions with endogenous SH-containing compounds to form, for example, disulphides. This process limits their duration of action and explains the short half-life of captopril (Drummer and Jarrott 1986; Duchin et al. 1982). It has been claimed that SH-containing ACE inhibitors act as scavengers of cytotoxic oxygen-derived free radicals and may thus have advantages in the treatment of myocardial ischaemia or reperfusion injuries (Chopra et al. 1990; Westlin and Mullane 1988). However, other studies did not support these observations and were not able to detect an inhibitory effect of SH-containing compounds, including captopril on superoxide radical generation when used in clinically relevant doses (Kukreja et al. 1990; Mehta et al. 1990). Therefore, the physiological significance of the claimed scavenging effect of captopril is still unclear. Furthermore, it has been suggested that SH-containing ACE inhibitors can potentiate the vasodilatory and antianginal effects of organic nitrates (Metelitsa et al. 1992; van Gilst et al. 1987), as has been shown for other SH donors such as *N*-acetylcysteine (Horowitz et al. 1983; Winniford et al. 1986).

The majority of ACE inhibitors such as enalapril and its analogues contain a carboxyl group (Fig. 1). Although the carboxyl group is a relatively weak zinc ligand when compared to the SH group, this drawback is overcome by the incorporation of additional side chains which bind to auxiliary binding sides within the active centre of ACE, thus improving the potency and the duration of

Sulfhydryl-containing ACE inhibitors

Captopril

Zofenopril

Carboxyl-containing ACE inhibitors

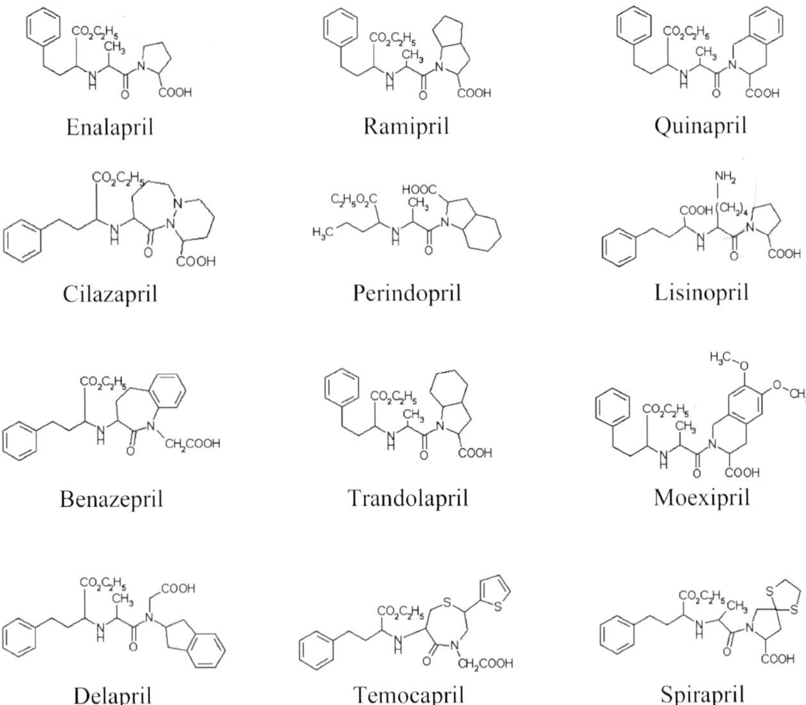

Phosphorus-containing ACE inhibitors

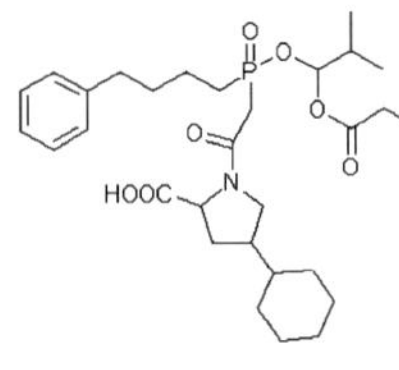

Fosinopril

Fig. 1 Chemical structures of sulfhydryl-containing, carboxyl-containing and phosphorus-containing angiotensin-converting enzyme inhibitors registered in European Union countries. With the exception of captopril and lisinopril, all ACE inhibitors are prodrugs which have to be converted to the active drug by enzymatic cleavage

action (Wyvratt 1988). Thus, carboxyl-containing ACE inhibitors are in general more potent than captopril when assessed by the IC_{50} values, i.e. the concentration necessary to inhibit 50% of the enzyme activity (usually from rabbit lung). However, the active diacid forms of this chemical class are poorly absorbed, which produces a limited bioavailability (Ulm 1983). This problem has been solved by the introduction of ethylester prodrugs which will be hydrolytically activated upon absorption (Todd and Heel 1986; Ulm 1983). The only exception is lisinopril, a lysin derivative of enalaprilat, which is orally active as a diacid (Lancaster and Todd 1988). Fosinopril is an ACE inhibitor that contains a phosphinyl group as zinc ligand (Fig. 1) and is characterized by an in vitro potency similar to captopril but by a longer duration of action in vivo (Ondetti 1988).

Earlier studies suggested that only one catalytically active site is present in the ACE molecule. Firstly, from the analysis of the zinc content of ACE, it was claimed that only a single zinc atom was bound per molecule of ACE (Bünning and Riordan 1985). Secondly, inhibitor binding studies with radiolabelled ACE inhibitors detected a single active site per ACE molecule (Cumin et al. 1989). However, recent findings from molecular biological studies and in vitro mutagenesis studies have demonstrated the existence of two catalytic sites within two homologous domains of ACE (Bernstein et al. 1989; Soubrier et al. 1988, 1993) each of which binds one zinc atom (Ehlers and Riordan 1991). The demonstration of two active sites in ACE is of particular interest with respect to ACE inhibitor binding and substrate selectivity. Firstly, the two active sites may differ in their selectivity to known substrates for ACE (e.g. Ang II, bradykinin, substance P). Secondly, ACE inhibitors of different structures and different zinc ligands may be differentiated by their interactions with the two putative active enzymatic sites.

2.2 Pharmacokinetics

Differences among ACE inhibitors with regard to their elimination half-lifes, potency, lipophilicity and route of elimination have been discussed intensively in recent reviews (Brown and Vaughan 1998; Kelly and O'Malley 1990; Leonetti and Cuspidi 1995; Salvetti 1990; Song and White 2002; Thind 1990; Unger and Gohlke 1994; Unger et al. 1990). Among the pharmacokinetic properties, the terminal half-life and the route of elimination appear to be the most relevant parameters. The terminal half-life reflects the affinity and the binding strength of an ACE inhibitor to ACE, which is related to the duration of the antihypertensive action of an ACE inhibitor. Therefore, drugs with a very short terminal half-life and weak binding to ACE such as captopril have to be administered two to three times a day, while drugs with a long terminal half-life and strong binding to ACE such as ramipril can be administered once daily. The route of elimination may be relevant in patients with kidney or liver disease. Most ACE inhibitors are eliminated by renal mechanism. However, some ACE inhibitors such as ramipril, spirapril, trandolapril and particularly fosinopril were eliminated by both the renal and hepatic mechanism.

2.3 Mechanism

Inhibition of ACE in the vascular endothelium of the lung and other organs and, to a minor extent, in the blood plasma accounts for the reduction in circulating Ang II levels (Fig. 2). Most of the experimental and clinical studies investigating the acute effects of ACE inhibitors on blood pressure demonstrated the expected changes in the parameters of the RAS. ACE activity was decreased, Ang II was lowered and Ang I was increased. Due to the withdrawal of the negative feedback of Ang II, renin activity was increased, leading to decreased angiotensinogen concentrations due to enhanced consumption (Unger and Gohlke 1994). The reduction of circulating Ang II levels in the blood leading to a diminution of the effects of the peptide on vascular tone, aldosterone release, and renal sodium handling, constitutes a principal mechanism of the cardiovascular action of ACE inhibitors (Fig. 2). Furthermore, the blockade of Ang II-mediated increases in plasma levels of AVP, catecholamines, endothelin-1 and PAI-1, as well as the inhibition of the Ang II-induced generation of growth factors and extracellular matrix components and prevention of Ang II-mediated oxidative stress also contribute to the overall cardiovascular effects of ACE inhibition (Fig. 2). Under chronic treatment conditions, the antihypertensive actions of ACE inhibitors were not always accompanied by the predicted changes in the parameters of the RAS and thus could not be explained exclusively by inhibition of the circulating RAS (Unger et al. 1990). Ang II can still be detected under chronic treatment conditions but virtually disappears after acute ACE inhibitor treatment. Furthermore, accumulated evidence from studies in animals with various types of experimental hypertension, as well as from clinical studies showing that blood pressure could be lowered by ACE inhibitors irrespective of whether the plasma RAS was stimulated, has cast some doubt on the initial idea that the antihypertensive actions of ACE inhibitors can be explained exclusively on the basis of reduced circulating Ang II levels. Therefore, additional pharmacological effects of these drugs have to be considered. Because ACE is identical with the bradykinin degrading enzyme, kininase II (Fig. 2), ACE inhibitors could, theoretically, potentiate the direct nitric oxide (NO)-mediated, or prostacyclin-mediated vasodepressor effects of endogenous kinins (Linz et al. 1995). Participation of kinins in the antihypertensive actions of ACE inhibitors has been shown under certain experimental conditions. This topic will be discussed in more detail in the chapter by Wohlfart and Wiemer, this volume.

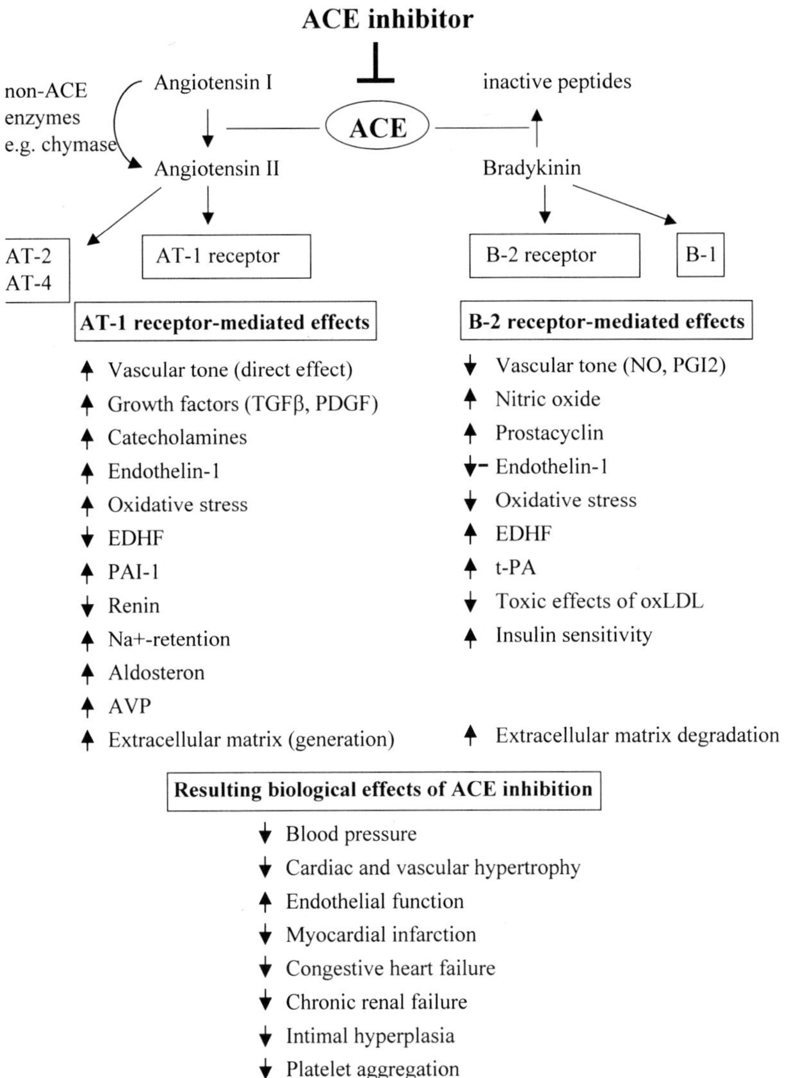

Fig. 2 Blockade of AT_1 receptor-mediated effects of angiotensin II and potentiation of bradykinin B_2-receptor-mediated effects of bradykinin by angiotensin-converting enzyme (*ACE*) inhibition and the resulting biological effects of ACE inhibition. *TGFβ*, transforming growth factor *β*; *PDGF*, platelet-derived growth factor; *EDHF*, endothelial-derived hyperpolarizing factor; *PAI-1*, plasminogen activator inhibitor 1; *AVP*, arginine vasopressin; *NO*, nitric oxide; *PGI2*, prostacyclin, *t-PA*, tissue plasminogen activator

2.4 Inhibition of Tissue ACE

The description of local RASs in various organs including those of cardiovascular regulation, i.e. heart, vascular wall, kidney, adrenal gland and brain has raised the possibility that ACE inhibitors might differ with respect to their ability to inhibit tissue ACE. It has been postulated that tissue penetration of ACE inhibitors could be a crucial determinant of these drugs ability to reach their target enzyme, i.e. ACE, within tissues. Tissue penetration mainly depends on the molecular size and lipophilicity of drugs and in particular on the presence of blood–tissue barriers, e.g. the blood–brain barrier or the blood–testis barrier. The access of ACE inhibitors to brain structures inside the blood–brain barrier is related to their lipid solubility (Gohlke et al. 1989). Ramipril and the more lipophilic ACE inhibitor Hoe 288 both inhibited ACE in the cerebrospinal fluid of rats following acute oral administration of doses between 10–30 mg/kg per day (ramipril) and 1–30 mg/kg per day (Hoe 288) (Gohlke et al. 1989). In contrast, enalapril, the least lipophilic ACE inhibitor used in this study, did not significantly inhibit ACE in the cerebrospinal fluid. An additional factor governing the access of prodrug ACE inhibitors such as enalapril, ramipril or perindopril to structures inside the blood–brain barrier may be the site and degree of metabolic activation of these drugs following oral application. Rat brain, for instance, appears to have only a limited capacity to hydrolyse prodrug ACE inhibitors to the active diacid compounds (Cohen et al. 1983; Gohlke et al. 1989; Unger et al. 1982). Therefore, degree and time course of brain ACE inhibition by a prodrug ACE inhibitor may depend on how much of the respective diacid compound can enter the brain from the blood. In contrast to the brain, testicular ACE seems to be effectively protected from ACE inhibitor access by the blood–testis barrier, since most ACE inhibitors tested thus far did not inhibit ACE activity in the testis after oral application (Jackson et al. 1988; Sakaguchi et al. 1988).

While the brain RAS is now generally accepted as a local RAS (Unger et al. 1988), the model of a tissue RAS still suffers from a number of conceptual problems with respect to peripheral tissues such as the heart or the vascular wall (Unger and Gohlke 1990). The term "tissue system" implies that a complete RAS exists within defined tissue compartments where all the components interact to function as an independent system. ACE is attached to the luminal surface of the vascular endothelial cell membrane by an anchor peptide. Thus the major part of the enzyme containing the active sites is localized outside the cell, protruding into the vascular lumen. In the kidney, a substantial portion of ACE activity is localized in epithelial brush border cells of the tubuli. Consequently, Ang II is always generated through ACE at the outer surface of endothelial cells or epithelial cells (Gohlke et al. 1992), irrespective of whether its direct precursor, Ang I, is produced locally or in the circulating blood. Therefore, it appears that a major task of the tissue RAS or better paracrine/autocrine system in the vascular wall is to provide the Ang II precursor, Ang I, which will then be converted to Ang II by endothelial ACE or by alternative pathways such as chymase.

Ang II thus generated may act locally on adjacent cells (paracrine), on endothelial cells of origin (autocrine) or may join the circulating pool of Ang II (endocrine). Therefore, ACE inhibitors—by inhibiting ACE at its strategic endothelial localization—attack the endocrine as well as the tissue or paracrine/autocrine RAS. Differences among ACE inhibitors in tissue ACE inhibition can be largely explained by their different enzyme binding properties, e.g. the velocity of dissociation of the enzyme inhibitor complex as mentioned above.

A more detailed discussion of the tissue RAS is provided in a different chapter of this handbook.

2.5 Adverse Effects

Side effects of ACE inhibitors include headache, cough, hyperkalaemia, hypotension and angioneurotic oedema. The incidence of these events has steadily declined over the years with the reduction of recommended doses.

Cough is probably the most common side effect of ACE inhibitors. However, the reported incidence of cough varies considerably from less than 1% to more than 30%. These variations may be largely explained by the fact that cough, although present, is frequently not recognized as a side effect associated with ACE inhibitor treatment. The frequency of cough is higher in women than in men. The mechanism of the ACE inhibitor-induced cough may involve increased levels of bradykinin and/or substance P, since both peptides were effectively degraded by ACE and were thus potentiated by ACE inhibition.

Profound first-dose hypotension has been reported in patients with hypertension and particularly in patients with congestive heart failure. This event is most often associated with previous volume depletion by aggressive treatment with diuretics and hyponatraemia, in addition to severe and secondary forms of hypertension. Commencing therapy at low doses and withdrawal of diuretics a few days before ACE inhibitor treatment minimizes the risk of first-dose hypotension. Special care concerning dosage has to be taken in patients with renal disease. Hyperkalaemia may occur as a result of a decrease in aldosterone, particularly in patients with impaired kidney function or those taking potassium-sparing diuretics. Evidence from animal studies and from clinical cases preclude the use of ACE inhibitors during pregnancy.

3 Therapeutic Use

In the following sections we will review more recent publications studying the effects of ACE inhibitor administration in different animal models of experimental hypertension, heart failure, myocardial infarction, endothelial dysfunction and atherosclerosis, diabetes mellitus and renal failure. Previous studies on these topics will only partly be included and the reader is referred to previous reviews on ACE inhibitors (Brown and Vaughan 1998; Leonetti and Cuspidi

1995; Salvetti 1990; Schölkens and Landgraf 2002; Taal and Brenner 2000; Unger and Gohlke 1994; Unger et al. 1990). Clinical trials have demonstrated that ACE inhibitor treatment can decrease mortality in patients with congestive heart failure (CONSENSUS, SAVE, SOLVD), in patients after myocardial infarction (AIRE, GISSI-3, ISIS-4, TRACE), in patients who had a previous stroke or transient ischaemic attack (PROGRESS) and in patients at high cardiovascular risk (HOPE). Furthermore, clinical studies have demonstrated to reduce the progression of diabetic nephropathy. These clinical studies are discussed in the chapter by Schulz, of this handbook, and will not be included in this review. Finally, studies investigating possible benefits of combination therapies of ACE inhibitors with, for example, diuretics, AT_1 receptor antagonists and others will not be included in this chapter because these topics are also discussed in the chapter by Azizi and Menard, in this volume.

3.1 Hypertension

ACE inhibitors are most effective in lowering blood pressure in experimental models of hypertension associated with a stimulated RAS such as 2K-1C-hypertensive rats, rats with aortic banding, TG(mRen2) 27 rats, transgenic rats made hypertensive by the insertion of an additional mouse renin gene or Tsukuba hypertensive mice carrying both human renin and angiotensinogen genes (Bao et al. 1992; Böhm et al. 1998; Demeilliers et al. 1998; Kai et al. 1999).

ACE inhibitors also reduce blood pressure in genetically hypertensive rats (SHRs, SHRSPs and Lyon hypertensive rats) and mice (Gillies et al. 1998; Gohlke et al. 1994; Leckie 2001; Unger et al. 1992), and also improve survival of normal SHRSPs (Linz et al. 1997; Linz et al. 1999a) as well as salt-loaded SHRSPs following lifelong treatment (Ogiku et al. 1993; Stier et al. 1989). Interestingly, in SHRs, the antihypertensive as well as the antihypertrophic effects of ACE inhibition persisted following drug withdrawal (Paull and Widdop 2001). In rats rendered hypertensive by extensive renal mass ablation (a model lacking genetic hypertensive determinants), ACE inhibition prevented the increase in blood pressure (Lopez-Hernandez et al. 1999). ACE inhibitors also reduced blood pressure in 1K-1C-hypertensive rats, rats made hypertensive by treatment with cyclosporine A or infusion of a nonselective antagonist of adenosine receptors (Lassila et al. 2000; Sousa et al. 2002; Sweet et al. 1981). In very old Wistar Kyoto rats with isolated systolic hypertension, ACE inhibition in combination with L-arginine ameliorated the hypertensive and associated cardiovascular changes that occurred with ageing. Finally, the hypertensive actions of chronic insulin infusion, as well as hypertension induced by feeding fructose-rich chow (a dietary model of insulin resistance) were both attenuated by ACE inhibitor treatment (Keen et al. 1998; Uchida et al. 2002). In contrast, ACE inhibitors had no or only weak effects on blood pressure in DOCA-salt hypertensive rats (a low-renin model of hypertension) (Brown et al. 1999; Pu et al. 2002). Clinically, ACE inhibitors decrease elevated systemic vascular resistance by the inhibition of the vasoconstrictive action of Ang II and the potentiation of the vasodilator action of bradykinin. ACE inhibitors did not induce reflex sympathetic activation and they maintain their effect without development of tolerance. All currently available ACE inhibitors are considered to be equally effective in lowering blood pressure after adequate dosing. Since ACE inhibitors do not negatively affect the metabolism of glucose and lipids, adverse metabolic consequences do not occur. ACE inhibitors were shown to be at least as effective in lowering blood pressure as beta-adrenoceptor blockers, diuretics or calcium antagonists. A number of studies have suggested that treatment with ACE inhibitors is less effective in black hypertensive patients than monotherapy with other antihypertensives. However, in combination with diuretics, ACE inhibitors are as effective as in other ethnic groups.

Reduction of dietary salt intake can improve the response to ACE inhibition. This potentiating effect is most likely due to the stimulatory action of a low-salt diet on the RAS.

3.2 Congestive Heart Failure

In recent years, a number of experimental studies in several animal species have been conducted to investigate the efficacy of ACE inhibitor treatment in congestive heart failure. Various animal models of heart failure have been employed, including myocardial infarction-induced heart failure in rats a well as pacing-induced heart failure in dogs and pigs. In general, most of these studies demonstrated that ACE inhibition can improve cardiac function and remodelling in animals with experimentally induced heart failure. These ACE inhibitor-induced cardiac changes include increases in left ventricular (LV) ejection fraction, LV myocyte shortening velocity, dP/dt and cardiac output, decreases in LV end-diastolic pressure and central venous pressure and decreases in LV mass, myocardial fibrosis, and myocyte cross sectional-area. Part of these effects may be explained by the ACE inhibitor-induced decrease in systemic and pulmonary vascular resistance. However, recent studies revealed some additional cardiac and systemic effects of ACE inhibitors which may be related to their beneficial actions in heart failure. Several detrimental changes associated with the development of heart failure were prevented or attenuated by ACE inhibitor treatment. For example in rats, ACE inhibitor treatment suppressed the increase in cardiac TGFβ1 mRNA expression and preserved cardiac high-energy phosphates (Brooks et al. 1997; Ma et al. 2001; Murakami et al. 2002; Podesser et al. 2002; Sanbe et al. 1995). The heart failure-induced increase in the expression of LV uncoupling protein 2 was suppressed (Murakami et al. 2002) and the increase in α myosin heavy-chain gene expression and the decrease in ANP and pro α1 (III) collagen gene expression was prevented by ACE inhibitor treatment (Brooks et al. 1997; Carraway et al. 1999). Moreover, ACE inhibition partially corrected the changes in the profile of heart sarcolemmal PLCs and reduced the mass of sarcolemmal phosphatidylinositol 4,5-biphosphate and the activities of enzymes responsible for its synthesis (Tappia et al. 1999). In heart failure-prone SHHF/Mcc-fa(cp) rats, ACE inhibitor treatment decreased the shift from V1 to V3 myosin isozymes, which was associated with heart failure (Carraway et al. 1999).

In dogs with heart failure induced by intracoronary microembolizations, the incidence of cardiomyocyte apoptosis at the border of the infarct region (evaluated by TUNEL staining) was reduced in ACE inhibitor-treated animals (Goussev et al. 1998). Li et al. (2001) showed that atrial Ang II concentrations and the expression of MAP kinases such as extracellular signal-regulated kinase (ERK) were increased in hearts from dogs with pacing-induced heart failure. ACE inhibitor treatment attenuated these heart failure-induced atrial changes and reduced atrial fibrosis and the mean duration of atrial fibrillation.

B_2-receptor antagonism in the presence of an ACE inhibitor suppressed the expression of eNOS and of sarcoplasmatic reticulum Ca^{2+}-ATPase mRNA and up-regulated the expression of collagen type I and III and also increased cardiac collagen content when compared to ACE inhibitor treatment alone, indicating the importance of bradykinin for the cardioprotective actions of ACE inhibitors (Fujii et al. 2002). ACE inhibition also increased nitrite release from coronary microvessels by a bradykinin-dependent mechanism, thus improving the impaired endothelial NO production in failing hearts (Zhang et al. 1999). Finally, in dogs with right-sided heart failure, ACE inhibitor treatment restored myocardial beta-adrenoceptor density and reduced the attenuation of cardiac norepinephrine uptake (Kawai et al. 1999). In pigs, the increases in plasma catecholamines and endothelins and the decreased β-adrenergic responses observed with heart failure were reduced by ACE inhibitor treatment (Krombach et al. 1998; Spinale et al. 1997a, 1997b). Moreover, the decreased density in L-type Ca^{2+} channels and sarcoplasmatic reticulum Ca^{2+}-ATPase in failing hearts were normalized by ACE inhibitor treatment when combined with an AT_1 receptor antagonist (Spinale et al. 1997b). Preservation of protein expression of sarcoplasmatic reticulum Ca^{2+}-ATPase as well as of phospholamban by ACE inhibitor treatment has also been demonstrated in guinea-pigs with pressure overload-induced heart failure (Takeishi et al. 1999). In these animals, ACE inhibitor treatment attenuated the translocation of protein kinase C-α and -ε. Most recently, Makino et al. (2003) demonstrated that ACE inhibitor treatment reversed several heart failure-induced alterations, including elevated plasma norepinephrine levels as well as reductions in β-adrenoceptor density and adenylate cyclase activity in rabbits with heart failure induced by myocardial infarction. Furthermore, ACE inhibition normalized the myocardial protein levels of β-adrenoceptor kinase 1 and Giα, which were found to be elevated in the failing heart (Makino et al. 2003). Taken together, several haemodynamic and humoural effects of ACE inhibitors in combination with improvements in cardiac function and remodelling are the basis for the beneficial actions of these drugs in the treatment of congestive heart failure.

3.3
Myocardial Ischaemia (Myocardial Infarction and Coronary Heart Disease)

Although ACE inhibitors are not considered antianginal drugs such as nitrates, beta-blockers or Ca^{2+} antagonists, they may be useful in the treatment of ischaemic heart disease by mechanisms which include reduction of myocardial oxygen demand, improvement of coronary blood flow, reduction of blood pressure and ventricular wall tension without reflex tachycardia and reduction of myocardial contractility. Some evidence for the clinical efficiency of these drugs has been provided by clinical studies suggesting an antianginal effect of ACE inhibitors in normotensive and hypertensive patients with coronary heart disease (Yusuf et al. 2000).

Experimental and clinical studies further indicate a protective role of ACE inhibitors after acute myocardial infarction (MI). Previous studies with isolated ischaemic rat hearts demonstrated that ACE inhibitor treatment can prevent or attenuate tissue damage and fatal arrhythmias during ischaemia and reperfusion (Li and Chen 1987; Linz et al. 1986; Westlin and Mullane 1988). The beneficial effect of ACE inhibitors in the isolated ischaemic rat heart could be attenuated by a bradykinin antagonist, indicating a role of ACE inhibitor-induced bradykinin potentiation. However, a reduction in both Ang II and catecholamine levels may also be involved (Linz et al. 1995).

The effect of ACE inhibition was also evaluated in animals with MI induced by coronary artery occlusion. These studies largely differ in the (a) time points of initiating drug therapy, (b) type of ACE inhibitors used, (c) dosage and (d) duration of drug treatment. In several studies, the effect of pretreatment with an ACE inhibitor on the consequences of MI was studied. In most studies, ACE inhibitor treatment was started after myocardial infarction. ACE inhibitor treatment was initiated early, that is up to 4 h after MI, delayed, that is 24 h after MI or late, that is 1–2 two weeks after MI.

3.3.1
Effect of Pretreatment with an ACE Inhibitor

Pretreatment of rats with ramipril for 7 days prior to MI improved the MI-induced increase in LVEDP and right atrial pressure and the decrease in cardiac index (Mori et al. 1998) and caused a decrease in infarct size and interstitial fibrosis (Sandmann et al. 2001b). Furthermore, ACE inhibitor treatment inhibited the up-regulation of the NA^+/H^+ exchanger and the Na^+/HCO_3^- symporter and decreased the up-regulation of calpain I and calpain II in the LV free wall of rats with MI (Sandmann et al. 2001a, 2001b,). In female rats, pretreatment with captopril for 10 weeks reduced infarct size and increased the threshold of ventricular fibrillation, both before occlusion and after reperfusion, and decreased the average episodes of ventricular tachycardia and fibrillation (Zhu et al. 2000). Pretreatment of SHR with captopril for 12 weeks prior to MI decreased the inducibility of ventricular arrhythmias and reduced the mortality 3 h after MI from 72% in control rats to 40% in ACE inhibitor-treated rats (Nguyen et al. 1998b). ACE inhibitors can also affect infarct size when applied shortly before induction of MI. In rats, the application of ramiprilat 5 min before ischaemia reduced the infarct size from 40% (vehicle-treated rats) to 18% (Wolfrum et al. 2001). Similar results were reported in dogs after pretreatment with cilazaprilat 10 min before MI. The infarct size was markedly reduced from 47% in control rats to 15% in ACE inhibitor-treated rats (Node et al. 1998). This effect of the ACE inhibitor was associated with an increase in bradykinin and NO measured in the coronary venous blood at 10 min after reperfusion. The beneficial effect of cilazaprilat on infarct size was abolished by a combined inhibition of NO synthase and blockade of Ca^{2+}-activated K^+ channel, while inhibition of cyclooxygenase had no effect (Node et al. 1998). In pigs with MI induced by low-flow ischaemia, infusion of ramiprilat for 30 min prior to MI reduced infarct size from 20% in controls to 9.8%. This effect was even more pronounced when ramiprilat was combined with the AT_1 receptor antagonist, candesartan (infarct size 6.7%) but was abolished by bradykinin B_2 receptor blockade.

3.3.2 Effect of Early Treatment

Treatment of rats with quinapril 30 min after MI decreased LVEDP and LV diastolic wall stress only in rats with large MI. The cardiac index and stroke volume index were both restored by ACE inhibitor treatment but not by AT_1 receptor antagonism (Hu et al. 1998). When treatment was started immediately after MI, the 1-year survival rate was similar in rats treated with captopril or an AT_1 receptor antagonist (Milavetz et al. 1996). Captopril inhibited endothelial cell proliferation and coronary vessel growth when treatment was initiated directly after MI, but not 3 weeks after MI. The increased survival in rats with MI treated for 8 weeks with captopril (treatment start 4 h after MI) was associated with an improved cardiac function and cardiac remodelling and a decrease in the mRNA expression of TGFβ1 and TNFα and decreased plasma concentrations of endothelin-1, Ang II and catecholamines (Lapointe et al. 2002). In SHRs, an early start of drug treatment (4 h after MI) also reduced mortality from 82% to 56% after 2 months and decreased the inducibility of ventricular arrhythmias (Nguyen et al. 1998a).

3.3.3 Effect of Delayed Treatment

Several studies started drug treatment in those animals which survived the first 24 h after MI. In an ex vivo study in isolated rat hearts, treatment with captopril prevented the changes in cardiac energy metabolism associated with MI. In this study, rats were treated with captopril for 8 weeks starting 24 h after MI (Hügel et al. 1999). In another study (Wei et al. 2002), cardiac haemodynamics and remodelling were improved after 4 weeks of ACE inhibitor treatment. These improvements were associated with an ACE inhibitor-induced prevention of the increase in the expression of the cytokines IL-1β and IL-6 and the reduced expression of TNFα and IL-5. Furthermore, a 4-week treatment with cilazapril prevented systolic and diastolic dysfunction and decreased the expression of β-myosin heavy chain, α-sceletal actin and ANP as well as the expression of collagen I and II in the noninfarcted ventricle (Yoshiyama et al. 1999). ACE inhibitor treatment after MI also interferes with the PKC signal transduction pathway. For example, captopril reversed the decreased expression of diacylglycerol (DAG) kinase ε, which terminated signalling from DAG and thereby attenuated PKC activity (Takeda et al. 2001). In another study, captopril inhibited the up-regulation of PKC in the infarcted heart (Namiuchi et al. 2000). In both studies, captopril was commenced 24 h after MI and continued for 3 weeks. Recently, Jin et al. (2001) reported on 37 genes which were differentially expressed between MI and sham-operated rats. During ACE inhibitor treatment, changes of ten of these genes were partially or completely inhibited.

3.3.4
Effects of Late Treatment

Several other studies also revealed beneficial effects of ACE inhibitors even when treatment was started late, that is more than 1 week after MI. Trandolapril improved the haemodynamic status by decreasing LVEDP and urinary cGMP at a low dose (0.1 mg/kg), which had no effect on blood pressure and cardiac fibrosis development (Richer et al. 1999). Low-dose treatment with enalapril improved the arterial elastic properties by increasing the elastin content and the elastin-to-collagen ratio, but did not alter infarct size (Driss et al. 1999). At a higher dose, enalapril reduced mortality after 312 days of treatment from 80% in controls to 56% (Levijoki et al. 2001). Trandolapril at a dose of 3 mg/kg attenuated cardiac sarcoplasmatic reticulum dysfunction by preventing the MI-induced down-regulation of Ca^{2+}-release channels but had no effect on infarct size (Yamaguchi et al. 1998). Besides the inhibition of Ang II generation, an attenuation of the MI-induced increase in cardiac AT_1 receptor expression by ACE inhibitor treatment may also contribute to the overall effects of these drugs on MI. When ACE inhibitor treatment was started 1 week before or 6 weeks after induction of myocardial infarction in rats (Stauss et al. 1994), a reduction in infarct size could be observed when treatment was started before MI but not when treatment was started 6 weeks after induction of myocardial infarction. However, either treatment caused a reduction in end-diastolic pressure. When treatment was started very late, that is, 3 months after the induction of MI in rats, cardiac hypertrophy and remodelling as well as ventricular dysfunction were reversed in rats treated for 6 months with lisinopril (Mulder et al. 1997b). These effects of the ACE inhibitor were associated with an increase in survival (28% in controls and 61% lisinopril-treated rats). The effect on survival was even more pronounced, when treatment was started earlier, that is 1 week after MI (90% survival) (Mulder et al. 1997b).

Therefore, in most experimental studies investigating the effect of ACE inhibition on MI, pretreatment of rats before MI as well as early or delayed initiation of ACE inhibitor treatment improved cardiac haemodynamics and remodelling, decreased the occurrence of cardiac arrhythmias, altered cardiac expression of several cytokines, decreased the circulating levels of catecholamines, Ang II and endothelin I and increased survival. In contrast, in a previous study (Schoemaker et al. 1991), treatment with captopril (s.c. infusion via osmotic minipumps) starting 24 h after MI for 3 weeks not only failed to improve cardiac function but appeared to have deleterious effects to the heart. However, cardiac function was improved in this study when captopril treatment was started late, that is 3 weeks after MI.

3.4
Cardiovascular Hypertrophy and Remodelling

Hypertension-induced pathophysiological changes in cardiac and vascular structure represent a compensatory response to the increased cardiac output and elevated wall stress. In addition, Ang II exerts trophic actions and may accelerate cardiovascular hypertrophy and chronic vascular disease by directly inducing cellular growth in addition to its systemic actions in hypertension. Regression of left ventricular mass after ACE inhibitor treatment was demonstrated in different forms of experimental hypertension. In rats with renal hypertension due to aortic banding (pressure overload hypertrophy) chronic treatment with a subantihypertensive dose of ramipril prevented left ventricular hypertrophy, suggesting that in this form of hypertension, early-onset treatment with ACE inhibitors can induce structural changes of the heart independently of the blood pressure-lowering actions of these drugs (Linz et al. 1989). In addition, low-dose ramipril treatment improved myocyte relaxation and improved both sarcoplasmatic reticulum (SR)-dependent and non-SR-dependent calcium cycling (Boateng et al. 2001). In contrast, in SHRs and stroke-prone SHRs (SHRSPs) the reduction in LVH by chronic ACE inhibitor treatment strongly correlates with the reduction in blood pressure (Gohlke et al. 1997; Unger et al. 1992). Similarly, in TG(mRen-2) rats, the ACE inhibitor induced decrease in LVH and interstitial and perivascular fibrosis was strongly dependent on blood pressure reduction (Bishop et al. 2000).

At the age of 6 months, SHRSPs had developed LVH and showed increases in LV myocyte cross-sectional area, LV interstitial and perivascular collagen content as well as increases in the relative amount of V3 myosin heavy chain. The development of LVH in these animals is associated with a reduction in coronary reserve maximum. Treatment with the ACE inhibitor captopril from 12 to 24 weeks of age prevented these cardiac changes (Ikeda et al. 2000). In SHRs, regression of LVH by ACE inhibition was also accompanied by a PKC-dependent normalization of the Na^+/H^+ and Na^+-independent Cl^-/HCO_3^--exchange activities (Ennis et al. 1998) as well as by a decrease in dihydropyridine-sensitive L-type calcium channels (Vulpis et al. 1998). Furthermore, ACE inhibitor treatment increased total beta-adrenoceptor density, restored beta1 and beta 2 adrenoceptor proportions and restored adenylyl-cyclase reactivity to normal (Laflamme et al. 1997). Similar changes in the beta-adrenergic signal transduction pathway by ACE inhibition have been demonstrated in transgenic TG(mRen-2)27 rats (Zolk et al. 1998). In addition, the depletion of tissue catecholamines and the desensitization of LV beta-1 adrenoceptors, which was evident in untreated TG(mRen-2) rats, was prevented by enalapril treatment (Witte et al. 2001). These results suggest that ACE inhibition can normalize sympathetic hyperreactivity and restore beta adrenergic signal pathway sensitivity in experimental hypertension. Normalization of LVH in TG(mRen-2)27 rats by captopril treatment was accompanied by normalization of the decreased expression of SR Ca^{2+}-ATPase and the increased expression of ANP-mRNA (Flesch et al. 1997).

Cardiac hypertrophy induced by chronic inhibition of NO synthase is associated with an activation of both 70-kDa S6 kinase and extracellular signal-regulated protein kinase (ERK) (Sanada et al. 2001) and a decrease in the number and the numerical density of cardiomyocytes (Pereira and Mandarim-De-Lacerda 2001). Cardiac hypertrophy was attenuated by ACE inhibition together with an inactivation of 70-kDa S6 kinase but not ERK. In contrast, AT_1 receptor blockade inactivated ERK but had no effect on 70-kDa S6 kinase, suggesting that different signalling pathways are modulated by ACE inhibitors and AT_1 receptor antagonists, respectively (Sanada et al. 2001). In addition, the loss of cardiomyocytes in L-NAME-treated rats was prevented by treatment with enalapril (Pereira and Mandarim-De-Lacerda 2001).

Cardiac benefits from ACE inhibition were most pronounced when treatment was started early, that is before the development of cardiac dysfunction. However, several studies have demonstrated that ACE inhibitor treatment is also effective in animals with long-term hypertension and related cardiovascular disorders. Treatment of 1-year-old SHRs for 2 months with trandolapril not only reduced blood pressure and LVH but also reduced the QT interval, which was found to be increased in older SHR compared to age-matched WKY rats (Baillard et al. 2000). In aged SHRs, collagen content markedly increased together with an increase in the ratio of collagen type I to type III. Treatment with captopril decreased collagen content and decreased the collagen type I-to-type III ratio, primarily by increasing type III collagen. The later effect was not observed with lisinopril (Yang et al. 1997). Treatment of very old SHRs (21 months old) did not restore contractile function or reduce cardiac fibrosis, but still reduced LVH and improved clinical signs of heart failure (Brooks et al. 1997). In LV papillary muscles from captopril-treated 2-year-old SHRs, the expression of alpha-myosin heavy chain was increased and the expression of the Na^+/Ca^{2+} exchanger was decreased. Furthermore, the LV inotropic responses to beta-adrenergic stimulation was improved by ACE inhibitor treatment (Brooks et al. 2000). Captopril treatment also reduced skeletal muscle actin transcripts and reversed the increase in beta-myosin heavy chain that occurred with ageing (Dalton et al. 2000).

In a model of right ventricular hypertrophy induced by pulmonary artery banding in rabbits, ramipril preserved papillary muscle contractility and prevented AT_1 receptor and G-protein up-regulation (Rouleau et al. 2001). These effects were partially attenuated by B_2 receptor blockade.

In Dahl salt-sensitive rats fed an 8% NaCl diet from 7 weeks of age, benazepril improved diastolic dysfunction, attenuated the up-regulation of LV ANP mRNA and improved survival. The effect was more pronounced when the ACE inhibitor was combined with an AT_1 receptor antagonist. The combination therapy additionally suppressed the increase in LV endothelin-1 levels and LV hydroxyproline contents (Kim et al. 2001). In the same model, imidapril increased the expression of eNOS, inhibited the expression of iNOS and decreased type-1 collagen expression in the LV and improved perivascular and myocardial fibrosis (Kobayashi et al. 1999). In contrast, in Dahl salt-sensitive rats fed an 8% NaCl diet from 12 weeks of age, ACE inhibitor treatment did not attenuate blood pressure and LVH, but significantly decreased both blood pressure and LVH, after switching from a high-salt to low-salt diet (Sugimoto et al. 1998).

In an ex-vivo study, LV myocytes isolated from SHRs which had been treated for 4 weeks with perindopril showed a decreased cell volume, cross-sectional area and cell length (Onodera et al. 2002). This effect was not observed after treatment with the beta blocker, bisoprolol. In cartilage and tendon tissue cultures, the conversion of procollagen to collagen was inhibited by captopril and enalapril. Because Ang II, bradykinin and the nonspecific angiotensin receptor antagonist saralasin did not affect this conversion, the authors speculated that ACE inhibitors may directly interact with the specific procollagen proteases (Mannistö et al. 2001).

3.5 Angiogenesis

Capillary rarefaction is a further detrimental morphological change associated with the development of LVH leading to relative ischaemia. Long-term treatment of SHRs and SHRSPs with ramipril not only prevented the development of LVH but also prevented capillary rarefaction by increasing the capillary length density (Gohlke et al. 1997; Unger et al. 1992). This effect of ramipril appeared to be independent of the antihypertensive and antihypertrophic effect of the ACE inhibitor, since it was also observed at a low dose which did not affect LVH. Thus, ACE inhibitor treatment, in addition to lowering blood pressure and reducing LVH, can improve myocardial oxygen and substrate supply via an increased myocardial capillarization. The angiogenic effect of ramipril was associated with an improved cardiac function, as demonstrated by an increase in left ventricular pressure, dp/dtmax and coronary flow, with no change in heart rate and increased myocardial tissue concentrations of glycogen and the energy-rich phosphates ATP and creatine phosphate (Gohlke et al. 1994). All ACE inhibitor-induced effects on capillary density, cardiac function and metabolism could be observed even at subantihypertensive doses and were abolished by chronic bradykinin receptor blockade (Gohlke et al. 1994, 1997).

ACE inhibitor-induced angiogenesis has also been shown in rat limb muscle and ischaemic hindlimb of rabbit and mouse (Cameron et al. 1992; Fabre et al. 1999; Silvestre et al. 2001). The ACE inhibitor-induced increase in vessel density and capillary number in the ischaemic hindlimb of wild type mice was associated with an increase in tissue eNOS protein level and was not observed in B_2 receptor-deficient mice (Silvestre et al. 2001). These results suggested that the proangiogenic effect of ACE inhibition involves a B_2 receptor-dependent and NO-related mechanism.

In contrast, ACE inhibition prevented inner retinal blood vessel growth in neonatal (mRen-2)27) rats, inhibited the induction of neovascularization in rat cornea and had antitumor effects (Moravski et al. 2000; Volpert et al. 1996).

3.6 Endothelial Dysfunction

The endothelium is an important regulatory organ maintaining vascular function and structure. ACE, which is mainly localized on the luminal site of the vascular endothelium, is critical for the local generation of Ang II from Ang I and the degradation and inactivation of bradykinin. Thus, the endothelium is a major target organ for ACE inhibitors. Endothelial dysfunction is associated with ageing as well as with several pathophysiological disorders such as hypertension, atherosclerosis, myocardial infarction, congestive heart failure, renal failure and diabetes. Dysfunction of the endothelium as well as endothelial cell destruction is not only a cause of such cardiovascular disorders but will also promote the progression of these diseases. Endothelial dysfunction can be defined as a decreased capability to produce and release endothelial-derived mediators such as NO and PGI_2, which are potent vasodilators as well as inhibitors of smooth muscle proliferation and thrombosis. Furthermore, interference with endothelial cell function, e.g. as a result of chronic hypertension or atherosclerosis will increase the risk of thrombus formation. Endothelial dysfunction induced by high blood pressure can be effectively prevented by early onset ACE inhibitor treatment (Gohlke et al. 1993; Keaton et al. 1998; Sharifi et al. 1998; Teisman et al. 1998). The ACE inhibitor-induced prevention of endothelial dysfunction persisted even after drug withdrawal (Keaton et al. 1998). Furthermore, existing endothelial dysfunction found in older SHRs can be regressed by ACE inhibitor treatment (Linz et al. 1999a; Rodrigo et al. 1997), resulting in an increase in survival of these animals (Linz et al. 1999a). ACE inhibition was also found to prevent diabetes-induced impairment of endothelium-dependent relaxation to acetylcholine (Pieper and Siebeneich 2000). ACE inhibitor treatment protected against the acute vascular toxicity of oxidized LDL (oxLDL) by attenuating the oxLDL-induced decrease in aortic cGMP and the decrease in acetylcholine-induced relaxation (Berkenboom et al. 1997; Fontaine et al. 2002). Endothelium-dependent relaxation was also restored in hypercholesterolaemic animals by treatment with an ACE inhibitor together with an inhibition of the development of atherosclerosis (Hernandez et al. 1998). In most of these studies, an ACE inhibitor-induced increase in the endothelium-dependent vascular relaxation to acetylcholine was measured as an indicator for an improved endothelial dysfunction. Vascular relaxation induced by endothelium-derived hyperpolarizing factor (EDHF) declines with ageing and hypertension. Long-term ACE inhibitor treatment improved the age-related as well as hypertension-related decline in EDHF-mediated vascular relaxation (Goto et al. 2000a, 2000b; Kansui et al. 2002). Vascular relaxation in response to acetylcholine and flow-mediated dilation of small arteries was found to be impaired in rats with MI-induced heart failure. Furthermore, tissue cyclic GMP content was lower in rats with heart failure compared to control rats. Treatment with different ACE inhibitors fully normalized the acetylcholine-induced dilatation of aortic segments (Buikema et al. 2000; Toyoshima et al. 1998) and small arteries (Mulder et al. 1997a), improved flow-dependent dilatation of small arteries (Varin et al. 2000) and partially restored tissue cyclic GMP levels (Toyoshima et al. 1998).

The attenuated increase in coronary flow in response to serotonin which was observed in ex vivo Langendorf preparations 4 weeks after MI, was prevented by ACE inhibitor treatment (Qi et al. 1999). In addition, ACE inhibition prevented the reduction of cardiac eNOS expression, which was observed in rats with heart failure (Qi et al. 1999; Varin et al. 2000) and in aged SHRs (Linz et al. 1999a). Furthermore, aortic eNOS expression was completely restored in rats chronically treated with L-NAME (De Gennaro Colonna et al. 2002).

In a porcine coronary artery balloon-injury model, ACE inhibition restored the vasodilator response of injured arteries to bradykinin. This ACE inhibitor-induced improvement of endothelial dysfunction was accompanied by the inhibition of neointima formation and the increase in the local expression of hepatocyte growth factor, a stimulator of endothelial cell growth (Matsumoto et al. 2001).

In contrast to the above studies, ACE inhibitor treatment had no effect or only slightly improved vascular endothelial function in DOCA-salt hypertensive rats, while dual NEP/ACE blockade effectively prevented endothelial dysfunction in these animals (Pu et al. 2002; Quaschning et al. 2001).

3.7 Atherosclerosis

Atherosclerosis is a chronic, inflammatory-activated condition of the vascular system and can, through the rupture of a vulnerable atherosclerotic plaque followed by thrombosis, escalate into an acute clinical event (Lusis 2000). Several factors are involved in the pathogenesis of atherosclerosis, including endothelial injury or dysfunction, enhanced arterial permeability to lipoproteins, local inflammation, platelet deposition and smooth muscle migration and proliferation (Bondjers et al. 1991). For years there has been increasing evidence that the RAS is also involved in this atherogenic process, particularly in the inflammatory process within the atherosclerotic plaques. Ang II formed by ACE has pronounced atherogenic and proinflammatory properties and promotes the risk of plaque rupture.

In the regions of the plaques where macrophages and foam cells accumulate, there is also an elevated concentration of tissue ACE which produces increased amounts of Ang II. Tissue ACE is now recognized as a key factor in cardiovascular diseases (Dzau et al. 2001). The involvement of the local RAS and its close association with the formation of inflammation mediators and interleukins point to a link between chronic inflammation and instability of the plaque.

Inhibition of the activity of tissue ACE and the local formation of proinflammatory Ang II has been shown to be a promising therapeutic approach to stabilizing the plaque and thus protecting it from rupturing, thereby avoiding the subsequent acute clinical events.

Preclinical studies showed that ACE inhibitors not only have blood pressure-lowering properties but also vasoprotective effects through the endothelial formation of NO and PGI_2 (Linz et al. 1999b; Wiemer et al. 1991 1994). This also included the prevention of coronary endothelial dysfunction (Martorana et al. 1999). At an early stage, studies and results of this kind suggested that there might be an effect on diet-induced experimental atherogenesis. Rabbits fed an atherogenic diet show a loss of endothelial function (and particularly of endothelium-dependent relaxation) and a marked reduction in cGMP, a biochemical marker of NO release. Chronic administration of the ACE inhibitor ramipril with the diet maintains endothelial function and vascular cGMP content (Becker et al. 1991). In a model for hereditary hyperlipidaemia, the Watanabe rabbit, trandolapril can markedly reduce the atherosclerotic involvement of the internal surface of the total aorta (Chobanian et al. 1992). Similarly, in the atherosclerosis model of the transgenic apolipoprotein-E-deficient mouse, ramipril produced a remarkable reduction in the atherosclerotic involvement of the internal surface of the aorta, together with antioxidant effects (Keidar et al. 2000). In rabbits, atherosclerotic changes can be induced in the femoral arteries through the combination of local endothelial damage and atherogenic diet. Quinapril prevented the arterial activation of NF-κB, expression of the monocyte chemoattractant protein-1 (MCP1) and the infiltration of macrophages. Ang II itself is capable of activating NF-κB and inducing the expression of MCP1, thereby contributing to the infiltration of macrophages (Hernandez-Presa et al. 1997). In cuff-induced thickening of the femoral neointima in mice, an inflammatory vascular lesion with increased ACE expression and up-regulation of the Ang II receptors, the administration of perindopril leads to a reduction in neointimal thickening (Akishita et al. 2000, 2001).

All these experimental observations with the most varied atherogenesis models made it almost certain that treatment with a suitable ACE inhibitor should be capable of favourably influencing the prognosis of cardiovascular events in patients with a high cardiovascular risk. The molecular biology and experimental pharmacology of ACE inhibitors suggested that they have a directly antiatherosclerotic and vasoprotective action quite separate from the lowering of blood pressure (Dzau et al. 2001).

Suppression of the increased ACE activity within the plaque can lead to the stabilization and deactivation of the plaque by reducing inflammation in the vascular wall, thus lessening the risk of rupture and thrombosis and the resultant acute clinical cardiovascular events. The remarkable improvement in the long-term prognosis of atherosclerotic patients with increased cardiovascular risk might be the clinical result of the contribution made by ACE inhibition in the vascular wall.

3.8 Nephroprotection

In experimental models of renal failure such as rats subjected to 5/6 nephrectomy (Anderson et al. 1985), streptozotocin-induced diabetes mellitus (Zatz et al. 1986) or immunologically induced glomerular damage (Hostetter et al. 1982), glomerular hydraulic pressure increases together with the occurrence of proteinuria and glomerulosclerosis. These changes appear to be partly attributable to intrarenal actions of Ang II (Hostetter et al. 1982) and usually go along with elevated systemic blood pressure. Treatment with ACE inhibitors lowered the intraglomerular pressure and consequently reduced albuminuria and retarded the progressive loss of renal function (Anderson et al. 1985; Meggs et al. 1988; Lafayette et al. 1992). This unique ACE inhibitor action has been related to specific renal actions, in particular to postglomerular vasodilatation. Most recently, the contribution of the haemodynamic and nonhaemodynamic effects of Ang II to the pathogenesis of progressive renal injury was reviewed (Taal and Brenner 2000). Besides elevating the glomerular hydraulic pressure, Ang II is involved in the accumulation of extracellular matrix. The later effect is in part the result of an enhanced synthesis and decreased degradation of extracellular matrix due to the Ang II-induced increase in the expression of profibrotic cytokine TGFβ and of plasminogen activator inhibitor-1 (PAI-1) (Taal and Brenner 2000). In addition, Ang II may be involved in inflammatory processes associated with chronic renal injury (Taal and Brenner 2000). ACE inhibitor treatment prevented the up-regulation of TGFβ in the remnant kidney model (Hamaguchi et al. 1997; Junaid et al. 1997; Wu et al. 1997). TGFβ and PAI-1 gene expression in the renal cortex was also reduced in rats with Thy-1.1-induced chronic glomerulonephritis after treatment with lisinopril (Shinosaki et al. 2002). This effect of the ACE inhibitor was associated with a decrease in proteinuria and a suppression of the progression of glomerulosclerosis and tubulointerstitial fibrosis (Shinosaki et al. 2002). Antiproteinuric effects of ACE inhibitors and prevention of glomerulosclerosis have been demonstrated in several experimental models of renal disease including 5/6 nephrectomy (Abbate et al. 1999; Flores et al. 1998; Hamaguchi et al. 1997; Mackie et al. 2002; Yoshida et al. 2002), passive Heymann nephritis (Zoja et al. 1998a), adriamycin nephrosis (Wapstra et al. 2001), Dahl-sensitive rats with spontaneous nephrosis (Yoneda et al. 1998) and PAN-induced chronic focal and segmental glomerulosclerosis (Suyama et al. 2000). In MWF rats, a genetic model of progressive proteinuria and renal injury, ACE inhibition progressively reduced proteinuria and prevented the progression of renal damage (Remuzzi et al. 1999). Further effects of ACE inhibitor treatment which may contribute to their renoprotective actions include an increase in the activity of the antioxidant enzymes superoxide dismutase and glutathione peroxidase in kidney homogenates (Verbeelen et al. 1998), prevention of the increase in ET-1 expression in kidney tissue (Ruiz-Ortega et al. 1997; Zoja et al. 1998b), and inhibition of interstitial macrophage infiltration (Abbate et al. 1999; Suyama et al. 2000; Uhlenius et al. 1999). ACE inhibitors were also effective in decreasing the age-related accumulation of extracellular matrix (collagen types I and IV and fibronectin) in the rat kidney (Cruz et al. 2000).

Long-term treatment with ACE inhibitors protected against hypertension-induced renal damage and prolonged survival in SHRSPs (Linz et al. 1999a). Chronic ACE inhibition also reduced mortality in rats subjected to 5/6 nephrectomy (Flores et al. 1998), Heymann nephritis rats chronically treated with L-NAME (Uhlenius et al. 1999) and transgenic mice (T26) resembling human HIV-associated nephropathy (Bird et al. 1998).

3.9 Diabetes Mellitus

Several experimental models of diabetes mellitus are available to study the effect of ACE inhibition in diabetic animals. In most experimental studies, diabetes was induced in rats and mice by treatment with streptozotocin.

Insulin resistance, reflected by a defective insulin action on lipid and glucose metabolism, is an important risk factor for coronary heart disease. Hypertension and diabetes mellitus are commonly associated with hyperinsulinaemia and insulin resistance. Thus, the effectiveness of antihypertensive treatment with regard to prevention of coronary heart disease will not only be determined by lowering high blood pressure, but also by improving insulin resistance.

In experimental studies in diabetic animals, plasma insulin concentration was shown to be either increased, decreased or unchanged following ACE inhibitor treatment (Duarte et al. 1999; Kim et al. 1997; Shiuchi et al. 2002). Most importantly, insulin resistance in diabetic animals was improved by ACE inhibition, as demonstrated by increased glucose uptake in diaphragms and skeletal muscle (Duarte et al. 1999; Shiuchi et al. 2002). Furthermore, ACE inhibition improved insulin sensitivity, as shown by a decrease in the time to produce a 50% fall in initial blood sugar levels (Mehta et al. 1999). Several studies using different models of experimental diabetes showed a decrease in blood glucose after ACE inhibitor treatment (Duarte et al. 1999; Ittner et al. 2000; Rosenthal et al. 1997; Shiuchi et al. 2002), while in other studies blood glucose remained unchanged (Kim et al. 1997; Reddi et al. 2000). In one study in the type 2 diabetic mouse KK-Ay, the decrease in plasma glucose and insulin and the improvement of insulin resistance was associated by an ACE inhibitor-induced enhanced translocation of glucose transporter 4 (GLUT4) to the plasma membrane (Shiuchi et al. 2002).

Microalbuminuria is an early marker of diabetic nephropathy. ACE inhibition reduced albuminuria and microalbuminuria in streptozotocin diabetic rats (de Cavanagh et al. 2001; Hill et al. 2001; Mifsud et al. 2002; Onozato et al. 2002; Osicka et al. 2000; Yavuz et al. 1999) and other animal models of diabetes (Fabris et al. 2001; Reddi et al. 2000). In some studies, the ACE inhibitor effect on albuminuria was independent of blood pressure reduction (Fabris et al. 2001; Onozato et al. 2002). Experimental diabetic nephropathy is associated with glomerulosclerosis as well as tubulointerstitial fibrosis. One of the key mediators which is implicated in the accumulation of extracellular matrix in diabetes appears to be TGF-β. Treatment of diabetic rats with ACE inhibitors not only prevented the development of glomerulosclerosis and tubulointerstitial lesions (Fabris et al. 1999; Gilbert et al. 1998; Mifsud et al. 2002; Reddi et al. 2000; Velasquez et al. 1997), but also prevented the up-regulation of TGF-β gene expression as well as glomerular TGF-β receptor expression in streptozotocin-treated rats (Gilbert et al. 1998; Hill et al. 2001; Kelly et al. 2002). TGF-β gene expression was also reduced in mesenteric arteries and the heart of ACE inhibitor-treated diabetic rats (Kim et al. 1997; Rumble et al. 1998). Besides increased synthesis of extracellular matrix components, a decreased activity of matrix metalloproteinases (MMP), which are responsible for matrix degradation, can contribute to glomerular and tubulointerstitial fibrosis in diabetes. ACE inhibitor treatment resulted in an attenuation of the diabetes-associated changes in MMP, such as decreased expression and activity of MMP9 (McLennan et al. 2002). Furthermore, the diabetes-induced increased expression of the tissue inhibitor of metalloproteinase was attenuated by ACE inhibition (McLennan et al. 2002). Therefore, ACE inhibitor treatment can modulate diabetes-associated accumulation of extracellular matrix by both, decreased synthesis and increased degradation.

In genetically obese diabetic mice, ACE inhibition blocked the development in coronary perivascular fibrosis (Zaman et al. 2001). In addition, ACE inhibitor treatment attenuated the increase in PAI-1 activity in blood and the increased cardiac expression of PAI-1 and tissue factor (Zaman et al. 2001). Therefore, ACE inhibition may influence microvascular remodelling and fibrinolytic activity in diabetic rats. In addition, in diabetic apolipoprotein E-deficient mice, perindopril treatment inhibited the development of atherosclerotic lesions and diabetes-induced ACE, connective tissue growth factor, and vascular cell adhesion molecule-1 overexpression in the aorta (Candido et al. 2002).

A further parameter which is implicated in the pathogenesis of diabetes mellitus is oxidative stress. Increased free radical reactions and oxidant tissue damage were associated with diabetes. In diabetic rats, treatment with enalapril improved the tissue oxidant/antioxidant status by increasing total glutathione in kidney and liver and by decreasing lipid oxidation and oxidation of glutathione and proteins in heart and kidney (de Cavanagh et al. 2001). The increased renal expression of the NADPH oxidase component p47phox but not gp91phox in diabetic rats was prevented by ACE inhibition (Forbes et al. 2002; Onozato et al. 2002). Furthermore, nitrotryosine, a marker of protein oxidation, was found to be increased in kidneys of diabetic rats and was attenuated in ACE inhibitor-treated animals.

In renal failure as well as in diabetes, levels of advanced glycation end products (AGE) are elevated and contribute to the progression of the disease. Furthermore, AGEs increase oxidative stress and induce the expression of TGF-β (Oldfield et al. 2001; Vlassara and Palace 2002). In streptozotocin-induced diabetes treatment with ramipril attenuated the accumulation of AGEs in the kidney to a similar degree as the AGE formation inhibitor aminoguanidine (Forbes et al. 2002).

3.10 Cognitive Enhancement

The role of angiotensin II in cognition and behaviour has been extensively reviewed recently (Gard 2002). Preclinical investigations have suggested that ACE inhibitors may improve cognition in animals. In these studies, ACE inhibitors increased basal performance in young and old animals and reversed the impairment in cognitive performance caused by the muscarinic antagonist scopolamine (Costall et al. 1989). The precise mechanism of action is still unknown. ACE has been localized in different regions of the rat brain and is involved in the metabolism of several neuropeptides such as Ang II, substance P and enkephalins, which may all participate in learning and memory processes. Therefore, ACE inhibition in the brain may lead to improved cognitive function via blockade of substance P degradation or inhibition of Ang II generation. Ang II has been shown to impair performance in learning and memory paradigms in animals (Barnes et al. 1989), possibly by interaction with the cholinergic system. This is supported by reports showing that Ang II can inhibit the release of ^{3}H-acetylcholine in vitro in slices of the rat entorhinal cortex (Barnes et al. 1989; Wiemer et al. 1989). Furthermore, ACE inhibitor treatment has been shown to enhance acetylcholine metabolism and to increase acetylcholine release in rat brain in vivo (Wiemer et al. 1989). Whether or not ACE inhibitors have to penetrate the blood–brain barrier to exert their central actions is currently not clear. If access to brain regions inside the blood–brain barrier is a prerequisite more lipophilic ACE inhibitors with a higher propensity to penetrate the blood–brain barrier such as ramipril (Gohlke et al. 1989) should be more efficient than more hydrophilic ACE inhibitors. Theoretically, ACE inhibitors can act on brain areas known to lack a blood–brain barrier such as the circumventricular organs to modify the activity of neurons involved in the control of mental processes. In this case, the lipophilic property of an ACE inhibitor will be less important for the central ACE inhibitor effect on cognition.

While the enhancing potencies of ACE inhibitors on cognitive functions have been assessed for the most part in animal studies, investigations in humans are scarce, and the results do not unequivocally support the assumption of a memory-improving capacity of ACE inhibitors (Frcka and Lader 1988; Lichter et al. 1986). Clearly more studies are necessary to establish a possible beneficial action of ACE inhibitors in the treatment of cognitive dysfunctions of various aetiologies such as Alzheimer's disease or senile dementia.

In summary, several cardiovascular risk factors such as hypertension, LVH and remodelling, endothelial dysfunction, atherosclerosis or insulin resistance were positively influenced by ACE inhibitor treatment. These effects of ACE inhibitors help to explain their clinical benefit in congestive heart failure, postmyocardial infarction, chronic renal insufficiency, atherosclerotic diseases and diabetes mellitus.

References

Abbate M, Zoja C, Rottoli D, Corna D, Perico N, Bertani T, Remuzzi G (1999) Antiproteinuric therapy while preventing the abnormal protein traffic in proximal tubule abrogates protein- and complement-dependent interstitial inflammation in experimental renal disease. J Am Soc Nephrol 10:804–13

Akishita M, Horiuchi M, Yamada H, Zhang L, Shirakami G, Tamura K, Ouchi Y, Dzau VJ (2000) Inflammation influences vascular remodeling through AT_2 receptor expression and signaling. Physiol Genomics 2:13–20

Akishita M, Shirakami G, Iwai M, Wu L, Aoki M, Zhang L, Toba K, Horiuchi M (2001) Angiotensin converting enzyme inhibitor restrains inflammation-induced vascular injury in mice. J Hypertens 19:1083–8

Anderson S, Meyer TW, Rennke HG, Brenner BM (1985) Control of glomerular hypertension limits glomerular injury in rats with reduced renal mass. J Clin Invest 76:612–619

Baillard C, Mansier P, Ennezat PV, Mangin L, Medigue C, Swynghedauw B, Chevalier B (2000) Converting enzyme inhibition normalizes QT interval in spontaneously hypertensive rats. Hypertension 36:350–354

Bao G, Gohlke P, Qadri F, Unger T (1992) Chronic kinin receptor blockade attenuates the antihypertensive effect of ramipril. Hypertension 20:74–79

Barnes JM, Barnes NM, Costall B, Horovitz ZP, Naylor RJ (1989) Angiotensin II inhibits the release of [3H]acetylcholine from rat entorhinal cortex in vitro. Brain Res 491:136–143

Becker RH, Wiemer G, Linz W (1991) Preservation of endothelial function by ramipril in rabbits on a long-term atherogenic diet. J Cardiovasc Pharmacol 18 [Suppl 2]:S110–S115

Berkenboom G, Langer I, Carpentier Y, Grosfils K, Fontaine J (1997) Ramipril prevents endothelial dysfunction induced by oxidized low-density lipoproteins: a bradykinin-dependent mechanism. Hypertension 30:371–376

Bernstein KE, Martin BM, Edwards AS, Bernstein EA (1989) Mouse angiotensin-converting enzyme is a protein composed of two homologous domains. J Biol Chem 264:11945–11951

Bird JE, Durham SK, Giancarli MR, Gitlitz PH, Pandya DG, Dambach DM, Mozes MM, Kopp JB (1998) Captopril prevents nephropathy in HIV-transgenic mice. J Am Soc Nephrol 9:1441–1447

Bishop JE, Kiernan LA, Montgomery HE, Gohlke P, McEwan JR (2000) Raised blood pressure, not renin-angiotensin systems, causes cardiac fibrosis in TGR m(Ren2)27 rats. Cardiovasc Res 47:57–67

Boateng SY, Naqvi RU, Koban MU, Yacoub MH, MacLeod KT, Boheler KR (2001) Low-dose ramipril treatment improves relaxation and calcium cycling after established cardiac hypertrophy. Am J Physiol Heart Circ Physiol 280:H1029–H1038

Böhm M, Zolk O, Flesch M, Schiffer F, Schnabel P, Stasch JP, Knorr A (1998) Effects of angiotensin II type 1 receptor blockade and angiotensin-converting enzyme inhibition on cardiac beta-adrenergic signal transduction. Hypertension 31:747–754

Bondjers G, Glukhova M, Hansson GK, Postnov YV, Reidy MA, Schwartz SM (1991) Hypertension and atherosclerosis. Cause and effect, or two effects with one unknown cause? Circulation 84:VI2–VI16

Brooks WW, Bing OH, Conrad CH, O'Neill L, Crow MT, Lakatta EG, Dostal DE, Baker KM, Boluyt MO (1997) Captopril modifies gene expression in hypertrophied and failing hearts of aged spontaneously hypertensive rats. Hypertension 30:1362–1368

Brooks WW, Bing OH, Boluyt MO, Malhotra A, Morgan JP, Satoh N, Colucci WS, Conrad CH (2000) Altered inotropic responsiveness and gene expression of hypertrophied myocardium with captopril. Hypertension 35:1203–1209

Brown L, Duce B, Miric G, Sernia C (1999) Reversal of cardiac fibrosis in deoxycorticosterone acetate-salt hypertensive rats by inhibition of the renin-angiotensin system. J Am Soc Nephrol 10 [Suppl 11]:S143–S148

Brown NJ, Vaughan DE (1998) Angiotensin-converting enzyme inhibitors. Circulation 97:1411–1420

Buikema H, Monnink SH, Tio RA, Crijns HJ, de Zeeuw D, van Gilst WH (2000) Comparison of zofenopril and lisinopril to study the role of the sulfhydryl-group in improvement of endothelial dysfunction with ACE-inhibitors in experimental heart failure. Br J Pharmacol 130:1999–2007

Bünning P (1987) Kinetic properties of the angiotensin converting enzyme inhibitor ramiprilat. J Cardiovasc Pharmacol 10 [Suppl 7]:S31–S35

Bünning P, Riordan JF (1985) The functional role of zinc in angiotensin converting enzyme: implications for the enzyme mechanism. J Inorg Biochem 24:183–198

Cameron NE, Cotter MA, Robertson S (1992) Angiotensin converting enzyme inhibition prevents development of muscle and nerve dysfunction and stimulates angiogenesis in streptozotocin-diabetic rats. Diabetologia 35:12–18

Candido R, Jandeleit-Dahm KA, Cao Z, Nesteroff SP, Burns WC, Twigg SM, Dilley RJ, Cooper ME, Allen TJ (2002) Prevention of accelerated atherosclerosis by angiotensin-converting enzyme inhibition in diabetic apolipoprotein E-deficient mice. Circulation 106:246–253

Carraway JW, Park S, McCune SA, Holycross BJ, Radin MJ (1999) Comparison of irbesartan with captopril effects on cardiac hypertrophy and gene expression in heart failure-prone male SHHF/Mcc-fa(cp) rats. J Cardiovasc Pharmacol 33:451–460

Chobanian AV, Haudenschild CC, Nickerson C, Hope S (1992) Trandolapril inhibits atherosclerosis in the Watanabe heritable hyperlipidemic rabbit. Hypertension 20:473–477

Chopra M, McMurray J, Stewart J, Dargie HJ, Smith WE (1990) Free radical scavenging: a potentially beneficial action of thiol-containing angiotensin converting enzyme inhibitors. Biochem Soc Trans 18:1184–1185

Cohen ML, Kurz KD, Schenck KW (1983) Tissue angiotensin converting enzyme inhibition as an index of the disposition of enalapril (MK-421) and metabolite MK-422. J Pharmacol Exp Ther 226:192–196

Costall B, Coughlan J, Horovitz ZP, Kelly ME, Naylor RJ, Tomkins DM (1989) The effects of ACE inhibitors captopril and SQ29,852 in rodent tests of cognition. Pharmacol Biochem Behav 33:573–579

Cruz CI, Ruiz-Torres P, del Moral RG, Rodriguez-Puyol M, Rodriguez-Puyol D (2000) Age-related progressive renal fibrosis in rats and its prevention with ACE inhibitors and taurine. Am J Physiol Renal Physiol 278:F122–F129

Cumin F, Vellaud V, Corvol P, Alhenc-Gelas F (1989) Evidence for a single active site in the human angiotensin I-converting enzyme from inhibitor binding studies with [3H] RU 44 403: role of chloride. Biochem Biophys Res Commun 163:718–725

Cushman DW, Cheung HS, Sabo EF, Ondetti MA (1977) Design of potent competitive inhibitors of angiotensin-converting enzyme. Carboxyalkanoyl and mercaptoalkanoyl amino acids. Biochemistry 16:5484–5491

Dalton GR, Jones JV, Levi AJ, Levy A (2000) Changes in contractile protein gene expression with ageing and with captopril-induced regression of hypertrophy in the spontaneously hypertensive rats. J Hypertens 18:1297–306

De Cavanagh EM, Inserra F, Toblli J, Stella I, Fraga CG, Ferder L (2001) Enalapril attenuates oxidative stress in diabetic rats. Hypertension 38:1130–1136

De Gennaro Colonna V, Rossoni G, Rigamonti A, Bonomo S, Manfredi B, Berti F, Muller E (2002) Enalapril and quinapril improve endothelial vasodilator function and aortic eNOS gene expression in L-NAME-treated rats. Eur J Pharmacol 450:61–66

Demeilliers B, Jover B, Mimran A (1998) Contrasting renal effects of chronic administrations of enalapril and losartan on one-kidney, one clip hypertensive rats. J Hypertens 16:1023–1029

Driss AB, Himbert C, Poitevin P, Duriez M, Michel JB, Levy BI (1999) Enalapril improves arterial elastic properties in rats with myocardial infarction. J Cardiovasc Pharmacol 34:102–107

Drummer OH, Jarrott B (1986) The disposition and metabolism of captopril. Med Res Rev 6:75–97

Duarte J, Martinez A, Bermejo A, Vera B, Gamez MJ, Cabo P, Zarzuelo A (1999) Cardiovascular effects of captopril and enalapril in obese Zucker rats. Eur J Pharmacol 365:225–232

Duchin KL, Singhvi SM, Willard DA, Migdalof BH, McKinstry DN (1982) Captopril kinetics. Clin Pharmacol Ther 31:452–458

Dzau VJ, Bernstein K, Celermajer D, Cohen J, Dahlof B, Deanfield J, Diez J, Drexler H, Ferrari R, van Gilst W, Hansson L, Hornig B, Husain A, Johnston C, Lazar H, Lonn E, Luscher T, Mancini J, Mimran A, Pepine C, Rabelink T, Remme W, Ruilope L, Ruzicka M, Schunkert H, Swedberg K, Unger T, Vaughan D, Weber M (2001) The relevance of tissue angiotensin-converting enzyme: manifestations in mechanistic and endpoint data. Am J Cardiol 88:1L–20L

Ehlers MR, Riordan JF (1989) Angiotensin-converting enzyme: new concepts concerning its biological role. Biochemistry 28:5311–5318

Ehlers MR, Riordan JF (1991) Angiotensin-converting enzyme: zinc- and inhibitor-binding stoichiometries of the somatic and testis isozymes. Biochemistry 30:7118–7126

Ennis IL, Alvarez BV, Camilion de Hurtado MC, Cingolani HE (1998) Enalapril induces regression of cardiac hypertrophy and normalization of pHi regulatory mechanisms. Hypertension 31:961–967

Erdös EG (1975) Angiotensin I converting enzyme. Circ Res 36:247–255

Erdös EG, Skidgel RA (1987) The angiotensin I-converting enzyme. Lab Invest 56:345–348

Fabre JE, Rivard A, Magner M, Silver M, Isner JM (1999) Tissue inhibition of angiotensin-converting enzyme activity stimulates angiogenesis in vivo. Circulation 99:3043–3049

Fabris B, Candido R, Armini L, Fischetti F, Calci M, Bardelli M, Fazio M, Campanacci L, Carretta R (1999) Control of glomerular hyperfiltration and renal hypertrophy by an angiotensin converting enzyme inhibitor prevents the progression of renal damage in hypertensive diabetic rats. J Hypertens 17:1925–1931

Fabris B, Candido R, Carraro M, Fior F, Artero M, Zennaro C, Cattin MR, Fiorotto A, Bortoletto M, Millevoi C, Bardelli M, Faccini L, Carretta R (2001) Modulation of incipient glomerular lesions in experimental diabetic nephropathy by hypotensive and subhypotensive dosages of an ACE inhibitor. Diabetes 50:2619–2624

Ferreira SH, Bartelt DC, Greene LJ (1970) Isolation of bradykinin-potentiating peptides from Bothrops jararaca venom. Biochemistry 9:2583–2593

Flesch M, Schiffer F, Zolk O, Pinto Y, Stasch JP, Knorr A, Ettelbruck S, Böhm M (1997) Angiotensin receptor antagonism and angiotensin converting enzyme inhibition improve diastolic dysfunction and Ca(2+)-ATPase expression in the sarcoplasmic reticulum in hypertensive cardiomyopathy. J Hypertens 15:1001–1009

Flores O, Arevalo M, Gallego B, Hidalgo F, Vidal S, Lopez-Novoa JM (1998) Beneficial effect of the long-term treatment with the combination of an ACE inhibitor and a calcium channel blocker on renal injury in rats with 5/6 nephrectomy. Exp Nephrol 6:39–49

Fontaine D, Fontaine J, Dupont I, Dessy C, Piech A, Carpentier Y, Berkenboom G (2002) Chronic hydroxymethylglutaryl coenzyme a reductase inhibition and endothelial function of the normocholesterolemic rat: comparison with angiotensin-converting enzyme inhibition. J Cardiovasc Pharmacol 40:172–180

Forbes JM, Cooper ME, Thallas V, Burns WC, Thomas MC, Brammar GC, Lee F, Grant SL, Burrell LA, Jerums G, Osicka TM (2002) Reduction of the accumulation of advanced glycation end products by ACE inhibition in experimental diabetic nephropathy. Diabetes 51:3274–3282

Frcka G, Lader M (1988) Psychotropic effects of repeated doses of enalapril, propranolol and atenolol in normal subjects. Br J Clin Pharmacol 25:67–73

Fujii M, Wada A, Tsutamoto T, Ohnishi M, Isono T, Kinoshita M (2002) Bradykinin improves left ventricular diastolic function under long-term angiotensin-converting enzyme inhibition in heart failure. Hypertension 39:952–957

Gard PR (2002) The role of angiotensin II in cognition and behaviour. Eur J Pharmacol 438:1–14

Gavras H, Brunner HR, Laragh JH, Sealey JE, Gavras I, Vukovich RA (1974) An angiotensin converting-enzyme inhibitor to identify and treat vasoconstrictor and volume factors in hypertensive patients. N Engl J Med 291:817–821

Gilbert RE, Cox A, Wu LL, Allen TJ, Hulthen UL, Jerums G, Cooper ME (1998) Expression of transforming growth factor-beta1 and type IV collagen in the renal tubulointerstitium in experimental diabetes: effects of ACE inhibition. Diabetes 47:414–422

Gillies LK, Werstiuk ES, Lee RM (1998) Cross-over study comparing effects of treatment with an angiotensin converting enzyme inhibitor and an angiotensin II type 1 receptor antagonist on cardiovascular changes in hypertension. J Hypertens 16:477–486

Gohlke P, Urbach H, Schölkens B, Unger T (1989) Inhibition of converting enzyme in the cerebrospinal fluid of rats after oral treatment with converting enzyme inhibitors. J Pharmacol Exp Ther 249:609–616

Gohlke P, Bünning P, Unger T (1992) Distribution and metabolism of angiotensin I and II in the blood vessel wall. Hypertension 20:151–157

Gohlke P, Lamberty V, Kuwer I, Bartenbach S, Schnell A, Linz W, Schölkens BA, Wiemer G, Unger T (1993) Long-term low-dose angiotensin converting enzyme inhibitor treatment increases vascular cyclic guanosine 3',5'-monophosphate. Hypertension 22:682–687

Gohlke P, Linz W, Schölkens BA, Kuwer I, Bartenbach S, Schnell A, Unger T (1994) Angiotensin-converting enzyme inhibition improves cardiac function. Role of bradykinin. Hypertension 23:411–418

Gohlke P, Kuwer I, Schnell A, Amann K, Mall G, Unger T (1997) Blockade of bradykinin B2 receptors prevents the increase in capillary density induced by chronic angiotensin-converting enzyme inhibitor treatment in stroke-prone spontaneously hypertensive rats. Hypertension 29:478–482

Goto K, Fujii K, Onaka U, Abe I, Fujishima M (2000a) Angiotensin-converting enzyme inhibitor prevents age-related endothelial dysfunction. Hypertension 36:581–587

Goto K, Fujii K, Onaka U, Abe I, Fujishima M (2000b) Renin-angiotensin system blockade improves endothelial dysfunction in hypertension. Hypertension 36:575–580

Goussev A, Sharov VG, Shimoyama H, Tanimura M, Lesch M, Goldstein S, Sabbah HN (1998) Effects of ACE inhibition on cardiomyocyte apoptosis in dogs with heart failure. Am J Physiol 275:H626–H631

Hamaguchi A, Kim S, Wanibuchi H, Iwao H (1997) Imidapril inhibits increased transforming growth factor-beta1 expression in remnant kidney model. Eur J Pharmacol 331:27–30

Hernandez A, Barberi L, Ballerio R, Testini A, Ferioli R, Bolla M, Natali M, Folco G, Catapano AL (1998) Delapril slows the progression of atherosclerosis and maintains endothelial function in cholesterol-fed rabbits. Atherosclerosis 137:71–76

Hernandez-Presa M, Bustos C, Ortego M, Tunon J, Renedo G, Ruiz-Ortega M, Egido J (1997) Angiotensin-converting enzyme inhibition prevents arterial nuclear factor-kappa B activation, monocyte chemoattractant protein-1 expression, and macrophage infiltration in a rabbit model of early accelerated atherosclerosis. Circulation 95:1532–1541

Hill C, Logan A, Smith C, Gronbaek H, Flyvbjerg A (2001) Angiotensin converting enzyme inhibitor suppresses glomerular transforming growth factor beta receptor expression in experimental diabetes in rats. Diabetologia 44:495–500

Horowitz JD, Antman EM, Lorell BH, Barry WH, Smith TW (1983) Potentiation of the cardiovascular effects of nitroglycerin by N-acetylcysteine. Circulation 68:1247–1253

Hostetter TH, Rennke HG, Brenner BM (1982) The case for intrarenal hypertension in the initiation and progression of diabetic and other glomerulopathies. Am J Med 72:375–380

Hu K, Gaudron P, Anders HJ, Weidemann F, Turschner O, Nahrendorf M, Ertl G (1998) Chronic effects of early started angiotensin converting enzyme inhibition and angiotensin AT_1-receptor subtype blockade in rats with myocardial infarction: role of bradykinin. Cardiovasc Res 39:401–412

Hügel S, Horn M, de Groot M, Remkes H, Dienesch C, Hu K, Ertl G, Neubauer S (1999) Effects of ACE inhibition and beta-receptor blockade on energy metabolism in rats postmyocardial infarction. Am J Physiol 277:H2167–H2175

Ikeda Y, Nakamura T, Takano H, Kimura H, Obata JE, Takeda S, Hata A, Shido K, Mochizuki S, Yoshida Y (2000) Angiotensin II-induced cardiomyocyte hypertrophy and cardiac fibrosis in stroke-prone spontaneously hypertensive rats. J Lab Clin Med 135:353–359

Ittner KP, Zimmermann M, Bucher M, Gessele W, Kees F, Kramer BK, Grobecker HF (2000) The effect of urapidil and ramipril on hyperglycemia in streptozotocin diabetic rats. Naunyn Schmiedebergs Arch Pharmacol 361:92–97

Jackson B, Cubela RB, Sakaguchi K, Johnston CI (1988) Characterization of angiotensin converting enzyme (ACE) in the testis and assessment of the in vivo effects of the ACE inhibitor perindopril. Endocrinology 123:50–55

Jin H, Yang R, Awad TA, Wang F, Li W, Williams SP, Ogasawara A, Shimada B, Williams PM, de Feo G, Paoni NF (2001) Effects of early angiotensin-converting enzyme inhibition on cardiac gene expression after acute myocardial infarction. Circulation 103:736–742

Junaid A, Rosenberg ME, Hostetter TH (1997) Interaction of angiotensin II and TGF-beta 1 in the rat remnant kidney. J Am Soc Nephrol 8:1732–1738

Kai T, Sugimura K, Shimada S, Kurooka A, Ishikawa K (1999) Renin-angiotensin system stimulates cardiac and renal disorders in Tsukuba hypertensive mice. Clin Exp Pharmacol Physiol 26:206–211

Kansui Y, Fujii K, Goto K, Abe I, Iida M (2002) Angiotensin II receptor antagonist improves age-related endothelial dysfunction. J Hypertens 20:439–446

Kawai H, Fan TH, Dong E, Siddiqui RA, Yatani A, Stevens SY, Liang CS (1999) ACE inhibition improves cardiac NE uptake and attenuates sympathetic nerve terminal abnormalities in heart failure. Am J Physiol 277:H1609–H1617

Keaton AK, White CR, Berecek KH (1998) Captopril treatment and its withdrawal prevents impairment of endothelium-dependent responses in the spontaneously hypertensive rat. Clin Exp Hypertens 20:847–866

Keen HL, Brands MW, Smith MJ Jr, Hall JE (1998) Maintenance of baseline angiotensin II potentiates insulin hypertension in rats. Hypertension 31:637–642

Keidar S, Attias J, Coleman R, Wirth K, Schölkens B, Hayek T (2000) Attenuation of atherosclerosis in apolipoprotein E-deficient mice by ramipril is dissociated from its antihypertensive effect and from potentiation of bradykinin. J Cardiovasc Pharmacol 35:64–72

Kelly DJ, Cox AJ, Tolcos M, Cooper ME, Wilkinson-Berka JL, Gilbert RE (2002) Attenuation of tubular apoptosis by blockade of the renin-angiotensin system in diabetic Ren-2 rats. Kidney Int 61:31–39

Kelly JG, O'Malley K (1990) Clinical pharmacokinetics of the newer ACE inhibitors. A review. Clin Pharmacokinet 19:177–196

Kim S, Wanibuchi H, Hamaguchi A, Miura K, Yamanaka S, Iwao H (1997) Angiotensin blockade improves cardiac and renal complications of type II diabetic rats. Hypertension 30:1054–1061

Kim S, Yoshiyama M, Izumi Y, Kawano H, Kimoto M, Zhan Y, Iwao H (2001) Effects of combination of ACE inhibitor and angiotensin receptor blocker on cardiac remodeling, cardiac function, and survival in rat heart failure. Circulation 103:148–154

Kobayashi N, Higashi T, Hara K, Shirataki H, Matsuoka H (1999) Effects of imidapril on NOS expression and myocardial remodelling in failing heart of Dahl salt-sensitive hypertensive rats. Cardiovasc Res 44:518–526

Kostis JB (1989) Angiotensin-converting enzyme inhibitors. Emerging differences and new compounds. Am J Hypertens 2:57–64

Krombach RS, Clair MJ, Hendrick JW, Houck WV, Zellner JL, Kribbs SB, Whitebread S, Mukherjee R, de Gasparo M, Spinale FG (1998) Angiotensin converting enzyme inhibition, AT1 receptor inhibition, and combination therapy with pacing induced heart failure: effects on left ventricular performance and regional blood flow patterns. Cardiovasc Res 38:631–645

Kukreja RC, Kontos HA, Hess ML (1990) Captopril and enalaprilat do not scavenge the superoxide anion. Am J Cardiol 65:24I-27I

Lafayette RA, Mayer G, Park SK, Meyer TW (1992) Angiotensin II receptor blockade limits glomerular injury in rats with reduced renal mass. J Clin Invest 90:766–771

Laflamme K, Oster L, Cardinal R, de Champlain J (1997) Effects of renin-angiotensin blockade on sympathetic reactivity and beta-adrenergic pathway in the spontaneously hypertensive rat. Hypertension 30:278–287

Lapointe N, Blais C Jr, Adam A, Parker T, Sirois MG, Gosselin H, Clement R, Rouleau JL (2002) Comparison of the effects of an angiotensin-converting enzyme inhibitor and a vasopeptidase inhibitor after myocardial infarction in the rat. J Am Coll Cardiol 39:1692–1698

Lassila M, Finckenberg P, Pere AK, Vapaatalo H, Nurminen ML (2000) Enalapril and valsartan improve cyclosporine A-induced vascular dysfunction in spontaneously hypertensive rats. Eur J Pharmacol 398:99–106

Leckie BJ (2001) The action of salt and captopril on blood pressure in mice with genetic hypertension. J Hypertens 19:1607–1613

Leonetti G, Cuspidi C (1995) Choosing the right ACE inhibitor. A guide to selection. Drugs 49:516–535

Levijoki J, Pollesello P, Kaheinen P, Haikala H (2001) Improved survival with simendan after experimental myocardial infarction in rats. Eur J Pharmacol 419:243–248

Li D, Shinagawa K, Pang L, Leung TK, Cardin S, Wang Z, Nattel S (2001) Effects of angiotensin-converting enzyme inhibition on the development of the atrial fibrillation substrate in dogs with ventricular tachypacing-induced congestive heart failure. Circulation 104:2608–2614

Li K, Chen X (1987) Protective effects of captopril and enalapril on myocardial ischemia and reperfusion damage of rat. J Mol Cell Cardiol 19:909–915

Lichter I, Richardson PJ, Wyke MA (1986) Differential effects of atenolol and enalapril on memory during treatment for essential hypertension. Br J Clin Pharmacol 21:641–645

Linz W, Jessen T, Becker RH, Schölkens BA, Wiemer G (1997) Long-term ACE inhibition doubles lifespan of hypertensive rats. Circulation 96:3164–3172

Linz W, Schölkens BA, Han YF (1986) Beneficial effects of the converting enzyme inhibitor, ramipril, in ischemic rat hearts. J Cardiovasc Pharmacol 8 [Suppl 10]:S91–S99

Linz W, Schölkens BA, Ganten D (1989) Converting enzyme inhibition specifically prevents the development and induces regression of cardiac hypertrophy in rats. Clin Exp Hypertens A 11:1325–50

Linz W, Wiemer G, Gohlke P, Unger T, Schölkens BA (1995) Contribution of kinins to the cardiovascular actions of angiotensin-converting enzyme inhibitors. Pharmacol Rev 47:25–49

Linz W, Wohlfart P, Schoelkens BA, Becker RH, Malinski T, Wiemer G (1999a) Late treatment with ramipril increases survival in old spontaneously hypertensive rats. Hypertension 34:291–295

Linz W, Wohlfart P, Schölkens BA, Malinski T, Wiemer G (1999b) Interactions among ACE, kinins and NO. Cardiovasc Res 43:549–561

Lopez-Hernandez FJ, Carron R, Montero MJ, Flores O, Lopez-Novoa JM, Arevalo MA (1999) Antihypertensive effect of trandolapril and verapamil in rats with induced hypertension. J Cardiovasc Pharmacol 33:748–755

Lusis AJ (2000) Atherosclerosis. Nature 407:233–241

Ma M, Watanabe K, Wahed MI, Inoue M, Sekiguchi T, Kouda T, Ohta Y, Nakazawa M, Yoshida Y, Yamamoto T, Hanawa H, Kodama M, Fuse K, Aizawa Y (2001) Inhibition of progression of heart failure and expression of TGF-beta 1 mRNA in rats with heart failure by the ACE inhibitor quinapril. J Cardiovasc Pharmacol 38 [Suppl 1]:S51–S54

Mackie FE, Meyer TW, Campbell DJ (2002) Effects of antihypertensive therapy on intrarenal angiotensin and bradykinin levels in experimental renal insufficiency. Kidney Int 61:555–563

Makino T, Hattori Y, Matsuda N, Onozuka H, Sakuma I, Kitabatake A (2003) Effects of angiotensin-converting enzyme inhibition and angiotensin II type 1 receptor blockade on beta-adrenoceptor signaling in heart failure produced by myocardial Infarction in rabbits: reversal of altered expression of beta-adrenoceptor kinase and G i alpha. J Pharmacol Exp Ther 304:370–379

Mannistö TK, Karvonen KE, Kerola TV, Ryhanen LJ (2001) Inhibitory effect of the angiotensin converting enzyme inhibitors captopril and enalapril on the conversion of procollagen to collagen. J Hypertens 19:1835–1839

Martorana PA, Ruetten H, Goebel B, Koehl D, Roegner B, Schoelkens BA, Keil M (1999) Ramiprilat prevents the development of acute coronary endothelial dysfunction in the dog. Basic Res Cardiol 94:238–245

Matsumoto K, Morishita R, Moriguchi A, Tomita N, Aoki M, Sakonjo H, Nakamura T, Higaki J, Ogihara T (2001) Inhibition of neointima by angiotensin-converting enzyme inhibitor in porcine coronary artery balloon-injury model. Hypertension 37:270–274

McLennan SV, Kelly DJ, Cox AJ, Cao Z, Lyons JG, Yue DK, Gilbert RE (2002) Decreased matrix degradation in diabetic nephropathy: effects of ACE inhibition on the expression and activities of matrix metalloproteinases. Diabetologia 45:268–275

Meggs LG, Garrick R, Chander P, Ben-Ari J, Gammon D, Goodman AI (1988) Amelioration of systemic hypertension by converting enzyme inhibition in the renal ablation model. Am J Hypertens 1:190–192

Mehta AA, Patel S, Santani DD, Goyal RK (1999) Effect of nifedipine and enalapril on insulin-induced glucose disposal in spontaneous hypertensive and diabetic rats. Clin Exp Hypertens 21:51–59

Mehta JL, Nicolini FA, Lawson DL (1990) Sulfhydryl group in angiotensin converting enzyme inhibitors and superoxide radical formation. J Cardiovasc Pharmacol 16:847–849

Metelitsa VI, Martsevich S, Kozyreva MP, Slastnikova ID (1992) Enhancement of the efficacy of isosorbide dinitrate by captopril in stable angina pectoris. Am J Cardiol 69:291–296

Mifsud SA, Skinner SL, Cooper ME, Kelly DJ, Wilkinson-Berka JL (2002) Effects of low-dose and early versus late perindopril treatment on the progression of severe diabetic nephropathy in (mREN-2)27 rats. J Am Soc Nephrol 13:684–692

Milavetz JJ, Raya TE, Johnson CS, Morkin E, Goldman S (1996) Survival after myocardial infarction in rats: captopril versus losartan. J Am Coll Cardiol 27:714–719

Moravski CJ, Kelly DJ, Cooper ME, Gilbert RE, Bertram JF, Shahinfar S, Skinner SL, Wilkinson-Berka JL (2000) Retinal neovascularization is prevented by blockade of the renin-angiotensin system. Hypertension 36:1099–1104

Mori T, Nishimura H, Okabe M, Ueyama M, Kubota J, Kawamura K (1998) Cardioprotective effects of quinapril after myocardial infarction in hypertensive rats. Eur J Pharmacol 348:229–234

Mulder P, Compagnon P, Devaux B, Richard V, Henry JP, Elfertak L, Wimart MC, Thibout E, Comoy E, Mace B, Thuillez C (1997a) Response of large and small vessels to alpha and beta adrenoceptor stimulation in heart failure: effect of angiotensin converting enzyme inhibition. Fundam Clin Pharmacol 11:221–230

Mulder P, Devaux B, Richard V, Henry JP, Wimart MC, Thibout E, Mace B, Thuillez C (1997b) Early versus delayed angiotensin-converting enzyme inhibition in experimental chronic heart failure. Effects on survival, hemodynamics, and cardiovascular remodeling. Circulation 95:1314–1319

Murakami K, Mizushige K, Noma T, Tsuji T, Kimura S, Kohno M (2002) Perindopril effect on uncoupling protein and energy metabolism in failing rat hearts. Hypertension 40:251–255

Namiuchi S, Kagaya Y, Chida M, Yamane Y, Takahashi C, Fukuchi M, Tezuka F, Watanabe J, Ido T, Shirato K (2000) Regional and temporal profiles of phorbol 12,13-dibutyrate binding after myocardial infarction in rats: effects of captopril treatment. J Cardiovasc Pharmacol 35:353–360

Nguyen T, El Salibi E, Rouleau JL (1998a) Postinfarction survival and inducibility of ventricular arrhythmias in the spontaneously hypertensive rat: effects of ramipril and hydralazine. Circulation 98:2074–2080

Nguyen T, El Salibi E, Rouleau JL (1998b) Reduced periinfarction mortality as a result of long-term therapy with captopril but not hydralazine or propranolol in spontaneously hypertensive rats. J Cardiovasc Pharmacol 32:884–895

Node K, Kitakaze M, Kosaka H, Minamino T, Mori H, Hori M (1998) Role of Ca2+-activated K+ channels in the protective effect of ACE inhibition against ischemic myocardial injury. Hypertension 31:1290–1298

Ogiku N, Sumikawa H, Minamide S, Ishida R (1993) Influence of imidapril on abnormal biochemical parameters in salt-loaded stroke-prone spontaneously hypertensive rats (SHRSP). Jpn J Pharmacol 61:69–73

Oldfield MD, Bach LA, Forbes JM, Nikolic-Paterson D, McRobert A, Thallas V, Atkins RC, Osicka T, Jerums G, Cooper ME (2001) Advanced glycation end products cause epithelial-myofibroblast transdifferentiation via the receptor for advanced glycation end products (RAGE). J Clin Invest 108:1853–1863

Ondetti MA (1988) Structural relationships of angiotensin converting-enzyme inhibitors to pharmacologic activity. Circulation 77:I74–I78

Ondetti MA, Williams NJ, Sabo EF, Pluscec J, Weaver ER, Kocy O (1971) Angiotensin-converting enzyme inhibitors from the venom of Bothrops jararaca. Isolation, elucidation of structure, and synthesis. Biochemistry 10:4033–4039

Ondetti MA, Rubin B, Cushman DW (1977) Design of specific inhibitors of angiotensin-converting enzyme: new class of orally active antihypertensive agents. Science 196:441–444

Onodera T, Okazaki F, Miyazaki H, Minami S, Ito T, Seki S, Taniguchi M, Taniguchi I, Mochizuki S (2002) Perindopril reverses myocyte remodeling in the hypertensive heart. Hypertens Res 25:85–90

Onozato ML, Tojo A, Goto A, Fujita T, Wilcox CS (2002) Oxidative stress and nitric oxide synthase in rat diabetic nephropathy: effects of ACEI and ARB. Kidney Int 61:186–194

Osicka TM, Yu Y, Panagiotopoulos S, Clavant SP, Kiriazis Z, Pike RN, Pratt LM, Russo LM, Kemp BE, Comper WD, Jerums G (2000) Prevention of albuminuria by aminoguanidine or ramipril in streptozotocin-induced diabetic rats is associated with the normalization of glomerular protein kinase C. Diabetes 49:87–93

Paull JR, Widdop RE (2001) Persistent cardiovascular effects of chronic renin-angiotensin system inhibition following withdrawal in adult spontaneously hypertensive rats. J Hypertens 19:1393–1402

Pereira LM, Mandarim-De-Lacerda CA (2001) The effect of enalapril and verapamil on the left ventricular hypertrophy and the left ventricular cardiomyocyte numerical density in rats submitted to nitric oxide inhibition. Int J Exp Pathol 82:115–122

Pieper GM, Siebeneich W (2000) Temocapril, an angiotensin converting enzyme inhibitor, protects against diabetes-induced endothelial dysfunction. Eur J Pharmacol 403:129–132

Podesser BK, Schirnhofer J, Bernecker OY, Kroner A, Franz M, Semsroth S, Fellner B, Neumüller J, Hallstrom S, Wolner E (2002) Optimizing ischemia/reperfusion in the failing rat heart—improved myocardial protection with acute ACE inhibition. Circulation 106:I277–I283

Pu Q, Touyz RM, Schiffrin EL (2002) Comparison of angiotensin-converting enzyme (ACE), neutral endopeptidase (NEP) and dual ACE/NEP inhibition on blood pressure and resistance arteries of deoxycorticosterone acetate-salt hypertensive rats. J Hypertens 20:899–907

Qi XL, Stewart DJ, Gosselin H, Azad A, Picard P, Andries L, Sys SU, Brutsaert DL, Rouleau JL (1999) Improvement of endocardial and vascular endothelial function on myocardial performance by captopril treatment in postinfarct rat hearts. Circulation 100:1338–1345

Quaschning T, d'Uscio LV, Shaw S, Lüscher TF (2001) Vasopeptidase inhibition exhibits endothelial protection in salt-induced hypertension. Hypertension 37:1108–1113

Reddi AS, Nimmagadda VR, Lefkowitz A, Kuo HR, Bollineni JS (2000) Effect of antihypertensive therapy on renal injury in type 2 diabetic rats with hypertension. Hypertension 36:233–238

Remuzzi A, Fassi A, Bertani T, Perico N, Remuzzi G (1999) ACE inhibition induces regression of proteinuria and halts progression of renal damage in a genetic model of progressive nephropathy. Am J Kidney Dis 34:626–632

Richer C, Gervais M, Fornes P, Giudicelli JF (1999) Combined selective angiotensin II AT1-receptor blockade and angiotensin I-converting enzyme inhibition on coronary flow reserve in postischemic heart failure in rats. J Cardiovasc Pharmacol 34:772–781

Rodrigo E, Maeso R, Munoz-Garcia R, Navarro-Cid J, Ruilope LM, Cachofeiro V, Lahera V (1997) Endothelial dysfunction in spontaneously hypertensive rats: consequences of chronic treatment with losartan or captopril. J Hypertens 15:613–618

Rosenthal T, Erlich Y, Rosenmann E, Cohen A (1997) Effects of enalapril, losartan, and verapamil on blood pressure and glucose metabolism in the Cohen-Rosenthal diabetic hypertensive rat. Hypertension 29:1260–1264

Rouleau JL, Kapuku G, Pelletier S, Gosselin H, Adam A, Gagnon C, Lambert C, Meloche S (2001) Cardioprotective effects of ramipril and losartan in right ventricular pressure overload in the rabbit: importance of kinins and influence on angiotensin II type 1 receptor signaling pathway. Circulation 104:939–944

Ruiz-Ortega M, Gomez-Garre D, Liu XH, Blanco J, Largo R, Egido J (1997) Quinapril decreases renal endothelin-1 expression and synthesis in a normotensive model of immune-complex nephritis. J Am Soc Nephrol 8:756–768

Rumble JR, Gilbert RE, Cox A, Wu L, Cooper ME (1998) Angiotensin converting enzyme inhibition reduces the expression of transforming growth factor-beta1 and type IV collagen in diabetic vasculopathy. J Hypertens 16:1603–1609

Sakaguchi K, Chai SY, Jackson B, Johnston CI, Mendelsohn FA (1988) Inhibition of tissue angiotensin converting enzyme. Quantitation by autoradiography. Hypertension 11:230–238

Salvetti A (1990) Newer ACE inhibitors. A look at the future. Drugs 40:800–828

Sanada S, Kitakaze M, Node K, Takashima S, Ogai A, Asanuma H, Sakata Y, Asakura M, Ogita H, Liao Y, Fukushima T, Yamada J, Minamino T, Kuzuya T, Hori M (2001) Differential subcellular actions of ACE inhibitors and AT(1) receptor antagonists on cardiac remodeling induced by chronic inhibition of NO synthesis in rats. Hypertension 38:404–411

Sanbe A, Tanonaka K, Kobayasi R, Takeo S (1995) Effects of long-term therapy with ACE inhibitors, captopril, enalapril and trandolapril, on myocardial energy metabolism in rats with heart failure following myocardial infarction. J Mol Cell Cardiol 27:2209–2222

Sandmann S, Yu M, Kaschina E, Blume A, Bouzinova E, Aalkjaer C, Unger T (2001a) Differential effects of angiotensin AT_1 and AT_2 receptors on the expression, translation and function of the Na+-H+ exchanger and Na+-HCO3- symporter in the rat heart after myocardial infarction. J Am Coll Cardiol 37:2154–2165

Sandmann S, Yu M, Unger T (2001b) Transcriptional and translational regulation of calpain in the rat heart after myocardial infarction—effects of AT(1) and AT(2) receptor antagonists and ACE inhibitor. Br J Pharmacol 132:767–777

Schoemaker RG, Debets JJ, Struyker-Boudier HA, Smits JF (1991) Delayed but not immediate captopril therapy improves cardiac function in conscious rats, following myocardial infarction. J Mol Cell Cardiol 23:187–197

Schölkens BA, Landgraf W (2002) ACE inhibition and atherogenesis. Can J Physiol Pharmacol 80:354–359

Sharifi AM, Li JS, Endemann D, Schiffrin EL (1998) Effects of enalapril and amlodipine on small-artery structure and composition, and on endothelial dysfunction in spontaneously hypertensive rats. J Hypertens 16:457–466

Shinosaki T, Miyai I, Nomura Y, Kobayashi T, Sunagawa N, Kurihara H (2002) Mechanisms underlying the ameliorative property of lisinopril in progressive mesangioproliferative nephritis. Nephron 91:719–729

Shiuchi T, Cui TX, Wu L, Nakagami H, Takeda-Matsubara Y, Iwai M, Horiuchi M (2002) ACE inhibitor improves insulin resistance in diabetic mouse via bradykinin and NO. Hypertension 40:329–334

Silvestre JS, Bergaya S, Tamarat R, Duriez M, Boulanger CM, Levy BI (2001) Proangiogenic effect of angiotensin-converting enzyme inhibition is mediated by the bradykinin B(2) receptor pathway. Circ Res 89:678–683

Song JC, White CM (2002) Clinical pharmacokinetics and selective pharmacodynamics of new angiotensin converting enzyme inhibitors: an update. Clin Pharmacokinet 41:207–224

Soubrier F, Alhenc-Gelas F, Hubert C, Allegrini J, John M, Tregear G, Corvol P (1988) Two putative active centers in human angiotensin I-converting enzyme revealed by molecular cloning. Proc Natl Acad Sci U S A 85:9386–9390

Soubrier F, Wei L, Hubert C, Clauser E, Alhenc-Gelas F, Corvol P (1993) Molecular biology of the angiotensin I converting enzyme: II. Structure-function. Gene polymorphism and clinical implications. J Hypertens 11:599–604

Sousa T, Morato M, Albino-Teixeira A (2002) Angiotensin converting enzyme inhibition prevents trophic and hypertensive effects of an antagonist of adenosine receptors. Eur J Pharmacol 441:99–104

Spinale FG, de Gasparo M, Whitebread S, Hebbar L, Clair MJ, Melton DM, Krombach RS, Mukherjee R, Iannini JP, O SJ (1997a) Modulation of the renin-angiotensin pathway through enzyme inhibition and specific receptor blockade in pacing-induced heart failure: I. Effects on left ventricular performance and neurohormonal systems. Circulation 96:2385–2396

Spinale FG, Mukherjee R, Iannini JP, Whitebread S, Hebbar L, Clair MJ, Melton DM, Cox MH, Thomas PB, de Gasparo M (1997b) Modulation of the renin-angiotensin pathway through enzyme inhibition and specific receptor blockade in pacing-induced heart failure: II. Effects on myocyte contractile processes. Circulation 96:2397–2406

Stauss HM, Zhu YC, Redlich T, Adamiak D, Mott A, Kregel KC, Unger T (1994) Angiotensin-converting enzyme inhibition in infarct-induced heart failure in rats: bradykinin versus angiotensin II. J Cardiovasc Risk 1:255–262

Stier CT Jr, Benter IF, Ahmad S, Zuo HL, Selig N, Roethel S, Levine S, Itskovitz HD (1989) Enalapril prevents stroke and kidney dysfunction in salt-loaded stroke-prone spontaneously hypertensive rats. Hypertension 13:115–121

Sugimoto K, Fujimura A, Takasaki I, Tokita Y, Iwamoto T, Takizawa T, Gotoh E, Shionoiri H, Ishii M (1998) Effects of renin-angiotensin system blockade and dietary salt intake on left ventricular hypertrophy in Dahl salt-sensitive rats. Hypertens Res 21:163–168

Suyama K, Muso E, Yashiro M, Sasayama S (2000) Significant suppressive effect of low-dose temocapril, an ACE inhibitor with biliary excretion, on FGS lesions in hypertensive rats. Nephron 86:491–498

Sweet CS, Gross DM, Arbegast PT, Gaul SL, Britt PM, Ludden CT, Weitz D, Stone CA (1981) Antihypertensive activity of N-[(S)-1-(ethoxycarbonyl)-3-phenylpropyl]-L-Ala-L-Pro (MK-421), an orally active converting enzyme inhibitor. J Pharmacol Exp Ther 216:558–566

Taal MW, Brenner BM (2000) Renoprotective benefits of RAS inhibition: from ACEI to angiotensin II antagonists. Kidney Int 57:1803–1817

Takeda M, Kagaya Y, Takahashi J, Sugie T, Ohta J, Watanabe J, Shirato K, Kondo H, Goto K (2001) Gene expression and in situ localization of diacylglycerol kinase isozymes in normal and infarcted rat hearts: effects of captopril treatment. Circ Res 89:265–272

Takeishi Y, Bhagwat A, Ball NA, Kirkpatrick DL, Periasamy M, Walsh RA (1999) Effect of angiotensin-converting enzyme inhibition on protein kinase C and SR proteins in heart failure. Am J Physiol 276:H53–H62

Tappia PS, Liu SY, Shatadal S, Takeda N, Dhalla NS, Panagia V (1999) Changes in sarcolemmal PLC isoenzymes in postinfarct congestive heart failure: partial correction by imidapril. Am J Physiol 277:H40–H49

Teisman AC, Pinto YM, Buikema H, Flesch M, Böhm M, Paul M, van Gilst WH (1998) Dissociation of blood pressure reduction from end-organ damage in TGR(mREN2)27 transgenic hypertensive rats. J Hypertens 16:1759–1765

Thind GS (1990) Angiotensin converting enzyme inhibitors: comparative structure, pharmacokinetics, and pharmacodynamics. Cardiovasc Drugs Ther 4:199–206

Todd PA, Heel RC (1986) Enalapril. A review of its pharmacodynamic and pharmacokinetic properties, and therapeutic use in hypertension and congestive heart failure. Drugs 31:198–248

Toyoshima H, Nasa Y, Kohsaka Y, Isayama Y, Yamaguchi F, Sanbe A, Takeo S (1998) The effect of chronic treatment with trandolapril on cyclic AMP-and cyclic GMP-dependent relaxations in aortic segments of rats with chronic heart failure. Br J Pharmacol 123:344–352

Uchida T, Okumura K, Ito T, Kamiya H, Nishimoto Y, Yamada M, Tomida T, Matsui H, Hayakawa T (2002) Quinapril treatment restores the vasodilator action of insulin in fructose-hypertensive rats. Clin Exp Pharmacol Physiol 29:381–385

Uhlenius N, Tikkanen T, Miettinen A, Holthofer H, Tornroth T, Eriksson A, Fyhrquist F, Tikkanen I (1999) Renoprotective effects of captopril in hypertension induced by nitric oxide synthase inhibition in experimental nephritis. Nephron 81:221–229

Ulm EH (1983) Enalapril maleate (MK-421), a potent, nonsulfhydryl angiotensin-converting enzyme inhibitor: absorption, disposition, and metabolism in man. Drug Metab Rev 14:99–110

Unger T, Gohlke P (1990) Tissue renin-angiotensin systems in the heart and vasculature: possible involvement in the cardiovascular actions of converting enzyme inhibitors. Am J Cardiol 65:3I–10I

Unger T, Gohlke P (1994) Converting enzyme inhibitors in cardiovascular therapy: current status and future potential. Cardiovasc Res 28:146–158

Unger T, Schüll B, Rascher W, Lang RE, Ganten D (1982) Selective activation of the converting enzyme inhibitor MK 421 and comparison of its active diacid form with captopril in different tissues of the rat. Biochem Pharmacol 31:3063–3070

Unger T, Badoer E, Ganten D, Lang RE, Rettig R (1988) Brain angiotensin: pathways and pharmacology. Circulation 77:I40–I54

Unger T, Gohlke P, Gruber G (1990) Converting enzyme inhibitors. In: Mulrow DGaPJ (ed) Handbook of experimental pharmacology, vol 93. Springer Verlag, Berlin Heidelberg New York, pp 379–481

Unger T, Mattfeldt T, Lamberty V, Bock P, Mall G, Linz W, Schölkens BA, Gohlke P (1992) Effect of early onset angiotensin converting enzyme inhibition on myocardial capillaries. Hypertension 20:478–482

Van Gilst WH, de Graeff PA, Scholtens E, de Langen CD, Wesseling H (1987) Potentiation of isosorbide dinitrate-induced coronary dilatation by captopril. J Cardiovasc Pharmacol 9:254–255

Varin R, Mulder P, Tamion F, Richard V, Henry JP, Lallemand F, Lerebours G, Thuillez C (2000) Improvement of endothelial function by chronic angiotensin-converting enzyme inhibition in heart failure: role of nitric oxide, prostanoids, oxidant stress, and bradykinin. Circulation 102:351–356

Velasquez MT, Striffler JS, Abraham AA, Michaelis OET, Scalbert E, Thibault N (1997) Perindopril ameliorates glomerular and renal tubulointerstitial injury in the SHR/N-corpulent rat. Hypertension 30:1232–1237

Verbeelen DL, De Craemer D, Peeters P, Vanden Houte K, Van den Branden C (1998) Enalapril increases antioxidant enzyme activity in renal cortical tissue of five-sixths-nephrectomized rats. Nephron 80:214–219

Vlassara H, Palace MR (2002) Diabetes and advanced glycation endproducts. J Intern Med 251:87–101

Volpert OV, Ward WF, Lingen MW, Chesler L, Solt DB, Johnson MD, Molteni A, Polverini PJ, Bouck NP (1996) Captopril inhibits angiogenesis and slows the growth of experimental tumors in rats. J Clin Invest 98:671–679

Vulpis V, Lograno MD, Seccia TM, Pirrelli A (1998) L-type calcium channels modulate the regression of left ventricular hypertrophy after ace-inhibition in genetic hypertension. Pharmacol Res 38:317–322

Wapstra FH, Navis GJ, van Goor H, van den Born J, Berden JH, de Jong PE, de Zeeuw D (2001) ACE inhibition preserves heparan sulfate proteoglycans in the glomerular basement membrane of rats with established adriamycin nephropathy. Exp Nephrol 9:21–27

Wei GC, Sirois MG, Qu R, Liu P, Rouleau JL (2002) Subacute and chronic effects of quinapril on cardiac cytokine expression, remodeling, and function after myocardial infarction in the rat. J Cardiovasc Pharmacol 39:842–850

Westlin W, Mullane K (1988) Does captopril attenuate reperfusion-induced myocardial dysfunction by scavenging free radicals? Circulation 77:I30–I39

Wiemer G, Becker R, Gerhards H, Hock F, Stechl J, Ruger W (1989) Effects of Hoe 065, a compound structurally related to inhibitors of angiotensin converting enzyme, on acetylcholine metabolism in rat brain. Eur J Pharmacol 166:31–39

Wiemer G, Schölkens BA, Becker RH, Busse R (1991) Ramiprilat enhances endothelial autacoid formation by inhibiting breakdown of endothelium-derived bradykinin. Hypertension 18:558–563

Wiemer G, Schölkens BA, Linz W (1994) Endothelial protection by converting enzyme inhibitors. Cardiovasc Res 28:166–172

Winniford MD, Wheelan KR, Kremers MS, Ugolini V, van den Berg E Jr, Niggemann EH, Jansen DE, Hillis LD (1986) Smoking-induced coronary vasoconstriction in patients with atherosclerotic coronary artery disease: evidence for adrenergically mediated alterations in coronary artery tone. Circulation 73:662–667

Witte K, Huser L, Knotter B, Heckmann M, Schiffer S, Lemmer B (2001) Normalisation of blood pressure in hypertensive TGR(mREN2)27 rats by amlodipine vs. enalapril: effects on cardiac hypertrophy and signal transduction pathways. Naunyn Schmiedebergs Arch Pharmacol 363:101–109

Wolfrum S, Richardt G, Dominiak P, Katus HA, Dendorfer A (2001) Apstatin, a selective inhibitor of aminopeptidase P, reduces myocardial infarct size by a kinin-dependent pathway. Br J Pharmacol 134:370–374

Wu LL, Cox A, Roe CJ, Dziadek M, Cooper ME, Gilbert RE (1997) Transforming growth factor beta 1 and renal injury following subtotal nephrectomy in the rat: role of the renin-angiotensin system. Kidney Int 51:1553–1567

Wyvratt MJ (1988) Evolution of angiotensin-converting enzyme inhibitors. Clin Physiol Biochem 6:217–229

Yamaguchi F, Sanbe A, Takeo S (1998) Effects of long-term treatment with trandolapril on sarcoplasmic reticulum function of cardiac muscle in rats with chronic heart failure following myocardial infarction. Br J Pharmacol 123:326–334

Yang CM, Kandaswamy V, Young D, Sen S (1997) Changes in collagen phenotypes during progression and regression of cardiac hypertrophy. Cardiovasc Res 36:236–245

Yavuz DG, Ersoz O, Kucukkaya B, Budak Y, Ahiskali R, Ekicioglu G, Emerk K, Akalin S (1999) The effect of losartan and captopril on glomerular basement membrane anionic charge in a diabetic rat model. J Hypertens 17:1217–1223

Yoneda H, Toriumi W, Ohmachi Y, Okumura F, Fujimura H, Nishiyama S (1998) Involvement of angiotensin II in development of spontaneous nephrosis in Dahl salt-sensitive rats. Eur J Pharmacol 362:213–219

Yoshida K, Xu HL, Kawamura T, Ji L, Kohzuki M (2002) Chronic angiotensin-converting enzyme inhibition and angiotensin II antagonism in rats with chronic renal failure. J Cardiovasc Pharmacol 40:533–542

Yoshiyama M, Takeuchi K, Omura T, Kim S, Yamagishi H, Toda I, Teragaki M, Akioka K, Iwao H, Yoshikawa J (1999) Effects of candesartan and cilazapril on rats with myocardial infarction assessed by echocardiography. Hypertension 33:961–968

Yusuf S, Sleight P, Pogue J, Bosch J, Davies R, Dagenais G (2000) Effects of an angiotensin-converting-enzyme inhibitor, ramipril, on cardiovascular events in high-risk patients. The Heart Outcomes Prevention Evaluation Study Investigators. N Engl J Med 342:145–153

Zaman AK, Fujii S, Sawa H, Goto D, Ishimori N, Watano K, Kaneko T, Furumoto T, Sugawara T, Sakuma I, Kitabatake A, Sobel BE (2001) Angiotensin-converting enzyme inhibition attenuates hypofibrinolysis and reduces cardiac perivascular fibrosis in genetically obese diabetic mice. Circulation 103:3123–3128

Zatz R, Dunn BR, Meyer TW, Anderson S, Rennke HG, Brenner BM (1986) Prevention of diabetic glomerulopathy by pharmacological amelioration of glomerular capillary hypertension. J Clin Invest 77:1925–1930

Zhang X, Recchia FA, Bernstein R, Xu X, Nasjletti A, Hintze TH (1999) Kinin-mediated coronary nitric oxide production contributes to the therapeutic action of angiotensin-converting enzyme and neutral endopeptidase inhibitors and amlodipine in the treatment in heart failure. J Pharmacol Exp Ther 288:742–751

Zhu B, Sun Y, Sievers RE, Browne AE, Pulukurthy S, Sudhir K, Lee RJ, Chou TM, Chatterjee K, Parmley WW (2000) Comparative effects of pretreatment with captopril and losartan on cardiovascular protection in a rat model of ischemia-reperfusion. J Am Coll Cardiol 35:787–795

Zoja C, Abbate M, Corna D, Capitanio M, Donadelli R, Bruzzi I, Oldroyd S, Benigni A, Remuzzi G (1998a) Pharmacologic control of angiotensin II ameliorates renal disease while reducing renal TGF-beta in experimental mesangioproliferative glomerulonephritis. Am J Kidney Dis 31:453–463

Zoja C, Liu XH, Abbate M, Corna D, Schiffrin EL, Remuzzi G, Benigni A (1998b) Angiotensin II blockade limits tubular protein overreabsorption and the consequent upregulation of endothelin 1 gene in experimental membranous nephropathy. Exp Nephrol 6:121–131

Zolk O, Flesch M, Schnabel P, Teisman AC, Pinto YM, van Gilst WH, Paul M, Böhm M (1998) Effects of quinapril, losartan and hydralazine on cardiac hypertrophy and beta-adrenergic neuroeffector mechanisms in transgenic (mREN2)27 rats. Br J Pharmacol 123:405–412

Part 5

Inhibition of the Renin-Angiotensin System

AT_1 Antagonists

AT_1 Receptor Antagonists: Pharmacology

M. de Gasparo

MG Consulting Co, Rue es Planches 123, 2842 Rossemaison, Switzerland
e-mail: m.de_gasparo@bluewin.ch

Abstract Innovative chemical modifications of the first nonpeptide imidazole antagonist of Ang II led to the synthesis of various new orally active agents with increased potency and improved bioavailability (13%–80%). They block specifically and selectively the angiotensin AT_1 receptor without intrinsic agonist properties. The angiotensin receptor blockers (ARB) can be classified as surmountable, (losartan, eprosartan, telmisartan), insurmountable (candesartan) or mixed (valsartan, irbesartan, olmesartan) antagonists depending on their degree of tight binding and their dissociation rate. Candesartan and olmesartan are administered as prodrug converted to the active compound upon absorption. The ARBs are excreted essentially in the bile and mainly unchanged. If biotransformed, it involves oxidative reaction and conjugation. The metabolism of irbesartan, losartan, and candesartan requires cytochrome P450 enzymes. There is no accumulation by repeated doses. Plasma concentrations are little influenced by mild-to-moderate renal impairment but caution may be required in patients with hepatic insufficiency due to the biliary mechanism of excretion. Losartan is unique by its uricosuric property. In general, the ARBs do not interfere with other drugs in a clinically significant way, but caution should be taken if prescribed with potassium-sparing agents or supplements, especially in elderly patients with reduced renal function. The ARBs are generally well tolerated with an incidence of adverse effects or withdrawals similar to the placebo. First-dose hypotension is uncommon and there is no rebound hypertension after withdrawal. Angio-oedema is rare. The ARBs are contra-indicated during pregnancy. The efficacy of the ARBs in hypertension is well documented in various population and age groups and better tolerated than other antihypertensive agents for similar efficacy. The first properly powered trial in an hypertensive population, LIFE with losartan, has demonstrated a beneficial effect on the primary composite endpoint, including cardiovascular death, myocardial infarction and stroke, even more impressive in a diabetic subgroup. Based on the result of Val-HeFT, valsartan is approved in heart failure patients intolerant to ACE inhibitor. The renoprotective effect of the ARBs was demonstrated in diabetic nephropathy with irbesartan, losartan and valsartan. Various clinical studies suggest a beneficial effect of the ARBs beyond the blood pressure fall. Several large trials are in progress to establish the efficacy of the ARBs in patients with LV dysfunction following recent myocardial infarction. As blockade of the AT_1 receptor is accompanied by increased plasma Ang II, the potential of the stimulation of the unblocked AT_2 receptor is discussed. Possible further indications for ARB are also briefly reported.

Keywords losartan · EXP3174 · Valsartan · Irbesartan · Candesartan · Eprosartan · Telmisartan · Olmesartan

In 1956, Skeggs and co-workers (1956) suggested that it should be possible to prevent the vasoconstrictive action of angiotensin II on smooth muscle cells by means of analogues of the peptide. This concept lead to the synthesis of a large number of compounds culminating in saralasin, an octopeptide, which differs from Ang II in having the first (Asp) and last (Phe) amino acids replaced by sarcosine and alanine, respectively (Pals et al. 1971). Its peptide nature necessitating parenteral administration, its short half-life and its partial agonistic properties limited its clinical usefulness to that of a diagnostic probe for renin-dependent forms of hypertension.

Ten years later, nonpeptide structures with specific but weak Ang II receptor binding activity were synthesized and patented by Furakawa et al. at Takeda, Japan (1982) These 1-benzylimidazole 5-acetic acid derivatives were further optimized by DuPont Pharmaceuticals, resulting in losartan (Table 1), the first of a new class of orally active agents with specific and selective affinity for the Ang II AT_1 receptor (Timmermans et al. 1993)

Table 1 Structure of some angiotensin receptor antagonists and their affinity for the AT$_1$ receptor

	Losartan	Exp 3174	Valsartan	Irbesartan
	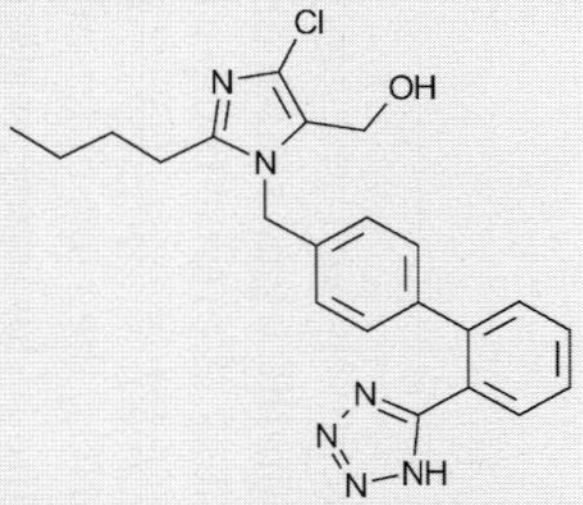	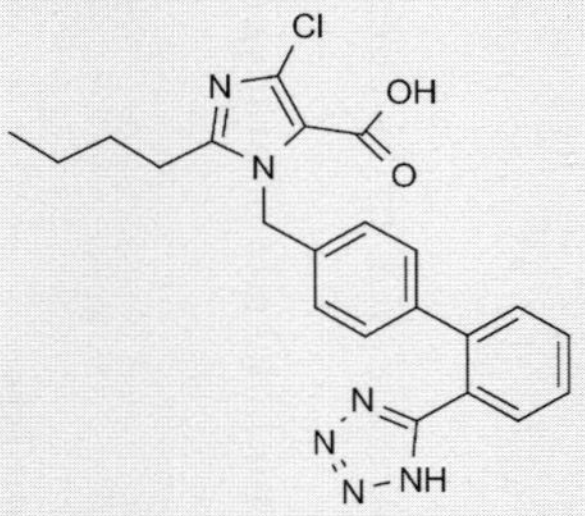	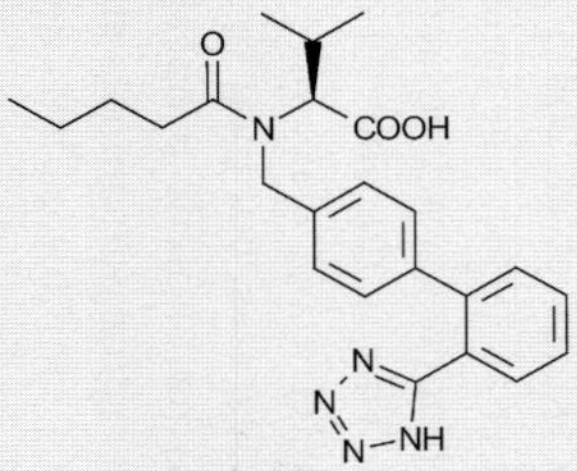	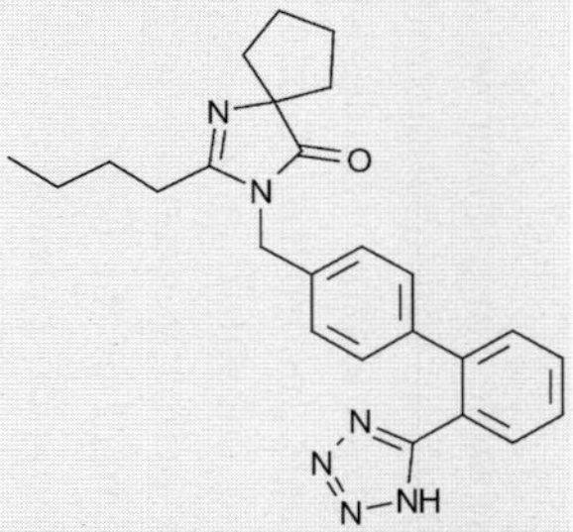
clog P	3.462	3.998	4.439	5.404
AT$_1$ affinity (nmol)	34±7[a]	0.45±0.1[a]	1.28±0.17[a]	1.9±0.19[a]
	Surmountable	Mixed (73/27)	Mixed (55/45)	Mixed (42/58)
K_{off} (min)	5.2±1.1	31±6	25±1	7.1±0.6
	Candesartan	Eprosartan	Telmisartan	Olmesartan
	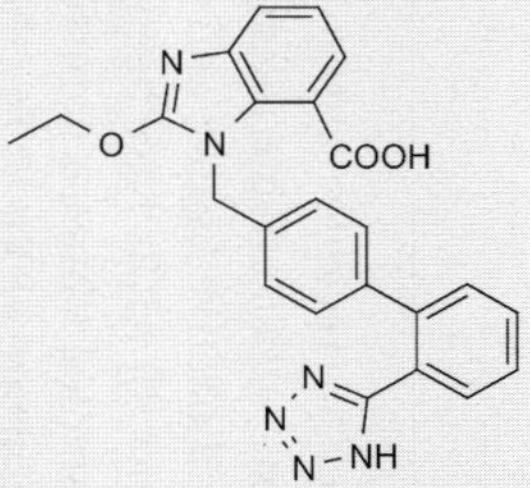	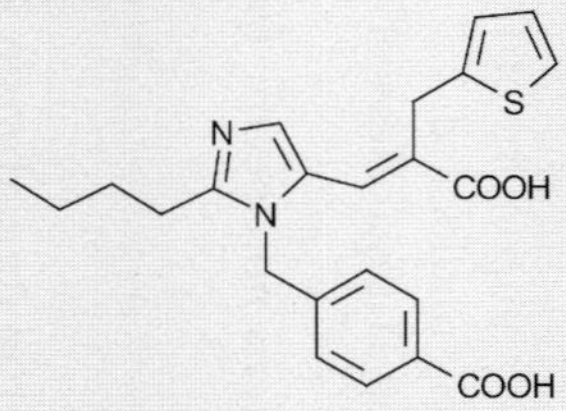	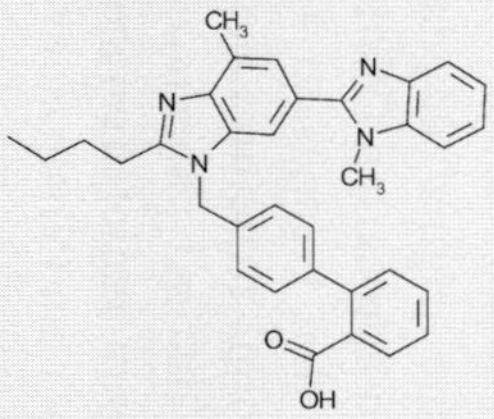	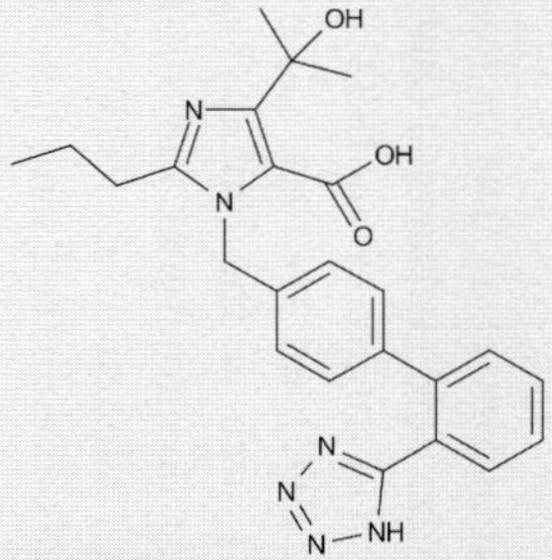
clog P	4.791	5.05	7.992	2.270
AT$_1$ affinity (nmol)	0.26±0.05[a]	1.4–3.9	3.7	8±0.8
	Insurmountable	Surmountable	Surmountable	Mixed
K_{off} (min)	152±58			

[a] From data published by Vauquelin's group.

An imidazole ring and an acidic biphenyltetrazole substituent characterize losartan (Furakawa et al. 1982). Replacing the imidazole ring, either with an acylated amino acid (valsartan) (Buhlmayer et al. 1994), an imidazolone moiety (irbesartan) (Bernhart et al. 1994) or a carboxy-benzimidazole (candesartan) (Noda et al. 1993) led to other AT_1 receptor antagonists. Incorporation of a carboxylic acid such as the biphenyl acidic group and substitution of a second phenylimidazole moiety at the six-position of the primary heterocycle gives Telmisartan (Wienen et al. 1993). Similarly, replacement of the imidazole ring by a fused six-membered ring heterocycle (dihydro-pyrido-pyrimidine-one) to form a quinazolinone derivative characterizes tasosartan (Ellingboe et al. 1994). Olmesartan is an imidazolecarboxylic acid derivative (Koike et al. 2001) and saprisartan is an imidazolecarboxylic acid amide (Judd et al. 1994). The imidazole-5-position of the original Takeda lead was also extended via a transacrylic acid group to yield eprosartan (Weinstock et al. 1991). This is the only AT_1 receptor antagonist that belongs to a nonbiphenyl, nontetrazole class of compounds. These various structures share some common features, but their differences may confer distinct pharmacodynamic and pharmacokinetic properties. By blocking the renin angiotensin system (RAS), AT_1 receptor antagonists share many properties of angiotensin-converting enzyme (ACE) inhibitors but can be differentiated in three ways:

1. Highly selective antagonism of the AT_1 receptor while the AT_2 receptor remains completely unblocked
2. Blockade of Ang II regardless of whether generated via ACE or by alternative pathways
3. Absence of actions nonspecific for the RAS (e.g. no blockade of bradykinin metabolism)

1 Pharmacology of the Angiotensin Receptor Antagonists

1.1 Binding

All the angiotensin receptor antagonists bind competitively and reversibly to the AT_1 receptor with variable affinities, depending on the tissue source of the receptor, the radioactive ligand, and the assay conditions. A precise comparison is difficult, as not all ARBs have been tested simultaneously.

In AT_1 transfected Chinese hamster ovary cells (CHO), the rank-order of potency is candesartan >EXP 3174 (the active metabolite of losartan) >valsartan >irbesartan >>losartan (Fierens et al. 1999; Vanderheyden et al. 1999; Verheijen et al. 2000) (Table 1). Eprosartan and telmisartan have higher affinities than losartan and are in the range of those of valsartan and irbesartan (Noda et al. 1993; Wienen et al. 1993). Olmesartan has a reported IC_{50} of 8 nM in the adrenal cortex membrane, whereas losartan and candesartan show an IC_{50} of 92 and 12 nM, respectively (Brunner 2002).

Preincubating the antagonists with the receptor before adding the radioactive and nonradioactive ligand allows differentiation into three types of antagonists: surmountable, insurmountable and mixed surmountable-insurmountable (Vanderheyden et al. 2000; Verheijen et al. 2000). This may explain the counterclockwise hysteresis observed between a decrease in blood pressure and plasma concentration (Delacretaz et al. 1995). Losartan, eprosartan and tasosartan exhibit surmountable antagonism: a dose-dependent parallel shift to the right of the Ang II dose-response but full expression of the agonist at high dosage of Ang II. These antagonists dissociate rapidly from the binding site.

In contrast, candesartan is fully insurmountable. There is a progressive nonparallel rightwards shift of the response to Ang II and a decrease in the maximal effect of Ang II with increasing concentration of candesartan. The very slow dissociation of candesartan from its binding site (T1/2 152±58 min) and the time required for the recovery of functional AT_1 receptor (6–8 h) suggest that there is a re-association of the released candesartan to its binding site. Olmesartan also appears to be an insurmountable antagonist (Koike et al. 2001)

Valsartan, irbesartan, telmisartan and EXP3174 are of the mixed type (Skeggs et al. 1956; Vanderheyden et al. 2000; Verheijen et al. 2000). These compounds bind to the receptor both in a high-affinity, slow-dissociating state and a low-affinity high-dissociating state. Therefore, as the antagonist dissociates from the receptor, there is a progressive restoration of the functional response to the agonist. The proportion of the labile and tight binding sites will affect the curve response of Ang II. The higher the dissociation constant of the tight binding sites, the greater the decrease in the maximal response. Thus, the insurmountable or mixed antagonists produce a form of pharmacological down-regulation of the AT_1 receptor. Increased circulating plasma Ang II levels can displace a competitive surmountable antagonist but not an insurmountable or mixed antagonist.

None of the nonpeptide antagonists has a significant affinity for the AT_2 receptor, indicating that at the concentration measured in the plasma of the patient, the AT_2 receptor remains free for binding of Ang II, whereas the AT_1 receptor is blocked. Intrinsic agonist properties of the nonpeptidic AT_1 antagonist have not been demonstrated in vivo either in animals or in humans.

The AT_1 receptor antagonists do not show affinity for other hormonal receptors or channels. However, tritiated losartan binds extensively to another site, which is not recognized by Sar^1-Ile^8 Ang II. The nature of this site is still conjectural (de Gasparo and Whitebread 1995).

Candesartan cilexetil and olmesartan medoxomil are inactive ester prodrugs which are rapidly hydrolysed in the gastrointestinal tract (Morimoto and Ogihara 1994; Warner and Jarvis 2002). Losartan is not a prodrug. It has moderate affinity for the AT_1 receptor. Fourteen percent of losartan, however, is converted to a metabolite, EXP3174, which has a 70-fold greater affinity for the receptor than losartan itself (Goa and Wagstaff 1996). The methyl hydroxyl group of the imidazole ring is oxidated to the carboxylated form.

1.2 Pharmacokinetics

The pharmacokinetic characteristics of the AT_1 receptor antagonists are summarized in Table 2 (Ellingboe et al. 1994; Morimoto and Ogihara 1994; Goa and Wagstaff 1996; Gillis and Markham 1997; Markham and Goa 1997; McClellan and Goa 1998; McClellan and Markham 1998; Sharpe et al. 2001; Brunner 2002; Easthope and Jarvis 2002; Warner and Jarvis 2002). The bioavailability of these agents varies from 13% for eprosartan to 80% for irbesartan. There is a clear reduction in bioavailability when valsartan is administered with food. However, this reduction occurs mainly in the time period 0–6 h after dosing and the mean plasma valsartan concentrations from 6–24 h are similar in the fed and fasted state. As valsartan is not extensively metabolized and has a low plasma clearance (2 l/h), the reduction in bioavailability is most likely to be caused by a slower rate of absorption. A negligible reduction of bioavailability is observed when eprosartan and telmisartan are taken with food.

The volume of distribution varies considerably between the antagonists depending on lipophilicity and/or plasma protein binding. Candesartan (9.1 l), EXP 3174 (12 l), eprosartan (13 l), valsartan (17 l), olmesartan (29 l), losartan (34 l) have significantly lower volumes of distribution than those of irbesartan (53–93 l) and telmisartan (500 l). A large volume of distribution may correlate with a higher plasma clearance: 10 l/h and 60 l/h for irbesartan and telmisartan, respectively, compared to 1.4 l/h for candesartan and olmesartan, 2 l/h for valsartan, 2.8 l/h for EXP 3174, 4.4 l/h for losartan and 7.8 l/h for eprosartan. Plasma clearance, however, is generally low compared to hepatic blood flow (90 l/h).

All the AT_1 receptor antagonists exhibit extensive binding to plasma protein (Morsing 1999). For example, the free plasma fractions of losartan and EXP 3174 do not exceed 1.3% and 0.2%, respectively. Large volumes of distribution does not appear to be limited by high-protein binding, however, although only the free fraction of the drug should be available for distribution outside the plasma compartment. Assuming a plasma volume of 6 l, an extracellular volume of 22 l and a total body water of 85 l, the high volume of distribution of irbesartan and telmisartan suggests that these substances concentrate at one or more sites and may gain access to the intracellular compartment, as a consequence of lipophilicity. Indeed, telmisartan has the highest lipophilicity of the AT_1 receptor antagonists, with a calculated log P of 7.992.

1.3 Metabolism and Elimination

The CYP450 isoenzymes 2C9 and 3A4 are responsible for the formation of the active metabolite of losartan, EXP3174 (Stearns et al. 1995; Yun et al. 1995). In this reaction, there is the formation of a highly reactive aldehyde intermediate EXP 3179. Irbesartan is metabolized negligibly by the isoenzyme 3A4. It is further oxidized and glycuronidated by the isoenzyme 2C9 (Marino and Vachharajani,

Table 2 Pharmacokinetic properties of some angiotensin receptor antagonists

	Losartan (Cozaar)	Exp 3174	Valsartan (Diovan)	Irbesartan (Aprovel Avapro)	Candesartan (Atacand, Biopress)	Eprosartan (Teveten)	Telmisartan (Micardis)	Olmesartan
Bioavailability	33%	14%	25%	60%–80%	Prodrug 42% Drug 15%	13%	40%–60%	Prodrug 28%
Food effect	Minimal	No	–46%	No	No	–25%	–6 to 19%	No
$T_{1/2}$	2 h	6–9 h	6–9 h	11–15 h	8–11 h	4.5–9 h	20–24 h	10–15 h
Plasma peak	1 h	2–4 h	2 h	1.5–2 h	3–4 h	1–3 h	0.5–1 h	1–3 h
Biliary excretion		58%	70%	80%	67%	90%	98%	60%
Kidney excretion	4%	30%–35%	30%	20%	33%	7%	1%	40%
Volume distribution	34 l	12 l	17 l	53–93 l	9.1 l	13 l	500 l	29 l
Protein binding	98.7%	99.8%	95%	99.5%	99.8%	98%	99.5%	
Plasma clearance	36 l/h	2.8 l/h	2 l/h	9.4–10.5 l/h	1.4 l/h	7.8 l/h	60 l/h	1.4 l/h
Renal clearance	4.4 l/h	1.6 l/h		0.2 l/h	0.8 l/h			0.5–0.7 l/h
Dosage (mg)	50–100		80–160	150–300	4–32	600	40–80	(10–20)
Maximum efficacy	3–6 weeks	3–6 weeks	2–4 weeks	4–6 weeks	4 weeks	2–3 weeks	4 weeks	

2001). A minor hepatic inactive metabolite of candesartan is formed trough the isoenzyme 2C9, whereas small percentages of valsartan, eprosartan and telmisartan are metabolized to inactive derivatives by a cytochrome P independent mechanism.

As all AT_1 receptor antagonists are excreted mainly unchanged in the bile, caution may be required in patients with hepatic insufficiency. Neither losartan, EXP 3174 nor candesartan is dialyzable.

1.4 Pharmacodynamics

Essentially three models have been used to compare the pharmacodynamic action of the AT_1 receptor antagonists in healthy volunteers: intravenous Ang II bolus injection, activation of endogenous RAS and direct measurement of AT_1 receptor blockade. Several antagonists have been compared in the same study either at the recommended starting doses for treatment of hypertension (Mazzolai et al. 1999) or at escalating oral doses (Belz et al. 2000).

Compared with placebo, a single dose of irbesartan (150 mg) caused a greater and more sustained blockade of the AT_1 receptor than that after a single dose of losartan (50 mg) or valsartan (80 mg) in healthy volunteers challenged by Ang II bolus injections. Losartan, valsartan and irbesartan were more effective than placebo at 4 h (43%, 51% and 88% blockade, respectively) and 30 h (21%, 26% and 45% blockade) after drug intake, but only irbesartan induced a significant residual blockade 30 h after dosing. Evaluation of in vitro receptor blockade capacity of plasma in the binding assay in smooth muscle cell membrane, an indirect measure of the plasma concentration, produces compatible results (Maillard et al. 1999). Plasma active renin and Ang II concentration increase proportionally to the extent of receptor blockade (Azizi et al. 1999a, 1999b; Mazzolai et al. 1999), but this is a less sensitive marker. Indeed, the percentage of change compared with baseline at peak or evaluated over 24 h (AUC) after single-dose administration of losartan, candesartan or valsartan indicates that losartan 50 mg, valsartan 80 mg and candesartan 8 mg are equivalent, whereas valsartan 160 mg and candesartan 16 mg are not significantly different (Azizi et al. 1999a, 1999b). These studies are limited by the use of only a single dose level of each AT_1 receptor antagonist.

Fig. 1 **A** Time-course of median dose ratios given as DR-1 following consecutive administration of oral doses of 4, 8, 16 mg candesartan cilexetil, 75, 150, 300 mg irbesartan, 25, 50, 100 mg losartan, 20, 40, 80 mg telmisartan, and 40, 80, 160 mg valsartan. A value of DR-1 of 1 represents 50% blockade of receptor blockade in vivo. D1, D2 and D3 represent the administration of the low, middle and high dose, respectively. **B–D** Schild regression plots (median DR-1 vs dose) for 4, 8 and 24 h after dosing. ▲ Candesartan cilexetil, ♦ irbesartan, ● losartan, ○ telmisartan, ■ valsartan. (Adapted from Belz et al. 2000)

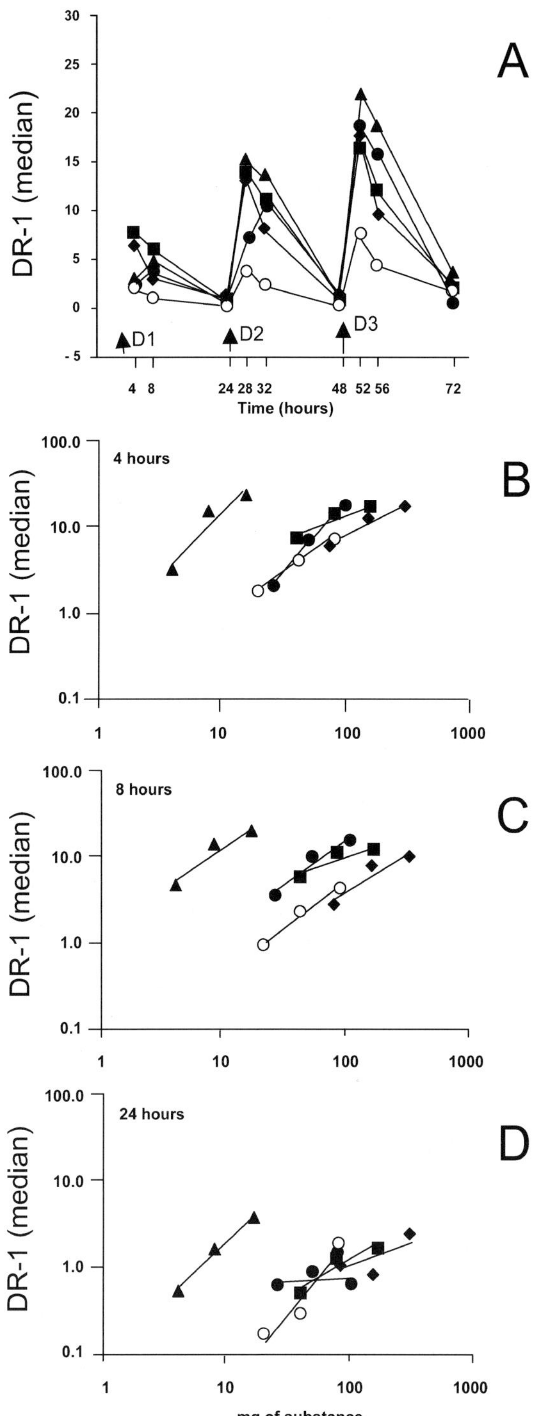
A
DR-1 (median)
D1
D2
D3
Time (hours)
B
4 hours
DR-1 (median)
C
8 hours
DR-1 (median)
D
24 hours
DR-1 (median)
mg of substance

An inconsistent reduction in plasma aldosterone has been reported. This observation supports modulation of aldosterone release by pathways other than Ang II such as potassium levels and changes in acid/base balance.

The pharmacological potency of increasing doses of various antagonists has also been assessed in vivo from the rightward shifts in Ang II dose-response curves for diastolic blood pressure in a randomized, double-blind parallel group design (Belz et al. 2000). The dose ratio, i.e. the extent by which the antagonist shifts the agonist response curve to the right, and the apparent K_i, i.e. the dose required to induce a twofold shift in the dose-response corresponding to 50% receptor blockade, were calculated (Belz et al. 1999, 2000; Belz 2001). Valsartan, telmisartan, losartan and candesartan inhibit dose-dependently the Ang II-induced increase in diastolic blood pressure. The median dose-ratios were not significantly different for candesartan (22) irbesartan (18), losartan (19) and valsartan (16), but the dose-ratio of telmisartan is significantly lower (8). Using the Schild regression technique, it appears that candesartan produces antagonistic effects at the lowest doses, indicating that it has the greatest pharmacological potency (Fig. 1). The apparent 24-h K_i-doses are 1, 123, 54, 93 mg for candesartan, irbesartan, telmisartan and valsartan, respectively, i.e. within the dose range recommended for clinical use in hypertensive patients. This model allows calculation of a functional plasma half-life in the range of 8–10 h for these drugs, whereas the value is less than 5 h for losartan. For irbesartan and telmisartan these results are slightly lower than expected from pharmacokinetics data, indicating that pharmacokinetic measures do not reliably predict pharmacodynamic responses. The Schild plot analysis also allows the detection of a second counter-regulatory vasodilatory mechanism, e.g. an increase in plasma renin activity as observed with irbesartan compared to candesartan (Belz 2003). This should be interpreted with caution as the increased circulating Ang II may favour the action of the AT_1 receptor blockade trough stimulation of the AT_2 (see below).

Ang II increases norepinephrine release through activation of prejunctional AT_1 receptor located on sympathetic nerve terminals and therefore enhances vasoconstriction through stimulation of the $\alpha 1$ adrenoceptors. Eprosartan, which does not contain the biphenyl-tetrazole moiety, has been reported to block the pressor response induced by activation of the sympathetic outflow in pithed rats (Ohlstein et al. 1997). Equivalent doses of losartan, valsartan and irbesartan were ineffective in this model, but this result has not been confirmed. The increased diastolic blood pressure following electrical stimulation of the thoracolumbar sympathetic outflow in pithed rats is blocked by all AT_1 antagonists tested, suggesting a class effect (Balt et al. 2001). Results in two other models confirm this observation. Similarly, AT_1 receptor antagonists inhibit Ang II concentration-dependent enhancement of the electrically stimulated efflux of ^{3}H-norepinephrine in isolated left rat atria but with different potency (Shetty and Delgrande 2000). Thus, AT_1 receptor antagonists inhibit vasoconstriction directly by acting on the vascular AT_1 receptor and indirectly in inhibiting the prejunctional AT_1 receptor.

1.5 Additional Properties of AT$_1$ Receptor Antagonists

Losartan has a dose-dependent uricosuric effect in healthy volunteers, in sodium-replete hypertensives and in patients with hypertension and renal diseases. This is a specific effect on urate transport, independent of Ang II blockade (Burnier et al. 1996); nor is a uricosuric effect observed with the metabolite EXP 3174 or with the other AT$_1$ receptor antagonists. The uricosuric effect of losartan may be beneficial in cyclosporin-treated patients having received transplants since cyclosporin induces hyperuricaemia (Minghelli et al. 1998). However, it may be a potential hazard to patients predisposed to urate stone formation, i.e. in nephrolithiasis, low urinary flow or aciduria. However, such an adverse effect has not yet been described, probably because the urinary pH tends to increase slightly during treatment with AT$_1$ receptor antagonists.

Losartan and irbesartan also reduce platelet aggregation in vitro and inhibit competitively TXA2/PGH2 platelet receptor binding (Li et al. 2000; Liu et al. 1992). The aldehyde EXP3179, an intermediate in the formation of the active metabolite EXP3174, may be responsible for the anti-inflammatory and antiaggregatory properties of losartan (Kramer et al. 2002). EXP 3174 itself, valsartan and candesartan have no such effect, indicating that the requirement for binding to AT$_1$ and platelet receptors is different (Lopez-Farre et al. 2001). Since a transient metabolite is responsible for this action, the antiplatelet activity of losartan is probably without major clinical significance.

1.6 Drug Interaction

The metabolism of losartan and irbesartan is CYP450 dependent and involves the 2C9 and 3A4 isoenzymes (Stearns et al. 1995). Inhibitors of 2C9 or of 3A4 decrease EXP3174 formation. Inhibition of CYP 2C19 with fluconazole increases both Cmax and AUC of losartan by 30% and 66%, respectively, and reduces these variables for EXP 3174 by 56% and 43% (Kazierad et al. 1997). The CYP 450 enzyme-inducer phenobarbital reduces the losartan:EXP3174 ratio by 20%, without affecting the half-life of EXP 3174 (Goldberg et al. 1996). A study in healthy individuals expressing different CYP450 2C9 genotypes did not demonstrate, however, a difference in the pharmacokinetics of losartan and its active metabolite (Lee et al. 2003). Although irbesartan does not require biotransformation for its pharmacological activity, its oxidative metabolism also involves CYP 2C9 and 3A4 isoenzymes, and fluconazole increases Cmax by 19% and the AUC by 63% (Marino and Vachharajani 2001). CYP 450 inhibitors or inducers do not affect the metabolism of other AT$_1$ receptor antagonists.

Therefore, for losartan and perhaps for irbesartan, there is a potential risk of interaction with other drugs, which induce these CYP450 isoenzymes. Competition at the enzymatic level between drugs metabolized through the CYP450 may alter disposition, leading to increased plasma levels and greater risk of adverse events. Such drugs include not only fluconazole, rifampicin, dexamethasone or carbamazepine, but also several agents often used concomitantly to treat hypertensive patients, e.g. statins, calcium channel blockers and oral antidiabetic drugs. Available clinical studies have demonstrated very little clinically significant interference. However, in rheumatoid arthritic patients in acute exacerbation and in remission, there is an inverse correlation between the C reactive protein levels, a marker of inflammation and endothelial dysfunction, and the formation of EXP 3174, suggesting that inflammation could alter the activation or metabolism of the CYP 450-dependent antagonist such as losartan, leading to changes in clinical efficacy (Lewanczuk, 2002). Telmisartan coadministered with digoxin is associated with an increase in plasma digoxin levels (49% and 20% at peak and trough, respectively), suggesting a need to monitor digoxin levels to avoid overdosing. Coadministration of valsartan, losartan and candesartan cilexetil with hydrochlorothiazide reduced the AUC of the diuretic by 31%, 17% and 14%, respectively, whereas hydrochlorothiazide increases plasma levels of candesartan by 31% (Jonkman et al. 1997). This is without clinical consequences.

In general, the AT_1 receptor antagonists do not interfere with other drugs in a clinically significant way. However, caution should be taken if prescribed with potassium sparing agents or supplements because of the risk of severe hyperkalaemia, especially in elderly patients with reduced renal function.

1.7 Safety and Tolerability

The AT_1 receptor antagonists do not directly interfere with any enzymatic processes of the RAS and do not affect either bradykinin or aminopeptidases metabolism. This is probably the reason why they are generally very well tolerated. The frequency of adverse events is not related to age, sex or dose and is indistinguishable from that with placebo. The overall incidence of any reported clinical adverse effects and the withdrawal rate in patients receiving AT_1 receptor antagonists were similar to that in placebo-treated patients (Fogari et al. 2000; Mazzolai and Burnier 1999). A recent retrospective analysis of prescription confirmed a higher persistence rate with AT_1 receptor antagonists in comparison with other antihypertensive drugs over a 3-year period (Bloom 1998; Degli et al. 2002). The most common adverse effects reported were dizziness, headache, asthenia and fatigue. The relationship of these symptoms with drug administration is uncertain. No significant and consistent adverse effects on haematology and clinical chemistry were noted. Clinically relevant hyperkalaemia is unusual but serum potassium levels should be monitored when AT_1 receptor antagonists are administered to patients with renal insufficiency, in diabetes mellitus or when potassium-sparing diuretics or potassium supplements are prescribed concomitantly. Changes in liver enzymes occur rarely and are usually transient. Tasosartan was uniquely associated with more significant liver damage, leading to its withdrawal.

A rapid fall in blood pressure is uncommon even in the elderly, although in patients who are volume-depleted or with blood pressure highly dependent on activation of the RAS, dramatic falls may be expected. Despite the high circulating Ang II plasma levels after AT_1 receptor blockade, there is no rebound hypertension after withdrawal of therapy. This may be due to down-regulation of the AT_1 receptor observed particularly with the insurmountable antagonists and the prolonged elimination half-life of AT_1 receptor antagonists compared with that of Ang II.

There is usually no significant change in glomerular filtration rate. A decreased creatinine clearance may be observed in patients cotreated with high doses of diuretics or those on a severely salt-restricted diet. The incidence of cough is similar to that with placebo and significantly less common than that with ACE inhibitors (Black et al. 1997; Lacourcière, 1999; Easthope and Jarvis 2002). Angio-oedema is occasionally observed with AT_1 receptor antagonists, mainly in patients with prior history of ACE inhibitor-associated angio-oedema (Frye and Pettigrew 1998; Van Rijnsoever et al. 1998). Sixty-three cases have been reported during postmarketing with irbesartan in Australia, the only AT_1 receptor antagonist commercially available in that country (Howes and Tran 2002).

AT_1 receptor antagonists should not be prescribed during pregnancy because of the risk of dysembryogenesis as well as severe hypotension and renal failure in the newborn. As these drugs are detected in the milk of the lactating rat, caution should be observed in the breast-feeding mother.

In conclusion, the Ang II antagonists appear to be the anthypertensive drugs with the lowest incidence of adverse effects.

2
Clinical Applications

2.1
Hypertension

The AT_1 receptor antagonists are effective antihypertensive drugs when the RAS is normal or increased (renovascular hypertension, genetic hypertension; negative sodium balance). These agents are less effective or even inactive when the RAS is suppressed or when there is a positive sodium balance (DOCA salt rat model, high-sodium diet). In theory, genetic factors such as polymorphism of the AT_1 gene (Vuagnat et al. 2001) could affect the consistency of the response to AT_1 receptor antagonists. In the general management of hypertension, where RAS phenotype and genotype are not assessed routinely, all AT_1 receptor antagonists exhibit similar antihypertensive response rates.

The antihypertensive effect of AT_1 receptor antagonists is linked to a direct and indirect inhibition of the vasoconstrictive action of Ang II and catecholamines, resulting in decreased systemic vascular resistance and improved hydroelectrolytic homeostasis (decreased tubular sodium and water reabsorption, reduced aldosterone secretion). The heart rate remains unchanged despite the fall in blood pressure following blockade of the catecholamine release. Furthermore, inhibition of the autocrine and paracrine effects of Ang II such as vascular smooth muscle hypertrophy and extracellular matrix stimulation by the AT_1 receptor antagonists probably contributes to the long-term beneficial outcome. Whereas blockade of the endocrine effect of Ang II is immediate, inhibition of autocrine/paracrine effects requires time to develop.

AT_1 receptor antagonists have a gradual onset of action and a smooth antihypertensive effect. First dose hypotension is therefore not a clinical problem. However, sodium and volume depletion, with consequent activation of the RAS, should be corrected to avoid excessive reduction in blood pressure. The maximal effect of AT_1 receptor antagonists usually appears between 2 and 5 weeks, depending on the drug and dose. Clinical studies have established the antihypertensive efficacy of AT_1 receptor antagonists in patients with mild to moderate or severe hypertension irrespective of age (Argenziano and Trimarco 1999; Neldam and Forsen 2001), race (Levine 1999; Weir et al. 2001) or gender (Gradman et al. 1999a) The responder rate defined as diastolic blood pressure $\leq$ 90 mmHg or fall in diastolic blood pressure $\geq$10 mmHg varies between 50% and 65% (Gradman et al. 1995; Holwerda et al. 1996; Gradman et al. 1999b; Hedner et al. 1999).

AT_1 receptor antagonist efficacy has been investigated in comparative trials against placebo, against other antihypertensive drugs (ACE inhibitors, calcium channel blockers, β blockers, diuretics) and against each other. There is a dose-related effect with all AT_1 receptor antagonists, although losartan appears to exhibit a shallow, flat dose-response curve (Christen et al. 1991; Neutel et al. 1998; Reif et al. 1998; Pool et al. 1999; Püchler et al. 2001). Except for irbesartan evaluated between 1 and 900 mg (Reeves et al. 1998), however, the dose-response relationships have been defined in cohorts of patients that were too small and over too narrow a dose range, neglecting the lower and upper extremes of the dose-response.

The calculated trough:peak ratio (i.e. the proportion of the peak effect remaining at trough) and 24-h blood pressure ambulatory monitoring have been used to evaluate the duration of action of AT_1 receptor antagonists. The trough:-peak ratio for different AT_1 receptor antagonists varies: losartan (50–100 mg): 50–75; valsartan (80–160 mg): 66; candesartan (8–16 mg) >80: irbesartan (150–300 mg): 58–74; eprosartan (400–800 mg): 67–88; telmisartan (100–200 mg): >50. The duration of action of AT_1 receptor antagonists appears to be dose related. Twenty-four-hour control of blood pressure seems to require twice-a-day dosing for losartan as its efficacy during the last 6 h of a 24-h cycle is not different from placebo (Mallion et al. 1999).

A meta-analysis of all the trials performed with AT_1 receptor antagonists up to October 1998 and including 11,281 patients suggests that all AT_1 receptor antagonists have equivalent antihypertensive effects (Conlin et al. 2000). The fall of diastolic and systolic blood pressure (without placebo substraction) was 8.2–8.9 and 10.4–11.8 mm Hg, respectively. There are several criticisms of this analysis: the included studies had very different designs, treatment durations, comparators and patient populations. Double-blind, randomized comparative studies with sufficient statistical power to compare AT_1 receptor antagonists are relatively rare. Furthermore, comparisons of valsartan with losartan or irbesartan gave different results in different direct comparative studies (Fogari et al. 1999; Hedner et al. 2000; Monterroso et al. 2000). Comparisons of candesartan and losartan in four parallel groups of about 80 patients with mild to moderate hypertension indicated a greater effect of candesartan at trough, whereas the drugs produced similar falls at peak effect (Andersson and Neldam 1998).

Multicenter, randomized, double-blind studies with sufficient subjects are therefore required. In 207 hypertensive patients, telmisartan (40 and 80 mg) reduced ambulatory blood pressure significantly more than losartan 50 mg during the 18- to 24-h postdose period (Mallion et al. 1999). These results suggest that losartan at a dose of 50 mg daily may be less effective than other AT_1 receptor antagonists, the difference being more marked at the end of the dosage interval. In a cross-over study including 426 patients with mild to moderate hypertension, irbesartan (150 mg) was significantly more effective than valsartan (80 mg) on ambulatory blood pressure monitoring, although differences were modest (Mancia et al. 2000). The calculated equipotent antihypertensive doses were 80.5 mg for losartan, 115.5 mg for valsartan, 216.6 mg for irbesartan and 13.5 mg for candesartan in a double-blind, cross-over study in a small number of mild to moderately hypertensive patients (Fogari et al. 2000) (Fig. 2).There is an outstanding need to compare AT_1 receptor antagonists across the dose ranges.

Overall, the AT_1 receptor antagonists are as effective as other antihypertensive drugs in various populations and age groups. Using treatment based on drugs of this class, it is therefore possible to obtain and maintain long-term and optimal reduction in blood pressure. Compared to calcium channel blockers (Chan et al. 1995; Kloner et al. 2001; Stumpe and Ludwig 2002), β blockers

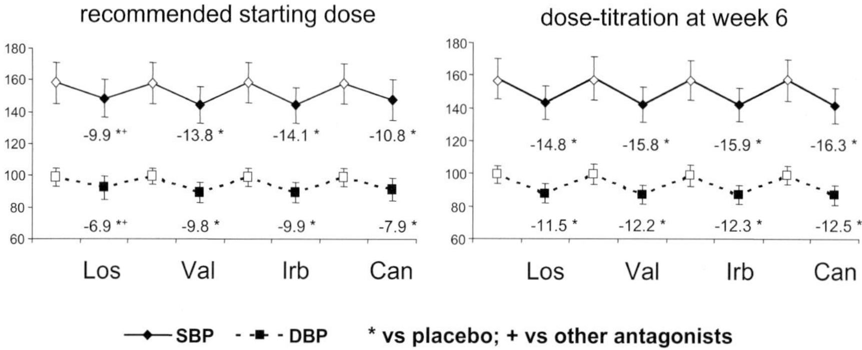

Fig. 2 Mean sitting blood pressure (± SD) after placebo run-in or washout period (◇❑) and active treatment with the recommended starting doses or with dose titration at the 6th week (♦■). (Adapted from Fogari et al. 2000)

(Dahlof et al. 1995; Stumpe et al. 1998; Ball et al. 2001) and ACE inhibitors (Black et al. 1997; Sever 1997; Larochelle et al. 1997; Gavras and Gavras 1999; Neutel et al. 1999), the AT_1 receptor antagonists are as effective in lowering blood pressure but better tolerated.

The efficacy of the AT_1 receptor antagonists, as of any other antihypertensive drugs, is usually defined with regard to a surrogate endpoint, i.e. blood pressure. However, the ultimate goal of the treatment is really to improve mortality and morbidity (Cohn 2002). The results of the first properly powered trial, LIFE (Losartan Intervention For Endpoint) are now available (Dahlof et al. 2002). LIFE compared once daily losartan-based and atenolol-based therapy in 9,193 patients with hypertension and ECG-determined left ventricular hypertrophy (LVH) in a double-blind, randomized, parallel-group design. Average doses of losartan and atenolol were around 80 mg daily. Diuretics and other antihypertensive drugs (except ACE inhibitors, β blockers and other AT_1 receptor antagonists) could be added as needed to normalize blood pressure. Blood pressure decreases were similar in both groups. However, losartan had an effect on the primary composite endpoint (all cardiovascular deaths, fatal and nonfatal myocardial infarction, fatal and nonfatal stroke) 13% greater than that of atenolol (p=0.021). The difference was mainly due to a reduction in stroke by 25% in the losartan group (p=0.001). Total mortality and rate of myocardial infarction were similar in both treatment groups.

The LIFE study included a diabetes subgroup of 1,195 patients in which the results of losartan compared to the β blocker were even more impressive (Lindholm et al. 2002). There was a 24% reduction in composite endpoints in the losartan group compared with the atenolol group (p=0.031) and a notable 39% reduction in all-cause mortality. Admission for heart failure was also reduced by 40%. There was also a 25% lower risk of developing diabetes amongst patients without diabetes at baseline.

Losartan induced a greater regression of LVH compare with atenolol. In a LIFE substudy (Devereux et al. 2002; Wachtell et al. 2002) using echocardiography, losartan reduces LV mass more than did atenolol (p=0.004) in 960 patients. The reversal of LVH was sustained through the mean follow-up of 4.8 years. The advantage of losartan could not be explained fully by greater reduction in LV mass or other differences between the groups in risk factors. This result clearly suggests a beneficial effect beyond the blood pressure fall.

It is probably fair to extrapolate these results with losartan to other AT_1 receptor antagonists. Indeed, similar results on LVH regression with valsartan, irbesartan, candesartan have been reported in smaller trials (Thurmann et al. 1998; Mitsunami et al. 1998; Malmqvist et al. 2003). The SILVER trial will further compare the effect of irbesartan and of the calcium channel blocker felodipine on the regression of the LVH in a large hypertensive population (Cohen et al. 1998).

SCOPE (Study on Cognition and Prognosis in the Elderly) compared candesartan with placebo in nearly 5,000 elderly patients with hypertension (Hansson et al. 1999). Active treatment with diuretic, calcium channel blocker or β blocker was allowed in both groups. Use of concomitant antihypertensive medication in the placebo group was much greater than anticipated and SCOPE was underpowered to compare active treatment regimens. There was no significant reduction in the primary endpoint combining cardiovascular morbidity, myocardial infarction and stroke, but there was a 28% reduction in nonfatal stroke (p=0.041) with candesartan. These differences were greater than might be expected for the difference in blood pressure between the groups. The trial failed to detect any significant effect of candesartan on cognitive function and dementia except in those with evidence of impaired cognitive function at randomization (Sever et al. 2002). However, in a substudy, comprehensive assessment of cognitive function using tests validated in the elderly and known to be more sensitive demonstrated than the candesartan group showed less decline in attention and episodic memory (Lithel et al. 2003). Overall, the results of SCOPE support the findings of LIFE.

VALUE (Valsartan Antihypertensive Long-Term Use Evaluation trial), a comparison of valsartan-based and amlodipine-based antihypertensive treatment in 15,314 hypertensive patients with high cardiovascular risk, should add substantially to the evidence for a potential advantage of AT_1 receptor antagonists. The primary endpoint is a composite of cardiovascular death, myocardial infarction, heart failure or urgent coronary revascularization (Mann and Julius 1998). The target primary endpoint of 1,450 events should be reached by 2004.

2.2 Heart Failure

Heart failure represents the final premorbid stage in the cardiovascular continuum from risk factors to death. Hypertension precedes heart failure in 91% of cases (Levy et al. 1996). In heart failure, ACE inhibitors improve haemodynamics by reducing afterload and increasing cardiac output. A potential advantage of AT_1 receptor antagonists over ACE inhibitors may reside in a more complete blockade of the RAS, as AT_1 antagonism is independent of the alternative pathways involved in Ang II synthesis and does not provoke systemic bradykinin accumulation. A change in the AT_1-AT_2 receptor balance (relative increase in AT_2 receptors) within the heart may also favor AT_1 receptor antagonists through stimulation of the AT_2 receptor since circulating Ang II is increased secondary to both heart failure and AT_1 receptor blockade of the negative feedback loop, which attenuates renin release (Rogg et al. 1996).

The first studies in heart failure patients have demonstrated a beneficial haemodynamic effect of the AT_1 receptor antagonists similar to that of ACE inhibitors but with better tolerance. There is a decrease in pulmonary capillary wedge pressure, pulmonary artery diastolic pressure and an increase in cardiac output (Crozier et al. 1995; Regitz-Zagrosek et al. 1995; Mazayev et al. 1998; McKelvie et al. 1999).

However, are AT_1 receptor antagonists superior to ACE inhibitors? The ELITE I trial (Evaluation of Losartan In The Elderly) in a small elderly population with heart failure suggested a 46% decrease in mortality in those treated with losartan (50 mg/d) compared with captopril (50 mg t.i.d) treated subjects (Pitt et al. 1997). A larger trial (ELITE II) with the same design and recruiting 3,152 patients failed to confirm the earlier result (Pitt et al. 1999). Caution is needed in interpreting these results, as the low dose of losartan (50 mg/d) employed might be insufficient to block efficiently the increased RAS activity in heart failure in contrast with full ACE inhibition with captopril. However, a meta-analysis of 17 randomized relatively small controlled trials that compared AT_1 receptor antagonists with placebo or ACE inhibitors in 12,469 patients with heart failure indicated that the AT_1 receptor antagonists were not superior to comparators in reducing mortality (Jong et al. 2002).

Subgroup analysis of patients included in the VAL-Heft trial and treated with valsartan with or without β-blockers but not receiving an ACE inhibitor appears more promising (Cohn and Tognoni 2002; Jong et al. 2002). In this population, there was a 44% reduction of mortality/morbidity ($p<0.001$) and a 33.1% reduction in all-cause mortality ($p=0.017$). The rate of hospitalization for heart failure was significantly reduced by 53% ($p<0.01$). These results support the use of AT_1 receptor antagonists with or without β blocker in patients intolerant to ACE inhibitor. Proof of the effectiveness of AT_1 receptor antagonists in heart failure was recently obtained with the results of the CHARM trial (Candesartan in Heart failure—Assessment of Reduction in Mortality and morbidity) (Swedberg et al. 1999). This study includes an arm that compares candesartan and placebo in approximately 2,000 patients intolerant to ACE inhibitors. This comparison indicates with sufficient power that candesartan significantly reduced mortality and morbidity in patients with symptomatic chronic heart failure (Granger et al. 2003).

As ACE inhibitors prescribed at the recommended dose do not fully inhibit circulating ACE in heart failure and Ang II levels may be further enhanced by alternative pathways (Jorde et al. 2000), combination therapy with AT_1 receptor antagonist and ACE inhibitor has potential advantages. This may lead to a more complete blockade of the RAS, while maintaining any beneficial effect of bradykinin (McDonald et al. 1995). In agreement with this hypothesis, a superior effect of this combination on ejection fraction, end diastolic and systolic volumes and brain natriuretic peptide (BNP) levels as observed in RESOLVD (Randomized Evaluation of Strategies for Left Ventricular Dysfunction) (McKelvie et al. 1999). The larger Val-Heft trial confirmed the beneficial effect of valsartan as an add-on therapy (to best conventional therapy including ACE inhibitor) on combined mortality/morbidity, with a risk reduction of 13.2% but with no significant effect on total mortality (Cohn and Tognoni 2002). CHARM (McMurray et al. 2003), has confirmed the beneficial effect of the combination therapy in significantly reducing cardiovascular death (adjusted hazard ratio, 0.83). VALIANT

(Valsartan In Acute myocardial infarction) (Pfeffer et al. 2000) and ONTARGET (ONgoing Telmisartan alone And in combination with Ramipril Global Endpoint Trial) (Yusuf 2002) will further address the add-on hypothesis of additional benefits of ACEI inhibitor combined with AT_1 receptor antagonist. These trials should allow a clearer conclusion in the near future.

2.3 Postmyocardial Infarction

Two outcome trials (OPTIMAAL and VALIANT) are evaluating the role of AT_1 receptor antagonists in patients with LV dysfunction following recent myocardial infarction. OPTIMAAL, which is powered to detect a difference between losartan 50 mg daily and captopril 50 mg thrice daily, has been completed (Dickstein et al. 1999). The much larger VALIANT trial is adequately powered not only to compare outcomes in patients treated with high doses of valsartan (up to 160 mg twice daily) and ACE inhibitor, but also to test the value of combination therapy of AT_1 receptor antagonism with ACE inhibitor and, crucially, the noninferiority of valsartan compared with an ACE inhibitor. The results of VALIANT were recently reported (Pfeffer et al. 2003). As equivalence of valsartan twice a day with captopril three times a day was confirmed, the superior tolerability of the AT_1 receptor antagonist appears decisive in guiding future clinical utility.

2.4 Renal Disease

The AT_1 receptor antagonist have effects on kidney function similar to those of ACEI inhibitors: renal vasodilation, increased renal blood flow without significant changes in glomerular filtration rate, resulting in a reduction of filtration fraction (Cooper et al. 2001). This reflects a preferential dilation of the efferent over the afferent arterioles (Fauvel et al. 1996; Price et al. 1997; MacKenzie et al. 1997; Pechere-Bertschi et al. 1998; Plum et al. 1998; Fricker et al. 1998). Better preservation of glomerular filtration rate with AT_1 receptor antagonists compared with ACE inhibitors should be beneficial in patients with renal impairment. There is a modest natriuretic effect following blockade of sodium reabsorption and aldosterone release through blockade of the AT_1 receptor at proximal tubular and adrenal zona glomerulosa sites (Burnier and Brunner 1994). This is amplified in the volume-depleted state (Burnier et al. 1996). The natriuretic effect of AT_1 receptor antagonists (like that of ACE inhibitors) is abolished by indomethacin, suggesting a non-AT_1 receptor-dependent mechanism (Fricker et al. 1998).

Like ACE inhibitors, AT_1 receptor antagonists attenuate microalbuminuria and proteinuria, especially in noninsulin-dependent diabetics (Muirhead et al. 1999). Valsartan decreases the glomerular sieving coefficient, especially of large pores, in patients with chronic renal failure (Plum et al. 1998). This improvement of structural membrane properties as well as influence on glomerular hydraulic pressure may explain the marked decrease in urinary protein excretion with only minor changes in glomerular filtration rate.

Several clinical trials have confirmed the beneficial effect of AT_1 receptor antagonists in diabetic nephropathy. RENAAL (losartan vs conventional antihypertensive treatment), IDNT (irbesartan vs amlodipine or conventional therapy) have demonstrated that AT_1 receptor antagonists decrease proteinuria and slow the progression of the renal disease to a greater extent than calcium channel blockers, diuretics or β blockers at equivalent levels of blood pressure control (Brenner et al. 2001; Lewis et al. 2001). Moreover, in incipient diabetic nephropathy characterized by microalbuminuria [IRMA 2 (irbesartan vs conventional therapy), MARVAL (valsartan vs amlodipine)], AT_1 receptor antagonists are superior in reducing urinary protein and in preserving kidney function (Parving et al. 2001; Viberti et al. 2002). As observed with ACEI inhibitors, the renoprotective effect of AT_1 receptor antagonists occurs independently of blood pressure lowering, probably due to a greater fall in glomerular capillary pressure and a better antifibrotic effect than observed with other antihypertensive drugs. Preclinical studies in various animal models have demonstrated a significant reduction in TGF-beta and osteonectin (SPARC) with an increased nephrin in the kidney of diabetic animals after AT_1 receptor blockade (Kelly et al. 2000; Cao et al. 2001; Cooper et al. 2001; Wilkinson-Berka et al. 2001).

3 Possible Mechanisms Beyond AT_1 Receptor Antagonism

There is accumulating evidence indicating that the AT_2 receptor counterbalances the effect of the AT_1 receptor. Usually expressed at low density in the adult, the AT_2 receptor is re-expressed in pathological circumstances such as congestive heart failure, myocardial infarction, kidney insufficiency, cerebral ischaemia or neuronal lesions by reactivation of a fetal gene program. (Nio et al. 1995; Makino et al. 1996; Rogg et al. 1996; Gallinat et al. 1998). Long-term treatment with AT_1 receptor antagonist results in an increase in plasma Ang II and therefore in a stimulation of the unblocked AT_2 receptor (de Gasparo and Siragy 1999; Horiuchi et al. 1999; Matsubara 1998). The AT_2 receptor has antiproliferative properties, it stimulates vasodilation through production of nitric oxide, induces transient apoptosis, and has antiangiogenic, neurotrophic and antifibrotic properties (Tsutsumi et al. 1999; Siragy et al. 2000; Henrion et al. 2001; Silvestre et al. 2002; Siragy et al. 2002). Stimulation of the AT_2 receptor has been reported to preserve left ventricular function after myocardial infarction (Yang et al. 2002).

The pioneering work of Carretero's group (Liu et al. 1997) has indicated a reduced beneficial effect of losartan in rats with ligated coronary arteries after blockade of the AT_2 and the bradykinin B2 receptor. The effect of ACE inhibitor as well as that of AT_1 receptor antagonist was similarly diminished in endothelial NO synthase knockout mice with heart failure compared to the wild type control animals, suggesting a kinin/NO-mediated mechanism (Liu et al. 2002). Stimulation of the AT_2 receptor is also accompanied by an increased of tissue interstitial bradykinin, NO and cGMP in kidney as well as in the heart (Tsutsumi et al.1999; Siragy et al. 2000, 2002). The AT_2 receptor could even be a physiological AT_1 receptor antagonist by forming a heterodimer with the AT_1 receptor, independently of any stimulation (Abdalla et al. 2001).

In contrast to Ichihara et al. (2001), Wu and co-workers (2002) recently reported that systolic blood pressure, cardiac function as well as remodelling did not differ between wild type and AT_2 knockout mice following myocardial infarction. However, the increase in ejection fraction and cardiac output and the decrease in LV diastolic dimension, myocyte cross-sectional area and interstitial collagen fraction were significantly reduced in AT_2 KO mice treated with valsartan compared to those treated with enalapril, which similarly improved cardiac function and remodelling in both strains. These results suggest that the AT_2 receptor does not play an important role in regulation of cardiac function under basal and post-MI remodelling conditions and in the beneficial effect of ACE inhibition but is essential for the therapeutic effect of the AT_1 receptor antagonist.

This counter-regulatory hypothesis has been challenged by conflicting data in other preclinical models (Mazzolai et al. 2000; Mifune et al. 2000; Senbonmatsu et al. 2000; Ichihara et al. 2001; Opie and Sack 2001; Schneider and Lorell, 2001; Wolf 2002). Furthermore, the results of the large clinical trials (Val-Heft, LIFE) can be explained without ascribing a major protective role to the unopposed stimulation of the AT_2 receptor. However, the role of AT_2 receptor stimulation in the action of AT_1 receptor antagonists is supported by recent observation of Malmqvist et al. (2003), who reported a negative correlation between the Ang II plasma levels and LV mass in irbesartan-treated patients with LVH. The greater the Ang II, the lower the cardiac hypertrophy. Unfortunately, the lack of pharmacological tools such as AT_2 receptor agonist and antagonist that can be used in the clinical setting does not allow the direct verification of such a finding in humans.

4
Perspective

Encouraged by the proven benefit of AT_1 receptor antagonists in hypertension, heart failure and kidney diseases, other indications for this new class of drug are under investigation.

4.1
Prevention of Diabetes Mellitus

The results of LIFE, RENAAL, IRMA, CHARM and CALM suggest that AT_1 receptor antagonists slow the progression or even delay the appearance of diabetes. Moreover, the observation that the RAS blockade increases adiponectin and improves insulin sensitivity in essential hypertension has some potential for the prevention of diabetes (Furuhashi et al. 2003). NAVIGATOR (Nateglimide and Valsartan in Impaired Glucose Tolerance Outcome Trial) is evaluating valsartan and nateglinide alone and in combination in patients with impaired glucose tolerance to further assess the effect of AT_1 receptor antagonism and the development of full blown diabetes in a prediabetic population (Califf et al. 2002).

4.2 Diabetic Proliferative Retinopathy

AT_1 receptor stimulation mediates VEFG and eNOS production and the AT_2 receptor stimulation appears to negatively modulate ischaemia-induced angiogenesis through activation of an apoptotic process (Silvestre et al. 2002). These mechanisms appear to be involved in the genesis of diabetic retinopathy. There is a close inter-relationship between diabetic nephropathy and proliferative retinopathy (Gilbert et al. 1998). Therefore, by extrapolating the results of RENAAL, IDNT and IRMA, a beneficial effect of the RAS blockade in the progression of the diabetic retinopathy might be anticipated. In transgenic m(ren-2)27 rats, which overexpress renin in tissues, losartan, like lisinopril, prevented inner retinal blood vessel growth induced by a sequence of hyperoxia and hypoxia (Moravski et al.2000). DIRECT (Diabetic Retinopathy Candesartan Trial) is a study investigating the effect of candesartan in diabetic retinopathy and should establish whether AT_1 receptor antagonists have a role in the management of this condition.

4.3 Glaucoma

The rabbit eye expresses the various components of the RAS (Ramirez et al. 1996). Captopril reduces intraocular pressure in normal volunteers and in patients with primary open-angle glaucoma (Costagliola et al. 1995). In a small pilot study, losartan administered orally decreased ocular pressure, probably by reducing the rate of aqueous humour production (Costagliola et al. 1999) More work is required.

4.4 Atherosclerosis

Experimental studies have shown a relationship between Ang II and lipid deposition in arteries, reactive oxygen species, monocyte activation, stimulation of adhesion molecules (ICAM, VCAM) and increased oxidized LDL-cholesterol deposition in monocytes (Strawn et al. 2000a; Dzau 2001). The AT_1 receptor mediates the inflammatory reaction to Ang II through stimulating nuclear factor NFκB, several proto-oncogenes (cfos, cjun, cmyc) and various growth factors such as TGFβ, PDGF. Although a beneficial effect of AT_1 receptor antagonism has been observed in various animal models of diet-induced atherogenesis, similar inhibition of the atherosclerotic process has not yet been documented in humans (De La Heras et al. 1999; Strawn et al. 2000b).

4.5 Cancer

Angiogenesis is essential for tumour progression and metastasis. In vitro as well as in vivo, the AT_1 receptor antagonists have clear antiangiogenic properties due to their inhibitory effect on proangiogenic growth factors such as VEGF, bFGF and PFGF. AT_1 receptor antagonists also inhibit growth by blocking the activity of various kinase pathways (Greco et al. 2003). Potentially, these drugs may therefore constitute a novel approach in prevention of cancer. ACE inhibition has been reported to reduced the incidence of cancer in a retrospective cohort study of hypertensive patients (Lever et al. 1998). There is a slight excess of cancer in hypertensive patients, mainly because of an association with renal cell cancer (Grossman et al. 2002). The possible role of AT_1 receptor antagonists in lowering cancer risk in hypertension merits further study.

4.6 Lung Diseases

Human lung expresses both AT_1 and AT_2 receptors. The AT_1 receptor is essentially localized on vascular smooth muscle cells, macrophages and in the stroma underlying the airway epithelium. In contrast, The AT_2 receptor is found in bronchial epithelial cell brush border and in mucous glands. In patients with chronic obstructive pulmonary disease (COPD), there is a more than fivefold increase in the AT_1/AT_2 ratio with a significant overexpression of the AT_1 receptor (Bullock et al. 2001). This observation suggests a potential role of Ang II in lung damage. In vitro, losartan inhibits Ang II-induced human lung fibroblast proliferation via inhibition of TGF-beta production (Marshall et al. 2000). Valsartan also inhibits in vivo fibrosis associated with bleomycin treatment not by inhibition of inflammatory cell infiltration into the airways, but by inhibition of the release of mediators such as TGF-beta and MCP-1 (Battram et al. 2002). AT_1 receptor antagonists may therefore be beneficial in the treatment of pulmonary fibrotic and airway diseases.

5 Conclusion

The pharmacology AT_1 receptor antagonists offers a novel approach to the management of hypertension and other diseases. These agents provide specific blockade of the RAS through a highly selective action in the AT_1 receptor which mediates the detrimental effects of Ang II. Unopposed stimulation of AT_2 receptors which modulate favourable influences of Ang II may further contribute to the overall action. The specificity of the mechanism of action is associated with unsurpassed tolerability. All available agents have prolonged duration of action allowing once daily administration, which further facilitates patient compliance.

The haemodynamic properties of AT_1 receptor antagonists mean that these drugs have favourable actions in hypertension and heart failure. Recent large-scale prospective trials have indicated that these effects are translated into outcome benefits. In heart failure, AT_1 receptor antagonists appear to have advantages equivalent to and additive to those of ACE inhibitors. High-risk hypertensive patients such as those with type 2 diabetes and nephropathy or LVH treated with AT_1 receptor antagonists gain morbidity and mortality benefits beyond those which can be explained by blood pressure reduction.

Several ongoing trials should further define the role of AT_1 receptor antagonists in patients with hypertension, heart failure, LV dysfunction postmyocardial infarction, impaired glucose tolerance and high cardiovascular risk. The future prospects for AT_1 receptor antagonists is good. Preliminary evidence supports the potential for preventing the progression of type 2 diabetes mellitus, atherosclerosis and even cancer, for attenuation of microvascular complications of diabetes and in the management of glaucoma and lung diseases. A full description of the pharmacology of AT_1 receptor antagonists must await evaluation in all these conditions.

Acknowledgements. The precious scientific comments and linguistic assistance of Gordon T. McInnes are greatly appreciated.

References

ABC (2000) Evaluation of candesartan cilexetil in black patients with systemic hypertension: the ABC Trial. Heart Dis 2:392–399

Abdalla S, Lother H, Abdel-Tawab AM, Quitterer U (2001) The angiotensin II AT_2 receptor is an AT_1 receptor antagonist. J Biol Chem 276:39721–39726

Andersson OK, Neldam S (1998) The antihypertensive effect and tolerability of candesartan cilexetil, a new generation angiotensin II antagonist, in comparison with losartan. Blood Press 7:53–59

Argenziano L, Trimarco B (1999) Effect of eprosartan and enalapril in the treatment of elderly hypertensive patients: subgroup analysis of a 26-week, double-blind, multicentre study. Eprosartan Multinational Study Group. Curr Med Res Opin 15:9–14

Azizi M, Bernard MC, Ménard J (1999a) Pharmacodynamic-pharmacokinetic study of renin release during angiotensin II blockade. J Hypertens 17 [Suppl 3]:S194

Azizi M, Chatellier G, Guyene TT, Ménard J (1999b) Pharmacokinetics-pharmacodynamic interactions of candesartan cilexetil and losartan. J Hypertens 17:561–568

Ball KJ, Williams PA, Stumpe KO (2001) Relative efficacy of an angiotensin II antagonist compared with other antihypertensive agents. Olmesartan medoxomil versus antihypertensives. J Hypertens Suppl 19 [Suppl 1]:S49–S56

Balt JC, Mathy MJ, Pfaffendorf M, van Zwieten PA (2001) Inhibition of facilitation of sympathetic neurotransmission and angiotensin II-induced pressor effects in the pithed rat: comparison between valsartan, candesartan, eprosartan and embusartan. J Hypertens. 19:2241–2250

Battram CH, Bayley D, Campbell EM, Giddings J, Jones G, Lewis C, Schindler F, Poll C (2002) Impact of valsartan (Diovan) in a murine model of pulmonary fibrosis. Am J Respir Crit Care Med 165 [Suppl 8,2]:A554

Belz GG (2001) Pharmacological differences among angiotensin II receptor antagonists. Blood Press Suppl 2:13–18

Belz GG (2003) Angiotensin II dose-effect curves and Schild regression plots for characterization of different angiotensin II AT_1 receptor antagonists in clinical pharmacology. Br J Clin Pharmacol 56:3–10

Belz GG, Butzer R, Kober S, Mang C, Mutschler E (1999) Time course and extent of angiotensin II antagonism after irbesartan, losartan, and valsartan in humans assessed by angiotensin II dose response and radioligand receptor assay. Clin Pharmacol Ther 66:367–373

Belz GG, Breithaupt-Grogler K, Butzer R, Fuchs W, Hausdorf C, Mang C (2000) The pharmacological potency of various AT_1 antagonists assessed by Schild regression technique in man. J Renin Angiotensin Aldosterone Syst 1:336–341

Bernhart CA, Perreaut PM, Ferrari BP et al (1994) A new series of imidazoleones: highly specific and potent nonpeptide AT_1 angiotensin II receptor antagonist. J Med Chem 36:3371–3380

Black HR, Graff A, Shute D, Stoltz R, Ruff D, Levine J, Shi Y, Mallows S (1997) Valsartan, a new angiotensin II antagonist for the treatment of essential hypertension: efficacy, tolerability and safety compared to an angiotensin-converting enzyme inhibitor, lisinopril. J Hum Hypertens 11:483–489

Bloom BS (1998) Continuation of initial antihypertensive medication after 1 year of therapy. Clin Ther 20:671–681

Brenner BM, Cooper ME, de Zeeuw D, Keane WF, Mitch WE, Parving HH, Remuzzi G, Snapinn SM, Zhang Zhonxin, Shahinfar S, RENAAL Study Investigators (2001) Effects of losartan on renal and cardiovascular outcomes in patients with type 2 diabetes and nephropathy. N Engl J Med 345:861–869

Brunner HR (2002) The new oral angiotensin II antagonist olmesartan medoxomil: a concise overview. J Hum Hypertens 16 [Suppl 2]:S13–S16

Buhlmayer P, Furet P, Criscione L, de Gasparo M, Whitebread S, Schmidlin T, Lattmann R, Wood J (1994) Valsartan, a potent, orally active angiotensin II antagonist developed from the structurally new amino acid series. Bioorg Med Chem Lett 4:29–34

Bullock GR, Steyaert I, Bilbe G, Carey RM, Kips J, dePaepe B, Pauwels R, Praet M, Siragy HM, de Gasparo M (2001) Distribution of type-1 and type-2 angiotensin receptors in the normal human lung and in lungs from patients with chronic obstructive pulmonary disease. Histochem Cell Biol 115:117–124

Burnier M, Brunner HR (1994) Angiotensin II receptor antagonists and the kidney. Curr Opin Nephrol Hypertens 3:537–545

Burnier M, Roch-Ramel F, Brunner HR (1996) Renal effects of angiotensin II receptor blockade in normotensive subjects. Kidney Int 49:1787–1790

Califf R, Haffner S, McMurray J, Picher E, Pratley R (2002) Nateglinide and valsartan in impaired glucose tolerance outcomes research: rationale and design of the NAVIGATOR trial. Diabetes 51 [Suppl 2]:470-P, A116

Cao Z, Bonnet F, Davis B, Allen TJ, Cooper ME (2001) Additive hypotensive and anti-albuminuric effects of angiotensin-converting enzyme inhibition and angiotensin receptor antagonism in diabetic spontaneously hypertensive rats. Clin Sci (Colch) 100:591–599

Chan JC, Critchley JA, Lappe JT, Raskin SJ, Snavely D, Goldberg AI, Sweet CS (1995) Randomised, double-blind, parallel study of the anti-hypertensive efficacy and safety of losartan potassium compared with felodipine ER in elderly patients with mild to moderate hypertension. J Hum Hypertens 9:765–771

Christen Y, Waeber B, Nussberger J, Lee RJ, Timmermans PBMWM, Brunner HR (1991) Dose-response relationships following oral administration of DuP 753 to normal humans. Am J Hypertens 4:350S–354S

Cohen A, Bregman B, Agabiti-Rosei E, Williams B, Dubourg O, Clairefond P, Brudi P, Gosse P, Gueret P (1998) Comparison of irbesartan vs felodipine in the regression after 1 year of left ventricular hypertrophy in hypertensive patients (the SILVER trial). Study of irbesartan in left ventricular hypertrophy regression. J Hum Hypertens 12:479–483

Cohn JN (2002) Hypertension therapeutic research: a plea for change in the new millennium. Am J Hypertens 15:97–98

Cohn JN, Tognoni G (2002) A randomized trial of the angiotensin-receptor blocker valsartan in chronic heart failure. N Engl J Med 345:1667–1675

Conlin PR, Spence JD, Williams B, Ribeiro AB, Saito I, Benedict C, Bunt AMG (2000) Angiotensin II antagonists for hypertension: are there differences in efficacy? J Hypertens 13:418–426

Cooper ME, Webb RL, de Gasparo M (2001) Angiotensin receptor blockers and the kidney: possible advantages over ACE inhibition? Cardiovasc Drug Rev 19:75–86

Costagliola C, Di Benedetto R, De Caprio L et al (1995) Effect of oral captopril (SQ14225) in intraocular pressure in man. Eur J Ophthalmo 5:19–25

Costagliola C, Verolino M, Iaccarino G, De Rosa ML, Ciancaglini M, Mastropasqua L (1999) losartan potassium oral administration decreased intraocular pressure in humans. A pilot study. Clin Drug Invest 17:329–332

Crozier I, Ikram H, Awan N, Cleland J, Stephen N, Dickstein K, Frey M, Young J, Klinger G, Makris L et al (1995) Losartan in heart failure. Hemodynamic effects and tolerability. Losartan Hemodynamic Study Group. Circulation 91:691–697

Dahlof B, Keller SE, Makris L, Goldberg AI, Sweet CS, Lim NY (1995) Efficacy and tolerability of losartan potassium and atenolol in patients with mild to moderate essential hypertension. Am J Hypertens 8:578–583

Dahlof B, Devereux RB, Kjeldsen SE, Julius S, Beevers G, Faire U, Fyhrquist F, Ibsen H, Kristiansson K, Lederballe-Pedersen O, Lindholm LH, Nieminen MS, Omvik P, Oparil S, Wedel H et al (2002) Cardiovascular morbidity and mortality in the Losartan Intervention For Endpoint reduction in hypertension study (LIFE): a randomised trial against atenolol. Lancet 359:995–1003

De Gasparo M, Whitebread S (1995) Binding of valsartan to mammalian angiotensin AT_1 receptors. Regul Pept 59:303–311

De Gasparo M, Siragy HM (1999) The AT_2 receptor: fact, fancy and fantasy. Regul Pept 81:11–24

De la Heras N et al (1999) AT(1) receptor antagonism reduces endothelial dysfunction and intimal thickening in atherosclerotic rabbits. Hypertension 34:969–975

Degli EE, Sturani A, Di Martino M, Falasca P, Novi MV, Baio G, Buda S, Volpe M (2002) Long-term persistence with antihypertensive drugs in new patients. J Hum Hypertens 16:439–444

Delacretaz E, Nussberger J, Biollaz J, Waeber B, Brunner HR (1995) Characterization of the angiotensin II receptor antagonist TCV-116 in healthy volunteers. Hypertension 25:14–21

Devereux RB, Palmieri V, Liu JE, Wachtell K, Bella JN, Boman K, Gerdts E, Nieminen MS, Papademetriou V, Dahlof B (2002) Progressive hypertrophy regression with sustained pressure reduction in hypertension: the Losartan Intervention For Endpoint Reduction study. J Hypertens 20:1445–1450

Dickstein K, Kjekshus J, for the OPTIMAAL Study group (1999) Comparison of the effects of losartan and captopril on mortality in patients after acute myocardial infarction: the OPTIMAAL trial design. Optimal therapy in myocardial infarction with the angiotensin II antagonist losartan. Am J Cardiol 83:477–481

Dzau VJ (2001) Tissue angiotensin and pathobiology of vascular disease. A unifying hypothesis. Hypertension 37:1047–1052

Easthope SE, Jarvis B (2002) Candesartan cilexetil. An update of its use in essential hypertension. Drugs 62:1253–1287

Ellingboe JW, Antane M, Nguyen TT, Collini MD, Antane S, Bender R, Hartupee D, White V, McCallum J, Park CH et al (1994) Pyrido[2,3-d] pyrimidine angiotensin II antagonists. J Med Chem 37:542–550

Fauvel JP, Velon S, Berra N, Pozet N, Madonna O, Zech P, Laville M (1996) Effects of losartan on renal function in patients with essential hypertension. J Cardiovasc Pharmacol 28:259–263

Fierens FLP, Vanderheyden PML, De Backer JP, Vauquelin G (1999) Binding of the antagonist [3H]candesartan to angiotensin II AT_1 receptor-transfected chinese hamster ovary cells. Eur J Pharmacol 367:413–422

Fogari R, Zoppi A, Mugellini A, Preti P, Banderali A, Pesce RM, Vanasia A (1999) Comparative efficacy of losartan and valsartan in mild-to-moderate hypertension: results of 24 hour ambulatory blood pressure monitoring. Curr Ther Res 60:195–206

Fogari R, Mugellini A, Zoppi A, Fogari E, Marasi G, Pesce RM, Banderali A (2000) A double-blind, crossover study of the antihypertensive efficacy of angiotensin II-receptor antagonists and their activation of the renin-angiotensin system. Curr Ther Res 61:669–679

Fricker AF, Nussberger J, Meilenbrock S, Brunner HR, Burnier M (1998) Effect of indomethacin on the renal response to angiotensin II receptor blockade in healthy subjects. Kidney Int 54:2089–2097

Frye CB, Pettigrew TJ (1998) Angioedema and photosensitive rash induced by valsartan. Pharmacotherapy 18:866–868

Furakawa Y, Kishimoto S, Nishikawa K (1982) Hypotensive imidazole derivatives and hypotensive imidazole-5-acetic acid derivatives. Patents issued to Takeda Chemical Industries Ltd. on July 20, 1982, and October 19, 1982, respectively. US Patents 4,340,598 and 4,355,040, Osaka, Japan, 1982

Furuhashi M, Ura N, Higashiura K, Murakami H, Tanaka M, Moniwa N, Yoshida D, Shimamoto K (2003)Blockade of the renin-angiotensin system increases adiponectin concentrations in patients with essential hypertension. Hypertension.42:76–81

Gallinat S, Yu M, Dorst A, Unger T, Herdegen T (1998) Sciatic nerve transection evokes lasting up-regulation of angiotensin AT_2 and AT_1 receptor mRNA in adult rat dorsal root ganglia and sciatic nerves. Brain Res 57:111–122

Gavras I, Gavras H (1999) Effects of eprosartan versus enalapril in hypertensive patients on the renin-angiotensin-aldosterone system and safety parameters: results from a 26-week, double-blind, multicenter study. Eprosartan Multinational Study group. Curr Med Res Opin 15:15–24

Gilbert RE, Tsalamandris C, Allen TJ, Colville D, Jerums G (1998) Early nephropathy predicts vision-threatening retinal disease in patients with type I diabetes mellitus. J Am Soc Nephrol 9:85–89

Gillis JC, Markham A (1997) Irbesartan: a review of its pharmacology and pharmakokinetic properties and therapeutic use in the management of hypertension. Drugs 54:885–902

Goa KL, Wagstaff AJ (1996) Losartan potassium: a review of its pharmacology, clinical efficacy and tolerability in the management of hypertension. Drugs 51:820–845

Goldberg MR, Lo MW, Deutsch PJ, Wilson SE, McWilliams EJ, McCrea JB (1996) Phenobarbital minimally alters plasma concentrations of losartan and its active metabolite E-3174. Clin Pharmacol Ther 59:268–274

Gradman AH, Arcuri KE, Goldberg AI, Ikeda LS, Nelson EB, Snavely DB, Sweet CS (1995) A randomized, placebo-controlled, double-blind, parallel study of various doses of losartan potassium compared with enalapril maleate in patients with essential hypertension. Hypertension 25:1345–1350

Gradman AH, Gray J, Maggiacomo F, Punzi H, White WB (1999a) Assessment of once-daily eprosartan, an angiotensin II antagonist, in patients with systemic hypertension. Eprosartan Study Group. Clin Ther 21:442–453

Gradman AH, Lewin A, Bowling BT, Tonkon M, Deedwania PC, Kezer AE, Hardison JD, Cushing DJ, Michelson EL (1999b) Comparative effects of candesartan cilexetil and losartan in patients with systemic hypertension. Candesartan Versus Losartan efficacy Comparison (CANDLE) Study group. Heart Dis 1:52–57

Granger CB, McMurray JJV, Yusuf S, Held P, Michelson EL, Olofsson B, Östergren J, Pfeffer MA, Swedberg K, for the CHARM Investigators and Committees (2003) Effects of candesartan in patients with chronic heart failure and reduced left-ventricular systolic function intolerant to angiotensin-converting-enzyme inhibitors: the CHARM-Alternative trial. Lancet 362:772–776

Greco S, Muscella A, Elia MG, Salvatore P, Storelli C, Mazzotta A, Manca C, Marsigliante S (2003) Angiotensin II activates extracellular signal regulated kinases via protein kinase C and epidermal growth factor receptor in breast cancer cells. J Cell Physiol 196:370–377

Grossman E, Messerli FH, Boyko V, Goldbourt U (2002) Is there an association between hypertension and cancer mortality? Am J Med 112:479–486

Hansson L, Lithell H, Skoog I, Baro F, Banki CM, Breteler M, Carbonin PU, Castaigne A, Correia M, Degaute JP, Elmfeldt D et al (1999) Study on COgnition and Prognosis in the Elderly (SCOPE). Blood Press 8:177–183

Hedner T, Oparil S, Rasmussen K, Rapelli A, Gatlin M, Kobi P, Sullivan J, Oddou-Stock P (1999) A comparison of the angiotensin II antagonists valsartan and losartan in the treatment of essential hypertension. Am J Hypertens 12:414–417

Henrion D, Kubis N, Levy BI (2001) Physiological and pathophysiological functions of the AT_2 subtype receptor of angiotensin II: from large arteries to the microcirculation. Hypertension 38:1150–1157

Holwerda NJ, Fogari R, Angeli P, Porcellati C, Hereng C, Oddou-Stock P, Heath R, Bodin F (1996) Valsartan, a new angiotensin II antagonist for the treatment of essential hypertension: efficacy and safety compared with placebo and enalapril. J Hypertens 14:1147–1151

Horiuchi M, Akishita M, Dzau VJ (1999) Recent progress in angiotensin II type 2 receptor research in the cardiovascular system. Hypertension 33:613–621

Howes LG, Tran D (2002) Can angiotensin receptor antagonists be used safely in patients with previous ACE inhibitor-induced angioedema? Drug Safety 25:73–76

Ichihara S, Senbonmatsu T, Price E, Ichiki T, Gaffney FA, Inagami T (2001) Angiotensin II type 2 receptor is essential for left ventricular hypertrophy and cardiac fibrosis in chronic angiotensin II-induced hypertension. Circulation 104:346–351

Jong P, Demers C, McKelvie RS (2002) Angiotensin receptor blockers in heart failure: meta-analysis of randomized controlled trials. J Am Coll Cardiol 39:463–470

Jonkman JHG, van Lier JJ, van Heiningen PNM, Lins R, Sennewald R, Högemann A (1997) Pharmacokinetic drug interaction studies with candesartan cilexetil. J Hum Hypertens 11 [Suppl 2]: S31–S35

Jorde UP, Ennezat PV, Lisker J, Suryadevara V, Infeld J, Cukon S, Hammer A, Sonnenblick EH, Le Jemtel TH (2000) Maximally recommended doses of angiotensin-converting enzyme (ACE) inhibitors do not completely prevent ACE-mediated formation of angiotensin II in chronic heart failure. Circulation 101:844–846

Judd DB, Dowle MD, Middlemiss D, Scopes DI, Ross BC, Jack TI, Pass M, Tranquillini E, Hobson JE, Panchal TA (1994) Bromobenzofuran-based non-peptide antagonists of angiotensin II: GR138950, a potent antihypertensive agent with high oral bioavailability. J Med Chem 37:3108–3120

Kazierad DJ, Martin DE, Blum RA, Tenero DM, Ilson B, Boike SC, Etheredge R, Jorkasky DK (1997) Effect of fluconazole on the pharmacokinetics of eprosartan and losartan in healthy male volunteers. Clin Pharmacol Ther 62:417–425

Kelly DJ, Skinner SL, Gilbert RE, Cox AJ, Cooper ME, Wilkinson-Berka JL (2000) Effects of endothelin or angiotensin II receptor blockade on diabetes in the transgenic (mRen-2)27 rat. Kidney Int 57:1882–1894

Kloner RA, Weinberger M, Pool JL, Chrysant SG, Prasad R, Harris SM, Zyczynski TM, Leidy NK, Michelson EL (2001) Comparative effects of candesartan cilexetil and amlodipine in patients with mild systemic hypertension. Comparison of Candesartan and Amlodipine for Safety, Tolerability and Efficacy (CASTLE) Study Investigators. Am J Cardiol 87:727–731

Koike H, Sada T, Mizuno M (2001) In vitro and in vivo pharmacology of olmesartan medoxomil, an angiotensin II type AT_1 receptor antagonist. J Hypertens Suppl 19 [Suppl 1]:S3–S14

Kramer C, Sunkomat, Witte J, Luchtefeld M, Walden M, Schmidt B, Boger RH, Forssmann WG, Drexler H, Schieffer B (2002) Angiotensin II receptor-independent antiinflammatory and antiaggregatory properties of losartan: role of the active metabolite EXP3179. Circ Res 90:770–776

Lacourcière Y (1999) The incidence of cough: a comparison of lisinopril, placebo and telmisartan, a novel angiotensin II antagonist. Telmisartan Cough Study Group. Int J Clin Pract 53:99–103

Larochelle P, Flack JM, Marbury TC, Sareli P, Krieger EM, Reeves RA (1997) Effects and tolerability of irbesartan vs enalapril in patients with severe hypertension. Am J Cardiol 80:1613–1615

Lee CR, Pieper JA, Hinderliter AL, Blaisdell JA, Goldstein JA (2003) Losartan and E3174 pharmacokinetics in cytochrome P450 2C9*1/*1, *1/*2, and *1/*3 individuals. Pharmacotherapy 23:720–725

Lever AF, Hole DJ, Gillis CR, McCallum IR, McInnes GT, MacKinnon PL, Meredith PA, Murray LS, Reid JL, Robertson JW (1998) Do inhibitors of angiotensin-I-converting enzyme protect against risk of cancer? Lancet 352:179–184

Levine B (1999) Effect of eprosartan and enalapril in the treatment of black hypertensive patients: subgroup analysis of a 26-week, double-blind, multicentre study. Eprosartan Multinational Study Group. Curr Med Res Opin 15:25–32

Levy D, Larson MG, Vasan RS, Kannel WB, Ho KK (1996) The progression from hypertension to congestive heart failure. JAMA 275:1557–1562

Lewanczuk RZ (2002) Effects of inflammation on a model of cytochrome P450 dependent versus independent antihypertensives: losartan versus valsartan. Am J Hypertens 15:P-539

Lewis EJ, Hunsicker LG, Clarke WR, Berl T, Pohl MA, Lewis JB, Ritz E, Atkins RC, Rohde R, Raz I (2001) Renoprotective effect of the angiotensin-receptor antagonist irbesartan in patients with nephropathy due to type 2 diabetes. N Engl J Med 345:851–860

Li P, Fukuhara M, Diz DI, Ferrario CM, Brosnihan KB (2000) Novel angiotensin II AT(1) receptor antagonist irbesartan prevents thromboxane A(2)-induced vasoconstriction in canine coronary arteries and human platelet aggregation. J Pharmacol Exp Ther 292:238–246

Lindholm LH, Ibsen H, Dahlof B, Devereux RB, Beevers G, de Faire U, Fyhrquist F, Julius S, Kjeldsen SE, Kristiansson K, Lederballe-Pedersen O, Nieminen MS, Omvik P, Oparil S, Wedel H, Aurup P, Edelman J, Snapinn S (2002) Cardiovascular morbidity and mortality in patients with diabetes in the Losartan Intervention For Endpoint reduction in hypertension study (LIFE): a randomised trial against atenolol. Lancet 359:1004–1010

Lithell H, Hansson L, Skoog I, Elmfeldt D, Hofman A, Olofsson B, Trenkwalder P, Zanchetti A; SCOPE Study Group (2003) The Study on Cognition and Prognosis in the Elderly (SCOPE): principal results of a randomized double-blind intervention trial. J Hypertens 21:875–886

Liu EC, Hedberg A, Goldenberg HJ, Harris DN, Webb ML (1992) DuP 753, the selective angiotensin II receptor blocker, is a competitive antagonist to human platelet thromboxane A2/prostaglandin H2 (TP) receptors. Prostaglandins 44:89–99

Liu YH, Yang XP, Sharov VG, Nass O, Sabbah HN, Peterson E, Carretero OA (1997) Effects of angiotensin-converting enzyme inhibitors and angiotensin II type 1 receptor antagonists in rats with heart failure. Role of kinins and angiotensin II type 2 receptors. J Clin Invest 99:1926–1935

Liu YH, Xu J, Yang XP, Yang F, Shedely E, Carretero OA (2002) Effect of ACE inhibitors and angiotensin II type 1 receptor antagonists on endothelial NO synthase knockout mice with heart failure. Hypertension 39:375–381

Lopez-Farre A, Sanchez DM, Monton M, Jimenez A, Lopez-Bloya A, Gomez J, Nunez A, Rico L, Casado S (2001) Angiotensin II AT(1) receptor antagonists and platelet activation. Nephrol Dial Transplant 16 [Suppl 1]:45–49

MacKenzie HS, Ots M, Ziai F, Lee KW, Kato S, Brenner BM (1997) Angiotensin receptor antagonists in experimental models of chronic renal failure. Kidney Int Suppl 63:S140–S143

Maillard MP, Mazzolai L, Daven V et al (1999) Assessment of angiotensin II receptor blockade in humans using a standardized angiotensin II receptor-binding assay. Am J Hypertens 12:1201–1208

Makino I, Shibata K, Ohgami Y, Fujiwara M, Furukawa T (1996) Transient upregulation of the AT_2 receptor mRNA level after global ischemia in the rat brain. Neuropeptides 30:596–601

Malacco E, Piazza S, Meroni R, Milanesi A (2000) Comparison of valsartan and irbesartan in the treatment of mild to moderate hypertension: a randomized, open-label, crossover study. Curr Ther Res Clin Exp 61:789–797

Mallion JM, Siche JP, Lacourcière Y (1999) ABPM comparison of the antihypertensive profiles of the selective angiotensin II receptor antagonists telmisartan and losartan in patients with mild-to-moderate hypertension. J Hum Hypertens 13:647–664

Malmqvist K, Ohman KP, Lind L, Nystrom F, Kahan T (2003) Long-term effects of irbesartan and atenolol on the renin-angiotensin-aldosterone system in human primary hypertension: The Swedish Irbesartan Left Ventricular Hypertrophy Investigation versus Atenolol (SILVHIA). J Cardiovasc Pharmacol 42:719–726

Mancia G, Korlipara K, Van Rossum P, Villa G, Silvert B, Gressin V (2000) Irbesartan results in superior blood pressure reduction versus valsartan. J Hypertens 18 [Suppl 2]:S208–S209

Mann J, Julius S (1998) The Valsartan Antihypertensive Long-term Use Evaluation (VALUE) trial of cardiovascular events in hypertension. Rationale and design. Blood Press 7:176–183

Marino MR, Vachharajani NN (2001) Drug interactions with irbesartan. Clin Pharmacokinet 40:605–614

Markham A, Goa KL (1997) Valsartan. A review of its pharmacology and therapeutic use in essential hypertension. Drugs 54:299–311

Marshall RP, McAnulty RJ, Laurent GJ (2000) Angiotensin II is mitogenic for human lung fibroblasts via activation of the type 1 receptor. Am J Respir Crit Care Med 161:1999–2004

Matsubara H (1998) Pathophysiological role of angiotensin II type 2 receptor in cardiovascular and renal diseases. Circ Res 83:1182–1191

Mazayev VP, Fomina IG, Kazakov EN, Sulimov VA, Zvereva TV, Lyusov VA, Orlov VA, Olbinskaya LI, Bolshakova TD, Sullivan J, Spormann DO (1998) Valsartan in heart failure patients previously untreated with an ACE inhibitor. Int J Cardiol 65:239–246

Mazzolai L, Burnier M (1999) Comparative safety and tolerability of angiotensin II receptor antagonists. Drug Safety 21:23–33

Mazzolai L, Maillard M, Rossat J, Nussberger J, Brunner HR, Burnier M (1999) Angiotensin II receptor blockade in normotensive subjects: a direct comparison of three AT_1 receptor antagonists. Hypertension 33:850–855

Mazzolai L, Pedrazzini T, Nicoud F, Gabbiani G, Brunner HR, Nussberger J (2000) Increased cardiac angiotensin II levels induce right and left ventricular hypertrophy in normotensive mice. Hypertension 35:985–991

McClellan KJ, Goa KL (1998) Candesartan cilexetil: a review of its use in essential hypertension. Drugs 56:847–869

McClellan KJ, Markham A (1998) Telmisartan. Drugs 56:1039–1044

McDonald KM, Mock J, D'Aloia A, Parrish T, Hauer K, Francis G, Stillman A, Cohn JN (1995) Bradykinin antagonism inhibits the antigrowth effect of converting enzyme inhibition in the dog myocardium after discrete transmural myocardial necrosis. Circulation 91:2043–2048

McKelvie RS, Yusuf S, Pericak D, Avezum A, Burns RJ, Probstfield J, Tsuyuki RT, White M, Rouleau J, Latini R, Maggioni A, Young J, Pogue J (1999) Comparison of candesartan, enalapril, and their combination in congestive heart failure: Randomized Evaluation of Strategies for Left Ventricular Dysfunction (RESOLVD) Pilot Study: The RESOLVD Pilot Study Investigators. Circulation 100:1056–1064

McMurray JJV, Östergren, Swedberg K, Granger CB, Held P, Michelson EL, Olofsson B, Yusuf S, Pfeffer MA, for the CHARM Investigators and Committees (2003) Effects of candesartan in patients with chronic heart failure and reduced left-ventricular systolic function taking angiotensin-converting-enzyme inhibitors: the CHARM-Added trial. Lancet 362:767–771

Mifune M, Sasamura H, Shimizu-Hirota R, Miyazaki H, Saruta T (2000) Angiotensin type 2 receptors stimulate collagen synthesis in cultured vascular smooth muscle cells. Hypertension 36:845–850

Minghelli G, Seydoux C, Goy JJ, Burnier M (1998) Uricosuric effect of the angiotensin II receptor antagonist losartan in heart transplant recipients. Transplantation 66:268–271

Mitsunami K, InoueS, Maeda K, Endoh S, Takahashi M, Okada M, Sugihara H, Kinoshita M (1998) Three-month effects of candesartan cilexetil, an angiotensin II type 1 (AT_1) receptor antagonist, on left ventricular mass and hemodynamics in patients with essential hypertension. Cardiovasc Drugs Ther 12:469–474

Mogensen CE, Neldam S, Tikkanen I, Oren S, Viskoper R, Watts RW, Cooper ME (2000) Randomised controlled trial of dual blockade of renin-angiotensin system in patients with hypertension, microalbuminuria, and non-insulin dependent diabetes: the candesartan and lisinopril microalbuminuria (CALM) study. BMJ 321:1440–1444

Monterroso VH, Rodriguez Chavez V, Carbajal ET, Vogel DR, Aroca Martinez GJ, Garcia LH, Cuevas JHB, Teran JL, Hitzenberger G, Smith RD (2000) Use of ambulatory blood pressure monitoring to compare antihypertensive efficacy and safety of two angiotensin II receptor antagonists, losartan and valsartan. Adv Ther 17:117–131

Moravski CJ, Kelly DJ, Cooper ME, Gilbert RE, Bertram JF, Shahinfar S, Skinner SL, Wilkinson-Berka JL (2000) Retinal neovascularization is prevented by blockade of the renin-angiotensin system. Hypertension 36:1099–1104

Morimoto S, Ogihara T (1994) TCV-116: a new angiotensin II type 1 receptor antagonist. Cardiovasc Drugs Rev 12:153–164

Muirhead N, Feagan BF, Mahon J, Lewanczuk RZ, Rodger NW, Botteri F, Oddou-Stock P, Pecher E, Cheung R (1999) The effects of valsartan and captopril on reducing microalbuminuria in patients with type 2 diabetes mellitus: a placebo-controlled trial. Curr Ther Res 60:650–660

Neldam S, Forsen B (2001) Antihypertensive treatment in elderly patients aged 75 years or over: a 24-week study of the tolerability of candesartan cilexetil in relation to hydrochlorothiazide. Drugs Aging 18:225–232

Neutel JM, Smith DHG (1998) Dose response and antihypertensive efficacy of the AT_1 receptor antagonist telmisartan in patients with mild to moderate hypertension. Adv Ther 15:207–217

Neutel JM, Frishman WH, Oparil S, Papademetriou V, Guthrie G (1999) Comparison of telmisartan with lisinopril in patients with mild-moderate hypertension. Am J Ther 6:161–166

Nio Y, Matsubara H, Murasawa S, Kanasaki M, Inada M (1995) Regulation of gene transcription of angiotensin II receptor subtypes in myocardial infarction. J Clin Invest 95:46–54

Noda M, Shibouta Y, Inada Y, Ojima M, Wada T, Sanada T, Kubo K, Kohara Y, Naka T, Nishikawa K (1993) Inhibition of rabbit aortic angiotensin II (AII) receptor by CV-11974, a new nonpeptide AII antagonist. Biochem Pharmacol 46:311–318

Ohlstein EH, Brooks DP, Feuerstein GZ, Ruffolo RRJ (1997) Inhibition of sympathetic outflow by the angiotensin II receptor antagonist, eprosartan, but not by losartan, valsartan or irbesartan: relationship to differences in prejunctional angiotensin II receptor blockade. Pharmacology 55:244–251

Opie LH, Sack MN (2001) Enhanced angiotensin II activity in heart failure: reevaluation of the counterregulatory hypothesis of receptor subtypes. Circ Res 88:654–658

Pals DT, Masucci FD, Sipos F, Denning GSJ (1971) A specific competitive antagonist of the vascular action of angiotensin. II. Circ Res 29:664–672

Parving HH, Lehnert H, Brochner-Mortensen J, Gomis R, Andersen S, Arner P (2001) The effect of irbesartan on the development of diabetic nephropathy in patients with type 2 diabetes. N Engl J Med 345:870–878

Pechere-Bertschi A, Nussberger J, Decosterd L, Armagnac C, Sissmann J, Bouroudian M, Brunner HR, Burnier M (1998) Renal response to the angiotensin II receptor subtype 1 antagonist irbesartan versus enalapril in hypertensive patients. J Hypertens 16:385–393

Pfeffer MA, Leizorovicz A, Maggioni AP, Rouleau JL, Van DeWerf F, Henis M, Neuhart E, Gallo P, Edwards S, Sellers MA, Velazquez E, Califf R (2000) Valsartan in Acute Myocardial Infarction Trial (VALIANT): rationale and design. Am Heart J 140:727–750

Pfeffer MA, McMurray JJ, Velazquez E, Rouleau JL, Kober L, Maggioni AP, Solomon SD, Swedberg K, Van DeWerf F, White H, Leimberger JD, Henis M, Edwards S, Zelenkofske S, Sellers MA, Califf R (2003) Valsartan, Captopril, or both in myocardial infarction complicated by heart failure, left ventricular dysfunction, or both. N Engl J Med 349:1893–1906

Pitt B, Segal R, Martinez FA, Meurers G, Cowley AJ, Thomas I, Deedwania PC, Ney DE, Snavely DB, ChangPI on behalf of ELITE Study Investigators (1997) Randomised trial of losartan versus captopril in patients over 65 with heart failure (Evaluation of Losartan in the Elderly Study, ELITE). Lancet 349:747–752

Pitt B, Poole-Wilson P, Segal R, Marinez FA, Dickstein K, Camm AJ, Konstam MA, Riegger G, Klinger GH, Neaton J, Sharma D, Thiyagarajan B (1999) Effects of losartan versus captopril on mortality in patients with symptomatic heart failure: rationale, design, and baseline characteristics of patients in the losartan heart failure survival study, ELITE II. J Card Fail 5:146–154

Plum J, Bunten B, Nemeth R, Grabensee B (1998) Effects of the angiotensin II antagonist valsartan on blood pressure, proteinuria, and renal hemodynamics in patients with chronic renal failure and hypertension. J Am Soc Nephrol 9:2223–2234

Pool JL, Glazer R, Chiang YT, Gatlin M (1999) Dose-response efficacy of valsartan, a new angiotensin II receptor blocker. J Hum Hypertens 14:275–281

Price DA, De'Oliveira JM, Fisher ND, Hollenberg NK (1997) Renal hemodynamic response to an angiotensin II antagonist, eprosartan, in healthy men. Hypertension 30:240–246

Püchler K, Laeis P, Stumpe KO (2001) Blood pressure response, but not adverse event incidence, correlates with dose of angiotensin II antagonist. J Hypertens 19 [Suppl 1]:S41–S48
Ramirez M, Davidson EA, Luttenauer L, Elena PP, Cumin F, Mathis G, de Gasparo M (1996) The renin-angiotensin system in rabbit eye. J Ocular Pharmacol Ther 12:299–312
Reeves RA, Lin CS, Kassler-Taub K, Pouleur H (1998) Dose-related efficacy of irbesartan for hypertension: an integrated analysis. Hypertension 31:1311–1316
Regitz-Zagrosek V, Neuss M, Fleck E (1995) Effects of angiotensin receptor antagonists in heart failure: clinical and experimental aspects. Eur Heart J 16 [Suppl N]:86–91
Reif M, White WB, Fagan TC, Oparil S, Flanagan TL, Edwards DT, Cushing DJ, Michelson EL (1998) Effects of candesartan cilexetil in patients with systemic hypertension. Candesartan Cilexetil Study Investigators. Am J Cardiol 82:961–965
Rogg H, de Gasparo M, Graedel E, Stulz P, Burkart F, Eberhard M, Erne P (1996) Angiotensin II-receptor subtypes in human atria and evidence for alterations in patients with cardiac dysfunction. Eur Heart J 17:1112–1120
Schneider MD, Lorell BH (2001) AT_2, Judgment Day. Which angiotensin receptor is the culprit in cardiac hypertrophy? Circulation 104:247–248
Senbonmatsu T, Ichihara S, Price E, Gaffney A, Inagami T (2000) Evidence for angiotensin II type 2 receptor-mediated cardiac myocyte enlargement during in vivo pressure overload. J Clin Invest 106:R25–R29
Sever P (1997) Candesartan cilexetil: a new, long acting, effective angiotensin II type 1 receptor blocker. J Hum Hypertens 11 [Suppl 2]:S91–S95
Sever P(2002)The SCOPE trial. Study on Cognition and Prognosis in the Elderly. J Renin Angiotensin Aldosterone Syst 3:61–62
Sharpe M, Jarvis B, Goa KL (2001) Telmisartan: a review of its use in hypertension. Drugs 61:1501–1529
Shetty SS, DelGrande D (2000) Differential inhibition of the prejunctional actions of angiotensin II in rat atria by valsartan, irbesartan, eprosartan, and losartan. J Pharmacol Exp Ther 294:179–186
Silvestre JS, Tamarat R, Senbonmatsu T, Icchiki T, Ebrahimian T, Iglarz M, Besnard S, Duriez M, Inagami T, Levy BI (2002) Antiangiogenic effect of angiotensin II type 2 receptor in ischemia-induced angiogenesis in mice hindlimb. Circ Res 90:1072–1079
Siragy HM, de Gasparo M, Carey RM (2000) Angiotensin type 2 receptor mediates valsartan-induced hypotension in conscious rats. Hypertension 35:1074–1077
Siragy HM, El-Kersh MA, DeGasparo M, Webb RL (2002) Differences in AT_2-receptor stimulation between AT_1-receptor blockers valsartan and losartan quantified by renal interstitial fluid cGMP. J Hypertens 20:1157–1163
Skeggs LT, Lentz KE, Kahn JR (1956) The amino acid sequence of hypertensin II. J Exp Med 104:193–197
Stearns RA, Chakravarty PK, Chen R, Chiu SH (1995) Biotransformation of losartan to its active carboxylic acid metabolite in human liver microsomes. Role of cytochrome P4502C and 3A subfamily members. Drug Metab Dispos 23:207–215
Strawn WB, Chappell MC, Dean RH, Kivlighn S, Ferrario CM (2000a) Inhibition of early atherogenesis by losartan in monkeys with diet-induced hypercholesterolemia. Circulation 101:1586–1593
Strawn WB, Dean RH, Ferrario CM (2000b) Novel mechanisms linking angiotensin II and early atherogenesis. J Renin Angiotensin Aldosterone Syst 1:11–17
Stumpe KO, Haworth D, Hoglund C, Kerwin L, Martin A, Simon T, Masson C, Kassler-Taub K, Osbakken M (1998) Comparison of the angiotensin II receptor antagonist irbesartan with atenolol for treatment of hypertension. Blood Press 7:31–37
Stumpe KO, Ludwig M (2002) Antihypertensive efficacy of olmesartan compared with other antihypertensive drugs. J Hum Hypertens 16 [Suppl 2]:S24–S28

Swedberg K, Pfeffer M, Granger C, Held P, McMurray J, Ohlin G, Olof B, Ostergren J, Yusuf S (1999) Candesartan in heart failure assessment of reduction in mortality and morbidity (CHARM): rationale and design. CHARM Programme Inveatigators. J Card Fail 5:276–282

Thurmann PA, Kenedi P, Schmidt A, Harder S, Rietbrock N (1998) Influence of the angiotensin II antagonist valsartan on left ventricular hypertrophy in patients with essential hypertension. Circulation 98:2037–2042

Timmermans PBMWM, Wong PC, Chiu AT, Herblin WF, Benfield P, Carini DJ, Lee RJ, Wexler RR, Saye JAM, Smith RD (1993) Angiotensin II receptors and angiotensin II receptor antagonists. Pharmacol Rev 45:205–251

Tsutsumi Y, Matsubara H, Masaki H, Kurihara H, Murasawa S, Takai S, Miyazaki M, Nozawa Y, Ozono R, Nakagawa K, Miwa T, Kawada N, Mori Y, Shibasaki Y, Tanaka Y, Fujiama S, Koyama Y, Fujiyama A, Takahashi H, Iwasaka T (1999) Angiotensin II type 2 receptor overexpression activates the vascular kinin system and causes vasodilation. J Clin Invest 104:925–935

Van Rijnsoever EW, Kwee-Zuiderwijk WJ, Feenstra J (1998) Angioneurotic edema attributed to the use of losartan. Arch Intern Med 158:2063–2065

Vanderheyden PML, Fierens FLP, De Backer JP, Frayman N, Vauquelin G (1999) Distinction between surmountable and insurmountable selective AT_1 receptor antagonists by use of CHO-K1 cells expressing human angiotensin II AT_1 receptors. Br J Pharmacol 126:1057–1065

Vanderheyden PM, Fierens FL, De Backer J, Vauquelin G (2000) Reversible and syntopic interaction between angiotensin receptor antagonists on Chinese hamster ovary cells expressing human angiotensin II type 1 receptors. Biochem Pharmacol 59:927–935

Verheijen I, Fierens FL, Debacker JP, Vauquelin G, Vanderheyden PM (2000) Interaction between the partially insurmountable antagonist valsartan and human recombinant angiotensin II type 1 receptors. Fundam Clin Pharmacol 14:577–585

Viberti G, Wheeldon NM, for the MicroAlbuminuria Reduction With VALsartan (MARVAL) Study Investigators (2002) Microalbuminuria reduction with valsartan in patients with type 2 diabetes mellitus: a blood pressure-independent effect. Circulation 106:672–678

Vuagnat A, Giacché M, Hopkins PN, Azizi M, Hunt SC, Vedie B, Corvol P, Jeunemaitre X (2001) Blood pressure response to angiotensin II, low-density lipoprotein cholesterol and polymorphisms of the angiotensin II type 1 receptor gene in hypertensive sibling pairs. J Mol Med 79:175–183

Wachtell K, Palmieri V, Olsen MH, Gerdts E, Papademetriou V, Nieminen MS, Smith G, Dahlof B, Aurigemma GP, Devereux RB (2002) Change in systolic left ventricular performance after 3 years of antihypertensive treatment: the Losartan Intervention for Endpoint (LIFE) Study. Circulation 106:227–232

Warner GT, Jarvis B (2002) Olmesartan medoxomil. Drugs 62:1345–1353

Weinstock J, Keenan R, Samanen J, Hempel J, Finkelstein JA, Franz RG, Gaitanopoulos DE, Girard GR, Gleason JG, Hill DT, Morgan TM, Peishoff CE, Aiyar N, Brooks DP, Fredrickson TA, Ohlstein EH, Ruffolo RR Jr, StackEJ, Sulpizio AC, Weidley EF, Edwards RM (1991) 1-(Carboxyl)imidazole-5-acrylic acids: potent and selective angiotensin II receptor antagonists. J Med Chem 34:1514–1517

Weir MR, Smith DH, Neutel JM, Bedigian MP (2001) Valsartan alone or with a diuretic or ACE inhibitor as treatment for African American hypertensives: relation to salt intake. Am J Hypertens 14:665–671

Wienen W, Hauel N, vanMeel JC et al (1993) Pharmacological characterization of the novel nonpeptide angiotensin II receptor antagonist BIBR 277. Br J Pharmacol 110:245–252

Wilkinson-Berka JL, Gibbs NJ, Cooper ME, Skinner SL, Kelly DJ (2001) Renoprotective and anti-hypertensive effects of combined valsartan and perindopril in progressive diabetic nephropathy in the transgenic (mRen-2)27 rat. Nephrol Dial Transplant 16:1343–1349

Wolf G (2002) "The road not taken": role of angiotensin II type 2 receptor in pathophysiology. Nephrol Dial Transplant 17:195–198

Wu L, Nakagami H, Chen R, Suzuki J, Akishita M, de Gasparo M, Horiuchi M (2002) Effect of angiotensin II type 1 receptor blockade on cardiac remodeling in angiotensin ii type 2 receptor null mice. Arterioscler Thromb Vasc Biol 22:49–54

Yang Z et al (2002) Angiotensin II type 2 receptor overexpression preserves left ventricular function after myocardial infarction. Circulation 106:106–111

Yun CH, Lee HS, Lee H, Rho JK, Jeong HG, Guengerich FP (1995) Oxidation of the angiotensin II receptor antagonist losartan (DuP 753) in human liver microsomes. Role of cytochrome P4503A(4) in formation of the active metabolite EXP3174. Drug Metab Dispos 23:285–289

Yusuf S (2002) From the HOPE to the ONTARGET and the TRANSCEND studies: challenges in improving prognosis. Am J Cardiol 89:18A–25A

Clinical Pharmacology of Angiotensin II Receptor Antagonists

M. Maillard · M. Burnier

Division of Hypertension and Vascular Medicine, CHUV, Av P. Decker,
1011 Lausanne, Switzerland
e-mail: Michel.Burnier@chuv.hospvd.ch

Abstract Ten years after the introduction for clinical use of losartan, the first orally active angiotensin II receptor antagonist, seven compounds—the pharmacological characteristics of which are described in this chapter—are registered by the US Food and Drug Administration and can be used in the United States and in various European countries for the treatment of hypertension, heart failure and for the prevention of type-2 diabetic nephropathy. These agents have a common mechanism of action—selective blockade of the binding of angiotensin II to the subtype 1 receptor—and an excellent tolerability profile, a real advantage as these agents are prescribed most frequently to asymptomatic patients. Within a rather short period of time, a large number of clinical studies has demonstrated the efficacy of angiotensin II receptor antagonists as antihypertensive agents. In addition, the completion of several large outcome trials have made it possible to define the potential benefits of these drugs to lower morbidity and mortality in various groups of patients, including hypertensive and heart failure patients and patients with type-2 diabetes nephropathy. While pharmacokinetic and pharmacodynamic differences clearly accounted for clinically significant differences, at least in terms of blood pressure control between antagonists, some unresolved questions still remain such as the right dose to use in order to achieve the optimal target organ protection, the role of the AT_2 receptors or the potential benefits of a combination with ACE inhibitors. These questions will probably find their answers in the next few years with the results of the ongoing studies.

Keywords Angiotensin II receptor antagonist · Pharmacology · Large outcome trials · Doses · Combination

1 Introduction

Since the beginning of the 1980s, angiotensin-converting enzyme (ACE) inhibitors have been used increasingly and successfully to block the renin-angiotensin system. Clinical trials have demonstrated that ACE inhibitors make it possible to reduce the cardiovascular morbidity and/or mortality in patients with hypertension (CAPP study: Hansson et al. 1999a; STOP2 study: Hansson et al. 1999b), congestive heart failure (the CONSENSUS Study: Kjekshus et al. 1992), myocardial infarction (Franzosi et al. 1998), type 1 and 2 diabetes (Lewis et al. 1993; UKPDS study: Holman et al. 1998), chronic renal failure (REIN study: Ruggenenti et al. 1999) and patients with a high cardiovascular risk (The HOPE study: Yusuf et al. 2000). However, despite their recognized clinical efficacy, ACE inhibitors still have some limitations. Thus, because the ACE enzyme is involved in the metabolism of several other peptides, ACE inhibitors are not devoid of side effects such as cough, which has been attributed to nonangiotensin-mediated mechanisms (Israili and Hall 1992; Morice et al. 1987). Since ACE also contributes to the metabolism of bradykinin, ACE inhibition results in an accumulation of bradykinin, which may lead to the development of angioedema (Nussberger et al. 1998). Finally, during long-term ACE inhibition, plasma angiotensin II levels decrease acutely, but some angiotensin II is still circulating at a measurable level during chronic treatment (Juillerat et al. 1990). These measurable angiotensin II levels are due to the reactive rise in plasma renin activity and plasma angiotensin I levels, which lead to the generation of the angiotensin II octapeptide, particularly if ACE activity is not fully inhibited around the clock. Angiotensin II could also be produced through non-ACE pathways that are not affected by ACE inhibitors (Dzau et al. 1993). Therefore it has been postulated that a more specific blockade of the renin-angiotensin cascade at the receptor level could lead to a more complete and better-tolerated inhibition of the renin-angiotensin system.

Based on these premises, blockade of angiotensin II AT_1 receptors with selective, nonpeptide, orally active angiotensin II antagonists became available in the beginning of the 1990s as an alternative to ACE inhibition. Since then, many antagonists have been synthesized and several of them (of which losartan was the first in 1995) received approval for the treatment of hypertension by the US Food and Drug Administration. Six antagonists are currently marketed and used in the United States and Europe to treat hypertensive patients and patients with congestive heart failure and type 2 diabetes (Burnier and Brunner 2000) (Fig. 1)).

2
Pharmacology of Angiotensin II Receptor Antagonists

All angiotensin II receptor antagonists share a common mechanism of action: they selectively block the angiotensin II subtype-1 (AT_1) receptors. In normotensive subjects, all antagonists have been shown to block the blood pressure response to exogenous angiotensin II dose dependently (Munafo et al. 1992; Christen et al. 1991; Müller et al. 1994; Delacrétaz et al. 1995; Hagmann et al. 1997; Maillard et al. 1999, 2000; Mazzolai et al. 1999; Ribstein et al. 2001). However, as shown in Table 1, these various antagonists differ in their pharmacological profile, and these differences may sometimes affect their clinical efficacy (Timmermans et al. 1993; Mazzolai et al. 1999; Maillard et al. 2002a). Based on in vitro binding studies, AT_1 receptor antagonists have been divided into two groups, surmountable and insurmountable antagonists (Timmermans et al. 1993; Morsing et al. 1999; Vanderheyden et al. 1999). Although both groups produce a rightward shift of the angiotensin II dose-response curve, the maximal response is unaffected by surmountable antagonists such as losartan (Mochizuki et al. 1995), whereas it is reduced by insurmountable antagonists (such as valsartan, irbesartan, candesartan, and telmisartan, for instance), leading to a nonparallel displacement of the angiotensin II response curve. Of note, surmountable/insurmountable antagonism only describes the ligand–antagonist interaction occurring when cells or tissue preparations are preincubated with the antagonist and thereafter exposed to the agonist. In contrast, the competitive or noncompetitive nature of a drug is related to experimental conditions in which the ligand and the antagonist are added simultaneously. Several recent studies have actually demonstrated that even though some AT_1 receptor antagonists are surmountable and others insurmountable, all are competitive antagonists, which means that they compete with angiotensin II at the receptor level according to the law of mass action (Fierens et al. 1999; Vanderheyden et al. 2000a).

The molecular basis for insurmountable antagonism is still a matter of debate and several potential mechanisms have been proposed (De Chaffoy de Courcelles et al. 1986; Timmermans et al. 1991; Liu et al. 1992; Wienen et al.

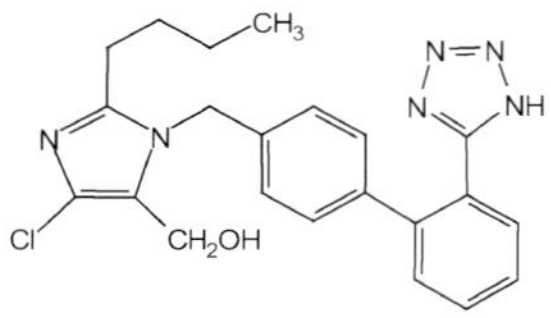

Losartan (Cozaar®)
Merck
Launched

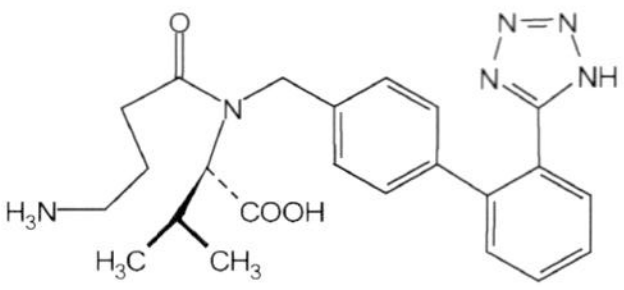

Valsartan (Diovan®, Tareg®, Nisis®)
Novartis
Launched

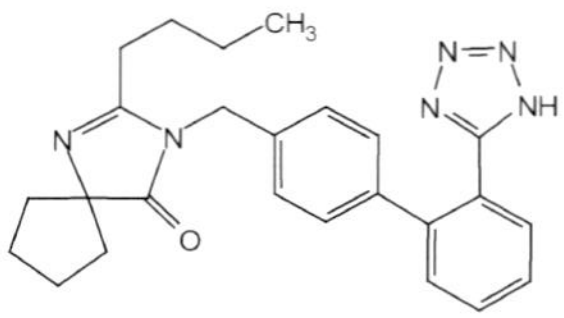

Irbesartan (Aprovel®, Avapro®, Karvea®)
Brsitol-Myers Squibb/Sanofi
Launched

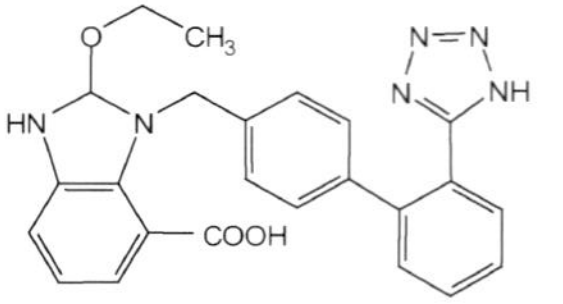

Candesartan (Atacand®, Kenzen®, Biopress®)
AstraZeneca/Takeda
Launched

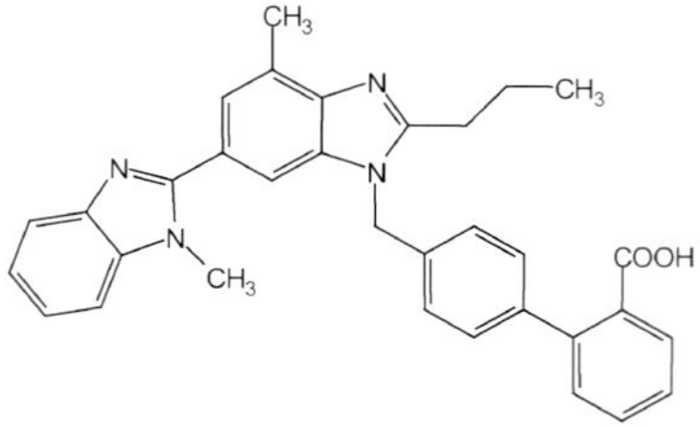

Telmisartan (Micardis®, Pritor®)
Boehringer Ingelheim/Glaxo Wellcome
Launched

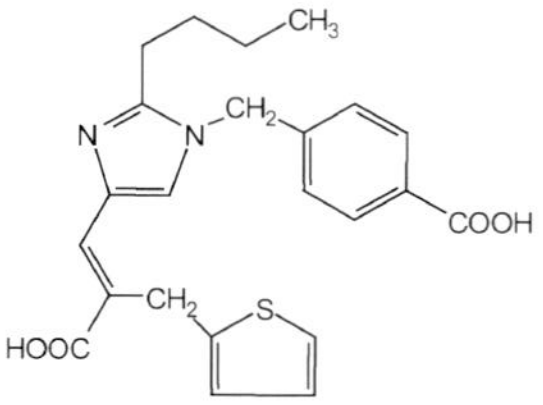

Eprosartan (Teveten®)
SmithKline Beecham
Launched

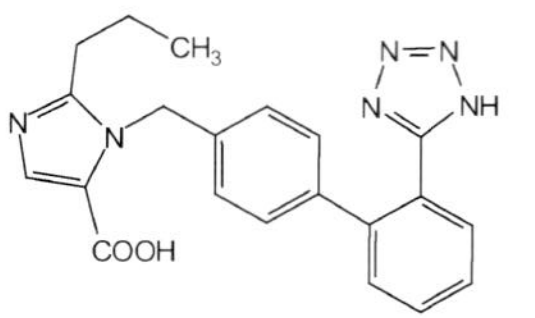

Olmesartan (Benicar®)
Sankyo
Approved by FDA

Fig. 1 Oral angiotensin II receptor antagonists

Table 1 Pharmacokinetic properties of angiotensin II receptor antagonists

Variable	Losartan	Valsartan	Irbesartan	Candesartan	Eprosartan	Telmisartan	Olmesartan
AT1 receptor affinity*	IC_{50}: 20 nmol/l	IC_{50}: 2.7 nmol/l	IC_{50}: 1.7 nmol/l	K_i: 0.6 nmol/l	IC_{50}: 1.5 nmol/l	K_i: 3.5 nmol/l	IC_{50}: 8.0 nmol/l
In vitro Ang II antagonism	Surmountable	Insurmountable	Insurmountable	Insurmountable	Surmountable	Insurmountable	Insurmountable
Bioavailability *F*(%)	33	25	60–80	40	15	43	26
Food interaction	10% decrease in *F* (clinically NS)	50% decrease in absorption	No	No	Delayed absorption (clinically NS)	No	No
Prodrug	No	No	No	Yes: candesartan cilexetil	No	No	Yes: olmesartan medoxomil
Active metabolite	Yes: EXP-3174	No	No	No	No	No	No
T_{max} (h)	1 (metabolite: 2–4)	2–4	1.5–2	2–5	1–2	0.5–1	≈2
Protein binding (%)	98.7 (95.8)	95.0	99	99.5	98	>99	>99
Vd (L)	28–50 (10–12)	17	50–100	8–10	15	500	35
Metabolism	Oxidation, mainly by CYP 2C9 and CYP 3A4	Unknown	Oxidation and glucuronide conjugation	*O*-demethylation	Glucuronide conjugation	Glucuronide conjugation	No
$t_{1/2}$ (h)	2 (6–9)	6–10	11–15	3–11	5–9	24	15
Daily dosage (mg)	50–100	80–320	150–300	8–32	400–800	40–80	20–40

IC_{50}, concentration displacing specifically 50% of the binding of angiotensin II, K_i = inhibition constant. NS, nonsignificant; T_{max}, time to peak plasma concentration; *V*d, apparent volume of distribution; $t_{1/2}$, elimination half-life.

1992; Panek et al. 1995; Fierens et al. 1999). Recently, increasing evidence has been provided suggesting that a slow dissociation from the receptor resulting in an increased longevity of the antagonist–receptor complex is one of the leading mechanisms of the insurmountable characteristic of angiotensin II receptor antagonists (Vanderheyden et al. 2000a, 2000b; Maillard et al. 2002b). Of course, we must note that the mode of AT_1 receptor antagonism probably does not play a major role in defining the antihypertensive effect of the antagonist (Timmermans 1999a), but it is likely that a slow off-rate from the AT_1 receptor may extend the time of occupancy of the receptor protein and lengthen the duration of antagonism.

Another common feature of all angiotensin II receptor antagonists is their high binding to plasma proteins (>95%), mainly albumin and α_1-acid glycoprotein. In general, the protein binding of antihypertensive drugs has little if any effect on its clinical efficacy. Yet we have found that some AT_1 receptor antagonists are so tightly bound to plasma proteins that they do not have any antagonistic effect in vivo (Dup 532). In additional studies, we have demonstrated that the strength of the binding between proteins and angiotensin II antagonists differs with the various compounds and may affect their efficacy. These data suggest that the qualitative interaction between AT_1 receptor antagonists and proteins, and not necessarily the quantitative aspect of the binding, is an important pharmacological characteristic of angiotensin II receptor antagonists (Maillard et al. 2001).

3
Losartan

Losartan was the first oral active AT_1 antagonist to be developed from the Takeda series of 1-benzylimidazole-5-acetic acid derivatives (Furukawa et al. 1982) and to be evaluated in humans (Christen et al. 1991). It has been extensively studied in both animals and human volunteers and its effectiveness as an antihypertensive agent was established in the early 1990s. In vitro, losartan selectively competes with the binding of angiotensin II to AT1 receptor with a median inhibitory concentration (IC_{50}) of 17–20 nmol/l (Chiu et al. 1991). Losartan undergoes first-pass hepatic metabolism via cytochrome P450 (CYP) isoenzymes 2C9 and 3A4 to its active carboxylic acid metabolic EXP-3174, which is 10–20 times more potent than losartan and has a longer duration of action than losartan itself. In fact, most of losartan's effects are due to EXP-3174 (Wong et al. 1990). However, since the oral bioavailability of EXP-3174 is very low, the drug on the market is losartan. On the isolated rabbit aorta, losartan produces a surmountable blockade of the contractile response induced by angiotensin II, whereas EXP-3174 causes an insurmountable blockade (Chiu et al. 1991; Wong et al. 1990). The main pharmacokinetic characteristics of losartan and EXP-3174 are presented in Table 1. Losartan and its metabolite are excreted by the kidney and bile. Neither compound is dialyzed. Food tends to slow the absorption of the drug, but this effect has only a minimal clinical impact. Finally, losartan is the only angiotensin II receptor antagonist that increases urinary uric acid excretion, thereby lowering plasma uric acid levels. This effect, which is due to the losartan molecule itself and not to its active metabolite, has been found in healthy subjects (Burnier et al. 1993) as well as in hypertensive subjects (Minghelli et al. 1998; Wuerzner et al. 2001). The ability of losartan to increase urinary uric acid excretion is mediated by an inhibitory action of losartan on the urate/anion exchanger found in the brush border of the renal proximal tubule (Roch-Ramel et al. 1994, 1997; Burnier et al. 1996).

The recommended doses of losartan are 50–100 mg once daily. Most clinical studies have been conducted with the 50-mg dose, but some recent trials have used losartan 100 mg once daily. Fixed combinations of losartan (50 and 100 mg) and hydrochlorothiazide (12.5 or 25 mg) are available in many countries. Losartan has received the indication for the treatment of hypertension and in some countries also for congestive heart failure.

4 Valsartan

Valsartan was the second nonpeptide angiotensin II receptor antagonist available for the treatment of hypertension. Valsartan is a nonheterocyclic compound in which the imidazole of losartan has been replaced with an acylated amino acid. In vitro it blocks the AT_1 receptor in an insurmountable fashion. Valsartan inhibits the binding of angiotensin II to rat aorta AT_1 receptors with an IC_{50} of 2.7 nmol/l (Criscione et al. 1995). In contrast to losartan, valsartan does not require conversion to an active metabolite (Markham and Goa 1997) and approximately 80%–90% of the drug is excreted unchanged in the bile (70%) and by the kidney (30%). Only one metabolite (valeryl-4-hydroxy valsartan), which is not active, accounting for 10% of recovered drug is also found in the feces. After oral application, valsartan is rapidly absorbed with a bioavailability of 25%, which is reduced by about 40% if food is taken concomitantly (Criscione et al. 1995).

The recommend starting doses of valsartan are 80–160 mg once daily with the possibly of increasing the dose to 320 mg daily). In heart failure, valsartan has been used safely at the dose of 160 mg b.i.d. Valsartan/hydrochlorothiazide fixed combinations are also available (Wellington and Faulds 2002)

5 Irbesartan

Like losartan, irbesartan is an imidazole derivative with a biphenyl-tetrazole side chain. The molecule has an imidazolinone ring in which the carbonyl group replaces the hydroxylmethyl group of losartan (or the carboxylic moieties of EXP-3174 or of valsartan) as the hydrogen bond acceptor. Irbesartan has a high affinity for the AT_1 receptor with an IC_{50} of 1.7 nmol/l in human vascular smooth muscle cells (Herbert et al. 1994). In vitro, irbesartan induces an insurmountable blockade of AT_1 receptors. Irbesartan has no active metabolite and food does not affect bioavailability, which, with an average of 60%–80%, is higher than that of other angiotensin II receptor antagonists (Brunner 1997). Irbesartan is longer acting than losartan and valsartan with an elimination half-life of about 11–15 h (Gillis and Markham 1997). There is some controversy about the binding of this drug to plasma proteins. This latter was originally claimed to be smaller than those usually observed within other angiotensin II antagonists (90%–92%) (Brunner 1997; Chung and Unger 1999; Timmermans 1999b), but in a more recent study irbesartan was found to have a protein binding (<99%) comparable to other antagonists (Morsing et al. 1999). Irbesartan is mainly excreted (60%–80%) in the feces, while about 25% appears in urine. As irbesartan is strongly metabolized via hepatic glucuronidation and oxidation (mainly CYP 2C9), only about 1% is excreted as the unchanged molecule (Perrier et al. 1994).

The recommended doses of irbesartan are 150 and 300 mg once daily. Fixed combinations with a small dose of hydrochlorothiazide are also available.

6 Candesartan

Candesartan is a potent long-lasting angiotensin II receptor antagonist that is rapidly produced in gastrointestinal tract by hydrolysis of an easily orally absorbed ester-prodrug candesartan-cilexetil (TCV-116). Candesartan inhibits the in vitro binding of angiotensin II to AT_1 receptors of rabbit aortic membrane with a inhibition constant (K_i) of 0.64 nmol/l (McClellan and Goa 1998). This drug produces an insurmountable antagonism due to a tight binding to and a slow dissociation from the AT_1 receptors (Ojima et al. 1997). Following oral administration, candesartan-cilexetil is quantitatively converted to candesartan, with an average absolute bioavailability of candesartan of about 40%. Candesartan has an elimination half-life of about 9 h, which might be extended in the elderly. It is mainly eliminated unchanged by the kidneys (30%) and the bile (70%). Of note, candesartan has a rather small apparent volume of distribution (0.1 l/kg) (McClellan and Goa 1998), which is probably the result of its high plasma protein binding (Delacrétaz et al. 1995). Another result of this high and tight binding to plasma proteins is that the mean extraction ratio for this drug from dialyzed blood is low (Burnier and Brunner 2000).

Due to the small volume of distribution of candesartan, the starting and maintenance doses of candesartan (8–16 mg once a day) is smaller than those of other antagonists. Titration to 32 mg has been used in some clinical trials and has been shown to provide greater antihypertensive efficacy (Andersson and Neldam 1988; Asmar and Lacourcière 2000). Candesartan/hydrochlorothiazide fixed combinations are available with different doses of candesartan.

7
Eprosartan

Eprosartan was the fourth selective nonpeptide angiotensin II receptor antagonist to gain approval for use in the treatment of hypertension in the United States. Eprosartan is the unique representative of an nonbiphenyl, nontetrazole AT_1 antagonist class (Timmermans 1999a). This compound is the latest development of an imidazole-5-acrylic acid series of nonpeptide angiotensin II receptor antagonists (Weinstock et al. 1991; Keenan et al. 1993). Introduction of a p-carboxylic acid on the N-benzyl ring resulted in nanomolar affinity for the AT_1 receptor and good oral activity, while the presence of a thienyl ring (sulfur-containing heterocycle) together with two acid groups were important to achieve good potency. Eprosartan requires no metabolic activation or transformation to produce an effective AT_1 receptor antagonism (McClellan and Balfour 1998). Eprosartan is a specific antagonist of the AT_1 receptor, which shows a true surmountable competitive antagonism in vitro (Timmermans 1999a). It inhibits [^{125}I]-angiotensin II binding to rat mesenteric artery membranes with an IC_{50} of 1.5 nmol/l, and It is also a potent inhibitor of labeled angiotensin binding to human liver membranes (IC_{50}=1.7 nmol/l) (Edwards et al. 1992). The bioavailability of eprosartan is smaller than those of other antagonists. It is limited by an incomplete oral absorption rather than high first-pass metabolism (Chung and Unger 1999). In addition, depending on the formulation, eprosartan absorption may be reduced by 25% and retarded by 1.5 h when the drug is administered with food. The bile represents the main excretory pathway (90%), while 7% of eprosartan is found as unchanged drug in urine. Several studies have investigated the ability of eprosartan to block presynaptic AT_1 receptors, located at the vascular neuroeffector junction but with contrasted results (Ohlstein et al. 1997; Shetty and DelGrande 2000; Hedner 2002). The usual maintenance dose eprosartan is 600 mg daily. Since eprosartan has a half-life of 5–7 h, most initial studies have been conducted using a twice-a-day regimen. However, the antihypertensive efficacy of one daily dose was compared with half-doses given twice daily in a double-blind, parallel group, placebo-controlled, multicenter study in 243 patients with mild to moderate hypertension (Hedner and Himmelmann 1999). This study showed that there was no significant difference in efficacy and tolerability of eprosartan between the once-daily or the twice-daily regimens, and that both induced significant blood pressure reductions. Thus eprosartan is now rather used in a once-daily regimen with a recommended dose of 400–800 mg per day.

8
Telmisartan

Telmisartan is a long-acting angiotensin II receptor antagonist. Chemically, telmisartan lacks the tetrazole unit and has a common benzimidazole group with candesartan. The substitution of this benzimidazole moiety with a basic heterocycle results in potent AT_1 antagonism and good absorption after oral application (Ries et al. 1993). Telmisartan selectively and insurmountably inhibits stimulation of the AT_1 receptor by angiotensin II without affecting other receptor systems involved in cardiovascular regulation. It inhibits the binding of labeled angiotensin II to AT_1 receptors in rat lung with a K_i of 3.7 nmol/l (Wienen et al. 1993). The absolute bioavailability of telmisartan is dose-dependent and varies from 30% to 60%. (Israili 2000). Telmisartan is more lipophilic than other antagonists. This feature, coupled with a high volume of distribution, could result in good tissue penetration (Wienen et al. 2000). Telmisartan is not a prodrug and has a long terminal elimination half-life (~24 h), making it suitable for once-daily dosing. The compound is not metabolized by cytochrome P450 isoenzymes, undergoes minimal glucurotransformation and is almost completely excreted in the feces (98%) (Wienen et al. 1999). In contrast to other angiotensin II receptor antagonists, some drug interactions with telmisartan have been described. In particular, telmisartan causes an increase in serum digoxin and may also decrease warfarin plasma levels during coadministration (Sharpe et al. 2001).

Telmisartan is used at the recommended starting dose of 40 mg once daily, and this dose may be increased to 80 mg in case of insufficient blood pressure lowering effect. The fixed dose combination of telmisartan and hydrochlorothiazide is now available in some countries.

9 Olmesartan

Olmesartan medoxomil is a new orally active angiotensin II receptor antagonist that gained FDA approval for use in the treatment of hypertension in 2002 and is being reviewed for registration in Europe. This compound is a prodrug containing, like candesartan-cilexetil, an ester moiety that after oral administration is rapidly converted in vivo to the active metabolite olmesartan. In vitro, the affinity of olmesartan for AT_1 bovine adrenal cortical receptors is comparable to other angiotensin II receptor antagonists (IC_{50} 8.0 nmol/l) (Brunner 2002). Olmesartan produces selective insurmountable inhibition of Ang ll-induced contractions of the guinea-pig aorta (Koike et al. 2001). Following oral administration, olmesartan has a faster onset but similar elimination half-life when compared with candesartan cilexetil. The absolute bioavailability of olmesartan after single oral dose of olmesartan medoxomil is 26% in healthy volunteers. The remainder unabsorbed drug is excreted without further metabolism. Feces is the major route of excretion of olmesartan, urinary excretion accounting for 5%–12% of the administered dose (Warner and Jarvis 2002).

The usual recommended doses of olmesartan are 20–40 mg once daily. Twice-daily dosing offers no advantage over the same total dose given once daily.

10
Clinical Uses of Angiotensin II Receptor Antagonists

10.1
Essential Hypertension

Numerous clinical studies have shown that angiotensin II receptor effectively lowers blood pressure in hypertensive patients, without affecting heart rate and regardless of gender and age (Goa and Wagstaff 1996; Gillis and Markham 1997; Markham and Goa 1997; McClellan and Balfour 1998; McClellan and Goa 1998; Reeves et al. 1998; Sharpe et al. 2001; Warner and Jarvis 2002). In addition, several short-term studies have demonstrated that the antihypertensive efficacy of angiotensin II receptor antagonists is comparable with that of other first-line antihypertensive agents such as ACE inhibitors, calcium-channel blockers, β-blocker drugs and diuretics (Burnier and Brunner 2000). Three large trials, of which two have been completed, were designed to assess the effects of angiotensin II receptor antagonists on morbidity and mortality in hypertensive patients (Table 2). The LIFE trial (Losartan Intervention For Endpoint Reduction in Hypertension) compared a losartan- and an atenolol-based regimen in 9,124 hypertensive patients with electrocardiogram-documented left ventricular hypertrophy (Dahlöf et al. 2002). Patients were followed for an average of 4.7 years. In this study, losartan and atenolol induced a comparable decrease in blood pressure. Yet the losartan-based treatment was associated with a 13% lower relative risk (p=0.021) of primary cardiovascular event (i.e., death, myocardial infarction or stroke) and with significantly fewer fatal and nonfatal strokes (–25%, p<0.001). In a subgroup of 1,195 diabetic patients, the protection was even more marked, with a 24% reduction of the combined risk (p=0.031) and a 39% decrease in mortality (p=0.002). Moreover, losartan decreased the relative risk of developing diabetes by 25% and was more effective than atenolol in reversing left ventricular hypertrophy. Thus, the LIFE study is the first to demonstrate in hypertensive patients that an angiotensin II receptor antagonist can prevent cardiovascular morbidity and death and that it can do it more effectively than a beta-blocker-based treatment. This study also suggests that losartan could confer additional cardiac and metabolic benefits beyond the reduction of blood pressure (Dahlöf et al. 2002).

The second study is entitled SCOPE (Study on Cognition and Prognosis in the Elderly). This trial has evaluated the effects of candesartan-cilexetil on cardiovascular mortality and morbidity and on cognitive performance in a population of elderly hypertensive patients (70–89 years) (Innocenti et al. 2002). The results of this placebo-controlled study were recently published (Lithell et al. 2003) and show that candesartan was associated with a modest, statistically nonsignificant reduction in major cardiovascular events. As in the LIFE study, candesartan is particularly effective in reducing the incidence of nonfatal stroke. In addition, candesartan has no deleterious effect on cognition despite a strict control of blood pressure.

Table 2 A selection of ongoing and completed clinical trials with angiotensin II receptor antagonists

Drug	Trial	Population	(*n*)	End points	Completion	Results
Losartan	ELITE	Elderly patients with heart failure	722	Effect on renal function, hospital admission for CHF and/or mortality	1997	Losartan is safe and effective in the treatment of CHF
	ELITE II	Elderly patients with heart failure	3,121	All-cause mortality	1999	Losartan is not better than captopril in the treatment of CHF
	RENAAL	NIDDM patients with nephropathy	1,520	Composite of ESRD, doubling of creatinine, mortality	2000	Losartan provides renal protection and has disease-retarding effect
	LIFE	Hypertensive with LVH	9,124	Mortality, MI, stroke	2001	Losartan is better than atenolol in reducing the risk of primary cardiovascular events
	OPTIMAAL	Post-MI with LV dysfunction	5,477	All-cause mortality	2002	Losartan is not better than captopril after MI
Valsartan	MARVAL	NIDDM patients with hypertension	332	Microalbuminuria	2001	Valsartan has antiproteinuric effect in type 2 diabetes
	Val-HeFT	Heart failure	5,010	All-cause mortality	2001	Valsartan is better than placebo to reduce morbidity and mortality in CHF patients
	VALUE	Hypertensives with high risk	14,400	Cardiovascular mortality	2004	–
	VALIANT	Post-MI with LV dysfunction	14,500	All-cause mortality	2003	Valsartan is not inferior to ACE inhibitor
Candesartan	RESOLVD	Heart failure	768	Improvement of the physical capacity	Stopped 1999	Increase in HF-related hospitalizations in candesartan groups
	SCOPE	Elderly hypertensives	4,964	Cardiovascular mortality, MI, stroke, prevention of cognitive impairment	2002	No significant effect of candesartan in reduction of cardiovascular events. No deleterious effect on cognition
	CHARM-I	Heart failure, ACEI-intolerant	2'028	All-cause mortality	2003	Candesartan reduces cardiovascular death or hospitalization for CHF
	CHARM-II	Heart failure (LVEF <40%)	2,548	All-cause mortality	2003	Candesartan reduces cardiovascular death and hospitalization for CHF
	CHARM-III	Heart failure (LVEF >40%)	3,025	All-cause mortality	2003	No significant effect of candesartan in patients with preserved LV systolic function

Table 2 (continued)

Drug	Trial	Population	(*n*)	End points	Completion	Results
Irbesartan	IDNT	NIDDM patients with nephropathy	1,650	Composite of ESRD, doubling of creatinine, mortality	2000	Irbesartan provides renal protection and has disease-retarding effect. Irbesartan is better than amlodipine
	IRMA-2	NIDDM patients with hypertension	611	Microalbuminuria	2000	Irbesartan decreases microalbuminuria and retards the progression toward proteinuria
Telmisartan	ONTARGET	High risk of cardiovascular event	30,000	All-cause mortality	2007	–

CHF, Cardiac heart failure; EF, ejection fraction; ESRD, end-stage renal disease; MI, myocardial infarction; NIDDM, non-insulin-dependent diabetes mellitus.

The third study is VALUE (Valsartan Antihypertensive Long-Term Use Evaluation) in which more than 14,000 hypertensive patients older than 50 years and with additional risks of cardiovascular events are enrolled. In this study, a valsartan-based regimen (80–160 mg o.d.) is compared to an amlodipine-based treatment (5–10 mg od) and cardiovascular mortality is the primary end point (Mann and Julius 1998).

Finally, the Ongoing Telmisartan Alone and in combination with Ramipril Global Endpoint Trial (ONTARGET) compares the efficacy of the angiotensin II receptor blocker telmisartan with that of the ACE inhibitor ramipril given either alone or in association with telmisartan to about 30,000 patients at high risk of cardiovascular disease with or without hypertension. This study is interesting in several regards: it will be the first large trial directly comparing an ACE inhibitor and an angiotensin II receptor antagonist. It should also provide important information on the potential benefits of combining an ACE inhibitor and an antagonist, particularly in diabetic patients (Yusuf 2002).

10.2 Safety and Tolerability Profile in Hypertension

Despite the wide range of antihypertensive agents on the market, the use of most drugs is often limited by side effects, tolerability and noncompliance. Angiotensin II receptor antagonists are unique in this regard since they have an excellent safety and tolerability profile. So far, several thousand patients have taken part in double-blind controlled studies evaluating the antihypertensive activity of various angiotensin II antagonists. A characteristic of this class of drugs is an adverse-effect profile comparable with that seen in the placebo groups (Mazzolai and Burnier 1999). None of the seven drugs reviewed here has a specific, dose-dependent adverse effect that can be attributed to the drug itself. All angiotensin II receptor antagonists are less likely than ACE inhibitors to cause cough (Lacourcière et al. 1994; Benz et al. 1997; Elliot 1999; Lacourcière 1999), and angioedema, another side-effect of ACE inhibitors, was only rarely observed in patients treated with angiotensin II receptor antagonists. Indeed, a recent retrospective chart review and review of the literature reported four cases of angioedema associated with the consumption of losartan (Chiu et al. 2001). However, because angioedema may occur with many substances, including drugs and some food products, it is difficult to ascertain whether these published cases of angioedema are really linked to the administration of losartan or simply a coincidence.

No consistent adverse effects on routine hematological parameters were noted with the use angiotensin II receptor antagonists. Only minor clinically nonsignificant decreases in hemoglobin levels have been reported in hypertensive patients. Some rare cases of anemia have been reported in dialyzed patients or patients with chronic renal failure treated with losartan (Schwartzberg et al. 1998). Occasional elevation of liver enzymes (particularly ALAT) occurred rarely and usually resolved with or without discontinuation of therapy (Mazzolai and

Burnier 1999). Slight increases in serum potassium levels have been reported during angiotensin II receptor blockade. Hyperkalemia is more likely to develop in patients with renal insufficiency, or with diabetes mellitus, or in those patients taking potassium-sparing diuretics or potassium supplementation.

The safety and tolerability of angiotensin II receptor antagonists as well as their efficacy were carefully evaluated as a function of patient gender, age and race. None of these factors was found to influence the incidence of secondary events. In particular, angiotensin II receptor antagonists are equally well tolerated by elderly (>65 years), younger (<65 years) and very old patients (>75 years). Since all antagonists are cleared through the bile, no dosage adjustment is recommended in patients with moderate to marked renal impairment (creatinine clearance <40 ml/min) or with moderate hepatic dysfunction (Mazzolai and Burnier 1999). In case of severe hepatic dysfunction, the antihypertensive efficacy may be prolonged. For patients with possible depletion of intravascular volume (e.g., patients treated with diuretics, particularly those with impaired renal function), renin-angiotensin system blockade should be initiated carefully and consideration should be given to use of a lower starting dose. In studies evaluating the use of ACE inhibitors in patients with unilateral or bilateral renal artery stenosis, increases in serum creatinine or blood urea nitrogen together with oliguria and/or progressive azotemia and (rarely) with acute renal failure and/or death have been reported (Dominiczak et al. 1988). So far, few studies have evaluated the safety of angiotensin II receptor antagonists in patients with renal artery stenosis, but these preliminary reports all suggest that similar results may be expected (Mazzolai and Burnier 1999).

10.3 Clinical Applications in Cardiac Diseases

Today, ACE inhibitors belong to the standard therapy of heart failure with proven benefits in terms of morbidity and mortality (Remme and Swedberg 2001). ACE inhibitors also reduce the morbidity and improve the survival of patients after acute myocardial infarction (Franzosi et al. 1998). Finally, ACE inhibition appears to produce greater reductions in left ventricular hypertrophy than beta-blockers, calcium antagonists or diuretics, despite similar reductions in blood pressure (Dahlöf et al. 1994). Because angiotensin II receptor antagonists share most of the pharmacological actions of ACE inhibitors but with a better tolerability profile, they were of course considered in these three clinical indications. Several short-term clinical studies conducted in patients with symptomatic congestive heart failure without ACE inhibitors have demonstrated the angiotensin II receptor antagonists exert beneficial hemodynamic and neurohormonal effects (Thürmann and Colette 2002). In addition, a number of short-term studies have found that AT_1 receptor blockade could lead to a further reduction in afterload and improvement of exercise capacity in heart failure patients already treated with an ACE inhibitor (Houghton et al. 2000; Baruch et al. 1999, Gremmler et al. 2000; Tonkon et al. 2000; Hamroff et al. 1997, 1999).

Heart failure trials were triggered by the astonishing results of the Evaluation of Losartan in The Elderly (ELITE) study, which compared the safety and efficacy of captopril 50 mg three times daily and losartan 50 mg daily in elderly patients with heart failure, renal safety being the primary end point. This study found that losartan is as safe and effective as captopril, with no difference in renal dysfunction between both treatment regimens, but after 1 year of follow-up, a secondary end point (combined mortality from and hospital admission for heart failure) was surprisingly lower in the losartan group. (Pitt et al. 1997). These positive results were not confirmed in the larger ELITE II follow-up trial, which involved more than 3,000 elderly patients with evidence of heart failure and left ventricular dysfunction (Pitt et al. 2000). The primary end point was all-cause mortality, and the secondary end point was the frequency of sudden death or cardiac arrest with resuscitation. Elite II confirmed that patients treated with losartan had significantly fewer side effects than those on captopril. However, losartan was not found to be superior to captopril in reducing morbidity and mortality. Unfortunately, since ELITE II was designed to demonstrate superiority of losartan and was not powered to establish equal efficacy of the two drug classes, the investigators concluded that ACE inhibitors should remain the standard of therapy in patients with heart failure due to systolic dysfunction (Pitt et al. 2000).

In the Randomized Evaluation Strategies for Left Ventricular Dysfunction (RESOLVD) trial, candesartan was compared with enalapril, and a combination of candesartan and enalapril in 768 patients with NYHA class II–IV heart failure and a 6-min walking test of less than 500 m. RESOLVD was not designed to assess morbidity and mortality, since the primary end point was the change in the 6-min walking distance. Nevertheless, this study was discontinued earlier than expected because of an increase, albeit nonsignificant, of heart failure-related hospitalizations and deaths in the candesartan and combination group compared with the enalapril group (McKelvie et al. 1999). There were no significant differences with respect to primary objective of the study. The relatively negative findings of ELITE II and RESOLVD have yet to be interpreted carefully. Thus, in the former study, the dose of losartan was probably too small, since losartan 50-mg-induced AT1 blockade is far from being complete (Mazzolai et al. 1999). In the latter, the candesartan–enalapril combination was very effective in suppressing aldosterone levels and BNP concentrations, particularly at the higher dose of candesartan (8 mg), and the increase in heart failure-induced hospitalization was only marginal. Clearly, after these two trials, further studies were required to evaluate the potential benefits of AT1-receptor antagonists in heart failure.

The Valsartan Heart-Failure Trial (VALHeFT) (Cohn et al. 1999) evaluated the long-term effects of the addition of the angiotensin II receptor antagonist valsartan to standard therapy for heart failure. A total of 5,010 heart failure patients were randomized to receive either valsartan 160 mg or a placebo twice daily. The primary end point was mortality, and the combined end point of mortality and morbidity was defined as the incidence of cardiac arrest, hospitalization for heart failure or receipt of intravenous inotropic or vasodilator therapy. In this study, valsartan did not affect overall mortality but significantly the incidence of the combined end point, mainly because of a lower number of hospitalizations for heart failure. The favorable impact of valsartan was particularly impressive in the small subgroup of patients who were not being treated with an ACE inhibitor, who had a 44% reduction in the combined end point of mortality and morbidity and a 33% decrease in mortality. Surprisingly, patients who were already receiving a beta-blocker did not derive the same benefits from valsartan or had even a worse outcome. This observation made in a subgroup of patients must be interpreted cautiously. Thus, whether concomitant beta-blockade and AT1 receptor blockade is beneficial in heart failure patients remains an open question that should be investigated further. Overall, the ValHeft study strongly emphasizes the benefits of blocking angiotensin II receptors in heart failure patients.

Other studies have been conducted in patients with heart failure that improve further our understanding of the place of AT1 receptor antagonists in the management of patients with chronic congestive heart failure, one of which, the Candesartan in Heart failure-Assessment of Reduction in Mortality and morbidity trials (CHARM-I, -II and -III) (Swedberg et al. 1999), was a large clinical trial of 7,600 patients evaluating the potential benefits of candesartan in heart failure. The results of this study involving three groups of patients with different cardiac function and concomitant therapies—one group with a left ventricular ejection fraction (LVEF) below 40% receiving an ACE inhibitor; a second group with a low LVEF but patients are ACE intolerant; and in the third group patients with a LVEF over 40% and no ACE inhibitor therapy—were recently published in The Lancet (Pfeffer et al. 2003; McMurray et al. 2003; Granger et al. 2003; Yusuf et al. 2003) and show that the long-acting angiotensin receptor antagonist candesartan reduces both cardiovascular mortality and hospital admissions for congestive heart failure in patients with low LVEF whatever the concomitant therapies, with the greatest benefits achieved in ACE-intolerant patients with a LVEF below 40%. Other studies are being conducted in patients with diastolic dysfunction.

Blockade of the renin-angiotensin system may also be useful in patients following myocardial infarction. The Optimal Therapy in Myocardial Infarction with the angiotensin II Antagonist Losartan (OPTIMAAL) study has compared losartan 50 mg q.d. and captopril 50 mg t.i.d. in patients with left ventricular dysfunction following myocardial infarction (n=5,477) (Dickstein and Kjekshus 2002). In this recently published study, losartan added to the standard care (except ACE inhibitors) was not found to be superior to captopril in reducing all-cause mortality. As in the ELITE II study, the main question is the dose of losartan, which was probably chosen too low. Consequently, we are now waiting for the results of the VALIANT study (VALsartan In Acute myocardial infarction Trial), which is comparing valsartan to captopril and to the their combination in a similar population (n=14,500) (Pfeffer et al. 2000). In contrast to OPTIMAAL, physicians will be allowed to use much higher doses of valsartan (up to 320 mg daily) and the study is powered to evaluate noninferiority of the angiotensin II antagonist. Since a large proportion of patients will be receiving a beta-blocker, this study will be also provide new insights on the interplay between AT_1 receptor blockade and beta-blockade. The recent presented results show that valsartan is not inferior to captopril and that there is no particular interaction in patients receiving a beta-blocker.

10.4 Diabetic Nephropathy

As already mentioned, angiotensin II plays an important role in controlling renal hemodynamics and sodium excretion. Hence, blockade of the renin-angiotensin system has several favorable impacts on renal function. Several clinical studies have demonstrated that in decreasing the tone of the efferent arterioles, ACE inhibitors lower intraglomerular pressure and thereby decrease the urinary albumin excretion rate and retard the progression of chronic renal failure (Lewis et al. 1993; McKenzie and Brenner et al. 1998) There are also several experimental and small clinical studies indicating that angiotensin II receptor antagonists have comparable effects on kidney function than those of ACE-inhibitors. They have no influence on the glomerular filtration rate but they increase renal blood flow; consequently, the filtration fraction decreases (Burnier et al. 1993, 1995; Pechere-Bertschi et al. 1998). In patients, angiotensin II antagonists have regularly been found to reduce urinary albumin excretion (Nielsen et al. 1997; Perico et al. 1998; Vervoort et al. 1998; Christensen et al. 2001; Parving et al. 2001; Tutuncu et al. 2001).

In patients with type 1 diabetes, Lewis and colleagues have demonstrated that ACE inhibition is an effective therapeutic approach to retard the deterioration of renal function (Lewis et al. 1993). These data obtained in type 1 diabetes had never been confirmed in type 2 diabetes, although several small studies have suggested that ACE inhibitors or angiotensin II receptor antagonists can lower urinary albumin excretion in this situation. In order to evaluate the role of angiotensin II receptor blockade in the management of patients with type 2 diabetes, two sets of studies have been conducted: the first set assessed the progression from microalbuminuria to overt proteinuria, whereas the second set was designed to investigate the ability of angiotensin II receptor antagonists to retard the progression towards end-stage renal failure.

The Irbesartan Microalbuminuria study (IRMA-2) and the Microalbuminuria Reduction with Valsartan (MARVAL) study are two large prospective studies designed to measure the ability of irbesartan and valsartan to lower microalbuminuria and to slow the progression to overt proteinuria in patients with type 2 diabetes and microalbuminuria (Parving et al. 2001; Viberti et al. 2002). Both trials found that the antagonist of angiotensin II receptors lowers microalbuminuria. Interestingly, in IRMA-2, only the 300-mg dose of irbesartan produced a significant reduction in microalbuminuria. In MARVAL, a significant proportion of patients who were microalbuminuric became normoalbuminuric. These studies therefore suggest that angiotensin receptor antagonists have an effect on microalbuminuria comparable to ACE inhibitors. Unfortunately, there was no direct comparison of an ACE inhibitor and an angiotensin II antagonist. In the CALM study, almost 200 patients with type 2 diabetes and microalbuminuria were randomized to receive either the angiotensin antagonist candesartan (16 mg o.d.) or the ACE inhibitor lisinopril (20 mg o.d.) for 12 weeks followed by a randomization to either a monotherapy or a combination of both drugs for another 12-week period (Mogensen et al. 2000). Candesartan and lisinopril induced comparable changes in blood pressure and microalbuminuria. With the combination of both drugs, the reduction of proteinuria was significantly greater than that obtained with candesartan alone but not significantly greater than that achieved with the ACE inhibitor. These results suggest that the combination of an ACE inhibitor and an AT1 receptor antagonist could provide additional benefits, but whether this is indeed the case remains to be determined.

To evaluate the impact of angiotensin II receptor antagonist on the progression of diabetic nephropathy toward end-stage renal disease, two large clinical trials have been recently conducted and published. The first is the RENAAL study (Reduction in End Points in NIDDM with the Angiotensin II Antagonist Losartan) and the second is the IDNT study (Irbesartan Diabetic Nephropathy Trial) (Brenner et al. 2001; Lewis et al. 2001). Both studies have included hypertensive patients with type 2 diabetes and overt diabetic nephropathy. In RENAAL, 1,513 patients were randomized to losartan (50–100 mg o.d.) or placebo, both being taken in addition to a conventional antihypertensive treatment including calcium channel blockers, beta blockers, diuretics and centrally acting agents. In IDNT, 1,715 patients were randomized to irbesartan 300 mg daily, amlodipine 10 mg daily or to a placebo in addition of a standard therapy including a beta blocker and a diuretic. The primary outcome was similar in both trials, i.e., the composite of a doubling of the baseline serum creatinine, end-stage renal disease or death. Of note, the highest recommended dose of each antagonist was used in these trials, i.e., 100 mg losartan in RENAAL and 300 mg irbesartan in IDNT. Both studies demonstrated a significant effect of the angiotensin II receptor antagonist to reduce the progression of the diabetic nephropathy. For the primary end point, the risk reduction ranged from 16% to 20%. In RENAAL, losartan reduced the rate of first hospitalization for heart failure by 32% and urinary protein excretion by 33%.

Taken together, the results of these studies indicate that angiotensin II receptor blockade can confer significant renal benefits in patients with type 2 diabetes and nephropathy. Unfortunately, none of these studies included a control group with an ACE inhibitor to evaluate the respective impact of ACE inhibition and angiotensin II receptor blockade in protecting against the progression of nephropathy due to type 2 diabetes. This interesting question will perhaps find its answer in the ONTARGET trial, in which a large group of diabetic patients will be randomized to ramipril or telmisartan or the combination of both.

11
Are There Differences in Clinical Efficacy Among the Angiotensin II Receptor Antagonists?

As discussed earlier in this chapter, angiotensin II receptor antagonists share the same mechanism of action but each compound has its specific pharmacokinetic and pharmacodynamic properties that may account for potential differences in efficacy. A meta-analysis of 43 randomized placebo-controlled clinical trials, assessing a total of 11,281 patients has recently suggested comparable antihypertensive efficacy within this class of drugs (Conlin et al. 2000), but this meta-analysis included only the initial studies conducted with four compounds, i.e., losartan, valsartan, irbesartan and candesartan. Thereafter, several studies were specially designed to compare drugs within the class and have revealed some significant differences between the antagonists. Thus, for example, valsartan and losartan were compared in a randomized double-blind, placebo-controlled trial with an up-titration of the active regimens. They randomized 1,369 adults with uncomplicated essential hypertension to receive 80 mg valsartan, 50 mg losartan, or placebo once daily. After 4 weeks, doses of active medication and placebo were doubled. A comparable decrease in blood pressure at trough were observed at 4 and 8 weeks for both antagonists compared to placebo. But a higher response rate was observed with valsartan 160 mg than losartan 100 mg (Hedner et al. 1999).

The biggest differences in head-to-head comparison were observed when losartan was compared to longer-acting drugs such as candesartan, irbesartan or telmisartan. Indeed, all these studies show that these latter drugs may be more effective than losartan, particularly at trough, thus providing better 24-h control of blood pressure. For example, several randomized, controlled trials have demonstrated greater antihypertensive effects of candesartan cilexetil over losartan. These studies either evaluated the starting doses of both drugs or used a response titration design for comparison of their maximum doses (Andersson and Neldam 1988; Asmar and Lacourcière 2000). Two other double-blind, randomized, forced titration studies (CLAIM Study I and II) were designed to provide direct comparison of the blood pressure lowering effects at once-daily maximum doses in 654 and 622 hypertensive patients, respectively (Bakris et al. 2001; Vidt et al. 2001). Both studies showed that candesartan 32 mg lowered blood pressure by a significantly greater amount than losartan 100 mg in this hypertensive population.

In a study involving 567 hypertensive patients, mean blood pressure reduction after 8 weeks of treatment with losartan 100 mg did not differ significantly from that of 150 mg irbesartan, but a significant difference was observed with irbesartan 300 mg (Kassler-Taub et al. 1998). Another study by Oparil et al. (1998) also showed the superiority of irbesartan in a randomized double-blind, elective-titration study comparing losartan (50–100 mg) and irbesartan (150–300 mg). In another study, telmisartan was compared to losartan (Mallion et al. 1999) or valsartan (Bakris 2002) in hypertensive patients. Both studies suggested greater efficacy for telmisartan than losartan or valsartan in controlling blood pressure throughout the 24-h dosing interval, including the last 6 h before dosing. Finally, a recent multicenter, randomized, double-blind study compared the antihypertensive efficacy of once-daily treatment with the recommended starting doses of olmesartan (20 mg), losartan (50 mg), valsartan (80 mg), and irbesartan (150 mg) in 588 patients with essential hypertension. The reduction of sitting cuff diastolic blood pressure with olmesartan, the primary efficacy variable of this study, was significantly greater than with losartan, valsartan, and irbesartan. This was also the case in 24-h measurement, where reduction in mean diastolic blood pressure recorded with olmesartan was significantly greater than reductions with losartan and valsartan and showed a trend toward significance when compared to the reduction with irbesartan. The reduction in mean 24-h systolic blood pressure with olmesartan was significantly greater than the reductions with losartan and valsartan and equivalent to the reduction with irbesartan (Oparil et al. 2001).

Taken together these comparative studies suggest that there are indeed clinically significant differences between the antagonists, at least for blood pressure control. Yet one must take into account that all studies have been sponsored by the pharmaceutical industry and came generally to the same conclusion, i.e., the new compound is better than the older one. One important issue in these studies is the choice of the comparator and most importantly the choice of the doses. Thus, a critical question still remains: are the blood pressure differences observed between the antagonists due to the selected doses or to the pharmacokinetic properties of the compounds. Moreover, do these blood pressure differences have a real impact on morbidity and mortality?

12
What Is the Right Dose of an Angiotensin II Receptor Antagonist?

As for ACE inhibitors, the dose recommendations for the use of angiotensin II receptor antagonists have been based on clinical observation, i.e., the ability for these drugs to lower blood pressure over 24 h when given once a day in patients with mild to moderate hypertension. However, only a few studies have been designed to evaluate the dose-dependent reductions in blood pressure in hypertensive patients using a large range of doses (Chiolléro and Burnier 1998; McClellan and Goa 1998; Reeves et al. 1998; Sharpe et al. 2001). This failure to explore fully the lower and the upper extremes of the dose-response curves, together with the lack of large study populations, and perhaps the inability to accurately measure blood pressure changes has lead to clearly inadequate choice of starting and maintenance doses (i.e., providing less than maximal blockade) for some of these compounds (Burnier and Brunner 1999; Maillard et al. 2002a).

We have evaluated the blockade of the renin-angiotensin system produced by a series of angiotensin II receptor antagonists using two different techniques, the inhibition of blood pressure increase to exogenous angiotensin II (Christen et al. 1991) and an in vitro binding assay that quantified the displacement of an-

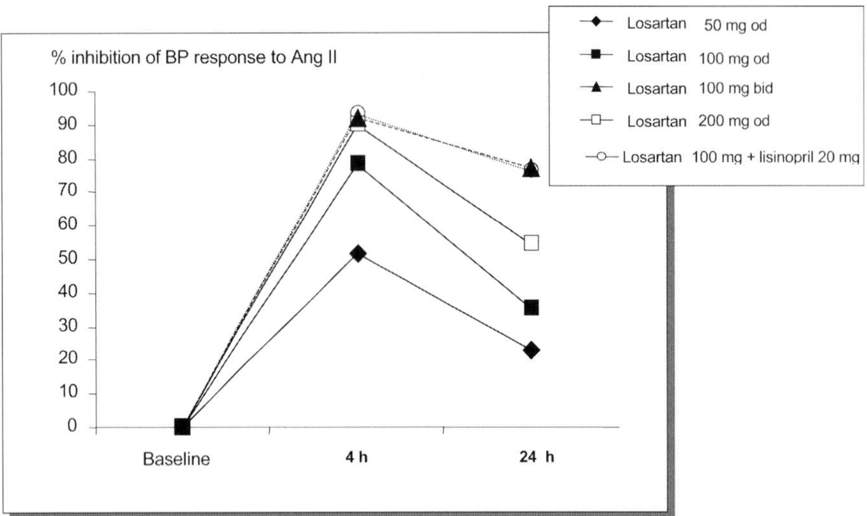

Fig. 2 Inhibition of the pressor response to exogenous angiotensin I or II. Values are based on mean blood pressure changes observed in healthy volunteers after 1 week treatment with different losartan regimens

giotensin II by the blocking agents (Maillard et al. 1999). With these techniques, we have found that the ability of the recommended starting doses of losartan (50 mg) valsartan (80 mg) and irbesartan (150 mg) to block the angiotensin II AT1 receptors in normotensive volunteers differ markedly and significantly, irbesartan showing the greatest inhibition at peak and trough (Mazzolai et al. 1999). In a separate study, we have reported that increasing the dose of valsartan from 80 to 320 mg had only a small effect on the peak inhibition but clearly enhanced the blockade at trough (Maillard et al. 2002a). The determinant impact of the dose or of the dose-regimen is illustrated with losartan in Fig. 2. The figure clearly shows that the initially recommended dose was too low to achieve a significant blockade of the angiotensin II receptor 4 h after drug intake and at trough. Once again, increasing the dose makes it possible to improve the inhibition at peak and prolong the duration of the blockade, but in the case of losartan, to obtain a blockade of the system around the clock, the ideal dosing regimen is probably 100 mg twice a day.

The other consequence of dose recommendations for the clinical use of these antagonists having been based on the antihypertensive efficacy rather on the capacity to fully inhibit the renin-angiotensin system is the relatively flat dose-response curves that is generally attributed to these compounds. However, the angiotensin II receptor antagonists have been developed in order to block the multiple effects of angiotensin II possibly throughout the day. Therefore, it may be necessary to add another agent to achieve adequate blood pressure control in some patients. As observed for ACE inhibitors, the addition of a low dose of a thiazide diuretic substantially potentiates the blood pressure lowering effect of

AT1 antagonists (Wellington and Faulds 2002). This resulted in greater reductions in blood pressure than with either agent alone. Thus, fixed-dose combinations of different angiotensin II receptor antagonists with hydrochlorothiazide have been marketed for the treatment of hypertension.

To achieve a more complete blockade of the renin-angiotensin system, some investigators have proposed combining an angiotensin II receptor antagonist and an ACE inhibitor, in order to act at two levels of the renin-angiotensin cascade (Azizi et al. 1995, 2000; Struckman and Rivey 2001; Ferrari 2002). With this approach, a greater decrease in blood pressure or urinary protein excretion has been reported, suggesting that the combination is indeed more effective than either drug alone (Mogensen et al. 2000). However, this approach also has some limitations. The first is that it does not obviate the side effects of ACE inhibitors. The second takes the form of a question: could the same blockade of the system be achieved with a higher dose of antagonists? Indeed, most of the positive studies combining an ACE inhibitor and an angiotensin II antagonist have used rather low doses of the antagonists, which we believe do not completely block the renin-angiotensin system. Hence, it is not surprising that the addition of an ACE inhibitor provides a greater blockade. To investigate this hypothesis, we have recently compared, in normotensive subjects, the degree of blockade of the renin-angiotensin system that can be obtained with different doses of losartan and telmisartan alone or in combination with lisinopril (Forclaz et al. 2003). In this study, we used doses of losartan and telmisartan beyond those generally recommended for the treatment of hypertension, i.e., losartan 100 mg o.d., 200 mg o.d. or 100 mg b.i.d. and telmisartan 80 mg, 160 mg o.d. and 80 mg b.i.d. When both drugs were administered at their maximal recommended doses, respectively 100 mg for losartan and 80 mg telmisartan, the addition of lisinopril provided a clear additional blockade of the receptor. However, the use of higher doses of the antagonist and in the case of losartan a twice-a-day regimen provided as much blockade as the ACE inhibitor/angiotensin II receptor blockade combination.

The results of these studies emphasize how difficult it is to chose the right dose of an angiotensin II receptor antagonist. Today, this choice is based almost exclusively on the antihypertensive efficacy. However, the recent morbidity and mortality trials have clearly shown that higher doses were the most effective in achieving target organ protection. This was the case in ValHeFT where an average of 240 mg of valsartan was used, but also in LIFE and RENAAL, in which 100 mg of losartan was used most frequently. In IRMA2, the 300 mg dose of irbesartan was found to be the most effective and this dose was chosen in IDNT. All these trials would therefore support the use of higher doses in clinical practice. Because of their excellent tolerability profile, the administration of high doses of angiotensin II antagonists should not represent a clinical problem other than cost.

13 Conclusions

Within 10 years, a tremendous number of clinical studies have demonstrated the efficacy of angiotensin II receptor antagonists as hypertensive agents. The development of angiotensin II antagonists has also been characterized by the completion of numerous large clinical trials which have shown within few years the ability of these agents to lower morbidity and mortality in various groups of patients, including hypertensive and heart failure patients and patients with incipient or overt type 2 diabetes nephropathy. Undoubtedly, the main advantage of angiotensin receptor antagonists is their excellent tolerability profile, which is a real advantage, as these agents are prescribed most frequently to asymptomatic patients. Yet a number of unresolved questions still remain, for example, the right dose to use in order to achieve the optimal target organ protection, the role of the AT_2 receptors, and the potential benefits of a combination with ACE inhibitors. These questions will probably find their answers in the next few years with the results of the ongoing studies and clinical trials.

References

Andersson OK, Neldam S (1988) The antihypertensive effect and tolerability of candesartan cilexetil, a new generation angiotensin II antagonist, in comparison with losartan. Blood Press 7:53–59

Asmar R, Lacourciere Y (2000) A new approach to assessing antihypertensive therapy: effect of treatment on pulse pressure. Candesartan cilexetil in Hypertension Ambulatory Measurement of Blood Pressure (CHAMP) Study Investigators. J Hypertens 18:1683–1890

Azizi M, Chatellier G, Guyene TT, Murieta-Geoffroy D, Menard J (1995) Additive effects of combined angiotensin-converting enzyme inhibition and angiotensin II antagonism on blood pressure and renin release in sodium-depleted normotensives. Circulation 92:825–834

Bakris G (2002) Comparison of telmisartan vs. valsartan in the treatment of mild to moderate hypertension using ambulatory blood pressure monitoring. J Clin Hypertens 4:26–31

Bakris G, Gradman A, Reif M, Wofford M, Munger M, Harris S, Vendetti J, Michelson EL, Wang R (2001) Antihypertensive efficacy of candesartan in comparison to losartan: the CLAIM study. J Clin Hypertens 3:16–21

Baruch L, Anand I, Cohen IS, Ziesche S, Judd D, Cohn JN (1999) Augmented short- and long-term hemodynamic and hormonal effects of an angiotensin receptor blocker added to angiotensin converting enzyme inhibitor therapy in patients with heart failure. Vasodilator Heart Failure Trial (V-HeFT) Study Group. Circulation 99:2658–2664

Benz J, Oshrain C, Henry D, Avery C, Chiang YT, Gatlin M (1997) Valsartan, a new angiotensin II receptor antagonist: a double-blind study comparing the incidence of cough with lisinopril and hydrochlorothiazide. J Clin Pharmacol 37:101–107

Brenner BM, Cooper ME, de Zeeuw D, Keane WF, Mitch WE, Parving HH, Remuzzi G, Snapinn SM, Zhang Z, Shahinfar S (2001) Effects of losartan on renal and cardiovascular outcomes in patients with type 2 diabetes and nephropathy. N Engl J Med 345:861–869

Brunner HR (1997) The new angiotensin II receptor antagonist, irbesartan: pharmacokinetic and pharmacodynamic considerations. Am J Hypertens 10:311S–317S

Brunner HR (2002) The new oral angiotensin II antagonist olmesartan medoxomil: a concise overview. J Hum Hypertens 16:S13–S16

Burnier M, Brunner HR (1999) Comparative antihypertensive effects of angiotensin II receptor antagonists. J Am Soc Nephrol 10:S278–S282

Burnier M, Brunner HR (2000) Angiotensin II receptor antagonists. The Lancet 355:637–645

Burnier M, Rutschmann B, Nussberger J, Versaggi J, Shahinfar S, Waeber B, Brunner HR (1993) Salt-dependent renal effects of an angiotensin II antagonist in healthy subjects. Hypertension 22:339–247

Burnier M, Roch-Ramel F, Brunner HR (1996) Renal effects of angiotensin II receptor blockade in normotensive subjects. Kidney Int 49:1787–1790

Chioléro A, Burnier M (1998) Pharmacology of valsartan, an angiotensin II receptor antagonist. Exp Opin Invest Drugs 7:1915–1925

Chiu AG, Krowiak EJ, Deeb ZE (2001) Angioedema associated with angiotensin II receptor antagonists: challenging our knowledge of angioedema and its etiology. Laryngoscope 111:1729–1731

Chiu AT, McCall DE, Price WAJ, Wong PC, Carini DJ, Duncia JV, Wexler RR, Yoo SE, Johnson AL, Timmermans PB (1991) In vitro pharmacology of DuP 753. Am J Hypertens 4:282S–287S

Christen Y, Waeber B, Nussberger J, Porchet M, Borland RM, Lee RJ, Maggon K, Shum LY, Timmermans PB, Brunner HR (1991) Oral administration of DuP 753, a specific angiotensin II receptor antagonist, to normal volunteers. Circ Res 83:1333–1342

Christensen PK, Lund S, Parving HH (2001) Autoregulated glomerular filtration rate during candesartan treatment in hypertensive type 2 diabetic patients. Kidney Int 60:1435–1442

Chung O, Unger T (1999) Pharmacology of angiotensin receptors and AT-1 receptor blockers. Basic Res Cardiol 93:15–23

Cohn JN, Tognoni G, Glazer RD, Spormann D, Hester A (1999) Rationale and design of the Valsartan Heart Failure Trial: a large multinational trial to assess the effects of valsartan, an angiotensin-receptor blocker, on morbidity and mortality in chronic congestive heart failure. J Card Fail 5:155–160

Conlin PR, Spence JD, Williams B, Ribeiro AB, Saito I, Benedict C, Bunt AMG (2000) Angiotensin II antagonists for hypertension: are there differences in efficacy? Am J Hypertens 13:418–426

Criscione L, Bradley W, Bühlmayer P (1995) Valsartan: preclinical and clinical profile of an antihypertensive angiotensin II antagonist. Cardiovasc Drug Rev 13:230–250

Dahlöf B (1994) Angiotensin-converting enzyme inhibitors and effects on left ventricular hypertrophy. Blood Press 2:35–40

Dahlöf B, Devereux RB, Kjeldsen SE, Julius S, Beevers G, Faire U, Fyhrquist F, Ibsen H, Kristiansson K, Lederballe-Pedersen O, Lindholm LH, Nieminen MS, Omvik P, Oparil S, Wedel H (2002) Cardiovascular morbidity and mortality in the Losartan Intervention For Endpoint reduction in hypertension study (LIFE): a randomized trial against atenolol. Lancet 359:995–1003

De Chaffoy de Courcelles D, Leysen, Roevens P, Van Belle H (1986) The serotonin S-2 receptor: a receptor-transducer coupling model to explain insurmountable effects. Drug Dev Res 8:173–178

Delacrétaz E, Nussberger J, Biollaz J, Waeber B, Brunner HR (1995) Characterization of the angiotensin II receptor antagonist TCV-116 in healthy volunteers. Hypertension 25:14–21

Dickstein K, Kjekshus J (2002) Effects of losartan and captopril on mortality and morbidity in high-risk patients after acute myocardial infarction: the OPTIMAAL randomised trial. Optimal Trial in Myocardial Infarction with Angiotensin II Antagonist Losartan. Lancet 360:752–760

Dominiczak A, Isles C, Gillen G, Brown JJ (1988) Angiotensin converting enzyme inhibition and renal insufficiency in patients with bilateral renovascular disease. J Hum Hypertens 2:53–56

Dzau VJ, Sasamura H, Lutz H (1993) Heterogeneity of angiotensin synthetic pathways and receptor subtypes: physiological and pharmacological implications. J Hypertens Suppl 11:S13–S18

Edwards RM, Aiyar N, Ohlstein EH (1992) Pharmacologic characterization of the nonpeptide angiotensin II receptor antagonist SKF 108566. J Pharmacol Exp Ther 260:175–181

Elliott WJ (1999) Double-blind comparison of eprosartan and enalapril on cough and blood pressure in unselected hypertensive patients. Eprosartan Study Group. J Hum Hypertens 13:413–417

Ferrari P, Marti HP, Pfister M, Frey FJ (2002) Additive antiproteinuric effect of combined ACE inhibition and angiotensin II receptor blockade. J Hypertens 20:125–130

Fierens FLP, Vanderheyden PML, de Backer JP, Vauquelin G (1999) Insurmountable angiotensin AT1 receptor antagonists: the role of tight antagonist binding. Eur J Pharmacol 372:199–206

Forclaz A, Maillard M, Nussberger J, Brunner HR, Burnier M (2003) Angiotensin II receptor blockade: is there truly a benefit of adding an ACE inhibitor? Hypertension 41:31–36

Franzosi MG, Santoro E, Zuanetti G, Baigent C, Collins R, Flather M, Kjekshus J, Latini R, Liu LS, Maggioni AP, Sleight P, Swedberg K, Tognoni G, Yusuf S for the ACE Inhibitor Myocardial Infarction Collaborative Group (1998) Indications for ACE inhibitors in the early treatment of acute myocardial infarction: systematic overview of individual data from 100,000 patients in randomised trials. Circulation 97:2202–2212

Furukawa Y, Kishimoto S, Nishikawa K (1982) Hypotensive imidazole-5-acetic acid derivatives. US Patent 4,35,040, 1982

Gillis JC, Markham A (1997) Irbesartan. A review of its pharmacodynamic and pharmacokinetic properties and therapeutic use in the management of hypertension. Drugs 54:885–902

Goa KL, Wagstaff AJ (1996) Losartan potassium: a review of its pharmacology, clinical efficacy and tolerability in the management of hypertension. Drugs 51:820–845

Granger CB, McMurray JJV, Yusuf S, Held P, Michelson EL, Olofsson B, Östergren J, Pfeffer MA, Swedberg K and CHARM Investigators and Committees (2003) Effects of candesartan in patients with chronic heart failure and reduced left-ventricular systolic function intolerant to angiotensin-converting-enzyme inhibitors: the CHARM-Alternative trial. Lancet 362:772–776

Gremmler B, Kunert M, Schleiting H, Ulbricht LJ (2000) Improvement of cardiac output in patients with severe heart failure by use of ACE-inhibitors combined with the AT1-antagonist eprosartan. Eur J Heart Fail 2:183–187

Hagmann M, Nussberger J, Naudin RB, Burns TS, Karim A, Waeber B, Brunner HR (1997) SC-52458, an orally active angiotensin II-receptor antagonist: inhibition of blood pressure response to angiotensin II challenges and pharmacokinetics in normal volunteers. J Cardiovasc Pharmacol 29:444–450

Hamroff G, Blaufarb I, Mancini D, Katz SD, Bijou R, Jondeau G, Olivari MT, Thomas S, LeJemtel TH (1997) Angiotensin II-receptor blockade further reduces afterload safely in patients maximally treated with angiotensin-converting enzyme inhibitors for heart failure. J Cardiovasc Pharmacol 30:533–536

Hamroff G, Katz SD, Mancini D, Blaufarb I, Bijou R, Patel R, Jondeau G, Olivari MT, Thomas S, Le Jemtel TH (1999) Addition of angiotensin II receptor blockade to maximal angiotensin-converting enzyme inhibition improves exercise capacity in patients with severe congestive heart failure. Circulation 99:990–992

Hansson L, Lindholm LH, Niskanen L, Lanke J, Hedner T, Niklason A, Luomanmäki K, Dahlof B, de Faire U, Mörlin C, Karlberg BE, Wester PO, Björk JE (1999a) Effect of angiotensin-converting-enzyme inhibition compared with conventional therapy on cardiovascular morbidity and mortality in hypertension: the Captopril Prevention Project (CAPPP) randomised trial. Lancet 353:611–616

Hansson L, Lindholm LH, EkbomT, Dahlof B, Lanke J, Schersten B, Wester PO, Hedner T, de Faire U (1999b) Randomised trial of old and new antihypertensive drugs in elderly patients: cardiovascular mortality and morbidity. The Swedish Trial in Old Patients with Hypertension-2 Study. Lancet 354:1751–1756

Hedner T (2002) The clinical profile of the angiotensin II receptor blocker eprosartan. J Hypertens 20:S33–S38

Hedner T, Himmelmann A (1999) The efficacy and tolerance of one or two daily doses of eprosartan in essential hypertension. The Eprosartan Multinational Study Group. J Hypertens 17:129–136

Hedner T, Oparil S, Rasmussen K, Rapelli A, Gatlin M, Kobi P, Sullivan J, Oddou-Stock P (1999) A comparison of the angiotensin II antagonists valsartan and losartan in the treatment of essential hypertension. Am J Hypertens 12:414–417

Herbert JM, Delisee C, Dol F, Cazaubon C, Nisato D, Chatelain P (1994) Effect of SR 47436, a novel angiotensin II AT1 receptor antagonist, on human vascular smooth muscle cells in vitro. Eur J Pharmacol 251:143–150

Holman R, Turner R, Stratton I, Cull C, Frighi V, Manley S, Matthews D, Neil A, Kohner E, Wright D, Hadden D, Fox C for the UK Prospective Diabetes Study Group (1998) Efficacy of atenolol and captopril in reducing risk of macrovascular and microvascular complications in type 2 diabetes: UKPDS 39. BMJ 317:713–720

Houghton AR, Harrison M, Cowley AJ, Hampton JR (2000) Combined treatment with losartan and an ACE inhibitor in mild to moderate heart failure: results of a double-blind, randomized, placebo-controlled trial. Am Heart J 140:791

Innocenti AD, Elmfeldt D, Hansson L, Breteler M, James O, Lithell H, Olofsson B, Skoog I, Trenkwalder P, Zanchetti A, Wiklund I (2002) Cognitive function and health-related quality of life in elderly patients with hypertension—baseline data from the study on cognition and prognosis in the elderly (SCOPE). Blood Press 11:157–163

Israili ZH (2000) Clinical pharmacokinetics of angiotensin II (AT1) receptor blockers in hypertension. J Hum Hypertens 14:S73–S86

Israili ZH, Hall WD (1992) Cough and angioneurotic edema associated with angiotensin-converting enzyme inhibitor therapy. A review of the literature and pathophysiology. Ann Intern Med 117:234–242

Juillerat L, Nussberger J, Ménard J, Mooser V, Christen Y, Waeber B, Graf P, Brunner HR (1990) Determinants of angiotensin II generation during converting enzyme inhibition. Hypertension 16:564–572

Kassler-Taub K, Littlejohn T, Elliott W, Ruddy T, Adler E (1998) Comparative efficacy of two angiotensin II receptor antagonists, irbesartan and losartan in mild-to-moderate hypertension. Irbesartan/Losartan Study Investigators. Am J Hypertens 11:445–453

Keenan RM, Weinstock J, Finkelstein JA, Franz RG, Gaitanopoulos DE, Girard GR, Hill DT, Morgan TM, Samanen JM, Peishoff CE (1993) Potent nonpeptide angiotensin II receptor antagonists. 1. (Carboxybenzyl)imidazole-5-acrylic acids. J Med Chem 36:1880–1892

Kjekshus J, Swedberg K, Snapinn S (1992) Effects of enalapril on long-term mortality in severe congestive failure. Am J Cardiol 62:103–107

Koike H, Sada T, Mizuno M (2001) In vitro and in vivo pharmacology of olmesartan medoxomil, an angiotensin II type AT1 receptor antagonist. J Hypertens 19:S3–S14

Lacourcière Y (1999) The incidence of cough: a comparison of lisinopril, placebo and telmisartan, a novel angiotensin II antagonist. Telmisartan Cough Study Group. Int J Clin Pract 53:99–103

Lacourcière Y, Brunner H, Irwin R, Karlberg BE, Ramsay LE, Snavely DB, Dobbins TW, Faison EP, Nelson EB (1994) Effects of modulators of the renin-angiotensin-aldosterone system on cough. Losartan Cough Study Group. J Hypertens 12:1387–1393

Lewis EJ, Hunsicker LG, Bain RP, Rohde RD (1993) The effect of angiotensin-converting-enzyme inhibition on diabetic nephropathy. The Collaborative Study Group. N Engl J Med 329:1456–1462

Lewis EJ, Hunsicker LG, Clarke WR, Berl T, Pohl MA, Lewis JB, Ritz E, Atkins RC, Rohde R, Raz I (2001) Renoprotective effect of the angiotensin-receptor antagonist irbesartan in patients with nephropathy due to type 2 diabetes. N Engl J Med 345:851–860

Lithell H, Hansson L, Skoog I, Elmfeldt D, Hofman A, Olofson B, Trenkwalder P, Zanchetti A, for the SCOPE Study Group (2003) The Study on Cognition and Prognosis in the Elderly (SCOPE): principal results of a randomized double-blind intervention trial. J Hypertens 21:875–886

Liu YJ, Shankley NP, Welsh NJ, Black JW (1992) Evidence that the apparent complexity of receptor antagonism by angiotensin II analogues is due to a reversible and syntopic action. Br J Pharmacol 106:233–241

Maillard M, Mazzolai L, Daven C, Centeno C, Nussberger J, Brunner HR, Burnier M (1999) Assessment of angiotensin II-receptor blockade in humans using a standardized angiotensin II-receptor binding assay. Am J Hypertens 12:1201–1208

Maillard M, Rossat J, Nussberger J, Ramis J, Pontes C, Burnier M, Brunner HR (2000) Pharmacological profile of UR-7247, an orally active angiotensin II receptor antagonist in healthy volunteers. J Cardiovasc Pharmacol 35:383–389

Maillard M, Centeno C, Frostell-Karlsson A, Brunner HR, Burnier M (2001) Does protein binding modulate the effect of angiotensin II receptor antagonists. J Renin Angiotensin Aldosterone Syst 2:S54–S58

Maillard M, Würzner G, Nussberger J, Centeno C, Burnier M, Brunner HR (2002a) Comparative angiotensin II receptor blockade in healthy volunteers: the importance of dosing. Clin Pharmacol Ther 71:68–76

Maillard M, Perregaux C, Centeno C, Stangier J, Wienen W, Brunner HR, Burnier M (2002b) In vitro and in vivo characterization of telmisartan: an insurmountable angiotensin II receptor antagonist. J Pharmacol Exp Ther 302:1089–1095

Mallion J, Siche J, Lacourciere Y (1999) ABPM comparison of the antihypertensive profiles of the selective angiotensin II receptor antagonists telmisartan and losartan in patients with mild-to-moderate hypertension. J Hum Hypertens 13:657–664

Mann J, Julius S (1998) The Valsartan Antihypertensive Long-Term Use Evaluation (VALUE) trial of cardiovascular events in hypertension. Rationale and design. Blood Press 7:176–183

Markham A, Goa KL (1997) Valsartan. A review of its pharmacology and therapeutic use in essential hypertension. Drugs 54:299–311

Mazzolai L, Burnier M (1999) Comparative safety and tolerability of angiotensin II receptor antagonists. Drug Safety 21:23–33

Mazzolai L, Maillard M, Rossat J, Nussberger J, Brunner HR, Burnier M (1999) Angiotensin II receptor blockade in normotensive subjects: A direct comparison of three AT1 receptor antagonists. Hypertension 33:850–855

McClellan KJ, Balfour JA (1998) Eprosartan. Drugs 55:713–718

McClellan KJ, Goa KL (1998) Candesartan cilexetil: a review of its use in essential hypertension. Drugs 56:847–869

McKelvie RS, Yusuf S, Pericak D, Avezum A, Burns RJ, Probstfield J, Tsuyuki RT, White M, Rouleau J, Latini R, Maggioni A, Young J, Pogue J (1999) Comparison of candesartan, enalapril, and their combination in congestive heart failure: randomized evaluation of strategies for left ventricular dysfunction (RESOLVD) pilot study. The RESOLVD Pilot Study Investigators. Circulation 100:1056–1064

McKenzie HS, Brenner BM (1998) Current strategies for retarding the progression of renal disease. Am J Kidney Dis 31:161–170

McMurray JJV, Östergren J, Swedberg K, Granger CB, Held P, Michelson EL, Olofsson B, Yusuf S, Pfeffer MA and CHARM Investigators and Committees (2003) Effects of candesartan in patients with chronic heart failure and reduced left-ventricular systolic function taking angiotensin-converting-enzyme inhibitors: the CHARM-Added trial. Lancet 362:767–771

Minghelli G, Seydoux C, Goy JJ, Burnier M (1998) Uricosuric effect of the angiotensin II receptor antagonist losartan in heart transplant recipients. Transplantation 66:268–271

Mochizuki S, Sato T, Furuta K, Hase K, Ohkura Y, Fukai C, Kosakai K, Wakabayashi S, Tomiyama A (1995) Pharmacological properties of KT3-671, a novel nonpeptide angiotensin II receptor antagonist. J Cardiovasc Pharmacol 25:22–29

Mogensen CE, Neldam S, Tikkanen I, Oren S, Viskoper R, Watts RW, Cooper ME (2000) Randomised controlled trial of dual blockade of renin-angiotensin system in patients with hypertension, microalbuminuria, and non-insulin dependent diabetes: the candesartan and lisinopril microalbuminuria (CALM) study. BMJ 321:1440–1444

Morice AH, Lowry RM, Brown MJ, Higenbottam T (1987) Angiotensin-converting enzyme and the cough reflex. Lancet 2:1116–1118

Morsing P, Adler G, Brandt-Eliasson U, Karp L, Ohlson K, Renberg L, Sjöquist P-O, Abrahamsson T (1999) Mechanistic differences of various AT-1 receptor blockers in isolated vessels of different origin. Hypertension 33:1406–1413

Müller P, Cohen T, de Gasparo M, Sioufi A, Racine-Poon A, Howald H (1994) Angiotensin II receptor blockade with single doses of valsartan in healthy, normotensive subjects. Eur J Clin Pharmacol 47:231–245

Munafo A, Christen Y, Nussberger J, Shum LY, Borland RM, Lee RJ, Waeber B, Biollaz J, Brunner HR (1992) Drug concentration response relationships in normal volunteers after oral administration of losartan, an angiotensin II receptor antagonist. Clin Pharmacol Ther 51:513–521

Nielsen S, Dollerup J, Nielsen B, Jensen HA, Mogensen CE (1997) Losartan reduces albuminuria in patients with essential hypertension. An enalapril controlled 3 months study. Nephrol Dial Transplant 12:19–23

Nussberger J, Cugno M, Amstutz C, Cicardi M, Pellacani A, Agostoni A (1998) Plasma bradykinin in angioedema. Lancet 351:1693–1697

Ohlstein EH, Brooks DP, Feuerstein GZ, Ruffolo RR Jr (1997) Inhibition of sympathetic outflow by the angiotensin II receptor antagonist, eprosartan, but not by losartan, valsartan or irbesartan: relationship to differences in prejunctional angiotensin II receptor blockade. Pharmacology 55:244–251

Ojima M, Inada Y, Shibouta Y, Sanada T, Kubo K, Nishikawa K (1997) Candesartan (CV-11974) dissociates slowly from the angiotensin AT1 receptor. Eur J Pharmacol 319:137–146

Oparil S, Guthrie R, Lewin AJ, Marbury T, Reilly K, Triscari J, Witcher JA (1998) An elective-titration study of the comparative effectiveness of two angiotensin II-receptor blockers, irbesartan and losartan. Irbesartan/Losartan Study Investigators. Clin Ther 20:398–409

Oparil S, Williams D, Chrysant SG, Marbury TC, Neutel J (2001) Comparative efficacy of olmesartan, losartan, valsartan, and irbesartan in the control of essential hypertension. J Clin Hypertens 3:283–291

Panek RL, Lu GH, Overhiser RW, Major TC, Hodges JC, Taylor DG (1995) Functional studies but not receptor binding can distinguish surmountable from insurmountable AT1 antagonism. J Pharmacol Exp Ther 273:753–761

Parving HH, Lehnert H, Brochner-Mortensen J, Gomis R, Andersen S, Arner P (2001) The effect of irbesartan on the development of diabetic nephropathy in patients with type 2 diabetes. N Engl J Med 345:870–878

Pechere-Bertschi A, Nussberger J, Decosterd L, Armagnac C, Sissmann J, Bouroudian M, Brunner HR, Burnier M (1998) Renal response to the angiotensin II receptor subtype

1 antagonist irbesartan versus enalapril in hypertensive patients. J Hypertens 16:385–393

Perico N, Remuzzi A, Sangalli F, Azzollini N, Mister M, Ruggenenti P, Remuzzi G (1998) The antiproteinuric effect of angiotensin antagonism in human IgA nephropathy is potentiated by indomethacin. J Am Soc Nephrol 9:2308–2317

Perrier L, Bourrié M, Marti E, Tronquet C, Massé D, Berger Y, Magdalou J, Fabre G (1994) In vitro N-glucuronidation of SB 47436 (BMS 186295), a new AT1 nonpeptide angiotensin II receptor antagonist, by rat, monkey and human hepatic microsomal fractions. J Pharmacol Exp Ther 271:91–99

Pfeffer MA, McMurray J, Leizorovicz A, Maggioni AP, Rouleau JL, Van De Werf F, Henis M, Neuhart E, Gallo P, Edwards S, Sellers MA, Velazquez E, Califf R (2000) Valsartan in acute myocardial infarction trial (VALIANT): rationale and design. Am Heart J 140:727–750

Pfeffer MA, Swedberg K, Granger CB, Held P, McMurray JJV, Michelson EL, Olofsson B, Östergren J, Yusuf S and CHARM Investigators and Committees (2003) Effects of candesartan on mortality and morbidity in patients with chronic heart failure: the CHARM-Overall programme. Lancet 362:759–766

Pitt B, Segal R, Martinez FA, Meurers G, Cowley AJ, Thomas I, Deedwania PC, Ney DE, Snavely DB, Chang PI (1997) Randomized trial of losartan versus captopril in patients over 65 with heart failure (Evaluation of Losartan in the Elderly Study, ELITE). Lancet 349:747–752

Pitt B, Poole-Wilson PA, Segal R, Martinez FA, Dickstein K, Camm AJ, Konstam MA, Riegger G, Klinger GH, Neaton J, Sharma D, Thiyagarajan B (2000) Effect of losartan compared with captopril on mortality in patients with symptomatic heart failure: randomised trial—the Losartan Heart Failure Survival Study ELITE II. Lancet 355:1582–1587

Reeves RA, Lin CS, Kassler-Taub K, Pouleur H (1998) Dose-related efficacy of irbesartan for hypertension: an integrated analysis. Hypertension 31:1311–1316

Remme WJ, Swedberg K (2001) Guidelines for the diagnosis and treatment of chronic heart failure. Eur Heart J 22:1529–1560

Ribstein J, Picard A, Armagnac C, Sissmann J, Mimran A (2001) Inhibition of the acute effects of angiotensin II by the receptor antagonist irbesartan in normotensive men. J Cardiovasc Pharmacol 37:449–460

Ries U, Mihm G, Narr B, Hasselbach KM, Wittneben H, Entzeroth M, Van Meel JCA, Wienen W, Hauel N (1993) 6-Substituted benzimidazoles as new nonpeptide angiotensin II receptor antagonists: synthesis, biological activity, and structure-activity relationships. J Med Chem 36:4040–4051

Roch-Ramel F, Werner D, Guisan B (1994) Urate transport in brush-border membrane of human kidney. Am J Physiol 266:F797–F805

Roch-Ramel F, Guisan B, Diezi J (1997) Effects of uricosuric and antiuricosuric agents on urate transport in human brush-border membrane vesicles. J Pharmacol Exp Ther 280:839–845

Ruggenenti P, Perna A, Gherardi G, Garini G, Zoccali C, Salvadori M, Scolari F, Schena FP, Remuzzi G (1999) Renoprotective properties of ACE inhibition in non-diabetic nephropathies with non-nephrotic proteinuria. Lancet 354:359–364

Schwarzberg A, Wittenmeier KW, Hallfritzsch U (1998) Anemia in dialysis patients as a side-effect of sartans. Lancet 352:286

Sharpe M, Jarvis B, Goa KL (2001) Telmisartan: a review of its use in hypertension. Drugs 61:1501–1529

Shetty SS, DelGrande D (2000) Differential inhibition of the prejunctional actions of angiotensin II in rat atria by valsartan, irbesartan, eprosartan, and losartan. J Pharmacol Exp Ther 294:179–186

Struckman DR, Rivey MP (2001) Combined therapy with an angiotensin II receptor blocker and an angiotensin-converting enzyme inhibitor in heart failure. Ann Pharmacother 35:242–248

Swedberg K, Pfeffer M, Granger C, Held P, McMurray J, Ohlin G, Olofsson B, Ostergren J, Yusuf S (1999) Candesartan in heart failure: assessment of reduction in mortality and morbidity (CHARM): rationale and design. J Card Fail 5:276–282

Thürmann PA, Colette D (2002) Angiotensin II type 1 receptor antagonists in chronic heart failure. Expert Opin Invest Drugs 11:705–716

Timmermans PB (1999a) Angiotensin II receptor antagonists: an emerging new class of cardiovascular therapeutics. Hpertens Res 22:147–153

Timmermans PB (1999b) Pharmacological properties of angiotensin II receptor antagonists. Can J Cardiol 15:26F–28F

Timmermans PB, Wong PC, Chiu AT, Herblin WF (1991) Nonpeptide angiotensin II receptor antagonists. Trends Pharmacol Sci 12:55–61

Timmermans PB, Wong PC, Chiu AT, Herblin WF, Benfield P, Carini DJ, Lee RJ, Wexler RR, Saye JA, Smith RD (1993) Angiotensin II receptors and angiotensin II receptor antagonists. Pharmacol Rev 45:205–251

Tonkon M, Awan N, Niazi I, Hanley P, Baruch L, Wolf RA, Block AJ (2000) A study of the efficacy and safety of irbesartan in combination with conventional therapy, including ACE inhibitors, in heart failure. Irbesartan Heart Failure Group. Int J Clin Pract 54:11–18

Tutuncu NB, Gurlek A, Gedik O (2001) Efficacy of ACE inhibitors and ATII receptor blockers in patients with microalbuminuria: a prospective study. Acta Diabetol 38:157–161

Vanderheyden PML, Fierens FLP, de Backer JP, Fraeyman N, Vauquelin G (1999) Distinction between surmountable and insurmountable angiotensin II type I antagonists by use of CHO-K1 cells expressing human angiotensin II AT1A receptors. Br J Pharmacol 126:1057–1065

Vanderheyden PML, Fierens FLP, de Backer JP, Vauquelin G (2000a) Reversible and syntopic interaction between angiotensin receptor antagonists on Chinese hamster ovary cells expressing human angiotensin II type 1 receptors. Biochem Pharmacol 59:927–935

Vanderheyden PML, Fierens FLP, Vauquelin G (2000b) Angiotensin II type 1 receptor antagonists. Why do some of them produce insurmountable inhibition? Biochem Pharmacol 60:1557–1563

Vervoort G, Hertenberg F, Wetzels JF, Smits P (1998) Influence of angiotensin converting enzyme inhibition and angiotensin II type 1 receptor antagonism on renal sodium and water handling and albuminuria during infusion of atrial natriuretic factor into healthy volunteers. J Hypertens 16:245–250

Viberti G, Wheeldon NM (2002) Microalbuminuria reduction with valsartan in patients with type 2 diabetes mellitus: a blood pressure-independent effect. Circulation 106:672–678

Vidt DG, White WB, Ridley E, Rahman M, Harris S, Vendetti J, Michelson EL, Wang R (2001) A forced titration study of antihypertensive efficacy of candesartan cilexetil in comparison to losartan: CLAIM Study II. J Hum Hypertens 15:475–480

Warner GT, Jarvis B (2002) Olmesartan medoxomil. Drugs 62:1345–1353

Weinstock J, Keenan RM, Samanen J, Hempel J, Finkelstein JA, Franz RG, Gaitanopoulos DE, Girard GR, Gleason JG, Hill DT, Morgan TM, Peishoff CE, Aiyar N, Brooks DP, Fredrickson TA, Ohlstein EH, Ruffolo RR Jr, Stack EJ, Sulpizio AC, Weidley EF, Edwards RM (1991) 1- (carboxybenzyl)imidazole-5-acrylic acids: potent and selective angiotensin II receptor antagonists. J Med Chem 34:1514–1517

Wellington K, Faulds D (2002). Valsartan/Hydrochlorothiazide. A review of its pharmacology, therapeutic efficacy and place in the management of hypertension. Drugs 62:1983–2005

Wienen W, Mauz AB, Van Meel JCA, Entzeroth M (1992) Different types of receptor interaction of peptide and nonpeptide angiotensin II antagonists revealed by receptor binding and functional studies. Mol Pharmacol 41:1081–1088

Wienen W, Hauel N, Van Meel JCA, Narr B, Ries U, Entzeroth M (1993) Pharmacological characterization of the novel nonpeptide angiotensin II receptor antagonist, BIBR 277. Br J Pharmacol 110:245–252

Wienen W, Hauel N, Busch U, Ebner T (1999) Characterization of the PK/PD profile of BIBR-277 1-O-acylglucuronide, the metabolite of telmisartan (abstract). J Hum Hypertens 13:S11

Wienen W, Entzeroth M, Van Meel JCA, Stangier S, Busch U, Ebner T, Schmid J, Lehmann H, Matzek K, Kempthorne-Rawson J, Gladigau V, Hauel NH (2000) A review on telmisartan: a novel, long-acting angiotensin II receptor antagonist. Cardiovasc Drug Rev 18:127–156

Wong PC, Price WAJ, Chiu AT, Duncia JV, Carini DJ, Wexler RR, Johnson AL, Timmermans PB (1990) Nonpeptide angiotensin II receptor antagonists. XI. Pharmacology of EXP3174: an active metabolite of DuP 753, an orally active antihypertensive agent. J Pharmacol Exp Ther 255:211–217

Wuerzner G, Gerster JC, Chioléro A, Maillard M, FalLab-Stubi CL, Brunner HR, Burnier M (2001) Comparative effects of losartan and irbesartan on serum uric acid in hypertensive patients with hyperuricemia and gout. J Hypertens 19:1855–1860

Yusuf S (2002) From the HOPE Study to ONTARGET and the TRANSCEND studies: challenges in improving prognosis. Am J Cardiol 89:18A–26A

Yusuf S, Sleight P, Pogue J, Bosh J, Davies R, Dagenais G (2000) Effects of an angiotensin-converting-enzyme inhibitor, ramipril, on cardiovascular events in high-risk patients. N Engl J Med 342:145–153

Yusuf S, Pfeffer MA, Swedberg K, Granger CB, Held P, McMurray JJV, Michelson EL, Olofsson B, Östergren J and CHARM Investigators and Committees (2003) Effects of candesartan in patients with chronic heart failure and preserved left-ventricular ejection fraction: the CHARM-Preserved Trial. Lancet 362:777–781

Combined Blockade of the Renin Angiotensin System with ACE Inhibitors and AT_1 Receptor Antagonists

M. Azizi · J. Ménard

Clinical Investigation Center 9201, Assistance Publique des Hôpitaux de Paris/INSERM, Hôpital Européen Georges Pompidou, 75908 Paris cedex 15, France
e-mail: michel.azizi@egp.ap-hop-paris.fr

Abstract The results from experimental models and randomized controlled clinical trials have shown that the more intense the blockade of the renin-angiotensin system (RAS) is during angiotensin I-converting enzyme (ACE) inhibition or AT_1 receptor blockade, the more effective the prevention of target organ damage is. Combined inhibition of the RAS is aimed at more complete blockade of the system through action at two different sites: ACE and AT_1 receptors. This is achieved either by neutralizing the rise in renin and angiotensin (Ang) I, which follows the interruption of the Ang II-renin negative feedback loop, or by directly antagonizing Ang II, whose synthesis is in part independent of the RAS, as nonrenin-dependent pathways have been identified in cardiac and vascular tissues. By comparison with higher doses of single-site blockers, a combination of an ACE inhibitor and an AT_1 receptor antagonist blocks the RAS more effectively. After demonstrating the synergistic or additive blood pressure effects of the combined RAS blockade in sodium-depleted normotensive subjects and animal models, the combined blockade of the RAS was shown to be more efficient than single-site blockade (a) in lowering BP in Caucasian hypertensive patients with predominantly low renin concentrations, with type II diabetes or with renal failure, (b) in lowering proteinuria and possibly retarding progression of renal failure in patients with diabetic and nondiabetic nephropathy, and (c) in finally improving left ventricular remodeling, cardiac function status and cardiovascular morbidity and mortality in patients with congestive heart failure. The advantage offered by combining two RAS blockers is to increase the beneficial effect of cardioprotection and nephroprotection, which are currently achieved with the highest doses of an ACE inhibitor or an AT_1 receptor antagonist.

Keywords Combination · Angiotensin-converting enzyme inhibitor · AT_1 receptor antagonist · Synergy

Blockade of the renin-angiotensin system (RAS) with angiotensin I-converting enzyme (ACE) inhibitors has become one of the most successful therapeutic approaches in medicine. Within 25 years, a high level of evidence has accumulated that attributes to this therapy a reduction in left ventricular mass (Dahlof et al. 1992; Schmieder et al. 1996), a decrease in overall mortality and death from chronic heart failure (Garg and Yusuf 1995; Flather et al. 2000), an improvement in the outcome for patients with ventricular systolic dysfunction and symptomatic heart failure (The SOLVD Investigators 1992), a reduction in postmyocardial infarction mortality (Pfeffer et al. 1988; ACE Inhibitor Myocardial Infarction Collaborative Group 1998), a retardation of the progression of renal insufficiency and an improvement of renal function in type 1 diabetes mellitus (Lewis et al. 1993), and a retardation of the progression of nondiabetic chronic renal disease (Maschio et al. 1996; The GISEN Group 1997; Jafar et al. 2001). ACE inhibitors have also been shown to reduce the rate of death, cardiac events and stroke in a broad range of patients with a high cardiovascular risk at baseline (Yusuf et al. 2000).

Research into mechanisms of ACE inhibition in preventing end-organ damage has focused on the respective roles of the circulating RAS and the tissue RAS (Dzau 2001). Local autocrine production of angiotensin II (Ang II) by either ACE or non-ACE-dependent pathways (Okunishi et al. 1984; Urata et al. 1990) modulates a tissue-specific response independent of the circulatory RAS, which may be an important participant in the pathogenesis of cardiovascular disease.

Multiple mechanisms may contribute to the beneficial effects of ACE inhibition in cardiovascular therapy. The hemodynamic consequences of Ang II suppression (Biollaz et al. 1981) and bradykinin accumulation (Linz et al. 1995; Gainer et al. 1998), and also the direct cellular effects of Ang II on heart and vascular smooth muscle cells (Sadoshima et al. 1993; Dzau 2001) are among the putative mechanisms explaining these beneficial effects. At first glance, the addition of two pharmacological agents, which both inhibit the RAS, such as an ACE inhibitor and an antagonist of type 1 Ang II receptors, receptors which are implicated in most of the biological effects of Ang II, should not offer an opportunity to increase the efficacy of a treatment based on one or the other, as shown by initial animal acute experiments (Wong et al. 1990). However, no ACE inhibitor, even when administered at the highest dose, is able to suppress plasma Ang II levels over 24 h at a level similar to that observed a few hours after the initial dose (Gadsbøll et al. 1990; Juillerat et al. 1990; Mooser et al. 1990; Van den Meiracker et al. 1992). This phenomenon results from the conjunction of the progressive drop in the competitive ACE inhibitor plasma level at the end of the dosing interval and the reactive rise in plasma active renin and Ang I secondary to the interruption of the Ang II feedback on renin release (Juillerat et al. 1990; Mooser et al. 1990). It may also be due to the existence of other Ang II-forming enzymatic pathways (Okunishi et al. 1984; Urata et al. 1990). The same reasoning can be applied to AT_1 receptor antagonists, which all have been proven to be in fact competitive antagonists, which also increase renin, Ang I and consequently Ang II concentrations (Doig et al. 1993).

1
First Demonstration of the Concept of Combined Blockade in Mildly Sodium-Depleted Normotensive Subjects

As a prerequisite to clinical studies in patients with hypertension or congestive heart failure, the biological effects of the combined blockade of the RAS were first studied in a human model of mild stimulation of renin release induced by the combination of a 40-mg furosemide-induced sodium depletion, followed by a 36-h low-sodium diet (Ménard et al. 1995). This model induces a two- to threefold increase in plasma active renin, aldosterone and angiotensins, and provides optimal conditions to unmask the renin dependency of blood pressure in normotensive subjects (McGregor et al. 1981). In contrast to the heterogeneity of the RAS of hypertensive patients, sodium-depleted normotensive volunteers have homogeneous levels of plasma active renin, and all have a diuretic-induced BP renin dependency (Doig et al. 1993; Azizi et al. 1995). The choice of the sodium depletion magnitude is crucial. A depletion that is too mild would not make subjects renin-dependent. A depletion that is too large would make them too dependent on the RAS, and therefore more prone to severe orthostatic hypotension and functional renal failure (Doig et al. 1993). The experimental condition of a threefold increase in plasma renin makes it possible to build dose-response curves on two indices of RAS blockade: the decrease in BP and the rise in plasma renin. Both indices change in parallel, although the precision in the assessment of their variation is not similar (Ménard and Guyene 1995). The in vitro renin measurement is much less sensitive to external factors than the assessment of a BP drop. In addition, the range of renin changes (one to ten times) is much larger than the range of BP changes (5–25 mmHg).

The initial work performed in sodium-depleted normotensives showed that the combined administration of single oral doses of 50 mg captopril and 50 mg losartan had a significant synergistic effect on plasma active renin and Ang I levels, an additive effect on BP decrease and no additive effect on plasma aldosterone decrease (Azizi et al. 1995). The synergistic effect of the combination on renin release indicated that a more complete and sustained RAS blockade was achieved during combined RAS blockade. Although the combination induced a higher renin release than each single-drug administration, ACE inhibition favorably influenced the competition between endogenous Ang II levels and the amount of AT_1 receptor antagonist present at the AT_1 receptor level (Azizi et al. 1995). Indeed, the additive BP decrease and the synergistic reactive renin rise with the combination were due (a) to neutralization by losartan of the AT_1 receptor-mediated effects of the persistent Ang II production during ACE inhibition and (b) to the complete neutralization by captopril of the losartan-induced rise in plasma Ang II. Full interpretation of the results of this first study was, however, complicated by the difference in the pharmacokinetics of captopril and losartan. Captopril was rapidly absorbed and was the first to block the RAS by a rapid ACE inhibition, whereas losartan needed to be metabolized before being fully active (Ohtawa et al. 1993). It was demonstrated that both the reduction in BP and the rise in plasma active renin concentrations were simultaneously dependent mainly on captopril for the first 4 h after drug intake, and mainly on losartan for the next several hours (Azizi et al. 1995). However, captopril 50 mg or losartan 50 mg were not the maximum doses necessary to obtain a complete and long-lasting blockade of the RAS, and presumably an increase in dosage of one or the other blocker would finally succeed in inducing similar effects to their combination.

In a second study using the same experimental design, the hemodynamic and hormonal effects of losartan 50 mg combined with enalapril 10 mg, a drug with a longer duration of ACE inhibition than captopril, were of greater magnitude than those obtained by doubling the dose of enalapril (20 mg) alone (Azizi et al. 1997). This second study confirmed the additive effects of ACE inhibition and Ang II antagonism. Both the BP drop and the plasma active renin rise were more pronounced after a single administration of a combination of 50 mg of losartan and 10 mg of enalapril, than after administration of 10 mg or 20 mg of enalapril given singly (Azizi et al. 1997). No additive effect of the combination was observed on the reduction in plasma aldosterone in this study. The combination of enalapril 10 mg and losartan 50 mg neutralized both the decrease in plasma Ang II induced by enalapril 10 mg and the rise in plasma Ang II induced by losartan 50 mg and limited the decrease in plasma Ang II induced by enalapril 10 mg (Azizi et al. 1997).

The single-dose administration of the combination of losartan and captopril (Azizi et al. 1995) and later of enalapril and losartan (Azizi et al. 1997) to a small number of sodium-depleted normal subjects demonstrated the original concept. Finally, even in conditions of high sodium intake and low basal renin status, two conditions known to reduce or neutralize the hypotensive effects of drugs interrupting the RAS at the level of ACE or AT_1 receptors, the dual RAS blockade was shown to be more effective than doubling the usual dose of an AT_1 receptor antagonist through a more efficient blockade of the RAS (Azizi et al. 2003). This may offer an alternative strategy for treating patients with various renin levels and may potentially increase the cardio- and nephroprotective benefits by a more complete RAS blockade.

The most clinically relevant subsequent question was whether these hemodynamic results and potentially beneficial effects would be observed in patients with hypertension, congestive heart failure (CHF), myocardial infarction or proteinuric nephropathy.

2 Effects of Combined RAS Blockade in Experimental Models of Hypertension

Before investigating the effects of combined RAS blockade in patients, we used a factorial design study to compare the hemodynamic, cardiac and hormonal effects of enalapril or losartan (1, 3, 10 and 30 mg/kg) or their combination administered to spontaneously hypertensive rats (SHR) (Ménard et al. 1997). The use of a factorial design in SHR made it possible to test enalapril and losartan over a wider dose range than can be investigated in humans and to analyze the effectiveness and consequences of RAS blockade on target organs at similar levels of inhibition of Ang II effects on its AT_1 receptor. In SHR on a normal sodium diet with intermediate renin concentrations (baseline PRA: 0.4±0.2 ng Ang I/ml/h), the combination of low doses of enalapril and losartan (3 mg/kg each) had a synergistic effect in reducing systolic blood pressure (SBP) and the left ventricular weight/body weight (LVW/BW) ratio, greater than each single site blocker administered at a higher doses (10 mg/kg) after 2–4 weeks (Ménard et al. 1997). The synergistic interaction between the effects of low doses of enalapril and losartan on BP and LVW/BW ratio were due to more effective inhibition of the RAS by their combination than by either agent alone, as attested by the magnitude of the synergistic reactive rise in renal and plasma renin (Ménard et al. 1997).

These results were confirmed by other studies in SHR (Nunez et al. 1997; Webb et al. 1998; Goto et al. 2000; Kim et al. 2000; Morgan et al. 2002) and in a renin-dependent model, the (mRen-2)27 transgenic rat (Richer et al. 1998). All these studies confirmed that the synergistic BP-lowering effect of the combined RAS blockade in SHR allows the individual dosages for each agent to be reduced from three- to tenfold (Nunez et al. 1997; Webb et al. 1998; Goto et al. 2000; Kim et al. 2000; Morgan et al. 2002) and still maintain the same BP-lowering effect as each drug given singly at a higher dose. Combined RAS blockade at low doses or single-site blockade at high doses induced a normalization of SBP (<140 mmHg) in most studies and could even cause profound hypotension with lethargy (Ménard et al. 1997).

The larger reduction in BP observed with the combined RAS blockade is due to a larger reduction in total peripheral resistance associated with a maintained cardiac output (Nunez et al. 1997; Richer et al. 1998). However, the synergistic BP response is finite. When the doses of each single-site blocker are increased toward a maximal inhibition of the RAS, the synergistic effect of combination therapy can no longer be observed (Morgan et al. 2002), but the combination remains more effective than single-site blockade.

3
Clinical Studies in Patients with Hypertension

With all the data that accumulated in both sodium-depleted healthy subjects and in SHR, there was a rationale for combining an ACE inhibitor with an AT_1 receptor antagonist for treating patients with essential hypertension. Indeed, a flat dose-response curve on BP measured at trough has been reported with most ACE inhibitors so far tested in essential hypertensive patients (Lees 1992). In order to increase the duration of action of these drugs up to the end of the dosing interval, increases in the once-daily administered dose or twice-daily administration have been proposed (Meredith et al. 1990).The attenuation of BP response to ACE inhibitors 24 h after drug intake is correlated with a return of plasma Ang II levels toward their initial values (Gadsbøll et al. 1990).

In a multicenter, randomized, double-blind, parallel-group, pilot study in 177 patients with mild-to-moderate hypertension, the BP effects of 6-week administration of combination therapy consisting of losartan 50 mg o.d. and enalapril 10 mg o.d. were compared to both losartan 50 mg o.d. and enalapril 10 mg o.d., each given singly (Azizi et al. 2000). The combination therapy was more effective on clinical diastolic blood pressure (DBP) measured at trough than was losartan, by 3.2 mmHg (95% CI, 0.7–5.7 mmHg, p=0.012), and more effective than enalapril by 4.0 mmHg (95% CI, 1.5–6.4 mmHg, p=0.002). Twenty-four-hour ambulatory mean DBP did not significantly differ between treatment groups, although the combination tended to lower BP more (Azizi et al. 2000). In a subgroup of 28 patients, higher plasma active renin and Ang I levels were measured during blockade by the combination therapy. These findings confirmed the induction of a more intensive inhibition of the RAS by the combination. The three treatment groups were similar with respect to the incidence of clinical adverse experiences during the double-blind treatment period (Azizi et al. 2000).

The results of this trial and other short-term trials (up to 16 weeks) subsequently reported show that combined RAS blockade administered at certain once-daily doses induces a supplementary drop in BP at trough, ranging from 3 to 6 mmHg, compared to single-site blockade in patients with either predominantly low basal renin hypertension (Azizi et al. 2000), type 2 diabetes (Mogensen et al. 2000) or in patients with progressive renal failure (Ruilope et al. 2000). In contrast to animal models, the range of doses of all the RAS blockers that have been studied in patients is limited. Various ACE inhibitors and AT_1 receptor antagonists at various standard daily doses were used and sodium diet was uncontrolled. Despite these limitations, in accordance with animal models (Ménard et al. 1997; Webb et al. 1998) and sodium-depleted normotensive subjects, combined RAS blockade at low doses (benazepril 5 mg o.d. + valsartan 80 mg o.d.) decreased trough BP more in hypertensive patients than doubling the dose of valsartan from 80 mg to 160 mg o.d. (Ruilope et al. 2000). With the exception of African-American hypertensive patients (Weir et al. 2001), combination of an ACE inhibitor with an AT_1 receptor antagonist has the same BP-lowering effect than combination of a single-site RAS blocker with a diuretic (Waeber et al. 2001) or a calcium channel blocker (Sakata et al. 2002), but it add the potential beneficial advantage of a more complete neutralization of Ang II at the tissue levels (Dzau 2001).

4
Hormonal Effects of Combined Blockade of the RAS by an ACE Inhibitor and an AT_1 Receptor Antagonist

When the RAS is blocked by an ACE inhibitor or an AT_1 receptor antagonist, there is competition between the inhibitor or the antagonist, progressively cleared from the body, and the available substrate (Ang I) or agonist (Ang II). Ang I or Ang II production is very dependent on the sensitive feedback between Ang II and renin release, which is permanently operational as suggested by the colocalization of AT_1 receptors and renin in the juxtaglomerular cells (Gasc et al. 1993; Nussberger 2000). When the inhibitor dissociates from the enzyme active sites, ACE starts again to generate Ang II in the presence of a high level of plasma and interstitial Ang I, which is always in competition with the inhibitor. At that time, the concurrent administration of an AT_1 receptor antagonist will protect the AT_1 receptor from newly produced agonist. Reciprocally, when less AT_1 receptor antagonist is present at the AT_1 receptors, an ACE inhibitor will reduce the amount of Ang II available to compete with the antagonist. An alternative to this internal counter-regulation of the RAS would be that an AT_1 receptor added to an ACE inhibitor blocks the effects of Ang II generated by pathways other than renin and ACE (Okunishi et al. 1984; Urata et al. 1990).

In contrast to the plasma active renin, prorenin and Ang I increases, no additive effect has been detected on the plasma aldosterone reduction (Azizi et al. 1997; Ménard et al. 1997; Azizi et al. 2000) in clinical settings other than CHF (Spinale et al. 1997a; McKelvie et al. 1999). In the context of CHF, the additional reduction in plasma aldosterone may provide additional benefit. In the absence of a highly activated RAS such as in CHF, the absence of an additive effect of the combination indicates that plasma aldosterone secretion is regulated through multiple pathways other than the RAS (Quinn and Williams 1988).

Table 1 Biochemical and hormonal consequences of combined RAS blockade

	ACE inhibitors	AT_1 receptor antagonists	ACE inhibitors + AT_1 receptor antagonists
AT_1 receptors	Not stimulated	Blocked	Not stimulated and blocked
AT_2 receptors	Not stimulated	Stimulated	Minor stimulation
AT (n) receptors	Not stimulated	Stimulated	Minor stimulation
Plasma and tissue ACE	Inhibited	Not inhibited	Inhibited
Tissue RAS	Inhibited	Blocked	Additive effect
Renin	Increased	Increased	Additive effect
Prorenin	Increased	Increased	Additive effect
Ang I	Increased	Increased	Additive effect
Ang II	Decreased	Increased	Decreased or normal
Angiotensinogen	Decreased	Decreased	Additive effect
Aldosterone	Decreased	Decreased	No additive effect except in CHF
Non–ACE-dependent Ang II	Present	Blocked	Blocked
Bradykinin and substance P	Increased	No change	Increased
Ang (1–7)	Increased	Increased	Additive effect
AcSDKP	Increased	No change	Increased
ACE induction	Increased	No change	Increased

This combined blockade of the RAS is also different from what has been clinically tested so far. Besides a more intensive blockade of AT_1 receptor functions, it induces accumulation of vasodilatory and natriuretic peptides such as bradykinin (Linz et al. 1995) and Ang (1–7) (Iyer et al. 1998) and the accumulation of the hematological peptide AcSDKP (Azizi et al. 1996), as do ACE inhibitors, and it stimulates potentially functioning AT_2 and other AT(n) receptors, less than the AT_1 receptor antagonists given singly (Table 1). The consequences of this long-term stimulation of AT_2 receptors are unknown and have equal chances to be beneficial, deleterious or neutral (Horiuchi et al. 1999; Opie and Sack 2001).

During long-term administration, ACE induction is another factor that may play a role in limiting the efficacy of ACE inhibitors (Boomsma et al. 1981); the combination of an ACE inhibitor and an AT_1 receptor antagonist, which would less increase plasma or tissue ACE, would be a worthwhile alternative to an increase in the dose of the ACE inhibitor (Costerousse et al. 1994).

5 Effect of the Combined Blockade on End-Organ Protection

5.1 Heart and Vessel Wall

5.1.1 Left Ventricular Hypertrophy Reversion

The synergistic BP-lowering effect of combined RAS blockade is associated with a synergistic reduction in left ventricular (LV) hypertrophy in rodent hypertensive models compared to single-site blockade (Ménard et al. 1997; Nunez et al. 1997; Kim et al. 2000 2001) with a correlation between left ventricular mass changes and BP changes (Nunez et al. 1997). However, beyond BP reduction, the more complete neutralization of Ang II trophic effects by the combined RAS blockade may have a positive impact on LV hypertrophy regression. In Stroke-prone SHR, a combination of low doses of perindopril and candesartan (0.2 mg/kg of each) or a tenfold higher dose of each drug given alone (2 mg/kg) reduced LVW and phenotypic changes to the same extent, and more than a high dose of the calcium channel blocker, amlodipine 3 mg/kg alone, even though the drop in BP achieved with the four different treatments was equivalent (Kim et al. 2000). The beneficial effect of combined RAS blockade on LV mass is associated with a larger improvement in coronary hemodynamics in SHR (Nunez et al. 1997) and in diastolic dysfunction (E/A wave ratio recorded by echocardiography) in the Dahl salt-sensitive rat fed with a high-sodium diet, a model of severe low-renin hypertension and diastolic heart failure (Kim et al. 2001).

5.1.2 Beneficial Effects of the Combined Blockade in Congestive Heart Failure

ACE inhibitors reduce mortality in patients with CHF by preventing new ischemic events and reducing progression of heart failure (Flather et al. 2000). However, despite receiving target doses of ACE inhibitors, a number of patients still have deterioration of LV function and a poor cardiac prognosis associated with a persistence of neurohormonal activation (Swedberg et al. 1990; Pouleur et al. 1993). Indeed, plasma Ang II and aldosterone concentrations are not suppressed in all patients treated with ACE inhibitors (MacFadyen et al. 1999). Even maximally recommended doses of ACE inhibitors (40 mg of long-acting drugs or 150 mg captopril) in patients with stable CHF do not fully prevent pressor response to ascending doses of Ang I and thus Ang II formation (Jorde et al. 1998). Alternative pathways for Ang II production (Urata et al. 1990) and the competitive nature of ACE inhibitors may also explain this phenomenon. The alternative of blocking AT_1 receptors to provide more benefits than ACE inhibitors in patients with CHF by achieving a more complete blockade of the Ang II effects generated both through ACE and non-ACE pathways has not been supported by the results of the ELITE II (Pitt et al. 2000) and OPTIMAAL trials (Dickstein and Kjekshus 2002) at the selected daily dose of losartan and by recent meta-analysis (Jong et al. 2002). There is, therefore, a rationale for combining ACE inhibitors and AT_1 receptor antagonists in patients with CHF.

The first evidence of the potential benefits of this combination came from experimental models of heart failure. In a pig model of pacing-induced heart failure, combining benazepril and valsartan, at doses that gave inhibition of Ang I or Ang II pressor responses at peak similar to each drug given alone, provided additional benefits for LV pump function and prevented neurohormonal activation at rest (Spinale et al. 1997b) and during exercise (Krombach et al. 1998). Acute synergistic hemodynamic improvements (reduction in total peripheral resistance and increase in cardiac output) with the combination of low doses of enalaprilat (1 mg/kg) and losartan (1 mg/kg) in comparison to a fourfold increase in the dose of either enalaprilat or losartan alone were subsequently demonstrated in a similar model (Shen et al. 1998). Over the long-term, combination of quinapril 10 mg o.d. and losartan 50 mg o.d., at doses that produce maximal inhibition of Ang I or Ang II pressor responses at peak similar to each drug given alone, delayed more LV dilatation than each drug given singly at the same dose in a model of rapid ventricular pacing in the New Zealand white rabbit (Kawai et al. 2000). The beneficial effects of combined RAS blockade on cardiac load and function and on neurohormonal activation is associated with an improvement in survival rate (Kim et al. 2001).

After demonstrating that combined blockade of the RAS was beneficial in animal models of CHF, adding an AT_1 receptor antagonist on top of a nonstandardized ACE inhibitor-based treatment in patients with stable CHF (class II–IV) in small-sample studies was shown to further decrease cardiac load and plasma aldosterone levels after 4 weeks, both significantly and safely (Baruch et al. 1999), to further increase cardiac output (Gremmler et al. 2000) and to improve peak exercise capacity and functional CHF class after 6 months in patients with severe CHF, despite treatment with maximally recommended or tolerated doses of ACE inhibitors (Hamroff et al. 1999). Subsequently, in 766 patients with stable CHF (class II–IV), a combination of candesartan 4–8 mg o.d. and enalapril 10 mg b.i.d. for 43 weeks was shown to be more effective in preventing LV remodeling and dilatation, in decreasing BP and in reducing both plasma aldosterone and BNP concentrations than either enalapril 10 mg b.i.d. or candesartan alone up-titrated to 16 mg o.d., with no improvement in mortality (McKelvie et al. 1999).

On the other hand, other small studies reported no additional benefits of adding an AT_1 receptor antagonist on top of a stable nonstandardized maximally tolerated ACE inhibitor treatment in terms of exercise capacity, ejection fraction or neurohormonal activation (Houghton et al. 2000; Murdoch et al. 2001; Ellis et al. 2002).

All these clinical studies have a number of limitations: short treatment periods, small group sizes, nonstandardized ACE inhibitor treatment that was not a clearly predetermined maximum dose, and use of only one dose of an AT_1 receptor antagonist.

The cardioprotective effects of the combined RAS blockade are demonstrated by the results of ValHeFT (Cohn and Tognoni 2001) and even more by CHARM-Added (Candesartan in Heart failure-Assessment of Reduction in Mortality and morbidity) (McMurray et al. 2003) trials, which both showed consistently that adding an AT_1 receptor blocker to an ACE inhibitor-based treatment improves cardiovascular outcome. In the CHARM-Added trial (McMurray et al. 2003), the administration of a high dose of candesartan (mean daily dose, 24 mg) to patients with class II–IV CHF and LV ejection fraction <40% who were being treated with a standard daily dose of an ACE inhibitor for at least 6 months resulted in 15% relative risk reduction [hazard ratio, 0.85, (95% CI, 0.75–0.96), $p<0.011$] in cardiovascular death and hospital admission for CHF, compared to a placebo. In other words, 23 patients have to be treated for 3 years to prevent one first event of cardiovascular death and admission of hospital for CHF. The beneficial effects of candesartan were consistent among all prespecified subgroups, including patients treated with beta-blockers.

5.1.3 Ischemic Heart Disease: Prevention of Left Ventricular Remodeling After Myocardial Infarction

Myocardial remodeling after acute myocardial infarction (MI) is accompanied by the accumulation of collagen in the noninfarcted region of the heart (Volders et al. 1993) and contributes to heart failure development and late mortality (Pfeffer et al. 1995). Both LV remodeling and collagen accumulation after MI are reversed experimentally by either an ACE inhibitor or an AT_1 receptor antagonist (Richer et al. 1992; Schieffer et al. 1994). In addition, ACE inhibitors have been shown to improve morbidity and mortality in patients with LV dysfunction with and without heart failure after MI (ACE Inhibitor Myocardial Infarction Collaborative Group 1998; Flather et al. 2000).

The effects of combined RAS blockade after MI have been studied in animal models in both acute or chronic studies. The acute IV administration of a combination of ramiprilat 50 µg/kg and candesartan 1 mg/kg enhances the reduction of infarct size following a 90-min low-flow ischemic episode and 120 min reperfusion in anesthetized pigs compared to a monotherapy by either drug alone at the same dose when started immediately (Weidenbach et al. 2000). This beneficial effect was prevented by the B2 receptor-antagonist, icatibant, and thus appeared to be mediated by bradykinin (Weidenbach et al. 2000). In contrast, when started later, e.g., 24 h after induction of MI in rats, an ACE inhibitor, an AT_1 receptor antagonist alone nor their combination could reduce the infarct size (Yu et al. 2001; Nakamura et al. 2003).

The studies comparing the preventive effects of ACE inhibitors, AT_1 receptor antagonists and their combination on post-MI LV remodeling and myocardial structural changes, such as collagen accumulation, macrophage and myofibroblast infiltration and myocardial expression of different genes, have yielded conflicting results, but none of them investigated either drug over its full dose range up to its maximum effect. Therefore, depending on the doses used, these studies have shown either the additive (Mankad et al. 2001; Yu et al. 2001; Nakamura et al. 2003) or a nonadditive (Taylor et al. 1998; Richer et al. 2001) effect of the combination of an ACE inhibitor and an AT_1 receptor antagonist. The interpretation of the results favored the bradykinin-mediated effects when the ACE inhibitor alone or in combination was more effective than the AT_1 receptor antagonist alone (Taylor et al. 1998) or the Ang II-mediated effects when the reverse was observed (Yu et al. 2001). However, Nakamura et al. (2003) have shown that combining low doses of the ACE inhibitor temocapril 1.5 mg/kg and the AT_1 receptor antagonist CS-866 0.5 mg/kg improved more LV phenotypic change, collagen accumulation and diastolic dysfunction after MI in rats than maximal effective doses of each RAS blocker (3 and 1 mg/kg, respectively), even though combination and single-site blockade were equipotent in reducing LV end-diastolic pressure and volume (Nakamura et al. 2003).

The critical issue of the most appropriate dose selection is highlighted by the results of the OPTIMAAL trial, where a nonsignificant difference in favor of a high dose of captopril (50 mg t.i.d.) was observed when compared to a rather low dose of losartan (50 mg o.d.) (Dickstein and Kjekshus 2002). The question whether combined RAS blockade will provide further benefits compared to single-site blockade in patients after MI will also be answered by the results of the VALIANT trial (Valsartan in Acute Myocardial Infarction) (Pfeffer et al. 2000), a multicenter, double-blind, randomized, active controlled parallel-group study comparing the efficacy and safety of valsartan titrated up to 160 mg b.i.d., captopril titrated up to 50 mg t.i.d., and their combination titrated up to captopril 50 mg t.i.d. + valsartan 80 mg b.i.d. in 14,500 patients with CHF and/or LV dysfunction after MI.

5.1.4 Vascular Wall

In normotensive rats, a combination of an ACE inhibitor and an AT_1 receptor antagonist at maximally effective doses prevented intimal thickening after balloon injury to a large extent, inhibited more potently PDGF-beta receptor tyrosyl phosphorylation and prevented intimal vascular smooth muscle cell proliferation more potently than single-site blockade at the same doses (Kim et al. 2002). Furthermore, a nonpeptide bradykinin B2 receptor antagonist or an NO synthase inhibitor reduced the prevention of intimal thickening by combined RAS blockade, suggesting that either bradykinin or NO partly contributes to its beneficial vascular effects. Bradykinin has also been reported to be involved in the reduction of infarct size after myocardial ischemia in pigs by combined RAS blockade (Weidenbach et al. 2000). Combined blockade of the RAS has also been shown to improve endothelial dysfunction (Goto et al. 2000). In L-NAME SHR, a model of severe hypertension and severe endothelial dysfunction and glomerulosclerosis, combined blockade of the RAS with a low dose of valsartan (5 mg/kg) and benazepril (5 mg/kg) improved survival, despite persistence of renal dysfunction and high BP, suggesting that survival improvement may be related to a direct effect of the combination on the endothelium (de Gasparo et al. 2000).

These results, combined with the results of the SECURE study, in which an effective dose of the ACE inhibitor, ramipril 2.5 mg, did not reduce intima-media thickness, is in favor of using either higher doses or a combined RAS blockade (Lonn et al. 2001).

5.2 Beneficial Effects of Combined RAS Blockade for Nephroprotection

The two main strategies for protection against loss of renal function in patients with chronic nephropathies have been to lower BP and to restrict dietary protein (Ruggenenti et al. 2001). Although lowering of BP results in lowering of urine protein excretion, a specific blockade of the RAS has additional effects and retards more the progressive course of chronic renal disease than other nonspecific antihypertensive treatments. Beyond their BP-lowering effects, ACE inhibitors are nephroprotective both in diabetic and nondiabetic renal disease (Lewis et al. 1993; Maschio et al. 1996; The GISEN Group 1997; Jafar et al. 2001) and AT_1 receptor antagonists are also nephroprotective, at least in type 2 diabetic nephropathy (Brenner et al. 2001; Lewis et al. 2001; Parving et al. 2001).

The nephroprotection afforded by RAS blockers is related to their antiproteinuric properties. Since the antiproteinuric response to treatment is the strongest predictor of long-term nephroprotective efficacy and the presence of a residual proteinuria during treatment is an independent risk factor for progression of renal disease (The GISEN Group 1997; Ruggenenti et al. 1998), the focus of nephroprotection has shifted toward maximal reduction of proteinuria, in addition to optimal control of BP (De Jong et al. 1999). However, many patients with chronic diabetic or nondiabetic nephropathy do not reach BP targets and have persistently elevated proteinuria, despite treatment with several antihypertensive agents, including recommended doses of ACE inhibitors. Since this insufficient response might be explained by incomplete blockade of the RAS (Vos et al. 1995), as in other clinical settings, there is also a rationale for either increasing the dosage of the RAS blockers more or combining an ACE inhibitor with an AT_1 receptor antagonist.

5.2.1 Experimental Models of Nephropathy

In different experimental models of renal failure associated with proteinuria and histological damage, combined RAS blockade has been shown to be either more effective than (Cao et al. 2000, 2001; Wilkinson-Berka et al. 2001), equivalent to (Burdmann et al. 1995; Kohzuki et al. 1995; Ots et al. 1998; Bos et al.

2002) or less effective (de Gasparo et al. 2000) than single-site blockade in terms of BP and proteinuria reduction, depending on the experimental model, the rat strain, the severity of the pretreatment histological damage, the doses of the drugs used alone or in combination and the duration of treatment. In the positive studies, the combination of low doses of an ACE inhibitor and an AT_1 receptor antagonist has been shown to induce similar reductions in BP and proteinuria and afford nephroprotection similar to higher doses of each RAS blocker alone (Wilkinson-Berka et al. 2001). In addition, a close correlation between proteinuria reduction and the drop in BP was observed, and the more the BP level was decreased the more proteinuria was reduced. In contrast, in Wistar rats with nonhypertensive adriamycin-induced proteinuric nephrosis that were resistant to a maximally effective dose of lisinopril (75 mg/l), Bos et al. (2002) were unable to show any additional hypotensive, antiproteinuric and nephroprotective effect of either doubling the lisinopril dose or adding an AT_1 receptor antagonist. The same observation was made in L-NAME SHR (de Gasparo et al. 2000).

Finally, all studies failed to demonstrate any short-term additional benefit of combined over single-site RAS blockade in terms of preventing glomerulosclerosis and tubulointerstitial injury and, consequently, in terms of preventing GFR decline. The mechanisms responsible for the dissociation between a reduction in BP and proteinuria and preservation of renal tissue are not fully understood and are probably multifactorial. Firstly, in contrast with humans with several operational non-ACE pathways, especially the chymase pathway, the contribution of such pathways in rodents seems limited (Hollenberg 2000), thus reducing the probability of detecting an additional effect of the combination of an AT_1 receptor antagonist and an ACE inhibitor. Secondly, improvement of renal structural changes requires not only a further reduction in BP and in glomerular perfusion pressure by RAS blockade, but also an intervention at the earliest stage of glomerular and tubular lesions (Ikoma et al. 1991; Burdmann et al. 1995; Remuzzi et al. 2002a), since once interstitial fibrosis is established, it is probably irreversible (Burdmann et al. 1995).

5.2.2
Patients with Diabetes Nephropathy

At the stage of incipient nephropathy, combining usual daily doses of lisinopril 20 mg and candesartan 16 mg in type 2 diabetic patients whose hypertension remained resistant to a 12-week treatment with either monotherapy, further reduced self-measured SBP/DBP by 8.6/5.6 mmHg and by 11.2/5.9 mmHg and microalbuminuria by 18% (NS) and 34% ($p<0.04$) compared to the continuation of either lisinopril 20 mg or candesartan 16 mg o.d. (Mogensen et al. 2000). At a later stage of disease progression, adding a high or standard usual dose of an AT_1 receptor antagonist (irbesartan 300 mg o.d. or candesartan 8 mg o.d., respectively) to a background of nonstandardized ACE inhibitor treatment in type 1 and 2 diabetic hypertensive patients, further safely reduced both 24-h ambulatory SBP/DBP (by 8/5 mmHg and 10/3 mmHg, respectively) and albuminuria (by 37% and 24%, respectively) compared to a placebo (Jacobsen et al. 2002; Rossing et al. 2002). The greatest decrease in BP was observed in patients with the highest residual plasma Ang II concentrations with ACE inhibitor treatment. Finally, the presence of a positive correlation between BP and albuminuria reduction in patients with diabetic nephropathy suggests that the nephroprotective effect of the combined RAS blockade is in part due to changes in systemic, local and glomerular capillary pressures, as observed in experimental models of diabetic nephropathy.

5.2.3
Patients with Chronic Nephropathies of Various Origins

The majority of the small sample size studies comparing the short-term antihypertensive and antiproteinuric effects (4–8 weeks) of combined RAS blockade with single-site blockade in hypertensive or normotensive patients with proteinuric nondiabetic nephropathies have shown a greater benefit of the combination over either ACE inhibition or AT_1 receptor blockade alone or an additional hypotensive and antiproteinuric effect of adding an AT_1 receptor antagonist on a background of ACE inhibition (Ruilope et al. 2000; Laverman et al. 2001; Perticucci et al. 2001; Russo et al. 2001; Berger et al. 2002; Ferrari et al. 2002; Nakao et al. 2003). In an elegant three-period crossover, prospective, randomized, open-blinded end-point study (PROBE), Ferrari et al. (2002) showed that a 6-week combination of fixed standard daily doses of irbesartan 150 mg and fosinopril 20 mg in ten Caucasian hypertensive patients with chronic proteinuric glomerulonephritis (mean basal proteinuria, 7.9 g/24 h) reduced more proteinuria (–58%) than either drug given alone (–37% and –32%, respectively) after controlling for confounding factors such as dietary sodium, protein intake and renal function. Fosinopril 20 mg and irbesartan 150 mg alone were equivalent in terms of BP reduction at trough (final SBP/DBP, 135/84 mmHg and 134/84 mmHg, respectively), and their combination was nonsignificantly additive (final SBP/DBP, 129/83 mmHg) (Ferrari et al. 2002). Using a double-blind placebo-controlled randomized two-period crossover design in a similar clinical situation, Berger et al. (2002) showed that adding a once-daily fixed low dose of 8 mg candesartan on a background nonstandardized ACE inhibitor treatment that normalized BP (<130/80 mmHg) for at least 3 months, further reduced significantly both ambulatory BP from 127/79 mmHg to 123/76 mmHg and proteinuria from 1.9 g/24 h to 1.3 g/24 h with no significant changes in GFR and renal plasma flow. Finally, combining low doses of valsartan 80 mg o.d. and benazepril 5–10 mg o.d. (Ruilope et al. 2000; Perticucci et al. 2001) or enalapril 10 mg o.d. and losartan 50 mg o.d. (Russo et al. 2001) has been shown to have either a greater or equivalent BP-lowering and antiproteinuric effect than doubling the dose of each single-site blocker.

Whether similar beneficial BP and renal effects would be achieved by increasing even more the dosage of each single-site blocker toward the top of its dose-response curve as well as twice-daily administrations remains controversial. After having determined the individual optimal antiproteinuric dose of losartan (50, 100 and 150 mg o.d.) and lisinopril (10, 20 and 40 mg o.d.) in an open label titration crossover protocol, Lavermann et al. (2001) have shown that the combination of maximal effective doses of losartan (100 mg o.d.) and lisinopril (40 mg o.d.) reduced proteinuria more than the optimal doses of losartan and lisinopril alone. BP was reduced more by the combination than by the optimal dose of losartan and not of lisinopril. On the other hand, in patients with hypertension and proteinuria resistant to a triple antihypertensive therapy including 3 months of lisinopril 40 mg o.d., the addition of losartan 50 mg o.d. failed to reduce BP and proteinuria compared to placebo (Agarwal 2001). Although losartan had no effect on BP and proteinuria, it significantly lowered urine TGF beta1 concentration, a sensitive marker of kidney damage, which was persistently increased despite maximal ACE inhibition with lisinopril 40 mg o.d. (Agarwal et al. 2002).

The short-term reduction in urine protein excretion demonstrated during combined RAS blockade is only a surrogate end-point. The next question is whether combination of ACE inhibitor and AT_1 receptor antagonist will slow the decline in renal function and postpone end-stage renal failure in these patients. In accordance with experimental models of nephrosis or renal failure, none of the short-term positive studies demonstrated an additional improvement in GFR with combined RAS blockade. However, the difference in GFR between groups over a short period may not be a sensitive marker of response to treatment. Indeed, the COOPERATE trial shows greater nephroprotection from the combined RAS blockade than from single-site RAS blockade after 3 years of treatment (Nakao et al. 2003). After screening and an 18-week run-in period to demonstrate that 3 mg o.d. of trandolapril was a maximally antiproteinuric dose, 263 of 336 patients with nondiabetic renal disease of different etiologies (mainly glomerulopathy, 65%) and persistent proteinuria greater than 0.3 g/24 h were randomly assigned to losartan 100 mg daily, trandolapril, 3 mg daily, or a combination of both drugs at equivalent doses (Nakao et al. 2003). The maximum fixed doses were achieved after a period of 9–12 weeks and all patients were requested to restrict sodium and protein intake. Patients were stratified by baseline GFR (< or >45 ml/min per 1.73 m^2), baseline proteinuria (<1 g/day, 1 to <3 g/day, <3 g/day) and reduction in proteinuria with trandolapril during the run-in period (< or $\geq$7% per day). Mean age of the patients was 44.8–45.9 years, mean creatinine clearance was 37.5–38.4 ml/min per 1.73 m^2, baseline proteinuria was 2.4–2.5 g/day. More than 90% of the patients were hypertensive at baseline and treated with a median of three (range, 1–3) antihypertensive drugs, excluding RAS blockers, and baseline BP was 130±10.5/76±5.6 mmHg. After a first interim analysis that showed a difference in end-point incidence in favor of the combination, the trial was stopped at year 3.

By year 3, the combination of trandolapril 3 mg and losartan 100 mg daily had a much greater nephroprotective effect than either losartan 100 mg or trandolapril 3 mg given alone. Ten (11%) patients on combination treatment reached the combined primary endpoint, i.e., time to doubling of serum creatinine concentration or end-stage renal disease, compared with 20 (23%) on trandolapril alone (hazard ratio, 0.38; 95% CI, 0.18–0.63, p=0·018) and 20 (23%) on losartan alone (hazard ratio, 0.40; 95%CI, 0.17–0.69, p=0.016). In other words, in comparison with single-site RAS blockade, 8–9 patients would have to be treated with the combination of trandolapril and losartan for 3 years to avoid the occurrence of one combined primary end-point (Nakao et al. 2003).

The benefits of combination treatment were shown independently of baseline urinary protein excretion and was mainly observed in patients with glomerulonephritis rather than with nephrosclerosis. The three treatment groups showed the same reductions in BP but the combination reduced more proteinuria during the whole trial duration (maximum median change, –75.6%) than either losartan (–42%) or trandolapril (–44.3%). The significant antiproteinuric effect of combination treatment was seen irrespective of both baseline proteinuria and GFR, and even patients classified as low-responders showed a greater response to combined RAS blockade than to single-site blockade.

These findings are in accordance with the results of the REIN follow-up study in patients with nondiabetic nephropathy (The GISEN Group 1997; Ruggenenti et al. 1998) and earlier studies in diabetic patients (Parving et al. 1987), where a progressive reduction in GFR decline was observed if the antihypertensive treatment was continued for a long enough period.

The renal effects of the combined blockade of the RAS require special attention. The intrarenal RAS is not regulated in the same way as the circulating RAS (Nussberger 2000). The huge amount of renin locally synthesized and released, the limited amount of endothelial ACE, the intrarenal consumption of angiotensinogen (Campbell 1997; Navar et al. 2001), the uptake of Ang II by the renal AT_1 receptors (Van Kats et al. 2001) influences intrarenal levels of Ang I and Ang II, which differ from their plasma levels. These observations explain why specific renal benefits are expected from the combined blockade at the level of the renal tissue under pathological conditions, and specific renal adverse effects can be predicted from renal hemodynamics.

The mechanisms by which combination of the ACE inhibitor and AT_1 receptor antagonist provide an additional antiproteinuric response are probably multifactorial, either due to systemic and glomerular capillary hemodynamic changes or to a more complete neutralization of Ang II effects, or both. An additional decrease in BP with combined RAS blockade correlated with an additional reduction in proteinuria was reported in several studies. However, the drop in BP was small, suggesting that the greater antiproteinuric effect might be attributed to more complete RAS blockade. For example, blocking the stimulating effects of Ang II on TGF beta1 production by both proximal tubular and glomerular mesangial cells, which may locally promote renal fibrosis by stimulating fibroblast proliferation and matrix accumulation (Kagami et al. 1994), could have beneficial nephroprotective effects (Sharma et al. 1999; Agarwal et al. 2002). An alteration in glomerular basement membrane pore selectivity (Morelli et al. 1990; Perna and Remuzzi 1996) or decreased mesangial matrix production and mesangial cell proliferation (Nakamura et al. 1999) could also mediate the reduction in proteinuria. Finally, the beneficial effect of combined RAS blockade could be due to the reinforcement of the blockade of the autocrine effects of Ang II on glomerular nephrin expression, since the down-regulation in both gene and protein expression of the transmembrane protein nephrin, which accompanies the development of albuminuria in experimental models of hypertension and diabetes, has been shown to be reversed by an AT_1 receptor antagonist or an ACE inhibitor (Hulthen 2003).

The recommended doses of ACE inhibitors and AT_1 receptor antagonists are derived form hypertension trials, and the dose-response for lowering proteinuria and retarding progression of renal failure may not be similar to the dose-response for BP reduction. There is both experimental evidence (Ikoma et al. 1991) and clinical evidence that RAS blockers should be up-titrated toward the maximal effective antiproteinuric or tolerated dose to afford the greatest nephroprotection, especially in patients with the highest baseline proteinuria (Jafar et al. 2001; Nakao et al. 2003). In the REIN study, the ramipril-induced reduction in proteinuria predicted the slower decline in GFR and lower incidence of end-stage renal failure (The GISEN Group 1997; De Jong et al. 1999). Forced titration or up-titration to maximally tolerated doses of either ACE inhibitors (Laverman et al. 2002; Pisoni et al. 2002) or AT_1 receptor antagonists (Laverman et al. 2001, 2002) increases their antiproteinuric effect in patients with chronic nondiabetic (Laverman et al. 2001) and diabetic (Andersen et al. 2002; Jacobsen et al. 2002; Rossing et al. 2002) nephropathies and persistent proteinuria. Moreover, the combination of maximal effective doses of losartan (100 mg o.d.) and lisinopril (40 mg o.d.) reduced proteinuria more than both optimal doses of losartan and lisinopril alone (Laverman et al. 2002). The IRMA II trial also convincingly showed the dose-dependent nephroprotective effects of 300 mg of irbesartan in reducing the incidence of overt nephropathy in patients with type II diabetes and microalbuminuria, whereas the dose of 150 mg could not be differentiated from placebo (Parving et al. 2001). At the stage of overt diabetic nephropathy, the same high dose of irbesartan 300 mg was also shown to be highly nephroprotective, whereas a high dose of amlodipine (10 mg) could not be differentiated from placebo (Lewis et al. 2001). In the RENAAL trial, the dose of losartan was up-titrated to 100 mg for achieving a target BP in 71% of the patients (Brenner et al. 2001). Finally, in the COOPERATE trial a greater nephroprotection was afforded by combination of trandolapril and losartan (Nakao et al. 2003).

In addition to the general advantages of the combined RAS blockade over single-site RAS blockade, what is its added value in patients with nephropathy of diabetic or nondiabetic origin?

1. Amplify the known beneficial effects of single-site RAS blockade.
2. Lower further BP and therefore reach the WHO recommended optimal BP goal more easily. This will probably be associated with greater prevention

in micro- and macrovascular complications of diabetes (Hansson et al. 1998; UK Prospective Diabetes Study Group 1998b).

3. Better prevent the progression to overt proteinuria and nephropathy of type 2 diabetic patients with microalbuminuria and even restore normoalbuminuria (Parving et al. 2001).
4. Reduce proteinuria further in patients with overt proteinuria and therefore provide more nephroprotection (Nakao et al. 2003) if the renal damage is not irreversible.
5. Achieve a more complete and permanent RAS blockade and overcome the counter-regulatory effects of renin release, which maintains active Ang II concentrations in the vicinity of AT_1 receptors. Beyond the BP-lowering effect, a more complete neutralization of the effects of Ang II on intrarenal hemodynamics (Anderson et al. 1985), glomerular size permeability (Remuzzi et al. 1999) or structure (Kagami et al. 1994) may also contribute to a greater reduction in proteinuria and finally to halting GFR decline (Nakao et al. 2003).

However, besides a more complete blockade of the RAS and the best BP control achievable, which will require combination with other antihypertensive drugs, a multifactorial intervention targeting lipid-lowering drugs, smoking cessation and, in addition, tight blood glucose control for diabetic patients is deemed necessary to optimize nephroprotection (Ruggenenti et al. 2001; Remuzzi et al. 2002b).

6
Safety of the Combined Blockade of the RAS

Theoretically, the price to be paid by combining an ACE inhibitor with an AT_1 receptor antagonist will be more side effects, especially those dependent on ACE inhibition such as cough (Mogensen et al. 2000; Waeber et al. 2001; Nakao et al. 2003), and the potential hazards of a complete RAS blockade, especially in situations where BP and renal functions are extremely renin-dependent. Indeed, hypotension, renal insufficiency, weight loss, lethargy and mortality were observed in SHR within 12 days following the onset of a combined enalapril 10 mg/kg and losartan, 10 mg/kg treatment (Ménard et al. 1997). Renal insufficiency, also described with the perindopril-losartan combination (Griffiths et al. 2001), is of functional origin, as no histological lesions can be detected (Ménard et al. 1997) and it can be prevented by a high-salt diet (Griffiths et al. 2001). In humans, the range of doses that can be used are much lower than in animals. In patients with baseline renal failure, no episodes of acute renal failure were reported at the dose of RAS blockers used (Ruilope et al. 2000; Nakao et al. 2003); on the contrary, a reduction in the incidence of the time to doubling serum creatinine and end-stage renal failure was even observed (Nakao et al. 2003). However, if treatments combining an ACE inhibitor and an AT_1 antagonist become more commonly used in patients, and especially in elderly patients and in patients with proteinuria, renal artery stenosis, renal failure or CHF, the experimental data predict a potential risk of functional renal failure, especially in patients who are sodium-restricted or are exposed to associated treatments such as diuretics or nonsteroidal anti-inflammatory drugs (Kamper et al. 2002) and during anesthetic induction (Brabant 1999).

To date, all clinical trials have reported a good tolerance profile of the combination of an ACE inhibitor and AT_1 receptor antagonist. There were a few reports of transient dizziness when the treatment was initiated (Russo et al. 2001; Ferrari et al. 2002; Laverman et al. 2002; Nakao et al. 2003) and an increase in serum potassium levels, especially when the baseline GFR was less than 35 ml/min (Ruilope et al. 2000; Jacobsen et al. 2002) or when spironolactone was part of the combination therapy in patients with CHF (McMurray et al. 2003). In the COOPERATE trial, the incidence of hyperkaliemia and dry cough was similar in the trandolapril (8/86 and 5/86, respectively) and the combination group (7/88 and 5/88, respectively) and both were higher than in the losartan group (4/89 and 1/89) (Nakao et al. 2003). However, during the run-in period under trandolapril alone, 24 patients were excluded for dry cough and one for angioedema (Nakao et al. 2003). In the CHARM-Added trial, 4.5% of the candesartan-treated patients with CHF compared to 3.1% of the placebo-treated patients (p=0.079) experienced a significant hypotension leading to treatment interruption (McMurray 2003). In addition, the incidence of significant creatinine increase was almost doubled in the candesartan group (7.8% vs 4.1% in the placebo group), as was the incidence of hyperkalemia.

In addition to monitoring serum potassium levels in patients with renal failure, hematocrit should be also monitored. Both ACE inhibitors (Hirakata et al. 1984) and AT_1 receptor antagonists (Schwarzbeck et al. 1998) have been reported to lower hematocrit, especially in patients with renal failure, and their combination has been reported to significantly lower hematocrit and hemoglobin levels in patients with either IgA nephropathy (Russo et al. 2001) or diabetic nephropathy (Jacobsen et al. 2002), as would be expected (Cole et al. 2000).

7 Discussion

The results from experimental models (Ikoma et al. 1991; Chobanian et al. 1992; Rakugi et al. 1994; Wollert et al. 1994; Chobanian et al. 1995) and randomized controlled clinical trials have shown that the more intense the blockade of the RAS is during ACE inhibition (Packer et al. 1999; Yusuf et al. 2000; Lonn et al. 2001), AT_1 receptor blockade (Parving et al. 2001) or combined RAS blockade (Nakao et al. 2003), the more effective the prevention of target organ damage is. Compared with higher doses of single-site blockers, a combination of an ACE inhibitor and an AT_1 receptor antagonist blocks the RAS more effectively. After the demonstration of the synergistic or additive blood pressure effects of the combined RAS blockade in sodium-depleted normotensive subjects (Azizi et al. 1995, 1997) and animal models (Ménard et al. 1997; Webb et al. 1998), the combined blockade of the RAS compared to single-site blockade was shown to be more efficient (a) in lowering BP in Caucasian hypertensive patients with predominantly low renin concentrations (Azizi et al. 2000), type II diabetes (Mogensen et al. 2000), or renal failure (Ruilope et al. 2000), (b) in lowering proteinuria and retarding progression of renal failure in patients with diabetic and overall nondiabetic nephropathy (Nakao et al. 2003), and (c) in finally improving LV remodeling and cardiac function status and cardiovascular morbidity and mortality in patients with CHF (McKelvie et al. 1999; Cohn and Tognoni 2001; McMurray et al. 2003).

However, the question of how dose selection of each RAS blocker influences the results of all studies reported to date remains a crucial one (Maillard et al. 2002). One advantage of the synergy or additivity of two antihypertensive drugs is ultimately reducing dosages of individual drugs to achieve the same therapeutic response. This is important when the side effects associated with a single-drug therapy are dose-dependent. The demonstration of additivity or synergism depends on the selected dosages of drugs. In animal models, the synergistic activity of the combined RAS blockade allows the individual dosages for each agent to be reduced from three- to tenfold (Ménard et al. 1997; Webb et al. 1998) and still maintain the same BP-lowering effect as each drug given singly at a higher dose, because it can overcome the counter-regulatory effects of renin release due to RAS inhibition. In sodium-depleted normotensive subjects (Azizi et al. 1995 1997) and patients with hypertension (Ruilope et al. 2000), combined RAS blockade was also shown to be more effective than doubling the dose of a single-site RAS blocker. Accordingly, Forclaz et al. (2003) have recently shown that a twice-daily dose of 100 mg losartan was required to achieve the same BP effect and the same RAS blockade over 24 h as a once-daily combination of losartan 100 mg and lisinopril 20 mg.

Thus, a combined blockade of RAS allows the use of lower doses of each of its components to achieve a more effective and more prolonged RAS blockade. This may reduce the dose-dependent adverse effects of one or both drugs, especially the ACE inhibitor, and makes possible a once-daily administration to achieve a permanent blockade of the RAS over 24 h, which may improve treatment compliance.

To summarize, the use of a combination of an ACE inhibitor and an AT_1 receptor antagonist at low doses:

- Improves 24-h blockade of the RAS especially at trough
- Improves compliance by maintaining a once-daily regimen
- Reduces the number of pills taken by the patient
- May reduce treatment costs
- Reduces the incidence of dose-dependent side effects
- Adds the specific effects of ACE inhibition to those of a more complete RAS tissue blockade (bradykinin, Ang 1–7, NO and AcSDKP accumulation)

- Adds the possibility of monitoring compliance to AT_1 receptor antagonists by measuring ACE inhibition-induced high concentrations of AcSDKP in urine (Azizi et al. 1999)

However, as shown by the results of the COOPERATE and the CHARM-Added trials, beyond BP lowering, there is clear evidence that blocking the RAS is more efficient using a combination of an ACE inhibitor and an AT_1 receptor antagonist at high doses to achieve the best nephroprotection in patients with chronic nephropathy and the best cardioprotection in patients with CHF, respectively.

8 Perspectives

The contrast between the positive results of the HOPE study (Yusuf et al. 2000), which used a high dose of 10 mg of the ACE inhibitor ramipril to prevent cardiovascular mortality and morbidity in patients with a high cardiovascular risk, and those of the SECURE study, in which an effective dose of the ACE inhibitor, ramipril 2.5 mg, did not reduce intima–media thickness (Lonn et al. 2001), and the positive results of the IRMA II study (Parving et al. 2001), the COOPERATE study (Nakao et al. 2003) and the CHARM-Added trial (McMurray 2003) highlight how important it is to block the RAS more strongly than that necessary to reduce only BP. The advantage offered by combining two RAS blockers is to increase the beneficial effect of cardioprotection (McMurray et al. 2003) and nephroprotection (Nakao et al. 2003), which are currently demonstrated with the highest doses of an ACE inhibitor (Yusuf et al. 2000) or an Ang II antagonist (Parving et al. 2001). Two other large on-going clinical trials, ON-TARGET (Yusuf 2002) and VALIANT (Pfeffer et al. 2000) will give additional answers on the cardioprotective effects of the combined blockade of the RAS in patients with a high-risk cardiovascular profile and those having had myocardial infarction.

References

ACE Inhibitor Myocardial Infarction Collaborative Group (1998) Indications for ACE inhibitors in the early treatment of acute myocardial infarction: systematic overview of individual data from 100,000 patients in randomized trials. Circulation 97:2202–2212

Agarwal R (2001) Add-on angiotensin receptor blockade with maximized ACE inhibition. Kidney Int 59:2282–2289

Agarwal R, Siva S, Dunn SR, Sharma K (2002) Add-on angiotensin II receptor blockade lowers urinary transforming growth factor-beta levels. Am J Kidney Dis 39:486–492

Andersen S, Rossing P, Juhl TR, Deinum J, Parving HH (2002) Optimal dose of losartan for renoprotection in diabetic nephropathy. Nephrol Dial Transplant 17:1413–1418

Anderson S, Meyer TW, Rennke HG, Brenner BM (1985) Control of glomerular hypertension limits glomerular injury in rats with reduced renal mass. J Clin Invest 76:612–619

Azizi M, Chatellier G, Guyene TT, Murieta-Geoffroy D, Ménard J (1995) Additive effects of combined angiotensin-converting enzyme inhibition and angiotensin II antagonism on blood pressure and renin release in sodium-depleted normotensives. Circulation 92:825–834

Azizi M, Rousseau A, Ezan E, Guyene TT, Michelet S, Grognet J-M, Lenfant M, Corvol P, Ménard J (1996) Acute angiotensin-converting enzyme inhibition increases the plasma level of the natural stem cell regulator N-acetyl-seryl-aspartyl-lysyl-proline. J Clin Invest 97:839–844

Azizi M, Guyene TT, Chatellier G, Wargon M, Ménard J (1997) Additive effects of losartan and enalapril on blood pressure and plasma active renin. Hypertension 29:634–640

Azizi M, Ezan E, Reny J-L, Junot C, Wdziecazk-Bakala J, Gerineau V, Ménard J (1999) Renal and metabolic clearance of AcSDKP during angiotensin converting enzyme inhibition in humans. Hypertension 33:879–886

Azizi M, Linhart A, Alexander J, Goldberg A, Menten J, Sweet C, Ménard J (2000) Pilot study of combined blockade of the renin-angiotensin system in essential hypertensive patients. J Hypertens 18:1139–1147

Azizi M, Lamarre-Cliche M, Labatide-Alanore A, Bissery A, Guyene TT, Ménard J (2002) Physiologic consequences of vasopeptidase inhibition in humans: effect of sodium intake. J Am Soc Nephrol 13:2454–2463

Azizi M, Bissery A, Bura-Rivi A, Ménard J (2003) Dual renin angiotensin system blockade restores blood pressure-renin dependency in low renin subjects. J Hypertens 21:1889–1897

Baruch L, Anand I, Cohen IS, Ziesche S, Judd D, Cohn JN (1999) Augmented short- and long-term hemodynamic and hormonal effects of an angiotensin receptor blocker added to angiotensin converting enzyme inhibitor therapy in patients with heart failure. Circulation 99:2658–2664

Berger ED, Bader BD, Ebert C, Risler T, Erley CM (2002) Reduction of proteinuria; combined effects of receptor blockade and low dose angiotensin-converting enzyme inhibition. J Hypertens 20:739–743

Biollaz J, Burnier M, Turini GA, Brunner DB, Porchet M, Gomez HJ, Jones KH, Ferber F, Abrams WB, Gavras H, Brunner HR (1981) Three long-acting converting enzyme inhibitors: relationship between plasma converting enzyme activity and response to angiotensin I. Clin Pharmacol Ther 29: 665–670

Boomsma F, De Bruyn JHB, Derkx FHM, Shalekamp MADH (1981) Opposite effects of captopril on angiotensin I converting enzyme "activity" and "concentration"; relation between enzyme inhibition and long -term blood pressure response. Clin Sci 60:4911–4918

Bos H, Henning RH, De Boer E, Tiebosch AT, De Jong PE, De Zeeuw D, Navis G (2002) Addition of AT_1 blocker fails to overcome resistance to ACE inhibition in adriamycin nephrosis. Kidney Int 61:473–480

Brabant SM, Bertrand M, Eyraud D, Darmon PL, Coriat P (1999) The hemodynamic effects of anesthetic induction in vascular surgical patients chronically treated with angiotensin II receptor antagonists. Anesth Analg 89:1388–1392

Brenner BM, Cooper ME, de Zeeuw D, Keane WF, Mitch WE, Parving HH, Remuzzi G, Snapinn SM, Zhang Z, Shahinfar S (2001) Effects of losartan on renal and cardiovascular outcomes in patients with type 2 diabetes and nephropathy. N Engl J Med 345:861–869

Burdmann EA, Andoh TF, Nast CC, Evan A, Connors BA, Coffman TM, Lindsley J, Bennett WM (1995) Prevention of experimental cyclosporin-induced interstitial fibrosis by losartan and enalapril. Am J Physiol 269: F491–F499

Campbell DJ (1997) Differential regulation of angiotensin peptides in plasma and kidney: effects of adrenalectomy and estrogen treatment. Clin Exp Hypertens 19:687–698

Cao Z, Cooper ME, Wu LL, Cox AJ, Jandeleit-Dahm K, Kelly DJ, Gilbert RE (2000) Blockade of the renin-angiotensin and endothelin systems on progressive renal injury. Hypertension 36:561–568

Cao Z, Bonnet F, Davis B, Allen TJ, Cooper ME (2001) Additive hypotensive and anti-albuminuric effects of angiotensin-converting enzyme inhibition and angiotensin receptor antagonism in diabetic spontaneously hypertensive rats. Clin Sci (Lond) 100:591–599

Chobanian AV, Haudenschild CC, Nickerson C, Hope S (1992) Trandolapril inhibits atherosclerosis in the Watanabe heritable hyperlipidemic rabbit. Hypertension 20:473–477

Chobanian AV, Hope S, Brecher P (1995) Dissociation between the antiatherosclerotic effect of trandolapril and suppression of serum and aortic angiotensin-converting enzyme activity in the Watanabe heritable hyperlipidemic rabbit. Hypertension 25:1306–1310

Cohn JN, Tognoni G (2001) A randomized trial of the angiotensin-receptor blocker valsartan in chronic heart failure. N Engl J Med 345:1667–1675

Cole J, Ertoy D, Lin H, Sutliff RL, Ezan E, Guyene TT, Capecchi M, Corvol P, Bernstein KE (2000) Lack of angiotensin II-facilitated erythropoiesis causes anemia in angiotensin-converting enzyme-deficient mice. J Clin Invest 106:1391–1398

Costerousse O, Allegrini J, Clozel JP, Ménard J, Alhenc-Gelas F (1994) Increased angiotensin-converting enzyme (ACE) gene expression during inhibitor treatment. Comparison with other blockers of the renin-angiotensin system. J Hypertens 12:S126

Dahlof B, Pennert K, Hansson L (1992) Reversal of left ventricular hypertrophy in hypertensive patients. A metaanalysis of 109 treatment studies. Am J Hypertens 5:95–110

De Gasparo M, Hess P, Nuesslein-Hildesheim B, Bruneval P, Clozel JP (2000) Combination of non-hypotensive doses of valsartan and enalapril improves survival of spontaneously hypertensive rats with endothelial dysfunction. J Renin Angiotensin Aldosterone Syst 1:151–158

De Jong PE, Navis G, De Zeeuw D (1999) Renoprotective therapy: titration against urine protein excretion. Lancet 354:352–353

Dickstein K, Kjekshus J (2002) Effects of losartan and captopril on mortality and morbidity in high-risk patients after acute myocardial infarction: the OPTIMAAL randomised trial. Optimal Trial in Myocardial Infarction with Angiotensin II Antagonist Losartan. Lancet 360:752–760

Doig JK, MacFadyen RJ, Sweet CS, Lees KR, Reid JL (1993) Dose-ranging study of the angiotensin type I receptor antagonist Losartan (DuP753/MK954), in salt-deplete normal man. J Cardiovasc Pharmacol 21:732–738

Dzau VJ (2001) Theodore Cooper Lecture. Tissue angiotensin and pathobiology of vascular disease: a unifying hypothesis. Hypertension 37:1047–1052

Ellis GR, Nightingale AK, Blackman DJ, Anderson RA, Mumford C, Timmins G, Lang D, Jackson SK, Penney MD, Lewis MJ, Frenneaux MP, Morris-Thurgood J (2002) Addition of candesartan to angiotensin converting enzyme inhibitor therapy in patients with chronic heart failure does not reduce levels of oxidative stress. Eur J Heart Fail 4:193–199

Ferrari P, Marti HP, Pfister M, Frey FJ (2002) Additive antiproteinuric effect of combined ACE inhibition and angiotensin II receptor blockade. J Hypertens 20:125–130

Flather MD, Yusuf S, Kober L, Pfeffer M, Hall A, Murray G, Torp-Pedersen C, Ball S, Pogue J, Moye L, Braunwald E (2000) Long-term ACE-inhibitor therapy in patients with heart failure or left-ventricular dysfunction: a systematic overview of data from individual patients. ACE-Inhibitor Myocardial Infarction Collaborative Group. Lancet 355:1575–1581

Forclaz A, Maillard M, Nussberger J, Brunner HR, Burnier M (2003) Angiotensin II receptor blockade: is there truly a benefit of adding an ACE inhibitor? Hypertension 41:31–36

Gadsbøll N, Nielsen MD, Giese J, Leth A, Lonborg-Jensen H (1990) Diurnal monitoring of blood pressure and the renin-angiotensin system in hypertensive patients on long-term angiotensin converting enzyme inhibition. J Hypertens 8:733–740

Gainer JV, Morrow JD, Loveland A, King DJ, Brown NJ (1998) Effect of bradykinin-receptor blockade on the response to angiotensin-converting-enzyme inhibitor in normotensive and hypertensive subjects. N Engl J Med 339:1285–1292

Garg R, Yusuf S (1995) Overview of randomized trials of angiotensin-converting enzyme inhibitors on mortality and morbidity in patients with heart failure. Collaborative Group on ACE Inhibitor Trials. JAMA 273:1450–1456

Gasc JM, Monnot C, Clauser E, Corvol P (1993) Co-expression of type 1 angiotensin II receptor (AT_1R) and renin mRNAs in juxtaglomerular cells of the rat kidney. Endocrinology 132:2723–2725

The GISEN Group (1997) Randomised placebo-controlled trial of effect of ramipril on decline in glomerular filtration rate and risk of terminal renal failure in proteinuric, non-diabetic nephropathy. Lancet 349:1857–1863

Goto K, Fujii K, Onaka U, Abe I, Fujishima M (2000) Renin-angiotensin system blockade improves endothelial dysfunction in hypertension. Hypertension 36:575–580

Gremmler B, Kunert M, Schleiting H, Ulbricht LJ (2000) Improvement of cardiac output in patients with severe heart failure by use of ACE-inhibitors combined with the AT_1-antagonist eprosartan. Eur J Heart Fail 2:183–187

Griffiths CD, Morgan TO, Delbridge LM (2001) Effects of combined administration of ACE inhibitor and angiotensin II receptor antagonist are prevented by a high NaCl intake. J Hypertens 19:2087–2095

Guyene TT, Bellet M, Sassano P, Serrurier D, Corvol P, Ménard J (1989) Crossover design for the dose determination of an angiotensin converting enzyme inhibitor in hypertension. J Hypertens 7:1005–1012

Hamroff G, Katz SD, Mancini D, Blaufarb I, Bijou R, Patel R, Jondeau G, Olivari MT, Thomas S, Le Jemtel TH (1999) Addition of angiotensin II receptor blockade to maximal angiotensin-converting enzyme inhibition improves exercise capacity in patients with severe congestive heart failure. Circulation 99:990–992

Hansson L, Zanchetti A, Carruthers SG, Dahlof B, Elmfeldt D, Julius S, Ménard J, Rahn KH, Wedel H, Westerling S (1998) Effects of intensive blood-pressure lowering and low-dose aspirin in patients with hypertension: principal results of the Hypertension Optimal Treatment (HOT) randomised trial. HOT Study Group. Lancet 351:1755–1762

Hirakata H, Onoyama K, Iseki K, Kumagai H, Fujimi S, Omae T (1984) Worsening of anemia induced by-term use of captopril in hemodialysis patients. Am J Nephrol 4:355–360

Hollenberg NK (2000) Implications of species difference for clinical investigation: studies on the renin-angiotensin system. Hypertension 35:150–154

Horiuchi M, Akishita M, Dzau VJ (1999) Recent progress in angiotensin II type 2 receptor research in the cardiovascular system. Hypertension 33:613–621

Houghton AR, Harrison M, Cowley AJ, Hampton JR (2000) Combined treatment with losartan and an ACE inhibitor in mild to moderate heart failure: results of a double-blind, randomized, placebo-controlled trial. Am Heart J 140:791

Hulthen UL (2003) Nephrin: a pivotal molecule in proteinuria influenced by angiotensin II. J Hypertens 21:39–40

Ikoma M, Kawamura T, Kakinuma Y, Fogo A, Ichikawa I (1991) Cause of variable therapeutic efficiency of angiotensin converting enzyme inhibitor on glomerular lesions. Kidney Int 40:195–202

Iyer SN, Chappell MC, Averill DB, Diz DI, Ferrario CM (1998) Vasodepressor actions of angiotensin-(1–7) unmasked during combined treatment with lisinopril and losartan. Hypertension 31:699–705

Jacobsen P, Andersen S, Rossing K, Hansen BV, Parving HH (2002) Dual blockade of the renin-angiotensin system in type 1 patients with diabetic nephropathy. Nephrol Dial Transplant 17:1019–1024

Jafar TH, Schmid CH, Landa M, Giatras I, Toto R, Remuzzi G, Maschio G, Brenner BM, Kamper A, Zucchelli P, Becker G, Himmelmann A, Bannister K, Landais P, Shahinfar S, de Jong PE, de Zeeuw D, Lau J, Levey AS (2001) Angiotensin-converting enzyme inhibitors and progression of nondiabetic renal disease. A meta-analysis of patient-level data. Ann Intern Med 135:73–87

Jong P, Demers C, McKelvie RS, Liu PP (2002) Angiotensin receptor blockers in heart failure: meta-analysis of randomized controlled trials. J Am Coll Cardiol 39:463–470

Jorde UP, Infeld J, Cukon S, Suryadevara VR, Hammer A, Lisker J, Sonnenblick EH, LeJemtel TH (1998) Incomplete suppression of Angiotensin-Converting-Enzyme (ACE) Inhibition in patients with chronic heart failure despite full dose ACE-Inhibitors. Circulation 98 [Suppl 1]: 1-82

Juillerat L, Nussberger J, Ménard J, Mooser V, Christen Y, Waeber B, Graf P, Brunner HR (1990) Determinants of angiotensin II generation during converting enzyme inhibition. Hypertension 16:564–572

Kagami S, Border WA, Miller DE, Noble NA (1994) Angiotensin II stimulates extracellular matrix protein synthesis through induction of transforming growth factor-beta expression in rat glomerular mesangial cells. J Clin Invest 93:2431–2437

Kamper AL, Nielsen AH, Baekgaard N, Just S (2002) Renal graft failure after addition of an angiotensin II receptor antagonist to an angiotensin-converting enzyme inhibitor: unmasking of an unknown iliac artery stenosis. J Renin Angiotensin Aldosterone Syst 3:135–137

Kawai H, Stevens SY, Liang CS (2000) Renin-angiotensin system inhibition on noradrenergic nerve terminal function in pacing-induced heart failure. Am J Physiol Heart Circ Physiol 279: H3012–H3019

Kim S, Zhan Y, Izumi Y, Iwao H (2000) Cardiovascular effects of combination of perindopril, candesartan, and amlodipine in hypertensive rats. Hypertension 35:769–774

Kim S, Yoshiyama M, Izumi Y, Kawano H, Kimoto M, Zhan Y, Iwao H (2001) Effects of combination of ACE inhibitor and angiotensin receptor blocker on cardiac remodeling, cardiac function, and survival in rat heart failure. Circulation 103:148–154

Kim S, Izumi Y, Izumiya Y, Zhan Y, Taniguchi M, Iwao H (2002) Beneficial effects of combined blockade of ACE and AT_1 receptor on intimal hyperplasia in balloon-injured rat artery. Arterioscler Thromb Vasc Biol 22:1299–1304

Kohzuki M, Yasujima M, Kanazawa M, Yoshida K, Fu LP, Obara K, Saito T, Abe K (1995) Antihypertensive and renal-protective effects of losartan in streptozotocin diabetic rats. J Hypertens 13:97–103

Krombach RS, Clair MJ, Hendrick JW, Houck WV, Zellner JL, Kribbs SB, Whitebread S, Mukherjee R, de Gasparo M, Spinale FG (1998) Angiotensin converting enzyme inhibition, AT_1 receptor inhibition, and combination therapy with pacing induced heart failure: effects on left ventricular performance and regional blood flow patterns. Cardiovasc Res 38:631–645

Laverman GD, Henning RH, de Jong PE, Navis G, de Zeeuw D (2001) Optimal antiproteinuric dose of losartan in nondiabetic patients with nephrotic range proteinuria. Am J Kidney Dis 38:1381–1384

Laverman GD, Navis G, Henning RH, De Jong PE, De Zeeuw D (2002) Dual renin-angiotensin system blockade at optimal doses for proteinuria. Kidney Int 62:1020–1025

Lees KR (1992) The dose-response relationship with angiotensin converting enzyme inhibitors: effects on blood pressure and biochemical parameters. J Hypertens 10 [Suppl 5]:S3–S11

Lewis EJ, Hunsicker LG, Bain RP, Rohde RD (1993) The effect of angiotensin-converting-enzyme inhibition on diabetic nephropathy. The Collaborative Study Group [see comments] [published erratum appears in N Engl J Med (1993) 330:152]. N Engl J Med 329:1456–1462

Lewis EJ, Hunsicker LG, Clarke WR, Berl T, Pohl MA, Lewis JB, Ritz E, Atkins RC, Rohde R, Raz I (2001) Renoprotective effect of the angiotensin-receptor antagonist irbesartan in patients with nephropathy due to type 2 diabetes. N Engl J Med 345:851–860

Linz W, Wiemer G, Gohlke P, Unger T, Scholkens BA (1995) Contribution of kinins to the cardiovascular actions of angiotensin-converting enzyme inhibitors. Pharmacol Rev 47:25–49

Lonn E, Yusuf S, Dzavik V, Doris C, Yi Q, Smith S, Moore-Cox A, Bosch J, Riley W, Teo K (2001) Effects of ramipril and vitamin E on atherosclerosis: the study to evaluate carotid ultrasound changes in patients treated with ramipril and vitamin E (SECURE). Circulation 103:919–925

MacFadyen RJ, Lee AF, Morton JJ, Pringle SD, Struthers AD (1999) How often are angiotensin II and aldosterone concentrations raised during chronic ACE inhibitor treatment in cardiac failure? Heart 82:57–61

Maillard MP, Wurzner G, Nussberger J, Centeno C, Burnier M, Brunner HR (2002) Comparative angiotensin II receptor blockade in healthy volunteers: the importance of dosing. Clin Pharmacol Ther 71:68–76

Mankad S, d'Amato TA, Reichek N, McGregor WE, Lin J, Singh D, Rogers WJ, Kramer CM (2001) Combined angiotensin II receptor antagonism and angiotensin-converting enzyme inhibition further attenuates postinfarction left ventricular remodeling. Circulation 103:2845–2850

Maschio G, Alberti D, Janin G, Locatelli F, Mann JF, Motolese M, Ponticelli C, Ritz E, Zucchelli P (1996) Effect of the angiotensin-converting-enzyme inhibitor benazepril on the progression of chronic renal insufficiency. The Angiotensin-Converting-Enzyme Inhibition in Progressive Renal Insufficiency Study Group. N Engl J Med 334:939–945

McGregor GA, Markandu ND, Smith SJ, Sagnella GA, Morton JJ (1981) Maintenance of blood pressure by the renin-angiotensin system in normal man. Nature 291:329–331

McKelvie RS, Yusuf S, Pericak D, Avezum A, Burns RJ, Probstfield J, Tsuyuki RT, White M, Rouleau J, Latini R, Maggioni A, Young J, Pogue J (1999) Comparison of candesartan, enalapril, and their combination in congestive heart failure: randomized evaluation of strategies for left ventricular dysfunction (RESOLVD) pilot study. The RESOLVD Pilot Study Investigators. Circulation 100:1056–1064

McMurray J, Ostergren J, Swedberg K, Granger CB, Held P, Michelson EL, Olofsson B, Yusuf S, Pfeffer, MA (2003) Effects of candesartan in patients with chronic heart failure and reducedleft-ventricular systolic function taking angiotensin-conveting-enzyme inhibitors: the CHARM-Added trial. Lancet 362:767–771

Ménard J, Guyene TT (1995) Commentary: renin assays: a debate for clinicians, not only for specialists. J Hypertens 13:367–369

Ménard J, Boger RS, Moyse DM, Guyene TT, Glassman HN, Kleinert HD (1995) Dose-dependent effects of the renin inhibitor zankiren HCl after a single oral dose in mildly sodium-depleted normotensive subjects. Circulation 91:330–338

Ménard J, Campbell DJ, Azizi M, Gonzales MF (1997) Synergistic effects of ACE inhibition and Ang II antagonism on blood pressure, cardiac weight and renin in spontaneously hypertensive rats. Circulation 96:3072–3078

Meredith PA, Donnelly R, Elliott HL, Howie CA, Reid JL (1990) Prediction of the antihypertensive response to enalapril. J Hypertens 8:1085–1090

Mogensen CE, Neldam S, Tikkanen I, Oren S, Viskoper R, Watts RW, Cooper ME (2000) Randomised controlled trial of dual blockade of renin-angiotensin system in patients with hypertension, microalbuminuria, and non-insulin dependent diabetes: the candesartan and lisinopril microalbuminuria (CALM) study. BMJ 321:1440–1444

Mooser V, Nussberger J, Juillerat L, Burnier M, Waeber B, Bidiville J, Pauly N, Brunner HR (1990) Reactive hyperreninemia is a major determinant of plasma angiotensin II during ACE inhibition. J Cardiovasc Pharmacol 15:276–282

Morelli E, Loon N, Meyer T, Peters W, Myers BD (1990) Effects of converting-enzyme inhibition on barrier function in diabetic glomerulopathy. Diabetes 39:76–82

Morgan T, Griffiths C, Delbridge L (2002) Low doses of angiotensin converting enzyme inhibitors and angiotensin type 1 blockers have a synergistic effect but high doses are less than additive. Am J Hypertens 15:1003–1005

Murdoch DR, McDonagh TA, Farmer R, Morton JJ, McMurray JJ, Dargie HJ (2001) ADEPT: addition of the AT_1 receptor antagonist eprosartan to ACE inhibitor therapy in chronic heart failure trial: hemodynamic and neurohormonal effects. Am Heart J 141:800–807

Nakamura T, Obata J, Kimura H, Ohno S, Yoshida Y, Kawachi H, Shimizu F (1999) Blocking angiotensin II ameliorates proteinuria and glomerular lesions in progressive mesangioproliferative glomerulonephritis. Kidney Int 55:877–889

Nakamura Y, Yoshiyama M, Omura T, Yoshida K, Izumi Y, Takeuchi K, Kim S, Iwao H, Yoshikawa J (2003) Beneficial effects of combination of ACE inhibitor and angiotensin II type 1 receptor blocker on cardiac remodeling in rat myocardial infarction. Cardiovasc Res 57:48–54

Nakao N, Yoshimura A, Morita H, Takada M, Kayano T, Ideura T (2003) Combination treatment of angiotensin-II receptor blocker and angiotensin-converting-enzyme inhibitor in non-diabetic renal disease (COOPERATE): a randomised controlled trial. Lancet 361:117–124

Nanas JN, Alexopoulos G, Anastasiou-Nana MI, Karidis K, Tirologos A, Zobolos S, Pirgakis V, Anthopoulos L, Sideris D, Stamatelopoulos SF, Moulopoulos SD (2000) Outcome of patients with congestive heart failure treated with standard versus high doses of enalapril: a multicenter study. High Enalapril Dose Study Group. J Am Coll Cardiol 36:2090–2095

Navar GL, Mitchell KD, Harrison-Bernard LM, Kobari H, Nishiyama A (2001) Intrarenal angiotensin II levels in normal and hypertensive states. J Renin Angiotensin Aldosterone Syst 2: S176–S184

Nunez E, Hosoya K, Susic D, Frohlich ED (1997) Enalapril and Losartan reduced cardiac mass and improved coronary hemodynamics in SHR. Hypertension 29:519–524

Nussberger J (2000) Circulating versus tissue angiotensin II. In: Epstein M, Brunner HR (eds) Angiotensin II receptor antagonists. . Philadelphia, Hanley and Belfus Inc., pp 69–78

Ohtawa M, Takayama F, Saitoh K, Yoshinaga T, Nakashima M (1993) Pharmacokinetics and biochemical efficacy after single and multiple oral administration of losartan, an orally active nonpeptide angiotensin II receptor antagonist, in humans. Br J Clin Pharmacol 35:290–297

Okunishi H, Miyazaki M, Toda N (1984) Evidence for a putatively new angiotensin II-generating enzyme in the vascular wall. J Hypertens 2:277–284

Opie LH, Sack MN (2001) Enhanced angiotensin II activity in heart failure: reevaluation of the counterregulatory hypothesis of receptor subtypes. Circ Res 88:654–658

Ots M, Mackenzie HS, Troy JL, Rennke HG, Brenner BM (1998) Effects of combination therapy with enalapril and losartan on the rate of progression of renal injury in rats with 5/6 renal mass ablation. J Am Soc Nephrol 9:224–230

Packer M, Poole-Wilson PA, Armstrong PW, Cleland JG, Horowitz JD, Massie BM, Ryden L, Thygesen K, Uretsky BF (1999) Comparative effects of low and high doses of the angiotensin-converting enzyme inhibitor, lisinopril, on morbidity and mortality in chronic heart failure. ATLAS Study Group. Circulation 100:2312–2318

Packer M, Califf RM, Konstam MA, Krum H, McMurray JJ, Rouleau JL, Swedberg K (2002) Comparison of omapatrilat and enalapril in patients with chronic heart failure: the Omapatrilat Versus Enalapril Randomized Trial of Utility in Reducing Events (OVERTURE). Circulation 106:920–926

Parving HH, Andersen AR, Smidt UM, Hommel E, Mathiesen ER, Svendsen PA (1987) Effect of antihypertensive treatment on kidney function in diabetic nephropathy. Br Med J (Clin Res Ed) 294:1443–1447

Parving HH, Lehnert H, Brochner-Mortensen J, Gomis R, Andersen S, Arner P (2001) The effect of irbesartan on the development of diabetic nephropathy in patients with type 2 diabetes. N Engl J Med 345:870–878

Perna A, Remuzzi G (1996) Abnormal permeability to proteins and glomerular lesions: a meta-analysis of experimental and human studies. Am J Kidney Dis 27:34–41

Perticucci E, Campbell R, Perna A, Ferrari P, Cattaneo D, Ruggenenti P, Koleva NZ, Aros G (2001) ACE inhibitors (ACEi), angiotensin II antagonists (ATA) or their combination: how to best renoprotect patients with nondiabetic chronic nephropathies? J Am Soc Nephrol 12:82A

Pfeffer JM, Fischer TA, Pfeffer MA (1995) Angiotensin-converting enzyme inhibition and ventricular remodeling after myocardial infarction. Annu Rev Physiol 57:805–826

Pfeffer MA, Lamas GA, Vaughan DE, Parisi AC, Braunwald E (1988) Effect of captopril on progressive ventricular dilatation after anterior myocardial infarction. N Engl J Med 319:80–86

Pfeffer MA, McMurray J, Leizorovicz A, Maggioni AP, Rouleau JL, Van De Werf F, Henis M, Neuhart E, Gallo P, Edwards S, Sellers MA, Velazquez E, Califf R (2000) Valsartan in acute myocardial infarction trial (VALIANT): rationale and design. Am Heart J 140:727–750

Pisoni R, Ruggenenti P, Sangalli F, Lepre MS, Remuzzi A, Remuzzi G (2002) Effect of high dose ramipril with or without indomethacin on glomerular selectivity. Kidney Int 62:1010–1019

Pitt B, Poole-Wilson PA, Segal R, Martinez FA, Dickstein K, Camm AJ, Konstam MA, Riegger G, Klinger GH, Neaton J, Sharma D, Thiyagarajan B (2000) Effect of losartan compared with captopril on mortality in patients with symptomatic heart failure: randomised trial–the Losartan Heart Failure Survival Study ELITE II. Lancet 355:1582–1587

Pouleur H, Konstam MA, Benedict CR, Donckier J, Galanti L, Melin J, Kinan D, Ahn S, Rousseau MF (1993) Progression of left ventricular dysfunction during enalapril therapy: relationship with neurohormonal activation (abstract). Circulation 88: I-293

Quinn SJ, Williams GH (1988) Regulation of aldosterone secretion. Ann Rev Physiol 50:409–426

Rakugi H, Wang DS, Dzau VJ, Pratt RE (1994) Potential importance of tissue angiotensin-converting enzyme inhibition in preventing neointima formation. Circulation 90:449–455

Remuzzi A, Perico N, Sangalli F, Vendramin G, Moriggi M, Ruggenenti P, Remuzzi G (1999) ACE inhibition and ANG II receptor blockade improve glomerular size - selectivity in IgA nephropathy. Am J Physiol 276: F457–F466

Remuzzi A, Gagliardini E, Donadoni C, Fassi A, Sangalli F, Lepre MS, Remuzzi G, Benigni A (2002a) Effect of angiotensin II antagonism on the regression of kidney disease in the rat. Kidney Int 62:885–894

Remuzzi G, Ruggenenti P, Perico N (2002b) Chronic renal diseases: renoprotective benefits of renin-angiotensin system inhibition. Ann Intern Med 136:604–615

Richer C, Mulder P, Fornes P, Domergue V, Heudes D, Giudicelli JF (1992) Long-term treatment with trandolapril opposes cardiac remodeling and prolongs survival after myocardial infarction in rats. J Cardiovasc Pharmacol 20:147–156

Richer C, Bruneval P, Ménard J, Giudicelli JF (1998) Additive effects of enalapril and losartan in (mREN-2)27 transgenic rats. Hypertension 31:692–698

Richer C, Fornes P, Domergue V, de Gasparo M, Giudicelli JF (2001) Combined angiotensin II AT_1-receptor blockade and angiotensin I-converting enzyme inhibition on survival and cardiac remodeling in chronic heart failure in rats. J Card Fail 7:269–276

Rossing K, Christensen PK, Jensen BR, Parving HH (2002) Dual blockade of the renin-angiotensin system in diabetic nephropathy: a randomized double-blind crossover study. Diabetes Care 25:95–100

Ruggenenti P, Perna A, Gherardi G, Gaspari F, Benini R, Remuzzi G (1998) Renal function and requirement for dialysis in chronic nephropathy patients on long-term ramipril: REIN follow-up trial. Gruppo Italiano di Studi Epidemiologici in Nefrologia (GISEN). Ramipril Efficacy in Nephropathy. Lancet 352:1252–1256

Ruggenenti P, Schieppati A, Remuzzi G (2001) Progression, remission, regression of chronic renal diseases. Lancet 357:1601–1608

Ruilope LM, Aldigier JC, Ponticelli C, Oddou-Stock P, Botteri F, Mann JF (2000) Safety of the combination of valsartan and benazepril in patients with chronic renal disease. European Group for the Investigation of Valsartan in Chronic Renal Disease. J Hypertens 18:89–95

Russo D, Minutolo R, Pisani A, Esposito R, Signoriello G, Andreucci M, Balletta MM (2001) Coadministration of losartan and enalapril exerts additive antiproteinuric effect in IgA nephropathy. Am J Kidney Dis 38:18–25

Sadoshima JU, Xu Y, Slayter HS, Izumo S (1993) Autocrine release of angiotensin II mediates stretch-induced hypertrophy of cardiac myocytes in vitro. Cell 75:977–984

Sakata K, Yoshida H, Obayashi K, Ishikawa J, Tamekiyo H, Nawada R, Doi O (2002) Effects of losartan and its combination with quinapril on the cardiac sympathetic nervous system and neurohormonal status in essential hypertension. J Hypertens 20:103–110

Schieffer B, Wirger A, Meybrunn M, Seitz S, Holtz J, Riede UN, Drexler H (1994) Comparative effects of chronic angiotensin-converting enzyme inhibition and angiotensin II type 1 receptor blockade on cardiac remodeling after myocardial infarction in the rat. Circulation 89:2273–2282

Schmieder RE, Martus P, Klingbeil A (1996) Reversal of left ventricular hypertrophy in essential hypertension. A meta-analysis of randomized double-blind studies. JAMA 275:1507–1513

Schwarzbeck A, Wittenmeier KW, Hallfritzsch U (1998) Anaemia in dialysis patients as a side-effect of sartanes. Lancet 352:286

Sharma K, Eltayeb BO, McGowan TA, Dunn SR, Alzahabi B, Rohde R, Ziyadeh FN, Lewis EJ (1999) Captopril-induced reduction of serum levels of transforming growth factor-beta1 correlates with long-term renoprotection in insulin-dependent diabetic patients. Am J Kidney Dis 34:818–823

Shen YT, Wiedmann RT, Greenland BD, Lynch JJ, Grossman W (1998) Combined effects of angiotensin converting enzyme inhibition and angiotensin II receptor antagonism in conscious pigs with congestive heart failure. Cardiovasc Res 39:413–422

Spinale FG, de Gasparo M, Whitebread S, Hebbar L, Clair MJ, Melton DM, Krombach RS, Mukherjee R, Iannini JP, O SJ (1997a) Modulation of the renin-angiotensin pathway through enzyme inhibition and specific receptor blockade in pacing-induced heart failure: I. Effects on left ventricular performance and neurohormonal systems. Circulation 96:2385–2396

Spinale FG, Mukherjee R, Iannini JP, Whitebread S, Hebbar L, Clair MJ, Melton DM, Cox MH, Thomas PB, de Gasparo M (1997b) Modulation of the renin-angiotensin pathway through enzyme inhibition and specific receptor blockade in pacing-induced heart failure: II. Effects on myocyte contractile processes. Circulation 96:2397–2406

Swedberg K, Eneroth P, Kjekshus J, Wilhelmsen L (1990) Hormones regulating cardiovascular function in patients with severe congestive heart failure and their relation to mortality. CONSENSUS Trial Study Group. Circulation 82:1730–1736

Swedberg K, Pfeffer M, Granger C, Held P, McMurray J, Ohlin G, Olofsson B, Ostergren J, Yusuf S (1999) Candesartan in heart failure—assessment of reduction in mortality and morbidity (CHARM): rationale and design. CHARM-Programme Investigators. J Card Fail 5:276–282

Taylor K, Patten RD, Smith JJ, Aronovitz MJ, Wight J, Salomon RN, Konstam MA (1998) Divergent effects of angiotensin-converting enzyme inhibition and angiotensin II-receptor antagonism on myocardial cellular proliferation and collagen deposition after myocardial infarction in rats. J Cardiovasc Pharmacol 31:654–660

The SOLVD Investigators (1992) Effect of enalapril on mortality and the development of heart failure in asymptomatic patients with reduced left ventricular ejection fractions. N Engl J Med 327:685–691

UK Prospective Diabetes Study Group (1998a) Efficacy of atenolol and captopril in reducing risk of macrovascular and microvascular complications in type 2 diabetes: UKPDS 39. BMJ 317:713–720

UK Prospective Diabetes Study Group (1998b) Tight blood pressure control and risk of macrovascular and microvascular complications in type 2 diabetes: UKPDS 38. BMJ 317:703–713

Urata H, Kinoshita A, Misono KS, Bumpus FM, Husain A (1990) Identification of a highly specific chymase as the major angiotensin II-forming enzyme in the human heart. J Biol Chem 265:22348–22357

Van Den Meiracker AH, Man in't veld AJ, Admiraal PJJ, Ritsema Van Eck HJ, Boomsma F, Derkx FHM, Schalekamp MADH (1992) Partial escape of angiotensin converting enzyme (ACE) inhibition during prolonged ACE inhibitor treatment: does it exist and does it affect the antihypertensive response? J Hypertens 10:803–812

Van Kats JP, Schalekamp MA, Verdouw PD, Duncker DJ, Danser AH (2001) Intrarenal angiotensin II: interstitial and cellular levels and site of production. Kidney Int 60:2311–2317

Volders PG, Willems IE, Cleutjens JP, Arends JW, Havenith MG, Daemen MJ (1993) Interstitial collagen is increased in the non-infarcted human myocardium after myocardial infarction. J Mol Cell Cardiol 25:1317–1323

Vos PF, Boer P, Braam B, Koomans HA (1995) Efficacy of intrarenal ACE-inhibition estimated from the renal response to angiotensin I and II in humans. Kidney Int 47:274–281

Waeber B, Aschwanden R, Sadecky L, Ferber P (2001) Combination of hydrochlorothiazide or benazepril with valsartan in hypertensive patients unresponsive to valsartan alone. J Hypertens 19:2097–2104

Webb RL, Navarrete AE, Davis S, de Gasparo M (1998) Synergistic effects of combined converting enzyme inhibition and angiotensin II antagonism on blood pressure in conscious telemetered spontaneously hypertensive rats. J Hypertens 16:843–852

Weidenbach R, Schulz R, Gres P, Behrends M, Post H, Heusch G (2000) Enhanced reduction of myocardial infarct size by combined ACE inhibition and AT(1)-receptor antagonism. Br J Pharmacol 131:138–144

Weir MR, Smith DH, Neutel JM, Bedigian MP (2001) Valsartan alone or with a diuretic or ACE inhibitor as treatment for African American hypertensives: relation to salt intake. Am J Hypertens 14:665–671

Wilkinson-Berka JL, Gibbs NJ, Cooper ME, Skinner SL, Kelly DJ (2001) Renoprotective and anti-hypertensive effects of combined valsartan and perindopril in progressive diabetic nephropathy in the transgenic (mRen-2)27 rat. Nephrol Dial Transplant 16:1343–1349

Wollert KC, Studer R, von Bulow B, Drexler H (1994) Survival after myocardial infarction in the rat. Role of tissue angiotensin-converting enzyme inhibition. Circulation 90:2457–2467

Wong PC, Price WA, Chiu AT, Duncia JV, Carini DJ, Wexler RR, Johnson AL, Timmermans PBMWM (1990) Nonpeptide angiotensin II receptor antagonist. IX. Antihypertensive activity in rats of DuP 753, an orally active antihypertensive agent. J Pharmacol Exp Ther 252:726–732

Yu CM, Tipoe GL, Wing-Hon Lai K, Lau CP (2001) Effects of combination of angiotensin-converting enzyme inhibitor and angiotensin receptor antagonist on inflammatory cellular infiltration and myocardial interstitial fibrosis after acute myocardial infarction. J Am Coll Cardiol 38:1207–1215

Yusuf S (2002) From the HOPE to the ONTARGET and the TRANSCEND studies: challenges in improving prognosis. Am J Cardiol 89:18A–25A

Yusuf S, Sleight P, Pogue J, Bosch J, Davies R, Dagenais G (2000) Effects of an angiotensin-converting-enzyme inhibitor, ramipril, on cardiovascular events in high-risk patients. The Heart Outcomes Prevention Evaluation Study Investigators. N Engl J Med 342:145–153

Part 5
Inhibition of the Renin-Angiotensin System
NEP/ACE Inhibitors

NEP/ACE Inhibitors: Experimental and Clinical Aspects

R. Corti[1] · F. Ruschitzka[1] · T. F. Lüscher[2]

[1] Cardio Vascular Center, Cardiology, University Hospital Zurich, Zurich, Switzerland
[2] Cardiology Department, University Hospital, Rämistrasse 100, 8091 Zurich, Switzerland
e-mail: cardiotfl@gmx.ch

Abstract The human cardiovascular system is regulated by hemodynamic and neurohumoral mechanisms. These regulatory systems play a key role in modulating cardiac function, vascular tone and structure as well as adhesion of blood cells by producing circulating and local vasoactive substances. Although neurohumoral systems are essential in vascular homeostasis, they become maladaptive in disease states such as hypertension, coronary disease and heart failure. The clinical success of blocking the renin-angiotensin-aldosterone system by angiotensin-converting enzyme inhibitors has led to efforts to block other humoral systems as well. Neutral endopeptidase (NEP) is an endothelial cell surface zinc metallopeptidase with a similar structure and catalytic site. NEP is the major enzymatic pathway for degradation of natriuretic peptides, a secondary enzymatic pathway for degradation of kinins, and the vasoactive and natriuretic peptide adrenomedullin. The natriuretic peptides can be viewed as endogenous inhibitors of the renin angiotensin system. Inhibition of NEP increases levels of natriuretic and vasodilatory peptides including atrial or A-type natriuretic peptide (ANP), B-type natriuretic peptide (BNP) of myocardial cell origin, and

C-type natriuretic peptide (CNP) of endothelial cell origin as well as bradykinin and adrenomedullin. By simultaneously inhibiting the renin-angiotensin-aldosterone system and potentiating the natriuretic peptide and kinin systems, vasopeptidase inhibitors reduce vasoconstriction, enhance vasodilation, improve sodium/water balance and in turn decrease peripheral vascular resistance and blood pressure and improve local blood flow. Within the blood vessel wall this leads to a reduction in vasoconstrictor and proliferative mediators such as angiotensin II and endothelin-1 and increased local levels of bradykinin (and in turn nitric oxide) and natriuretic peptides. In hypertension, vasopeptidase inhibitors such as omapatrilat are more potent in lowering blood pressure than any other compound. In heart failure, the difference compared to ACE inhibitors is less. Nevertheless, combined inhibition of angiotensin-converting enzyme and neutral endopeptidase is a new and promising approach to treat patients with hypertension, atherosclerosis or heart failure.

Keywords Hypertension · Heart failure · Therapy · Natriuretic peptide

Abbreviations

ACE	Angiotensin-converting enzyme
ANP	Atrial natriuretic peptide
BNP	Brain natriuretic peptide
CNP	C-type natriuretic peptide
CHF	Congestive heart failure
NEP	Neutral endopeptidase
RAAS	Renin-angiotensin-aldosterone system
EDHF	Endothelial-derived hyperpolarizing factor
NO	Nitric oxide
eNOS	Endothelial NO synthase

Structural, humoral and neuronal factors are involved in cardiovascular regulation, among them the sympathetic and parasympathetic nervous systems and the renin-angiotensin-aldosterone system (RAAS). The endothelium is a source of paracrine mediators such as nitric oxide (NO), endothelial-derived hyperpolarizing factor (EDHF) and endothelin (Fig. 1). Circulating and local regulatory mediators exhibit complex synergisms and interactions; the sympathetic nervous system stimulates secretion of renin and angiotensin II which centrally and at the presynaptic level increase sympathetic nerve activity and enhance endothelin and vasopressin production. The natriuretic peptide system consisting of A-type natriuretic peptide (ANP) and B-type natriuretic peptide (BNP) of myocardial origin and C-type natriuretic peptide (CNP) of endothelial origin, on the other hand, counteract the RAAS and endothelin. Endothelial substances act primarily locally and exhibit vasoconstrictor, vasodilating, and mitogenic effects. Some endothelial substances stimulate the production of cytokines and

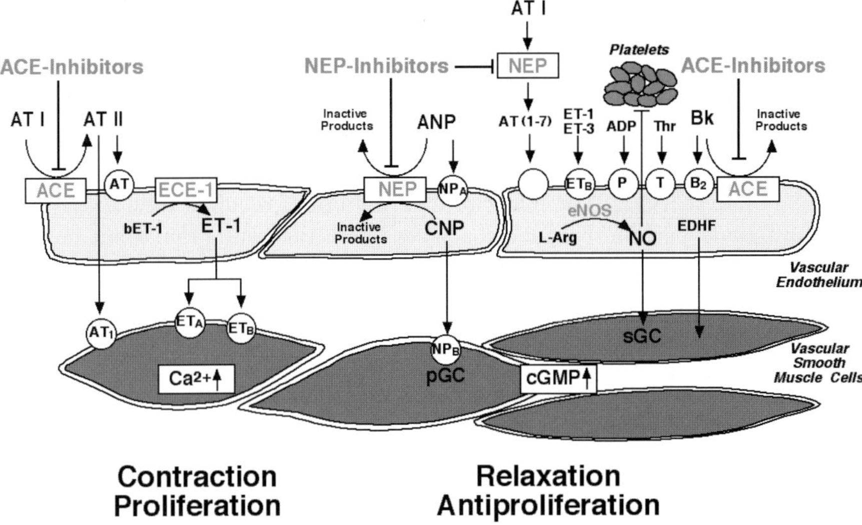

Fig. 1 The synergistic effects resulting from combined ACE and NEP inhibition are due to a similar mechanism, leading to the blockade of angiotensin (*AT*) synthesis and concomitant potentiation of the effects of natriuretic peptides and bradykinin, leading to vasodilatation, natriuresis and improvement in myocardial function

growth factors, leading to vascular smooth muscle cell proliferation. All these regulatory systems are crucial for proper circulatory homeostasis and for structural vascular and myocardial regulation.

1
Vascular Effects of Angiotensin-Converting Enzyme Inhibitors

The angiotensin-converting enzyme (ACE) is mainly located on endothelial cells where it transforms angiotensin I into angiotensin II and degrades bradykinin, a potent stimulator of the L-arginine and cyclooxygenase pathways (Palmer et al. 1987) (Fig. 1). Therefore, ACE inhibitors not only prevent the formation of a potent vasoconstrictor with proliferative properties, but also increase local concentrations of bradykinin and, in turn the production of NO (Zhang et al. 1997b) and prostacyclin (Wiemer et al. 1991). The latter may participate in the vascular protective effects of ACE inhibitors by improving local blood flow and preventing platelet activation. Accordingly, pretreatment of human saphenous vein and coronary artery with ACE inhibitors enhances endothelium-dependent relaxation to bradykinin (Yang et al. 1993; Auch-Schwelk et al. 1992). The decreased degradation of bradykinin could therefore explain the improved endothelial function observed with ACE inhibitors in normotensive and particularly in hypertensive rats (Kahonen et al. 1995; Dohi et al. 1994; Bossaller et al. 1992). However, the improvement of endothelial function by ACE inhibitors in L-NAME-induced hypertension suggests that they also enhance endothelium-dependent mediators other than NO (i.e., EDHF), since the activity of NO synthase (eNOS) remains suppressed in this model (Takase et al. 1996; Kung et al. 1995). Furthermore, ACE inhibitors stabilize the activity of B_2-bradykininergic receptors independent of effects on bradykinin metabolism, since stable analogs of bradykinin that are not inactivated by chymase II or ACE also exhibit enhanced endothelium-dependent relaxation in the presence of an ACE inhibitor (Mombouli et al. 1992). The effect of ACE inhibitors requires time to develop and can only in part be reproduced acutely, again suggesting that other mechanisms than ACE inhibition, which are rapid, are involved (Takase et al. 1996). Indeed, in Dahl salt-induced hypertension and in postinfarction rats, diminished eNOS expression is restored during chronic ACE-inhibition (Quaschning et al. 2001; Qi et al. 1999).

In contrast to striking improvements in experimental hypertension, studies in hypertensives revealed controversial results. ACE inhibitors somewhat improved endothelial function in subcutaneous arteries (Schiffrin and Deng 1995) and in the renal circulation (Mimran et al. 1995). In the forearm circulation, however, captopril and enalapril (Creager and Roddy 1994) or cilazapril (Kiowski et al. 1996) failed to improve vasodilation to a muscarinic agonist, while lisinopril selectively improved vasodilation to bradykinin without restoring NO bioavailability (Taddei et al. 1998). The reasons for this discrepancy are unclear. Endothelial dysfunction certainly is treated at later stages in patients than in experimental hypertension. Alternatively, duration of therapy and differences in tissue selectivity may be important. Indeed, in normal subjects, ACE inhibitors with high tissue selectivity such as quinaprilat have vascular effects, which are not shared by enalapril (Hornig et al. 1998; Haefeli et al. 1997). In patients with coronary artery disease, treatment with quinapril for 6 months improved endothelium-dependent vasomotion to acetylcholine in epicardial coronary arteries (Mancini et al. 1996) and in part in the coronary microcirculation (Schlaifer et al. 1997). Quinaprilat also improves flow-dependent dilation in congestive heart failure (CHF) as the result of increased availability of NO, whereas enalaprilat does not (Hornig et al. 1998).

The mechanisms involved may be related to inhibition of angiotensin formation and/or stimulation of the L-arginine/NO pathway. Many studies have documented the existence of a kallikrein-kinin system in myocardial and vascular tissue (Linz et al. 1995; Schölkens 1996; Zhang et al. 1997a). The antihypertensive and cardioprotective effect of ACE inhibitors has been explained in part as a consequence of diminished kinin degradation, resulting in the increase of endothelial NO production (Fig. 1) (Zhang et al. 1999; Blais et al. 2001). Indeed, studies of recombinant full-length ACE have shown that the K_m of ACE for bradykinin is substantially lower than for angiotensin I, reflecting a greater affinity for metabolism of bradykinin than for the production of angiotensin II (Jaspard et al. 1993). ACE is the major enzyme responsible for the metabolism of bradykinin, the exact percentage varying according to the tissue and species being evaluated (Blais et al. 1997). In endothelium and cardiac tissues, ACE is the major enzyme involved in the degradation of bradykinin, regardless of species. The other enzymes involved in the metabolism of bradykinin include carboxypeptidases, neutral endopeptidases and aminopeptidase P. ACE is also a major metabolic pathway for the degradation of one of the metabolites of bradykinin, des-arg^9-bradykinin, which in certain situations where its receptors are expressed (inflammatory situations), the B_1 receptor can have the same effects as bradykinin (Blais et al. 2001).

2 Clinical Effects of ACE Inhibitors

ACE inhibitors decrease systemic vascular resistance without increasing heart rate and promote natriuresis. The favorable effect of ACE inhibition has been documented in many large randomized trials in hypertension (Cheung and Lau 1999; Hansson et al. 1999), after myocardial infarction (GISSI-3 1994; Pfeffer et al. 1992; Ambrosioni et al. 1995; Swedberg et al. 1992) and in CHF (Yusuf et al. 1992). More recently, they have been shown to decrease clinical events in high-risk patients with atherosclerosis with and without ventricular dysfunction, prior to and after myocardial infarction (AIRE 1993; Kober et al. 1995; Yusuf et al. 2000). Moreover, the HOPE study confirmed that ACE inhibitors are vascular protective independent of their effects on blood pressure and ventricular remodeling (Yusuf et al. 2000).

The clinical effect of ACE inhibitors in the treatment of hypertension and CHF underlines the importance of neurohumoral blockade. Since the introduction of captopril in 1975, many long-acting molecules have been developed. However, in spite of their clinical efficacy, a substantial number of hypertensives are not adequately controlled with ACE inhibitors and require combination therapy with diuretics, beta-blockers and/or calcium antagonists. Furthermore, clinical studies in early stages of CHF demonstrated that ACE inhibitors (Richardson et al. 1987) are less effective in patients with high levels of ANP adrenaline and renin activity (Remes et al. 1991; Bayliss et al. 1986). Also, morbidity and mortality remains high in patients with CHF on ACE. Thus, the development of new drugs that act on the other neurohumoral systems than the RAAS may be advantageous.

3 Natriuretic Peptides

The family of natriuretic peptides consists of three forms, i.e., ANP, BNP and CNP. Both ANP and BNP are synthesized in the atrium of the heart under physiological conditions and in the ventricles (BNP) in the presence of ventricular hypertrophy, and in endothelial cells (CNP). BNP is a cardiac neurohormone specifically secreted from the cardiac ventricles as a response to ventricular volume expansion, pressure overload, and resultant increased wall tension. Measurement of BNP appears to be a particularly attractive parameter for diagnostic and prognostic evaluation of patients with CHF. Recent reports have determined

Table 1 Natriuretic peptides have contrasting biological effects to angiotensin II

Angiotensin II		Natriuretic peptide
↑	Blood pressure	↓
↓	Renal sodium secretion	↑
↑	Aldosterone	↓
↓	Renin secretion	↓
↑	Cell proliferation	↓
↑	Hypertrophy	↓

the role of BNP in the evaluation of acute dyspnea, in the emergency diagnosis of CHF and have defined BNP as a marker of risk stratification in CHF and acute coronary syndromes (Berger et al. 2002; Maisel et al. 2002; McCullough et al. 2002; Sabatine et al. 2002).

Natriuretic peptides, in addition to their diagnostic and prognostic value, appear as an intriguing target for pharmacological intervention in CHF. ANP infusion reduces blood pressure while increasing urine volume and urinary excretion of sodium; cyclic GMP inhibits renin and aldosterone secretion (Burnett et al. 1984; Janssen et al. 1989) and increases the hypotensive effect of BNP (Seymour et al. 1995). Moreover, ANP inhibits endothelin production and proliferation of vascular smooth cells and myocardial hypertrophy, and ANP has been shown to have significant sympatholytic effects as well (Azevedo et al. 2000). Because of its biological effect as an antagonist to angiotensin II, ANP is an endogenous inhibitor of the RAAS (Johnston et al. 1989) (Table 1). ANP and BNP production in the myocardium is induced by increased atrial pressure, as may occur with increased sodium intake and by decreased left ventricle function (systolic and diastolic left ventricular dysfunction) (Struthers 1994; Lang et al. 1994).

A hallmark of ventricular remodeling secondary to heart failure or left ventricular hypertrophy is the increase in plasma ANP and BNP (Burnett et al. 1986). Although the increased circulating natriuretic peptides may prevent water and sodium retention, progressive CHF is associated with a relative decrease in ANP production in association with an escape phenomenon of the RAAS, leading to increased water and sodium retention.

Circulating ANP BNP and CNP are quickly metabolized and inactivated by the specific enzyme, the widely located neutral endopeptidases (Fig. 1) as well as by cell-surface clearance receptor. The short half-life of the natriuretic peptides as well as the difficulty of administering and cost of producing a peptide limit the option of an exogenous application of the peptide as a possible therapeutic strategy. It should be noted that BNP has emerged as an efficacious intravenous agent for the treatment of CHF (Colucci et al. 2000). Therefore, pharmacological inhibition of the metabolism of natriuretic peptides is an attractive alternative therapeutic target.

4
Neutral Endopeptidases

Neutral endopeptidase (NEP) is an endothelial, membrane-bound metallopeptidase with zinc at its active site that cleaves endogenous peptides at the amino side of hydrophilic residues (Margulies et al. 1995) (Fig. 1). The membrane-bound metalloproteinase has a similar catalytic unit to ACE. NEP is widely distributed in endothelial, smooth muscle cells, cardiac myocytes, renal epithelial cells and fibroblasts (Erdos and Skidgel 1989; Graf et al. 1995; Dussaule et al. 1993). NEP is also found in the lung, gut, adrenal, brain, and heart. It catalyzes the degradation of vasodilator peptides, including ANP, BNP, CNP, substance P, and bradykinin as well as vasoconstrictor peptides, including endothelin-1 and angiotensin II (Stephenson and Kenny 1987; Lang et al. 1992; Kenny et al. 1993; Skidgel et al. 1984; Erdos and Skidgel 1989). Recently it was shown that the potent vasodilating and natriuretic peptide adrenomedullin is also a substrate for NEP (Lisy et al. 1998).

Selective NEP inhibitors prevent the degradation of natriuretic peptides in vitro and in vivo and increase their biological activity. In addition to degrading vasoactive peptides to inactive products, NEP is also involved in the enzymatic conversion of big endothelin to its active form, the vasoconstrictor peptide endothelin-1. Hence, the balance of NEP inhibition effects on vascular tone will depend on whether the predominant substrates degraded are vasodilators or vasoconstrictors and on the extent of NEP involvement in the processing of big endothelin-1 (Murphy et al. 1994) (Fig. 1). Indeed, in the human forearm circulation, certain NEP inhibitors cause vasoconstriction rather than vasodilatation, indicating that vasoconstrictor peptides such as angiotensin II and endothelin-1 can be substrates for NEP (Ferro et al. 1998). This explains why NEP inhibitors such as candoxatril, thiorphan and phosphoramidon increase circulating ANP concentrations in humans and induce natriuresis (Fig. 2), but do not lower (Murphy et al. 1994; Danilewicz et al. 1989; Gros et al. 1989; Schwartz et al. 1990; Northridge et al. 1989; Richards et al. 1991; Bevan et al. 1992; O'Connell et al. 1993) or even increase blood pressure in normotensive subjects (Ando et al. 1995). In essential hypertension, certain NEP inhibitors lower blood pressure (Richards et al. 1993a; Ogihara et al. 1994; Fettner et al. 1995), whereas others increase it (Bevan et al. 1992; Singer et al. 1991). Chronic treatment with NEP inhibitors augments the effects of ANP and lowers blood pressure in hypertension. However, the antihypertensive effects may be offset by an increased activity of the RAAS and sympathetic nervous system and/or by down-regulation of ANP receptors. The blood pressure response to endopeptidase inhibition in hypertension depends on the relative effects on vasodilator (including ANP) and vasoconstrictor (including the RAAS and sympathetic) systems (Richards et al. 1993b).

NEP is also involved in the metabolism of kinins. In most tissues, NEP accounts for only a small portion of the metabolism of kinins, but in human cardiac tissue, NEP accounts for nearly half of the metabolism of bradykinin (Blais

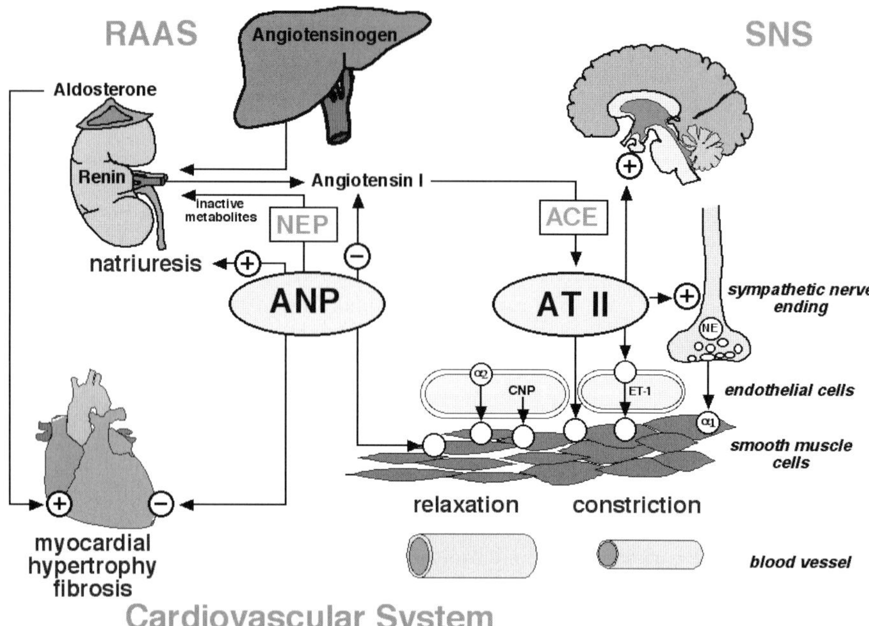

Fig. 2 Cardiovascular regulation by the sympathetic nervous system (*SNS*), the renin angiotensin aldosterone system (*RAAS*), atrial natriuretic peptide (*ANP*) and local mediators *ET-1*, endothelin; *CNP*, C-type natriuretic peptide; *AT II*, angiotensin II

et al. 2000). However, when ACE is inhibited, NEP becomes a major pathway for bradykinin metabolism. In experimental studies, the reduction of ischemia and reperfusion damage after NEP inhibition is kinin mediated (Zhang et al. 1998). Interestingly, in hypertension the selective NEP inhibitor candoxatril led only to minimal blood pressure reduction, whereas combination with an ACE inhibitor caused a marked decrease in blood pressure (Richards et al. 1993b).

Accordingly, in patients with CHF NEP inhibitors do not reduce afterload, although they do reduce pulmonary capillary wedge pressure, presumably due to their natriuretic effect and vasodilating properties (Kahn et al. 1990; Northridge et al. 1989). In moderate to severe CHF, acute NEP inhibition induces dose-dependent diuresis (Good et al. 1995), while chronic treatment does not provide this benefit. In dogs with evolving CHF, long-term NEP inhibition causes modest improvement in sodium excretion and enhances the renal response to exogenous ANP, suggesting up-regulation of NEP in CHF. Thus, the enzymatic degradation by NEP limits renal responses to increased ANP in chronic CHF independently of changes in systemic hemodynamic and augmented plasma concentrations of ANP (Margulies et al. 1995). Other possible explanations for the insufficiency of NEP inhibitors in CHF are tolerance to ANP, most likely due to down-regulation of ANP receptors and/or activation of the RAAS.

5
Combined ACE and NEP Inhibition

In many cardiovascular diseases, an array of regulatory mechanisms is involved, making drugs with multiple modes of action promising. The philosophy behind combined ACE and NEP inhibition is to simultaneously suppress a harmful neurohumoral system, the renin-angiotensin-aldosterone system, while augmenting a beneficial one (the NP system). As ACE inhibition leads to normalization of the physiological effect of ANP (Perrella et al. 1991; Margulies et al. 1991) and NEP inhibitors lower blood pressure more effectively in salt- and volume-dependent than in renin-dependent forms of hypertension (Pham et al. 1993), the combination of ACE and NEP inhibition may be especially useful in hypertension and CHF.

Indeed, in hypertension and CHF the hemodynamic and renal effect achieved after simultaneous inhibition of ACE and NEP is more pronounced than after selective inhibition (Trippodo et al. 1995b; Fink et al. 1996; Fournie-Zaluski et al. 1996; Gonzalez et al. 1996).

The synergistic effect of combined NEP and ACE inhibition is based on similar modes of action (Fig. 1), including blockade of angiotensin synthesis, simultaneous unmasking and potentiation of the effects of peptides, such as ANP BNP and bradykinin (by preventing their degradation), and in turn inducing vasodilatation, diuresis and improving myocardial function. The earliest dual metalloprotease inhibitors had limitations because of low potency, short duration of action, or limited oral bioavailability (Seymour et al. 1991; Gros et al. 1991;

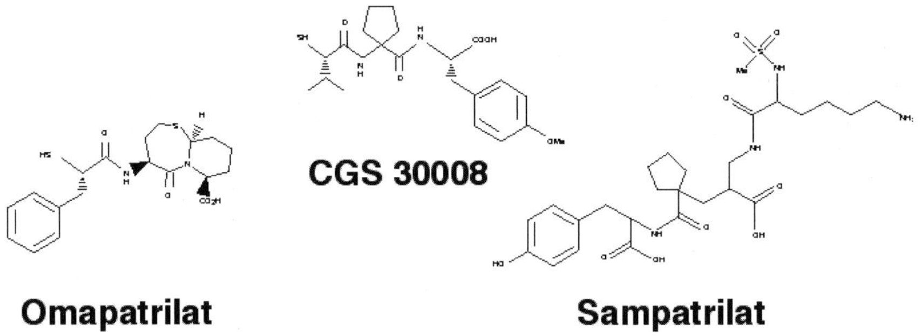

CGS 30008

Omapatrilat

Sampatrilat

Enzyme	Omapatrilat IC_{50} (nM)	CGS 30008 IC_{50} (nM)	Sampatrilat IC_{50} (nM)
Neutral endopeptidase (NEP)	8.9	2.2	20
Angiotensin-converting enzyme (ACE)	6.0	19	7

Fig. 3 Dual vasopeptidase inhibitors. In animal models, the various molecules display different selective inhibitory activity against NEP and ACE. Dual metalloproteinase inhibitors are more potent in hypertension or treatment of CHF than selective inhibition of ACE or NEP alone

Gonzalez Vera et al. 1995; Fournie-Zaluski et al. 1996). The new vasopeptidase inhibitors (Fig. 3) exhibit a long-lasting and potent effect in the cardiovascular system.

6
ACE/NEP Inhibition in Experimental Models of Hypertension

Coinhibition of ACE and NEP was expected to lower blood pressure in a broader range of conditions than inhibition of ACE or NEP alone, independent of the ac-

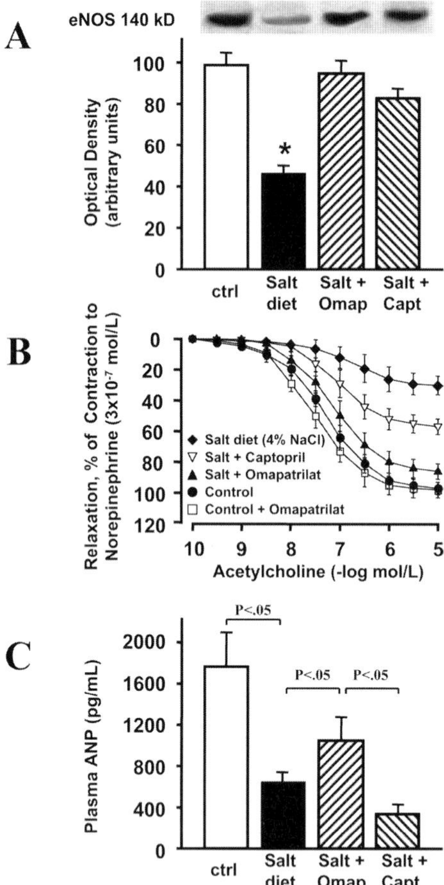

Fig. 4A–C Effects of omapatrilat or captopril on aortic eNOS expression (**A**), endothelium-dependent relaxations to acetylcholine (**B**) and plasma levels of ANP (**C**) in salt-sensitive hypertensive Dahl rats. While both drugs similarly increased eNOS, only omapatrilat elevated ANP and normalized endothelium-dependent relaxations. Results are shown as mean±SEM. *$p<0.05$ vs control rats and vs treatment groups. ‡ $p<0.05$ vs control. † $p<0.05$ vs salt-treated group. # $p<0.05$ vs salt-plus-omapatrilat (Omap) group. *Capt*, captopril

tivity of the RAAS or the degree of salt retention (Fournie-Zaluski et al. 1994; Seymour et al. 1996). Indeed, in an animal model of diabetes and hypertension treatment with a dual NEP/ACE inhibitor, lowered blood pressure and attenuated cardiac hypertrophy together with a reduction in albuminuria, featuring both angiotensin-dependent vasoconstriction and salt retention were reported (Tikkanen et al. 1998). Interestingly, omapatrilat—a potent oral dual NEP/ACE inhibitor—induced long-lasting, hypotensive effects in low-, normal-, and high-renin models of hypertension greater than those elicited by selective inhibition of either enzyme alone. Thus, combined NEP/ACE inhibition may be an effective and broad-spectrum antihypertensive principle (Trippodo et al. 1998).

Omapatrilat administered once daily to spontaneously hypertensive rats on low sodium (high-renin model of hypertension) or deoxycorticosterone acetate salt hypertensive rats (low-renin model) markedly reduced blood pressure up to 24 h (Trippodo et al. 1998). In stroke-prone spontaneously hypertensive rats, a model of malignant hypertension, chronic treatment with omapatrilat decreases systolic blood pressure, while endothelium-dependent relaxation of resistance arteries improved. Media width and media/lumen ratio decreased and lumen diameter tended to increase, while vascular stiffness was unaltered, suggesting that omapatrilat improves structure and endothelial function of resistance arteries in this model (Intengan and Schiffrin 2000). Similar observations were made in salt-induced hypertension of the rat where omapatrilat was more effective in reversing structural changes and endothelial dysfunction than captopril (D'uscio et al. 2001). In the aorta of the same model, both omapatrilat and captopril similarly increased eNOS expression, while only omapatrilat increased ANF levels and normalized endothelium-dependent relaxations to acetylcholine (Fig. 4) (Quaschning et al. 2001; D'Uscio et al. 2001).

7
ACE/NEP Inhibition in Hypertension

In normotensive patients, oral administration of omapatrilat leads to long-lasting (>24 h) and dose-dependent ACE inhibition and increases in urinary ANP levels (Liao et al. 1997; Vesterqvist et al. 1997). In a randomized, double-blind, placebo-controlled study on 36 normotensive patients, omapatrilat potently lowered blood pressure in a dose-dependent manner. The peak effect was registered in the first 3–8 h and was sustained for 24 h (Liao et al. 1999). Comparison with other antihypertensive drugs such as lisinopril, losartan and amlodipine revealed more pronounced antihypertensive effects of omapatrilat, particularly in the systolic range (Fig. 5) (Ruilope et al. 2000; Asmar et al. 2000). The pronounced effects of omapatrilat on systolic pressure are intriguing and suggest that large artery compliance and structure may be favorably affected. As systolic hypertension is difficult to treat, these new drugs may address unmet needs in the management of hypertension.

In patients with systolic hypertension, a 12-week double-blind, randomized clinical trial that compared monotherapy with the ACE inhibitor enalapril 40 mg

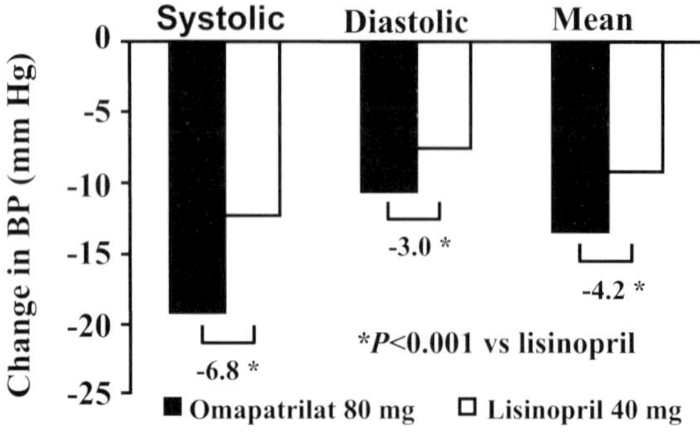

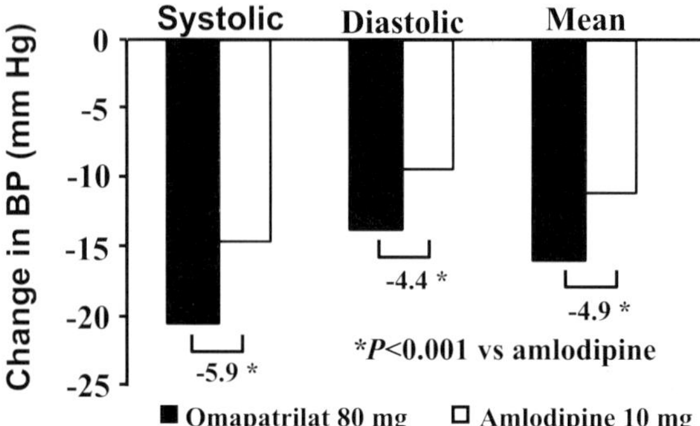

Fig. 5 Comparison of the antihypertensive effect of omapatrilat vs lisinopril or amlodipine. At 10 weeks, omapatrilat 80 mg had produced greater reductions in ambulatory blood pressure parameter than did lisinopril 40 mg (*upper panel*) or amlodipine 10 mg (*lower panel*)

daily (n=87) vs omapatrilat 80 mg daily (n=80) revealed a greater reduction in pulse pressure (peripheral pulse pressure, −8.2±12.2 mmHg vs −4.0±12.2 mmHg, and central pulse pressure −10.2±16.2 mmHg vs −3.2±16.9 mmHg) and characteristic aortic impedance in hypertensive subjects when compared with enalapril. These results suggest that aortic stiffness is maintained by specific, partially reversible mechanisms and underscore a potential role for pharmacological modulation of natriuretic peptides in the treatment of hypertension (Mitchell et al. 2002). Increased pulse pressure, an indicator of conduit vessel stiffness, is a strong independent predictor of cardiovascular events in hypertensive cohorts, which suggests that reduction of conduit vessel stiffness may be desirable in hypertension.

More recently, the OCTAVE (Omapatrilat Cardiovascular Treatment Assessment vs Enalapril) trial that enrolled 25,000 untreated or poorly controlled hypertensive patients and compared force titration of omapatrilat from 10 mg to 20 mg (with elective up-titration up to 80 mg) has been reported. The study included patients initiating therapy, replacing current therapy or adding on therapy to other antihypertensive drugs. The study was designed to allow physicians to electively increase dosage of omapatrilat or add other antihypertensive therapies as needed to control blood pressure. The study design allowed for the possibility that both treatment groups would yield equivalent blood pressure results at the end of the trial. Omapatrilat was more effective in reducing blood pressure. Despite more frequent increases in dosage and addition of other antihypertensive therapy in the enalapril group, omapatrilat resulted in consistently greater systolic blood pressure reductions (average of 3 mmHg whether alone or in combination with existing antihypertensive therapies). Consistently, the proportion of patients who reached blood pressure goals of less than 140 mmHg systolic and under 90 mmHg diastolic was about nine percentage points higher with omapatrilat than enalapril. Greater systolic blood pressure reduction was also consistently observed across a broad range of patient types, including those with diabetes, renal failure, severe hypertension, isolated systolic hypertension, and prior coronary or cerebrovascular events.

The beneficial effect of omapatrilat in reducing blood pressure was, however, counterbalanced by a higher incidence of angioedema. More than half of all cases of angioedema required no treatment or treatment with antihistamines only (1.28% with omapatrilat and 0.52% with enalapril). Severe cases, in which patients had to be treated with epinephrine or steroids, were 0.89% with omapatrilat and 0.17% with enalapril. Two cases of airway compromise occurred, both in omapatrilat-treated patients. One of these patients experienced an anaphylactic reaction that responded to treatment with epinephrine and did not require mechanical ventilation, whereas the other patient required mechanical airway protection prior to resolution. All patients with angioedema fully recovered. The overall incidence over 24 weeks was 2.17% with omapatrilat and 0.68% with enalapril. With both drugs, the risk of developing angioedema was higher in black patients (5.54% with omapatrilat and 1.62% with enalapril) than in non-black patients (1.78% with omapatrilat versus 0.55% with enalapril).

Sampatrilat, another ACE/NEP inhibitor, has been tested in hypertensive patients. Increasing dosages of sampatrilat (50, 100, 200 mg) administered for 10 days lowered clinical and ambulatory blood pressure, with a trend toward a dose response for systolic ambulatory blood pressure. Sampatrilat inhibited plasma ACE in a dose-dependent fashion but less so than lisinopril (20 mg daily) (Wallis et al. 1998). Lisinopril but not sampatrilat increased plasma renin activity, whereas sampatrilat but not lisinopril increased urinary cGMP excretion. A study performed in 58 black hypertensive subjects, known to be poorly responsive to ACE monotherapy, confirmed these results (Norton et al. 1999).

Fasidotril, the third combined ACE/NEP inhibitor reaching the clinical level has also been proven to efficiently reduce blood pressure in humans (Laurent et al. 2000).

8
ACE/NEP Inhibition in Experimental Models of Congestive Heart Failure

In cardiomyopathic hamsters with CHF, acute administration of omapatrilat reduces left ventricular systolic and end-diastolic pressure (Trippodo et al. 1995b). These changes were associated with a 40% increase in cardiac output, a 47% decrease in peripheral vascular resistance, and were significantly greater after administration of omapatrilat compared to SQ-28603 (a selective NEP inhibitor) or the ACE inhibitor enalapril (Trippodo et al. 1995a). In cardiomyopathic hamsters, chronic vasopeptidase inhibition with omapatrilat improves cardiac geometry and survival more than captopril (Trippodo et al. 1999).

In an experimental canine model of CHF, omapatrilat was superior to ACE inhibition alone in inducing an increase in sodium excretion and glomerular filtration rate in addition to a greater decrease in pulmonary capillary wedge pressure (Chen et al. 2001). Most importantly, these cardiorenal actions were markedly attenuated by a natriuretic peptide receptor antagonist, underscoring that the endogenous natriuretic peptides in part mediate the actions of omapatrilat.

9
ACE/NEP Inhibition in Congestive Heart Failure

In 48 patients with CHF (NYHA Class II-IV), treatment with omapatrilat for 3 months reduced afterload, improved cardiac function and in turn clinical status. Ejection fraction increased from 24% to 28%, while myocardial wall stress and heart rate decreased. Moreover, natriuresis increased and norepinephrine levels decreased (McClean et al. 2000). In a randomized, double-blind study in 369 patients with CHF, omapatrilat decreased blood pressure in a dose-dependent manner, increased left ventricular function and reduced pulmonary capillary wedge pressure. Plasma BNP, an important prognostic factor if increased, was lower after 12 weeks of treatment with 40 mg daily and was reflected by a reduced incidence of death and hospitalization for CHF (Ikram et al. 1999).

In the IMPRESS (Inhibition of Metalloproteinase in a Randomized Exercise and Symptoms Study in Heart Failure) trial, 573 patients with CHF (63% NYHA II and 37% NYHA III/IV) were randomized to either omapatrilat (40 mg daily) or lisinopril (20 mg daily) (Rouleau et al. 2000). After 12 weeks exercise, tolerance similarly increased in both groups. However, omapatrilat led to a better clinical status and lower incidence of the combined mortality/morbidity endpoint (i.e., hospitalization and discontinuation of study medication for worsening heart failure) compared to lisinopril (Fig. 6). Both drugs were well tolerated, but serious adverse events and marked elevations of creatinine were less frequent with omapatrilat. Omapatrilat increased ANP and resulted in lower plasma norepinephrine levels than lisinopril. It also did not increase endothelin-1 levels, suggesting that the combined effects of ACE inhibition and NEP inhibition prevented the rise of endothelin-1 in this setting. In the postinfarction rat

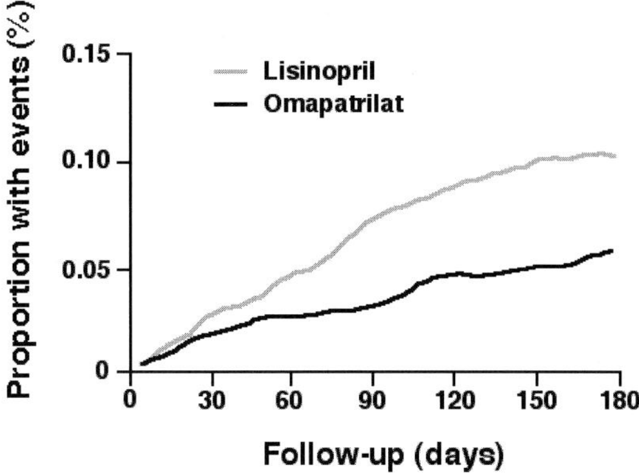

Fig. 6 In the IMPRESS study, omapatrilat 80 mg in CHF was more effective on composite end-points (i.e., death, admission or discontinuation of the treatment because of worsening CHF) than lisinopril 20 mg

model omapatrilat prevented the expected rise in endothelin-1 (unpublished data). In the IMPRESS study, omapatrilat also had a positive influence on conduit vessel stiffness as compared with lisinopril, reducing pulsatile load on the heart without compromising a potentially tenuous mean arterial pressure (Mitchell et al. 1999).

More recently, omapatrilat was compared with enalapril in patients with CHF in the OVERTURE trial (Omapatrilat Versus Enalapril Randomized Trial of Utility in Reducing Events). This study enrolled 5,770 patients in NYHA class II–IV, who were randomly assigned to double-blind treatment with either enalapril (10 mg b.i.d., n=2,884) or omapatrilat (40 mg once daily, n=2,886) for a mean of 14.5 months. The primary end-point—the combined risk of death or hospitalization for heart failure requiring intravenous treatment—was used prospectively to test both a superiority and noninferiority hypothesis (based on the effect of enalapril in the Study of Left Ventricular Dysfunction (SOLVD) Treatment Trial. A primary end-point was achieved in 973 patients in the enalapril group and in 914 patients in the omapatrilat group, fulfilling the criteria for noninferiority but not for superiority. The omapatrilat group had a 9% lower risk of cardiovascular death or hospitalization (p=0.024) and a 6% lower risk of death (p=0.34). Post hoc analysis of the primary end-point with the definition used in the SOLVD Treatment Trial (which included all hospitalization for HF) showed an 11% lower risk in patients treated with omapatrilat. Angioedema was reported in 24 (0.8%) omapatrilat-treated and 14 (0.5%) enalapril-treated patients, of whom three (two on enalapril and one on omapatrilat) required hospitalization for management: none required intubation or died. In a similar number of patients, the medication was withdrawn because of an adverse event (in 17.9% of omapatrilat group and in 17.0% of the enalapril group). Thus, in this large trial, omapatrilat reduced morbidity and mortality of patients with moderate to severe CHF but was not more effective than ACE inhibition alone in decreasing the risk of a primary clinical event.

10 Vasopeptidase Inhibition and Angioedema

Angioedema is a serious and potentially fatal complication of ACE inhibitors, which is relatively rare in the general population but more common amongst blacks and Afro-Caribbeans (Gibbs et al. 1999). ACE inhibitor-induced angioedema occurs with an incidence of 0.1%–0.5% (Warner et al. 2000). Less than 20% of angioedema attacks are potentially life-threatening (i.e., affecting the larynx or the upper respiratory tract), but may reach 50% among patients with hereditary angioedema (Agostoni et al. 1999). Surprisingly, angiotensin II receptor antagonists also have an increased risk of angioedema, particularly in patients with a prior episode of angioedema attributed to ACE inhibitors (Warner et al. 2000).

Symptoms of angioedema range from mild gastrointestinal disturbance (i.e., colic, nausea, vomiting, and diarrhea) to severe dyspnea due to larynx edema. The mechanism remains unclear, but bradykinin and its metabolite des-arg^9-bradykinin have been implicated in ACE-induced angioedema (Blais et al. 1999). Plasma bradykinin concentrations can rise more than tenfold during acute attacks of angioedema associated with ACE inhibitor therapy (Nussberger et al. 1998). Recently an enzyme defect involved in the des-arg9-bradykinin metabolism, aminopeptidase P, leading to bradykinin and to an even greater extent to des-arg^9-bradykinin accumulation, was reported (Blais et al. 1999).

Vasopeptidase inhibitors acting simultaneously on two enzymes that inactivate bradykinin, i.e., ACE and NEP, may increase the risk of angioedema. Company statements refer to a rate of angioedema associated with omapatrilat similar to that reported for ACE inhibitors. However, in data submitted to the New Drug Application, the incidence of angioedema was more than three times as common when the starting dose was 20 mg or more than it was with lower doses, which suggests a pharmacodynamic rather than allergic effect (Messerli and Nussberger 2000). In this report, 44 instances of angioedema occurred among more than 6,000 patients, and four cases were severe enough to require intubation. A possible explanation for the relatively high incidence could be that the omapatrilat trial program included significant numbers of Afro-Americans known to have a higher rate of angioedema compared to Caucasians. Recent clinical trials that were supposed to clarify this important issue have given controversial results. The OCTAVE trial investigated in 25,000 untreated or poorly controlled hypertensive cases and showed that force titration of omapatrilat from 10 mg to 20 mg (with elective up-titration up to 80 mg) is associated with a higher incidence of angioedema than enalapril. On the other hand, the OVERTURE trial, enrolling 5,770 patients with moderate to severe CHF showed that enalapril (10 mg b.i.d.) and omapatrilat (40 mg once daily) revealed similar incidence and severity of angioedema (0.5% and 0.8%, respectively). The discrepancy between these two trials could be explained through a possible resistance to the ability of bradykinin to produce cutaneous exudation in patients with CHF.

11 Conclusion

Vasopeptidase inhibitors are a promising new class of drugs. Additional adequately powered mortality/morbidity trials with combined ACE/NEP inhibitors in CHF and in isolated systolic hypertension (OPERA, The Omapatrilat in Persons with Enhanced Risk of Atherosclerotic events trial) are still underway. The role of ACE/NEP inhibitors in atherosclerosis is also to be defined. These trials will provide the definite answer whether this class of drugs confers clinical advantage over ACE inhibition alone, particularly in view of the reports of increased rate of angioedema. These side effects are of great clinical concern and question whether the encouraging experimental results obtained with vasopeptidase inhibitors will translate into long-term clinical benefit.

Acknowledgements. Original research of the authors was supported by grants of the Swiss National Research Foundation (3200–065447.01 to FR and 32–51069.97/1 to TFL).

References

Agostoni A, Cicardi M, Cugno M, Zingale LC Gioffre D, Nussberger J (1999) Angioedema due to angiotensin-converting enzyme inhibitors. Immunopharmacology 44:21–25

AIRE (1993) Effect of ramipril on mortality and morbidity of survivors of acute myocardial infarction with clinical evidence of heart failure. The Acute Infarction Ramipril Efficacy (AIRE) Study Investigators. Lancet 342:821–828

Ambrosioni E, Borghi C, Magnani B (1995) The effect of the angiotensin-converting-enzyme inhibitor zofenopril on mortality and morbidity after anterior myocardial infarction. The Survival of Myocardial Infarction Long-Term Evaluation (SMILE) Study Investigators [see comments]. N Engl J Med 332:80–85

Ando S, Rahman MA, Butler GC, Senn BL, Floras JS (1995) Comparison of candoxatril and atrial natriuretic factor in healthy men. Effects on hemodynamics, sympathetic activity, heart rate variability, and endothelin. Hypertension 26:1160–1166

Asmar R, Fredebohm WIS, Chang PI, Gressin V, Saini RK (2000) Omapatrilat compared with lisinopril in treatment of hypertension as assessed by ambulatory blood pressure monitoring. Am J Hypertens 13:S143

Auch-Schwelk W, Bossaller C, Claus M, Graf K, Grafe M, Fleck E (1992) Local potentiation of bradykinin-induced vasodilation by converting-enzyme inhibition in isolated coronary arteries. J Cardiovasc Pharmacol 20 [Suppl 9]:S62–S67

Azevedo ER, Newton GE, Parker AB, Floras JS, Parker JD (2000) Sympathetic responses to atrial natriuretic peptide in patients with congestive heart failure. J Cardiovasc Pharmacol 35:129–135

Bayliss J, Canepa-Anson R, Norell M, Poole-Wilson P, Sutton G (1986) The renal response to neuroendocrine inhibition in chronic heart failure: double-blind comparison of captopril and prazosin. Eur Heart J 7:877–884

Berger R, Huelsman M, Strecker K, Bojic A, Moser P, Stanek B, Pacher R (2002) B-type natriuretic peptide predicts sudden death in patients with chronic heart failure. Circulation 105:2392–2397

Bevan EG, Connell JM, Doyle J, Carmichael HA, Davies DL, Lorimer AR, McInnes GT (1992) Candoxatril, a neutral endopeptidase inhibitor: efficacy and tolerability in essential hypertension. J Hypertens 10:607–613

Blais C Jr, Drapeau G, Raymond P, Lamontagne D, Gervais N, Venneman I, Adam A (1997) Contribution of angiotensin-converting enzyme to the cardiac metabolism of bradykinin: an interspecies study. Am J Physiol 273:H2263–H2271

Blais CJ, Rouleau JL, Brown NJ, Lepage Y, Spence D, Munoz C, Friborg J, Geadah D, Gervais N, Adam A (1999) Serum metabolism of bradykinin and des-Arg9-bradykinin in patients with angiotensin-converting enzyme inhibitor-associated angioedema. Immunopharmacology 43:293–302

Blais C Jr, Fortin D, Rouleau JL, Molinaro G, Adam A (2000) Protective effect of omapatrilat, a vasopeptidase inhibitor, on the metabolism of bradykinin in normal and failing human hearts. J Pharmacol Exp Ther 295:621–626

Blais C Jr, Marceau F, Rouleau J et al (2001) The kallikrein-kininogen-kinin system: lessons from the quantification of endogenous kinins. Peptides 21:1903–1940

Bossaller C, Auch-Schwelk W, Weber F, Gotze S, Grafe M, Graf K, Fleck E (1992) Endothelium-dependent relaxations are augmented in rats chronically treated with the angiotensin-converting enzyme inhibitor enalapril. J Cardiovasc Pharmacol 20 [Suppl 9]:S91–S95

Burnett JC, Granger JP, Opgenorth TJ (1984) Effects of synthetic atrial natriuretic factor on renal function and renin release. Am J Physiol 247:F863–F866

Burnett JC Jr, Kao PC, Hu DC, Heser DW, Heublein D, Granger JP, Opgenorth TJ, Reeder GS (1986) Atrial natriuretic peptide elevation in congestive heart failure in the human. Science 231:1145–1147

Chen HH, Lainchbury JG, Matsuda Y, Harty GJ, Burnett JC (2001) Endogenous natriuretic peptides participate in the renal and humoral actions of acute vasopeptidase inhibition in experimental mild heart failure. Hypertension 38:187–191

Cheung BM, Lau CP (1999) Fosinopril reduces left ventricular mass in untreated hypertensive patients: a controlled trial. Br J Clin Pharmacol 47:179–187

Colucci WS, Elkayam U, Horton DP, Abraham WT, Bourge RC, Johnson AD, Wagoner LE, Givertz MM, Liang CS, Neibaur M, Haught WH, LeJemtel TH (2000) Intravenous nesiritide, a natriuretic peptide, in the treatment of decompensated congestive heart failure. Nesiritide Study Group. N Engl J Med 343:246–253

Creager MA Roddy MA (1994) Effect of captopril and enalapril on endothelial function in hypertensive patients. Hypertension 24:499–505

D'Uscio LV, Quaschning T, Burnett JCJ, Lüscher TF (2001) Vasopeptidase inhibition prevents endothelial dysfunction of resistance arteries in salt-sensitive hypertension in comparison with single ACE inhibition. Hypertension 37:28–33

Danilewicz JC, Barclay PL, Barnish IT, Brown D, Campbell SF, James K, Samuels GM, Terrett NK, Wythes MJ (1989) UK-69,578, a novel inhibitor of EC 3.4.24.11 which increases endogenous ANF levels and is natriuretic and diuretic. Biochem Biophys Res Commun 164:58–65

Dohi Y, Criscione L, Pfeiffer K, Luscher TF (1994) Angiotensin blockade or calcium antagonists improve endothelial dysfunction in hypertension: studies in perfused mesenteric resistance arteries. J Cardiovasc Pharmacol 24:372–329

Dussaule JC, Stefanski A, Bea ML, Ronço P, Ardaillou R (1993) Characterization of neutral endopeptidase in vascular smooth muscle cells of rabbit renal cortex. Am J Physiol 264:F45–F52

Erdos EG, Skidgel RA (1989) Neutral endopeptidase 24.11 (enkephalinase) and related regulators of peptide hormones. FASEB J 3:145–151

Ferro CJ, Spratt JC, Haynes WG, Webb DJ (1998) Inhibition of neutral endopeptidase causes vasoconstriction of human resistance vessels in vivo. Circulation 97:2323–2330

Fettner SH, Pai S, Zhu GR, Kosoglou T, Banfield CR, Batra V (1995) Pharmacokinetic-pharmacodynamic (PK-PD) modeling for a new antihypertensive agent (neutral metalloendopeptidase inhibitor SCH 42354) in patients with mild to moderate hypertension. Eur J Clin Pharmacol 48:351–359

Fink CA, Carlson JE, McTaggart PA, Qiao Y, Webb R, Chatelain R, Jeng AY, Trapani AJ (1996) Mercaptoacyl dipeptides as orally active dual inhibitors of angiotensin-converting enzyme and neutral endopeptidase. J Med Chem 39:3158–3168

Fournie-Zaluski MC, Gonzalez W, Turcaud S, Pham I, Roques BP, Michel JB (1994) Dual inhibition of angiotensin-converting enzyme and neutral endopeptidase by the orally active inhibitor mixanpril: a potential therapeutic approach in hypertension. Proc Natl Acad Sci USA 91:4072–4076

Fournie-Zaluski MC, Coric P, Thery V, Gonzalez W, Meudal H, Turcaud S, Michel JB, Roques BP (1996) Design of orally active dual inhibitors of neutral endopeptidase and angiotensin-converting enzyme with long duration of action. J Med Chem 39:2594–2608

Gibbs CR, Lip GY, Beevers DG (1999) Angioedema due to ACE inhibitors: increased risk in patients of African origin. Br J Clin Pharmacol 48:861–865

GISSI-3 (1994) Effects of lisinopril and transdermal glyceryl trinitrate singly and together on 6-week mortality and ventricular function after acute myocardial infarction. Gruppo Italiano per lo Studio della Sopravvivenza nell'infarto Miocardico. Lancet 343:1115–1122

Gonzalez Vera W, Fournie-Zaluski MC, Pham I, Laboulandine I, Roques BP, Michel JB (1995) Hypotensive and natriuretic effects of RB 105, a new dual inhibitor of angiotensin converting enzyme and neutral endopeptidase in hypertensive rats. J Pharmacol Exp Ther 272:343–351

Gonzalez W, Beslot F, Laboulandine I, Fournie-Zaluski MC, Roques BP, Michel JB (1996) Inhibition of both angiotensin-converting enzyme and neutral endopeptidase by S21402 (RB105) in rats with experimental myocardial infarction. J Pharmacol Exp Ther 278:573–581

Good JM, Peters M, Wilkins M, Jackson N, Oakley CM, Cleland JG (1995) Renal response to candoxatrilat in patients with heart failure. J Am Coll Cardiol 25:1273–1281

Graf K, Koehne P, Grafe M, Zhang M, Auch-Schwelk W, Fleck E (1995) Regulation and differential expression of neutral endopeptidase 24.11 in human endothelial cells. Hypertension 26:230–235

Gros C, Souque A, Schwartz JC, Duchier J, Cournot A, Baumer P, Lecomte JM (1989) Protection of atrial natriuretic factor against degradation: diuretic and natriuretic responses after in vivo inhibition of enkephalinase (EC 3.4.24.11) by acetorphan. Proc Natl Acad Sci USA 86:7580–7584

Gros C, Noel N, Souque A, Schwartz JC, Danvy D, Plaquevent JC, Duhamel L, Duhamel P, Lecomte JM, Bralet J (1991) Mixed inhibitors of angiotensin-converting enzyme (EC 3.4.15.1) and enkephalinase (EC 3.4.24.11): rational design, properties, and potential cardiovascular applications of glycopril and alatriopril. Proc Natl Acad Sci USA 88:4210–4214

Haefeli WE, Linder L, Luscher TF (1997) Quinaprilat induces arterial vasodilation mediated by nitric oxide in humans. Hypertension 30:912–917

Hansson L, Lindholm LH, Niskanen L, Lanke J, Hedner T, Niklason A, Luomanmaki K, Dahlof B, de Faire U, Morlin C, Karlberg BE, Wester PO, Bjorck JE (1999) Effect of angiotensin-converting-enzyme inhibition compared with conventional therapy on cardiovascular morbidity and mortality in hypertension: the Captopril Prevention Project (CAPPP) randomised trial [see comments]. Lancet 353:611–616

Hornig B, Arakawa N, Haussmann D, Drexler H (1998) Differential effects of quinaprilat and enalaprilat on endothelial function of conduit arteries in patients with chronic heart failure. Circulation 98:2842–2848

Ikram H, McClean DR, Mehta S, Heywood JT, Rousseau MF, Niederman AL, Sequeira RF, Fleck E, Singh SN, Coutu B, Brand R, Komajda M, Bryson CC, Qian C, Hanyok JJ (1999) Long-term beneficial hemodynamic and neurohormonal effects of vasopeptidase inhibition with omapatrilat in heart failure. J Am Coll Cardiol 33:185A

Intengan HD, Schiffrin EL (2000) Vasopeptidase inhibition has potent effects on blood pressure and resistance arteries in stroke-prone spontaneously hypertensive rats. Hypertension 35:1221–1225

Janssen WM, de Zeeuw D, van der Hem GK, de Jong PE (1989) Antihypertensive effect of a 5-day infusion of atrial natriuretic factor in humans. Hypertension 13:640–646

Jaspard E, Wei L, Alhenc-Gelas F (1993) Differences in the properties and enzymatic specificities of the two active sites of angiotensin I-converting enzyme (kininase II). Studies with bradykinin and other natural peptides. J Biol Chem 268:9496–9503

Johnston CI, Hodsman PG, Kohzuki M, Casley DJ, Fabris B, Phillips PA (1989) Interaction between atrial natriuretic peptide and the renin angiotensin aldosterone system. Endogenous antagonists. Am J Med 87:24S–28S

Kahn JC, Patey M, Dubois-Rande JL, Merlet P, Castaigne A, Lim-Alexandre C, Lecomte JM, Duboc D, Gros C, Schwartz JC (1990) Effect of sinorphan on plasma atrial natriuretic factor in congestive heart failure [letter]. Lancet 335:118–119

Kahonen M, Makynen H, Wu X, Arvola P, Porsti I (1995) Endothelial function in spontaneously hypertensive rats: influence of quinapril treatment. Br J Pharmacol 115:859–867

Kenny AJ, Bourne A, Ingram J (1993) Hydrolysis of human and pig brain natriuretic peptides, urodilatin, C-type natriuretic peptide and some C-receptor ligands by endopeptidase-24.11. Biochem J 291:83–88

Kiowski W, Linder L, Nuesch R, Martina B (1996) Effects of cilazapril on vascular structure and function in essential hypertension. Hypertension 27:371–376

Kober L, Torp-Pedersen C, Carlsen JE, Bagger H, Eliasen P, Lyngborg K, Videbaek J, Cole DS, Auclert L, Pauly NC (1995) A clinical trial of the angiotensin-converting-enzyme inhibitor trandolapril in patients with left ventricular dysfunction after myocardial infarction. Trandolapril Cardiac Evaluation (TRACE) Study Group [see comments]. N Engl J Med 333:1670–1676

Kostis JB, Cobbe S, Johnston C, Ford I, Murphy M, Weber MA, Black HR, Plouin PF, Levy D, Mancia G, Larochelle P, Kolloch RE, Alderman M, Ruilope LM, Dahlof B, Flack JM, Wolf R (2002) Design of the Omapatrilat in Persons with Enhanced Risk of Atherosclerotic events (OPERA) trial. Am J Hypertens 15:193–198

Kung CF, Moreau P, Takase H, Luscher TF (1995) L-NAME hypertension alters endothelial and smooth muscle function in rat aorta. Prevention by trandolapril and verapamil. Hypertension 26:744–751

Lang CC, Motwani JG, Coutie WJ, Struthers AD (1992) Clearance of brain natriuretic peptide in patients with chronic heart failure: indirect evidence for a neutral endopeptidase mechanism but against an atrial natriuretic peptide clearance receptor mechanism. Clin Sci (Colch) 82:619–623

Lang CC, Prasad N, McAlpine HM, Macleod C, Lipworth BJ, MacDonald TM, Struthers AD (1994) Increased plasma levels of brain natriuretic peptide in patients with isolated diastolic dysfunction. Am Heart J 127:1635–1636

Laurent S, Boutouyrie P, Azizi M et al (2000) Antihypertensive effects of fasidotril, a dual inhibitor of neprilysin and angiotensin-converting enzyme, in rats and humans. Hypertension 35:1148–1153

Liao W, Delaney C, Smith R (1997) Supine mean arterial blood pressure (MAP) lowering and oral tolerance of BMS-186716, a new dual metalloprotease inhibitor of angiotensin converting enzyme (ACE) and neutral endopeptidase (NEP), in healthy male subjects (abstract). Clin Pharmacol Ther 61:229

Liao W, Delaney CL, Beierle FA, Ferreira IM, Davis KD, Meier A, Vesterqvist O, Ford NF, Uderman HD (1999) Omapatrilat, a novel vasopeptidase inhibitor, reduces blood pressure in normotensive subjects. J Am Coll Cardiol 33:309A

Linz W, Wiemer G, Gohlke P, Unger T, Scholkens BA (1995) Contribution of kinins to the cardiovascular actions of angiotensin-converting enzyme inhibitors. Pharmacol Rev 47:25–49

Lisy O, Jougasaki M, Schirger JA, Chen HH, Barclay PT, Burnett JC (1998) Neutral endopeptidase inhibition potentiates the natriuretic actions of adrenomedullin. Am J Physiol 275:F410–F414

Maisel AS, Krishnaswamy P, Nowak RM, McCord J, Hollander JE, Duc P, Omland T, Storrow AB, Abraham WT, Wu AH, Clopton P, Steg PG, Westheim A, Knudsen CW, Perez A, Kazanegra R, Herrmann HC, McCullough PA (2002) Rapid measurement of B-type natriuretic peptide in the emergency diagnosis of heart failure. N Engl J Med 347:161–167

Mancini GB, Henry GC, Macaya C, O'Neill BJ, Pucillo AL, Carere RG, Wargovich TJ, Mudra H, Luscher TF, Klibaner MI, Haber HE, Uprichard AC, Pepine CJ, Pitt B (1996) Angiotensin-converting enzyme inhibition with quinapril improves endothelial vasomotor dysfunction in patients with coronary artery disease. The TREND (Trial on Reversing ENdothelial Dysfunction) Study. Circulation 94:258–265

Margulies KB, Perrella MA, McKinley LJ, Burnett JC Jr (1991) Angiotensin inhibition potentiates the renal responses to neutral endopeptidase inhibition in dogs with congestive heart failure. J Clin Invest 88:1636–1642

Margulies KB, Barclay PL, Burnett JC Jr (1995) The role of neutral endopeptidase in dogs with evolving congestive heart failure. Circulation 91:2036–2042

McClean DR, Ikram H, Garlick AH, Richards AM, Nicholls MG, Crozier IG (2000) The clinical, cardiac, renal, arterial and neurohormonal effects of omapatrilat, a vasopeptidase inhibitor, in patients with chronic heart failure. J Am Coll Cardiol 36:479–486

McCullough PA, Nowak RM, McCord J, Hollander JE, Herrmann HC, Steg PG, Duc P, Westheim A, Omland T, Knudsen CW, Storrow AB, Abraham WT, Lamba S, Wu AH, Perez A, Clopton P, Krishnaswamy P, Kazanegra R, Maisel AS (2002) B-type natriuretic peptide and clinical judgment in emergency diagnosis of heart failure: analysis from Breathing Not Properly (BNP) Multinational Study. Circulation 106:416–422

Messerli FH, Nussberger J (2000) Vasopeptidase inhibition and angio-oedema. Lancet 356:608–609

Mimran A, Ribstein J, DuCailar G (1995) Contrasting effect of antihypertensive treatment on the renal response to L-arginine. Hypertension 26:937–941

Mitchell GF, Block AJ, Hartley LH, Tardif JC, Rouleau J-L, Pfeffer MA (1999) The vasopeptidase inhibitor, omapatrilat, has a favorable, pressure-independent effect on conduit vessel stiffness in patients with congestive heart failure. Circulation 100:I-646

Mitchell GF, Izzo JL Jr, Lacourciere Y, Ouellet JP, Neutel J, Qian C, Kerwin LJ, Block AJ, Pfeffer MA (2002) Omapatrilat reduces pulse pressure and proximal aortic stiffness in patients with systolic hypertension: results of the conduit hemodynamics of omapatrilat international research study. Circulation 105:2955–2961

Mombouli JV, Illiano S, Vanhoutte PM (1992) Local production of kinins contributes to the endothelium dependent relaxations evoked by converting enzyme inhibitors in isolated arteries. Agents Actions Suppl 38:186–195

Murphy LJ, Corder R, Mallet AI, Turner AJ (1994) Generation by the phosphoramidon-sensitive peptidases, endopeptidase-24.11 and thermolysin, of endothelin-1 and c-terminal fragment from big endothelin-1. Br J Pharmacol 113:137–142

Northridge DB, Jardine AG, Alabaster CT, Barclay PL, Connell JM, Dargie HJ, Dilly SG, Findlay IN, Lever AF, Samuels GM (1989) Effects of UK 69 578: a novel atriopeptidase inhibitor. Lancet 2:591–593

Norton GR, Woodiwiss AJ, Hartford C, Trifunovic B, Middlemost S, Lee A, Allen MJ (1999) Sustained antihypertensive actions of a dual angiotensin-converting enzyme neutral endopeptidase inhibitor, sampatrilat, in black hypertensive subjects. Am J Hypertens 12:563–571

Nussberger J, Cugno M, Amstutz C, Cicardi M, Pellacani A, Agostoni A (1998) Plasma bradykinin in angio-oedema. Lancet 351:1693–1697

O'Connell JE, Jardine AG, Davies DL, McQueen J, Connell JM (1993) Renal and hormonal effects of chronic inhibition of neutral endopeptidase (EC 3.4.24.11) in normal man. Clin Sci (Colch) 85:19–26

Ogihara T, Rakugi H, Masuo K, Yu H, Nagano M, Mikami H (1994) Antihypertensive effects of the neutral endopeptidase inhibitor SCH 42495 in essential hypertension. Am J Hypertens 7:943–947

Palmer RM, Ferrige AG, Moncada S (1987) Nitric oxide release accounts for the biological activity of endothelium-derived relaxing factor. Nature 327:524–526

Perrella MA, Margulies KB, Burnett JC Jr (1991) Pathophysiology of congestive heart failure: role of atrial natriuretic factor and therapeutic implications. Can J Physiol Pharmacol 69:1576–1581

Pfeffer MA, Braunwald E, Moye LA, Basta L, Brown EJ Jr, Cuddy TE, Davis BR, Geltman EM, Goldman S, Flaker GC et al (1992) Effect of captopril on mortality and morbidity in patients with left ventricular dysfunction after myocardial infarction. Results of the survival and ventricular enlargement trial. The SAVE, Investigators [see comments]. N Engl J Med 327:669–677

Pham I, Gonzalez W, el Amrani AI, Fournie-Zaluski MC, Philippe M, Laboulandine I, Roques BP, Michel JB (1993) Effects of converting enzyme inhibitor and neutral endopeptidase inhibitor on blood pressure and renal function in experimental hypertension. J Pharmacol Exp Ther 265:1339–1347

Qi XL, Stewart DJ, Gosselin H, Azad A, Picard P, Andries L, Sys SU, Brutsaert DL, Rouleau JL (1999) Improvement of endocardial and vascular endothelial function on myocardial performance by captopril treatment in postinfarct rat hearts. Circulation 100:1338–1345

Quaschning T, d'Uscio LV, Shaw S, Luscher TF (2001) Vasopeptidase inhibition exhibits endothelial protection in salt-induced hypertension. Hypertension 37:1108–1113

Remes J, Tikkanen I, Fyhrquist F, Pyorala K (1991) Neuroendocrine activity in untreated heart failure. Br Heart J 65:249–255

Richards AM, Wittert G, Espiner EA, Yandle TG, Frampton C, Ikram H (1991) Prolonged inhibition of endopeptidase 24.11 in normal man: renal, endocrine and haemodynamic effects. J Hypertens 9:955–962

Richards AM, Crozier IG, Espiner EA, Yandle TG, Nicholls MG (1993a) Plasma brain natriuretic peptide and endopeptidase 24.11 inhibition in hypertension. Hypertension 22:231–236

Richards AM, Wittert GA, Crozier IG, Espiner EA, Yandle TG, Ikram H, Frampton C (1993b) Chronic inhibition of endopeptidase 24.11 in essential hypertension: evidence for enhanced atrial natriuretic peptide and angiotensin II. J Hypertens 11:407–416

Richardson A, Bayliss J, Scriven AJ, Parameshwar J, Poole-Wilson PA, Sutton GC (1987) Double-blind comparison of captopril alone against frusemide plus amiloride in mild heart failure. Lancet 2:709–711

Rouleau JL, Pfeffer MA, Stewart DJ, Isaac D, Sestier F, Kerut EK, Porter CB, Proulx G, Qian C, Block AJ (2000) Comparison of vasopeptidase inhibitor, omapatrilat, and lisinopril on exercise tolerance and morbidity in patients with heart failure: IMPRESS randomised trial. Lancet 356:615–620

Ruilope LM, Palatini P, Grossman E, Esposti D, Speier U, Chang PI, Gressin V, Plat F (2000) Randomized double-blind comparison of omapatrilat with amlodipine in mild-to-moderate hypertension. Am J Hypertens 13:S134–S135

Sabatine MS, Morrow DA, de Lemos JA, Gibson CM, Murphy SA, Rifai N, McCabe C, Antman EM, Cannon CP, Braunwald E (2002) Multimarker approach to risk stratification in non-ST elevation acute coronary syndromes: simultaneous assessment of troponin I, C-reactive protein, and B-type natriuretic peptide. Circulation 105:1760–1763

Schiffrin EL, Deng LY (1995) Comparison of effects of angiotensin I-converting enzyme inhibition and beta-blockade for 2 years on function of small arteries from hypertensive patients. Hypertension 25:699–703

Schlaifer JD, Wargovich TJ, O'Neill B, Mancini GB, Haber HE, Pitt B, Pepine CJ (1997) Effects of quinapril on coronary blood flow in coronary artery disease patients with endothelial dysfunction. TREND Investigators. Trial on Reversing Endothelial Dysfunction. Am J Cardiol 80:1594–1597

Schölkens BA (1996) Kinins in the cardiovascular system. Immunopharmacology 33:209–216

Schwartz JC, Gros C, Lecomte JM, Bralet J (1990) Enkephalinase (EC 3.4.24.11) inhibitors: protection of endogenous ANF against inactivation and potential therapeutic applications. Life Sci 47:1279–1297

Seymour AA, Abboa-Offei BE, Smith PL, Mathers PD, Asaad MM, Rogers WL (1995) Potentiation of natriuretic peptides by neutral endopeptidase inhibitors. Clin Exp Pharmacol Physiol 22:63–69

Seymour AA, Asaad MM, Abboa-Offei BE, Smith PL, Rogers WL, Dorso CR (1996) In vivo pharmacology of dual neutral endopeptidase/angiotensin-converting enzyme inhibitors. J Cardiovasc Pharmacol 28:672–678

Seymour AA, Swerdel JN, Abboa-Offei B (1991) Antihypertensive activity during inhibition of neutral endopeptidase and angiotensin converting enzyme. J Cardiovasc Pharmacol 17:456–465

Singer DR, Markandu ND, Buckley MG, Miller MA, Sagnella GA, MacGregor GA (1991) Dietary sodium and inhibition of neutral endopeptidase 24.11 in essential hypertension. Hypertension 18:798–804

Skidgel RA, Engelbrecht S, Johnson AR, Erdos EG (1984) Hydrolysis of substance P and neurotensin by converting enzyme and neutral endopeptidase. Peptides 5:769–776

SOLVD (1991) Effect of enalapril on survival in patients with reduced left ventricular ejection fractions and congestive heart failure. N Engl J Med 325:293–302

Stephenson SL, Kenny AJ (1987) The hydrolysis of alpha-human atrial natriuretic peptide by pig kidney microvillar membranes is initiated by endopeptidase-24.11. Biochem J 243:183–187

Struthers AD (1994) Ten years of natriuretic peptide research: a new dawn for their diagnostic and therapeutic use? BMJ 308:1615–1619

Swedberg K, Held P, Kjekshus J, Rasmussen K, Ryden L, Wedel H (1992) Effects of the early administration of enalapril on mortality in patients with acute myocardial infarction. Results of the Cooperative New Scandinavian Enalapril Survival Study II (CONSENSUS II) [see comments]. N Engl J Med 327:678–684

Taddei S, Virdis A, Ghiadoni L, Mattei P, Salvetti A (1998) Effects of angiotensin converting enzyme inhibition on endothelium-dependent vasodilatation in essential hypertensive patients. J Hypertens 16:447–456

Takase H, Moreau P, Kung CF, Nava E, Luscher TF (1996) Antihypertensive therapy prevents endothelial dysfunction in chronic nitric oxide deficiency. Effect of verapamil and trandolapril. Hypertension 27:25–31

Tikkanen T, Tikkanen I, Rockell MD, Allen TJ, Johnston CI, Cooper ME, Burrell LM (1998) Dual inhibition of neutral endopeptidase and angiotensin-converting enzyme in rats with hypertension and diabetes mellitus. Hypertension 32:778–785

Trippodo NC, Panchal BC, Fox M (1995a) Repression of angiotensin II and potentiation of bradykinin contribute to the synergistic effects of dual metalloprotease inhibition in heart failure. J Pharmacol Exp Ther 272:619–627

Trippodo NC, Robl JA, Asaad MM, Bird JE, Panchal BC, Schaeffer TR, Fox M, Giancarli MR, Cheung HS (1995b) Cardiovascular effects of the novel dual inhibitor of neutral endopeptidase and angiotensin-converting enzyme BMS-182657 in experimental hypertension and heart failure. J Pharmacol Exp Ther 275:745–752

Trippodo NC, Robl JA, Asaad MM, Fox M, Panchal BC, Schaeffer TR (1998) Effects of omapatrilat in low, normal, and high renin experimental hypertension. Am J Hypertens 11:363–372

Trippodo NC, Fox M, Monticello TM, Panchal BC, Asaad MM (1999) Vasopeptidase inhibition with omapatrilat improves cardiac geometry and survival in cardiomyopathic hamsters more than does ACE inhibition with captopril. J Cardiovasc Pharmacol 34:782–790

Vesterqvist O, Liao W, Manning JA (1997) Effects ode BMS-186716, a new dual metalloprotease inhibitor, on pharmacodynamic markers of neutral endopeptidase (NEP) and angiotensin converting enzyme (ACE) activity in healthy men (abstract). Clin Pharmacol Ther 61:230

Wallis EJ, Ramsay LE, Hettiarachchi J (1998) Combined inhibition of neutral endopeptidase and angiotensin-converting enzyme by sampatrilat in essential hypertension. Clin Pharmacol Ther 64:439–449

Warner KK, Visconti JA, Tschampel MM (2000) Angiotensin II receptor blockers in patients with ACE inhibitor-induced angioedema. Ann Pharmacother 34:526–528

Wiemer G, Scholkens BA, Becker RH, Busse R (1991) Ramiprilat enhances endothelial autacoid formation by inhibiting breakdown of endothelium-derived bradykinin. Hypertension 18:558–563

Yang Z, Arnet U, von Segesser L, Siebenmann R, Turina M, Luscher TF (1993) Different effects of angiotensin-converting enzyme inhibition in human arteries and veins. J Cardiovasc Pharmacol 22 [Suppl 5]:S17–S22

Yusuf S, Pepine CJ, Garces C, Pouleur H, Salem D, Kostis J, Benedict C, Rousseau M, Bourassa M, Pitt B (1992) Effect of enalapril on myocardial infarction and unstable angina in patients with low ejection fractions [see comments]. Lancet 340:1173–1178

Yusuf S, Sleight P, Pogue J, Bosch J, Davies R, Dagenais G (2000) Effects of an angiotensin-converting-enzyme inhibitor, ramipril, on cardiovascular events in high-risk patients. The Heart Outcomes Prevention Evaluation Study Investigators. N Engl J Med 342:145–153

Zhang X, Scicli GA, Xu X, Nasjletti A, Hintze TH (1997a) Role of endothelial kinins in control of coronary nitric oxide production. Hypertension 30:1105–1111

Zhang X, Xie YW, Nasjletti A, Xu X, Wolin MS, Hintze TH (1997b) ACE inhibitors promote nitric oxide accumulation to modulate myocardial oxygen consumption. Circulation 95:176–182

Zhang X, Nasjletti A, Xu X, Hintze TH (1998) Neutral endopeptidase and angiotensin-converting enzyme inhibitors increase nitric oxide production in isolated canine coronary microvessels by a kinin-dependent mechanism. J Cardiovasc Pharmacol 31:623–629

Zhang X, Recchia FA, Bernstein R, Xu X, Nasjletti A, Hintze TH (1999) Kinin-mediated coronary nitric oxide production contributes to the therapeutic action of angiotensin-converting enzyme and neutral endopeptidase inhibitors and amlodipine in the treatment in heart failure. J Pharmacol Exp Ther 288:742–751

Part 6
Clinical Qutlook

Inhibitors of the RAS: Evidence-Based Medicine

W. Schulz

Fuchshohl 23, 60431 Frankfurt, Germany
e-mail: Schulzwolf@aol.com

Abstract This chapter provides an overview of clinical evidence for angiotensin converting enzyme inhibitors (ACEi) and angiotensin II receptor blockers (ARBs or A2As). Only those data are presented which are relevant and fulfill the criteria of evidence-based medicine. The chapter is a snapshot of the current evidence as many trials are still ongoing. Given that the RAS is a ubiquitous system, it is conceivable that both drug classes are interacting with many organs. Evidence for treatment and prevention of cardiovascular (myocardial infarction, congestive heart failure) and cerebrovascular diseases (stroke), diabetes mellitus, and glomerular renal disease is extensively described. The position of both drug classes in relation to each other as well as in relation to other drug classes has been clarified for many indications. The clinical development of both substance classes was interesting: ARBs had a worse starting position because they entered the market later than ACE inhibitors. Unequivocal proof of efficacy could be shown with ACE inhibitors in placebo-controlled trials for the major indications. Thereafter, those indications were occupied by ACE inhibitors; repetition of placebo-controlled studies with ARBs was ethically unconscionable.

These major indications were almost lost for ARBs. Strenuous head-to-head comparisons with the aim to demonstrate at least non-inferiority were done; add-on trials were performed, or studies in patients intolerant to ACE inhibitors. As major indications were difficult to develop with ARBs, a fruitful evolution has focused on the more minor indications such as nephropathy. ACE inhibitors and ARBs are both suitable for first-line therapy of hypertension. In unconditional CHF, ACE inhibitors are the first-line therapy; ARBs are good in patients intolerant to ACE inhibitors, or may serve as add-on therapy to ACE inhibitors. In acute MI, ACE inhibitors are effective from the first day onwards, while no data exist for ARBs on day 30. In unselected patients after MI, ACE inhibitors are effective, while no data exist for ARBs. In patients after MI with poor LV function, ACE inhibitors are effective and ARBs are at best equally effective. In primary and secondary prevention of stroke, ACE inhibitors are effective and ARBs are likely equally effective. In type 1 diabetic and nondiabetic nephropathy, ACE inhibitors have demonstrated efficacy; in type 2 diabetic nephropathy, ARBs were shown to be effective. Are ACE inhibitors and ARBs alike or not alike? Do the effects of ACE inhibitors on the bradykinin system play a significant clinical role? One puzzling observation was seen in several trials: myocardial infarctions and cardiovascular mortality were apparently not as much reduced with ARBs as with ACE inhibitors. This last mystery is addressed in a prospective clinical trial—the ongoing ONTARGET trial—with about 30,000 patients randomized.

Keywords Renin-angiotensin system · Angiotensin converting enzyme inhibitors · Angiotensin II receptor blockers · Angiotensin II · Bradykinin · Evidence-based medicine

1 Introduction

The renin-angiotensin system has been extensively investigated under experimental conditions. Amongst the many options available for intervention within the renin-angiotensin cascade, a clinical breakthrough was finally achieved for two drug classes:

1. Inhibitors of the angiotensin I converting enzyme (ACE inhibitors)
2. Blockers of the angiotensin II receptor—mostly the AT1 subtype (AT1 receptor blockers or A2As)

The first ACE inhibitor to be introduced into the market was captopril in the early 1980s, while losartan as the first A2A was introduced only 15 years later, in 1995. Regulatory approval was obtained largely on the basis of blood pressure data and the first indication granted for losartan was treatment of essential hypertension in those patients experiencing side effects (cough) with ACE inhibitors.

Large-scale clinical trials were not available for quite some time after the introduction of ACE inhibitors and A2As. Meanwhile, however, clinical research has shifted from the measurement of surrogate markers such as blood pressure to evidence-based medicine (EBM), which essentially means the tally of relevant clinical events. The latter must consist—at least in part—of hard and irreversible end points such as fatal and nonfatal myocardial infarction and stroke. Hard and special softer clinical end points (such as hospitalization for heart failure) are the cornerstone of evidence-based medicine in cardiology.

The debate continues as to whether A2As are superior or, at a minimum, are equal in their effect to other cardiovascular agents, in particular ACE inhibitors. That debate is a long-standing one. A2As interfere with the end-stage of a biological cascade and block the generation of angiotensin II completely. ACE inhibitors interact somewhat earlier along the renin-angiotensin cascade and may, unlike A2As, potentiate bradykinin and follow-up reactions. The clinical value of these experimental findings has not yet been fully explored. However, some A2As have been tested in head-to-head comparisons against ACE inhibitors and answers have been found to at least some of the long-standing questions.

Still, the database is much broader for ACE inhibitors. They have been investigated as a first-line treatment in patients with special forms of hypertension, those with chronic heart failure (CHF), those with a high cardiovascular risk, and as an add-on treatment in patients after myocardial infarction. As A2As came later, the environment in which they were tested was a difficult one. There is now evidence that A2As are effective as a first-line treatment of type 2 diabetic nephropathy and as an add-on treatment in chronic heart failure, and that they may prove similarly effective to ACE inhibitors in hypertensive patients with additional risk factors. Outcomes vary for the indications investigated and these will all be discussed separately.

Many more large-scale trials are currently ongoing and new indications are under investigation.

2
Evidence from Large Interventional Trials

In evidence-based medicine, only clinical outcomes are regarded as suitable end points. These are parameters reflecting mortality and morbidity. When testing new drugs, they can be investigated in three different standard situations:

1. Initial, first-line therapy in comparison with placebo. This is possible for indications without an established therapy. As ACE inhibitors were present on the market earlier, it was possible to explore their effect, for example, in CHF (e.g., CONSENSUS). In contrast, A2As were investigated as first-line treatment in such left-over indications as type 2 diabetic nephropathy (e.g., IDNT).
2. Add-on therapy in comparison with add-on placebo. This is mandatory in indications for which an established therapy already exists and for which

add-on therapy is the only ethical way to test new substances. Add-on therapy with the A2A valsartan was tested, for instance, in CHF (ValHeFT). Testing add-on therapy is similar to testing combination therapy, at least in populations in which a very high percentage is receiving standardized treatment at baseline.

3. Head-to-head comparison with active drug. This is ethically permissible only if strong evidence of at least equal efficacy of a new therapy in comparison with established therapy is probable. Such trials have been done in hypertensive patients with LVH (LIFE) or following acute myocardial infarction (OPTIMAAL).

Proof of the evidence in interventional trials often requires an enormous sample size, which may pose challenges both logistical and monetary in nature. The determinants for a sample size calculation are, most importantly, the desired and clinically relevant effect, and further include the alpha-error, the beta-error, and the variability. The number of patients needed may equal or exceed 10,000 and they may require treatment over a long period of time, e.g., 4 years in a double-blinded randomized fashion.

Even with such large sample sizes, these trials are usually planned with only *one* active dose of the investigational drug. In light of the tremendous resources required, dose-finding is not affordable. Furthermore, repetitions of interventional studies with other dosages of an active substance are difficult to conduct once a beneficial effect on mortality has already been shown with one active dose. Conversely, the choice of a wrong dose may also lead to the failure of an study eventually leading to a verdict against an otherwise effective drug.

2.1 Unselected and Special Patient Groups with Hypertension

In former times, antihypertensive drugs were tested in trials in which hypertensive patients were included on the basis of their blood pressure values. They were not further selected with respect to additional risk factors such as diabetes mellitus. In these largely *unselected* hypertensive populations, active antihypertensive treatments (e.g., diuretics and beta-blockers) were frequently pooled for evaluation purposes in order to achieve sufficient statistical power against the placebo control (e.g., Australian Study, MRC Study, MRC-2 Study). The statistical power, however, was insufficient to evaluate the individual contribution of the diuretics or the beta-blockers alone with regard to the event reduction. This fact merits mention because diuretics and beta-blockers are acclaimed as gold standards even though the evidence cited, at least in the case of beta-blockers, is not at all that strong, claimed as it usually is on the basis of findings in *unselected* hypertensive patients. Furthermore, there are no flawless studies available investigating newer drug classes in populations that have no other risk factors.

Meanwhile, a different means than poding is being used to achieve sufficient statistical power: only those hypertensive patients are randomized who have additional risk factors and thus exhibit a higher cardiovascular risk. The likelihood that more events will occur is greater. Fewer patients are needed for recruitment and/or the observation period can be shortened.

The major studies with A2As in these "enriched" hypertensive populations—some of which are still ongoing—are:

1. LIFE study in patients with left ventricular hypertrophy
2. SCOPE study in patients of advanced age of 70–89 years with mild hypertension
3. VALUE study in patients with at least one additional risk factor (study is still ongoing)

ACE inhibitors have not been well investigated in outcome trials with hypertensive patients. Limited evidence in hypertensive patients was provided in the following studies, the conduct and outcomes of which were more or less convincing:

- CAPPP study
- CAPPP study in diabetic patients (preplanned subgroup analysis)
- UKPDS-39 study in diabetic patients
- ABCD study in diabetic patients
- FACET study in diabetic patients
- STOP-2 study in patients of advanced age
- AASK study in patients of Afro-American descent
- ALLHAT study

The focus of the development of ACE inhibitors, however, was centered on patients with CHF, postmyocardial infarction, or with increased cardiovascular risk (see following sections).

2.1.1 LIFE

The LIFE Study is the only head-to-head comparison of an A2A with another drug in hypertensive patients. It showed a better outcome with losartan than with atenolol (Dahlöf et al. 2002). The study randomized 9,194 patients with hypertension and left-ventricular hypertrophy to losartan or atenolol. A forced titration was performed with both drugs to achieve target blood pressures of <140/<90 mmHg. The primary end point was a composite of cardiovascular mortality, MI, or stroke. Mean doses for losartan were 82 mg, and for atenolol, 79 mg. Blood pressure effects were identical in both groups. However, after a mean follow-up of 4.8 years, losartan was significantly better than atenolol with respect to the primary end point (508 events in 4,606 patients with losartan vs 588/4,588 with atenolol; adjusted RR, −13%; p=0.021). In the prospectively defined stratum of 1,195 diabetic patients (Lindholm et al. 2002b), the risk reduction was even more pronounced (RR, –24%, *p*=0.031). In the LIFE study, the benefit of losartan over atenolol was largely carried by a reduction in stroke (10.8 vs. 14.5 per 1,000 patient years follow-up) but not by a reduction in myocardial infarction (9.2 vs 8.7 per 1,000 patients years follow-up).

Furthermore, losartan was compared to a beta-blocker the clinical benefit of which had not in fact been proven. The only outcome study with atenolol is the MRC study in elderly patients. In the MRC Study, 4,396 elderly patients were randomized to daily doses of atenolol 50 mg or HCT 25 mg or placebo (Medical Research Council Working Party 1992). Actively treated subjects (diuretic and beta-blocker groups combined) showed a 25% reduction in stroke and a 19% reduction in coronary events. The diuretic group showed a significantly reduced risk of stroke (-31%) and of cardiovascular events (–44%) compared with the placebo group. The atenolol group did not show significant reductions in these end points, i.e., the evidence for beneficial effects of atenolol has not in fact been proven.

2.1.2 SCOPE

In the placebo-controlled SCOPE study (Study on Cognition and Prognosis in the Elderly), candesartan cilexetil was investigated in 4,964 elderly patients aged 70–79 with mild hypertension (SBP, 160–179 or DBP, 90–99 mmHg). Candesartan was given in doses of 8 mg, doubled to 16 mg after 3 months where BP response was insufficient. After 4 years of treatment, candesartan had not reduced the risk of major cardiovascular events, which was the primary end point (–11%; *p*=0.19). However, it did significantly reduce the risk of nonfatal strokes, which was a secondary end point (–28%; *p*=0.041) in these patients. SCOPE was the first outcome study in this particular population (Hansson et al. 1999c, 2000; Sever 2002). Previous placebo-controlled trials involved elderly patients with higher blood pressures, for example, the STOP-Hypertension study (Hansson et al. 1990).

2.1.3 VALUE

The double-blind, randomized prospective, parallel group VALUE trial is ongoing. The effects of valsartan on the reduction of cardiac morbidity and mortality are compared to those of amlodipine. Patients with essential hypertension, aged 50 years and older, and at a particularly high risk of coronary events were enrolled. A total of 18,119 patients were screened and 15,314 patients in 31 countries were randomized. More than 92% of the randomized participants had been treated for high blood pressure for at least 6 months when screened for the study. The randomized population is now being treated (goal: blood pressure <140/90 mmHg) and treatment will continue until at least 1,450 patients experience a primary cardiac end point, defined as clinically evident or aborted myocardial infarction, hospitalization for heart failure or death caused by coronary heart disease. Results are expected soon (Kjeldsen et al. 2001).

2.1.4 CAPPP

The Captopril Prevention Project (CAPPP) was an interventional trial to compare the effects of ACE inhibition (captopril) with conventional therapy (diuretics, beta-blockers) on cardiovascular morbidity and mortality in patients with hypertension. CAPPP was an open, randomized trial with blinded end point evaluation in 10,985 patients aged 25–66 years with a baseline diastolic blood pressure of 100 mmHg or more. The primary end point was a composite of fatal and nonfatal myocardial infarction, stroke, and other cardiovascular deaths. A total of 5,492 patients were assigned captopril and 5,493 assigned conventional therapy. Primary end point events occurred in 363 patients in the captopril group (11.1 per 1,000 patient-years) and 335 in the conventional-treatment group (10.2 per 1,000 patient-years; relative risk 1.05, p=0.52). Thus, captopril and conventional treatment did not differ in their overall composite end point. Unfortunately, the influence from combination therapy (which was used frequently) was not borne out. Furthermore, captopril is an ACE inhibitor that is not considered to be as potent as the newer, more lipophilic ACE inhibitors such as quinapril, perindopril, or ramipril.

The results for some monocomponents of the primary composite end point were:

- Cardiovascular mortality: 76 events with captopril vs 95 events with conventional treatment (RR, 0.77; p=0.092)
- Rate of fatal and nonfatal myocardial infarction: 162 vs 161 events
- Fatal and nonfatal stroke: 189 vs 148 (RR, 1.25; p=0.044).

Due to the limited number of events, the results are difficult to interpret. Moreover, the difference in stroke risk, for example, was thought to be due to lower levels of blood pressure obtained initially in previously treated patients randomized to conventional therapy (Hansson et al. 1999b).

2.1.5 UKPDS-39

UKPDS-39 (UK Prospective Diabetes Study Group 1998a) was a substudy of UKPDS-38 (UK Prospective Diabetes Study Group 1998c), which in turn was a substudy of UKPDS-33 (UK Prospective Diabetes Study Group 1998b) in the UKPDS program. The purpose of UKPDS-38 was to determine whether tight control of blood pressure with either a beta-blocker or an angiotensin-converting enzyme inhibitor would show better effects than looser BP control in preventing the macrovascular and microvascular complications of patients with new type 2 diabetes (UKPDS-38 study). A total of 390 patients were allocated to looser control of blood pressure. Only the subgroup with tight blood pressure control (aiming at a blood pressure of <150/<85 mmHg) was further randomized to captopril or atenolol in patients with type 2 diabetes (UKPDS-39). Of those 758 patients allocated to tight control of blood pressure, 400 were allocated to captopril and 358 to atenolol. Predefined clinical end points were diabetes-related fatal and nonfatal events and all-cause mortality. Surrogate measures of microvascular and macrovascular disease included urinary albumin excretion and retinopathy assessed by retinal photography. Captopril and atenolol were equally effective in reducing blood pressure to a mean of 144/83 mmHg and 143/81 mmHg respectively, with a similar proportion of patients (27% and 31%) requiring three or more antihypertensive treatments. Captopril and atenolol were equally effective in reducing the risk of macrovascular end points.

Thus, blood pressure lowering with captopril or atenolol was similarly effective in reducing the incidence of diabetic complications under the special conditions of the trial. However, it is difficult to conclude that both monotherapies (captopril or beta-blockers) are equally effective overall. The population randomized was a subgroup (UKPDS-39) of a subgroup (UKPDS-38) of a large study population (UKPDS-33). Furthermore, the sample size was small, and most patients were on double or triple combinations. In essence, two types of not very well-defined combination therapies were compared.

2.1.6 CAPPP in Diabetic Patients

Captopril is evidently more effective than diuretics or beta-blockers in those 572 diabetic patients of the CAPPP study (Niskanen et al. 2001) in whom a preplanned subgroup analysis was performed. The primary end point, fatal and nonfatal myocardial infarction and stroke as well as other cardiovascular deaths, was markedly lower in the captopril than in the conventional therapy group (RRR 41%; p=0.018). Some attempts were made to explain the differing results of CAPPP and UKPDS39: diabetic patients in CAPPP were probably at a higher risk than in UKPDS, as the latter had milder disturbances in glucose metabolism. Furthermore, the treatment goal was lower in the UKPDS (<150/<85 mmHg) than in CAPPP (<90 mmHg).

There is other evidence from smaller studies (see the ABCD and FACET studies) underscoring that ACE inhibitors may enjoy an advantage in diabetic patients—at least over dihydropyridine calcium channel blockers (DHP-CCB).

2.1.7 ABCD

Enalapril was more effective than nisoldipine in hypertensive patients in the ABCD Study (Appropriate Blood Pressure Control in Diabetes) (Estacio et al. 1998). The primary hypothesis was that intense BP control (target DBP,

<75 mmHg), as compared with moderate control (target DBP, 80–89 mmHg), would prevent or slow the progression of diabetic nephropathy, neuropathy, retinopathy, and cardiovascular events in normotensive (480 patients) and hypertensive (470 patients) patients with NIDDM. The prespecified secondary hypothesis was that a long-acting DHP-CCB would have an equivalent effect to that of an ACE inhibitor in preventing these complications. The DSMB, however, had to stop treatment after 5 years in the subgroup of 470 hypertensive patients because the latter showed a significantly higher rate of fatal and nonfatal myocardial infarctions with nisoldipine (25 patients) than with enalapril (5 patients).

2.1.8 FACET

Fosinopril was more effective than amlodipine in the FACET Study in which 380 diabetic hypertensive patients were tested over 3.5 years (Tatti et al. 1998). The primary aim of the study was to assess treatment-related differences in serum lipids and diabetes control. The secondary end point was the rate of prospectively defined cardiovascular complications (myocardial infarction, stroke or hospitalization for angina pectoris). With fosinopril, the secondary end point was significantly lower in comparison with amlodipine (14/189 vs 27/191 events).

2.1.9 STOP-2

In the open, prospective, randomized STOP-2 trial, the effects of new antihypertensive drugs (ACE inhibitors and CCBs) were compared to the effects of old antihypertensive drugs (beta-blockers and diuretics) on cardiovascular mortality and morbidity. A total of 6,614 patients aged 70–84 years with hypertension (blood pressure ≥180 mmHg systolic, ≥105 mmHg diastolic, or both) were randomized to conventional antihypertensive drugs (atenolol 50 mg, metoprolol 100 mg, pindolol 5 mg, or hydrochlorothiazide 25 mg plus amiloride 2.5 mg daily) or to newer drugs (enalapril 10 mg or lisinopril 10 mg, or felodipine 2.5 mg or isradipine 2–5 mg, daily). Fatal stroke, fatal myocardial infarction, and other fatal cardiovascular diseases were assessed. Blood pressure was decreased to a similar extent in all treatment groups. The primary combined end point of fatal stroke, fatal myocardial infarction, and other fatal cardiovascular disease occurred in 221 of 2,213 patients in the conventional drugs group (19.8 events per 1,000 patient-years) and in 438 of 4,401 in the newer drugs group (19.8 per 1,000; RR, 0.99, p=0.89). The combined end point of fatal and nonfatal stroke, fatal and nonfatal myocardial infarction, and other cardiovascular mortality occurred in 460 patients taking conventional drugs and in 887 taking newer drugs (RR, 0.96, p=0.49). It was concluded that old and new antihypertensive drugs were similar in prevention of cardiovascular mortality or major nonfatal events (Hansson et al. 1999a).

The STOP-2 trial was not designed to determine the role of either new monotherapy alone in comparison to conventional antihypertensive treatment. Nevertheless, in the comparison of ACE inhibitors with CCBs, it was shown that ACE inhibitors were significantly better than CCBs with regard to myocardial infarction and the frequency of CHF.

2.1.10 ALLHAT

The NIH-sponsored ALLHAT study is a large-scale comparison of four antihypertensive drugs, with 42,516 patients randomized to amlodipine, chlorthalidone, doxazosin, and lisinopril (Davis et al. 1996).

Treatment with doxazosin was discontinued after 4 years (9,067 randomized patients), because a significantly higher incidence of heart failure was demonstrated in comparison with chlorthalidone (15,268 randomized patients) at a relative risk of 2.04 (2% with doxazosin, 1% with chlorthalidone). A greater rate of stroke (at a relative risk of 1.19, $p=0.04$) was also shown. Furthermore, a 25% increased risk of combined cardiovascular events was found, which may be due to a slight systolic blood (but no diastolic) pressure difference (Messerli 2000, 2001).

Final results of ALLHAT for the remaining patients treated with chlorthalidone, amlodipine, or lisinopril were published after a mean follow-up of 4.9 years. The primary outcome (combined fatal CHD or nonfatal myocardial infarction) occurred in 2,956 patients, with no difference between treatments. Likewise, the secondary end point, all-cause mortality, did not differ between groups. Other secondary end points were better with chlorthalidone (less CHF than with amlodipine, and less combined CVD, stroke, and CHF than with lisinopril). However, 5-year systolic blood pressures were significantly higher in the amlodipine (0.8 mmHg) and lisinopril (2 mmHg) groups compared with chlorthalidone. The conclusion that "thiazide-type diuretics are superior in preventing 1 or more major forms of CVD" seems questionable, for example, in light of the differences in blood pressures (ALLHAT Collaborative Research Group 2002).

2.1.11 ANBP-2

The prospective, randomized, open-label Australian ANBP-2 study compared the outcomes with ACE inhibitors and diuretics for hypertension in the elderly. A total of 6,083 patients with hypertension who were 65–84 years of age were followed for a median of 4.1 years at 1,594 family practices. By the end of the study, blood pressure had decreased to a similar extent in both groups. There were 695 cardiovascular events or deaths in the ACE inhibitor group and 736 cardiovascular events or deaths in the diuretic group (hazard ratio, 0.89; p=0.05). Effects were more pronounced in men than in women. It was concluded that initiation of antihypertensive treatment involving ACE inhibitors appeared to lead to better outcomes than treatment with diuretic agents.

2.1.12 Conclusions, Open Questions, Hypotheses

From the perspective of evidence-based medicine, the old indication of hypertension was not well investigated with ACE inhibitors and A2As. Why is that so? Placebo controls were used in studies with conventional therapy (diuretics and beta-blocker) and could not be used again in trials with the new drugs. Only trials with active comparators were permitted, for ethical reasons. Unfortunately, different "new" and "conventional" therapies were grouped together in the past. This practice has not been helpful in clearly establishing the benefits of ACE inhibitors alone. Only ALLHAT was a direct comparison of four different drugs as first-line treatment of hypertension (to which steps 2 and 3 were added as needed). However, even ALLHAT has not really answered the question of whether modern ACE inhibitors are equal or superior to other antihypertensive agents. A2As were not part of the ALLHAT trial.

The evidence for the effects of drugs within a substance class is even worse. There are no outcome data available from head-to-head comparisons. Therefore, the credit should only be given to those drugs for which benefit has been proven.

It must be emphasized that surrogate markers do not generate the evidence, although such studies are often used to claim the advantages of the one drug over the other. Recently, a better blood pressure lowering at maximally tolerated doses was demonstrated in the CLAIM studies with candesartan in comparison with losartan (Bakris et al. 2001; Vidt et al. 2001). The true clinical meaning of the findings is unclear. Furthermore, a larger LVH reduction has been shown for ACE inhibitors such as enalapril (SCAT Study; Teo et al. 2000), and ramipril (PART II Study; MacMahon et al. 2000) in comparison to placebo, and with ramipril (RACE Study; Agabiti-Rosei et al. 1995) in comparison to atenolol. A similar reduction in LVH has been found for candesartan in comparison with enalapril in the CATCH trial (Cuspidi et al. 2002). Again, the true clinical meaning of all these findings will remain unclear in the absence of outcome trials.

From the evidence that is currently available for ACE inhibitors, some assumptions may be derived:

- ACE inhibitors are equal to conventional drugs (STOP-2, CAPPP, ALLHAT) in unselected patients.
- ACE inhibitors are likely superior to CCBs in unselected patients (STOP-2).
- ACE inhibitors are possibly superior to other hypertensives in diabetic patients (CAPPP, ABCD, FACET, but not in UKPDS-39).

- An ACE inhibitor was shown to be superior to a DHP-CCB and to a beta-blocker with regard to nephroprotection in hypertensive Afro-Americans (see Sect. 2.6 in this chapter; Wright et al. 2002a).

For A2As, the data situation may be summarized as follows:

- An A2A was superior to a beta-blocker (LIFE).
- An A2A was superior to a beta-blocker in diabetic patients (LIFE).
- An A2A was not really superior to placebo (SCOPE) with respect to cardiovascular events and was likely superior with respect to cerebrovascular events in elderly patients with mild hypertension.

In the absence of direct comparisons, it is difficult to speculate on whether ACE inhibitors are better than A2As. The opposite conclusion, in any event, seems highly unlikely. In the following sections, in which the effects of the two drug classes in other important indications are described, the impression prevails that cardiac risk reduction is consistently shown with ACEi but not with A2As. A2As, on the other hand, reduce at least stroke reliably.

It is intriguing to speculate that ACE inhibitors could resolve the old dilemma in hypertension research posed by the fact that stroke was consistently reduced, whereas fatal and nonfatal myocardial infarction was not convincingly reduced with other antihypertensives.

2.2
Unselected and Special Patient Groups After Myocardial Infarction

Tremendous progress has been made in recent years in the treatment of acute myocardial infarction. In addition to the immediate goal of therapy, which is a rapid revascularization with either fibrinolysis or percutaneous coronary intervention (PCI), a whole battery of concomitant treatments has proven effective and has become state of the art.

Apart from antithrombotic drugs enhancing the likelihood of revascularization and preventing reinfarctions (with ASA, clopidogrel, heparins or LMWHs, fibrinogen receptor antagonists), other drugs with a different mode of action such as beta-blockers have proven effective in the long-term follow-up of myocardial infarction. In addition to beta-blockers, ACE inhibitors and, recently, A2As have been investigated and have also proven to be effective. They prevent remodeling and preserve myocardial function, which is thought to translate into better survival. Survival is the one and only primary outcome measure for drug approval.

There are three major types of studies with ACE inhibitors and A2As:

1. Placebo-controlled short-term studies with ACE inhibitors in largely *unselected* patients with AMI. The initiation of therapy is early after onset of MI

and the end point evaluation is also early, e.g., after 5 weeks (CONSENSUS II, ISIS-4, GISSI-3, CCS-1, SMILE, MITRA registry).

2. Placebo-controlled long-term studies with ACE inhibitors in *selected* patients with AMI and impaired LV function. The initiation of therapy after onset of MI was delayed, and the end point evaluation was performed after several years (SAVE, AIRE, TRACE).
3. Head-to-head comparisons of A2As with ACE inhibitors have been performed or are ongoing in selected patients with AMI and impaired LV function (OPTIMAAL). (VALIANT) The initiation of therapy is within 10 days after onset of the MI and the end point is evaluated after several years.

2.2.1 CONSENSUS II

CONSENSUS II was carried out in a group of 6,090 unselected patients representing almost every type of acute myocardial infarction (above all, irrespective of the extent of left ventricular damage). The result was a slight nonsignificant trend toward excessive mortality with 20 mg enalapril once daily (enalapril 10.2%, placebo 9.4%). Initially in this study, 1 mg enalapril was first infused within 2 h, followed by a 6-h pause. Thereafter, enalapril was administered orally in doses that were staggered from 2×2.5 mg to 20 mg up to the 5th day. Strong decreases in blood pressure (systolic to <90 mmHg) developed at a higher rate (12%) in the enalapril group than in the placebo group (3%). It was assumed that the titration scheme was too aggressive in CONSENSUS II (Swedberg et al. 1992).

2.2.2 ISIS-4

In ISIS-4 study, 58,050 patients were randomized to captopril (up to 25 mg b.i.d.) or placebo. Treatment was also started early up to 24 h after infarction onset. Captopril was tolerated well. After 35 days, all-cause mortality was 7.2% in the captopril group, and 7.7% in the placebo group (ISIS-4 Collaborative Group 1995).

2.2.3 GISSI-3

In the GISSI-3 study, 18,895 patients were randomized to lisinopril (up to 10 mg once daily) or placebo within 24 h after infarction onset. All-cause mortality in the lisinopril group was 6.3%, in the placebo group 7.1% after 42 days (Gruppo Italiano per lo Studio della Sopravivenza nell'Infarto Miocardio 1994).

2.2.4 CCS-1

In the Chinese CCS-1 study, 13,634 patients were randomized to captopril (up to 12.5 mg t.i.d.) or placebo within 36 h after infarction onset. All-cause mortality in the captopril group was 9.1%, in the placebo group was 9.6% after 4 weeks (Chinese Cardiac Study Collaborative Group 1995).

2.2.5 SMILE

In the SMILE study, 1,556 patients were enrolled within 24 h after the onset of symptoms of acute anterior myocardial infarction, and they were randomly assigned in a double-blind fashion to receive either placebo (784 patients) or zofenopril (772 patients) for 6 weeks. The primary end point was the incidence of death or severe congestive heart failure. The patients were reexamined after 1 year to assess survival.

The incidence of death or severe congestive heart failure at 6 weeks was significantly reduced in the zofenopril group (55 patients, 7.1%), as compared with the placebo group (83 patients, 10.6%); the cumulative reduction in the risk of death or severe congestive heart failure was 34% (95% confidence interval, 8%–54%; p=0.018). After 1 year of observation, the mortality rate was significantly lower in the zofenopril group (10.0%) than in the placebo group (14.1%); the reduction in risk was 29% (p=0.011). Treatment with zofenopril significantly improved both short-term and long-term outcome when this drug was started within 24 h after the onset of acute anterior myocardial infarction and continued for 6 weeks (Ambrosioni et al. 1995).

2.2.6 SAVE

In the SAVE study, 2,231 patients with reduced LV function were randomized to captopril (50 mg t.i.d.) or placebo within 3–16 days after the onset of infarction. All-cause mortality was 20% in the captopril group and 25% in the placebo group after 42 months (Pfeffer et al. 1992).

2.2.7 AIRE

In the AIRE study, 2,006 patients with clinical signs and/or symptoms of heart failure were randomized to ramipril (up to 5 mg b.i.d.) or to placebo within 2–11 days after infarction onset. All-cause mortality was 17% in the ramipril group and 23% in the placebo group after only 15 months (The AIRE Study Investigators 1993).

The AIREX follow-up study revealed that an additional survival benefit accrued thereafter and was maintained for several years (Hall et al. 1997).

2.2.8 TRACE

In the TRACE study, 1,749 patients with impaired LV function were randomized to trandolapril (up to 2 mg q.d.) or placebo within 3–7 days after infarction onset. All-cause mortality was 35% in the trandolapril group and 43% in the placebo group after 24–50 months (Køber et al. 1995). After patients had been followed up for a minimum of 6 years, median survival was improved by 15.3 months (Torp-Pedersen et al. 1999).

2.2.9 MITRA PLUS Registry

MITRA PLUS is a German prospective multicenter registry of current treatment of AMI with 14,608 consecutive patients with ST-elevation acute myocardial infarction. Treatment with ACE inhibitors was documented. Their effects on all-cause hospital mortality were evaluated. A total of 685 patients (4.7%) received ramipril, 39% received other ACE inhibitor therapy, and 56.3% no ACE inhibitor therapy. Hospital mortality was 5.8% with ramipril, 8.9% with other ACE inhibitor therapy, and 12.3% in patients receiving no ramipril treatment (Wienbergen et al. 2002).

2.2.10 OPTIMAAL

OPTIMAAL was recently completed in 5,477 patients with confirmed AMI. Only those patients who had heart failure during the acute phase or a new Q-wave anterior infarction or reinfarction were randomized. Patients were randomly assigned and titrated to a target dose of losartan (50 mg once daily) or captopril (50 mg t.i.d.). The primary end point was all-cause mortality. There were 499 (18%) deaths in the losartan group and 447 (16%) in the captopril group after 2.7 years of follow-up (Dickstein et al. 1998, 2002).

2.2.11 VALIANT

VALIANT is a study in 14,500 high-risk patients after acute myocardial infarction. Patients must exhibit clinical signs and symptoms of CHF or LV dysfunction. Patients are randomized to valsartan (160 mg b.i.d.), or captopril (50 mg t.i.d.), or its combination (captopril 50 mg t.i.d. + valsartan 80 mg b.i.d.) between 12 h and 10 days after the onset of acute infarction. The trial is powered to detect a 15% reduction in all-cause mortality rate with either use of valsartan compared with captopril. The study is designed as an event-driven trial (until 2,700 deaths have accrued) unless the duration of follow-up exceeds 6 years (Pfeffer et al. 2000; Pfeffer et al. 2003). Results were released recently: After a median follow-up of 24,7 months, 997 patients died in the valsartan group, 958 in the captopril group, and 941 in the combination group.

2.2.12
Conclusions, Open Questions, Hypotheses

Acute treatment of myocardial infarction and its long-term management thereafter are both very important. In addition to rapid reperfusion and prevention of rethrombosis, other modes of action were thought to be helpful. The hypothesis—always the basis for a new study in EBM—was that vascular and myocardial effects could be beneficial. Such evidence was generated in myocardial infarction, both for ACE inhibitors and A2As.

ACE inhibitors have been investigated in all kinds of different situations and populations following acute myocardial infarction. It has been proven that:

- ACE inhibitors reduce acute in-hospital mortality in largely unselected populations.
- ACE inhibitors reduce long-term mortality in populations with impaired LV function or clinical signs and symptoms of heart failure. Median survival is prolonged by 1 or more years.
- ACE inhibitors need to be given early in AMI because 40% of the 30-day survival advantage is observed in days 0–1 (ACE Inhibitor Myocardial Infarction Collaborative Group 1998).

The data basis for A2As, on the other hand, is not broad and their real value in comparison with ACE inhibitors is not entirely clear.

- An A2A was not inferior to an ACE inhibitor, although there was a trend for a worse outcome in some end points (for losartan in comparison with captopril). No relevant difference, however, was seen for valsartan in comparison with captopril.

Patients with impaired LV function following AMI have much in common with patients who suffer from unconditional CHF. In those patients, again, an A2A was formally not inferior to an ACE inhibitor; again, however, a trend for a worse outcome was seen with losartan in comparison with captopril (see Sect. 2.3).

2.3 Chronic Heart Failure

In chronic heart failure (CHF), it took quite some time for evidence of prognostic benefit from drug treatment to accrue. Many unsuccessful attempts, for example, with positive inotropes, were made. Alone a vasodilator combination comprising hydralazine-isosorbide dinitrate had shown a borderline significant prognostic effect in the VA Cooperative Study (Cohn et al. 1986). Then clinical breakthrough was obtained with ACE inhibitors, which were the first drugs for which unequivocal proof of prognostic benefit in CHF was provided.

In the interim, ACE inhibitors have been investigated in all stages of symptomatic and asymptomatic heart failure. Only thereafter has the presence of a benefit from other therapies such as beta-blockers and spironolactones been shown. These drugs must be given in addition to ACE inhibitors. The major studies with ACE inhibitors are:

- CONSENSUS I
- VHeFT II
- SOLVD treatment
- SOLVD prevention
- ATLAS

Considerable hope prevailed that A2As could substitute for ACE inhibitors because their performance would equal or even exceed that of ACE inhibitors. Two major studies have been completed with A2As, one of which involves a direct head-to-head comparison study (ELITE II) and the other one an add-on study (ValHeFT). The evidence culled from the following studies will be discussed:

- ELITE I
- ELITE II
- RESOLVD
- ValHeFT
- CHARM

2.3.1 CONSENSUS I

The placebo-controlled CONSENSUS Study was the landmark trial in which enalapril showed overwhelming benefit after a short period of time. A total of 253 patients with New York Heart Association class IV (NYHA IV) disease were randomized to enalapril or placebo. After only 6 months, the rate of mortality in the placebo group was 44%; in the enalapril group, the rate was 26%, and the study was prematurely stopped (The CONSENSUS Trial Study Group 1987).

2.3.2 VHeFT

In the active-control V-HeFT II Study, 804 men with reduced exercise tolerance were randomized to enalapril or a combination of hydralazine with ISDN. The latter combination was in use at that time for patients with CHF. After 2 years, the mortality was lower in the enalapril arm than in the hydralazine + isosorbide dinitrate arm (18% vs 25%). Reduction in mortality was mainly attributable to a reduction of sudden cardiac death (Cohn et al. 1991).

2.3.3 SOLVD Treatment and SOLVD Prevention

The aim of the SOLVD trials (The SOLVD Investigators 1990) was to explore whether ACE inhibitors are also effective in patients with NYHA III (SOLVD treatment study) or even in asymptomatic patients with impaired LV function (SOLVD prevention study). A total of 2,569 patients were randomized in the double-blind SOLVD treatment study to enalapril or placebo. After an average follow-up of 41.4 months, the rate of total mortality in the placebo group was 39.7%; in the enalapril group, the rate was 35.2% (The SOLVD Investigators 1991).

In the double-blind SOLVD prevention study (The SOLVD Investigators 1992), 2,111 patients were treated with enalapril and 2,117 patients with placebo for 37.4 months. At 8% (RRR), the reduction in total mortality was not significant. The hospitalization rate due to heart failure, however, was reduced significantly by 37% (RRR), and the rate of development of symptomatic heart failure was also reduced significantly by 44% (RRR). Symptomatic drops in blood pressure were infrequent in the SOLVD collective (Hood et al. 1991).

2.3.4 ATLAS

In the ATLAS Study, 3,164 patients were investigated over an average period of 46 months; they had been titrated to a low (2.5–5.0 mg daily) or to a high (35 mg daily) dose of lisinopril. Survival alone (primary endpoint) did not differ significantly between the two groups. However, hospital admission for heart failure and the combined end point of death or hospital admission was significantly reduced with the higher doses. Thus, it is currently believed that ACE inhibitors, if tolerated, show better effects with higher doses. However, there is no flawless dose-finding study available addressing this question properly in patients with CHF (Packer 1996; Packer et al. 1999).

2.3.5
ELITE I

A first pilot study (ELITE I study) showed prognostic benefit of 50 mg losartan once-daily in comparison to captopril (50 mg t.i.d.) in 722 heart-failure patients of advanced age (at least 65 years): 17 deaths with losartan vs 32 deaths for captopril. The study, however, was not planned as a survival study and mortality was not the primary end point (Pitt et al. 1997).

2.3.6
ELITE II

In the pivotal ELITE II trial, a total of 3,152 patients over the age of 60 with an LVEF less than 40% and NYHA symptoms II–IV were randomized to 50 mg losartan or 3×50 mg captopril. Mortality in the captopril group was somewhat lower (250 out of 1,574 patients) than in the losartan group (280 out of 1,578 patients). Also, an almost significant trend in favor of captopril was observed in conjunction with sudden cardiac death (captopril group 115 patients, losartan group 142 patients; p=0.08). In combination with beta-blockers, which a total of 24% of all patients received (well distributed in both groups), captopril's performance was significantly superior to that of losartan (Pitt et al. 2000).

2.3.7
RESOLVD

The RESOLVD study was planned as a study to investigate exercise tolerance, ventricular function, quality of life (QoL), neurohormone levels, and tolerability in congestive heart failure (CHF). A total of 768 patients with NYHA II–IV and LVEF less than 0.40 and a 6-min walking distance under 500 m were randomized to candesartan (4, 8, or 16 mg), enalapril (4 or 8 mg), or the combination of both. The study was not planned as a survival study. However, the study was halted after 43 weeks because the External Safety and Efficacy Monitoring Committee (ESEMC) voiced concern about the use of candesartan: mortality, the hospitalization rate for heart failure, and their combined end point were higher with candesartan and its combinations (candesartan alone: 14.6%, candesartan plus enalapril 15.1%, enalapril alone 6.4%). These results were published in the fine-print method section of the RESOLVD study (McKelvie et al. 1999). It is worth mentioning that the combination of candesartan and enalapril showed an improvement in several surrogate markers (ventricular volumes, LV ejection fraction, concentrations of aldosterone and BNP) in comparison to the individual monosubstances, yet a survival benefit was not shown. The lesson to be learned may be that these surrogate markers *cannot* reliably predict the prognostic benefit of an intervention in CHF patients. Interestingly, the randomized placebo-controlled addition of the beta-blocker metoprolol improved both surrogate markers and survival in all groups (The RESOLVD Investigators 2000).

2.3.8 ValHeFT

The Val-HeFT study was a double-blind placebo-controlled randomized trial in 5,010 patients with CHF and NYHA II–IV and EF under 40%. The active drug was valsartan, which was titrated up from 40 mg to a relatively high dose of 160 mg in addition to standard therapy (92% with ACE inhibitors, 86% with diuretics, 93% with digoxin, 35% with beta-blockers). There were two primary end points: overall mortality and the combination of overall mortality with morbidity criteria. Morbidity was defined as sudden cardiac death with resuscitation, hospitalization for heart failure, or the need for parenteral inotropic or vasodilator therapy). The rate of overall mortality after 2 years did not differ between the groups (19.7% valsartan vs 19.4% placebo), whereas a significant difference was discernible in the combined morbidity/mortality end point (28.8% valsartan vs 32.1% placebo; RRR, 13%, p=0.009). This difference was attributable only to a significant improvement in the hospitalization rate (13.8% valsartan vs 18.2% placebo; RRR, 28%). Thus, survival benefit is not proven, while improvement in morbidity (shown as reduced hospitalization rate) was shown. A small subgroup of 7% of the patients was not treated with an ACE inhibitor. A nonsignificant reduction of 33.1% in all-cause mortality and 44% in the combined end point of mortality and morbidity was seen (Cohn and Tognoni 2001). Were these patients to be excluded from the overall group, then the observed overall reduction in the combined end point would no longer be significant.

2.3.9 CHARM

CHARM was a program of three placebo-controlled studies designed to define the clinical benefits of candesartan in a broad spectrum of patients with systolic and diastolic heart failure (Swedberg et al. 1999). The CHARM program randomized 7,801 patients with:

- EF <40%, as add-on treatment in patients already on ACE-inhibitor treatment (CHARM-Added trial with 2,548 patients enrolled)
- EF <40%, who were ACE-inhibitor intolerant (CHARM-Alternative trial with 2,028 patients enrolled)
- EF >40%, who were not treated with ACE inhibitors (CHARM-Preserved trial with 3,023 patients randomized)

The three studies were combined to evaluate all-cause mortality as the primary outcome. The outcomes were released only recently. All-cause mortality, the primary end point of the overall program, was reduced by 10% (RRR) (Pfeffer et al. 2003). Furthermore, the combined end points of cardiovascular mortality or CHF hospitalization were evaluated for the component trials.

In the CHARM-Added trial (McMurray et al. 2003), the composite outcome (cardiovascular death or hospital admission for CHF) was reduced after 41 months follow-up (38% vs 42% with placebo).

In the placebo-controlled CHARM-Alternative trial (Granger et al. 2003), the same combined end point was significantly reduced after a median follow-up of 33.7 months (33% vs 40% on placebo; RRR, 33%; p=0.0004).

In the CHARM-Preserved trial (Yusuf et al. 2003), however, the combined end point was not significantly reduced after 36.6 months follow-up (22% vs 24% on placebo). Cardiovascular mortality did not differ between groups (170 vs 170 deaths), but fewer patients in the candesartan group than in the placebo group were admitted to hospital once for CHF (230 vs 279 on placebo).

2.3.10 Conclusions, Open Questions, Hypotheses

Patients with CHF have a poor prognosis with 5-year mortality rates of 50% and higher (Ho et al. 1993). CHF is also the most frequent cause for hospitalization. Modern drug treatment ameliorates this situation: considerable progress has been made in recent years with the combination of ACE inhibitors, beta-blockers, and spironolactone.

ACE inhibitors can be regarded as the breakthrough to life-saving drug therapy in CHF. The available evidence was summarized in a large meta-analysis (Flather et al. 2000). It can now be safely stated that:

- ACE inhibitors are effective at all stages of heart failure, in asymptomatic (NYHA I) and symptomatic patients (NYHA II, III, and IV). The effects are not only significant, they are clinically relevant.

A2As have been investigated in three major studies so far:

- An A2A was not inferior to an ACE inhibitor in the ELITE II study. However, there was a negative trend with losartan in comparison with captopril in most end points.
- An A2A did not show better survival if given in addition to ACE inhibitors in the ValHeFT study. However, the rate of hospitalization for heart failure was reduced.
- An A2A showed survival benefit over placebo in those patients who were not on ACE inhibitors in ValHeFT. This subgroup was small and evaluation not prespecified.

It was speculated that losartan may have been underdosed. On the other hand, a high dose of valsartan was given in the ValHeFT and still there was no additive effect on mortality. Thus, the likelihood that A2As are superior to ACE inhibitors with regard to total or cardiovascular mortality is very low.

The situation is different and not fully understood with regard to hospitalization for heart failure. Why then does valsartan show additional effects with respect to hospitalization for heart failure? It may be that baseline therapy with captopril was not sufficiently dosed for prevention of hospitalization.

The CHARM program brought only some clarification:

- A2As are highly effective in CHF patients who do not tolerate ACE inhibitors.
- Additive effects (in addition to baseline therapy with ACE inhibitors) are, again, only seen with regard to hospitalization for CHF but not with regard to cardiovascular mortality.

Apart from the above considerations, different dosages are apparently needed for prolongation of life and for a reduction in hospitalization. This assumption fits another observation from the ATLAS study: a higher dose of the ACE inhibitor lisinopril did not provide an additional reduction in mortality, but there was an additional effect on hospitalization.

Furthermore, all these assumptions (and uncertainties) demonstrate again the disadvantage of not having dose-finding in outcome trials, a task which, however, is almost impossible to fulfill (Dickstein 2001).

2.4 Patient Groups with a High Cardiovascular Risk

The concept of isolated risk factor intervention is well established. The next step in the evolution of preventive medicine is to apply preventive measures on the basis of total risk (British Cardiac Society et al. 1998). The reason is that priority should be given to those patients who are at highest risk of developing CHD. Consequently, the budgets of health care providers are preferably directed towards patients with the highest overall risks. Risk equations and computer programs such as the Cardiac Risk Assessor have been developed based on the Framingham data set (Andersen et al. 1991). Furthermore, coronary risk charts have been developed (Wood et al. 1998) or coronary risk can easily be assessed on the internet (e.g., www.chd-taskforce.de).

The definitions of primary and secondary prevention were termed, for example, to form groups with lower and higher risk. These definitions may be superseded in the future because they do not serve this purpose precisely. It became evident that asymptomatic patients in the highest risk decile of primary prevention may have a higher risk than patients following MI in secondary prevention (Assmann et al. 2002).

Large-scale intervention studies have been successfully conducted with platelet inhibitors and lipid-lowering agents such as statins. The effects are substantial but not sufficient. New concepts were derived from research in atherosclerosis, which was fostered by a better understanding of endothelial changes, plaque formation, etc. A particular role was seen for the renin-angiotensin system.

With ACE inhibitors, the following studies are or have been conducted:

- HOPE
- DIABHYCAR
- EUROPA
- PEACE (ongoing)

With A2As, the following studies are ongoing:

- ONTARGET
- TRANSCEND

2.4.1 HOPE

In the HOPE study, ramipril was assessed in a broad spectrum of patients who were at high risk for cardiovascular events but who did not have left ventricular dysfunction or heart failure. A total of 9,297 high-risk patients (55 years of age or older) who had evidence of vascular disease or diabetes plus one other cardiovascular risk factor and who were not known to have a low ejection fraction or heart failure were randomly assigned to receive ramipril (10 mg once per day orally) or matching placebo for a mean of 5 years. The primary outcome was a composite of myocardial infarction, stroke, or death from cardiovascular causes. Of those who were assigned to receive ramipril, 14.0% reached the primary endpoint, as compared with 17.8% who were assigned to receive placebo (RR, 0.78; $p<0.001$).

Treatment with ramipril reduced the rates of death from cardiovascular causes (6.1% as compared with 8.1% in the placebo group; RR, 0.74; $p<0.001$), myocardial infarction (9.9% vs 12.3%; RR, 0.80; $p<0.001$), stroke (3.4% vs 4.9%; RR, 0.68; $p<0.001$), death from any cause (10.4% vs 12.2%; RR, 0.84; $p=0.005$), revascularization procedures (16.3% vs 18.8%; RR, 0.85; $p<0.001$), cardiac arrest (0.8% vs 1.3%; RR, 0.62; $p=0.02$), and complications related to diabetes (6.4% vs 7.6%; RR, 0.84; $p=0.03$). In conclusion, ramipril significantly reduced the rates of death, myocardial infarction, and stroke in a broad range of high-risk patients who were not known to have a low ejection fraction or heart failure (The Heart Outcomes Prevention Evaluation Study Investigators 2000a).

2.4.2 HOPE in Diabetic Patients

Diabetic patients were a prespecified stratum in the HOPE study. The primary evaluation of these patients was identical to those of the entire HOPE population. Ramipril lowered the risk of the combined primary outcome by 25% ($p=0.0004$), myocardial infarction by 22%, stroke by 33%, cardiovascular death by 37%, and total mortality by 24% (The Heart Outcomes Prevention Evaluation Study Investigators 2000b).

2.4.3 DIABHYCAR

In the double-blind placebo-controlled randomized DIABHYCAR study, 4,912 hypertensive and nonhypertensive patients with type 2 diabetes and microalbuminuria were included. They received 1.25 mg ramipril or placebo consecutively with the usual treatment. The primary end point was a composite of cardiovascular death, myocardial infarction, stroke, acute heart failure requiring hospitalization, or end-stage renal failure. After a median follow-up of 4.5 years, 793 primary end points were observed (3.8/100 patient years). There were no significant treatment effects (RR, 0.97).

The population investigated in DIABHYCAR is very similar to the prespecified diabetic population of HOPE and its Micro-HOPE subpopulation. The outcomes in Micro-HOPE in which 10 mg of ramipril were given were very positive. The authors of DIABHYCAR hence concluded that the high dose of 10 mg ramipril is important and needed for the demonstration of a clinical effect (Gueret et al. 2002).

2.4.4 EUROPA

The EUROPA Study is a double-blind, placebo-controlled, multicenter study in 24 countries in Europe. A total of 10,500 patients (>18 years) with coronary artery disease without clinical signs of heart failure were randomized to either placebo or perindopril (titrated up to 8 mg per day) for at least 36 months. The primary end point was a composite of cardiovascular death, nonfatal MI, or cardiac arrest with successful resuscitation (Fox et al. 1998). Results were released only recently (The European trial On Reduction of Cardiac Events with Perindopril in Stable Coronary Artery Disease Investigators 2003). They were largely similar to those of the HOPE study. Effects on stroke, however, were much smaller. EUROPA did not cover as broad a study population as was investigated in the HOPE study.

2.4.5 PEACE

The PEACE trial is an 8,100-patient, randomized, double-blind, placebo-controlled trial in patients (>50 years) with proven coronary artery disease and preserved left-ventricular function (LVEF >40%). Patients are randomized to either trandolapril (up-titrated to 4 mg per day) or placebo. All patients will be followed up for 4 years. The primary end point is a composite of cardiovascular death, nonfatal MI, or need for revascularization (Pfeffer et al. 1998).

2.4.6
ONTARGET

Currently, the three-arm ONTARGET study is investigating the effects of an A2A (telmisartan), of an ACE inhibitor (ramipril), and of their combination. ONTARGET is a long-term study over 5.5 years in 23,400 patients. Patients must have established coronary artery disease, stroke, peripheral vascular disease, or diabetes with end-organ damage. Patients with congestive heart failure will be excluded. The primary end point is a composite of cardiovascular death, MI, stroke, and hospitalization for heart failure. Secondary end points will investigate reductions in the development of diabetes mellitus, nephropathy, dementia, and atrial fibrillation (Yusuf 2002).

2.4.7
TRANSCEND

In a parallel study, patients unable to tolerate an ACE inhibitor will be randomized to receive telmisartan or placebo. The end points are as in ONTARGET, with 8,000 patients recruited.

2.4.8
Conclusions, Open Questions, Hypotheses

The ACE inhibitors ramipril and perindopril are the only ACE inhibitors for which the results of a large-scale clinical trial have become available. These angiotensin-converting enzyme inhibitors improve the clinical outcome among patients with apparently normal left ventricular function. It can be stated that:

- The ACE inhibitor ramipril reduces cardiovascular events (fatal and nonfatal MI and stroke) in a broad spectrum of high-risk patients.
- The ACE inhibitor ramipril reduces cardiovascular events also in diabetic patients. However, dose matters: a high dose of 10 mg is effective while a low dose of 1.25 mg is not effective.
- The ACE inhibitor perindopril reduced cardiovascular events (but not stroke) in a smaller spectrum of patients with CHD.

A direct comparison of an A2A (telmisartan) with an ACE inhibitor (ramipril) is ongoing (ONTARGET study). Currently—in the absence of results from clinical trials—it can only be stated that:

- A2As have not shown that they are better or worse than, or are equal to, the ACE inhibitor ramipril in high-risk patients.

2.5 Stroke

Stroke is the second largest cause of death in many societies. Disability in survivors is frequent and often irreversible. Therefore, primary and secondary prevention is of great importance. Intervention with platelet inhibitors such as ASA, ticlopidine, and clopidogrel has been proven beneficial in primary and/or secondary prevention trials. Still, event rates are high and deserve further reduction.

Blood pressure is a major determinant for the risk of stroke among both hypertensive and nonhypertensive individuals. It has been shown that antihypertensive treatment consistently reduces the occurrence of stroke. The effect is apparently related to the amount of blood pressure lowering; the choice of antihypertensive drug appears to be of little importance. Since the first major antihypertensive trials were performed (e.g., the VA studies in 1967 and 1970) (VA Cooperative Group on Antihypertensive Agents 1967, 1970), the reduction in stroke was always much easier to prove than the reduction in myocardial infarction. Therefore, the primary target of many antihypertensive studies has been the occurrence of stroke, for example, when calcium channel blockers were studied in the elderly (SHEP, Syst-Eur, Syst-China).

Atherosclerosis, plaque rupture, and vascular occlusion are promoted by angiotensin II (Lonn et al. 1994) and, thus, the renin-angiotensin system (RAS) may also play an important role in the pathogenesis of stroke. Conceivably, drugs interfering with the RAS are of interest not only in hypertensive but also in other populations with a high cardiovascular risk. No study dedicated to primary stroke prevention alone in high-risk populations and involving these substances is currently available. For secondary prevention after stroke, only the PROGRESS study has been completed so far. Therefore, this overview also encompasses those studies in which stroke contributed to a primary composite end point and prespecifying stroke as a secondary end point.

With ACE inhibitors, add-on therapy has been tested in two large-scale outcome trials:

- HOPE (in a broad spectrum of high-risk patients)
- PROGRESS (in secondary prevention after stroke).

With A2As, there is no study particularly tailored to stroke prevention but stroke was a monocomponent of the primary composite end point in the following studies:

- LIFE (in hypertensive patients with LVH)
- SCOPE (in elderly patients with mild hypertension).

2.5.1 HOPE

In the placebo-controlled HOPE study, the effects of the ACE inhibitor ramipril on stroke prevention were investigated. Stroke was part of the primary composite end point and was a prespecified secondary end point. HOPE was performed as a randomized controlled trial with a 2×2 factorial design in 267 hospitals in 19 countries in 9,297 patients with vascular disease or diabetes plus an additional risk factor. Follow-up was 4.5 years. Stroke was confirmed by computed tomography or magnetic resonance imaging when available. Blood pressure was recorded at entry to the study, after 2 years, and at the end of the study. Reduction in blood pressure was modest (3.8 mmHg systolic and 2.8 mmHg diastolic). The relative risk of any stroke was reduced by 32% (156 vs 226) in the ramipril group compared with the placebo group, and the relative risk of fatal stroke was reduced by 61% (17 vs 44). Benefits were consistent across baseline blood pressures, drugs used, and subgroups defined by the presence or absence of previous stroke, coronary artery disease, peripheral arterial disease, diabetes, or hypertension. It was concluded that ramipril reduces the incidence of stroke in patients at high risk, despite a modest reduction in blood pressure (Bosch et al. 2002).

2.5.2 PROGRESS

The study on perindopril protection against recurrent stroke (PROGRESS) was designed to determine the effects of a blood-pressure-lowering regimen in hypertensive and nonhypertensive patients with a history of stroke or transient ischemic attack. A total of 6,105 patients were randomly assigned active treatment (n=3,051) or placebo (n=3,054). Active treatment comprised a flexible regimen based on perindopril (4 mg daily), with the addition of the diuretic indapamide at the discretion of treating physicians. The primary outcome was total stroke (fatal or nonfatal). Over 4 years of follow-up, active treatment reduced blood pressure by 9/4 mmHg. A total of 307 (10%) individuals assigned active treatment suffered a stroke, compared with 420 (14%) assigned placebo (RRR, 28%, $p<0.0001$). Active treatment also reduced the risk of total major vascular events (26%). There were similar reductions in the risk of stroke in hypertensive and nonhypertensive subgroups (all $p<0.01$). Combination therapy with perindopril plus indapamide reduced blood pressure by 12/5 mmHg and stroke risk by 43% (30%–54%). Single-drug therapy reduced blood pressure by 5/3 mmHg and produced no significant reduction in the risk of stroke. This blood-pressure-lowering regimen reduced the risk of stroke among both hypertensive and nonhypertensive individuals with a history of stroke or transient ischemic attack. Combination therapy with perindopril and indapamide produced a greater reduction in both blood pressure and risk than did single-drug therapy with perindopril alone (PROGRESS Collaborative Group 2001).

2.5.3 LIFE

In the LIFE study, a total of 9,193 participants aged 55–80 years with hypertension and LVH were randomized to losartan or atenolol (see Sect. 2.1 in this chapter). Stroke was significantly reduced with losartan in comparison to atenolol: a total of 232 and 309 patients, respectively, experienced fatal or nonfatal stroke (RR, 0.75; p=0.001). Patients in the LIFE study represent a hypertensive population that is quite different from those in HOPE and in PROGRESS (Dahlöf et al. 2002).

2.5.4 SCOPE

In the placebo-controlled SCOPE study (Study on Cognition and Prognosis in the Elderly), candesartan was investigated in 4,964 elderly patients aged 70–79 with mild hypertension (SBP, 160–179 or DBP, 90–99 mmHg). Candesartan was given as 8 mg and doubled to 16 mg after 3 months if BP response was not sufficient. After 4 years of treatment, candesartan had not reduced the risk of major cardiovascular events, which was the primary end point (–11%; p=0.19). However, it did significantly reduce the risk of nonfatal strokes, which was a secondary end point (–28%; p=0.041) in these patients. SCOPE was the first outcome study in the elderly with mild hypertension (Sever 2002). This population differed from those in HOPE, PROGRESS, and LIFE.

2.5.5 Conclusions, Open Questions, Hypotheses

It has been known for quite some time that blood pressure reduction is an effective tool for stroke prevention, apparently independent of the type of the antihypertensive used. In recent years, stroke prevention was extended to populations with a risk greater than that of hypertension alone. ACE inhibitors and A2As were deemed particularly suited for this task.

For ACE inhibitors, an interesting data situation was generated in three large-scale trials:

- The ACE inhibitor ramipril has shown to have a strong preventive effect that is largely independent from blood pressure lowering.
- The ACE inhibitor perindopril, however, showed only a small effect in primary prevention of stroke.
- The ACE inhibitor perindopril has shown to have a secondary preventive effect that is pronounced only in combination with indapamide.

This situation gives rise to some questions that could only be answered in direct head-to-head comparison studies:

- Are the differences in outcomes between ramipril and perindopril genuine?
- Are methodological issues responsible for the disparity in results? Secondary prevention, for example, may well be much more difficult to attain or the type of concomitant treatment given may play an important role.

For A2As, data are available from two large-scale clinical trials:

- A2As prevented stroke in two different populations (LIFE, SCOPE).

A2As are likely equally effective than ACE inhibitors.

2.6 Nephropathy

The incidence and prevalence of nephropathy is on the increase. A pragmatic classification is based on the presence or absence of diabetes:

- Nondiabetic nephropathy
- Type 1 diabetic nephropathy (T1D)
- Type 2 diabetic nephropathy (T2D)

The most important cause for nephropathy is type 2 diabetes, which accounts for about 75% of all nephropathies. Most importantly, type 2 diabetes is on the rise. Consequently, nephropathy and end-stage renal failure (ESRF) are also increasing. Interestingly, patients with type 2 diabetes may have a worse prognosis than those with type 1 diabetes because the former often have additional risk factors such as hypertension, dyslipidemia, etc.

A brief mention of the various end points summarizing the outcome of the fierce regulatory debates of recent years follows.

- The ultimate goal of drug treatment is prevention of end-stage renal failure, which means a need for dialysis or a need for transplantation. This end point of ESRF *must* be investigated, either alone or as a major contributor to a composite end point. Otherwise, health authorities will not grant regulatory approval for the indication nephroprotection. Doubling of serum creatinine concentrations (DOC) has frequently been used as another component of a composite end point. Patients whose creatinine level doubles show ESRF usually 9–15 months thereafter.
- Other end points such as a halt in the decline of renal function (delta GFR) or in the decline in creatinine clearance are not accepted by health authorities, a circumstance which is regretful. Stopping nephropathy at earlier stages is desirable from a clinical point of view.
- The same is true for albuminuria and proteinuria as an end point. Proteinuria is not only a marker but also a modifiable risk factor for renal and cardiovascular disease, for instance, in type 2 diabetes. Reduction in proteinur-

ia can be measured as well as prevention of progression from one to another stage (e.g., from microalbuminuria to macroalbuminuria). From a medical standpoint, there is a good reason to do so and to intervene early. For example, patient, with type 1 diabetes developing microalbuminuria will inevitably develop ESRF in approximately 15 years barring death in the interim from heart attack or stroke. As it is not possible to conduct a double-blind controlled clinical trial over 15 years, it seems reasonable to evaluate and stop progression of proteinuria as an intermediate step. Such evidence, however, is not accepted by health authorities for drug approval.

At baseline, the severity of glomerular nephropathy may be expressed by creatinine serum values, creatinine clearance, or protein or albumin excretion. For the same level of creatinine clearance, it is protein excretion that predicts outcome. Therefore, patients with impaired creatinine clearance are often stratified further according to proteinuria levels at baseline.

Until a decade ago, no reliable proof was available of any preventive effects with pharmaceutical drugs. Only in 1993 was the first landmark study (ACE Inhibitor I study) published showing prevention of DOC with the ACE inhibitor captopril in type 1 diabetic patients. Thereafter, evidence was provided with composite end points (DOC, ESRF, and death) in nondiabetic patients (AIPRI, REIN). Surprisingly, patients with T2D nephropathy were not initially tested, thus leaving the door open for placebo-controlled trials testing first-line treatment with A2As.

The relevant studies with the following outcome parameters (according to the strength of regulatory acceptance in descending order) are described below:

- Progression from impaired renal function with macroalbuminuria to ESRF
- Decline in renal function
- Progression from microalbuminuria to macroalbuminuria
- Progression from normalbuminuria to microalbuminuria

2.6.1 Progression from Impaired Renal Function to ESRF

RENAAL. In the placebo-controlled randomized RENAAL study, a total of 1,513 patients were randomized to 50–100 mg once daily losartan or placebo. Mean follow-up was 3.4 years (Brenner et al. 2001). Patients had to have type 2 diabetes, a serum creatinine between 1.3 and 3.0 mg/dl, and signs of protein excretion such as proteinuria over 500 mg. The primary outcome was a composite of a doubling of the base-line serum creatinine, end-stage renal disease, or death. The prespecified secondary end point, morbidity and mortality from cardiovascular causes was a composite of myocardial infarction, stroke, first hospitalization for heart failure or unstable angina, coronary or peripheral revascularization, or death from cardiovascular causes. The primary end point was significantly reduced by a RR of 16% (327 vs 359 events; p=0.02); the secondary end

point was insignificantly reduced by 10% (247 vs 268 events; $p=0.26$). Hospitalization for heart failure was significantly reduced by 32% (89 vs 127, $p<0.005$); myocardial infarction was insignificantly reduced by 28% (50 vs 68, $p=0.08$). Death rate was not reduced (21% vs 20.3%).
IDNT. In the three-arm IDNT study, a total of 1,715 patients were randomly assigned 300 mg daily irbesartan, 10 mg daily amlodipine, or placebo (Lewis et al. 2001). Patients had to have type 2 diabetes, hypertension (>135 mmHg SBP or >85 mmHg DBP), proteinuria (>900 mg per 24 h), and a serum creatinine concentration between 1.0 and 3.0 mg/dl in women and 1.2 and 3.0 mg/dl in men. Mean treatment duration was 2.6 years. The primary end point was a composite of DOC, ESRF, or death from any cause. The secondary, cardiovascular end point was the composite of death from cardiovascular causes, nonfatal myocardial infarction, heart failure resulting in hospitalization, a permanent neurological deficit caused by a cerebrovascular event, or lower limb amputation above the ankle. With irbesartan, the primary end point was 20% lower than that in the placebo group ($p=0.02$) and 23% lower than that in the amlodipine group ($p=0.006$). The secondary cardiovascular end point was not significantly altered with irbesartan (23.8% irbesartan, 22.6 amlodipine, 25.3% placebo). Death rates were not significantly different (15.0% irbesartan, 14.6% amlodipine, 16.3% placebo).

2.6.2 Deterioration in Renal Function

ACE Inhibitor I and II Studies. In the placebo-controlled ACE Inhibitor I study, 207 patients were randomized to captopril and 202 to placebo. In the placebo group, antihypertensive drugs other than ACE inhibitors were given in order to achieve the same blood pressure targets. Patients with type 1 diabetes were included if their urinary protein excretion was greater than 500 mg per day and the serum creatinine concentration was less than 2.5 mg/dl. The primary end point was doubling of creatinine (to at least 2.0 mg/dl); one of the secondary end points was a combination of death, dialysis, and transplantation. The primary end point was reduced by 48% with captopril (25 vs 43 patients on placebo had DOC; $p=0.007$); the secondary end point was reduced by 50% (23 vs 42 patients; $p=0.006$). Eight patients died with captopril, 14 with placebo. The effect of the ACE inhibitor was regarded as independent of a small disparity in blood pressure between the groups (Lewis et al. 1993).

Thereafter, those 129 patients remaining after the ACE Inhibitor I study with kidney survival were randomized in the ACE Inhibitor II study to an intensive blood pressure goal (group I, MAP <92 mmHg) or to a normal blood pressure goal (group II, MAP 100–107 mmHg) with low or high doses of the ACE inhibitor ramipril. After a 2-year follow-up, there was a significant difference in total urinary protein excretion between group I (535 mg/24 h) and group II (1,723 mg/24 h). Other exploratory outcome parameters did not differ significantly, possibly because the patients included represented the good risks and their follow-up period was too short (Lewis et al. 1999).

REIN. The REIN study was a placebo-controlled study in 352 patients with glomerular nephropathy (creatinine clearance of 20–70 ml/min/1.73 m^2 and persistent proteinuria at enrollment). The vast majority of the subjects were nondiabetic. Before randomization, subjects were stratified according to baseline 24-h proteinuria values (stratum 1, proteinuria >1 and <3 g/24 h; stratum 2, proteinuria ≥3 g/24 h). Subjects were randomized within strata to receive either placebo or ramipril during a 27-month double-blind phase. Thereafter, ramipril subjects were to remain on open therapy for a further 33 months; placebo subjects were to remain in the study without further study medication.

The primary efficacy variable was the rate of change in glomerular filtration rate (GFR) per month. The most important secondary variable was the combined clinical end point of end-stage renal failure or doubling of the baseline serum creatinine level (DOC). During the overall study period, the mean rate of delta GFR per month was significantly less in the ramipril than in the conventional-treatment group (0.37 vs 0.51 ml/min/1.73m^2). Furthermore, ramipril decreased the risk for ESRD by 48% (28 vs 53 patients) (Ruggenenti et al. 2000). Effects were more pronounced in patients of stratum 2 with a higher protein excretion at baseline (The GISEN Group 1997) than in patients of stratum 1 (Ruggenenti et al. 1999).

AIPRI. In the double-blind AIPRI study, 583 patients with chronic renal insufficiency caused by various underlying renal diseases were randomized to benazepril or placebo. The primary outcome measure was defined as a composite end point of DOC or need for dialysis. Actually, however, the primary end point was driven by DOC and not by ESRF: at 3 years, 31 patients in the benazepril and 57 in the placebo group had reached the primary end point ($p<0.001$). A total of 86 patients had DOC and only 2 had ESRF. Eight patients in the benazepril group and one in the placebo group died (Maschio et al. 1996).

AASK. The AASK study was a randomized, double-blind, 3×2 factorial trial conducted in 1,094 African-Americans aged 18–70 years with hypertensive renal disease (GFR 20–65 ml/min/1.73 m^2) and proteinuria. Two hundred seventeen participants were randomly assigned to receive amlodipine, 5–10 mg/d, 441 patients to receive ramipril, 2.5–10 mg/d (n=436), and 441 patients to receive metoprolol, 50–200 mg/d. Other agents were added in a stratum for which tight blood pressure control was foreseen. The primary outcome measure was the rate of change in GFR; the main clinical outcome was a composite end point of reduction in GFR of more than 50% or 25 ml/min/1.73 m^2, end-stage renal disease, or death. Following an interim analysis after 3 years, treatment with amlodipine was terminated by the DSMB because ramipril was significantly superior to amlodipine. After adjustment for baseline covariates, the ramipril group had a 38% reduced risk of clinical endpoints, a 36% slower mean decline in GFR after 3 months (p=0.002), and less proteinuria ($p<0.001$) (Agodoa et al. 2001). Furthermore, it was shown after the end of the scheduled follow-up of

48 months, that ramipril also significantly reduced the risk by 22% (p=0.042) in comparison to metoprolol (Wright et al. 2002b).

2.6.3 Progression from Microalbuminuria to Macroalbuminuria

EUCLID. In the EUCLID study, 530 normotensive patients with IDDM and normoalbuminuria (n=440), microalbuminuria (n=79), or macroalbuminuria (n=6) were randomized to lisinopril or placebo. Patients were not on medication for hypertension. The rate of change in AER was the primary end point of the study. In the overall population, albumin excretion rate after 2 years was 2.2 μg/min lower (RRR, 18.8%) in the lisinopril group compared to placebo in an ITT analysis. For patients who completed 24 months of treatment, the final treatment difference in AER was 38.5 μg/min in those with microalbuminuria, and 0.23 μg/min in those with normoalbuminuria at baseline (The EUCLID Study Group 1997). Furthermore, retinopathy progressed less with lisinopril than with placebo.

STENO Type 2. STENO type 2 was an open randomized study in 160 patients with type 2 diabetes mellitus and microalbuminuria. Patients were allocated standard treatment (n=80) or intensive treatment (n=80). Intensive treatment was a stepwise implementation of behavior modification, pharmacological therapy targeting hyperglycemia, hypertension, dyslipidemia, and microalbuminuria. In the intense group, ACE inhibitors were prescribed independent of blood pressure. The primary end point was the development of overt nephropathy (median albumin excretion rate >300 mg/24 h in at least one of the two yearly examinations). Secondary end points were the incidence or progression of diabetic retinopathy and neuropathy. Patients in the intensive group had significantly lower rates of progression to nephropathy (odds ratio, 0.27), progression of retinopathy (OD, 0.45), and progression of autonomic neuropathy (OD, 0.32) than those in the standard group. It was concluded that the effect was attributable, at least in part, to ACE inhibitor treatment (Gaede et al. 1999).
Micro-HOPE. Micro-HOPE was a prespecified substudy of the 3,577 diabetic patients of the HOPE study. In Micro-HOPE, diabetic patients had to have microalbuminuria. At baseline, 31% in the ramipril group and 33% in the placebo group had microalbuminuria. Development of macroalbuminuria (which is regarded as a reliable marker of clinical overt nephropathy) was the prespecified main outcome in Micro-HOPE. The effects of 10 mg ramipril per day were compared to those of placebo. Macroalbuminuria (>300 mg/24 h urinary albumin, >500 mg/24 h urinary protein, or >36 mg/mmol albumin/creatinine ratio in the absence of 24-h urine) developed in 6.5% of *all* diabetic patients with ramipril and 8.4% with placebo, and thus was significantly reduced by 24% in the diabetic subgroup (p=0.0027) of HOPE. A total of 225 (20%) participants with microalbuminuria and 41 (2%) without microalbuminuria developed macroalbuminuria (RR, 14.0; p<0.0001). New macroalbuminuria was reduced by 24% with

ramipril in the total population of HOPE as well. HOPE and Micro-HOPE are the world's largest long-term observations of nonalbuminuric and microalbuminuric patients (Heart Outcomes Prevention Evaluation Study Investigators 2000).

IRMA II. In the IRMA II study, 590 hypertensive patients with type 2 diabetes were randomized to 150 mg daily irbesartan, 300 mg daily irbesartan, or placebo. The primary outcome was the time to onset of diabetic nephropathy, defined by persistent albuminuria in overnight specimens, with a urinary albumin excretion rate that was greater than 200 μg per minute and at least 30% higher than the base-line level. The primary end point was reached by 5.2%, 9.7%, and 14.7% of the 300 mg, 150 mg, and placebo group, respectively (Parving et al. 2001).

MARVAL. The MARVAL study was designed to evaluate the BP-independent effect of valsartan on the urinary albumin excretion rate (UAER) in type 2 diabetic patients with microalbuminuria. A total of 332 patients, with or without hypertension, were randomly assigned to 80 mg/d valsartan or 5 mg/d amlodipine for 24 weeks. A BP of 135/85 mmHg was targeted by means of dose doubling followed by addition of bendrofluazide and doxazosin whenever needed. The primary end point was the percent change in UAER from baseline to 24 weeks. The UAER at 24 weeks was 56% of baseline with valsartan and 92% of baseline with amlodipine, an effect that was highly significant between the groups. Valsartan lowered UAER similarly in both the hypertensive and normotensive subgroups. More patients reversed to normoalbuminuria with valsartan than with amlodipine (29.9% vs 14.5%; p=0.001). It was concluded that for the same level of attained BP and the same degree of BP reduction, valsartan lowered UAER more effectively than amlodipine in patients with type 2 diabetes and microalbuminuria, including the subgroup with baseline normotension. This indicates a BP-independent antiproteinuric effect of valsartan (Viberti and Wheeldon 2002).

Comparison of Losartan and Enalapril. Losartan and enalapril were compared with regard to changes in UAE in a small pilot study in 103 hypertensive type 2 diabetics with early nephropathy. The study was a one-year prospective, double-blind trial. Other antihypertensive drugs to achieve target blood pressure were given in 59.6% of the losartan group and 51% of the enalapril group. Both losartan and enalapril administered alone or in combination with other agents induced similar decreases in UAE: with losartan from 64.1 to 41.5 μg/min and with enalapril from 73.9 to 33.5 μg/min. Furthermore, a similar stabilization of the decline in glomerular filtration rate (GFR) was observed (Lacourciere et al. 2000).

2.6.4 Progression from Normalbuminuria to Microalbuminuria

The large-scale DCCT trial was performed in 1,441 patients with IDDM. This study successfully showed that prevention of progression of diabetic nephropathy, retinopathy, and neuropathy is possible with intensified insulin therapy in type 1 diabetic patients (The Diabetes Control and Complications Trial Research Group 1993). Furthermore, there was considerable hope that effects can be magnified further with substances inhibiting the renin-angiotensin system.
EUCLID. In the EUCLID study (see also "EUCLID" in Sect. 2.6.3), 530 normotensive patients with IDDM, of whom 440 were normalbuminuric, were randomized to lisinopril or placebo. Patients were not on medication for hypertension. The rate of change in AER was the primary end point of the study. In the overall population, albumin excretion rate after 2 years was 2.2 µg/min lower (RRR, 18.8%) in the lisinopril group compared to placebo in an ITT analysis. For patients who completed 24 months of treatment, the final treatment difference in AER was 0.23 µg/min in those with normoalbuminuria at baseline (Wright et al. 2002b).

Ravid et al.. The Ravid et al. study was a randomized, double-blind, placebo-controlled trial with 6-year follow-up in 156 normotensive, normoalbuminuric patients with type 2 diabetes. The aim was to evaluate the effect of prolonged ACE inhibition on renal function and albuminuria in normotensive, normalbuminuric patients in whom type 2 diabetes was diagnosed after 40 years of age who had a baseline mean blood pressure less than 107 mmHg and albuminuria (albumin excretion $\leq$30 mg/24 h). Patients were randomized to 10 mg/d enalapril or placebo. Initially, enalapril decreased albumin excretion from a mean 11.6 mg/24 h to 9.7 mg/24 h at 2 years; thereafter, a gradual increase to 15.8 mg/24 h at 6 years occurred. With placebo, albumin excretion increased from 10.8 mg/24 h to 26.5 mg/24 h at 6 years. Transition to microalbuminuria occurred in 6.5% patients with enalapril and 19% with placebo. After 6 years, creatinine clearance decreased with enalapril by 0.025 ml/s per year and by 0.04 ml/s per year with placebo (p=0.040). It was concluded that enalapril attenuated the decline in renal function and reduced the extent of albuminuria in normotensive, normoalbuminuric patients with type 2 diabetes. (Ravid et al. 1998).
ABCD-2V. The ongoing ABCD-2V study is an extension of the ABCD trial (see "ABCD" in Sect. 2.1.7) in 772 hypertensive or normotensive patients with type 2 diabetes to evaluate the effects of moderate vs intensive PB control. Patients are randomized to valsartan or placebo with add-on hydrochlorothiazide and metoprolol as needed. The primary objective is nephropathy; the secondary objective is the effects on cardiovascular, retinal, and neurological complications after 5 years of follow-up are expected soon.

2.6.5 Conclusions, Open Questions, Hypotheses

Nephropathy is an increasing burden for society, especially in patients with type 2 diabetes. The effects of diet, glucose control, and, most importantly, blood pressure lowering, are well established. In addition, blood pressure-independent effects were postulated for substances blocking the renin-angiotensin system.

ACE inhibitors have been thoroughly investigated with respect to different outcomes:

- ACE inhibitors have improved outcomes, which included ESRF in patients with type 1 diabetic nephropathy and in nondiabetic nephropathy.
- ACE inhibitors reduce decline in renal function.
- ACE inhibitors prevent progression from micro- to macroalbuminuria.
- ACE inhibitors prevent progression from normo- to microalbuminuria.
- An ACE inhibitor was better than amlodipine and metoprolol in Afro-Americans.

These effects of ACE inhibitors occur independently from blood pressure lowering. Furthermore, in a meta-analysis also including other smaller studies, these effects of ACE inhibitors were confirmed (Giatras et al. 1997).

ACE inhibitors, however, have not been well investigated in patients with type 2 diabetes, which has opened a window of opportunity for A2As. With A2As, the following evidence is established:

- A2As have improved composite end points, which included ESRF in patients with type 2 diabetic nephropathy.
- A2As prevent progression from micro- to macroalbuminuria.
- A2As prevent progression from normo- to microalbuminuria.

With ACE inhibitors and A2As, all these outcomes occur to a great extent independently from blood pressure lowering.

Still, an important question remains: are ACE inhibitors more effective than A2As in treating nephroprotection? In ACE Inhibitor I, REIN, and AIPRI, the reduction in end points is mostly 50%, in RENAAL and IDNT only 20%. Is this a true difference or an artifact? Only in a direct head-to-head comparison study could this question be answered.

Even more important is the question whether combination treatment of A2As with ACE inhibitors is superior to either treatment alone. The outcome of a recently published, relatively small Japanese trial, the COOPERATE study, points toward a benefit of combination treatment.

2.7
Other Potential Indications

There are many more ideas with respect to new indications for ACE inhibitors and A2As such as:

- Prevention of type 2 diabetes (DREAM, NAVIGATOR)
- Prevention of microvascular diabetic complications (DIRECT)
- Prevention of restenosis following stent implantation (ValPREST)
- Protection following revascularization (APRES)

2.7.1
Prevention of Type 2 Diabetes

Prevention of type 2 diabetes has come into the focus of recent clinical research. It has been shown that different types of intervention can successfully suppress new diabetes, including diet and exercise (Tuomilehto et al. 2001; Pan et al. 1977; Diabetes Prevention Program Research Group 2002), acarbose (The STOP-NIDDM Trial Research Group 1998; Chiasson et al. 2002), orlistat (Keating and Jarvis 2001), metformin (Diabetes Prevention Program Research Group 2002), troglitazone (Buchanan et al. 2002), statins (Freeman et al. 2001).

In addition, substances affecting the renin-angiotensin system showed similar effects in post-hoc analysis of major trials such as:

- The HOPE study, in which ramipril reduced the new onset of type 2 diabetes by 34%.
- The LIFE study, in which losartan reduced the new onset of type 2 diabetes by 25% (Lindholm et al. 2002a).

These post-hoc data yielded good reasons to proceed with two major prospective large-scale trials.

DREAM. The post-hoc finding in the HOPE study that ramipril treatment was associated with a significant 34% reduction in new diagnoses of diabetes is now being prospectively evaluated in the large-scale DREAM study. A total of 4,000 patients with impaired glucose tolerance (IGT) will be randomized, requiring an enormous screening effort. The Diabetes Reduction Assessment with ramipril and rosiglitazone (DREAM) is a trial with a 2×2 factorial design with four treatment groups: ramipril (up to 15 mg/day), rosiglitazone (8 mg/day), ramipril/rosiglitazone combination, and placebo. The primary composite end point is the occurrence of new type 2 DM or death. Participants will be followed up for at least 3 years after randomization (Gerstein 2002).

NAVIGATOR. NAVIGATOR is a randomized, double-blind, placebo-controlled trial with a 2×2 factorial design in 7,500 patients with impaired glucose tolerance (IGT) at high risk for cardiovascular events. Study participants will receive valsartan, nateglinide, both valsartan and nateglinide, or placebo. There are two primary end points: delay or prevention of progression to diabetes (as assessed 3 years after enrollment of the last patient) and, in an extension phase, prevention of cardiovascular morbidity and mortality (which will be assessed after 1,200 events approximately 5–6 years after enrollment).

2.7.2 Prevention of Microvascular Diabetic Complications

EUCLID. In the EUCLID study (see also "EUCLID" in Sects. 2.6.3 and 2.6.4), 530 normotensive patients with IDDM and normoalbuminuria (n=440), microalbuminuria (n=79), or macroalbuminuria (n=6) were randomized to lisinopril or placebo. After 24 months, retinopathy had progressed by at least one level in 13.2% of 159 patients on lisinopril and 23.4% of 166 patients on placebo. However, patients on lisinopril had a significantly lower HbA1c at baseline than those on placebo (6.9% vs 7.3%; p=0.05). This means that the findings on retinopathy need to be confirmed in a larger trial (Chaturvedi et al. 1998).

DIRECT. Diabetic retinopathy is, indeed, investigated in a large-scale ongoing clinical trial. In the placebo-controlled DIRECT study, 4,500 patients with type 2 diabetes are being investigated with candesartan titrated up to 32 mg candesartan. The primary end point is the occurrence of diabetic retinopathy.

2.7.3 Prevention of Restenosis Following PCI

ValPREST. The open-label, placebo-controlled ValPREST study was conducted in 250 patients to evaluate the effects of 6-month administration of valsartan on restenosis rate after stenting of type B2/C lesions (Peters et al. 2001). In-stent restenosis rates (ISR) calculated based on quantitative coronary angiography (QCA) and the need for reintervention as primary and secondary end points were analyzed after a repeat angiogram at 6 months in 99 patients with 80 mg valsartan and 101 patients with placebo. The ISR was 19.2% (n=19/99) with valsartan and 38.6% (n=39/101) with placebo (p<0.005). The reintervention rate was 28.7% (n=29/101) in the placebo group and 12.1% (n=12) in the valsartan group (p<0.005). These results need to be confirmed in a larger double-blind trial with clinical outcome as the primary end point. Previously, several attempts with ACE inhibitors in, for example, the MERCATOR, MARCATOR, and PARIS studies (MERCATOR Study Group 1992; Berger et al. 1996; Meurice et al. 2001), had failed to show efficacy. It must be emphasized that, in the past, many attempts to inhibit restenosis after angioplasty or stenting had yielded negative results with other drugs as well (Paranandi and Topol 1994). In future, it is unlikely that drugs affecting the renin-angiotensin system will demonstrate stronger effects than those arising from new therapies such as brachytherapy

(Seabra-Gomes 2002) or stents coated with rapamycin (Morice et al. 2002) or paclitaxel (Liistro et al. 2002).

2.7.4
Cardiovascular Prognosis After Revascularization

APRES. The double-blind APRES study was an interesting pilot study (Kjoller-Hansen et al. 2000). APRES was conducted in 159 asymptomatic patients with moderate left ventricular dysfunction who underwent revascularization with coronary artery bypass graft (CABG) (n=130) or PTCA (n=29). Patients were randomized 6 days after revascularization to ramipril or placebo; they were followed up for 33 months. The primary composite end point (cardiac death, AMI, development of clinical heart failure, or recurrent angina pectoris) was not significantly altered (36 vs 41 events), because of the event rates for angina pectoris (34 vs 30 events). The triple composite end point (without angina pectoris) was significantly lower with ramipril (8 events) vs placebo (18 events). Patients in the APRES study reflect a high-risk population such as those in the HOPE study. Whether additional protection is needed in patients following bypass surgery remains an open question.

3
Summary

This overview is simply a snapshot of the current situation, which is likely to change in the light of the many ongoing large-scale trials.

In daily therapeutic use, ACE inhibitors cover a broad spectrum of clinical indications. They have been introduced for:

- Hypertension

Thereafter, they have provided the breakthrough for several indications and may be regarded as:

- First-line treatment of congestive heart failure to which other drugs are added
- After fibrinolysis or PCI, first-line treatment in acute myocardial infarction
- First-line treatment of patients with a high cardiovascular risk
- First-line treatment of patients with nondiabetic and type 1 diabetic nephropathy
- Secondary prevention of stroke

This broad spectrum of indications does not apply in the case of A2As. However, patients on A2As demonstrate fewer side effects, especially cough. Undoubtedly, A2As are suitable as:

- First-line treatment for patients with type 2 diabetic nephropathy
- Treatment of CHF in patients who do not tolerate ACE inhibitors
- Add-on treatment of CHF in addition to ACE inhibitors

There is no indication so far for which evidence of superiority of A2As over ACE inhibitors has been provided. Evidence is also accumulating which demonstrates that both drug classes are superior to other antihypertensives in patients with diabetes, an increasing burden to mankind. Furthermore potential for prevention of new type 2 diabetes is on the horizon.

References

ACE Inhibitor Myocardial Infarction Collaborative Group (1998) Indications for ACE inhibitors in the early treatment of acute myocardial infarction. Circulation 97:2202–2212

The Acute Infarction Ramipril Efficacy (AIRE) Study Investigators (1993) Effect of ramipril on mortality and morbidity of survivors of acute myocardial infarction with clinical evidence of heart failure. Lancet 342:821–828

Agabiti-Rosei E, Ambrosioni E, Dal Palu C et al (1995) ACE inhibitor ramipril is more effective than the beta-blocker atenolol in reducing left ventricular mass in hypertension. Results of the RACE (ramipril cardioprotective evaluation) study on behalf of the RACE study group. J Hypertens 13:1325–1334

Agodoa LY, Appel L, Bakris GL et al (2001) Effect of ramipril vs amlodipine on renal outcomes in hypertensive nephrosclerosis: a randomized controlled trial. JAMA 285:2719–2728

The ALLHAT Officers and Coordinators for the ALLHAT Collaborative Research Group (2002) Major outcomes in high-risk hypertensive patients randomized to angiotensin-converting enzyme inhibitor or calcium channel blocker vs diuretic. JAMA 288:2981–2997

Ambrosioni E, Borghi C, Magnani B (1995) The effect of the angiotensin-converting-enzyme inhibitor zofenopril on mortality and morbidity after anterior myocardial infarction. The Survival of Myocardial Infarction Long-Term Evaluation (SMILE) Study Investigators. N Engl J Med 332:80–85

Andersen KM, Wilson PWF, Odell PM et al (1991) An updated coronary risk profile: a statement for health professionals. Circulation 83:356–362

Assmann G, Cullen P, Schulte H (2002) Simple scoring scheme for calculating the risk of acute coronary events based on the 10-Year follow-up of the Prospective Cardiovascular Münster (PROCAM) Study. Circulation 105:310–315

Bakris G, Gradman A, Reif M et al (2001) Antihypertensive efficacy of candesartan in comparison to losartan: the CLAIM study. J Clin Hypertens 3:16–21

Berger PB, Holmes DR Jr, Ohman EM et al (1996) Restenosis, reocclusion and adverse cardiovascular events after successful balloon angioplasty of occluded versus nonoccluded coronary arteries. Results from the Multicenter American Research Trial With Cilazapril After Angioplasty to Prevent Transluminal Coronary Obstruction and Restenosis (MARCATOR). J Am Coll Cardiol 27:1–7

Bosch J, Yusuf S, Pogue J et al (2002) Use of ramipril in preventing stroke: double blind randomised trial. BMJ 324:699–702

Brenner BM, Cooper ME, de Zeeuw D et al (2001) Effects of losartan on renal and cardiovascular outcomes in patients with type 2 diabetes and nephropathy. N Engl J Med 345:861–869

British Cardiac Society, British Hyperlipidaemia Association, British Hypertension Society, endorsed by the British Diabetic Association (1998) Joint British recommendations on prevention of coronary heart disease in clinical practice. Heart 80 [Suppl 2]:S1–S29

Buchanan TA, Xiang AH, Peters RK et al (2002) Preservation of pancreatic β-cell function and prevention of type 2 diabetes by pharmacological treatment of insulin resistance in high-risk Hispanic women. Diabetes 51:2796–2803

Chaturvedi N, Sjolie AK, Stephenson JM et al (1998) Effect of lisinopril on progression of retinopathy in normotensive people with type 1 diabetes. The EUCLID Study Group. EURODIAB controlled trial of lisinopril in insulin-dependent diabetes mellitus. N Engl J Med 351, 28–3

Chiasson J-L, Josse RG, Gomis R et al (2002) Acarbose for prevention of type 2 diabetes mellitus: the STOP-NIDDM randomised trial. Lancet 359; 2072–2077

Chinese Cardiac Study Collaborative Group (1995) Oral captopril versus placebo among 13,634 patients with suspected acute myocardial infarction: interim report from the Chinese Cardiac Study (CCS-1). Lancet 345:686–687

Cohn JN, Tognoni G (2001) Valsartan Heart Failure Trial Investigators. A randomized trial of the angiotensin-receptor blocker valsartan in chronic heart failure. N Engl J Med 345:1667–1675

Cohn JN, Archibald DG, Ziesche S et al (1986) Effect of vasodilator therapy on mortality in chronic congestive heart failure. N Engl J Med 314:1547–1552

Cohn JN, Johnson G, Ziesche S et al (1991) A comparison of enalapril with hydralazine-isosorbide dinitrate in the treatment of chronic congestive heart failure. N Engl J Med 325:303–310

The CONSENSUS Trial Study Group (1987) Effects of enalapril on mortality in severe congestive heart failure. Results of the Cooperative North Scandinavian Enalapril Survival Study (CONSENSUS). The CONSENSUS Trial Study Group. N Engl J Med 316:1429–1435

Cuspidi C, Muiesan ML, Valagussa L et al (2002) Comparative effects of candesartan and enalapril on left ventricular hypertrophy in patients with essential hypertension: the CATCH study. Eur Heart J 23 (abstract P1530), 277

Dahlöf B, Devereux RB, Kjeldsen SE et al (2002) Cardiovascular morbidity and mortality in the losartan intervention for end point reduction in hypertension study (LIFE): a randomised trial against atenolol. Lancet 359:995–1003

Davis BR, Cutler JA, Gordon DJ et al (1996) Rationale and Design for the Antihypertensive and Lipid Lowering Treatment to Prevent Heart Attack Trial (ALLHAT). Am J Hypertens 9:342–360

The Diabetes Control and Complications Trial Research Group (1993) The effect of intensive treatment of diabetes on the development and progression of long-term complications in insulin-dependent diabetes mellitus. The Diabetes Control and Complications Trial Research Group. N Engl J Med 329:977–986

Diabetes Prevention Program Research Group (2002) Reduction in the incidence of type 2 diabetes with lifestyle intervention or metformin. N Engl J Med 346:393–403

Dickstein K (2001) ELITE II and Val-HeFT are different trials: together what do they tell us? Curr Control Trials Cardiovasc Med 2:240–243

Dickstein K, Kjekshus J for the OPTIMAAL Study Group (1998) Comparison of the effects of losartan and captopril on mortality in patients after acute myocardial infarction: The OPTIMAAL Trial. Design Exp Opin Invest Drugs 7:1897–1914

Dickstein K, Kjekshus J and the OPTIMAAL Steering Committee (2002) Effects of losartan and captopril on mortality and morbidity in high-risk patients after acute myocardial infarction: the OPTIMAAL randomised trial. Lancet 360:752–760

Estacio RO, Jeffers BW, Hiatt WR et al (1998) The effect of nisoldipine as compared with enalapril on cardiovascular outcomes in patients with non-insulin-dependent diabetes and hypertension. N Engl J Med 338:645–662

The EUCLID study group (1997) Randomised placebo-controlled trial of lisinopril in normotensive patients with insulin-dependent diabetes and normoalbuminuria or microalbuminuria. The EUCLID Study Group. Lancet 349:1787–1792

The European trial On Reduction of Cardiac Events with Perindopril in Stable Coronary Artery Disease Investigators (2003) Efficacy of perindopril in reduction of cardiovascular events among patients with stable coronary artery disease: randomised, double-blind, placebo-controlled, multicentre trial (the EUROPA study). Lancet 362:782–788

Flather MD, Yusuf S, Kober L et al (2000) Long-term ACE-inhibitor therapy in patients with heart failure or left-ventricular dysfunction: a systematic overview of data from individual patients. Lancet 355:1575–1581

Fox KM, Henderson JR, Bertrand ME et al (1998) The European trial on reduction of cardiac events with perindopril in stable coronary artery disease (EUROPA). Eur Heart J 19: J52–J55

Freeman DJ, Norrie J, Sattar N et al (2001) Pravastatin and the development of diabetes mellitus: evidence for a protective treatment effect in the west of Scotland Coronary Prevention Study. Circulation 103:357–362

Gaede P, Vedel P, Parving HH et al (1999) Intensified multifactorial intervention in patients with type 2 diabetes mellitus and microalbuminuria: the Steno type 2 randomised study. Lancet 353:617–622

Gerstein HC (2002) Reduction of cardiovascular events and microvascular complications in diabetes with ACE inhibitor treatment: HOPE and MICRO-HOPE. Diabetes Metab Res Rev 18: S82–S85

Giatras I, Lau J, Levey A (1997) Effect of angiotensin-converting enzyme inhibitors on the progression of nondiabetic renal disease: a meta-analysis of randomized trials. Angiotensin-Converting-Enzyme Inhibition and Progressive Renal Disease Study Group. Ann Intern Med 127:337–345

The GISEN Group (1997) Randomised placebo-controlled trial of effect of ramipril on decline in glomerular filtration rate and risk of terminal renal failure in proteinuric, non-diabetic nephropathy. The GISEN Group (Gruppo Italiano di Studi Epidemiologici in Nefrologia). Lancet 349:1857–1863

Granger CB, McMurray JJV, Yusuf S et al (2003) Effects of candesartan in patients with chronic heart failure and reduced left-ventricular systolic function intolerant to angiotensin-converting-enzyme inhibitors: the CHARM-Alternative trial. Lancet 362:772–776

Gruppo Italiano per lo Studio della Sopravivenza nell'Infarto Miocardio (1994) GISSI-3: effects of lisinopril and transdermal glyceryl trinitrate singly and together on 6-week mortality and ventricular function after acute myocardial infarction. Lancet 343:1115–1122

Gueret P, Marr M, Lievre M et al (2002) Effects of low-dose ramipril on cardiovascular events in type 2 diabetes patients with microalbuminuria/proteinuria: the DIABHYCAR study (abstract 169). Eur Heart J 4:8

Hall A, Murray GD, Ball SG (1997) Follow-up study of patients randomly allocated ramipril or placebo for heart failure after acute myocardial infarction: AIRE Extension (AIREX) Study. Lancet 349:1493–1497

Hansson L, Dahlöf B, Ekbom T et al (1990) Key Learnings from the STOP-hypertension study: an update on the progress of the ongoing Swedish study of antihypertensive treatment in the elderly. Cardiovasc Drugs Ther 4:1253–1256

Hansson L, Lindholm LH, Ekbom T et al (1999a) Randomised trial of old and new antihypertensive drugs in elderly patients: cardiovascular mortality and morbidity the Swedish Trial in Old Patients with Hypertension-2 study. Lancet 354:1751–1756

Hansson L, Lindholm LH, Niskanen L et al (1999b) Effect of angiotensin-converting-enzyme inhibition compared with conventional therapy on cardiovascular morbidity and mortality in hypertension: the Captopril Prevention Project (CAPPP) randomised trial. Lancet 353:611–616

Hansson L, Lithell H, Skoog I et al (1999c) Study on Cognition and Prognosis in the Elderly (SCOPE). Blood Press 8:177–183

Hansson L, Lithell H, Skoog I et al (2000) Study on Cognition and Prognosis in the Elderly (SCOPE): baseline characteristics. Blood Press 9:146–151

Heart Outcomes Prevention Evaluation Study Investigators (2000a) Effects of an angiotensin-converting-enzyme inhibitor, ramipril, on cardiovascular events in high-risk patients. N Engl J Med 342:145–153

Heart Outcomes Prevention Evaluation Study Investigators (2000b) Effects of ramipril on cardiovascular and microvascular outcomes in people with diabetes mellitus: results of the HOPE study and MICRO-HOPE substudy. Lancet 355:253–259

Ho KKL, Anderson KM, Kannel WB et al (1993) Survival after the onset of congestive heart failure in Framingham Heart Study subjects. Circulation 88:107–115

Hood WB, Youngblood M, Ghali JK et al (1991) Initial blood pressure response to enalapril in hospitalized patients (Studies of Left Ventricular Dysfunction [SOLVD]). Am J Cardiol 68:1465–1468

ISIS-4 (Fourth International Study of Infarct Survival) Collaborative Group (1995) ISIS-4: a randomised factorial trial assessing early oral captopril, oral mononitrate, and intravenous magnesium sulphate in 58050 patients with suspected acute myocardial infarction. Lancet 345:669–685

Keating GM, Jarvis B (2001) Orlistat in the prevention and treatment of type 2 diabetes mellitus. Drugs 61:2107–2119

Kjeldsen SE, Julius S, Brunner H et al (2001) Characteristics of 15,314 hypertensive patients at high coronary risk. The VALUE Trial. The Valsartan Antihypertensive Long-term Use Evaluation. Blood Press 10:83–91

Kjoller-Hansen L, Steffensen R, Grande P (2000) The Angiotensin-converting Enzyme Inhibition Post Revascularization Study (APRES). J Am Coll Cardiol 35:881–888

Køber L, Torp-Pedersen C, Carlsen JE et al (1995) A clinical trial of the angiotensin-converting-enzyme inhibitor trandolapril in patients with left ventricular dysfunction after myocardial infarction. N Engl J Med 333:1670–1676

Lacourciere Y, Belanger A, Godin C et al (2000) Long-term comparison of losartan and enalapril on kidney function in hypertensive type 2 diabetics with early nephropathy. Kidney Int 58:762–769

Lewis EJ, Hunsicker LG, Bain RP et al (1993) The effect of angiotensin-converting-enzyme inhibition on diabetic nephropathy. The Collaborative Study Group. N Engl J Med 329:1456–1462

Lewis EJ, Hunsicker LG, Clarke WR et al (2001) Renoprotective effect of the angiotensin-receptor antagonist irbesartan in patients with nephropathy due to type 2 diabetes. N Engl J Med 345:851–860

Lewis JB, Berl T, Bain RP et al (1999) Effect of intensive blood pressure control on the course of type 1 diabetic nephropathy. Collaborative Study Group. Am J Kidney Dis 34:809–817

Liistro F, Stankovic G, Di Mario C et al (2002) First clinical experience with a paclitaxel derivate-eluting polymer stent system implantation for in-stent restenosis: immediate and long-term clinical and angiographic outcome. Circulation 105:1883–1886

Lindholm LH, Ibsen H, Borch-Johnsen K et al (2002a) Risk of new-onset diabetes in the Losartan Intervention For End point reduction in hypertension study. J Hypertens 20:1879–1886

Lindholm LH, Ibsen H, Dahlöf B et al (2002b) Cardiovascular morbidity and mortality in patients with diabetes in the losartan intervention for end point reduction in hypertension study (LIFE): a randomised trial against atenolol. Lancet 359:1004–1010

Lonn EM, Yusuf S, Jha P et al (1994) Emerging role of angiotensin-converting enzyme inhibitors in cardiac and vascular protection. Circulation 90:2056–2069

MacMahon S, Sharpe N, Gamble G et al (2000) Randomized, placebo-controlled trial of the angiotensin-converting enzyme inhibitor, ramipril, in patients with coronary or other occlusive arterial disease. PART-2 Collaborative Research Group. Prevention of Atherosclerosis with Ramipril. J Am Coll Cardiol 36:438–443

Maschio G, Alberti D, Janin G et al (1996) Effect of the angiotensin-converting-enzyme inhibitor benazepril on the progression of chronic renal insufficiency. The Angiotensin-Converting-Enzyme Inhibition in Progressive Renal Insufficiency Study Group. N Engl J Med 334:939–945

McKelvie RS, Yusuf S, Pericak D et al (1999) Comparison of candesartan, enalapril, and their combination in congestive heart failure: Randomized Evaluation of Strategies for Left Ventricular Dysfunction (RESOLVD) pilot study: the RESOLVD Pilot Study Investigators. Circulation 100:1056–1064

McMurray JVJ, Östergren J, Swedberg K et al (2003) Effects of candesartan in patients with chronic heart failure and reduced left-ventricular systolic function taking angiotensin-converting-enzyme inhibitors: the CHARM-Added trial. Lancet 362:767–771

Medical Research Council Working Party (1992) Medical Research Council trial of treatment of hypertension in older adults: principal results. MRC Working Party. BMJ 304:405–412

MERCATOR Study Group (1992) Does the new angiotensin converting enzyme inhibitor cilazapril prevent restenosis after percutaneous transluminal coronary angioplasty? Results of the MERCATOR study: a multicenter, randomized, double-blind placebo-controlled trial. Multicenter European Research Trial with cilazapril after angioplasty to prevent transluminal coronary obstruction and restenosis. Circulation 86:100–110

Messerli FH (2000) Implications of discontinuation of doxazosin arm of ALLHAT. Lancet 355:863–864

Messerli FH (2001) Doxazosin and congestive heart failure. J Am Coll Cardiol 38:1295–1296

Meurice T, Bauters C, Hermant X et al (2001) Effect of ACE inhibitors on angiographic restenosis after coronary stenting (PARIS): a randomised, double-blind, placebo-controlled trial. Lancet 357:1321–1324

Morice MC, Serruys PW, Sousa JE et al (2002) RAVEL Study Group. A randomized comparison of a sirolimus-eluting stent with a standard stent for coronary revascularization. N Engl J Med 346:1773–1780

Nakao N, Yoshimura A, Morita H et a (2003) Combination treatment of angiotensin-II receptor blocker and angiotensin-converting-enzyme inhibitor in non-diabetic renal disease (COOPERATE): a randomised controlled trial. Lancet 361:117–124

Niskanen L, Hedner T, Hansson L et al (2001) Reduced cardiovascular morbidity and mortality in hypertensive diabetic patients on first-line therapy with an ACE inhibitor compared with a diuretic/beta-blocker-based treatment regimen: a subanalysis of the Captopril Prevention Project. Diabetes Care 24:2091–2096

Packer M (1996) Do angiotensin-converting enzyme inhibitors prolong life in patients with heart failure treated in clinical practice? J Am Coll Cardiol 28:1323–1327

Packer M, Poole-Wilson PA, Armstrong PW et al (1999) Comparative effects of low and high doses of the angiotensin-converting enzyme inhibitor, lisinopril, on morbidity and mortality in chronic heart failure. ATLAS Study Group. Circulation 100:2312–2318

Pan XR, Li GW, Hu YH et al (1977) Effects of diet and exercise in preventing NIDDM in people with impaired glucose tolerance. The Da Qing IGT and Diabetes Study. Diabetes Care 20:537–544

Paranandi SN, Topol EJ (1994) Contemporary clinical trials of restenosis. J Invasive Cardiol 6:109–124

Parving H-H, Lehnert H, Bröchner-Mortensen J et al (2001) The effect of irbesartan on the development of diabetic nephropathy in patients with type 2 diabetes. N Engl J Med 345:870–878

Peters S, Gotting B, Trummel M et al (2001) Valsartan for prevention of restenosis after stenting of type B2/C lesions: the VAL-PREST trial. J Invasive Cardiol 13:93–97
Pitt B, Segal R, Martinez FA et al (1997) Randomised trial of losartan versus captopril in patients over 65 with heart failure (Evaluation of Losartan in the Elderly Study, ELITE) Lancet 349:747–752
Pitt B, Poole-Wilson PA, Segal R et al (2000) Effect of losartan compared with captopril on mortality in patients with symptomatic heart failure: randomised trial – the Losartan Heart Failure Survival Study ELITE II. Lancet 355:1582–1587
Pfeffer MA, Braunwald E, Moyé LA et al (1992) Effect of captopril on mortality and morbidity in patients with left ventricular dysfunction after myocardial infarction. N Engl J Med 327:669–677
Pfeffer MA, Domanski M, Rosenberg Y et al (1998) Prevention of events with angiotensin-converting enzyme inhibition (The PEACE Study Design). Am J Cardiol 82:25H–30H
Pfeffer MA, McMurray J, Leizorovics A et al (2000) Valsartan in Acute Myocardial Infarction Trial (VALIANT): rationale and design. Am Heart J 140:727–734
Pfeffer MA, McMurray JJV, Velazquez EJ et al (2003) Valsartan, Captopril, or both in myocardial infarction complicated by heart failure. Left ventricular dysfunction, or both. N Engl J Med 349:1893–1906
Pfeffer MA, Swedberg K, Granger CB et al (2003) Effects of candesartan on mortality and morbidity in patients with chronic heart failure: the CHARM-Overall programme. Lancet 362:759–766
PROGRESS Collaborative Group (2001) Randomised trial of a perindopril-based blood-pressure-lowering regimen among 6,105 individuals with previous stroke or transient ischaemic attack. Lancet 358:1033–1041
Ravid M, Brosh D, Levi Z et al (1998) Use of enalapril to attenuate decline in renal function in normotensive, normoalbuminuric patients with type 2 diabetes mellitus. A randomized, controlled trial. Ann Intern Med 128:982–988
The RESOLVD Investigators (2000) Effects of metoprolol CR in patients with ischemic and dilated cardiomyopathy. The Randomized Evaluation of Strategies for Left Ventricular Dysfunction Pilot Study. Circulation 101:378–384
Ruggenenti P, Perna A, Gherardi G et al (1999) Renoprotective properties of ACE-inhibition in non-diabetic nephropathies with non-nephrotic proteinuria. Lancet 354:359–364
Ruggenenti P, Perna A, Gherardi G et al (2000) Chronic proteinuric nephropathies: outcomes and response to treatment in a prospective cohort of 352 patients with different patterns of renal injury. Am J Kidney Dis 35:1155–1165
Seabra-Gomes R (2002) Intracoronary brachytherapy for restenosis: an efficient technique in the struggle for survival? Eur Heart J 23:1319–1321
Sever P (2002) The SCOPE trial. J Renin Angiotensin Aldosterone Syst 3:61–62
The SOLVD Investigators (1990) Studies of left ventricular dysfunction (SOLVD) – rationale, design and methods: two trials that evaluate the effect of enalapril in patients with reduced ejection fraction. Am J Cardiol 66:315–322
The SOLVD Investigators (1991) Effect of enalapril on survival in patients with reduced left ventricular ejection fractions and congestive heart failure. N Engl J Med 325:293–302
The SOLVD Investigators (1992) Effect of enalapril on mortality and the development of heart failure in asymptomatic patients with reduced left ventricular ejection fractions. N Engl J Med 327:685–691
The STOP-NIDDM Trial Research Group (1998) The STOP-NIDDM trial: an international study on the efficacy of an alpha-glucosidase inhibitor to prevent type 2 diabetes in a population with impaired glucose tolerance: rationale, design, and preliminary screening data. Study to Prevent Non-Insulin-Dependent Diabetes Mellitus. Diabetes Care 21:1720–1725

Swedberg K, Held P, Kjekshus J et al (1992) Effects of the early administration of enalapril on mortality in patients with acute myocardial infarction. N Engl J Med 327:678–684

Swedberg K, Pfeffer M, Granger C et al (1999) For the CHARM-programme investigators. Candesartan in heart failure—assessment of Reduction in Mortality and morbidity (CHARM): rationale and design. J Cardiac Failure 5:276–282

Tatti P, Pahor M, Byington RP et al (1998) Outcome results of the Fosinopril Versus Amlodipine Cardiovascular Events Randomized Trial (FACET) in patients with hypertension and NIDDM. Diabetes Care 21:597–603

Teo KK, Burton JR, Buller CE et al (2000) Long-term effects of cholesterol lowering and angiotensin-converting enzyme inhibition on coronary atherosclerosis: the Simvastatin/Enalapril Coronary Atherosclerosis Trial (SCAT). Circulation 102:1748–1754

Torp-Pedersen C, Køber L for the TRACE Study Group (1999) Effect of ACE inhibitor trandolapril on life expectancy of patients with reduced left-ventricular function after acute myocardial infarction. Lancet 354:9–12

Tuomilehto J, Lindström J, Eriksson JG et al (2001) Prevention of type 2 diabetes mellitus by changes in lifestyle among subjects with impaired glucose tolerance. N Engl J Med 344:1343–1350

UK Prospective Diabetes Study Group (1998a) Efficacy of atenolol and captopril in reducing risk of macrovascular and microvascular complications in type 2 diabetes: UKPDS 39. BMJ 317:713–720

UK Prospective Diabetes Study Group (1998b) Intensive blood-glucose control with sulphonylureas or insulin compared with conventional treatment and risk of complications in patients with type 2 diabetes (UKPDS 33). Lancet 352:837–853

UK Prospective Diabetes Study Group (1998c) Tight blood pressure control and risk of macrovascular and microvascular complications in type 2 diabetes: UKPDS 38. BMJ 317:703–713

VA Cooperative Study Group on Antihypertensive Agents (1967) Effects of Treatment on morbidity in hypertension. results in patients with diastolic blood pressures averaging 115 through 129 mmHg. JAMA 202:116–122

VA Cooperative Study Group on Antihypertensive Agents (1970) Effects of treatment on morbidity in hypertension. II. Results in patients with diastolic blood pressure averaging 90 through 114 mmHg. JAMA 213:1143–1152

Viberti G, Wheeldon NM for the MicroAlbuminuria Reduction with VALsartan (MARVAL) Study Investigators (2002) Microalbuminuria reduction with valsartan in patients with type 2 diabetes mellitus: a blood pressure-independent effect. Circulation 106:672–678

Vidt DG, White WB, Ridley E et al (2001) A forced titration study of antihypertensive efficacy of candesartan cilexetil in comparison to losartan: CLAIM study II. J Hum Hypertens 15:475–480

Wienbergen H, Schiele R, Gitt AK et al (2002) Impact of ramipril versus other angiotensin-converting enzyme inhibitors on outcome of unselected patients with st-elevation acute myocardial infarction. Am J Cardiology 90:1045–1049

Wing LMH, Reid CM, Ryan P et al (2003) A comparison of outcomes with angiotensin-converting-enzyme inhibitors and diuretics for hypertension in the elderly. N Engl J Med 348:583–592

Wood DA, De Backer, Faergeman O et al (1998) Prevention of coronary heart disease in clinical practice. Recommendations of the second joint task force of the European Society of Cardiology, European Atherosclerosis Society and European Society of Hypertension. Eur Heart J 19:1434–1503

Wright JT Jr, Agodoa L, Contreras G et al (2002a) African American Study of Kidney Disease and Hypertension Study Group. Successful blood pressure control in the African American Study of Kidney Disease and Hypertension. Arch Intern Med 162:1636–1643

Wright JT Jr, Bakris G, Greene T et al, from the AASK Study Group (2002b) Effect of blood pressure lowering and antihypertensive drug class on progression of hypertensive kidney disease: results from the AASK trial. JAMA 288:2421:2431

Yusuf S (2002) From the HOPE to the ONTARGET and the TRANSCEND studies: challenges in improving prognosis. Am J Cardiol 89:18A–25A

Yusuf S, Pfeffer MA, Swedberg K et al (2003) Effects of candesartan in patients with chronic heart failure and preserved left-ventricular ejection fraction: the CHARM-Preserved trial. Lancet 362:777–781

Subject Index

Printing: Saladruck, Berlin
Binding: Stein+Lehmann, Berlin